Second Edition

SAFETY, HEALTH, and ASSET PROTECTION

Management Essentials

■ ■ ■

Edited by

Richard W. Lack

Taylor & Francis
Taylor & Francis Group

Boca Raton   London   New York   Singapore

A CRC title, part of the Taylor & Francis imprint, a member of the
Taylor & Francis Group, the academic division of T&F Informa plc.

Library of Congress Cataloging-in-Publication Data

Safety, health, and asset protection : management essentials / edited by Richard W. Lack.-- 2nd ed.
    p. cm.
Includes bibliographical references and index.
ISBN 1-56670-370-0 (alk, paper)
   1. Industrial safety. 2. Industrial hygiene. 3. Asset-liabilty management. I. Lack, Richard W.

T55 .S2152 2001
658.3′82--dc21
                                              2001038456

Visit the CRC Press Web site at www.crcpress.com

Foreword, Second Edition

"Communication occurs when the message has been received and understood." This is a quote from one of my professors in graduate school a number of years ago. It is as true today as it was then. As the safety and health management professions have moved into prominence in the past twenty years, the prime goal of each of us has become imparting, sharing, or teaching others we serve, the management and state-of-the-art technical skills we possess. Our professions have become exceedingly more complex and require, among other capabilities, the realization that no one practitioner has the answers to all the questions. The dilemma then becomes where do we go to locate all those answers? Do we buy a book on just safety engineering standards or modern management principles? Or do we look for a comprehensive, authoritative volume that gives the reader or investigator a sound glimpse into the subject of inquiry and then follows through with lists of other recommended books and research publications for more detailed reading.

Richard W. Lack, the editor, has organized such a comprehensive authoritative volume. This volume presents a group of safety and health professionals who are also recognized authors and authorities. The second edition of *Essentials of Safety and Health Management* gives the reader an unparalleled spectrum of theory, facts, current research, and a glimpse into the future that is found nowhere else. I highly recommend the addition of this valuable professional tool to your library.

Edwin P. Granberry, Jr.
Winter Park, Florida

Foreword, First Edition

Many books address the topic of safety and health management. Few of them accumulated the outstanding group of authors whose work and experience are contained in *Essentials of Safety and Health Management*. The painstaking effort made to develop an authoritative book will soon become apparent to you as you explore new dimensions of these old topics. The welcome addition of international developments is particularly noteworthy. You hold an education in your hands. The influence of this book on your work in safety and health will be considerable.

Joseph LaDou, M.D.
Division of Occupational and Environmental Medicine
University of California
San Francisco, California

Foreword, First Edition

The profession and practice of occupational safety is very different than it was just a few years ago. The workforce makeup and psyche is very different as well. The practice of safety, founded in sound technical principles, is today carried out in an environment of rapidly changing management approaches.

Safety practitioners find their responsibilities broadening, while the clean lines delineating the various disciplines in occupational safety and health are blurring. The need to be competent in the broad practice of occupational safety spurs the professional to learn new technologies and products, and new applications of familiar technology.

The rapidly developing alliances and exchanges among world nations and marketplaces has expanded the safety practitioner's world so that as global cooperation develops, so does global competition. The safety professional is in need of at least a working knowledge of the technical standards and cultural approaches of other countries to occupational safety and health.

Fast-paced change and broadening responsibilities dictate that the professional remain competent through professional development. This volume is encyclopedic in breadth, providing an update of some fundamental topics, while introducing new and innovative subjects. This book prepares the professional to meet the challenges outlined above and can also serve the manager in need of an overview of occupational safety and health management.

This collection of topics and authors explores new approaches to the broadly accepted elements of accident prevention, while providing an overview of additional information necessary for the professional to maintain competency and confidence in a rapidly evolving profession.

Essentials of Safety and Health Management may be the *only* book that professionals will need to add to their bookcases in order to remain current and meet the challenges of the changing workplace, workforce, and management environment. This remarkable collection of theory, application, and fact makes a significant contribution to the body of knowledge in the Science of Safety.

Margaret Mock Carroll, CSP, P.E.
Sandia National Laboratory
Albuquerque, New Mexico

Preface, Second Edition

Even though the first edition is still basically current, to keep pace with this ever changing world of ours, our publisher and I felt it was time to update the information where necessary. We decided to expand the scope of our text to include chapters about selected technical aspects of safety and health management, and related aspects of security and fire protection. The authors of these two chapters, Jim Cawood, Peter Schontag, and Robert Panero are top professionals in their respective fields.

I am indebted to fellow safety professionals, Dale Gray and Jerry Williams, for their new chapter topics on negotiation and new approaches to performance measurement.

Finally, at the suggestion of my esteemed colleague Professor David Fender at Murray State University, we have included a chapter on ethics. The author is Phil Ulmer, a much-respected speaker on this topic.

Again, on behalf of the publisher and all contributing authors, we invite your comments and suggestions for the book's improvement in future editions.

Richard W. Lack
Cheyenne, Wyoming

Preface, First Edition

This book was born in the minds of two outstanding educators and professionals, Marion Gillen, RN, MPH, and Barbara Plog, MPH, CIH, CSP. Both are associated with the Center for Occupational and Environmental Health (COEH) of the University of California at Berkeley. At the time, Gillen was serving as continuing education coordinator responsible for developing occupational health and safety and environmental courses. Plog is an instructor for the School of Public Health at the University of California at Berkeley, where she teaches courses in industrial hygiene and safety. More recently, Gillen has commenced studies for her Ph.D., and Plog has taken over duties as director of continuing education at COEH.

During this time, while Gillen and Plog were discussing the book with Lewis Publishers, Gillen was also working with me on a new course to be called, *Essentials of Safety Management*.

Some time later, Gillen and Plog decided that their work and personal study commitments were such that a book project was no longer feasible. Accordingly, Gillen invited me to take over the project. It sounded like such an exciting and challenging opportunity that I immediately decided to investigate the possibilities.

First, I approached the members of our faculty for the *Essentials of Safety Management* course. All were enthusiastically willing to join me on this project! The pioneering group of co-authors who first joined this project were: Jim Arnold, Gabe Gillotti, Tom Hanley, Kathleen Kahler, Bob Lapidus, Steve McConnell, and Herman Woessner.

Subsequently, I decided that the scope of our book needed expanding to include certain topics not covered in our existing two-day course. A number of other professionals were therefore invited to join the group, and the complete panel may be found in the list of contributing authors.

Each of the co-authors has had wide experience in the various chapter subjects he or she is contributing. Several have already published books and or papers and magazine articles. Many are also frequent speakers and instructors at professional seminars and courses.

I designed this book with the objective of serving a wide range of readers. First, the book is intended to serve as a desk reference source, with primary emphasis on the management aspects of each subject area. An overview of the subject elements is provided, together with examples of their practical application in the workplace. In addition, each subject chapter includes a Further Reading section. Here are listed selected recommended books, articles, etc., to which readers may go for more in-depth information.

Second, the book is intended to serve both the needs of academia in preparing career professionals and those already in practice. It will also serve as a useful reference for managers who have responsibilities in the safety and health field.

Some readers may question why there are so many chapters devoted to management- and communications-related subjects. It is the opinion of the faculty of the *Essentials of Safety Management* course that, if safety and health practitioners are to perform their job functions effectively, it is essential they possess a comprehensive knowledge of the state of the art in the fields of management and communications, and continuously develop their skills in these areas.

Finally, it is intended that the audience for this book should extend beyond the shores and borders of the United States. Obviously, we cannot address the many thousands of technical, legal, and regulatory aspects for all the countries of this globe. We have, however, included a chapter on international aspects. The author of this chapter, Kathy Seabrook, has provided many valuable insights, particularly with regard to European countries. Furthermore, the contents of the entire book has been designed so that its scope is virtually universal, and the principles and procedures can be applied worldwide.

On behalf of the publisher and co-authors, we look forward to hearing from you. Your comments and suggestions for the book's improvement will be a most welcome aid in planning future editions.

Richard W. Lack
San Francisco, California

The Editor

Richard W. Lack, PE, CSP, CPP, CHCM, CSHM, RSP (UK) is a consultant for Safety & Protection Services Consulting, Cheyenne, Wyoming. Prior to starting his consulting business in 1997, Lack pursued a 40-year career in safety management with a wide variety of industries and organizations, both in the U.S. and internationally. Assignments included the San Francisco International Airport Commission; Castle & Cooke, Inc.; the Great Plains Coal Gasification project in Beulah, ND; and Reynolds Metals Company and Kaiser Aluminum & Chemical Corporation in Jamaica, West Indies.

Lack is a registered professional engineer in safety engineering (California). He has certifications as a safety professional, protection professional, hazard control manager, safety manager, and is a registered safety professional in the United Kingdom.

He has served twice on the board of directors of the American Society of Safety Engineers, and was recently elected to the honor of Fellow of the Society. He has served on the governing board of the National Safety Management Society (NSMS) and as chapter president of the NSMS Golden Gate chapter. In 1980, Lack served as founding chairman of the Arkansas chapter of the American Society for Industrial Security (ASIS) and chairman of the ASIS North Dakota chapter in 1986.

Lack is a member of the American Society of Safety Engineers, the National Safety Council, the National Safety Management Society, the Institution of Occupational Safety and Health (UK), and the American Society for Industrial Security, among others. He is a regular speaker at professional conferences and an author of numerous papers. He was editor and contributor for the book *Security Management,* published by the National Safety Council in 1997, and *Dictionary of Terms Used in the Safety Profession, fourth edition,* published by the American Society of Safety Engineers in 2001.

Acknowledgments, Second Edition

Just as in the first edition, it is not possible for me to list every person who has helped with the production of this second edition. Let me therefore mention a few individuals who have made very significant contributions.

My wife, Phillippa, who despite her work and artistic commitments has proofread all the new and revised manuscripts, and has provided valuable input on grammar, style, layout, etc.

Sincere thanks go to Randi Gonzalez, associate acquisitions editor, Lewis Publishers, for her assistance and tireless support of this project. Without her dedicated efforts, a project of this complexity would never have had cohesion.

It goes without saying that the publisher and I owe a huge debt of gratitude to all our contributing authors; to repeat my first edition statement, the book stands as a true monument to their work.

Acknowledgments, First Edition

It would be impossible for me to list every person who has contributed to the production of this book. Nevertheless, I do want to mention a few who have made extraordinarily significant contributions.

First of all, my wife, Phillippa, who in spite of her work assignments, craft projects, and the numerous affairs of our household, took the time to proofread my chapters and also provided invaluable counsel on grammar, layout, etc.

My thanks also to my employer, the Airports Commission, San Francisco International Airport, and in particular, my supervisor, Assistant Deputy Director, Mel Leong, who provided much needed support and encouragement for me to take on this project. Two Airport staffers, Imelda Quesada and Leticia Aguilar, contributed outstanding support with all my correspondence to the publisher and co-authors, and typing of the book outline and table of contents. My thanks especially to Leticia Aguilar, who types all my chapters and spent many hours working up the diagrams and example forms on her computer.

My contribution to this book would not have been possible had it not been for the guidance and teaching gained over many years from my foremost mentor, Homer K. Lambie. It is my principal goal to capture and preserve Homer's unique philosophies on safety management systems. They have inspired and guided several generations of safety professionals and countless managers during his 40 years of work, primarily with the Kaiser Aluminum and Chemical Corporation.

On a personal basis, my sincere thanks to another of my mentors, Dr. John Grimaldi, for his support and advice which has been invaluable particularly in developing my material for the book. My thanks also to Margaret Carroll for her support of this project and her thoughtful review and comments on my chapters. I have already mentioned Marion Gillen and Barbara Plog in the Preface. It goes without saying that this book would not have happened had it not been for their efforts, perception, and support.

My sincere thanks go to all the contributing authors. Their time and effort cannot be measured in value, the book is the true monument to their work.

I would like to specially acknowledge John Northey and others concerned on the editorial staff of Paramount Publishing Limited. Their support and acceptance of my earliest writings gave me a much needed opportunity to get started as a writer in our professional field. In this book, I have drawn from several of these papers, and this is so indicated in the relevant chapters.

My grateful acknowledgment to the staff of Lewis Publishers for their enthusiastic and unflagging support of this project. Their guidance and patience extended to a complete newcomer to the world of publishing was greatly appreciated.

Finally, my thanks to many fellow professionals in the American Society of Safety Engineers, National Safety Management Society, and the Institution of Occupational Safety and Health for their guidance, support, and encouragement.

Contributors

John D. Adams, Ph.D.
Director, Organizational Systems
Saybrook Graduate School
San Francisco, California

Yvonne F. Alexander, M.A.
Alexander Communications
San Francisco, California

Melissa M. Allain, J.D.
Vice President, Environmental &
  Real Property Law Group
AutoNation, Inc.
Fort Lauderdale, Florida

Laura Benjamin, B.A.
Laura Benjamin International
Colorado Springs, Colorado

Michael-Laurie Bishow, Ph.D.
Contra Costa County Administrator's
  Office
Policy & Innovation Institute
Martinez, California

Marc Bowman, B.S.
Woodward-Clyde Consultants
Oakland, California

Ken Braly, M.S.
San Jose, California

Roger L. Brauer, Ph.D., P.E., CSP
Executive Director
Board of Certified Safety Professionals
Tolono, Illinois

Rosa Antonia Carillo, M.S.
Carillo & Associates, Inc.
Seal Beach, California

Margaret Mock Carroll, CSP, P.E.
Safety Engineering Manager
Sandia National Labs
Albuquerque, New Mexico

James S. Cawood, CPP
President
Factor One, Inc.
San Leandro, California

Gerard C. Coletta, M.S.
Vice President and Practice Leader
  Business Continuity Consulting
Palmer and Cay
Boston, Massachusetts

Kenneth M. Colonna, CSP
CNA Risk Management
Naperville, Illinois

Barbara K. Cooper, MSPH, CIH, CSP
Davis, California

Ellen E. Dehr, CIH
Industrial Hygienist
Port of San Francisco
San Francisco, California

Judith A. Erickson, Ph.D.
President
Erickson Associates
Irvine, California

David L. Fender, Ed.D., CSP
Assitant Professor
Department of Occupational Safety &
  Health
Murray State University
Murray, Kentucky

Michael L. Fischman, M.D., MPH
Fischman Occupational and
  Environmental Medical Group
Walnut Creek, California

Robert S. Fish, Ph.D.
San Jose, California

E. Scott Geller, Ph.D.
Professor, Department of Psychology
Virginia Polytechnic Institute and
  State University
Blacksburg, Virginia

Gabriel J. Gillotti
Director
Voluntary Programs and Outreach
United States Dept. of Labor – OSHA
San Francisco, California

Dale A. Gray, P.E., CSP
Sanibel Island, Florida

Edwin P. Granberry, Jr.
President
Granberry and Associates, Inc.
Winter Park, Florida

Thomas A. Hanley, M.A.
Regional Manager
Division of Occupational Safety &
  Health
Anaheim, California

Mark D. Hansen, CSP, CPE, P.E.
Manager, Health, Safety, &
  Environmental
Austin International
Houston, Texas

J. Michelle Hickey, J.D.
Partner
Holland & Knight LLP
Los Angeles, California

Joseph LaDou, M.D., M.S.
Division of Occupational and
  Environmental Medicine
University of California
San Francisco, California

Robert A. Lapidus, CSP
Safety Management Consultant
Lapidus Consulting
Fredericksburg, Texas

Audrey Lawrence, MPH, CIH
Safety & Health Manager
Airport Commission
San Francisco International Airport
San Francisco, California

Steven M. McConnell, CSP, REA
Safety Engineer
Lawrence Livermore National
  Laboratory
Livermore, California

Donald L. Morelli, M.S., CPE
San Carlos, California

Barbara Newman, M.A., M.F.C.C.
Los Angeles, California

Robert A. Panero
Principal Loss Control Engineer
P.G.&E. Corporation
San Francisco, California

Neva Nishio Petersen, MPH, CIH
Safety Manager
Integrated Device Technology, Inc.
Salinas, California

Lucy O. Reinke, M.D., MPH
Fischman Occupational and
  Environmental Medical Group
Walnut Creek, California

Peter B. Rice, CSP, CIH
Director, EHS
ClickSafety
Novato, California

Anne Durrum Robinson, M.A.
Consultant Human Resource
  Development
Creativity, Communication, Common
  Sense
Austin, Texas

Peter K. Schontag, P.E.
Aptos, California

Kathy A. Seabrook, CSP, RSP (UK)
Global Business Solutions
Mendham, New Jersey

Michal F. Settles, Ed.D.
Department Manager, Human Resources
Bay Area Rapid Transit District
Oakland, California

James P. Seward, M.D., MPP
Medical Director
University of California
Lawrence Livermore National
 Laboratory
Livermore, California

Carolyn R. Shaffer, MA
Growing Community Associates
Berkeley, California

Steven I. Simon, Ph.D.
President
Cultural Change Consultants, Inc.
Larchmont, New York

Robin Spencer, B.A., CHMM, REA
Orinda, California

George E. Swartz, CSP
Davenport, Florida

Paula R. Taylor, M.A.
Paula Taylor & Associates
Oakland, California

Sondra Thiederman, Ph.D.
Cross-Cultural Communications
San Diego, California

Jan Thomas, Ph.D., CSP
Circle Safety and Health Consultants,
 LLC
Richmond, Virginia

Michael D. Topf, B.S., M.A.
President
Topf Organization
King of Prussia, Pennsylvania

Philip E. Ulmer, P.E.
Principal Consultant
Northwest Safety Management, Inc.
Eagle River, Alaska

Jerry L. Williams, CSP
Assistant Professor
Central Missouri State University
Warrensburg, Missouri

Joan F. Woerner, REA, REHS, CSP
Edy's Grand Ice Cream
Ft. Wayne, Indiana

Herman Woessner, MSS, CSP, REA
EH&S Project Analyst
Progress Energy
St. Petersburg, Florida

Holland A. Young
Department of Aviation
City of Austin
Austin, Texas

Bonita R. Zahara
Zahara and Associates
Spokane, Washington

Contents

PART I

Introduction

## CHAPTER 1

Where Have We Been?

Richard W. Lack

CONTENTS

INTRODUCTION

From my perspective, this question goes back over a professional experience spanning more than 40 years both in the United States and overseas. Historically, we have seen a world recovering from a series of major conflicts, starting with World War II in the 1940s. This was followed by the Korean War in the 1950s and the Vietnam War in the 1960s. These disastrous chapters in our history have since been followed by continued tensions and outbreaks of hostility that have erupted in many different parts of our globe. Such situations continue today, when we consider the troubles in the Balkans and the Middle East.

Politically speaking, in recent years we have seen some dramatic changes from the Cold War, Iron Curtain, and Berlin Wall days.

The earlier effect on the economies of the United States and other developed nations was an emphasis on heavy engineering and manufacturing. This was necessary to produce the arms and technology for a massive war effort.

Many of us began our careers in military service and followed this experience by entering heavy industry. It might have been a steel mill, a shipbuilding yard, a chemical refinery, or a major engineering works. In my case, it was mining bauxite, the raw material for aluminum.

The technologies developed originally for military purposes have in many cases been turned toward something much more beneficial for mankind. Witness the development of the wartime rocket, which was the predecessor of our space program.

Developments in communications and computer technologies have helped to bring us rapidly into the Information Age.

To adjust to all these changes, industries in the developed nations have been forced to make major shifts. Many have disappeared altogether; others have been radically restructured. Another more recent trend has been the public's concern for preserving and protecting our environment.

Against this backdrop of political, historical, and economic influences, where do we stand in terms of losses, both of people and of property? According to the National Safety Council, the 1999 unintentional-injury deaths total was 96,900 — the highest total since 1988.

The breakdown for all unintentional injuries in 1999 is as follows:

Injuries	Deaths	Deaths per 100,000 Persons
All classes	96,900	35.5
Motor vehicle	41,300	15.1
Public non-motor vehicle	24,100	8.8
Work	5,100	1.9
Home	28,800	10.6

(From National Safety Council (NSC), *Injury Facts*, Itasca, IL: NSC, 2000.)

These injuries were estimated to cost $225 billion in medical expenses, insurance, lost wages, and property damage.

Dealing with work injuries in 1999, highlights from *Injury Facts* are:

- Work injury costs were $122.6 billion, which breaks down to a cost per worker of $910 and cost per disabling injury of $28,000.
- Total time lost due to injuries in 1999 was 80,000,000 days.
- For a worker population of 134,688,000, there were 3,800,000 disabling injuries.

The Bureau of Labor Statistics National Consensus of Fatal Occupational Injuries 1999 is reproduced on the following pages.

In the book *Violence in America*, it is noted that in 1987 there were 20,000 homicides and 31,000 suicides in the United States. Assaultive violence was reported to cause from 19,000 to 23,000 deaths a year.

According to the *UC Berkeley Wellness Letter*, almost 70 million Americans are injured seriously enough every year to require medical attention, or at least a day of restricted activity. Around the world, 16,000 people die from injuries every day. Unintentional injuries are a leading cause of death and disability.

The previous figures show that although overall injury rates are on a gradual downward trend, they represent a staggering burden in terms of economic loss, inestimable human suffering, and misery. The loss far exceeds any losses sustained by war and other forms of hostilities.

Occasionally, some catastrophic event such as an explosion at a mine or industrial plant will attract attention. Such events have been a driver for increased legislation, especially during the past 20 years, and these developments are discussed in another chapter of this book. More stringent safety regulations have certainly had some impact, but opinions seem to vary widely as to the degree.

Dan Petersen provides an excellent historical review in his article *The Barriers to Safety Excellence* published in the December 2000 issue of *Occupational Hazards*. This article includes an analysis of what is standing in the way of companies achieving safety excellence and outlines six criteria for attaining it.

As mentioned earlier, total injuries may be down, but time lost is up. Allied to this are skyrocketing medical costs. According to a 1990 study for the California Legislature, America's national health spending has increased from a total of $248 billion in 1980 to $647 billion in 1990. Quoted in this report from a U.S. Department of Commerce source, which is based on 1988 expenditures, health care is the largest industry in the U.S. economy at 10% of gross national product (GNP). This exceeds defense (9%), food and beverage (5%), and education (4%). In closing this introduction, here is a quote from this report to ponder:

> In the past, occupational health and safety focused on injuries and illnesses with a direct association to the occupational work environment. Nowadays, the distinction between general health problems which have their origins elsewhere are brought into the workplace. They reflect health problems endemic in modern society at large — abuse of drugs and alcohol, cigarette smoking, poor nutritional habits, lack of aerobic exercise, traumatic injury, obesity, psychological and unhealthy stress associated with problems of coping with life in general.

News

United States
Department
of Labor

Bureau of Labor Statistics   Washington, D.C. 20212

Technical information: (202) 691-6175
Media information: (202) 691-5902
Internet address: http://stats.bls.gov/oshhome.htm

USDL 00-236
FOR RELEASE: 10 a.m. EDT
Thursday, August 17, 2000

NATIONAL CENSUS OF FATAL OCCUPATIONAL INJURIES, 1999

The number of fatal work injuries that occurred during 1999 was 6,023, nearly the same as the previous year's total despite an increase in employment, according to the Census of Fatal Occupational Injuries, conducted by the Bureau of Labor Statistics, U.S. Department of Labor. Decreases in job-related deaths from homicides and electrocutions in 1999 were offset by increases from workers struck by falling objects or caught in running machinery. Homicides fell from the second-leading cause of fatal work injuries to the third, behind highway fatalities and falls. Construction reported the largest number of fatal work injuries for any industry and accounted for one-fifth of the fatality total.

Profiles of 1999 Fatal Work Injuries

Highway crashes continued as the leading cause of on-the-job fatalities during 1999, accounting for one-fourth of the fatal work injury total. (See Table 1 and Chart 1.) The number of these fatalities increased slightly over 1998 to reach the highest level since the BLS fatality census began in 1992. Slightly over two-fifths of the 1491 victims of job-related highway fatalities were employed as truck drivers.

In contrast to fatalities resulting from crashes that occurred on public roadways, the number of workers killed in nonhighway crashes and overturnings or killed after being struck by a vehicle declined from the previous year. The number of workers killed in air, water, and rail vehicle incidents during 1999 was about the same as in 1998.

In 1999, deaths resulting from on-the-job falls increased slightly to 717. This increase, coupled with a decline in homicides, made falls the second-leading cause of fatal work injuries for the first time since the fatality census began in 1992. (See Chart 2.) About half of the fatal falls were from a roof, ladder, or scaffold, and slightly over half of the fatal falls occurred in the construction industry.

Now the third-leading cause of on-the-job deaths, workplace homicides fell to the lowest level since the fatality census' inception in 1992. Job-related homicides totaled 645 in 1999, a 10 percent drop from the 1998 total and a 40 percent decline from the 1080 homicides that occurred in 1994, which had the highest count in the 8-year period. The drop in homicides at work was most pronounced in retail trade, where homicides

Chart 1: The manner in which workplace fatalities occured, 1999

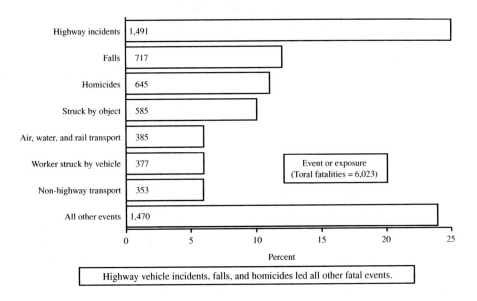

Highway incidents	1,491
Falls	717
Homicides	645
Struck by object	585
Air, water, and rail transport	385
Worker struck by vehicle	377
Non-highway transport	353
All other events	1,470

Event or exposure
(Toral fatalities = 6,023)

Percent

Highway vehicle incidents, falls, and homicides led all other fatal events.

Chart 2: Homicides and Falls 1992-1999

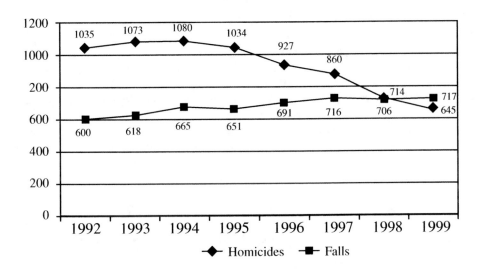

1035 1073 1080 1034 927 860 714 717

600 618 665 651 691 716 706 645

◆ Homicides ■ Falls

fell by 51 percent from 1994. The following table shows counts of workplace homicides for selected industries during 1994 to 99.

Industry	1994	1995	1996	1997	1998	1999
Total homicides................................	1,080	1,034	927	860	714	645
Retail trade...	530	422	437	395	286	260
Grocery stores.................................	196	152	146	141	95	76
Eating and drinking places..............	135	121	135	109	69	94
Gasoline service stations	41	36	23	34	24	17
Taxicab	87	68	50	74	48	51
Detective and armored car services....	49	27	29	21	18	17
Police protection.................................	65	61	45	61	50	41

Among the job-related homicides for which a motive could be ascertained from the source documents, robbery continued to be the primary motive, followed by violence by co-workers and customers or clients. Occupations with high numbers of homicides include those that typically engage in cash transactions or have valuables on hand, including managers of food and lodging establishments, sales supervisors and proprietors, cashiers, and taxicab drivers.

Workers struck by objects or equipment accounted for 10 percent of the fatal work injuries in 1999. These types of injuries increased from the previous year. Electrocutions accounted for 5 percent of the fatal injuries and decreased by 17 percent from 1998. Contact with overhead power lines accounted for about two-fifths of the electrocutions.

On average, about 17 workers were fatally injured each day during 1999. Eighty-three percent of fatally injured workers died the day they were injured; 97 percent died within 30 days. There were 235 multiple-fatality incidents (incidents that resulted in two or more worker deaths), resulting in 617 job-related deaths. Although this was a slight increase over the 227 multiple-fatality events reported for 1998, there was a more substantial increase in the number of deaths resulting from these types of incidents in 1999 than in the previous year, when 555 worker deaths occurred.

Occupation Highlights (Table 2 and Chart 3):

* Occupations with large numbers of fatal injuries included truck drivers, construction trades, and farm occupations.
* Fatal injuries to truck drivers were at their highest level in the 8-year period.
* Mechanics and repairers also reported a noticeable increase in fatal work injuries over the previous year, reaching its highest level in the 8-year period.
* In contrast, the number of fatalities in sales occupations fell to its lowest level during the same period, primarily because of the drop in homicides.

Relative Risk (Tables 3 and 4):

A comparison of percent distributions of fatalities and employment can be used to evaluate the relative risk of a job-related fatality for a given industry or worker characteristic. For example, the construction industry accounted for 20 percent of the fatality

Chart 3: Occupation with large numbers of worker fatalities and the leading event, 1999

Truck drivers	Highway crash, jack-knifing — 898
Farm occupations	Tractor-related — 557
Sales occupations	Homicide — 356
Construction laborers	Falls — 341
Laborers, except construction	Contact with objects & equipment — 193
Police and detectives, public	Varied — 132
Timber cutters	Falling trees — 114
Groundskeepers and gardeners	Vehicle — 106
Electricians	Contact with elec. current — 105
Carpenters	Falls — 103

Leading fatal event shown for each occupation

0   100   200   300   400   500   600   700   800   900

(Total fatalities = 6,023)

Truck driver fatalities, primarily involving highway crashes and jack-knifings, accounted for 15 percent of the job-related fatalities.

total, 3 times its 6 percent share of total employment. While employment can be used to evaluate the relative risk of a fatal work injury, other measures, such as hours worked, also can be used.

Industry Highlights (Table 3):

* Industry divisions with large numbers of fatalities relative to their employment include agriculture, forestry, and fishing; construction; transportation, and public utilities; and mining.
* Retail trade posted a substantial decline in the number of fatal work injuries in 1999 over the previous year; transportation and public utilities posted a substantial increase.
* A decline in fatal work injuries among government workers resulted in the lowest levels during the 8-year period.

Demographic Highlights (Table 4):

* Men, the self-employed, and older workers suffered fatal injuries more often than their employment shares would suggest. Differences in the industries and occupations of these worker groups explain in part their high relative risk of fatal injury on the job.
* Highway-related incidents were the leading cause of job-related fatalities among both men and women. Homicides, which had been the leading cause for women, were the second-leading cause in 1999. Falls ranked second for men.
* Two-fifths of fatally injured workers under 18 years of age were killed while doing farm work; another one-fifth were killed while working for a retail trade establishment.

State Highlights by Major Regional Area (Table 5)

* In general, the states with the largest number of persons employed have the largest number of work-related fatalities. Four of the largest states — California, Texas, Florida, and New York — accounted for over one-fourth of the total fatalities in the U.S. Each state's and region's industry mix, geographical features, age of population, and other characteristics of the workforce must be considered when evaluating state and region fatality profiles.

* In all four regions of the U.S. — Northeast, Midwest, South, and West — highway motor vehicle incidents were the leading event for occupational fatalities. In the Southern and Western states, the second leading event was homicide. In the Northeast and Midwest states, the second-leading events were falls to lower level and being struck by an object, respectively.

* About 44 percent of the fatal occupational highway incidents and almost half of the work-related homicides occurred in the South, which has 35 percent of total employment.

Background of the Program

The Census of Fatal Occupational Injuries, part of the BLS occupational safety and health statistics program, provides the most complete count of fatal work injuries available because it uses diverse state and federal data sources to identify, verify, and profile fatal work injuries. Information about each workplace fatality (occupation and other worker characteristics, equipment being used, and circumstances of the event) is obtained by cross-referencing source documents, such as death certificates, workers' compensation records, and reports to federal and state agencies. This method assures counts are as complete and accurate as possible.

This is the eighth year that the fatality census has been conducted in all 50 states and the District of Columbia. The BLS fatality census is a federal/state cooperative venture in which costs are shared equally. Additional state-specific data are available from the participating state agencies listed in Table 6.

Another BLS program, the Survey of Occupational Injuries and Illnesses, profiles worker and case characteristics of nonfatal workplace injuries and illnesses that result in lost worktime and presents frequency counts and incidence rates by industry. Copies of the 1998 news release on nonfatal injuries and illnesses are available from BLS by calling (202) 691-6179 or by accessing the Web site listed below. Incidence rates for 1999 by industry will be published in December 2000, and information on 1999 worker and case characteristics will be available in April 2001. For additional data, access the BLS Internet site: http://www.bls.gov/oshhome.htm. To request a copy of BLS Report 934, which includes several articles and highlights 1997 fatality data, e-mail your address to CFOIstaff@bls.gov or write to Bureau of Labor Statistics, 2 Massachusetts Avenue, NE, Room 3180, Washington, D.C. 20212.

Table 1   Fatal Occupational Injuries by Event or Exposure, 1994–99

Event or Exposure[a]	Fatalities			
	1994–98 average	1998[b] Number	1999 Number	1999 Percent
Total	6,280	6,055	6,023	100
Transportation incidents	2,640	2,645	2,613	43
Highway	1,374	1,442	1,491	25
Collision between vehicles, mobile equipment	662	707	711	12
Moving in same direction	113	120	129	2
Moving in opposite directions, oncoming	240	272	269	4
Moving in intersection	136	143	160	3
Vehicle struck stationary object or equipment	272	307	334	6
Noncollision	368	375	388	6
Jack-knifed or overturned — no collision	280	302	321	5
Nonhighway (farm, industrial premises)	387	388	353	6
Overturned	215	217	206	3
Aircraft	304	224	227	4
Worker struck by a vehicle	382	413	377	6
Water vehicle	104	112	102	2
Rail vehicle	78	60	56	1
Assaults and violent acts	1,168	962	893	15
Homicides	923	714	645	11
Shooting	748	574	506	8
Stabbing	68	61	60	1
Other	107	79	79	1
Self-inflicted injuries	215	221	208	3
Contact with objects and equipment	984	944	1,029	17
Struck by object	564	520	585	10
Struck by falling object	364	319	358	6
Struck by flying object	60	59	55	1
Caught in or compressed by equipment or objects	281	266	302	5
Caught in running equipment or machinery	148	129	163	3
Caught in or crushed in collapsing materials	124	140	128	2
Falls	686	706	717	12
Fall to lower level	609	625	634	11
Fall from ladder	101	111	96	2
Fall from roof	146	157	153	3
Fall from scaffold	89	98	92	2
Fall on same level	53	51	66	1
Exposure to harmful substances or environments	583	576	529	9
Contact with electric current	322	334	278	5
Contact with overhead powerlines	136	153	124	2
Contact with temperature extremes 45	48	50	1	
Exposure to caustic, noxious, or allergenic substances	118	105	106	2
Inhalation of substance	66	48	55	1
Oxygen deficiency	96	87	93	2
Drowning, submersion	77	75	75	1
Fires and explosions	199	206	216	4
Other events or exposures[c]	21	16	26	-

[a] Based on the 1992 BLS Occupational Injury and Illness Classification Structures.
[b] The BLS news release issued Aug. 4, 1999, reported a total of 6026 fatal work injuries for calendar year 1998. Since then, an additional 29 job-related fatalities were identified, bringing the total job-related fatality count for 1998 to 6055.
[c] Includes the category "bodily reaction and exertion."
Note: Totals for major categories may include subcategories not shown separately. Percentages may not add to totals because of rounding. Dashes indicate less than 0.5 percent or data that are not available or that do not meet publication criteria.
Source: Bureau of Labor Statistics, U.S. Department of Labor, in cooperation with state and federal agencies, Census of Fatal Occupational Injuries, 1994–99.

Table 2 Fatal Occupational Injuries by Occupation and Major Event or Exposure, 1999

Occupation[a]	Fatalities		Major Event or Exposure[b] (percent of total for occupation)			
	Number	Percent	Highway[c]	Homicide	Struck by object	Fall to lower level
Total..	6,023	100	25	11	10	11
Managerial and professional specialty	597	10	24	19	4	7
Executive, administrative, and managerial	371	6	22	26	5	8
Professional specialty	226	4	27	8	3	6
Technical, sales, and administrative support..............	610	10	27	32	2	3
Technicians and related support occupations.	158	3	16	—	3	3
Airplane pilots and navigators....................................	94	2	—	—	—	—
Sales occupations	356	6	28	49	2	2
Supervisors and proprietors, sales occupations.........	140	2	13	62	4	2
Sales workers, retail and personal services	144	2	27	51	—	—
Cashiers..	55	1	—	80	—	—
Administrative support occupations, including clerical ...	96	2	44	19	—	4
Service occupations..	468	8	20	33	1	9
Protective service occupations	261	4	26	32	1	3
Firefighting and fire prevention occupations, including supervisors	57	1	18	—	—	5
Police and detectives, including supervisors	132	2	39	36	—	—
Guards, including supervisors......................................	72	1	10	50	—	4
Farming, forestry, and fishing...................................	897	15	13	2	21	5
Farming operators and managers	362	6	13	1	15	3
Farmers, except horticultural.....................................	233	4	13	—	17	3
Managers, farms, except horitcultural	118	2	13	—	10	4
Other agricultural and related occupations....................	335	6	18	4	14	8
Farm workers, including supervisors	206	3	22	4	9	4
Forestry and logging occupations.................................	122	2	6	—	67	3
Timber cutting and logging occupations	102	2	4	—	73	3
Fishers, hunters, and trappers.....................................	78	1	—	—	—	—
Fishers, including vessel captains and officers	78	1	—	—	—	—
Precision production, craft, and repair	1,142	19	11	3	12	28
Mechanics and repairers	353	6	12	7	19	13
Construction trades...	633	11	11	1	7	39
Carpenters and apprentices..	103	2	6	—	13	48
Electricians and apprentices	105	2	15	—	4	12
Painters...	38	1	—	—	—	68
Roofers...	59	1	5	—	—	85
Structural metal workers	43	1	—	—	19	77
Operators, fabricators, and laborers	2,194	36	37	5	10	8
Machine operators, assemblers, and inspectors...........	216	4	5	4	14	15
Transportation and material moving occupations...........	1,320	22	56	6	7	2
Motor vehicle operators..	1,063	18	67	8	5	2
Truck drivers...	898	15	70	2	6	2
Driver-sales workers	42	1	79	10	—	—
Taxicab drivers and chauffeurs................................	74	1	28	69	—	—
Material moving equipment operators.........................	205	3	14	—	14	5
Handlers, equipment cleaners, helpers, and laborers....	658	11	11	4	14	17
Construction laborers	341	6	11	—	14	25
Laborers, except construction	193	3	11	6	16	10
Military[d]...	80	1	24	—	4	—

[a] Based on the 1990 Occupational Classification System developed by the Bureau of the Census.
[b] The figure shown is the percent of the total fatalities for that occupational group.
[c] "Highway" includes deaths to vehicle occupants resulting from traffic incidents that occur on the public roadway, shoulder, or surrounding area. It excludes incidents occurring entirely off the roadway, such as in parking lots and on farms; incidents involving trains; and deaths to pedestrians or other nonpassengers.
[d] Resident armed forces.
Note: Totals for major categories may include subcategories not shown separately. Percentages may not add to totals because of rounding. There were 35 fatalities for which there was insufficient information to determine an occupation classification. Dashes indicate less than 0.5 percent or data that are not available or that do not meet publication criteria.
Source: Bureau of Labor Statistics, U.S. Department of Labor, in cooperation with state and federal agencies, Census of Fatal Occupational Injuries, 1999.

Table 3 Fatal Occupational Injuries and Employment by Industry, 1999

Industry	SIC Code[a]	Fatalities				Employment[b] (in thousands)	
		1994–98 average	1998 (revised)	1999			
		Number	Number	Number	Percent	Number	Percent
Total		6,280	6,055	6,023	100	134,666	100
Private industry		5,625	5,457	5,461	91	114,570	85
Agriculture, forestry and fishing		826	840	807	13	3,349	2
Agricultural production – crops	01	379	380	350	6	955	1
Agricultural production – livestock	02	170	174	163	3	993	1
Agricultural services	07	168	170	164	3	1,317	1
Mining		159	147	121	2	563	—
Coal mining	12	37	30	35	1	84	—
Oil and gas extraction	13	84	76	50	1	329	—
Construction		1,082	1,174	1,190	20	8,479	6
General building contractors	15	191	213	183	3	—	—
Heavy construction, except building	16	253	272	280	5	—	—
Special trades contractors	17	629	680	709	12	—	—
Manufacturing		733	698	719	12	19,994	15
Food and kindred products	20	75	72	83	1	1,643	1
Lumber and wood products	24	191	172	190	3	824	1
Transportation and public utilities		948	911	1,006	17	7,947	6
Local and interurban passenger transportation	41	100	85	102	2	593	—
Trucking and warehousing	42	528	564	605	10	2,679	2
Transportation by air	45	90	74	74	1	864	1
Electric, gas, and sanitary services	49	88	83	86	1	1,029	1
Wholesale trade		253	229	237	4	5,173	4
Retail trade		683	570	507	8	22,300	17
Food stores	54	187	135	115	2	3,511	3
Automotive dealers and service stations	55	116	120	82	1	2,238	2
Eating and drinking places	58	155	107	145	2	6,718	5
Finance, insurance, and real estate		109	92	105	2	8,610	6
Services		773	763	732	12	38,240	28
Business services	73	203	196	161	3	6,756	5
Automotive repair, services, and parking	75	111	133	132	2	1,576	2
Government[c]		656	598	562	9	20,096	15
Federal (including resident armed forces)		204	162	147	2	4,427	3
State		126	136	108	2	5,237	4
Local		319	296	301	5	10,433	8
Police protection	9221	104	102	91	2	—	—

[a] Standard Industrial Classification Manual, 1987 Edition.
[b] Employment is an annual average of employed civilians 16 years of age and older from the Current Population Survey, 1999, adjusted to include data for resident armed forces from the Department of Defense.
[c] Includes fatalities to workers employed by government organizations regardless of industry.
Note: Totals for major categories may include subcategories not shown separately. Percentages may not add to totals because of rounding. There were 37 fatalities for which there was insufficient information to determine a specific industry classification, though a distinction between private sector and government was made for each. Dashes indicate less than 0.5 percent or data that are not available or that do not meet publication criteria.
Source: Bureau of Labor Statistics, U.S. Department of Labor, in cooperation with state and federal agencies, Census of Fatal Occupational Injuries, 1994–99.

FURTHER READING

Books

Baker, S. P., O'Neil, B., Ginsburge, M. J. et al. *The Injury Fact Book.* New York: Oxford University Press, 1992.

Foege, W. H., Baker, S. P., Davis, J. H. et al. *Injury in America: A Continuing Public Health Problem.* Washington, D.C.: National Academy Press, 1985.

Mellius, J. M., Althafer, C. A., Perry, W. H. et al. *Proposed National Strategies for the Prevention of Leading Work-Related Diseases and Injuries*, Part I. Washington, D.C.: Association of Schools of Public Health, 1988.

National Safety Council (NSC). *Injury Facts.* 2000 edition. Itasca, IL: National Safety Council, 2000.

Polakoff, P. I., O'Rourke, P. F. et al. *Healthy Worker–Healthy Workplace. The Productivity Connection.* Sacramento, CA: California Legislature, 1990.

Rice, D. P., MacKenzie, E. J. et al. *Cost of Injury in the United States: A Report to Congress, 1989.* Atlanta, GA: Atlanta Centers for Disease Control, 1989.

Resenberg, M. L., Feniey, M. A. et al. *Violence in America — A Public Health Approach.* New York: Oxford University Press, 1991.

Sauter, S. L., Althafer, C. A., Cahill, L. M. et al. *Proposed National Strategies for the Prevention of Leading Work-Related Diseases and Injuries,* Part II L. Washington, D.C.: Association of Schools of Public Health, 1988.

Other Publications

Casteil J. Safety and health watch. *Saf. Health* 148(2): 143–156, August 1993.

Hoskin A. F. Council updates work–death statistics. *Saf. Health.* 147(5): 80–81, May 1993.

Petersen, D. The barriers to safety excellence, *Occup. Hazards* 37–42, December 2000.

University of California. How accidental are accidents? *Berkeley Wellness Letter* 17: 8, 1–2. May 2001.

## CHAPTER 2

What Works Today? A Safety Professional's Viewpoint

Richard W. Lack

CONTENTS

OVERVIEW

Against the background previously discussed, let us now turn to our profession and the organizations in which we work to discuss what issues we face and what works today. In attempting to address this question, I decided to involve the right brain in helping to focus on the issues. To do this, I used the "mind mapping" approach. See Figure 2.1 for what I developed.

Most safety professionals, I believe, would agree that the past few years have been an exciting and challenging experience. We have had to master many new skills along the way. As an example, in the experience, after starting in a traditional safety field, subsequent responsibilities have been in the areas of security, fire protection, and health, including industrial hygiene and hazardous materials. Other added responsibility areas have included ergonomics, indoor air quality, asbestos, lead, bloodborne pathogens, and the list goes on.

The mind map demonstrates the complexity of the issue facing our profession today. Despite these complex, ever changing issues, I believe prevention must still be our watchword. Prevention of injury, illness, and loss is accomplished by establishing a basic systematic approach that all levels in your organization will understand and effectively implement.

1-56670-370-0/02/$0.00+$1.50
© 2002 by CRC Press LLC

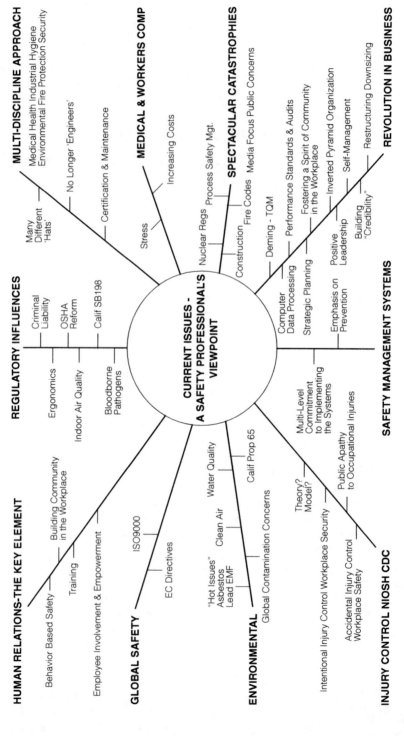

Figure 2.1 Current Issues — a safety professional's viewpoint.

To achieve effective prevention, every aspect of your safety and health management systems must be designed to find and control hazards. This will in turn help to protect your organization's assets — both people and property — from the various risks to which they may be exposed.

From my viewpoint, despite all the tremendous changes in technology, the principal challenges facing members of the safety and health profession rarely change. In fact, they have been with us since the beginning of time. "They," of course, are *people*.

We should never lose sight of our need to continuously improve our people-relationship skills. This includes both communication and management skills. The overall mission, therefore, of the authors involved in the production of this book is dedicated to providing you, the reader, with the information and techniques you need to help you successfully meet these challenges.

FURTHER READING

Books

Grose, V. L. *Managing Risk — Systematic Loss Prevention for Executives.* Englewood Cliffs, NJ: Prentice-Hall, 272–366, 1987.

Brauer, R. L. *Safety and Health for Engineers.* New York: Van Nostrand Reinhold, 79–91, 1990.

Manuele, F. A. *On the Practice of Safety.* 2nd ed. New York: Van Nostrand Reinhold, 175–187, 1997.

Other Publications

Lack, R. W. Systips: getting back to the basics — or learning from history so we are not forced to repeat our mistakes. *National Safety Management Society Insights into Management* 2: 5/6, 6, November/December 1990.

PART II

Safety Program Management Aspects

## CHAPTER 3

Safety Program Organization

Richard W. Lack

CONTENTS

LEGISLATIVE ASPECTS

The continuing loss experience of a wide spectrum of organizations and industries has resulted in an increasing stream of legislative initiatives requiring employers to establish a minimum safety and health organization.

Naturally, as readers are well aware, many employers have outstanding professional staffs and have established and maintained highly effective programs that produce superior results. Unfortunately, these employers are in the minority when one looks at the broad picture.

In the United States, an early initiative was the Occupational Safety and Health Administration's (OSHA) Safety and Health Program Management Guidelines of 1989. These guidelines described the elements of an effective safety program including management and employee responsibilities. In the original draft of these guidelines, there was a reference to a competent person for conducting work site analysis. In the final version, this term was removed. However, the commentary includes a discussion of the relative competence for the various approaches to work site analysis.

The California State Law SB 198 of 1991 requires all employers to nominate an injury and illness prevention program administrator who has the authority and responsibility for implementing and maintaining the employer's Injury and Illness Prevention Plan.

1-56670-370-0/02/$0.00+$1.50

In Europe, the European Community (EC) directive *Management of Health and Safety at Work,* as adopted by the British Health and Safety Commission in 1992, contains some more specific references.

Regulation 6 in the HSE Regulations states in part:

> (1) Every employer shall, subject to paragraphs (6) and (7) (self-employed or business partnerships), appoint one or more competent persons to assist him in undertaking the measures he needs to take to comply with the requirements and prohibitions imposed upon him by or under the relevant statutory provisions.

The approved Code of Practice relevant to this regulation discusses various options open to employers, such as appointing a health and safety department for large employers or hiring external services for smaller employers. There is also reference to the degree of competence needed for persons appointed to carry out the requirements of the regulations.

This trend can be expected to continue, especially in the area of competence, where several bills are pending in various state legislatures requiring licensing and registration of safety and health professionals.

ECONOMIC ASPECTS

Despite these legislative drivers, during recent years the business world has been suffering from the combined impacts of recession and increasing global competition. These pressures have resulted in a veritable tidal wave of reorganization in one form or another.

Familiar terms summarizing the response of businesses both large and small are downsizing, restructuring, or the latest one, reengineering. Added to this is the compelling need to control costs, especially costs not directly related to production, such as accidental losses or overheads.

Because of these economic drivers, the trend has been for the larger organizations to reduce corporate or divisional safety and health staffs and/or combine responsibilities with other functions such as environmental, security, and fire protection.

On the positive side, there has been a trend among smaller organizations toward establishing a safety and health function. This trend is expected to continue, and there should also be an increasing trend of firms seeking assistance from outside consultants.

MANAGEMENT RESPONSIBILITIES OF THE SAFETY AND HEALTH POSITION

The technical duties and responsibilities of this position will be many and varied and, as such, are clearly beyond the scope of this publication. The administrative or management responsibilities of this position are of critical concern.

In 1966, the American Society of Safety Engineers (ASSE) published a paper, *The Scope and Functions of the Safety Professional,* which summarized the functional activities of safety and health professionals. The paper is periodically reviewed, and no doubt it may be expanded or amended in the future in light of changing conditions.

During 1992 and 1993, the ASSE Government Affairs Committee prepared a working draft position paper, *Management Responsibilities of the Safety Professional.* In this paper, the principal management responsibilities for the safety and health professional were established as follows:

- Assist the line leadership in assessing the effectiveness of the unit safety and health programs.
- Provide guidance to line leadership and employees so that they understand the programs and how to implement them.
- Assist line leadership in the identification and evaluation of high-risk hazards and develop measures for their control.
- Provide staff engineering services on the safety and health aspects of engineering.
- Maintain working relationships with regulatory agencies.
- Attain and maintain a high level of competence in all related aspects of the profession.
- Represent the organization in the community on matters of safety and health.

Expanding this list could be an endless task; however, these seven broad responsibility areas are a good starting point.

The key to an effective management system is to ensure that the person in this position functions as a system or program administrator, not an implementor. When problems exist, it is usually found that the organization's line management has failed to understand the philosophy and role of the safety and health function. Conversely, those involved in the safety and health functions must be sure they are not immersing themselves in technical details at the expense of administering and providing support for the management system.

A particular event has stuck in my memory: When I was a young safety professional, I attended the National Safety Congress in Chicago for the first time in 1964. Concurrent with this Congress, Homer Lambie, Corporate Safety Director had assembled a large group of Kaiser Aluminum safety professionals. One of the special sessions that Mr. Lambie had organized for our group was a luncheon meeting with John Grimaldi, then safety director of the General Electric Company. From my notes on his talk, I quote Dr. Grimaldi's remarks as follows: "Insofar as the safety professional at GE is concerned, an effective safety professional is one that persuades management action rather than promotes employee awareness. The safety professional should develop decision-making information for managers to apply rather than assume responsibility for corrective action." In other words, the safety professional teaches methods of solving problems rather than providing answers. This meeting was held on October 26, 1964, and the principles are as current today as they were 30 years ago. Certainly, I have done my best to consistently live by them throughout my professional career.

To clarify this point, let us consider a typical day in the life of a full-time safety and health professional. A normal day when nothing unusual occurs should include some or all of the following activities:

- Meeting with one manager or supervisor to discuss the status and progress of his or her safety and health programs
- Conducting an audit or survey with or without members of management to assess the effectiveness of the unit's programs
- Reviewing any loss event that may have occurred and assisting line management with the investigation of those considered significant
- Working up programs such as training presentations, new or revised procedures, committee meetings, and assignments
- Conducting training programs
- Attending committee and other meetings
- Conducting research and providing professional consultation
- Developing and guiding staff members, if they are in a supervisory role

This discussion now leads us to the final element of this chapter, which is to look at actually fitting the safety and health function into the management organization.

Obviously, the variables in any organization are immense: the unit's nature of business, overall mission, leadership style, and the employee relations climate will all contribute to differences in the management structure of the organization. Despite these differences, it is still highly desirable from an effectiveness standpoint for the unit leadership to position their staff support functions in a direct reporting relationship with the chief executive or principal leader of the organization. An example ideal organization chart is shown in Figure 3.1.

In smaller units, the safety function may be grouped together with several others. The key is that safety and health is a management system, and to be effective it must be managed from the highest level in any unit organization.

The responsibility of the individual in this position is to bring to that organization a system that is designed to prevent unplanned (or, in the case of security, deliberate) accidents or incidents that can produce adversity of one kind or another for the organization. Systems necessary to achieve this objective are discussed in other chapters.

Many businesses are going though a process of reengineering in an effort to streamline their organization and related systems. The benefits of this process are cost reductions, quality improvements, time to market advantage, and greater employee job satisfaction and empowerment. These benefits lead in turn to a wide host of opportunities for improvement in the unit's day-to-day operations. Examples could be increased revenues, increased productions, improved service, increased growth potential, enhanced communications and decision making, and — above all — reduced risks.

When all these factors are considered, it is clear that the safety and health professionals will need to demonstrate superior qualities in the fields of professional competence, systems design and administration, management, and communications skills. Professional competence and skills are discussed in more detail by Dr. Roger Brauer in Chapter 52. Management and communications skills are discussed in several other chapters.

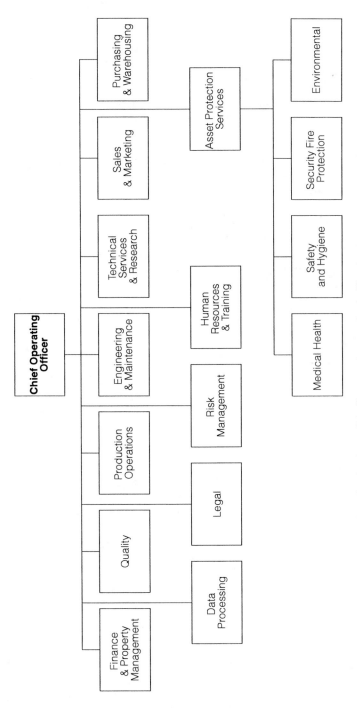

Figure 3.1   This chart indicates the position of safety in an "ideal" corporate or major facility structure.

FURTHER READING

Books

Brid, F. E., Jr. and Germain George, L. *Practical Loss Control Leadership.* Loganville, GA:
    Det Norske Veritas, 1996.
Kohn, J. P., Friend, M. A., and Winterberger, C. A. *Fundamentals of Occupational Safety and
    Health.* Rockville, MD: Government Institutes, 1996.
Hammer, M. and Champy, J. *Re-Engineering the Corporation.* New York: Harper Business,
    1993.
Petersen, D. *Techniques of Safety Management.* New York: Aloray, 1989.
Thomen, J. R. *Leadership in Safety Management.* New York: John Wiley & Sons. 1991.

Other Publications

Health and Safety Commission. *Management of Health and Safety at Work Regulations 1992:
    Approved Code of Practice.* London: British Health and Safety Commission, 1992.
U.S. Department of Labor, Occupational Safety and Health Administration (OSHA). *Safety
    and Health Program Management Guidelines: Issuance of Voluntary Guidelines. Fed.
    Regist.* 54(16), January 26, 1989.

CHAPTER **4**

Safety Standards

Peter B. Rice

CONTENTS

WHAT IS A STANDARD?

The National Fire Protection Association defines a standard as "a document, the main text of which contains only mandatory provisions using the word shall to indicate requirements and which is in a form generally suitable for mandatory reference by another standard or code or for adoption into law."

Literally thousands of occupational safety and health standards exist. Employers tend to focus on standards adopted by the Occupational Safety and Health Administration (OSHA), however, other standards do exist. Safety standards are generally those that must be followed by employers and employees in the workplace. A number of agencies and groups are involved in safety and health standards, as well as specifications that are used in the American workplace and in other countries around the world. These include the federal Occupational Safety and Health Administration

(OSHA), the National Institute of Occupational Safety and Health (NIOSH), the American National Standards Institute (ANSI), the American Society of Testing Materials (ASTM), the National Fire Protection Association (NFPA) and others.

What is the difference between a safety standard and a health standard? In the past, the delineation between safety and health standards was clearer or more clean-cut than today. Many of the new standards adopted by OSHA are actually a blend of safety, health, and even environmental issues (i.e., OSHA's Hazardous Waste Operations and Emergency Response standard found in Title 29 of the Code of Federal Regulations, Part 1910 Section 120).

OSHA AND OSHA STANDARDS

The U.S. Congress created OSHA under the federal Occupational Safety and Health Act of 1970. The Act established the legal and regulatory framework for occupational safety and health. This federal law established a national policy, agency (i.e., OSHA), and mechanisms "to assure safe and healthful working conditions for working men and women."

Since its inception, OSHA's main goal has been to protect American workers, and to ensure that every worker returns home whole and healthy every day. Because OSHA's mission is to protect American workers, it has been given a number of tasks, directives, and tools including:

- Establishing programs to encourage employers and employees to reduce workplace hazards
- Establishing responsibilities and rights for employers and employees to achieve better safety and health conditions
- Conducting research to develop innovative ways of dealing with workplace hazards
- Maintaining a reporting and record keeping system to monitor job-related injuries and illnesses
- Establishing training programs
- Developing, analyzing, evaluating, and approving state occupational safety and health programs
- Providing technical and compliance assistance, training and education, and cooperative programs and partnerships to help employers reduce worker accidents and injuries
- Developing mandatory job safety and health standards and enforcing them through worksite inspections, employer assistance, and sometimes, by imposing citations or penalties or both.

The Department of Labor promulgates safety and health standards under OSHAct, with technical advice from NIOSH and other organizations.

The National Institute of Occupational Safety and Health, is part of the Health and Human Services Department. It conducts research related to standards development, development of occupational exposure and laboratory methods, and respiratory device testing and approval.

One of the most important aspects of OSHA is the provision for approval and funding of state programs. Federal OSHA has established criteria for state plans and, after application and evaluation, may approve state plans. Once approved, the state is provided funding to implement the federal law and regulations, as well as state-specific nuances approved by federal OSHA.

When it comes to safety standards, state OSHA safety standards must be either identical to or at least as effective as federal OSHA standards and must keep pace with federal standards. State plans must adopt standards comparable to federal standards within six months of a federal standard's promulgation. Most state plan standards are very similar to federal standards, but states with approved plans may have different and independent standards.

Most of the safety and health standards now in force under OSHAct for general industry were promulgated soon after OSHAct went into effect on April 28, 1971. Those standards represented a compilation of material authorized by the Act from existing federal, state, and consensus standards (i.e., ANSI, NFPA, and ASTM). These, with some amendments, deletions, and additions, remain the body of standards under OSHA.

Categories

The OSHA standards consist of four major categories:

- Design standards that specify a particular design (e.g., guardrails, ventilation)
- Performance standards that specify a level of performance or objective that must be obtained and leaves the method for achieving it up to the employer
- Vertical standards that apply to a particular industry such as petrochemical, pulp and paper, or agriculture
- Horizontal standards that apply to all workplaces and relate to broad areas such as emergency action plans, walking and working surfaces, and sanitation

OSHA STANDARDS DEVELOPMENT

The development of standards is a continuing process. NIOSH and other organizations provide information and data about health and safety hazards, but the final authority for promulgation of the standards remains with the secretary of the Department of Labor.

In the recent years, following the adoption of OSHA, the Department of Labor has made a significant effort to increase the rate at which health standards are promulgated. These have included Subpart G of Part 1910, Code of Federal Regulations, which address OSHA Standards and Subpart Z, Toxic and Hazardous Substances.

Safety standards were the foundation of the early OSHA regulations, yet many have been amended, and some have even been repealed. OSHA does continue to promulgate and adopt new safety standards as the need arises.

Kinds of hazards that OSHA safety standards address include: harmful physical agents (i.e., noise, microwave radiation, and ionizing radiation); electrical hazards; fall hazards; hazardous waste; infectious diseases; fire and explosion hazards; machine hazards; and more.

Table 4.1   Part 1910 — Occupational Safety and Health Standards

Subpart	Safety and Health Standards
D	Walking-working surfaces
E	Means of egress
F	Powered platforms, manlifts, and vehicle-mounted work platforms
H	Hazardous materials
I	Personal protective equipment
J	General environmental controls
K	Medical and first aid
L	Fire protection
M	Compressed gas and compressed air equipment
N	Materials handling and storage
O	Machinery and machine guarding
P	Hand and portable powered tools and other hand-held equipment
Q	Welding, cutting, and brazing
R	Special industries
S	Electrical
T	Commercial diving operations

Table 4.2   29 CFR Safety Standards

Part	CFR Safety Standards
1915	Occupational safety and health standards for shipyard employment
1917	Marine terminals
1918	Safety and health regulations for longshoring
1925	Safety and health standards for federal service contracts
1926	Safety and health regulations for construction
1928	Occupational safety and health standards for agriculture

Tables 4.1 and 4.2 provide listings of OSHA's Safety Standards.

As noted earlier, literally thousands of occupational safety and health standards exist. Some of these standards apply to all employers, and they include the requirements for an emergency action plan, fire prevention plan, work space, walking-working surfaces, and means of egress. Also, hazard communication standards apply in all places of employment in which hazardous chemicals (as defined by the regulation) are found.

- The emergency action plan standard (29 CFR, 1910.38a) requires every employer to plan appropriate actions for employee safety in advance of foreseeable emergencies. It applies to all employers, private and public, irrespective of business activity, industry, or number of employees. The sole exemption is relief from the written program requirement if the employer has ten or fewer employees.
- The fire prevention plan standard (29 CFR 1910.38b) requires every employer to evaluate potential fire hazards and establish fire prevention requirements, including safe work practices and preventive maintenance on fire suppression systems and portable equipment. It applies to all employers, private and public, irrespective of business activity, industry, or number of employees. The sole exemption is relief from the written program requirement if the employer has ten or fewer employees.

- Walking-working surfaces (29 CFR 1910, Subpart D) is applicable for employee-occupied areas. This standard applies to all permanent places of employment, except where only domestic, mining, or agricultural work is performed, and requires employers to ensure safety with respect to: housekeeping; worksurfaces; aisles and passageways; covers and/or guardrails to protect personnel from the hazards of open pits, tanks, vats, ditches, etc.; and floor loading protection.
- Means of egress (29 CFR 1910, Subpart E) requires that every building or structure, new or old, designed for human occupancy be provided with exits sufficient to permit the prompt escape of occupants in case of fire or other emergency.

Federal OSHA standards and, in some cases, state plan standards can be found in several locations including public libraries, the Government Printing Office, or on the web at www.osha.gov or state plan Web sites. You can contact OSHA directly at: U.S. Department of Labor, Occupational Safety and Health Administration (OSHA), Directorate of Safety Standards Programs, 200 Constitution Avenue, N.W., Washington, D.C. 20210, (202) 693-2277.

As noted earlier, a number of other parties assist OSHA in the standards-setting process. These include NIOSH, state and local governments, employer or labor representatives, other interested parties, and nationally recognized standards-producing organizations. Standards and specifications organizations produce standards and basic reference materials for safety professionals who will frequently be confronted with hazards and safe work practices and conditions. They are also used by OSHA and other agencies in developing required standards.

American National Standards Institute (ANSI)

Two of the most important, nationally recognized ANSI standards and specifications-producing organizations, which have a large impact in the adoption, amending, or repeal of OSHA's standards, are:

1. Washington, D.C. Headquarters
    1819 L Street, N.W., 6th Floor
    Washington, D.C. 20036
    Tel: 202-293-8020
    Fax: 202-293-9287

2. New York City Office
    25 West 43d Street, 4th Floor
    New York, NY 10036
    Tel: 212-642-4900
    Fax: 212-398-0023

The Institute's Web site is www.ansi.org.

ANSI has served in its capacity as administrator and coordinator of the United States private sector voluntary standardization system for more than eighty years. Founded in 1918 by five engineering societies and three government agencies, the Institute remains a private, nonprofit membership organization supported by a diverse constituency of private and public sector organizations. Throughout its history, the

ANSI Federation has maintained as its primary goal the enhancement of global competitiveness of U.S. businesses and the American quality of life by promoting and facilitating voluntary consensus standards and conformity assessment systems while promoting their integrity. The Institute represents the interests of its nearly 1000 company, organization, government agency, institutional, and international members through its office in New York City and its headquarters in Washington, D.C.

ANSI does not itself develop American National Standards (ANS), but it facilitates development by establishing consensus among qualified groups. The Institute ensures that its guiding principles — consensus, due process, and openness — are followed by the more than 175 distinct entities currently accredited under one of the Federation's three methods of accreditation (organization, committee, or canvass). In 1999 alone, the number of American National Standards increased by nearly 5.5% to a new, approved total of 14,650. ANSI-accredited developers are committed to supporting the development of national and, in many cases, international standards addressing the critical trends of technological innovation, marketplace globalization, and regulatory reform.

ANSI promotes the use of U.S. standards internationally, advocates U.S. policy and technical positions in international and regional standards organizations, and encourages the adoption of international standards as national standards where they meet the needs of the user community.

ANSI has dozens of occupational safety specification standards including personal protective equipment, machine tool safety, respiratory protection, and many others.

American Society for Testing and Materials (ASTM)

ASTM is the world's largest source of voluntary consensus standards for materials, products, systems, and services. Currently, more than 8000 ASTM standards exist. ASTM membership is drawn from a broad spectrum of individuals, agencies, and industries concerned with materials, products, and systems. The 30,000 members include engineers, scientists, researchers, educators, testing experts, companies, associations and research institutes, governmental agencies, as well as departments from federal, state and municipal groups, educational institutions, consumers, and libraries. Many ASTM standards are published for occupational safety and health.

ASTM sponsors committees on geothermal resources and energy, quality control, food service equipment, and protective coatings. The committee on consumer product safety, for example, has helped develop standards that will assist in protecting the public by reducing the risk of injury associated with the use of consumer products such as cigarette lighters, bathtubs, shower structures, children's furniture, and other products.

ASTM standards are of interest to the safety professional because they identify hazard areas and establish guidelines for safe performance, similar to ANSI standards. The society also publishes standards for atmospheric sampling and analysis, fire tests of materials and construction, as well as methods of testing building construction, fatigue testing, radiation effects, pavement skid resistance, protective equipment for electrical workers, and others.

The ASTM address is 100 Barr Harbor Drive, West Conshohocken, PA 19428-2959. ASTM's Web site is www.astm.org.

Fire Protection Organizations

Fire Protection Organizations also contribute helpful information in the standards development process.

Factory Mutual Research

At Factory Mutual Research, a not-for-profit scientific research and testing organization managed by Factory Mutual Global, the members' common goal is preventing and controlling property loss. Factory Mutual Research works with commercial and industrial clients to solve problems through the unique interaction of its three operating divisions: Standards, Research, and Product Certification (approvals).

Research results set new standards that advance loss prevention practices and help develop new products. In addition, manufacturers use Factory Mutual Research's independent, third-party testing laboratory services to obtain approval certifying the reliability of their products and services.

Guided by more than 165 years of loss prevention experience, Factory Mutual Research can help businesses stay competitive on a local, national, or even international scale.

Factory Mutual Research is located at 1301 Atwood Avenue, Johnston, RI 02919, phone: (401) 275-3000, fax: (401) 275-3029. The Web site is www.fmglobal.com.

National Fire Protection Association (NFPA)

For more than 100 years, the NFPA has been developing and updating codes and standards concerning all areas of fire safety. An international, nonprofit organization with more than 65,000 members from seventy nations, NFPA's mission is to reduce the burden of fire on the quality of life by advocating scientifically based consensus codes and standards, research, and education for fire and related safety issues. While the NFPA is involved with extensive fire research and produces numerous fire safety educational programs and materials, its lifeblood is its codes and standards making system.

Currently, more than 300 NFPA fire codes and standards are used throughout the world. Some examples of these documents include:

- NFPA 1, Fire Prevention Code addresses basic fire prevention requirements necessary to establish a reasonable level of fire safety and property protection from the hazards created by fire and explosion.
- NFPA 54, National Fuel Gas Code provides safety requirements for fuel gas equipment installations, piping, and venting.
- NFPA 70, National Electrical Code addresses proper installation of electrical systems and equipment.
- NFPA 101, Life Safety Code provides minimum building design, construction, operation, and maintenance requirements needed to protect building occupants from the dangers of the effects of fire.

Today, virtually all of NFPA's codes and standards today affect every building, process, service, design, and installation in society. One reason these documents have been so widely accepted and adopted is because of the unique, open process under which they are developed and updated.

The NFPA is located at 1 Batterymarch Park, P.O. Box 9101, Quincy, MA 02269-9101, phone: (617) 770-3000, fax: (617) 770-0700. NFPA's Web site is www.nfpa.org.

Underwriters' Laboratories Inc. (UL)

UL is an independent, not-for-profit product safety testing and certification organization. UL has tested products for public safety for more than a century. Each year, more than 16 billion UL Marks are applied to products worldwide.

Since UL's founding in 1894, it has the reputation as the leader in U.S. product safety and certification. Building on its household name in the United States, UL is becoming one of the most recognized, reputable conformity assessment providers in the world. Today, UL's services extend to helping companies achieve global acceptance, whether it is an electrical device, a programmable system, or a company's quality process.

UL is located at 333 Pfingsten Road, Northbrook, IL 60062-2096, phone: 847-272-8800, fax: 847-272-8129. The Web site is www.ul.com.

## CHAPTER 5

Engineering Design and Construction: Safety Management Aspects

Steven M. McConnell

CONTENTS

1-56670-370-0/02/$0.00+$1.50
© 2002 by CRC Press LLC

INTRODUCTION

Too many injuries (and fatalities) occur in construction. The construction industry is second only to the mining industry in the number of fatalities per 100,000 workers. Frequency and severity rates in construction far exceed those of most other industries as well. It is not surprising that many people have come to accept that injuries, even fatalities, are a part of doing construction work. That concept represents a paradigm — and it is time for us to realize the need for a shift in this paradigm. Most safety professionals would agree that the hazards faced by workers in construction are more numerous and are often more serious than those faced by workers in most other industries. That doesn't mean we have to accept a higher rate of injuries. Our goal remains the prevention of injuries and illnesses. Our challenge is to take the term prevention beyond reduction; we should be focused on elimination.

The term "Construction" represents a broad scope of activity including new building, renovation, demolition, and tenant improvement work. Contractors run projects that vary in size from very small to extremely large. The amount and effectiveness of training and education for individual construction workers varies greatly — sometimes more significantly in nonunion environments. Workers come and go with regularity on a typical construction project. Payment to the contractor for work performed on a project is often based on adherence to strict project schedules, which sometimes tempts the contractor to shortcut safety measures in an attempt to gain time in the schedule. Combine these issues with the many inherently dangerous tasks performed on a construction project, and it is easy to conclude that managing safety in the construction industry is a difficult and dynamic process. But the safety professional can be successful by employing sound management techniques.

Safety professionals share a common frustration associated with reacting to events that are often the result of a design error or oversight. This premise is encountered in the construction industry as well as other industries, and it illustrates the importance of safety being associated with the design phase of a job.

Architects and engineers play an important part in a job's safety potential both during and after construction. Input from the safety department, in the form of comments from design reviews, etc., is necessary to help identify and correct potential hazards before actual construction takes place. However, because of the volume of drawings and submittals associated with most projects, the safety professional often misses the opportunity to provide valuable input.

When time (or ability) constraints prevent your full involvement in the review process, the conceptual/design stage of a job provides the best opportunity to identify and control many of the serious, potential hazards that may exist. For example, this is the perfect time to consider how fall protection requirements can be met on a job. When associated needs are identified up front and added to the plan, the provision of fall protection can be built into the job. This is far more desirable and effective than trying to address these concerns on a haphazard basis during the construction phase of a job.

Each project is unique and may present a different set of concerns, but basic questions should be asked of the design/engineering firm and/or the contractor to

Table 5.1   Questions to be Asked during the Conceptual/Design Stage of a Job

Is trenching or excavation work in excess of five feet required for the job? How can workers be protected?

What types/classes of soil are found on the site?

Where is fall protection going to be required, and how will the requirements be met?

Will lifting equipment such as boomtrucks, cherry pickers, or cranes be operated in near proximity to overhead electrical lines? What about buried utilities?

When is it safe to close in a building, and how will emergency rescue/evacuation be performed?

Is the work scheduled in a manner that will avoid stacking of contractors?

Has the need to use hazardous chemicals been evaluated and their use eliminated where possible or less hazardous materials substituted where possible?

Is there a potential for employee exposure to ACM, lead, silica, etc., and, if so, can the exposure be eliminated or reduced?

Has adequate space been designated for a material storage area or boneyard?

Are stairs, ramps, building paraphets, etc., designed to code?

Where will confined space work activity take place and what type of monitoring is needed?

ensure that those aspects of the job are assessed. Examples of the type of questions to be asked are given in Table 5.1.

It is recognized that the method for managing the safety program may vary depending on whether you represent the client or the contractor. This chapter will attempt to speak generally with respect to construction safety management techniques. Some specific guidance will be given representing the viewpoint of the safety professional providing services for the client and the contractor.

MAKING THE SAFETY PROGRAM WORK

Before moving to a discussion of safety management from the perspectives of the client and contractor, it is important to establish what makes a safety program successful in a construction environment. In the author's view, three important elements must occur before the safety program can be successful on a project:

1. Pre-job planning
   - comprehensive written safety program
   - plan key work activities
   - develop an alcohol and substance abuse program
   - employee orientation program
2. Defined accountability and responsibility
   - field supervisors through project management
   - workforce
3. Strong, sincere support from the superintendent

Each of these components is worthy of additional discussion.

Pre-Job Planning

The need to preplan the job is obvious to safety professionals. This is the timeframe to write the site-specific safety program that will govern the work activity during the job. The program must be comprehensive and based, at minimum, on the applicable regulations. The key to a good program is to achieve balance between making the program comprehensive while still being a usable document. The requirements and rules must be stated clearly and concisely, brevity will promote understanding by anyone using the program.

The preplanning phase is the opportune time to develop your plans to safeguard workers from hazards associated with various facets of the job. Consider where fall protection will be needed during different phases of the job, and plan for it. Consider excavation and trenching requirements and how you will protect the workers. Consider scaffolding requirements and work in confined spaces. It is your opportunity to identify these types of complex hazards before being confronted with them on the job. Preplanning in these areas is essential to job site safety.

Many contractors have programs established to control alcohol and substance abuse on the job. The details of the program should be reviewed during the preplanning phase of the job. The program should be written, and the contractors coming onto the job must be required to adopt it or have their own programs that are at least equivalent. Points to consider include pre-job screenings of workers, random and postaccident testing, site inspections for illegal drugs and alcohol and disciplinary measures levied against those who violate the policy. Legal counsel must review the alcohol and substance abuse program prior to its implementation.

Every employee coming onto the job requires an orientation. The safety professional needs to be involved in the development of the orientation program. Indeed, in many cases the safety professional is given the task of developing and presenting the orientation. This is your opportunity to speak. It is your chance to tell the craft and project management what is expected of them on the project site. It is your opportunity to highlight key expectations and areas of concern. Most important, it is your opportunity to impart your philosophy about injury prevention. This is where you effectively tell people what your safety program is all about. Avoid leaving anything on the table — let people know what is expected.

Defined Accountability and Responsibility

The safety professional must work with project management to establish clear roles and responsibilities. It must be clear, from the beginning, that the responsibility for safety on the job site belongs to project management. The safety professional does not own that responsibility. The safety professional facilitates the safety program; project management owns it. The adage rings true, "What the boss pays attention to is what gets managed." If the superintendent does not own safety, project management will not give safety the needed amount of attention. They will manage what their boss pays attention to instead. Safety professionals are in a position of convincing the top manager of this fact. Management doesn't take ownership of safety until safety is the first thought in their minds. Take an example of a very tedious, complex lift to be made on the job. The manager who has truly taken ownership of safety asks, "How can we make this lift safely?" while the manager who hasn't taken on ownership simply asks, "How can we make this lift?" When management owns safety, the manager will react to an injury by stopping whatever he or she is doing in order to respond to the scene and to begin an immediate investigation into what happened. Those who have taken ownership care. Clearly defined roles and responsibilities will help establish this culture. Real accountability will sustain it.

Strong, Sincere Support from the Superintendent

You know that you have the superintendent's support when the crafts understand that, when the safety professional is talking, it is really the superintendent talking. The safety program will ultimately fail without strong support from the superintendent. If the superintendent and the safety professional are not on the same page, the result will be inconsistent messages in the field. The safety professional cannot succeed on a job if the superintendent contradicts his or her directions. This same page relationship is not easy to develop, especially on short-term jobs. Still, the safety professional must make the effort to make it happen. The pre-job phase of a project is a stressful, busy time for the superintendent. Still, you must get time on the superintendent's calendar to discuss the philosophy of the safety program for the job. You have to make it happen. It may be in the parking lot after hours or at 5:30 a.m. before others get to work. If you want to improve your chances of success on the job, make the extra effort to make it happen.

Safety professionals fail miserably when the workforce views them as the safety cops, whether by the craft or management. Avoiding that perception is achieved through a combination of the three elements discussed previously in this section. Credibility is established with project management during the pre-job planning phases of a job. You earn respect at this stage by supplying quality input to the job planning effort. You also earn respect by demonstrating your willingness to be a flexible, team player. The workforce will recognize whether you have the respect of management. When you do, it is much easier to build that respect with the crafts. When you do not, it is an uphill battle in the field.

The employee orientation program helps avoid the safety cop syndrome by serving as a vehicle to highlight your responsibilities. The orientation should inform personnel that you are responsible for identifying hazards and establishing appropriate controls, as well as enforcing adherence to those controls. This message must be stated clearly so that everyone will know the job duties of the safety professional. Being professional on the job site and addressing workers appropriately when non-compliant behaviors are discovered also helps the safety professional to be perceived as someone who is doing his or her job instead of someone who is acting as the safety cop.

The third element, strong and sincere support from the superintendent, if achieved will help you avoid the perception of being the safety cop because people will have respect for the superintendent who owns the safety program. Most people will not transfer that respect to you until your own behavior changes their opinions.

The safety professional should strive to build a positive rapport with people on the job site. Make compliments about the good practices you encounter and enforce the rules in a consistent manner. Be up front with those you encounter on the job site. Never smile unwittingly and tell someone that your inspection went okay — and then blindside him or her with a negative report. If it didn't go well, take the time to explain the issues and offer some suggestions to help them improve the work activity. Even better, invite them to accompany you on the next inspection.

OWNER CONTROLLED INSURANCE PROGRAM (OCIP) VERSUS CONVENTIONAL INSURANCE PROGRAMS

The primary reason that the OCIP method is selected is cost savings. Most of the savings realized from the OCIP occur because the project owner buys the required insurance in bulk, eliminating the premium redundancy and duplications in coverage. It also eliminates contractor mark-ups associated with insurance coverage (often 5 to 10 percent) and reduces the number of worker's compensation and third-party liability claims because the owner can implement an effective safety program. The owner has direct control over the selection of the insurer and, typically, only one insurer is selected for the workers' compensation package. The owner is in a good position to monitor the performance of the insurer and its financial solvency. This is an obvious advantage compared to the project insured under the conventional method.

Under an OCIP, the general contractor is usually responsible for project safety. The related impact of an OCIP on the safety professional depends on whether the individual works for the general contractor or a supporting subcontractor. In the first case, the individual has a more authoritative role while, in the latter, he or she is in the position of ensuring their company is complying with the requirements of the OCIP.

Under an OCIP, the general contractor creates a safety program that includes:

- Formalized safety program
- Written, site-specific safety plan
- Contractor prequalification program
- Worker safety training
- Hazard identification and control program
- Work monitoring
- Periodic toolbox meetings
- Site safety staffing requirements
- Criteria for response to safety deficiencies
- Safety orientation
- Safety audits
- Drug and alcohol testing program

The requirements of a contractor's safety program should be stated in the contract. The safety professional, whether working for the general contractor or a subcontractor, needs to know (in detail) what the requirements are. At the same time, the owner usually retains the right to make the job site safe.

Although a contractor may claim that an OCIP safety program disrupts the contractor's own safety program, in reality the opposite is true. In many cases, the contractor's program is improved. The general contractor performs extra work associated with monitoring its subcontractor's safety activity, and that can lead to improved safety on the job.

The owner must manage all aspects of an OCIP effectively in order to realize the advantages, and that includes project safety.

FROM THE CLIENT'S PERSPECTIVE

If you manage the safety function for the client, you are interested in completing the job on time and within budget while minimizing job-related accidents and injuries. It is important to stress the safety-related aspects of the job with the contractor(s) from the beginning (the bid package stage). Safety-related requirements expected of the contractor(s) should be given to those bidding the project at the pre-bid meeting. Project management for your company must support you in this process by emphasizing project safety concerns when discussing project details. Project management's efforts to impart the importance of safety on the job are vital to the overall success of the program. This is the appropriate time to go beyond the boilerplate message and specify what is expected in terms of safety on the job and the safety-related documentation or data that will be required from the contractor, including the information you want included in the contractor's bid for the project. Typical items to request and discuss are illustrated in Table 5.2. This is the time for project management to outline clear roles and responsibilities for safety on the job.

It is generally recommended that a safety program be managed using only one set of safety rules for the project — yours. The contractor(s) must be made aware of this concept and any requirements that exceed OSHA requirements should be highlighted in the bid process. This practice will eliminate surprises later and allow the contractor to accurately bid the project. This may represent your best opportunity to convince the contractor that your company is serious about job site safety and health. Provide copies of specific requirements to the contractor; post them on site once the job begins.

As the safety professional representing the client, it is important to establish how you will assess the safety efforts of the contractor(s). You are responsible to provide oversight of their efforts as they relate to job site safety. Develop and communicate your plan to oversee their efforts, explain to them what reports you will generate and distribute, and explain the response mechanism required of the contractor. Set the tone for how job site safety will be managed.

FROM THE CONTRACTOR'S PERSPECTIVE

The safety professional for the contractor also wants the job completed on schedule, within budget, and without accidents and injuries. A good safety record makes it easier for the company to obtain work and obviously translates into higher profits. Safety performance has become a key job qualification for contractors and it is usually the safety professional who is responsible for providing safety performance-related

Table 5.2 Items Requested from Contractor

Safety professional's general expectations regarding project safety
Contractor's EMR for past three years
Copy of the contractor's OSHA citations for past three years
Information regarding fatalities or serious injuries for the past three years, including control
  measures implemented
Contractor's written safety program, if one exists
Qualifications of the safety professional and the amount of coverage anticipated for the job
Methods to be used to control anticipated hazards:
  • Job site inspection program
  • Accident investigation program
  • Fall protection
  • Lockout, tagout program
  • General electrical safety
  • Hazard communication program
  • Confined space program
  • Respiratory protection program
  • Scaffold plan
  • Trenching/excavation plan
  • Employee safety orientation program
  • Employee training/education program
  • Subcontractor orientation program
  • Names of key project personnel
  • Name of competent person(s)
  • Disciplinary action program
  • Client oversight/participation in safety plan

information requested by the client; he or she should be well versed with regard to the company's safety performance and programs. The safety professional can best serve the company by anticipating client needs and developing a site-specific program for the subject project. While the site-specific program may include many generic components, it is important for it to be directed toward, and address the needs of, the subject project. The safety professional should be prepared to provide and discuss the things requested by the client. Typical items to be presented by the contractor are listed in Table 5.3.

Table 5.3 Items Presented by the Contractor

Company's safety policy statement
Written safety program/site-specific program to follow
Safety performance history, including accident statistics
Company's OSHA record
Job site hazard identification/control program
Hazards anticipated on the job
Methods used to control anticipated hazards
Oversight of subcontractor safety
Disciplinary action program
Employee orientation/education program
Methods to document safety program efforts

The contractor's safety professional should be sincere in his or her attempt to explain the company's safety program and performance record, and cooperation with the client's representative should be stressed. This relationship will last throughout the project, and it helps to get started on the right foot. The contractor's safety professional should build positive rapport with the client's safety representative and indicate a willingness to achieve compliance with the site safety expectations.

THE MULTIPLE EMPLOYER SITE

It is common for a prime contractor to use several subcontractors to complete a project. Clients sometimes bid projects in phases, which creates a multiple employer site. The need for one set of rules becomes even more important in these conditions.

Managing safety on these projects becomes even more difficult, at least administratively. Establishing an effective paper trail is necessary to ensure that appropriate documentation exists and that appropriate people are informed. Guidelines must be developed to ensure that safety-related documentation is maintained accurately and is appropriately distributed. It is a good practice to maintain safety performance-related statistics (both for the project overall and by individual contractor) that are updated on a regular basis, perhaps monthly. Figure 5.1 illustrates a form that can be used to summarize safety performance for both the job and individual contractor on a monthly basis.

A method must be developed to ensure the consistent enforcement of safety requirements and the fair, uniform use of disciplinary action. Intentions regarding enforcement (or oversight) efforts should be clearly communicated to the contractor(s) before actual work begins on the project. Again, there should be no surprises once the work begins.

It is important for the safety professional to recognize that responsibility, and subsequent liability, for job site safety can be shared by more than one contractor or company, at least in terms of an OSHA inspection. For example, on a particular job site, one contractor is assigned responsibility to install and maintain handrails (on stairs, overhead work platforms, etc.) regardless of who employs the individual(s) using the work area. When OSHA visits the site, the inspector observes a pipe fitter, who is not using fall protection, welding a pipe approximately twenty feet above the ground while standing on a job-built wooden platform that is missing part of the required handrails. In this case, there may be three citations issued. Potential citations would be issued to the:

- Prime or general contractor because they are viewed to have control of the job
- Subcontractor assigned responsibility for work platforms because the omission of a portion of the handrails created a potentially serious hazard
- Subcontractor employing the pipe fitter because they allowed the individual to work in the area where the unsafe condition existed

In other scenarios, the owner/client may share in liability, particularly in third party lawsuits. Indemnifications are not always viewed as viable in the courtroom.

Monthly Safety Performance Review

Project Name: _____

Client Contact: _____

Project Manager: _____   Month: _____

ES&H Manager: _____   Year: _____

Job Site Telephone: _____   T.C. Rate: _____

OSHA Telephone: _____   LWD Rate: _____

Contractor Performance

Contractor	Site Contact/Phone	T.C. Rate/LWD Rate
_____	_____	_____
_____	_____	_____
_____	_____	_____
_____	_____	_____
_____	_____	_____
_____	_____	_____

TOTALS _____

RATES

$$\text{T.C.} = \frac{\text{Total Recordable OSHA Cases} \times 200,000}{\text{Total Hours Worked}}$$

$$\text{LWD} = \frac{\text{\# LWD Cases} \times 200,000}{\text{Total Hours Worked}}$$

Figure 5.1   Sample monthly safety performance review.

It is important for the safety professional to discuss the liability issue with the company's senior management and legal counsel in order to understand exactly where the company fits into the situation and how oversight efforts will be handled.

General Site Safety

Whether you represent the client or the contractor, certain things need to occur in order to implement an effective site safety program. The next logical step following contractor selection is the pre-construction meeting. This type of meeting has been common for years, and it is held primarily to discuss the scope of work in detail with each contractor. In recent years, safety has become a highly visible agenda item for these meetings. Potential contractors were informed in the pre-bid stage that certain safety-related information and programs would be required of them — this is the time for delivery. The scope of work as it relates to safety is usually addressed in detail at the pre-construction meeting; specific attention is given to anticipated hazards that may create the potential for serious or fatal injury. These hazards include those items listed in Table 5.1 and other project-specific tasks that may be identified. The method(s) the contractor will use to control these hazards should be discussed in detail and agreed upon. These will be the standards to which the contractor is held when work activity begins.

It is necessary to reach agreement on compliance efforts at this stage. In many cases it is possible to achieve compliance in a number of ways; whether you represent the client or the contractor, you want to establish the method(s) for ensuring compliance on a particular job.

Companies use a variety of techniques to establish performance criteria on a job. An effective strategy used by the Brown and Root Building Company requires a contractor to provide a model for compliance. For example, the subcontractor with the most scaffold work on a job is required to build a scaffold model (i.e., full size — two tiers high by 16 feet long), which is erected adjacent to the construction office. The model serves as the standard for scaffolding used on that job. Subcontractors have no excuse for erecting scaffolding that does not meet the job standard. The same system could be used to establish job site standards in other areas including job-built stairs, ladders, and trenching operations. Everyone benefits when such standards are established for the job *before* breaking ground on the project, and personnel safety is enhanced.

The pre-construction meeting is also an effective opportunity to address safety-related documentation requirements, client–contractor interface, and common safety concerns such as housekeeping and related requirements. This meeting provides the opportunity to review specific documentation requirements. Figure 5.2 illustrates the type of documentation that may be introduced at this stage.

The pre-construction meeting is also used to establish the contractor responsible for adequately orienting its subcontractors to site safety requirements before work begins on the site. The objective is to minimize accidents and injuries on the project; subcontractors can not be overlooked in the process.

Progress Review Meetings

The safety progress review meeting is an effective tool for improving safety once the project is underway. These meetings occur on a regular schedule, usually

---

TRENCH & EXCAVATION INSPECTION LOG

Date: _____ Inspector: _____ Job: _____

Contractor: _____

Trench/Excavation Description: _____

Soil Description: _____

Visual Soil Test Made: Yes ☐ No ☐ Type: _____

Protective System Used: Sloping ☐ Box ☐

      Wood Shoring ☐ Hydraulic Shoring ☐

Conditions: Wet ☐ Dry ☐ Submerged ☐

Hazardous Atmosphere? Yes ☐ No ☐

(Confined Space Procedure Mandatory if Yes)
Trench / Excavation Dimensions:

Length _____ Width _____ Depth _____

GENERAL ITEMS – WRITE COMMENTS ON BACK

Item	Yes	No
Adequate access	☐	☐
Spoil pile away from edge	☐	☐
Utilities or structures secured	☐	☐
Materials stockpiled away from edge	☐	☐
DOT vests worn as needed	☐	☐
Barricades used where needed	☐	☐

Figure 5.2 Sample trench and excavation inspection log.

monthly, and are attended by contractor site managers and their safety professionals. The owner (in the case of an OCIP) or the general contractor typically conducts the meetings in the absence of an OCIP. The purpose is simple — to review project safety and recommend enhancements.

These meetings provide a forum for the owner or general contractor to critique a contractor's safety effort and/or project safety as a whole. Contractors may be asked to comment on the following:

- Injuries and incidents
- Project accident statistics
- Specific project performance measures
- Scheduled significant activities

This type of meeting also helps to improve consistency in project safety. Management representatives and site safety professionals are present to hear and discuss project safety. Project review meetings are a mechanism for two-way communication.

All of these efforts combine to establish the mechanics and criteria of the site safety program. When addressed at or before the pre-construction meeting, the job can begin without confusion regarding safety requirements for the job.

THE DISCIPLINARY ACTION PROGRAM

Behavior-based safety and positive recognition programs continue to gain popularity in construction, but the implementation of a fair, progressive disciplinary action program should not be overlooked. The typical construction worker booms from one job to the next with regularity; most workers never align themselves with one company long enough to develop loyalty, whether they want to do so or not. For them, violating a safety requirement may seem trivial. In fact, some may do so purposely because they view it as the macho thing to do. Walking the steel rather than cooning a beam is a classic example. Such behavior, of course, can result in catastrophic consequences for the worker. But even short of death or serious injury, this type of behavior can be detrimental to the company. When an OSHA inspector observes noncompliance they usually look to cite the employer and possibly others, as previously discussed.

An effective, consistently implemented disciplinary program can be beneficial in two ways. First, the action (though viewed negatively) may reduce the potential for future noncompliances by the employee and co-workers. Second, documentation related to the program and its enforcement site-wide may persuade OSHA not to issue a particular citation. If a citation is issued, this documentation may sometimes be used effectively during an appeal process. It may help get the violation removed or the monetary fine reduced.

The disciplinary program should be communicated to all contractors before they begin work on the project. The more successful programs include progressive disciplinary action. In such programs the employee receives more significant discipline for each violation. A typical progression of discipline would be as follows:

- First offense: verbal warning, documented
- Second offense: written warning
- Third offense: suspension of employee
- Fourth offense: termination

In practice, disciplinary programs are implemented in a variety of ways. Some companies eliminate the verbal warning, choosing instead to issue a written warning on the first offense. Depending on the severity of the noncompliance, some companies suspend or terminate employees on the first offense. Regardless of how such a program is implemented, it is important that all parties understand it from the beginning of the job. The primary key to success is having the program written, communicated, and enforced uniformly on the job. Enforcement actions should be documented in writing.

PROGRAM GOALS AND MEASUREMENTS

Similar to managing the function of safety and health in other industries, successful construction safety programs have measurable goals and performance measures. In many cases, program goals may come from the home office, but on-site managers may also have input or may even develop additional goals themselves. It is helpful to get your job superintendent or top official involved in the formation of a site-specific safety goal. Strive to get them to own the goal.

The importance of goals and measurements is well documented, but it has not always been common to establish them in the construction industry. With the realization that workers' compensation costs were reducing profits and that higher EMRs meant less work coming in, companies have escalated their loss-control efforts. Attention to meaningful goals and performance measures is a part of that escalation.

At Bechtel, for example, a zero accidents program was implemented in January 1993. Not every company chooses to implement goals this ambitious, but positive results can be realized nonetheless. Employees pay attention to things that get measured.

Much can be written about the zero-injury philosophy. Some people are not convinced that the achievement of zero injuries on a project is possible. Most safety professionals accept the concept that injuries are preventable. It is not a far stretch to adopt the philosophy of zero injuries. When we establish goals relative to safety performance, we need to remember what those goals really mean. For example, if 50 people were injured during a given year, to the extent that they were off the job or working under medical restrictions, and we establish a goal to reduce the number of those types of injuries by 20% the next year, aren't we really saying it is okay to injure forty people seriously enough that they become a L/RWD statistic? Do we really want to start the year by stating we hope to injure forty people?

It is more proactive to set the goal at zero and to treat any subsequent injury as an anomaly. Investigate any injury that does occur and develop controls to prevent reoccurrence. Communicate what is learned in the investigation, implement the necessary controls, and reset the goal for zero.

Program measurements should be established early and communicated to all contractors at the start of the job. Contractors should be responsible for the safety performance of their short-term subcontractors because they oversee their work daily and are responsible to orient them to the safety requirements of the project site. All parties must understand this relationship.

It is common to see job site statistics, in the form of frequency and severity rates, reported as both total job performance and by individual contractor (see Figure 5.1).

In many cases, other meaningful data are measured as well. These may include performance measurements such as total injuries, a company's or supervisor's compliance with conducting safety meetings, training/educating employees, safety sampling, conducting required site safety inspections, and hazard abatement. A number of measurements, in addition to injury/illness rates can be used to assess safety performance, for both management and nonmanagement employees (see Figure 5.3).

Whatever the measurement(s), specific criteria should be established so that accountability for safety performance can be appropriately assigned. Timely feedback, which details experience on the job including a comparison to project goals, should be given to employees and management. Evaluation of the safety and health program is recognized as the final process of hazard control and is important to individuals interested in enhancing the program's effectiveness.

HAZARD IDENTIFICATION AND CONTROL

A basic component of the program, hazard identification and control, is at the heart of the loss-control effort. Activities on construction projects occur and change rapidly and can result in potentially serious hazards at any time. It is common to have several hazardous operations underway at one time on a construction site. Identifying hazards and keeping track of abatement efforts can be a difficult task. Systems may be established to run either manually or by computer to record and track site hazards (see Figure 5.4). Desired data can be obtained and tracked for the site as a whole, by building, by contractor, by supervisor, etc. The information to be gathered is determined by you, the safety professional, but the reader is cautioned to determine the uses ahead of time in order to be sure the needed data is obtained. Figure 5.4 illustrates a hazard-tracking log. The sample log in Figure 5.4 can be modified as necessary to better fit the needs of your specific job. For example, you may want to add a column to assign responsibility for corrective action. Computerized tracking systems offer obvious advantages to you.

The tracking system provides a closed loop mechanism for managing the disposition of identified job site hazards. If inspected by external agencies, this information can be used to illustrate your hazard control efforts. An inspector will usually be favorably impressed by the fact that potential hazards and corrective measures are routinely identified and monitored until acceptable corrective action is achieved.

The system can also help identify problem areas or problem contractors on the job. With this knowledge, specific action can be taken to address such problems on a proactive basis.

The purpose of tracking hazard identification and abatement efforts is three-fold:

- Potential problem areas (work activity) can be identified.
- The data provide a method of "trending" site performance.
- Documentation serves as a "closed loop" approach to hazard control.

Inherently, someone is identified as being responsible for locating job site hazards. It is often you, the safety professional. It may be appropriate to provide yourself

CONTRACTOR _____ MANAGEMENT REP _____ MONTH/YEAR _____

INJURY DATA

Measure	Month	Year-to-date
# Reported injuries		
# L/RWOs		
# Open cases		
# Effort		
Frequency rate		
Severity rate		

JOB HAZARD ANALYSES

	Yes	No
Written when required		
Proper signature chain		
Adequate detail		
Posted at work activity		
Reviewed by employees		
Submitted timely		

GENERAL CONTRACTOR SAFETY INSPECTIONS

# of deficiencies		
# of deficiencies mitigated		
# open deficiencies		
Timely response		
Proactive program		

CONTRACTOR SAFETY PROGRAM

Measure	Yes	No
Daily prework briefings completed		
Weekly toolbox meetings documented		
Accidents investigated/controls implemented		
Documented safety inspections by contractor		
Equipment certifications current		
Required permits obtained/current		
Confined space procedure followed		
Timely reporting of incidents to GC		
Enforces program requirements with subs		
New hire orientation complete and timely		
Reps/workers adhere to safety requirements		
Management promotes safety effectively		

GENERAL COMMENTS

Completed by _____ Title _____ Date _____

Figure 5.3 Sample safety performance status sheet.

Project: _____

Date Found	Reference	Contractor	General Description	Action(s) Taken	Date Closed

Figure 5.4 Sample hazard identification and tracking log.

with assistance. Teams can be established (labor/management safety committees, foremen, etc.) to inspect sites. Supervisors and even the job superintendent can also participate in site safety inspections. This activity is encouraged, but when it occurs, make sure the purpose of the walk-through is to inspect the site for safety. Too often, these efforts fail when the key individual gets hung up or sidetracked with a schedule or material-related problem because the focus is removed from site safety.

Inspection results should usually be documented; it will provide you with a history of the job and documentation that may later prove useful if OSHA visits the site. Consideration should be given to developing a system, which requires contractors to respond, in writing, relative to corrective actions taken to mitigate identified hazards.

Many safety professionals question whether a checklist should be used during the inspection process. Whereas this author generally discourages the use of checklists, it is acknowledged that they may be of use in some circumstances, particularly when the individuals performing the inspection are not trained safety professionals or are new to the field of hazard control. In either case the checklist should be viewed as a tool to help complete the inspection process but not as the final inspection report. Checklists offer some degree of value to remind inspectors of the various topics they need to consider on the job site during the inspection, but they can be difficult to complete, especially when set up in a yes/no format. For example, a typical checklist might include the question "Are *all* electrical cords maintained in good condition?" If the inspector observed fifty or sixty cords on the site and found only one in disrepair, he would still need to answer the question no. It may be more useful if the checklist provides space for written comment. However, many such examples could be used to illustrate the difficulties associated with the use of checklists, which supports the view that their value is highest when they are used as a tool to both help the inspector remember items to observe and when writing the inspection report. An example of a construction-oriented checklist is located at the end of this chapter (see Appendix A).

The use of checklists may also result in field workers doubting the expertise of the safety professional. Generally, the safety professional is viewed as the expert when it comes to safety and health issues, and it works to our advantage to provide our services from this approach. It is imperative for us to realize our limits and acknowledge (up front) when we need to do further research or obtain information from others before providing direction. Seeking consultation or conducting additional research does not tarnish one's image of being the expert. In the author's opinion, the use of checklists may be viewed as a crutch by the crafts, and that may lead to suspicion with regard to the direction you provide. If a checklist is used, it must be used in a professional manner.

KEY PROGRAM COMPONENTS

To this point, our focus has been on the management aspects of construction safety. What about specific components of the overall safety program? Safety professionals are familiar with OSHA requirements, whether impacted by federal or state regulations, and generally understand the need for a set of rules (known as a "code of safe practices" in some states) that dictate compliance with applicable standards for the job. These requirements should be addressed in your site safety and health manual. But avoid becoming complacent; stay current with OSHA trends and revisions to standards. Keep your management team informed about regulatory activity. Keep your written program up to date as well.

One way to be prepared for your job is to research the standards and develop checklists addressing key areas. These checklists can become part of your files and may be taken from job to job. They can be used as necessary to address key stages or components of a job before they actually occur. In other words, they will help

Table 5.4   Checklists for Construction Safety

Job site medical provisions
Emergency plan
Contractor's accident reporting requirements
Fire protection
Temporary power for the site
Fuels storage
Chemicals storage
Traffic routes
Sites for contractors field offices
Sites for materials storage areas
Job site security needs
Capping/covering of impalement hazards
Assured grounding program
Requirements for reporting to client
Fall protection program
(LOTO) program
Personnel manbaskets
Steel erection

you be proactive on your job. When something comes up on your site, you will be prepared. Table 5.4 illustrates some topics to consider for the development of a checklist.

This practice may sound inconsistent with the opinions previously expressed about the use of checklists. However, these checklists are not intended for your use in the field. These checklists can be placed into a reference binder and consulted during different phases of the job. They will serve as a quick tool for your use to remind you of key requirements associated with work activity occurring on the job.

Model safety and health programs reach far beyond a basic site safety plan and are often customized for a specific job. For example, if you are involved in a project requiring a lot of deep trenches then you may launch a safe trench program focusing on proper shoring methods and sloping requirements. You may implement a requirement for a permit to be obtained prior to personnel entering an excavation and working below grade — that way you can ensure that a trench is inspected prior to authorizing personnel to enter it. You may also arrange to have competent person training available on site. The program can engage employees by having them calculate statistics such as the number of yards of soil moved without injury, the total distance of trenching completed without incident, or the number of man-hours spent working in trenches without mishap. Be sure to use this information when promoting your program.

The most successful programs usually place emphasis on those areas where serious accident potential exists. Although construction sites are viewed as dangerous in general, many other operations present an even greater potential for serious injury. These operations should be identified and given your attention. Table 5.5 illustrates many of the operations or processes that present serious accident potential; be proactive — anticipate these operations and plan for them. Your time and energy is well spent focusing on those that apply on your job.

Table 5.5   Areas for Special Emphasis

Cranes and hoisting equipment
Rigging materials and practices
Electrical safety (assured grounding, LOTO)
Confined space work
Trenching/excavations
Exposure to lead or asbestos
Heavy equipment operation
Blasting
Steel erection
Roof work
Scaffolding (particularly two-point suspended)
Weather extremes

These areas and other identified high-risk processes can be assessed using the job hazard analysis (JHA) technique. This well-known process is used to break down the components or steps of a task to identify where potential for injury exists and to establish acceptable control measures to reduce the injury potential. Figure 5.5 illustrates an example of a typical JHA form. This accident-prevention method provides an opportunity to get workers involved in your accident-prevention efforts and has been used successfully for years. Completed JHAs may be used to train and educate workers; they also make good safety meeting material.

Accident/incident investigation procedures are an important element of your program. As mentioned previously, management ownership of safety means the cognizant manager will stop what he or she is doing when an accident occurs and will respond to the scene to ensure that an investigation is initiated. The manager takes control of the investigation. This practice is in sharp contrast to the usual practice — the safety professional conducts the investigation and provides management with the report. It is also common to see management involved in the investigation but not to the extent desirable.

The depth of an investigation is often determined by the nature of the incident. In all cases, the investigation should be sufficiently detailed to yield a root cause and one or more contributory causes. The report should include at least one recommendation to prevent a similar occurrence. The results of the investigation should be shared with management and workers so that everyone on the project learns from what happened and understands how to prevent the same type of accident. The process of sharing the details of accident causation and preventive measures can be referred to as a "lessons learned" program. This type of program can be effectively implemented. Lessons learned should be distributed in an easily recognized, unique format to supervisors and management. They should be posted conspicuously around the job site to promote worker awareness. Finally, your program should require that each lessons learned be discussed at group safety meetings to ensure the workforce understands how to prevent the incident.

JOB HAZARD ANALYSIS

Contractor: _____   Date: _____

Job Number: _____   Location: _____

Phase Number: _____   Page _____ of _____

Activity or Operation	Unsafe Condition or Process	Preventive Measure or Corrective Action

Reviewed and Approved

Author: _____   Date: _____

Supervisor: _____   Date: _____

Site Safety: _____   Date: _____

Contractor Rep: _____   Date: _____

Figure 5.5   Sample job hazard analysis sheet.

SHARPEN YOUR SAW

Knowledge of safety and health issues in construction is a key to presenting yourself as the safety and health expert on a job site. Safety professionals must make an effort to learn about the many activities performed on a typical construction project. The individual can increase his or her knowledge by establishing a network of peers, attending professional development conferences, seminars, or classes, and participating in professional societies or associations. Appendix B provides a sample of professional development courses available to the safety professional. Appendix C provides resources for further reading.

CONCLUSION

The following are ten points for success in construction safety management:

1. Obtain full support from project management.
2. Know your project, anticipate hazards, and preplan the job adequately.
3. Define and communicate clear roles and responsibilities.
4. Understand applicable regulations and agency interpretations.
5. Strive to manage your program proactively.
6. Make yourself and your program visible to the workforce.
7. Communicate project safety requirements early in the pre-bid and pre-construction phases.
8. Establish time-bound, measurable goals for safety performance, and provide feedback to management and workers.
9. Enforce safety program requirements in a fair, uniform manner.
10. Involve others in the program (workers and supervisors), implement JHAs, and establish yourself as the expert regarding safety and health.

Managing safety in construction is a dynamic process. People and activities change on a regular basis; it is common to have multiple high-risk operations occurring at the same time. In many cases a safety professional may manage safety efforts on multiple projects simultaneously. The high accident rates in the construction industry have prompted OSHA to establish a strong presence, and thousands of citations are issued annually. Many health hazards associated with construction work were previously ignored or not even recognized. The safety professional is often the industrial hygienist as well and is expected to provide leadership in both areas.

Despite these challenges, the safety professional can succeed and be successful in the construction arena. A daily presence on the job site is fundamental to success. Early communication of project safety requirements is essential to proper job setup. A strong presence in the pre-bid and pre-construction stages will yield positive results. Time-bound, measurable goals, combined with periodic feedback to employees, contractors, and management is a necessary component of the program.

All of these efforts can be referred to as safety management. The secret to a successful construction safety program is to employ good management techniques. The safety professional must plan, organize, lead, and control. Indeed, managing safety involves the same principles as managing other functions in an organization. Good management techniques will integrate safety into the company culture and set the tone for safety performance on the project.

Not unlike safety professionals in other industries, the construction safety professional must strive to stay current with the regulatory community, safe practices in the construction field, and management techniques. The safety professional must establish himself or herself as the project authority, or expert, regarding safety and health. Make it your ambition to have the workforce and management recognize your expertise and seek your advice regarding safety and health issues. Keep yourself current with regard to safety and health issues. Professionalism must be practiced daily.

The Board of Certified Safety Professionals (BCSP) offers a Certified Safety Professional (CSP) specialty in construction safety. The achievement of this designation can solidify the expertise of the holder and will likely grow to become the norm for practitioners in this field. Of course, the designation itself does not make the individual a success — it establishes his or her knowledge base. It is the implementation of sound management techniques that leads to success in the field.

Many components of a safety program, including some of those discussed within this chapter, can be considered controversial. Therefore, plans should be reviewed by the company's legal counsel prior to implementation. The approach to safety enforcement may vary depending on whether you represent the client or the contractor, particularly in the areas of drug testing and giving direction to workers.

Whereas enforcement of safety programs in construction continues to be accomplished by methods emphasizing the negative, the proactive use of positive recognition should not be overlooked. The negative techniques seem driven primarily by a fear of OSHA, or perhaps out of respect for the OSHA inspection and the dynamic nature of a construction project. Still, the use of positive approaches can be included in your program to maximize your overall effectiveness.

Many companies manage a recognition program and reward safe performance. Typically, incentive programs are based on cents per hour worked without injuries, milestone cash awards, and spot cash awards, or group performance awards. The criteria chosen must be defined and communicated to all program participants. Some in the profession condemn this type of incentive program because employees or contractors may be influenced to under report accident/injury experience in order to meet objectives. Others suggest that such programs increase worker interest in safety and help to improve overall safety performance. Still others believe this type of program may have a positive psychological effect on employees, which results in improved safety. If you choose to employ this tactic, be sure the criterion is fair and clearly communicated to all those involved. Monitor the results and provide feedback in terms of performance to all interested parties. Do not consider this type of program to be your savior; attach the program to the overall project or company goal. Make it meaningful and make it interesting, but never encourage under reporting of injuries or illnesses. Many safety professionals have had success with incentive programs when they have been used as short-term stimuli for the overall safety program, and most agree that the value of the award(s) offered does not determine the overall success of the incentive program. Employees simply like to achieve the awards and be recognized by management and their peers.

APPENDIX A
COMPREHENSIVE CONSTRUCTION PROJECT SURVEY

Comprehensive Construction Project Survey

Project:	_____
Location:	_____
Prime Contractor:	_____
Inspector:	_____
Date of Inspection:	_____

1. Records, Administration, General

	Yes	No	N/A
1.1 Copy of OSHA regulations on site			
1.2 Required signage, postings in place			
1.3 Company safety plan or prejob safety plan on site			
1.4 Contractor identifying and correcting hazards			
1.5 Tool box safety meetings held weekly			
1.6 Required permits obtained and posted			
1.7 Confined space entry procedure followed			

2. Housekeeping Subpart C 1926.25

2.1 Stairs, work areas, walkways free of hazards			
2.2 Proper containers provided for trash, rags, etc.			
2.3 Scrap and debris removed daily			
2.4 Nails bent over or removed from scrap lumber			

3. First Aid, Sanitation & Illumination
Subpart D 1926.50-51 & 56

3.1 First aid kit available and complete			
3.2 Drinking water, cups and receptacle provided			
3.3 Adequate number portable toilets provided			
3.4 Washing facilities provided			
3.5 Work area lighting adequate			

4. Occupational Health & Environmental
Controls Subpart D 1926.52-59

4.1 Hazard assessments completed where needed			
4.2 Hazard communication program satisfactory			
4.3 MSDSs available for employee use			
4.4 Hearing protection provided and worn			
4.5 Exposures to gases, vapors, fumes, dusts, controlled			
4.6 Asbestos work conforms to regulations			
4.7 Hazardous materials properly labeled			
4.8 Employees trained to work with hazardous materials			

Construction Survey

5. Personal Protective Equipment Subpart E
1926.100 - 107

5.1 Hazard assessments completed where necessary
5.2 PPE provided and used where necessary
5.3 Hardhats worn by all personnel in construction zone
5.4 Safety harnesses, belts, lanyards inspected and used
5.5 Safety nets provided where needed
5.6 Employees trained in use of PPE

6. Fire Protection and Prevention
Subpart F 1926.150-151

6.1 Flammable & combustible liquids in safety cans
6.2 Extinguishers properly placed and inspected
6.3 Hydrants clear and access open
6.4 Proper signage posted where required

7. Handling, Storage of Materials
Subpart H 1926.251

7.1 Materials properly stored to prevent toppling
7.2 Maximum safe loads for floors posted and observed
7.3 Aisles kept clear and adequately sized
7.4 Mechanical aids used when possible

8. Rigging Equipment for Material Handling
Subpart H 1926.251

8.1 Safe working load limits not exceeded
8.2 Wire rope not secured by knots
8.3 Wire rope clips not used to form eyes of slings
8.4 Custom lifting devices labeled for maximum load
8.5 Rigging equipment properly stored
8.6 Rigging equipment in good condition

9. Tools, Hand and Power Subpart I
1926.300-303

9.1 Tools maintained in good condition
9.2 Damaged tools repaired or replaced, not used
9.3 All mechanical safeguards in use

Construction Survey

9. Tools, Hand and Power (continued)

9.4 Power tools grounded or double insulated
9.5 Pneumatic tools equipped with attachment restraints and restraints on air line connections
9.6 Compressed air used at 30 psi or less
9.7 Powder actuated tool operators properly trained
9.8 Grinders provided with safety guards
9.9 Work rests and wheel guards on bench mounted grinders adjusted properly

10. Woodworking Tools Subpart I 1926.304 & Subpart O 1910.213

10.1 Portable circular saws equipped with upper and lower blade guards, lower guard returns automatically
10.2 Radial saws - upper hood enclosed, lower portion of blade guarded to full diameter of blade, non-kickback fingers or dogs, adjustable stops or other effective device to eliminate table over-run and automatic return provided
10.3 Belts, pulleys, gears, shafts & moving parts guarded
10.4 Hand fed circular rip and crosscut table saws - saw above table guarded by hood, provided with non-kickback fingers or dogs and spreader

11. Welding & Cutting Subpart J 1926.350-354

11.1 Gas cylinders secured in upright position
11.2 Valve caps in place when cylinders not in use
11.3 Special wrench available when required by cylinder
11.4 Cylinders transported properly
11.5 Fire wall or 20' clearance for oxygen & acetylene
11.6 Fire extinguisher immediately available
11.7 Valves closed, equipment purged when not in use
11.8 Welding curtains used where needed
11.9 Welding cables in good condition & properly insulated
11.10 Proper ventilation and/or respirators used

12. Electrical Subpart K 1926.400-417

12.1 All exposed live parts guarded
12.2 All conductors and equipment "approved"
12.3 Tools and cords in good condition
12.4 Proper working clearances provided
12.5 Ground fault circuit interrupters provided
12.6 Portable and/or cord plug connected equipment grounded

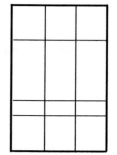

Construction Survey

12. Electrical (continued)

12.7 Temporary lights provided guards
12.8 Cables & cords protected from damage
12.9 Outlet boxes covered
12.10 Flexible cords used in continuous lengths without
       splicing
12.11 Portable hand lights provided with guards
12.12 Lock-out, tag-out procedures followed

13. Scaffolds Subpart L 1926.451

13.1 Guardrails and toeboards on all open sides and ends of
       scaffold platforms 10 feet or more above the ground
13.2 Guardrails installed on all open sides and ends of
       scaffolding from 4 feet to 10 feet high with minimum
       horizontal dimension of 45 inches
13.3 Tube and coupler scaffold posts accurately spaced,
       erected on suitable base and maintained plumb, brace
       connections secure
13.4 Planks secured or not less than 6 inches nor more
       than 12 inches over the end support
13.5 Ladder or equivalent safe access provided
13.6 Scaffold tied into structure when required
13.7 Lifeline and harness provided and used by each
       worker on swinging and single-point adjustable
       suspension scaffolds

14. Floor, Wall Openings Subpart M 1926.500

14.1 Floor, roof & wall openings guarded by a standard
       railing & toeboards or a secured cover
14.2 Open sided floors & platforms 6 feet or more above
       ground guarded with standard railing & toeboard
       where necessary
14.3 Runways 4 feet or more above ground properly
       guarded
14.4 Standard railings consist of top rail, toeboard &
       posts, & have a vertical height of approximately
       42 inches
14.5 Anchor points & framing members for railings of
       all types capable of withstanding 200 pound load
       in any direction

Construction Survey

15. Roofing Subpart M 1926.500

15.1 During built-up roof work employees protected against falling from unprotected sides by motion stopping safety system, or warning line 6 feet from edge and safety monitor

15.2 Guardrails 4 feet on either side of material access point at roof

15.3 Long sleeved clothing and adequate foot protection

15.4 Class BC fire extinguisher near kettle

15.5 Materials stored a minimum of 6 feet from edge of roof

16. Cranes & Derricks Subpart N 1926.550

16.1 Crane certified annually

16.2 Daily visual inspection, mace logs available

16.3 Signals and load capacities posted

16.4 Fire extinguisher provided and maintained

16.5 Swing radius of rear of rotating super-structure of crane barricaded

16.6 Power lines de-energized or safe distance maintained (groundman provided)

16.7 Safety latch provided on hook

16.8 Tag lines used to control loads

16.9 Personnel prohibited from riding on hook or loads

17. Aerial Lifts, Elevating & Rotating Work Platforms Subpart N 1926.556

17.1 Safety belt worn by each occupant of an aerial lift with lanyards attached to the boom or basket

17.2 Safety rails on all open sides of elevating work platforms

18. Earthmoving Equipment Subpart O 1926.602

18.1 Scissor points guarded on front-end loaders which constitute hazard to operator

18.2 Braking system, horn, seat belts (unless designed for stand-up operation) back-up alarm, overhead protection in good condition

Construction Survey

19. Excavating, Trenching & Shoring
Subpart P 1926.650-652

19.1 Name of competent person provided
19.2 Employees in excavations protected from cave-ins
19.3 Excavations 5 feet or greater in depth guarded by sloping, benching, shoring, shielding, or other equivalent means
19.4 Slopes excavated to angle of repose
19.5 Banks benched to proper heights
19.6 Spoil material pulled back at least 2 feet
19.7 Adjacent structure adequately shored / supported
19.8 Roads and sidewalks shored / supported
19.9 Ladder or ramp provided in trenches 4 feet or more in depth with no more than 25 feet lateral travel required for employees
19.10 Excavation adequately barricaded & lighted
19.11 Equipment a safe distance from excavation
19.12 Bridges over excavations equipped with safety rail
19.13 Daily excavation records available
19.14 Location of underground utilities complete

20. Concrete Work, Forms & Shoring Subpart
O 1926.700-701

20.1 Forms propely installed & braced
20.2 Adequate shoring, plumbed & cross braced
20.3 Shoring in place until strength attained
20.4 Protruding re-bar guarded, capped or covered
20.5 Pipelines, hoses and tremies used as a part of concrete pumping systems secured by wire or equivalent means in addition to the regular couplings or connections
20.6 Form lumber organized once stripped

21. Steel Erection Subpart R 1926.750-752

21.1 Safety railing installed around floor periphery, 1/2 inch wire rope or equal
21.2 Ladder or other acceptable means of access provided
21.3 Safety harnesses (belts for positioning) used
21.4 "Christmas treeing" of steel prohibited
21.5 Safety nets provided where required
21.6 Means provided to keep bolts and drift pins from falling when being knocked out (during bolting)

Construction Survey

22. Demolition Subpart T 1926.850-860

22.1 Engineering survey of structure to be demolished made and discussed prior to operations and approved by competent person
22.2 Public and employee protection provided
22.3 Floor openings covered
22.4 Material chutes enclosed or guarded
22.5 Adequate access ladders or stairs provided
22.6 Dust controlled adequately
22.7 Utilities cut off or controlled
22.8 Clear operating space for trucks and equipment

23. Power Transmission and Distribution Subpart V 1926.950-960

23.1 Existing conditions including energized lines & their voltages determined prior to starting work
23.2 Safe working & clearance distances maintained
23.3 Procedures followed to de-energize lines
23.4 Employees trained in first aid & CPR
23.5 PPE inspected and tested prior to use
23.6 Live-line tools certified, inspected before use
23.7 Portable metal or conductive ladders not used near energized lines or equipment
23.8 Aerial lift trucks near energized lines or equipment are grounded or barricaded, or insulated
23.9 Conductors and equipment treated as energized until tested or determined to be de-energized or grounded
23.10 Grounding procedures followed
23.11 Open manholes or vaults posted by warning signs & protected by barricades, temporary covers, or other guards
23.12 Manholes or vaults provided with forced ventilation or found to be safe by testing for oxygen & flammable gases or fumes prior to entry
23.13 Authorization obtained prior to starting work in energized substations
23.14 Temporary fence provided when existing substation fence is removed or expanded for construction

24. Stairways Subpart X 1926.1050-1060

24.1 Stairway or ladder provided at all personnel points of access where there is a break in elevation of 19 inches or more & no other means of access provided

Construction Survey

24. Stairways (continued)

24.2 Riser height and tread depth are uniform
24.3 Stairways free of slippery conditions or hazardous
      projections
24.4 Stairs with 4 or more risers or rising more than
      30 inches equipped with at least one handrail &
      one stair rail system along each unprotected side
      or edge
24.5 Stair rails not less than 36 inches from upper
      surface of the stair rail system to the surface of
      the tread
24.6 Midrails, screens, mesh or equivalent between top
      of stair rail system & top of stairs
24.7 Handrails and the top of stair rail systems capable
      of withstanding a force of at least 200 pounds
24.8 Unprotected sides and edges of stairway landings
      provided with guardrails
24.9 Electrical cords & cables collected on one side of
      stairs to reduce trip / fall hazard

25. Ladders Subpart X 1926.1050-1060

25.1 Extension ladders extend 3 feet above landing (or grab
      rail provided) & are secured against displacement
25.2 Non self supporting ladders used at an angle of
      1:4
25.3 Personnel face ladder and use side rail(s)
25.4 Personnel avoid carrying objects that could come
      loose and fall
25.5 Metal spreader or locking device on step ladders
      to hold ladder open when in use
25.6 Ladders with broken or missing rungs, cleats,
      steps, broken or split rails, corroded components
      or other faulty or defective components tagged &
      removed from service
25.7 Safety feet provided on extension ladders

NOTE: This checklist does not address all OSHA requirements for construction activities. For
      additional requirements, and conditions not included such as blasting and explosives,
      underground construction, ionizing radiation, roll-over protection, use of helicopters,
      etc. refer directly to the applicable OSHA / ANSI standard.

APPENDIX B
PROFESSIONAL DEVELOPMENT COURSES — SAMPLE LIST

- Crane/Inspection Certification Bureau, Orlando, Florida
  - Inspections, mobile, overhead, rigging, train-the-trainer
  - http://cicb.com/default.asp
- Crane Institute of America, Orlando, Florida
  - Inspections, mobile, overhead, rigging, train-the-trainer
  - http://www.craneinstitute.com/
- Scaffold Industry Association
  - Scaffold user hazard awareness
  - Primary access training
  - Professional scaffolder training
  - http://www.scaffold.org/coming/index.html
- OSHA Training Institute
  - www.oshalgov-outreach, training
  - #301 Excavation, trenching, and soil mechanics
  - #308 Principles of scaffolding
  - #311 Fall arrest systems
  - #326 Health hazards in the construction industry for safety personnel
  - #500 Trainer course in occupational safety and health standards for the construction industry
  - #502 Update for construction industry outreach trainers
  - #510 Occupational safety and health standards for the construction industry
- American Society of Safety Engineers (ASSE)
  - www.asse.org
- Local construction associations

APPENDIX C
FURTHER READING LIST

Many articles and books have been written to address safety in the construction industry. Following is a recommended list of further reading material.

Books

Ellis, J. *Introduction To Fall Protection,* 3rd ed. American Society of Safety Engineers, Des Plaines, IL 2001.

Finkel, M. *Guidelines for Hot Work in Confined Spaces: Recommended Practices for Industrial Hygienists and Safety Professionals.* American Society of Safety Engineers, Des Plaines, IL 2000.

MacCollum, D. *Crane Hazards and Their Prevention.* American Society of Safety Engineers, Des Plaines, IL 1999.

Hislop, R. *Construction Site Safety: A Guide for Managing Contractors.* Boca Raton, FL: CRC Press, 1999.

Petersen, D. *Safety Management: A Human Approach,* 2nd ed., Aloray, Inc., Goshen, NY, 1988.

Hinze, J. *Construction Safety.* Englewood Cliffs, NJ: Prentice-Hall, John Wiley & Sons, NY, 1998.

Kohn, J. and Terry, T. Safety and Health Management Planning, Government Institutes, Rockville, MD, 1999.

Levitt/Samelson. *Construction Safety Management.* 2nd ed., 1994.

National Safety Council (NSC). *Accident Prevention Manual for Business and Industry, Engineering and Technology,* 10th ed., Washington, D.C.: NSC, 1992.

MacCollum, D. *Construction Safety Planning.* New York: John Wiley & Sons, 1997.

Petersen, D. *Human Error Reduction and Safety Management.* Aloray, Inc., Deer Park, NY, 1984.

Other Publications

Petersen, D. The barriers to safety excellence. *Occup. Hazards.* 62,12,37–42, 2000.

Willen, J. Safety incentive programs — problem solver or troublemaker? *Saf. and Health.* 162, 3, 50–53, 2000.

Burkhammer, S. Construction safety. *Prof. Saf.* 45, 9, 49–51, 2000.

Brasser, G. Financial Risk Considerations for CM's and Owners. *CMAA's 2000 National Construction Management Conference,* September 2000.

Bennett, B. Orientation program for casual contractors. *Prof. Saf.* 45, 8, 24–28, 2000.

Winn/Frederick/Becker Adding construction to the academic safety curriculum. *Prof. Saf.* 45, 7, 16–18, 2000.

Adams, S. Today's safety professional: manager or engineer? *Prof. Saf.* 45, 6, 24–27, 2000.

Hansen, M. Engineering design for safety: petrochemical process plant design. *Prof. Saf.* 45, 1, 20–25, 2000.

Petersen, D. Safety management 2000: our strengths and weaknesses. *Prof. Saf.* 45, 1, 16–19, 2000.

Carnevate, L. More on OCIPs, Part 1: Deducting insurance costs from contractor bids. *The Risk Management Letter.* 20(6): 1999.

Smith, S. Under construction: safety at multi-employer sites. *Occup. Hazards.* 60, 5, 56–61, 1998.

Bukowski, J. Best practices for owner controlled insurance programs — Part 1. *The Risk Management Letter.* 17(6): 1996.

Nelson, E. Remarkable zero-injury safety performance. *Prof. Saf.* 41, 1, 22–25, 1996.

Minter, S. Building safety into construction. *Occup. Hazards.* September 1993.

Roughton, J. Managing a safety program through job hazard analysis. *Prof. Saf.* 37, 1, 28–31, 1992.

CHAPTER **6**

Overview of the Key Elements of a Systematic Safety Management Program

Richard W. Lack

CONTENTS

1-56670-370-0/02/$0.00+$1.50
© 2002 by CRC Press LLC

INTRODUCTION

A number of learned fellow members of the safety profession have recently been debating the so-called paradigm shift from the unsafe act/unsafe condition world of the mid-20th century, to quote my colleague Dan Petersen, a management consultant, author and educator who specializes in safety and organizational behavior. In fact, an intriguing article on this issue was published in the American Society of Safety Engineers journal, *Professional Safety*, in December 1992. The author, Gary Winn, provided a fascinating review of trends, including research background on the meaning of the word paradigm. He reluctantly concluded that although the safety profession appears to be getting close to a shift, there is no certainty as to when this will actually occur; "We're waiting impatiently," as he says.

From my perspective, safety practitioners owe a great deal to Dan Petersen for his outstanding research and pioneering work in the human behavior aspects of safety management. Many individuals have made significant contributions in this field, and my list of all those whom I have studied is too long to mention here. I will, therefore, limit my recognition to those teachers who have had a major influence on my development and understanding of the humanistic element in safety. These teachers are Frank Bird, John Grimaldi, Homer Lambie, Dan Petersen, and Bill Pope.

The Dupont Company and Technica DNV (formerly the International Loss Control Institute) have developed safety management systems that I have observed to be very effective. However, I should point out that, like any management system, they require careful tailoring to your organization's traditions, climate, and nature of operations. Strong management support and comprehensive training of key personnel are essential if these systems are to be implemented effectively.

The behavior-based safety process is not new; in fact, Dan Petersens's studies trace it back to the 1930s. It has, of course, gone through many stages of change and development.

For safety practitioners considering programs of this type, I recommend a careful review of Scott Geller's chapter, *Addressing the Human Dynamics of Occupational Safety and Health* (see Chapter 12). This will provide the reader with valuable insights into the psychological and motivational aspects of human work performance, and it also emphasizes how complex and confusing this field can be to the unwary. Steven Simon, Rosa Carillo and Michael Topf also discuss cultural aspects of the accident prevention process in their chapters (see Chapters 13 and 14, respectively).

In summary, these systems deserve very thorough investigation and observation, preferably including trial on a pilot basis, before a decision is made on unit-wide adoption. In the meantime, while this experimentation and debate continues, this chapter will, for practical reasons, focus on the more traditional elements of a sound safety management program. These elements that have stood the test of time have proven themselves to be effective in the continuous reduction of unintentional losses. Furthermore, these programs are very simple and flexible and can be implemented in a variety of ways. Thus, they can fit equally well into an organization such as a municipality, which may be structured with multiple layers of hierarchy, versus, for example, a hi-tech firm organized on a team basis with maximum employee decision-making empowerment.

In selecting the key elements of a safety management system, one could come up with a very long list if one wished to include every conceivable aspect. Frank Bird and the International Loss Control Institute have identified the following twenty program elements that are considered essential for a successful loss-control effort:

- Leadership and administration
- Management training
- Planned inspections
- Task analysis and procedures
- Accident/incident investigation
- Task observation
- Emergency preparedness
- Organization rules
- Accident/incident analysis
- Employee training
- Personal protective equipment
- Health control and services
- Program evaluations systems
- Engineering controls
- Personal communications
- Group meetings
- General promotion
- Hiring and placement

- Purchasing controls
- Off-the-job safety

In the interest of simplicity, I have distilled these twenty programs down to an eight-point program in the following section.

Eight-Point Safety and Health Management System

Administrative Elements

- Manual (procedures and guidelines)
- Committees and coordinators
- Training, interest, and motivation

Action Elements

- Inspections
- Hazard control
- Job hazard analysis
- Safety meetings
- Accident investigation

These key elements will be discussed in detail in the balance of this chapter. Many other related programs such as industrial hygiene, medical health, engineering controls, construction, and risk management are discussed in other chapters of this book.

EVALUATING THE EFFECTIVENESS OF THE SAFETY MANAGEMENT SYSTEM

Many organizations still evaluate the effectiveness of their safety and health systems by their injury experience. This is one form of measurement, but it is not a reliable approach because it does not measure system failures, only the chance products of these failures.

A more reliable system, which is discussed in this and other chapters of this book, is to measure the quantity and quality of activity in each element of the safety system. This approach provides management with information as to potential weak links in the system, substandard performance, and training needs for those individuals who are implementing the system.

Primary Elements of an Effective Safety Management System

Safety and Health Manual (Procedures and Guidelines)

A manual serves as the foundation of an effective safety and health management system. The manual is to the system what a blueprint is to construction. Without some basic procedures and guidelines, the loss control effort will be uncoordinated and haphazard in its emphasis. Issues will be addressed as they come up, rather than on a systematic prevention-oriented basis.

Obviously, a manual can be organized in many different ways, and each unit's needs will be different. Important criteria include:

- The manual must be users friendly. That is, there should be a logical layout of the contents so people can easily find procedures.
- The index and numbering system must make the job of filing new or revised procedures an easy process.
- The index and numbering system must be expandable into a total classification system so future procedures can be fitted into the system.
- The index system must be such that it can be translated to a reference filing system where additional materials can be kept and located easily.

Taking these criteria into consideration, a system adapted from one originally developed by Homer Lambie, former safety director at Kaiser Aluminum and Chemical Corporation, has been found to work very satisfactorily. The core of this system is a classification index (see Figure 6.1). The three general sections for safety procedures include:

- Administration (100 Series Procedures) — This section includes procedures primarily related to the planning, coordination, and evaluation of the unit's safety activities.
- Safety Engineering Standards (200 Series Procedures) — This section includes procedures primarily related to the adoption of standards for a safe work environment.
- Service Programs (300 Series Procedures) — This section includes procedures primarily related to the human element and the adoption of procedures aimed at the behavioral aspects of safety.

For the sake of consistency and control, there should be a standard layout method for each manual procedure. Again, every organization has different needs. A typical procedure layout is shown in Figure 6.2.

The Safety Engineering Standards sections is subdivided into nine major sections that relate to the source areas of most typical workplace hazards. They are self-explanatory; however, a few guidelines are called for in order to help with selection of the appropriate classification.

First, it is important when using this system to keep in mind the question, "What is the center of attraction in this procedure?" For example, a mobile crane could be classified as either equipment or vehicle. Where you classify it will depend on the center of attraction. Is your procedure concerned with the hoisting aspects or the mobile equipment aspects?

The source category descriptions, as previously noted, are self-explanatory. However, experience has shown that there is a need to emphasize the distinctions between the categories of Machines and Equipment and between the categories of Materials — Health and Materials — Traumatic. These categories are defined as follows:

Safety and Health Manual
Classification Index

Issue Date:

Revised Date:

Tab No.	Section	Procedure Classif. Range	Assigned Number	Subject	Issue Date	Revised Date	Auto Review Date
1	Administration		101	OSHA Inspections			
		110		**Union Safety**			
		120		Accident Investigation			
			121	Occupational Injury & Illness Investigation			
			122	Vehicle Accident Investigation (No Injury) Investigation			
		130		**Safety Reports**			
			131	Safety & Health Program Responsibilities			
			132	Statistical Injury Experience Reports			
			133	Safety Activity Reports			
		140		**Safety Manuals**			
		150		**Purchasing Safety**			
			151	Control of Purchasing Hazardous Materials			
		160		**Contractor Safety & Health**			
			161	Contractor Selection			
			162	Contractor Safety Program			
		170		**Professional Associations**			
2	Safety Engineering Standards		201	Regulatory Safety Standards			
			202	National Safety Standards			
			203	Safety & Health Design Review & Engineering Projects			
		210		WORK AREA			
		200		Floors			
			201	Excavation Procedure			
			202	Standard for Scaffolds			
			203	Standard for Portable Ladders			
			204	Standard for Traffic Signs & Controls for Work in Roadways			
		210		Temperature			
		215		Illumination			

Figure 6.1 A classification index layout for a safety and health program manual.

Machines — This category includes all types of machines that produce or finish materials or objects. Examples are lathes, drill presses, milling machines, grinders, and extrusion presses.

Equipment — This category includes all types of systems, facilities, or equipment that convey or process materials or objects. Examples are cranes, hoists, escalators, elevators, chemical process plants, conveyors, ovens, and robotic equipment.

Tab No.	Section	Procedure Classif. Range	Assigned Number	Subject	Issue Date	Revised Date	Auto Review Date
		220		MACHINES			
		221		Mechanical Power & Transmission			
		222		Point of Operation			
		223		Nip Points			
		224		Production Machines			
		225		Metal Working Machines Woodworking Machines			
		230		PORTABLE TOOLS			
		231		Hand Tools Manual			
		232		Hand Tools Power (Except Electrical)			
		240		EQUIPMENT			
			241	Inspection and Maintenance of Cranes & Hoisting Equipment			
			242	Inspection and Maintenance of Elevators			
			243	Inspection and Maintenance of Boilers and Pressure Vessels			
		244		Fans			
		245		Pumps, Pipes, Valves, Tanks			
		246		Conveyors			
		247		Gas Welding & Burning			
		250		MATERIALS — HEALTH			
		251		Toxic Materials			
			255	Standards for Chlorine Handling and Transportation			
		260		MATERIALS — TRAUMATIC			
		261		Compressed Gasses & Liquids			
			261.1	Compressed Gas Storage & Transportation			
		262		Acids and Caustics			
			262.1	Standard for Nitric Acid Handling and Transportation			
			262.2	Emergency Eyewash/Showers			
			263	Flammables & Explosives			
			264	Molten Metals			
			265	Hot Materials			
			266	Flying Particles			
			267	Sharp Edges			
			268	Weight & Shape, Stacking			

Figure 6.1 (continued) A classification index layout for a safety and heatth program manual.

Tab No.	Section	Procedure Classif. Range	Assigned Number	Subject	Issue Date	Revised Date	Auto Review Date
		270		ELECTRICAL			
			271	Motors & Generators			
			272	Transformers & Rectifiers			
			272.1	High Voltage Switching			
			273	Wiring Connectors			
			273.1	Electrical Grounding			
			273.2	Ground Fault Circuit Interrupters			
			274	Electrical Hand Tools			
			275	Electrical Welding & Burning			
		280		VEHICLE			
			281	Transporting Passengers Safety Regulations			
			282	Seat Belts			
			283	Power Driven Trucks			
			284	Forklifts			
			285	Railroads			
			286	Marine Vessels			
			287	Aircraft			
		290		MULTI-CATEGORY			
			291	Confined Space Entry			
			292	Lock Out — Tag Out			
			293	Hazardous Area Entry			
			294	Safe Work Permits			
3	Safety Service Programs	300		INSPECTIONS			
			301	Safety Inspection Program			
		310		HAZARD CONTROL			
			311	Hazard Control Program			
			311.1	Safety Recommendations			
		320		JOB HAZARD ANALYSIS			
			321	Job Hazard Analysis Program			
		330		SAFETY MEETINGS			
			331	Safety Meeting Program			
		340		ACCIDENT EXPERIENCE REVIEWS			
			341	Significant Incident — Damage Control Program			

Figure 6.1 (continued)   A classification index layout for a safety and heatth program manual.

Tab No.	Section	Procedure Classif. Range	Assigned Number	Subject	Issue Date	Revised Date	Auto Review Date
		350		SAFETY COMMITTEES & COORDINATORS			
			351	Executive Safety Committee			
			352	Labor-Management Safety Committee			
			353	Technical Safety Committees (Traffic, Electrical, Chemical, etc.)			
			354	Interest Motivation Safety Committee			
			355	Departmental Safety Committees			
		360		SAFETY TRAINING			
			361	New Employee Safety Orientation			
			362	New Supervisor Safety Orientation			
			363	Monthly Safety Training Meetings for Supervisors			
			364	Supervisory Training			
			365	Lineworker/Team Training			
			366	Safety Visuals Library			
		370		PERSONAL PROTECTIVE EQUIPMENT			
			371	Personal Protective Equipment Policy			
			372	Safety Shoe Policy			
			373	Prescription Eyewear Policy			
		380		AWARD AND RECOGNITION PROGRAMS			
		390		INTEREST AND INFORMATION PROGRAMS			
			391	Accident Experience Information			
			392	Hazard Control Ideas			
			393	Safety Idea Exchange			
			394	Safety Bulletins			
			395	Posters & Signs			
			397	Off-The-Job Safety Program			
4	Medical Health		401	Medical Surveillance Program			
			402	Bloodborne Pathogen Exposure Control Program			
			403	Cumulative Trauma Disorders Control Program			

Figure 6.1 (continued)   A classification index layout for a safety and health program manual.

Tab No.	Section	Procedure Classif. Range	Procedure Assigned Number	Subject	Issue Date	Revised Date	Auto Review Date
5	Industrial Hygiene		501	Respiratory Protection Program			
			511	Hearing Conservation Program			
			521	Employee Exposure Monitoring			
		530		Ventilation			
		540		Radiation			
			551	Asbestos Control Program			
			561	Lead Exposure Control Program			
6	Hazardous Materials		601	Hazardous Communication Program			
			611	Chemical Work Practices Guidelines			
			621	Process Safety Management Program			
			631	Laboratory Hazard Control Program			
7	Emergency Procedures		701	Disaster Plan			
			702	Earthquake Emergency Plan			
			703	Fire			
			704	Tornado			
			705	Flood			
8	Fire Protection	900		Fire Protection Standards			
			911	Fire Suppression Equipment Inspection & Maintenance			
			912	Fire Extinguisher Inspection & Maintenance			
			921	Fire Insurance Inspections			
			931	Fire Prevention Plan			
9	Security	900		Security Standards			
			911	Security Surveys			
			921	Access Control Procedure			
			931	Threats and Violence Control Plan			
10	Workers Compensation		1001	Insurance			
			1002	Claims Management			
			1003	Alternative Duty Policy			

Figure 6.1 (continued) A classification index layout for a safety and health program manual.

ORGANIZATION NAME	DATE:
SAFETY AND HEALTH MANUAL	NUMBER:
POLICY PROCEDURES	MANUAL SECTION:
SUBJECT:	REFERENCE:
	APPROVAL

I. **PURPOSE**

   Description of what the unit wants to be done.

II. **OBJECTIVE**

   Description of the reasons why the procedure is necessary.

III. **DEFINITION**

   Description of the principal terms referred to in the procedure. May also contain Policy Statements.

IV. **PROCEDURE**

   Description how the procedure will be implemented.

Figure 6.2   Layout of a typical safety and health manual procedure.

   Materials — Health — This category includes all types of materials that may cause "health injuries or illnesses" resulting from an acute or chronic exposure effect. Examples include asbestos, cyanide, benzene, lead, carbon monoxide, and hydrogen sulfide.

   Materials — Traumatic — This category includes all types of materials that may cause "traumatic injury" from the effect of material or impact contact. Examples include sodium hydroxide, sulfuric acid, flammables, welding sparks, blasting materials, oxygen, refractory materials and other abrasives, and materials with sharp edges.

   This manual classification system can be easily adapted as a checklist for system audits. The Safety Engineering Standards section can also serve as a checklist for safety inspections and for hazard recognition.

   In summary, some form of a safety and health manual is already a mandatory document in several countries. In the United States, some states require it, and most likely it will be required under the proposed federal OSHA Safety and Health Program Standard.

   The manual should be widely distributed in every organization. All members of the management staff, including first-line supervisors, should be included in the distribution. In organizations operating with work teams where the teams are responsible for conducting their own safety programs, each team should have a copy.

   Certain procedures, such as hazard control or accident investigation, may require the development of supplementary training guidelines for supervision. This type of document can be inserted into the manual. Alternatively, many organizations find it

ATM

Figure 6.2a A camel is a horse designed by a committee.

more effective to develop a separate supervisor training manual in which supervisors can keep all such material as it may be developed from time to time.

Development of the contents of the safety manual should be the responsibility of top management, with professional advice and input provided by in-house safety personnel or outsourced to consultants.

Many of the procedures in the Engineering Standards section will be selected from those prepared by professional organizations such as the National Safety Council (NSC) and the American National Standards Institute (ANSI), trade associations, state and federal agencies, product manufacturers, etc. They can be added to or modified as necessary to meet the unit's needs.

All procedures should be regularly reviewed and audited to ensure that they are both current and are being effectively implemented by the organization.

Once a manual is set up (following the guidelines previously discussed) and efficiently maintained, it will provide the organization with a positive focus and a solid building block to help ensure continuously improving safety results.

Safety Committees and Coordinators

Safety Committees

The size of an organization, its management structure, and its overall employee relations "climate" will all be factors in determining the number and type of safety committees that best fit its needs. "Top-down" autocratic-style organizations will likely have very few committees of any kind. Others that are run in a more participatory or consensus style may have many different committees with a wide variety of responsibilities.

Committees have been the subject of jokes for many years. My favorite is "a camel is a horse designed by a committee" (see Figure 6.2a). Most of these jokes are probably a little exaggerated, but they do reflect some of the potential problems that can plague committees.

First, what are the advantages of a safety committee? Obviously, the overall purpose of a systematic safety and health program is to prevent accidents. To achieve this purpose, the system must be focused on the target of finding and controlling hazards. The key advantages of the committee meeting approach in hazard discovery and control are:

- The pooling of a group's experience and expertise; this, together with the opportunity to brainstorm problems, results in the development of innovative and practical solutions
- The opportunity for groups of people to work together in committee meetings results in improved communications
- A committee's recommendations, especially on controversial issues, are usually accepted more positively by others in the organizations.

Some of the pitfalls that must be avoided if committees are going to be effective are:

- Committees may try to take over the job of line operations.
- Committee members are unclear as to their duties and responsibilities.
- The purpose and objectives of the committee have not been clearly defined.
- The flow of problems to committees and their recommended solutions are undefined.

Committee Organization — The following are recommended guidelines for committee operations.

- Membership should be between three and twelve. Anything over twelve can result in the meeting discussions being dominated by the more aggressive members. The bigger the group, the more difficult it is to "draw out" quieter members who may have important ideas to contribute.
- The committee should report to one individual.
- Committee members are either appointed or selected by the group.
- Subcommittees are responsible to the chairperson of the committee or some member designated by the chairperson.
- The chairperson of the committee either is appointed or is nominated and selected by majority vote of the committee.
- Committees should have the responsibility to recommend only. Their recommendations must be implemented by those persons having the necessary authority.
- Committees should operate in a business-like fashion. Minutes should be recorded and distributed promptly after meetings. Agendas should be established, and the committee should meet at a regularly scheduled date and time.

See Appendix B for a sample Committee Method of Operation.

Types of Committees — There obviously can be a wide variety of safety committees. Broad categories for safety committees are as follows:

- Executive Safety Committee. This should be the principal unit policy advisory committee to the chief executive of the unit or organization (see Appendix A).
- Labor–Management Safety Committee. This committee, when appointed, should report to the unit's top manager. It will provide a forum for joint discussion on safety problems and recommendations for improvements.
- Safety Program Committees. These committees will report to the executive safety committee and will submit recommendations for the improvement of specific program elements such as
  - Training
  - Recognition and awards
  - Hazard control

- Departmental Safety Committees. For large departments, it may be desirable for the department manager to appoint a committee that will provide recommendations for the improvement of departmental programs.
- Technical Safety Committees. Technical committees should be appointed as needs arise for the consideration of special technical problems. Examples are:
  - Electrical hazard control
  - Chemical hazard control
  - Traffic safety
  - Crane safety
  - Ergonomics

Safety committees, if provided with the proper direction and structure, can be a very effective element in your overall system. Their effectiveness, as with any committee, is measured by the action taken as a result of their meetings to solve specific safety problems.

Safety Coordinators

The safety coordinator program is designed to provide support and assistance to department management. The position will normally be a part-time assignment. Duties of a safety coordinator will include assisting the department manager with the administrative aspects of the various programs, as well as defining program weak points and making recommendations for their improvement.

The unit safety professional should meet regularly with all coordinators to provide them with specialized training and guidance.

A guideline procedure should be developed so that all concerned will understand the purpose, objectives, and procedures of this program. This guideline should state that safety coordinators will not relieve line supervisors of this basic responsibility for the safety of their assigned employees. Their primary function is to assist the unit manager with the administration of the unit's safety and health programs.

The duties and responsibilities of safety coordinators will vary with the unit's needs. Some examples include:

- Audit unit safety manuals to ensure they are kept up to date
- Ensure that supervisors complete required documentation for new employee safety orientation, hazard communication, and other mandatory safety training
- Assist with investigation of serious accidents and incidents
- Review unit hazard control logs
- Track control action on hazards pending correction
- Assist supervisors with administration of the job hazard analysis program
- Audit employee compliance with safety procedures

Safety coordinators can provide vital support to their unit management. In addition, those assigned to these positions have an invaluable opportunity to increase their knowledge of the unit's operations and also of safety program management techniques. Development of their leadership and communications skills as safety coordinators will help to prepare them for future promotions.

APPENDIX A: SAMPLE POLICY PROCEDURE
FOR EXECUTIVE SAFETY COMMITTEE

Purpose

The Executive Safety Committee will provide the director with recommendations as to unit-wide programs and major problems.

Objectives

- To ensure continuous improvement of unit safety and health systems and programs
- To provide a means for the efficient and effective resolution of major potential problems that could adversely affect the success of the enterprise

Committee Duties and Responsibilities

- A written method of operation (MO) should be developed to outline how the committee will receive problems and other items for discussion.
- Problems for discussion should be ranked by priority, assigned a title, document control number, and date.
- Wherever possible, committee members should receive the meeting agenda ahead of time so they can come prepared to discuss the issues and contribute to developing their solutions.

APPENDIX B: SAMPLE SAFETY COMMITTEE METHOD OF OPERATION

Procedure

Committee Organization

- The director is the control chairperson of the executive committee.
- The operations manager is chairperson of the meeting.
- The committee meetings will be held each month on the third Wednesday at 10:00 a.m.

Committee Duties and Responsibilities

- The committee will establish basic unit-wide programs.
- Any deviations from standard policy will be reviewed by the committee.
- The committee will normally deal only with multidepartment problems.
- Problems for discussion will be circulated as agenda items so that members are prepared to discuss them and reach a conclusion.
- Individual committee members or small subcommittees will be appointed by the chairperson to conduct periodic audits and surveys to assess the effectiveness of the unit safety and health programs.

Committee Meeting Agenda

Meeting agendas will include the following:

Agenda Item	Person(s) Responsible
• Review of overall unit safety program activity and results	Safety manager
• Discussion on significant accidents	Department managers
• Discussions on identified problems	Committee members
• Reports from subcommittees	Subcommittee chairpersons
• Review new or revised policies and procedures submitted for approval	Committee members

FURTHER READING

Boylston, R. P. *Managing Safety and Health Programs*. New York: Van Nostrand Reinhold, 1990. (This entire book has extensive references to committee activities and functions.)

Colvin, R. J. *The Guidebook to Successful Safety Programming*. Boca Raton, FL: Lewis Publishers, 51–66, 1992.

Thomen, J. R. *Leadership in Safety Management*. New York: John Wiley & Sons, 61–88, 1991.

Kohn, J. P. *Safety and Health Management Planning*. Rockville, MD: Government Institutes, Inc. 384–389, 1999.

Ridley, J. et al. *Safety at Work*. London: Butterworths, 137–138, 294–295, 1983.

Grimaldi, J. and Simonds, R. *Safety Management*. Boston, MA: Irwin, 118–125, 1989.

Senecal, P. and Snyder, J. J. *Safety Committees that Work*. Cleveland, OH: Penton Publishing Inc., 1997.

SAFETY AND HEALTH TRAINING

Training in this element of the system is usually provided by specialist personnel or employees who have received special instructor training. Much of this training is legally mandated by Occupational Safety and Health Administration (OSHA) and other regulations. The following is a list of the more common training requirements for health-related training:

- Hazard communication
- Hazardous waste operator
- Asbestos awareness
- Hearing conservation
- Infection exposure control
- Respiratory protection
- Ergonomic (proposed regulations)

The specific content and requirements of these training programs will be discussed in more detail in the chapter on health.

Safety-Related Training

Program	Scope	Requirements
Excavations	Employees who enter excavations	Initial training for competent persons and awareness training for all employees who enter and work in excavations
Confined space	Entry supervisors Confined space attendants and authorized entrants	Training, initial and annual; Basic first aid and CPR certificate for confined space attendants
Lockout/tagout	Employees who work with/around energized equipment	Training (initial and annual)
Motor vehicle accident prevention	Employees who drive for employer on business	Training, behind the wheel every 2 years for frequent drivers; Preventability workshops for employees who have accidents
Emergency response	All employees	Annual training and retraining

This is by no means a complete list because, depending on the industry, there are numerous other specialized training requirements.

The following occupations require legally mandated safety training:

- Window cleaning operations
- Aerial lift operation
- Forklift truck operation
- Powered industrial trucks
- Mechanical power presses

- Welding and cutting
- Crane operation
- Scaffold erection and dismantling
- Power actuated tools
- Fire extinguisher operation

The Federal Mine Safety and Health Act of 1977 went even further than OSHA and actually contains specific mandatory safety training requirements in the Act itself. Examples (quoted in part) are below.

Sec. 115(a)

Each operator of a coal or other mine shall have a health and safety training program which shall be approved by the secretary.

Each training program approved by the secretary shall provide as a minimum that

- New miners having no underground experience shall receive no less than 40 hours of training if they are to work underground. Such training shall include instruction in the statutory rights of miners and their representatives under this Act, use of the self-rescue device and use of respiratory devices, hazard recognition, escape-ways, walk-around training, emergency procedures, basic ventilation, basic roof control, electrical hazards, first aid, and the health and safety aspects of the task to which he will be assigned.
- All miners shall receive no less than 8 hours of refresher training no less frequently than once each 12 months.
- Any miner reassigned to a new task in which he has had no previous experience shall receive training in accordance with a training plan approved by the secretary.

Keeping track of this training has become something of a paperwork nightmare. Fortunately, computer systems are now available, which not only satisfy regulatory documentation requirements, but also track when employees are due for training or retraining. (see Figure 6.3) for a sample format for tracking safety training completed.) Interactive-type programs are now available, and these are a big help with scheduling, especially in multishift operations. For more in-depth information on training systems and techniques, refer to Chapters 47 and 48.

Much of this training, especially new employee and supervisor orientation, requires close coordination between safety professionals and the human resource function. A sign-off checklist is commonly used to ensure that all elements are covered. See Figures 6.4 and 6.5 for sample formats.

Successfully completing this array of training always presents scheduling problems and other production and timing conflicts. To help minimize these problems and spread the training throughout the year, one system that works well is to establish a schedule of monthly meetings. All the required topics can thus be covered, plus other subjects as needs are identified. The monthly training session can be repeated a sufficient number of times to allow for shift schedules and coverage.

Safety Training Record

Division	Section	Shift	Time Started	Date Started	MO.	DAY	YR
				Date Completed			
				Time Ended	Total Time		
					Hours	Minutes	

Check Training Course Completed

New Employee Training		Annual Refresher Training	
Course No.	Title	Course No.	Title
	Materials Hazard Control		HAZMAT Part I
	Hazard Communications		HAZMAT Part II
	MSDS		Asbestos
	HAZMAT Emergency Response		Respiratory
	Hazardous Waste		Hearing Conservation
	Asbestos Awareness		First Aid-CPR
	Respirator Protection		Defensive Driver
	Hearing Conservation		Bloodborne Pathogens
	First Aid-CPR		Fire Extinguisher
	Defensive Driver		Earthquake Plan
	Bloodborne Pathogens		Emergency/Evacuation
	Fire Extinguisher		Confined Space Entry
	Earthquake Plan		Lock-Out Tag-Out
	Emergency/Evacuation		

	Confined Space Entry				
	Lock-Out Tag-Out				
	HAZ Control Proced. Gen. Safe. Pract.				
JHA No.	JHA Title				

	JHA Training (Check One)				
	New JHA				
	Review of Old JHA				
	On Spot JHA				
JHA No.	JHA Title				

Attendance List

I acknowledge by my signature below that I have received the training stated on this report.

Name	Signature	Name	Signature

Instructor's Name: _____ Position Title: _____

Signature: _____ Section: _____ Date: _____

Figure 6.3 Sample format for safety training completed.

New Employee Orientation Checklist

Date Employed		
MO.	DAY	YR.

Name   Job Classification   Department

Item Covered	Initial	Item Covered	Initial
Human Resources		**Security**	
Introduction to Facility/Plant/Organization		Procedures for Entering and Leaving Plant	
Date Schedule		Parking Regulations	
Food Service		ID Badge Regulations	
Probation Period		Loss Reporting	
Role of Supervisor			
Performance Reviews		**Safety**	
Pay Rates and Increases		The Unit Safety and Health Program	
Overtime Pay		Purpose and function of Safety & Health Dept.	
Payday		In-Plant Medical Center	
Promotions		Accident Reporting and Investigations	
Pay Deductions (Union NSI)		Workers Compensation Procedures	
		Special Plant Hazards	
Emergency Messages		Special Programs (Asbestos, Infection Control)	
Reporting Absences		Emergency Procedures	
Rules of Conduct		Fire Prevention and Protection System	
Holidays		Fire Reporting and Fire Fighting Techniques	
Vacation		General Safety Rules	
Sick Leave		Respiratory Protection	
Credit Union		Hearing Conservation	
Change House Facilities & Regulations		Safe Work Clothing and Practice	
Change House Locker Assigned		"Employees Guide to Safety" issued	
Identification Badge Issued			
Issued all Personal Protective Equipment		Shown 5-Point Emergency Care Film	

The items appearing above were discussed with the employee on _____

Employee Signature   **Human Resources Supervisor**

_____   _____

Human Resources Superintendent   **Safety Supervisor**

_____   _____

Figure 6.4   Sample new employee orientation checklist.

Besides organizing and/or conducting these training programs, the safety professional, or designated person having this responsibility, will also need to establish a reference/resource library to support these activities. The training reference/resource library should include:

- An audiovisual library so that supervisors and team leaders can borrow videos, tapes, slides, etc., for their training needs
- A reference library of books, pamphlets, and bulletins

New Employees Orientation Flow Sheet

Instructions:
After you have completed your phase, sign in the space provided. After completing Section 4, return form to Personnel Office.

_____
Employee's Name

_____
Classification

1. Employment

General information about Company
Character of employment
Schedule (working)
Rates of pay, hours, and overtime
Equal Employment Opportunities
Vacations, holidays
Importance of good attendance record
Union
Probationary period
Job-bidding
Group insurance-life, hospital-surgical
Bond purchase
Proper clothing, safety shoes
Pension plan, S.U.B.
Employee's Association
Issue Security badge

3. Safety and Health

a. MSHA Safety Training (Total 3 hours)
   Statutory rights
   Employee's handbook, safety manual
   Introduction to work environment Part I
   Hazard recognition
   Health/Safety of tasks assigned Part I
b. Other Items
   Cafeteria
   First-aid facilities
   Safety equipment
   Safety shoes, how to order
   Visit to storeroom
   Visit to Medical Department
   Medical Surveillance Program
   Point out high risk potential hazards and measures

_____
Personnel Representative

_____
Safety Representative

2. Plant Protection

Locker room facilities
Master bulletin board
Telephone calls
Parking facilities
Security procedures

4. Department

a. MSHA Safety Training (Total 5 hours)
   Introduction to Work Environment Part II
      General Nature of work in department
      Relation of dept. to other depts.
      Location of lockers and rest rooms
      Brief tour of department
   Health/Safety of Tasks Assigned Part II
      Departmental safety practices
      Proper clothing for work
      Importance of good housekeeping
      Report injuries, treatment, & record
b. Other items
   Nature of work assigned
   Work schedule, hours, lunch arrangements
   Overtime work and procedures
   Importance of good attendance record
   Call in when absent
   Tool checks
   Requisition of storeroom items
   Public address system
   Introduce to other employees

_____
Safety/Security Representative

_____
Supervisor

Figure 6.5 Sample new employee orientation checklist to comply with MSHA requirements.

- A library of professional journals, magazines, and periodicals
- A reference file of articles, accident experience information, hazard control ideas, and other safety information; the person maintaining this file should circulate new items to the proper individuals in the organization for their information and action as appropriate

Preferably, the reference and audio-visual libraries should be established in a learning center, training room, conference room, or similar facility. In this complex, all the visual aid projection equipment and other training materials will be stored as needed.

To summarize, the training phase must as a minimum include:

- New employee orientation
- Supervisory/team leader training
- Backup management training
- Line employee training
- Specialized training and retraining according to task and operations needs
- Learning center and training equipment
- Training information and reference library

Training, both the specialized type and the specific, are absolutely vital factors for the success of the safety and health effort. Unit management will need to commit increasing time and effort to emphasize training and ensure that it is built into the unit's goals and strategic planning process.

Interest and Motivation

As mentioned earlier, after training has been provided, successful loss prevention programs must include ways and means to promote interest among the employees and to motivate them to put their knowledge into action.

Interest Programs

For the innovative professional, methods of creating and maintaining employee interest in safety and health issues, and the unit's programs, are almost limitless. Some common approaches are:

- Bulletin boards
- Posters and signs
- Distribution of magazines, bulletins, etc. (e.g., the National Safety Council's Family Safety bulletin and numerous other subscription bulletins)
- Special in-house flyers or handouts
- Speeches and presentations

It also promotes employee acceptance when you encourage employee involvement. Go out and find employees with graphic or artistic skills to help illustrate your handouts. Seek out those with writing skills who would be willing to contribute short articles. Publicize accident and incident experiences, especially information on near misses. Some organizations have, for example, a brown bag lunch program in which they arrange for in-house and outside invited speakers to speak on a variety of topics. Finally, in order to encourage employees, top management must support these programs and use every opportunity to demonstrate their commitment.

Motivation Programs

Much has been written about employee motivation, and this aspect of human behavior is explored in depth in later chapters.

A highly recommended reading on this subject is Homer Lambie's article *Accident Control Through Motivation*. It is as current today as when it was written 30 years ago. It really crystallizes the spirit of the human soul, its desires, and needs as they relate to accident prevention.

Safety motivation programs, for the purpose of this chapter, will include:

- Award and recognition for outstanding safety results
- Award and recognition for outstanding safety program performance
- Incentive programs
- Award and recognition for outstanding safety service and/or heroism

Recognition of outstanding safety results is the most common of all motivation programs. These programs provide recognition and awards for injury-free records. The unit award can be on an individual basis, by section, or department, or for the total organization.

A word of caution: experience has shown that when the award is based on all injuries, there is a tendency for injuries to go underground (i.e., to go into the nonoccupational care system rather than risk spoiling the group's chance for an award). To avoid this problem, many organizations base their records on lost workday cases. In this way, the measurement is based on cases that are inherently more severe in terms of injury and, as a result, they are less likely to be concealed.

In the writer's experience, a more effective recognition system is one that is based not only on the program results, but also on the program performance. This type of award is more positive because it emphasizes the concept that a performance- or results-based award recognizes the group's accident prevention efforts as well as their superior injury-free results. Results alone may be just the product of luck.

Incentive programs come in such a variety of forms that it is clearly beyond the scope of this chapter to discuss them in detail. A well-planned incentive program that is properly promoted can be a useful aid to an effective safety program. For this reason, presented here is a consolidated list of hints and pitfalls to avoid when designing either a contest or a recognition program:

- Organization contests are often geared toward the best individual or group. This can result in others not being positively reinforced. One or a few winners and many losers results in a very negative impact.
- Contests or incentive programs deteriorate over time, and if the goal is too long range, the groups lose interest.
- Contests should be based on rewarding positive effort by all the contestants. Too many contests are based on the luck of the draw.
- Contests should be short term (about three to six months in duration). The results are also more positive when all participants are rewarded with some tangible recognition.
- The reward should be a token, not a prize. People like something that they can associate with their organization, the contest, and their groups. Poll your group to obtain their preferences.
- People love to see themselves in a picture. A plant where I worked some years ago established a world safety record for the industry. The company photographer went around the plant and took dozens of pictures of people at work. This was made into a huge, two-page collage and published in the local newspaper. A special

reprint was also provided for employees, and all employees were given an inscribed wall clock with a picture of the plant as a background.

- If you are going to select a results-only recognition program, make sure the accomplishment is recognizing a superior record. For example, recognize the first 100 injury-free days for a new plant, or a million hours without a lost workday due to injury. For individuals, reward one, three, and five years without a lost workday due to injury.

To summarize, the value of interest and motivation programs is measured by the extent to which they assist every employee in the organization to develop a strong sense of responsibility and the desire to continuously improve their skills in finding and controlling hazards.

FURTHER READING

Books

Balge, M. Z. and Krieger, G. R. *Occupational Health & Safety,* 3rd ed. Itasca, IL: National Safety Council, 2000.

Bird, F. E. and Germain, G. L. *Practical Loss Control Leadership.* Loganville, GA: Det Norske Veritas (USA) Inc., 45–68. (Motivation), 1996.

Broadwell, M. M. *The Supervisor On-The-Job Training.* Reading, MA: Addison-Wesley, 1989.

Colvin, R. J. *The Guidebook to Successful Safety Programming.* Boca Raton, FL: Lewis Publishers, 113–131, 1992.

Eckenfelder, D. J. *Values-Driven Safety.* Rockville, MD: Government Institutes Inc., 1996.

Fettig, A. *Winning the Safety Commitment,* Battlecreek, MI: Growth Unlimited, 1998.

Gellerman, S. W. *Motivation in the Real World.* New York: Penguin Books, 1993.

Grimaldi, J. V. and Simonds, R. H. *Safety Management.* Boston, MA: Irwin, 185–191 (Motivation), 478–493 (Training), 1989.

Grund, E. V. *Lockout/Tagout.* Itasca, IL: National Safety Council, 1995.

Heath, E. D. and Ferry, T. S. *Training in the Work Place.* Goshen, NY: Aloray 1990. (This book contains extensive information on safety training.)

Kohn, J. P. and Ferry, T. S. *Safety and Health Management Planning.* Rockville, MD: Government Institutes Inc., 1999.

Krause, T. R. Hidley, J. H., and Hodson, S. J. *The Behavior Based Safety Process.* New York: Van Nostrand Reinhold, 1990. (This book contains extensive research and information on motivation as it affects employee behavior.)

Krause, T. R. *Employee-Driven Systems for Safe Behavior,* New York: Van Nostrand Reinhold, 1995.

Lambie, H. K. *Accident Control through Motivation: Selected Readings in Safety.* Macon, GA: Academy Press, 1973.

MacCollum, D. V. *Construction Safety Planning.* New York: Van Nostrand Reinhold, 1995.

McSween, T. E. *The Values-Based Safety Process.* New York: Van Nostrand Reinhold, 1995.

Petersen, D. *Authentic Involvement,* Itasca, IL: National Safety Council, 2001.

Petersen, D. *Safe Behavior Reinforcement.* Goshen, NY: Aloray, 1989. (This book contains extensive research and information on motivation as it affects employee behavior.)

Petersen, D. *Safety Management: A Human Approach.* Goshen, NY: Aloray, 59–63 (Training), 139–150 (Motivation), 205–215 (Training), 217–271 (Motivation), 1988.

Petersen, D. *Techniques of Safety Management: A Systems Approach.* Goshen, NY: Aloray, 141–158, 1989.

Pike, R. W. *Creative Training Techniques Handbook*. Minneapolis, MN: Lakewood Publications, 1989.

Robinson, A. D. Incentives and rewards. In *Human Resources Management and Development Handbook*, Tracey, W. R., Ed. New York: AMACON (American Management Association), Chapter 42, 1994.

Swartz, G. Safety *Culture and Effective Safety Management*. Itasca, IL: National Safety Council, 2000.

Other Publications

Jones, S. E. The key issues of safety and health. Occupational hazards. *NSMS Focus*, 87–90, May 1991.

Lack, R. W. Industrial safety training — coping with the unsafe act problem in today's world. *Prof. Saf.* 24(2): 33–37, February 1979.

Lack, R. W. Is your safety program a paper tiger? *Am. Soc. Saf. Eng. Manage. Div. Newsl.* 9: 6–9, March 1985.

Lapidus, R. The psychology of safety: how to get people to perform in a safe manner. Issues in perspective. *Nat. Saf. Manage. Soc.* 1(4): 1–6, December 1986.

Sarkus, D. J. Safety and psychology — Where do we go from here? *Prof. Saf.* 46(1): 18–25, January 2001.

Swartz, G. Safety in supervisors salary review — a formal approach. *Prof. Saf.* 36(5): 21–24, May 1991.

Winn, G. L. In the crucible: testing for a real paradigm shift. *Prof. Saf.* 37(12): 30–33, December 1992.

Winn, G. L. Total quality? The "new" paradigm seems out of reach for safety managers. *Occup. Health Saf.* 63(10): 53–54, October 1994.

SAFETY INSPECTIONS

Introduction

On January 26, 1989, the U.S. Department of Labor and OSHA issued their document *Safety and Health Program Management Guidelines; Issuance of Voluntary Guidelines*. Among the recommended actions, there is a section dealing with inspections. This is quoted (in part) as follows:

(C)(2) Worksite analysis (i) so that all hazards are identified:

(A) Conduct comprehensive baseline worksite surveys for safety and health and periodic comprehensive update surveys

(B) Analyze planned and new facilities, processes, materials, and equipment; and

(C) Perform routine job hazard analyses

(ii) Provide for regular site safety and health inspections, so that new or previously missed hazards and failures in hazard controls are identified

Comment: Identification at a worksite of those safety and health hazards which are recognized in its industry is a critical foundation for safety and health protection. It is the general duty of the employer under the Occupational Safety and Health Act of 1970.

Personnel performing regular inspections should, however, possess a degree of experience and competence that is adequate to recognize hazards in the areas they review, and to identify reasonable means for the correction or control of these hazards. Such competence should normally be expected of ordinary employees who are capable of safely supervising the performance of the operations of the specific workplace.

The frequency and scope of these routine inspections depends on the nature and severity of the hazards that could be present, and the relative stability and complexity of work site operations.

The State of California Title 8, General Industry Safety Orders, Article 1, Section 3203, *Injury and Illness Prevention Program*, went into effect in July 1991 and requires employers to establish, implement, and maintain an effective injury and illness prevention program. This standard is quoted (in part) as follows:

> (a)(4) Include procedures for identifying and evaluating workplace hazards including scheduled periodic inspections to identify unsafe conditions and work practices.

> (b) Records of the steps taken to implement and maintain the Program shall include:

> (I) Records of scheduled and periodic inspections required by Sub-section (a)(4) to identify unsafe conditions and work practices, including person(s) conducting the inspection, the unsafe conditions and work practices that have been identified and action taken to correct the identified unsafe conditions and work practices.

The material that follows is based on a paper on a related subject published in 1980 by *Protection*, the official journal of the Institution of Industrial Safety Officers, located in the United Kingdom (now the Institution for Occupational Safety and Health). It is printed with permission from Paramount Publishing Ltd.

> The traditional approach to industrial safety and housekeeping, probably copied from the military, was the routine inspection. This approach required that senior members of management, usually second level supervision or higher, made regular inspections and pointed out to first-line supervisors what must be done. This method had some benefits in that it was a form of training for the supervisors in what sort of problems they should be looking for, but it is inefficient insofar as the second-line supervisor's time is concerned and very often ineffective because little effort is made to find out why the problems existed in the first place.

The following section will, therefore, describe a more systematic approach to safety inspections, which in one form or another is being practiced by many organizations around the world.

The Key Parts of an Effective Safety Inspection System

An effective safety inspection system is built around three key elements: assignment of responsibility, the inspector's self-survey, and follow-up controls.

Assignment of Responsibility

The first step in establishing a sound safety inspection system is based on the principle of total single responsibility. This requires that the entire work site, including fence, gates, yards, and grounds, be divided up and assigned to the lowest level of management.

To establish this part of the program, it is recommended that, first, a work site map be drawn up showing major department areas of responsibility, preferably subdivided down to subdepartment areas usually under the direction of second-line supervisors. This map should have the approval of department heads and top management and be kept current with unit expansions, management changes, etc.

Based on the plant safety inspection area responsibility map, second-line supervisors should then develop maps showing how their respective areas are subdivided among their first-line supervisors or work team leaders.

Confusion often arises here because of the difference between supervisors' program responsibilities versus their job responsibilities for safety and housekeeping. Let us take a simple example to see how this approach can be misunderstood. In this sketch, we show a typical production area subdivided among four supervisors:

SMITH	JONES
BROWN	GREEN

This department operates on a 40-hour, 7-days-a-week basis, and only one of the supervisors is on duty at any particular time. Supervisor Smith may, therefore, be tempted while on duty to polish his or her area and ignore problems in the other supervisors' areas. To overcome this problem, supervisors must be instructed that, on a daily basis, they have a job responsibility to maintain good safety and housekeeping in the total operating area under their supervision. Conversely, their program responsibility calls for them to periodically survey their assigned area to detect and solve safety and housekeeping problems. If such problems are not systematically eliminated, the result will be steady deterioration of the level of safety and housekeeping in the total department.

On a daily basis, supervisors must ensure job safety and housekeeping. A few examples include:

- Compliance with safety, health, and environmental procedures and job hazard analyses
- Cleanup of production spills
- Collection and proper storage of tools and equipment
- Proper disposal of waste and scrap

On a periodic basis (recommended minimum of once a month), supervisors should survey their assigned areas of responsibility for safety and housekeeping problems. Checklists can be used to assist inspectors in their survey for safety problems.

An example of this type is provided by Cal/OSHA Consultation and is illustrated, in part, in Appendix C: Hazard Assessment Checklist at the end of this chapter.

Naturally, housekeeping problems will be many and varied, but for simplicity they may be divided into four major areas:

Cleanliness — Includes floors and waste collection in all areas.

Orderliness — Includes material and equipment storage, aisles and exist, fire protection, and emergency equipment.

Appearance — Includes windows, walls, doors, furniture, signs, and bulletin boards.

Lighting — Includes area lighting, emergency lighting, and warning lights.

Two other principles that require emphasis are:

- Area responsibility must be assigned to one person only on a 24-hour basis. This is most important, for the success of the system depends on each area being the responsibility of one person. Thus, any given area will reflect the ability and interest of one person to define and solve the safety and housekeeping problems that may exist in that area. Likewise, if problems do exist, it is clear who is responsible to act on the problems. Supervisors often object to this concept, pointing out that they have no authority over other people messing up their area when they are not at work. Nevertheless, experience has shown that this approach actually helps supervisors to develop as managers because they have to practice leadership abilities in working with others to get their cooperation.
- Production problems such as spills, piles of scrap, and stacks of parts or equipment are not necessarily housekeeping problems, especially when they are placed in a predictable area for a predictable length of time. Obviously, these require attention, but they are not the kind of problems that supervisors should be concerned with when searching for substandard housekeeping conditions.

Area Self-Survey

The self-survey approach is designed to teach the inspector to do what managers do — that is, go around and find their own problems. The system helps train inspectors to discover their own problems instead of having them pointed out by their supervisors.

Naturally, supervisors and team leaders will survey their areas frequently, but at some regular interval, such as once a month, they should be required to evidence their surveys in writing. This then is the self-survey, and in order to be an effective training tool it should incorporate the features illustrated in Figure 6.6.

Source–Cause

As mentioned earlier, one of the most common problems with safety programs is that people tend to treat the symptom rather than get at the root of the problem. The result of this approach is that the problem tends to keep recurring. A typical example would be:

Problem — Tools are scattered on the tool storage floor (tripping hazard).

Action — Employees were instructed to tidy them up. The following week, the supervisor checks again — same problem. He berates the employees and once again they tidy them up, and for a few weeks the tool store stays reasonably tidy, but before long the problem is back again!

Safety Inspection

Department	Inspection Responsibility Area/Equipment Assigned	Inspected By	Date

Problem		Control Action		Follow-Up
Describe WHAT is wrong and where it is located	Give reasons WHY it exists (source/cause)	Priority Code	Describe actions taken or planned to control the problem and its source/cause	Code	Name/Date

			Reviewer's Comments	Problem Priority Code
Problem	Is the problem clearly described and located? Are the reasons why the problem exists identified?	10 30		**E** Emergency
Priority	Do the priorities reflect and understanding of the potential effects of the problems listed?	10		**A** 1 Week
				B 1 Month
Control	Did the action control the problems and the reasons for their existence?	40		**C** 3 Months
				D 6 Months
Follow-Up	Does the follow-up specify who will do what, when, to ensure action completed?	10		**Follow-Up Codes**
				C Corrected (Date)
	Total	100		**S** Scheduled (Date)
			Signature: Date:	**P** Passed to (Name)

Distribution   Safety   Originator   Department

Figure 6.6   Sample format for safety inspection report.

Applying the source–cause approach, the supervisor must investigate why the problem exists. He considers what thing might produce the problem (source) or what is the reason why the people do it the wrong way (cause).

Let us now develop the previous example:

Problem — Tools are scattered on the tool store floor.

Source — Insufficient storage shelving and racks; no labeling for each tool item.

Cause — Employees were not trained to store similar items together.

Action — Source: Install shelves and racks and label storage areas. Cause: Train employees how and where to store tools.

The important thing is for the supervisors to apply the source–cause approach to all problems. By considering the whys as well as the whats, their control actions will be more permanent and/or long range.

Problem	Control through
Source (origin of the problem)	Engineering
Cause (incorrect action of the people)	Training

Priority

Part of the training process for first-line supervisors is the evaluation of identified problems in order to determine their priority for correction. The system indicated on the sample format above is similar to that used in many typical hazard control programs. By using the same approach in both hazard control and housekeeping, supervisors do not have to learn and apply two different systems. Furthermore, many housekeeping problems can also be a hazard.

A simple evaluation system, which can be used for both programs, is as follows:

Criteria	PRIORITY		
	High	Moderate	Low
Frequency — Number of persons exposed to the problem	3	1	0
Severity — Potential result of the problem in terms of injury, damage, or adverse employee or public relations	6	3	1
Probability — Is the problem *likely* to produce injury or adverse reaction?	3	1	0

Priority Rating		
E	Today	10–12 points
A	7 Days	8–10 points
B	1 Month	6–8 points
C	3 Months	4–6 points
D	6 Months	0–4 points

It should be pointed out that in considering control of any problem, and certainly one involving a hazard, the supervisor must first consider what immediate temporary control action can be taken pending a permanent solution. For example:

Problem — Hose was left lying on the floor.
Source — No rack provided.
Cause — Employees fail to coil hose up after use.
Immediate Temporary Control — Coil up hose.
Priority Evaluation — 7 Points = 1 month.
Permanent Control — Write maintenance work ticket to install hose rack.
Long-Range Control — Train employees to coil up hose on rack.

Control

As pointed out previously, for control of all problems, effective action must be taken to get at the source and/or the cause of the problem. For this reason, the self-survey report is laid out so that the reporter must indicate what will be done to control the whys.

Follow-Up

Here the inspector should indicate what action has been taken, at the time the report is submitted, to process each of the stated control actions. Supervisors and team leaders should be trained that there are only three ways to process control of an identified problem:

- Corrected — Action has already been completed.
- Scheduled — Action such as job ticket or engineering request submitted but pending scheduling.
- Passed — Action is beyond the authority or ability of the inspector to resolve and has been passed to a higher level or to the proper person responsible for their action.

The letters C, S, or P should be indicated in the survey report, along with the date and name in the case of the passed category.

This completes the description of the first-line supervisor's part in the self-survey system. However, to assist supervisors in improving their effectiveness, it is recommended that second-line supervisors be required to evaluate these reports and give their supervisors necessary feedback.

To help second-line supervisors in this evaluation process, some organizations include a points rating system as well as written comments. See Figure 6.7 for a sample format of this evaluation process.

Follow-Up Control Systems

The third key element of an effective industrial safety inspection system consists of follow-up controls. These are primarily management controls established at levels above the first-line supervisor or team leader. Management controls can be divided into those actions that are departmental and those that are plant-wide or multidepartment.

Hazard Survey Checklist

Hazard = Unsafe Act or Unsafe Condition

Hazard Source/Cause Category	Items in Areas Surveyed

1. Work Areas — Floors, aisles, yards and grounds, roadways, ladders, platforms, scaffolds, temperature, illumination.

2. Machines — All *production* or *finishing* machines: lathes, drill-presses, milling machines, grinders, Point of Operation and Nip points.

3. Equipment — All equipment that *conveys* or *processes*: Fans, boilers and pressure vessels, ovens and heaters, compressors, and air-receivers, pumps, pipes, valves, tanks, hoses and connections, cranes, hoists, slings, escalators, elevators, conveyors, robots, gas welding and burning. *Rotating* equipment.

4. Portable Tools — All portable manual hand tools, all portable compressed air operated hand tools.

5. Materials — Health — All types of materials that may produce *health* injury resulting from an acute or chronic toxic effect: Asbestos, lead, carbon monoxide, hydrogen sulfide, benzene, cyanide, arsenic.

6. Materials — Traumatic — All types of materials that may produce a traumatic injury resulting from corrosive, explosive or impact contact. Acids and caustics, flammables and explosives, compressed gases and liquids, molten metals, hot materials, flying sparks, sharp edges, abrasive materials, weight-shape, stacked materials, radiation, noise. Also include bodily movement in handling materials.

7. Electrical — Motors and generators, transformers and rectifiers, wiring and conductors, electrical hand tools, electrical welding and burning. Electrical guards — insulation, isolation, over-current protection and grounding. Ground-fault protection.

8. Vehicles — Manual hand trucks and dollies, automobiles, power trucks, forklifts, railroad equipment and rolling stock, marine vessels, aircraft.

9. Multi-Category — Critical procedures relating to more than one category:

   Excavations

   Confined Space Entry

   Lockout-Tagout

   Personal Protective Equipment

Figure 6.7   A typical format for a supervisor's hazard control notebook.

Departmental Controls

Second-line supervisors, in addition to evaluating the survey reports of their supervisors, obviously need to ensure that their programs are producing the desired results. Here are some of the actions that should be taken by the department levels above the first line.

- Keep a department hazard log or follow-up file. Either all items listed in survey reports as pending should be transferred to a hazard log or the report should be kept in a pending file. This file should be regularly reviewed by the second-line supervisor to ensure that action has been taken as scheduled.
- Perform department safety and housekeeping audits. Second-line supervisors should periodically inspect their department areas and compare their findings with what is reported in their supervisors' self-surveys. Exceptions should be discussed with the supervisors concerned to help them upgrade their abilities in the program.

Unit Controls

Just as in many other aspects of the total enterprise, management must establish controls to measure the results of the plan and ensure that the program is effective. The following list is by no means comprehensive, but will outline those elements considered to be the minimum needed for a successful program:

Goals and Objectives — Each department and subdepartment should establish goals and objectives for the continuous improvement of their safety inspection programs. Progress on these goals should be reviewed by upper levels of management.

Audits — Periodic audits should be conducted to measure the results of the program. This can be done in any number of ways. In some organizations it is done by the safety department, in others by special committees. The audit group randomly checks self-surveys for action on listed undesirable conditions and notifies the management if and where problems exist. They will also make random inspections of areas to measure program effectiveness and determine training needs. Primarily, the audit group should be a program assessment committee rather than a problem discovery committee.

Appearance Standards

The following is a list of some recommended items for establishing standards for appearance of the work site:

- Standard paint scheme for buildings and offices
- Routine maintenance painting schedule for all buildings and facilities
- Schedule and performance standards for custodial service
- Standard layout for all signs
- Beautification program for grounds and surrounding areas
- Easy maintenance materials selected for floors (nonskid) and walls in heavy traffic areas
- Scheduled roof maintenance plan
- Scheduled engineering inspection of all buildings and facilities
- Recycling and waste collection program and schedule, with disposal containers well identified and distributed around the facility

Adoption of these kinds of programs automatically results in improved appearance and takes care of many problems that usually are beyond the authority of first-line supervisors to resolve.

Interest and Motivation

Even with the best of procedures, there is still a need to provide ways to promote employee interest and motivate them to use the available program. It is beyond the scope of this chapter to describe all of the numerous programs that have been tried in the past. Suffice it to say that experience has shown that the most outstanding programs often fail without motivation. It is a human need to seek recognition, so it is recommended that after installing a sound system, some simple motivation program should be implemented. This might include quarterly plant inspections and awards for good safety and housekeeping, with appropriate publicity in the organization's newsletter.

Summary and Conclusions

The purpose and objective of this section was to outline the major elements of an effective safety inspection system. The concept behind the system described is based on delegating the responsibility for safety and housekeeping down to the lowest level of management and then helping them do it. Supervisors systematically search for and control hazards as well as substandard housekeeping conditions in their areas. At the same time, supervisors are taught to consider the source and cause of the problems they discover so that the problems are less likely to recur.

FURTHER READING

Books

Boylston, R. P. *Managing Safety and Health Programs*. New York: Van Nostrand Reinhold, 61–71, 142–143, 167–188, 208–219, 1990.

Bird, F. E. and Germain, G. L. *Practical Loss Control Leadership*. Loganville, GA: Det Norske Veritas (USA) Inc. 123-148, 1996.

Peterson, D. *Techniques of Safety Management*. Goshen, NY: Aloray, 159–169, 1989.

Colvin, R. J. *The Guidebook to Successful Safety Programming*. Boca Raton, FL: Lewis Publishers, 97–110, 1992.

Holt, A. St. J. and Andrews, H. *Principles of Health and Safety at Work*. Leicester, U.K.: IOSH Publishing, 105–111, 1993.

Grimaldi, J. V. and Simonds, R. H. *Safety Management,* Boston, MA: Irwin, 144–146, 1989.

Other Publications

Cal/OSHA Consultation Service. *Guide to Developing Your Workplace Injury and Illness Prevention Program*. CS-1, Cal/OSHA Consultation Service, San Francisco, CA, 1995.

Lack, R.W. The frustrating problem of industrial housekeeping. *Protection* 17(11): 12–16, November 1980.

U.S. Department of Labor, Occupational Safety and Health Administration. OSHA Safety and Health Program Management Guidelines: Issuance of Voluntary Guidelines. *Fed. Regist.* 54(16), January 26, 1989.

APPENDIX C:
HAZARD ASSESSMENT CHECKLIST

GENERAL WORK ENVIRONMENT

☐ Are all worksites clean and orderly?

☐ Are work surfaces kept dry or appropriate means taken to assure the surfaces are slip-resistant?

☐ Are all spilled materials or liquids cleaned up immediately?

☐ Is combustible scrap, debris and waste stored safely and removed from the worksite promptly?

☐ Is accumulated combustible dust routinely removed from elevated surfaces, including the overhead structure of buildings?

☐ Is combustible dust cleaned up with a vacuum system to prevent the dust going into suspension?

☐ Is metallic or conductive dust prevented from entering or accumulation on or around electrical enclosures or equipment?

☐ Are covered metal waste cans used for oily and paint-soaked waste?

☐ Are all oil and gas fired devices equipped with flame failure controls that will prevent flow of fuel if pilots or main burners are not working?

☐ Are paint spray booths, dip tanks and the like+ wcleaned regularly?

☐ Are the minimum number of toilets and washing facilities provided?

☐ Are all toilets and washing facilities clean and sanitary?

☐ Are all work areas adequately illuminated?

☐ Are pits and floor openings covered or otherwise guarded?

PERSONAL PROTECTIVE EQUIPMENT & CLOTHING

☐ Are protective gaggles or face shields provided and worn where there is any danger of flying particles or corrosive materials?

☐ Are approved safety glasses required to be worn at all times in areas where there is a risk of eye injuries such as punctures, abrasions, contusions or burns?

☐ Are workers who need corrective lenses (glasses or contacts lenses) in working environments with harmful exposures, required to wear only approved safety glasses, protective goggles, or use other medically approved precautionary procedures?

☐ Are protective gloves, aprons, shields, or other means provided against cuts, corrosive liquids and chemicals?

☐ Are hard hats provided and worn where danger of failing objects exists?

☐ Are hard hats inspected periodically for damage to the shell and suspension system?

☐ Is appropriate foot protection required where there is the risk of foot injuries from hot, corrosive, poisonous substances, falling objects, crushing or penetrating actions?

☐ Are approved respirators provided for regular or emergency use where needed?

☐ Is all protective equipment maintained in a sanitary condition and ready for use?

☐ Do you have eye wash facilities and a quick drench shower within the work area where workers are exposed to injurious corrosive materials?

☐ Where special equipment is needed for electrical workers, is it available?

☐ When lunches are eaten on the premises, are they eaten in areas where there is no exposure to toxic materials or other health hazards?

☐ Is protection against the effects of occupational noise exposure provided when sound levels exceed those of the Cal/OSHA noise standard?

WALKWAYS

☐ Are aisles and passageways kept clear?

☐ Are aisles and walkways marked as appropriate?

□ Are wet surfaces covered with non-slip materials?

□ Are holes in the floor, sidewalk or other walking surface repaired properly, covered or otherwise made safe?

□ Is there safe clearance for walking in aisles where motorized or mechanical handling equipment is operating.

□ Are spilled materials cleaned up immediately?

□ Are materials or equipment stored in such a way that sharp projectiles will not interfere with the walkway?

□ Are changes of direction or elevations readily identifiable?

□ Are aisles or walkways that pass near moving or operating machinery, welding operations or similar operations arranged so workers will not be subjected to potential hazards?

□ Is adequate headroom provided for the entire length of any aisle or walkway?

□ Are standard guardrails provided wherever aisle or walkway surfaces are elevated more than 30 inches above any adjacent floor or the ground?

□ Are bridges provided over conveyors and similar hazards?

FLOOR & WALL OPENINGS

□ Are floor openings guarded by a cover, guardrail, or equivalent-on all sides (except at entrance to stairways or ladders)?

□ Are toeboards installed around the edges of a permanent floor opening (where persons may pass below the opening)?

□ Are skylight screens of such construction and mounting that they will withstand a load of at least 200 pounds?

□ Is the glass in windows, doors, glass walls which are subject to human impact, of sufficient thickness and type for the condition of use?

□ Are grates or similar type covers over floor openings such as floor drains, of such design that foot traffic or rolling equipment will not be affected by the grate spacing?

□ Are unused portions of service pits and pits not actually in use either covered or protected by guardrails or equivalent?

□ Are manhole covers, trench covers and similar covers, plus their supports, designed to carry a truck Year axle load of at least 20,000 pounds when located in roadways and subject to vehicle traffic?

□ Are floor or wall openings in fire resistive construction provided with doors or covers compatible with the fire rating of the structure and provided with self-closing feature when appropriate?

STAIRS & STAIRWAYS

□ Are standard stair rails or handrails on all stairways having four or more risers?

□ Are all stairways at least 22 inches wide?

□ Do stairs have at least a S'S' overhead clearance?

□ Do stairs angle no more than 50 and no less than 30 degrees?

□ Are stairs of hollow-pan type treads and landings filled to noising level with solid material?

□ Are stop risers an stairs uniform from top to bottom, with no riser spacing greater than 7–1/2 inches?

□ Are steps on stairs and stairways designed or provided with a surface that renders them slip resistant?

□ Are stairway handrails located between 30 and 34 inches above the leading edge of stair treads?

□ Do stairway handrails have a least 1-1/2 inches of clearance between the handrails and the wall or surface they are mounted on?

□ Are stairway handrails capable of withstanding a load of 200 pounds, applied in any direction?

□ Where stairs or stairways exit directly into any area where vehicles may be operated, are adequate barriers and warnings provided to prevent workers stepping into the path of traffic?

□ Do stairway landings have a dimension measured in the direction of travel, at least equal to width of the stairway?

□ Is the vertical distance between stairway landings limited to 12 feet or less?

ELEVATED SURFACES

□ Are signs posted, when appropriate, showing the elevated surface load capacity?

□ Are surfaces elevated more than 30 inches above the floor or ground provided with standard guardrails?

□ Are all elevated surfaces (beneath which people or machinery could be exposed to falling objects) provided with standard 4-inch toeboards?

□ Is a permanent means of access and egress provided to elevated storage and work surfaces?

□ Is required headroom provided when a necessary?

□ Is material on elevated surfaces piled, stacked or racked in a manner to prevent it from tipping, failing, collapsing, rolling or spreading?

□ Are dock boards or bridge plates used when transferring materials between docks and trucks or rail cars?

EXITING OR EGRESS

□ Are all exits marked with an exit sign and illuminated by a reliable light source?

□ Are the directions to exits. when not immediately apparent, marked with visible signs?

□ Are doors, passageways or stairways, that are neither exits nor access to exits and which could be mistaken for exits, appropriately marked "NOT AN EXIT", "TO BASEMENT", "STOREROOM", and the like?

□ Are exit signs provided with the word "EXIT" in lettering at least 5 inches high and the stroke of the lettering at least 1/2 inch wide?

□ Are exit doors side-hinged?

□ Are all exits kept free of obstructions?

□ Are at least two means of egress provided from elevated platforms, pits or roams where the absence of a second exit would increase the risk of injury from hot, poisonous, corrosive, suffocating, flammable, or explosive substances?

□ Are there sufficient exits to permit prompt escape in case of emergency?

□ Are special precautions taken to protect workers during construction and repair operations?

□ Is the number of exits from each floor of a building, and the number of exits from the building itself, appropriate for the building occupancy load?

□ Are exit stairways which are required to be separated from other parts of a building enclosed by at least two hour fire-resistive construction in buildings more than four stories in height, and not less than one-hour fire resistive construction elsewhere?

□ When ramps are used as part of required exiting from a building, is the ramp slope limited to I-foot vertical and 12 feet horizontal?

□ Where exiting will be through frameless glass doors, glass exit doors. storm doors, and such, are the doors fully tempered and meet the safety requirements for human impact?

EXIT DOORS

□ Are doors which are required to serve as exits designed and constructed so that the way of exit travel is obvious and direct?

□ Are windows which could be mistaken for exit doors, made inaccessible by means of barriers or railings?

□ Are exit doors openable from the direction of exit travel without the use of a key or any special knowledge or effort, when the building is occupied?

□ Is a revolving, sliding or overhead door prohibited from serving as a required exit door?

□ Where panic hardware is installed on a required exit door, will it allow the door to open by applying a force of 15 pounds or less in the direction of the exit traffic?

□ Are doors an cold storage rooms provided with an inside release mechanism which will release the latch and open the door even if it's padlocked or otherwise locked on the outside?

☐ Where exit doors open directly onto any street, alley or other area where vehicles may be operated, are adequate barriers and warnings provided to prevent workers stepping into the path of traffic?

☐ Are doors that swing in both directions and are located between rooms where there is frequent traffic, provided with viewing panels in each door?

PORTABLE LADDERS

☐ Are all ladders maintained in good condition joints between steps and side rails tight, al; hardware and fittings securely attached, and moveable parts operating freely without binding or undue play?

☐ Are non-slip safety feet provided an each ladder?

☐ Are non-slip safety feet provided on each metal or rung ladder?

☐ Are ladder rungs and steps free of grease and oil'

☐ Is it prohibited to place a ladder in front of doors opening toward the ladder except when the door is blocked open, locked or guarded?

☐ Is it prohibited to place ladders on boxes, barrels, or other unstable bases to obtain additional height?

☐ Are workers instructed to face the ladder when ascending or descending?

☐ Are workers prohibited from using ladders that are broken, missing steps, rungs, or cleats, broken side rails or other faulty equipment?

☐ Are workers instructed not to use the top 2 steps of ordinary stepladders as a step?

☐ When portable rung ladders are used to gain access to elevated platforms, roofs, and the like., does the ladder always extend at least 3 feet above the elevated surface?

☐ Is it required that when portable rung or cleat type ladders are used the base is so placed that slipping will not occur, or it is lashed or otherwise held in place?

☐ Are portable metal ladders legibly marked with signs reading "CAUTION" "Do Not Use Around Electrical Equipment" or equivalent wording?

☐ Are workers prohibited from using ladders as guys, braces, skids, gin poles, or for other than their intended purposes?

☐ Are workers instructed to only adjust extension ladders while standing at a base (not while standing an the ladder or from a position above the ladder)?

☐ Are metal ladders inspected for damage?

☐ Are the rungs of ladders uniformly spaced at 12 inches, center to center?

HAND TOOLS & EQUIPMENT

☐ Are all tools and equipment (both, company and worker-owned) used by workers at their work-place in good condition?

☐ Are hand tools such as chisels, punches, which develop mushroomed heads during use, recon-ditioned or replaced as necessary?

☐ Are broken or fractured handles on hammers, axes and similar equipment replaced promptly?

☐ Are worn or bent wrenches replaced regularly?

☐ Are appropriate handles used on files and sim-ilar tools?

☐ Are workers made aware of the hazards caused by faulty or improperly used hand tools?

☐ Are appropriate safety glasses, face shields, and similar equipment used while using hand tools or equipment which might produce flying materials or be subject to breakage?

☐ Are jacks checked periodically to assure they are in good operating condition?

☐ Are tool handles wedged tightly in the head of all tools?

☐ Are tool cutting edges kept sharp so the tool will move smoothly without binding or skipping?

☐ Are tools stared in dry, secure location where they won't be tampered with?

☐ Is eye and face protection used when driving hardened or tempered spuds or nails?

PORTABLE (POWER OPERATED) TOOLS & EQUIPMENT

☐ Are grinders, saws, and similar equipment provided with appropriate safety guards?

☐ Are power tools used with the correct shield, guard or attachment recommended by the manufacturer?

☐ Are portable circular saws equipped with guards above and below the base shoe?

☐ Are circular saw guards checked to assure they are not wedged up, thus leaving the lower portion of the blade unguarded?

☐ Are rotating or moving parts of equipment guarded to prevent physical contact?

☐ Are all cord-connected, electrically-operated tools and equipment effectively grounded or of the approved double insulated type?

☐ Are effective guards in place over belts, pulleys, chains, sprockets, an equipment such as concrete mixers, air compressors, and the like?

☐ Are portable fans provided with full guards or screens having openings 1/2 inch or less?

☐ Is hoisting equipment available and used for lifting heavy objects, and are hoist ratings and characteristics appropriate for the task?

☐ Are ground-fault circuit interrupters provided an all temporary electrical 1 5 and 20 ampere circuits, used during periods of construction?

☐ Are pneumatic and hydraulic hoses an power-operated tools checked regularly for deterioration or damage?

ABRASIVE WHEEL EQUIPMENT GRINDERS

☐ Is the work rest used and kept adjusted to within 1/8 inch of the wheel?

☐ Is the adjustable tongue an the top side of the grinder used and kept adjusted to within 1/4 inch of the wheel?

☐ Do side guards cover the spindle, nut, and flange and 75 percent of the wheel diameter?

☐ Are bench and pedestal grinders permanently mounted?

☐ Are goggles or face shields always worn when grinding?

☐ Is the maximum RPM rating of each abrasive wheel compatible with the RPM rating of the grinder motor?

☐ Are fixed or permanently mounted grinders connected to their electrical supply system with metallic conduit or other permanent wiring method?

☐ Does each grinder have an individual an and off control switch?

☐ Is each electrically operated grinder effectively grounded?

☐ Before new abrasive wheels are mounted, are they visually inspected and ring tested?

☐ Are dust collectors and powered exhausts provided on grinders used in operations that produce large amounts of dust?

☐ Are splash guards mounted on grinders that use coolant, to prevent the coolant reaching workers?

☐ Is cleanliness maintained around grinder?

POWDER ACTUATED TOOLS

☐ Are workers who operate powder-actuated tools trained in their use and carry a valid operators card?

☐ Do the powder-actuated tools being used have written approval of the Division of Occupational Safety and Health?

☐ Is each powder-actuated tool stored in its own locked container when not being used?

☐ Is a sign at least 7" by 10" with bold type reading "POWDER-ACTUATED TOOL IN USE" conspicuously posted when the tool is being used?

☐ Are powder-actuated tools left unloaded until they are actually ready to be used?

☐ Are powder-actuated tools inspected for obstructions or defects each day before use?

☐ Do powder-actuated tools operators have and use appropriate personal protective equipment such as hard hats, safety goggles, safety shoes and ear protectors?

MACHINE GUARDING

☐ Is there a training program to instruct workers an safe methods of machine operation?

☐ Is there adequate supervision to ensure that workers are following safe machine operating procedures?

☐ Is there a regular program of safety inspection of machinery and equipment?

☐ Is all machinery and equipment kept clean and properly maintained?

☐ Is sufficient clearance provided around and between machines to allow for safe operations, set up and servicing, material handling and waste removal?

☐ Is equipment and machinery securely placed and anchored, when necessary to prevent lipping or other movement that could result in personal injury?

☐ Is there a power shut-off switch within reach of the operator's position at each machine?

☐ Can electric power to each machine be locked out for maintenance, repair, or security?

☐ Are the noncurrent-carrying metal parts of electrically operated machines bonded and grounded?

☐ Are foot-operated switches guarded or arranged to prevent accidental actuation by personnel or failing objects?

☐ Are manually operated valves and switches controlling the operation of equipment and machines clearly identified and readily accessible?

☐ Are all emergency stop buttons colored red?

☐ Are all pulleys and belts that are within 7 feet of the floor or working level properly guarded?

☐ Are all moving chains and gears properly guarded?

☐ Are splash guards mounted an machines that use coolant, to prevent the coolant from reaching workers?

☐ Are methods provided to protect the operator and other workers in the machine area from hazards created at the point of operation, ingoing nip points, rotating parts, flying chips, and sparks?

☐ Are machinery guards secure and so arranged that they do not offer a hazard in their use?

☐ If special hand tools are used for placing and removing material, do they protect the operator's hands?

☐ Are revolving drums, barrels, and containers required to be guarded by an enclosure that is interlocked with the drive mechanism, so that revolution cannot occur unless the guard enclosure is in place, so guarded?

☐ Do arbors and mandrels have firm and secure bearings and are they free from play?

☐ Are provisions made to prevent machines from automatically starting when power is restored after a power failure or shutdown?

☐ Are machines constructed so as to be free from excessive vibration when the largest size tool is mounted and run at full speed?

☐ If machinery is cleaned with compressed air, is air pressure controlled and personal protective equipment or other safeguards used to protect operators and other workers from eye and body injury?

☐ Are fan blades protected with a guard having openings no larger than 1/2 inch, when operating within 7 feet of the floor?

☐ Are saws used for ripping, equipped with anti-kick back devices and spreaders?

☐ Are radial arm saws so arranged that the cutting head will gently return to the back of the table when released?

LOCKOUT BLOCKOUT PROCEDURES

☐ Is all machinery or equipment capable of movement. required to be de-energized or disengaged and blocked or locked out during cleaning, servicing, adjusting or setting up operations, whenever required?

□ Is the locking-out of control circuits in lieu of locking-out main power disconnects prohibited?

□ Are all equipment control valve handles provided with a means for locking-out?

□ Does the lock-out procedure require that stared energy (i.e. mechanical, hydraulic, air,) be released or blocked before equipment is locked-out for repairs?

□ Are appropriate workers provided with individually keyed personal safety locks?

□ Are workers required to keep personal control of their key(s) while they have safety locks in use?

□ Is it required that workers check the safety of the lock out by attempting a start up after making sure no one is exposed?

□ Where the power disconnecting means for equipment does not also disconnect the electrical control circuit:

□ Are the appropriate electrical enclosures identified?

□ Is means provide to assure the control circuit can also be disconnected and locked out?

WELDING, CUTTING & BRAZING

□ Are only authorized and trained personnel permitted to use welding, cutting or brazing equipment?

□ Do all operator have a copy of the appropriate operating instructions and are they directed to follow them?

□ Are compressed gas cylinders regularly examined for obvious signs of defects, deep rusting, or leakage?

□ Is care used in handling and storage of cylinders, safety valves, relief valves, and the like, to prevent damage?

□ Are precautions taken to prevent the mixture of air or oxygen with flammable gases, except at a burner or in a standard torch?

□ Are only approved apparatus (torches, regulators, pressure-reducing valves. acetylene generators, manifolds) used?

□ Are cylinders kept away from sources of heat?

□ Is it prohibited to use cylinders as rollers or supports?

□ Are empty cylinders appropriately marked, their valves closed and valve-protection caps on?

□ Are signs reading: DANGER NO-SMOKING, MATCHES, OR OPEN LIGHTS, or the equivalent posted?

□ Are cylinders, cylinder valves, couplings, regulators, hoses, and apparatus keep free of oily or greasy substances?

□ Is care taken not to drop or strike cylinders?

□ Unless secured on special trucks, are regulators removed and valve-protection caps put in place before moving cylinders?

□ Do cylinders without fixed hand wheels have keys, handles, or non-adjustable wrenches on stem valves when in service?

□ Are liquefied gases stared and shipped valve-end up with valve covers in place?

□ Are workers instructed to never crack a fuel-gas cylinder valve near sources of ignition?

□ Before a regulator is removed, is the valve closed and gas released form the regulator?

□ Is red used to identify the acetylene (and other fuel-gas) hose, green for oxygen hose, and black for inert gas and air hose?

□ Are pressure-reducing regulators used only for the gas and pressures for which they are intended?

□ Is open circuit (No Load) voltage of arc welding and cutting machines as low as possible and not in excess of the recommended limits?

□ Under wet conditions, are automatic controls for reducing no-load voltage used?

□ Is grounding of the machine frame and safety ground connections of portable machines checked periodically?

□ Are electrodes removed from the holders when not in use?

□ Is it required that electric power to the welder be shut off when no one is in attendance?

□ Is suitable fire extinguishing equipment available for immediate use?

□ Is the welder forbidden to coil or loop welding electrode cable around his body?

□ Are wet machines thoroughly dried and tested before being used?

□ Are work and electrode lead cables frequently inspected for wear and damage, and replaced when needed?

□ Do means for connecting cables' lengths have adequate insulation?

□ When the object to be welded cannot be moved and fire hazards cannot be removed, are shields used to confine heat, sparks, and slag?

□ Are fire watchers assigned when welding or cutting is performed, in locations where a serious fire might develop?

□ Are combustible floors kept wet, covered by damp sand, or protected by fire-resistant shields?

□ When floors are wet down, are personnel protected from possible electrical shock?

□ When welding is done an metal walls, are precautions taken to protect combustibles on the other side?

□ Before hot work is begun, are used drums, barrels, tanks, and other containers so thoroughly cleaned that no substances remain that could explode, ignite, or produce toxic vapors?

□ Is it required that eye protection helmets, hand shields and gaggles meet appropriate standards?

□ Are workers exposed to the hazards created by welding, cutting, or bracing operations protected with personal protective equipment and clothing?

□ Is a check made for adequate ventilation in and where welding or cutting is preformed?

□ When working in confined places are environmental monitoring tests taken and means provided for quick removal of welders in case of an emergency?

COMPRESSORS & COMPRESSED AIR

□ Are compressors equipped with pressure relief valves, and pressure gauges?

□ Are compressor air intakes installed and equipped to ensure that only clean uncontaminated air enters the compressor?

□ Are air filters installed an the compressor intake?

□ Are compressors operated and lubricated-in accordance with the manufacturer's recommendations?

□ Are safety devices an compressed air systems checked frequently?

□ Before any repair work is done on the pressure system of a compressor, is the pressure bled off and the system locked-out?

□ Are signs posted to warn of the automatic starting feature of the compressors?

□ Is the belt driven system totally enclosed to provide protection for the front, back, top, and sides?

□ Is it strictly prohibited to direct compressed air towards a person?

□ Are workers prohibited from using highly compressed air for cleaning purposes?

□ If compressed air is used for cleaning off clothing, is the pressure reduced to less than 10 psi?

□ When using compressed air for cleaning, do workers use personal protective equipment?

□ Are safety chains or other suitable locking devices used at couplings of high pressure hose lines where a connection failure would create a hazard?

□ Before compressed air is used to empty containers of liquid, is the safe working pressure of the container checked?

□ When compressed air is used with abrasive blast cleaning equipment, is the operating valve a type that must be hold open manually?

□ When compressed air is used to inflate auto tires, is a clip-on chuck and an inline regulator preset to 40 psi required?

□ Is it prohibited to use compressed air to clean up or move combustible dust if such action could cause the dust to be suspended in the air and cause a fire or explosion hazard?

COMPRESSED AIR RECEIVERS

□ Is every receiver equipped with a pressure gauge and with one or more automatic, spring-loaded safety valves?

□ Is the total relieving capacity of the safety valve capable of preventing pressure in the receiver from exceeding the maximum allowable working pressure of the receiver by more than 10 percent?

□ Is every air receiver provided with a drainpipe and valve at the lowest point for the removal of accumulated oil and water?

□ Are compressed air receivers periodically drained of moisture and oil?

□ Are all safety valves tested frequently and at regular intervals to determine whether they are in good operating condition?

□ Is there a current operating permit issued by the Division of Occupational Safety and Health?

□ Is the inlet of air receivers and piping systems kept free of accumulated oil and carbonaceous materials?

COMPRESSED GAS & CYLINDERS

□ Are cylinders with a water weight capacity over 30 pounds equipped with means for connecting a valve protector device, or with a collar or recess to protect the valve?

□ Are cylinders legibly marked to clearly identify the gas contained?

□ Are compressed gas cylinders stored in areas which are protected from external heat sources such as flame impingement, intense radiant heat, electric arcs, or high temperature lines?

□ Are cylinders located or stared in areas where they will not be damaged by passing or failing objects, or subject to tampering by unauthorized persons?

□ Are cylinders stored or transported in a manner to prevent them creating a hazard by tipping, falling or rolling?

□ Are cylinders containing liquefied fuel gas, stored or transported in a position so that the safety relief device is always in direct contact with the vapor space in the cylinder?

□ Are valve protectors always placed an cylinders when the cylinders are not in use or connected for use?

□ Are all valves closed off before a cylinder is moved, when the cylinder is empty, and at the completion of each job?

□ Are low pressure fuel-gas cylinders checked periodically for corrosion, general distortion, cracks, or any other defect that might indicate a weakness or render it unfit for service?

□ Does the periodic check of low pressure fuel-gas cylinders include a close inspection of the cylinders' bottom?

HOIST & AUXILIARY EQUIPMENT

□ Is each overhead electric hoist equipped with a limit device to stop the hook travel at its highest and lowest point of safe travel?

□ Will each hoist automatically stop and hold any load up to 125 percent of its rated load, if its actuating force is removed?

□ Is the rated load of each hoist legibly marked and visible to the operator?

□ Are stops provided at the safe limits of travel for trolley hoist?

□ Are the controls of hoists plainly marked to indicate the direction of travel or motion?

□ Is each cage-controlled hoist equipped with an effective warning device?

□ Are close-fitting guards or other suitable devices installed an hoist to assure hoist ropes will be maintained in the sheave groves?

□ Are all hoist chains or rapes of sufficient length to handle the full range of movement for the application while still maintaining two full wraps on the drum at all times?

☐ Are nip points or contact points between hoist rapes and sheaves which are permanently located within 7 feet of the floor, ground or working platform, guarded?

☐ Is it prohibited to use chains or rope slings that are kinked or twisted?

☐ Is it prohibited to use the hoist rope or chain wrapped around the load as a substitute, for a sting?

☐ Is the operator instructed to avoid carrying loads over people?

☐ Are only workers who have been trained in the proper use of hoists allowed to operate them?

INDUSTRIAL TRUCKS — FORKLIFTS

☐ Are only trained personnel allowed to operate industrial trucks?

☐ Is substantial overhead protective equipment provided on high lift rider equipment?

☐ Are the required lift truck operating rules posted and enforced?

☐ Is directional lighting provided on each industrial truck that operates in an area with less than 2 foot candles per square fact of general lighting?

☐ Does each industrial truck have a warning horn, whistle, gong or other device which can be clearly heard above the normal noise in the areas where operated?

☐ Are the brakes an each industrial truck capable of bringing the vehicle to a complete and safe stop when fully loaded?

☐ Will the industrial truck's parking brake effectively prevent the vehicle from moving when unattended?

☐ Are industrial trucks operating in areas where flammable gases or vapors, or combustible dust or ignitable fibers may be present in the atmosphere, approved for such locations?

☐ Are motorized hand and hand/rider trucks so designed that the brakes are applied, and power to the drive motor shuts off when the operator releases his/her grip on the device that controls the travel?

☐ Are industrial trucks with internal combustion engine operated in buildings or enclosed areas, carefully checked to ensure such operations do not cause harmful concentration of dangerous gases or fumes?

SPRAYING OPERATIONS

☐ Is adequate ventilation assured before spray operations are started?

☐ Is mechanical ventilation provided when spraying operation is done in enclosed areas?

☐ When mechanical ventilation is provided during spraying operations, is it so arranged that it will not circulate the contaminated air?

☐ Is the spray area free of hot surfaces?

☐ Is the spray area at least 20 feet from flames, sparks, operating electrical motors and other ignition sources?

☐ Are portable lamps used to illuminate spray areas suitable for use in a hazardous location?

☐ Is approved respiratory equipment provided and used when appropriate during spraying operations?

☐ Do solvents used for cleaning have a flash point of 100°F or more?

☐ Are fire control sprinkler heads kept clean?

☐ Are "NO SMOKING" signs posted in spray areas, paint rooms, paint booths, and paint storage areas?

☐ Is the spray area kept clean of combustible residue?

☐ Are spray booths constructed of metal, masonry, or other substantial noncombustible material?

☐ Are spray booth floors and baffles noncombustible and easily cleaned?

☐ Is infrared drying apparatus kept out of the spray area during spraying operations?

☐ Is the spray booth completely ventilated before using the drying apparatus?

□ Is the electric drying apparatus properly grounded?

□ Are lighting fixtures for spray booths located outside of the booth and the interior lighted through sealed clear panels?

□ Are the electric motors for exhaust fans placed outside booths or ducts?

□ Are belts and pulleys inside the booth fully enclosed?

□ Do ducts have access doors to allow cleaning?

□ Do all drying spaces have adequate ventilation?

ENTERING CONFINED SPACES

□ Are confined spaces thoroughly emptied of any corrosive or hazardous substances, such as acids or caustics, before entry?

□ Before entry, are all lines to a confined space, containing inert, toxic, flammable, or corrosive materials valved off and blanked or disconnected and separated?

□ Is it required that all impellers, agitators, or other moving equipment inside confined spaces be locked-out if they present a hazard?

□ Is either natural or mechanical ventilation provided prior to confined space entry?

□ Before entry, are appropriate atmospheric tests performed to check for oxygen deficiency, toxic substance and explosive concentrations in the confined space before entry?

□ Is adequate illumination provided for the work to be performed in the confined space?

□ Is the atmosphere inside the confined space frequently tested or continuously monitor during conduct of work?

□ Is there an assigned safety standby worker outside of the confined space, whose sale responsibility is to watch the work in progress, sound an alarm if necessary, and tender assistance?

□ Is the standby worker or other workers prohibited from entering the confined space

without lifelines and respiratory equipment if there is any questions as to the cause of an emergency?

□ In addition to the standby worker, is there at least one other trained rescuer in the vicinity?

□ Are all rescuers appropriately trained and using approved, recently inspected equipment?

□ Does all rescue equipment allow for lifting workers vertically from a top opening?

□ Are there trained personnel in First Aid and CPR immediately available?

□ Is there an effective communication system in place whenever respiratory equipment is used and the worker in the confined space is out of sight of the standby person?

□ Is approved respiratory equipment required if the atmosphere inside the confined space cannot be made acceptable?

□ Is all portable electrical equipment used inside confined spaces either grounded and insulated, or equipped with ground fault protection?

□ Before gas welding or burning is started in a confined space, are hoses checked for leaks, compressed gas bottles forbidden inside of the confined space, torches lighted only outside of the confined area and the confined area tested for an explosive atmosphere each time before a lighted torch is to be taken into the confined space?

□ If workers will be using oxygen-consuming equipment--such as salamanders. torches, furnaces,-in a confined space, is sufficient air provided to assure combustion without reducing the oxygen concentration of the atmosphere below 19.5 percent by volume?

□ Whenever combustion-type equipment is used in confined space, are provisions made to ensure the exhaust gases are vented outside of the enclosure?

□ Is each confined space checked for decaying vegetation or animal matter which may produce methane?

☐ Is the confined space checked for possible industrial waste which could contain toxic properties?

☐ If the confined space is below the ground and near areas where motor vehicles will be operating, is it possible for vehicle exhaust or carbon monoxide to enter the space?

ENVIRONMENTAL CONTROLS

☐ Are all work areas properly illuminated?

☐ Are workers instructed in proper first aid and other emergency procedures?

☐ Are hazardous substances identified which may cause harm by inhalation, ingestion, skin absorption or contact?

☐ Are workers aware of the hazards involved with the various chemicals they may be exposed to in their work environment, such as ammonia, chlorine, epoxies, caustics?

☐ Is worker exposure to chemicals in the workplace kept within acceptable levels?

☐ Can a less harmful method or product be used?

☐ Is the work area's ventilation system appropriate for the work being performed?

☐ Are spray painting operations done in spray rooms or booths equipped with an appropriate exhaust system?

☐ Is worker exposure to welding fumes controlled by ventilation, use of respirators, exposure time, or other means?

☐ Are welders and other workers nearby provided with flash shields during welding operations?

☐ If forklifts and other vehicles are used in buildings or other enclosed areas, are the carbon monoxide levels kept below maximum acceptable concentration?

☐ Has there been a determination that noise levels in the facilities are within acceptable levels?

☐ Are steps being taken to use engineering controls to reduce excessive noise levels?

☐ Are proper precautions being taken when handling asbestos and other fibrous materials?

☐ Are caution labels and signs used to warn of asbestos?

☐ Are wet methods used, when practicable, to prevent the emission of airborne asbestos fibers, silica dust and similar hazardous materials?

☐ Is vacuuming with appropriate equipment used whenever possible rather than blowing or sweeping dust?

☐ Are grinders, saws, and other machines that produce respirable dusts vented to an industrial collector or central exhaust system?

☐ Are all local exhaust ventilation systems designed and operating properly such as airflow and volume necessary for the application? Are the ducts free of obstructions or the belts slipping?

☐ Is personal protective equipment provided, used and maintained wherever required?

☐ Are there written standard operating procedures for the selection and use of respirators where needed?

☐ Are restrooms and washrooms kept clean and sanitary?

☐ Is all water provided for drinking, washing, and cooking potable?

☐ Are all outlets for water not suitable for drinking clearly identified?

☐ Are workers' physical capacities assessed before being assigned to jobs requiring heavy work?

☐ Are workers instructed in the proper manner of lifting heavy objects?

☐ Where heat is a problem, have all fixed work areas been provided with spot cooling or air conditioning?

☐ Are workers screened before assignment to areas of high heat to determine if their health condition might make them more susceptible to having an adverse reaction?

☐ Are workers working on streets and roadways where they are exposed to the hazards of traffic, required to wear bright colored (traffic orange) warning vest?

☐ Are exhaust stacks and air intakes located that contaminated air will not be recirculated within a building or other enclosed area?

☐ Is equipment producing ultra-violet radiation property shielded?

FLAMMABLE & COMBUSTIBLE MATERIALS

☐ Are combustible scrap, debris and waste materials (i.e. oily rags) stored in covered metal receptacles and removed from the worksite promptly?

☐ Is proper storage practiced to minimize the risk of fire including spontaneous combustion?

☐ Are approved containers and tanks used for the storage and handling of flammable and combustible liquids?

☐ Are all connections on drums and combustible liquid piping, vapor and liquid tight?

☐ Are all flammable liquids kept in closed containers when not in use (e.g. parts cleaning tanks, pans)?

☐ Are bulk drums of flammable liquids grounded and bonded to containers during dispensing?

☐ Do storage rooms for flammable and combustible liquids have explosion-proof lights?

☐ Do storage rooms for flammable and combustible liquids have mechanical or gravity ventilation?

☐ Is liquefied petroleum gas stored, handled, and used in accordance with safe practices and standards?

☐ Are liquefied petroleum storage tanks guarded to prevent damage from vehicles?

☐ Are all solvent wastes, and flammable liquids kept in fire-resistant covered containers until they are removed from the worksite?

☐ Is vacuuming used whenever possible rather than blowing or sweeping combustible dust?

☐ Are fire separators placed between containers of combustibles or flammables, when stacked one upon another, to assure their support and stability?

☐ Are fuel gas cylinders and oxygen cylinders separated by distance, fire resistant barriers or other means while in storage?

☐ Are fire extinguishers selected and provided for the types of materials in areas where they are to be used?
Class A: Ordinary combustible material fires.
Class B: Flammable liquid, gas or grease fires.
Class C: Energized-electrical equipment fires.

☐ If a Halon 1301 fire extinguisher is used, can workers evacuate within the specified time for that extinguisher?

☐ Are appropriate fire extinguishers mounted within 75 feet of outside areas containing flammable liquids, and within 10 feet of any inside storage area for such materials?

☐ Is the transfer/withdrawal of flammable or combustible liquids performed by trained personnel?

☐ Are fire extinguishers mounted so that workers do not have to travel more than 75 feet for a class "A" fire or 50 feet for a class "B" fire?

☐ Are workers trained in the use of fire extinguishers?

☐ Are extinguishers free from obstructions or blockage?

☐ Are all extinguishers serviced, maintained and tagged at intervals not to exceed one year?

☐ Are all extinguishers fully charged and in their designated places?

☐ Is a record maintained of required monthly checks of extinguishers?

☐ Where sprinkler systems are permanently installed, are the nozzle heads directed or arranged so that water will not be sprayed into operating electrical switch boards and equipment?

☐ Are "NO SMOKING" signs posted where appropriate in areas where flammable or combustible materials are used or stored?

☐ Are "NO SMOKING" signs posted on liquefied petroleum gas tanks?

☐ Are "NO SMOKING" rules enforced in areas involving storage and use of flammable materials?

☐ Are safety cans used for dispensing flammable or combustible liquids at a point of use?

☐ Are all spills of flammable or combustible liquids cleaned up promptly?

☐ Are storage tanks adequately vented to prevent the development of excessive vacuum or pressure as a result of filling, emptying, or atmosphere temperature changes?

☐ Are storage tanks equipped with emergency venting that will relieve excessive internal pressure caused by fire exposure?

☐ Are spare portable or butane tanks which are sued by industrial trucks stored in accord with regulations?

FIRE PROTECTION

☐ Do you have a fire prevention plan?

☐ Does your plan describe the type of fire protection equipment and/or systems?

☐ Have you established practices and procedures to control potential fire hazards and ignition sources?

☐ Are workers aware of the fire hazards of the material and processes to which they are exposed?

☐ Is your local fire department well acquainted with your facilities, location and specific hazards?

☐ If you have a fire alarm system, is it tested at least annually?

☐ If you have a fire alarm system, is it certified as required?

☐ If you have interior stand pipes and valves, are they inspected regularly?

☐ If you have outside private fire hydrants, are they flushed at least once a year and on a routine preventive maintenance schedule?

☐ Are fire doors and shutters in good operating condition?

☐ Are fire doors and shutters unobstructed and protected against obstructions. including their counterweights?

☐ Are fire door and shutter fusible links in place?

☐ Are automatic sprinkler system water control valves, air and water pressures checked weekly/periodically as required?

☐ Is maintenance of automatic sprinkler system assigned to responsible persons or to a sprinkler contractor?

☐ Are sprinkler heads protected by metal guards, when exposed to physical damage?

☐ Is proper clearance maintained below sprinkler heads?

☐ Are portable fire extinguishers provided in adequate number and type?

☐ Are fire extinguishers mounted in readily accessible locations?

☐ Are fire extinguishers recharged regularly and noted on the inspection tag?

☐ Are workers periodically instructed in the use of extinguishers and fire protection procedures?

HAZARDOUS CHEMICAL EXPOSURES

☐ Are workers trained in the safe handling practices of hazardous chemicals such as acids, caustics, and the like?

☐ Are workers aware of the potential hazards involving various chemicals stored or used in the workplace — such as acids, bases, caustics, epoxies, phenols?

☐ Is worker exposure to chemicals kept within acceptable levels?

☐ Are eye wash fountains and safety showers provided in areas where corrosive chemicals are handled?

☐ Are all containers, such as vats and storage tanks labeled as to their contents — e.g. "CAUSTICS"?

☐ Are all workers required to use personal protective clothing and equipment when handling chemicals (i.e. gloves, eye protection, respirators)?

☐ Are flammable or toxic chemicals kept in closed containers when not in use?

☐ Are chemical piping systems clearly marked as to their content?

☐ Where corrosive liquids are frequently handled in open containers or drawn from storage vessels or pipe lines, is adequate means readily available for neutralizing or disposing of spills or overflows properly and safety?

☐ Have standard operating procedures been established and are they being followed when cleaning up chemical spills?

☐ Where needed for emergency use, are respirators stored in a convenient. clean and sanitary location?

☐ Are respirators intended for emergency use adequate for the various uses for which they may be needed?

☐ Are workers prohibited from eating in areas where hazardous chemicals are present?

☐ Is personal protective equipment provided, used and maintained whenever necessary?

☐ Are there written standard operating procedures for the selection and use of respirators where needed?

☐ If you have a respirator protection program, are your workers instructed on the correct usage and limitations of the respirators?

☐ Are the respirators NIOSH approved for this particular application?

☐ Are they regularly inspected and cleaned sanitized and maintained?

☐ If hazardous substances are used in your processes, do you have a medical or biological monitoring system in operation?

☐ Are you familiar with the Threshold Limit Values or Permissible Exposure Limits of airborne contaminants and physical agents used in your workplace?

☐ Have control procedures been instituted for hazardous materials, where appropriate, such as respirators, ventilation systems, handling practices, and the like?

☐ Whenever possible, are hazardous substances handled in properly designed and exhausted booths or similar locations?

☐ Do you use general dilution or local exhaust ventilation systems to control dusts, vapors, gases, fumes, smoke, solvents or mists which may be generated in your workplace?

☐ Is ventilation equipment provided for removal of contaminants from such operations as production grinding, buffing, spray painting, and/or vapor decreasing, and is it operating properly?

☐ Do workers complain about dizziness, headaches, nausea, irritation, or other factors of discomfort when they use solvents or other chemicals?

☐ Is there a dermatitis problem — do workers complain about skin dryness, irritation, or sensitization?

☐ Have you considered the use of an industrial hygienist or environmental health specialist to evaluate your operation?

☐ If internal combustion engines are used, is carbon monoxide kept within acceptable levels?

☐ Is vacuuming used, rather than blowing or sweeping dusts whenever possible for clean-up?

☐ Are materials which give off toxic asphyxiant, suffocating or anesthetic fumes. stored in remote or isolated locations when not in use?

HAZARDOUS SUBSTANCES COMMUNICATION

☐ Is there a list of hazardous substances used in your workplace?

☐ Is there a written hazard communication program dealing with Material Safety Data Shoots (MSDS) labeling, and worker training?

☐ Who is responsible for MSDS, container labeling, worker training?

☐ Is each container for a hazardous substance (i.e. vats, bottles, storage tanks,) labeled with product identity and a hazard warning (communication of the specific health hazards and physical hazards)?

☐ Is there a Material Safety Data Sheet readily available for each hazardous substance used?

☐ How will you inform other employers whose workers share the same work area where the hazardous substances are used?

☐ Is there an worker training program for hazardous substances?

☐ Does this program include:

☐ An explanation of what an MSDS is and how to use and obtain one?

☐ SDS contents for each hazardous substance or class of substances?

☐ Explanation of "Right to Know"?

☐ Identification of where workers can see the employer's written hazard communication program and where hazardous substances are present in their work area?

☐ The physical and health hazards of substances in the work area, how to detect their presence, and specific protective measures to be used?

☐ Details of the hazard communication program, including how to use the labeling system and MSDS'S?

☐ How workers will be informed of hazards of non-routine tasks, and hazards of unlabeled pipes?

ELECTRICAL

☐ Are your workplace electricians familiar with the Cal/OSHA Electrical Safety Orders?

☐ Do you specify compliance with Cal/OSHA for all contract electrical work?

☐ Are all workers required to report as soon as practicable any obvious hazard to life or property observed in connection with electrical equipment or lines?

☐ Are workers instructed to make preliminary inspections and/or appropriate tests to determine what conditions exist before starting work on electrical equipment or lines?

☐ When electrical equipment or lines are to be serviced, maintained or adjusted, are necessary switches opened, locked-out and tagged whenever possible?

☐ Are portable electrical tools and equipment grounded or of the double insulated type?

☐ Are electrical appliances such as vacuum cleaners, polishers, vending machines grounded?

☐ Do extension cords being used have a grounding conductor?

☐ Are multiple plug adapters prohibited?

☐ Are ground-fault circuit interrupters installed an each temporary 15 or 20 ampere, 120 volt AC circuit at locations where construction, demolition, modifications, alterations or excavations are being performed?

☐ Are all temporary circuits protected by suitable disconnecting switches of plug connectors at the junction with permanent wiring?

☐ Is exposed wiring and cards with frayed or deteriorated insulation repaired or replaced promptly?

☐ Are flexible cards and cables free of splices or taps?

☐ Are clamps or other securing means provided on flexible cords or cables at plugs. receptacles, tools, equipment and is the card jacket securely hold in place?

☐ Are all card, cable and raceway connections intact and secure?

☐ In wet or damp locations, are electrical tools and equipment appropriate for the use or location or otherwise protected?

☐ Is the location of electrical power lines and cables (overhead, underground, under-floor, other side of walls) determined before digging, drilling or similar work is begun?

☐ Are metal measuring tapes, ropes, handlines or similar devices with metallic thread woven into the fabric prohibited where they could come in contact with energized parts of equipment or circuit conductors?

☐ Is the use of metal ladders prohibited in area where the ladder or the person using the ladder could come in contact with energized parts of equipment, fixtures or circuit conductors?

☐ Are all disconnecting switches and circuit breakers labeled to indicate their use or equipment served?

☐ Are disconnecting means always opened before fuses are replaced?

☐ Do all interior wiring systems include provisions for grounding metal parts of electrical raceways, equipment and enclosures?

☐ Are all electrical raceways and enclosures securely fastened in place?

☐ Are all energized parts of electrical circuits and equipment guarded against accidental contact by approved cabinets or enclosures?

☐ Is sufficient access and working space provided and maintained about all electrical equipment to permit ready and safe operations and maintenance?

☐ Are all unused openings (including conduit knockouts) in electrical enclosures and fittings closed with appropriate covers, plugs or plates?

☐ Are electrical enclosures such as switches, receptacles, junction boxes, etc., provided with tight-fitting covers or plates?

☐ Are disconnecting switches for electrical motors in excess of two horsepower, capable of opening the circuit when the motor is in a stalled condition, without exploding? (Switches must be horsepower rated equal to or in excess of the motor hp rating).

☐ Is low voltage protection provided in the control device of motors driving machines or equipment which could cause probably injury from inadvertent starting?

☐ Is each motor disconnecting switch or circuit breaker located within sight of the motor control device?

☐ Is each motor located within sight of its controller or the controller disconnecting means capable of being lacked in the open position or is a separate disconnecting means ins-tailed in the circuit within sight of the motor?

☐ Is the controller for each motor in excess of two horsepower, rated in horsepower equal to or in excess of the rating of the motor is serves?

☐ Are workers who regularly work an or around energized electrical equipment or lines instructed in the cardiopulmonary resuscitation (CPR) methods?

☐ Are workers prohibited from working alone an energized lines or equipment over 600 volts?

NOISE

☐ Are there areas in the workplace where continuous noise levels exceed 85 dBA? (To determine maximum allowable levels for intermittent or impact noise, see Title 8, Section 5097.)

☐ Are noise levels being measured using a sound level meter or an octave band analyzer and records being kept?

☐ Have you tried isolating noisy machinery from the rest of your operation?

☐ Have engineering controls been used to reduce excessive noise levels?

☐ Where engineering controls are determined not feasible, are administrative controls (i.e. worker rotation) being used to minimize individual worker exposure to noise?

☐ Is there an ongoing preventive health program to educate workers in safe levels of noise and exposure, effects of noise on their health, and use of personal protection?

☐ Is the training repeated annually for workers exposed to continuous noise above 85 dBA?

☐ Have work areas where noise levels make voice communication between workers difficult been identified and posted?

☐ Is approved hearing protective equipment (noise attenuating devices) available to every worker working in areas where continuous noise levels exceed 85 dBA?

☐ If you use ear protectors, are workers properly fitted and instructed in their use and care?

☐ Are workers exposed to continuous noise above 85 dBA give periodic audiometric testing to ensure that you have an effective hearing protection system?

FUELING

☐ Is it prohibited to fuel an internal combustion engine with a flammable liquid while the engine is running?

☐ Are fueling operations done in such a manner that likelihood of spillage will be minimal?

☐ When spillage occurs during fueling operations, is the spilled fuel cleaned up completely, evaporated, or other measures taken to control vapors before restarting the engine?

☐ Are fuel tank caps replaced and secured before starting the engine?

☐ In fueling operations is there always metal contact between the container and fuel tank?

☐ Are fueling hoses of a type designed to handle the specific type of fuel?

☐ Is it prohibited to handle or transfer gasoline in open containers?

☐ Are open lights, open flames, or sparking or arcing equipment prohibited near fueling or transfer of fuel operations?

☐ Is smoking prohibited in the vicinity of fueling operations?

☐ Are fueling operations prohibited in building or other enclosed areas that are not specifically, ventilated for this purpose?

☐ Where fueling or transfer of fuel is done through a gravity flow system, are the nozzles of the self-closing type?

IDENTIFICATION OF PIPING SYSTEMS

☐ When nonpotable water is piped through a facility, are outlets or laps posted to alert workers that it is unsafe and not to be used for drinking, washing or other personal use?

☐ When hazardous substances are transported through above ground piping, is each pipeline identified at points where confusion could introduce hazards to workers?

☐ When pipelines are identified by color painting, are all visible parts of the line so identified?

☐ When pipelines are identified by color painted bands or tapes, are the bands or tapes located at reasonable intervals and at each outlet, valve or connection?

☐ When pipelines are identified by color. is the color code posted at all locations where confusion could introduce hazards to workers?

☐ When the contents of pipelines are identified by name or name abbreviation, is the information readily visible on the pipe near each valve or outlet?

☐ When pipelines carrying hazardous substances are identified by tags, are the tags constructed of durable materials, the message carried clearly ad permanently distinguishable and are tags installed at each valve or outlet?

☐ When pipelines are heated by electricity, steam or other external source, are suitable warning signs or tags placed at unions, valves, or other serviceable parts of the system?

MATERIAL HANDLING

☐ Is there safe clearance for equipment through aisles and doorways?

☐ Are aisleways designated, permanently marked, and kept clear to allow unhindered passage?

☐ Are motorized vehicles and mechanized equipment inspected daily or prior to use?

☐ Are vehicles shut off and brakes set prior to loading or unloading?

☐ Are containers or combustibles or flammables, when stacked while being moved, always separated by dunnage sufficient to provide stability?

□ Are dock boards (bridge plates) used when loading or unloading operations are taking place between vehicles and docks?

□ Are trucks and trailers secured from movement during loading and unloading operations?

□ Are dock plates and loading ramps constructed and maintained with sufficient strength to support imposed loading?

□ Are hand trucks maintained in safe operating condition?

□ Are chutes equipped with sideboards of sufficient height to prevent the materials being handled from failing off?

□ Are chutes and gravity roller sections firmly placed or secured to prevent displacement?

□ At the delivery and of rollers or chutes, are provisions made to brake the movement of the handled materials.

□ Are pallets usually inspected before being loaded or moved?

□ Are hooks with safety latches or other arrangements used when hoisting materials so that slings or load attachments won't accidentally slip off the hoist hooks?

□ Are securing chains, ropes, chockers or slings adequate for the job to be performed?

□ When hoisting material or equipment, are provisions made to assure no one will be passing under the suspended loads?

□ Are Material Safety Data Sheets available to workers handling hazardous substances?

TRANSPORTING WORKERS & MATERIALS

□ Do workers who operate vehicles on public thoroughfares have valid operator's licenses?

□ When seven or more workers are regularly transported in a van, bus or truck, is the operator's license appropriate for the class of vehicle being driven?

□ Is each van, bus or truck used regularly to transport workers, equipped with an adequate number of seats?

□ When workers are transported by truck, are provision provided to prevent their failing from the vehicle?

□ Are vehicles used to transport workers, equipped with lamps, brakes, horns, mirrors, windshields and turn signals in good repair?

□ Are transport vehicles provided with handrails, steps, stirrups or similar devices, so placed and arranged that workers can safely mount or dismount?

□ Are worker transport vehicles equipped at all times with at least two reflective type flares?

□ Is a full charged fire extinguisher, in good condition, with at least 4 B:C rating maintained in each worker transport vehicle?

□ When cutting tools with sharp edges are carried in passenger compartments of worker transport vehicles, are they placed in closed boxes or containers which are secured in place?

□ Are workers prohibited from riding on top of any load which can shift, topple, or otherwise become unstable?

CONTROL OF HARMFUL SUBSTANCES BY VENTILATION

□ Is the volume and velocity of air in each exhaust system sufficient to gather the dusts, fumes, mists, vapors or gases to be controlled, and to convey them to a suitable point of disposal?

□ Are exhaust inlets, ducts and plenums designed, constructed, and supported to prevent collapse or failure of any part of the system?

□ Are clean-out parts or doors provided at intervals not to exceed 12 feet in all horizontal runs of exhaust ducts?

□ Where two or more different type of operations are being controlled through the same exhaust system, will the combination of substances being controlled, constitute a fire, explosion or chemical reaction hazard in the duct?

☐ Is adequate makeup air provided to areas where exhaust systems are operating?

☐ Is the intake for makeup air located so that only clean, fresh air, which is free of contaminates, will enter the work environment?

☐ Where two or more ventilation systems are serving a work area, is their operation such that one will not offset the functions of the other?

SANITIZING EQUIPMENT & CLOTHING

☐ Is personal protective clothing or equipment, that workers are required to wear or use, of a type capable of being easily cleaned and disinfected?

☐ Are workers prohibited from interchanging personal protective clothing or equipment, unless it has been properly cleaned?

☐ Are machines and equipment, which processes, handle or apply materials that could be injurious to workers, cleaned and/or decontaminated before being overhauled or placed in storage?

☐ Are workers prohibited from smoking or eating in any area where contaminates are present that could be injurious if ingested?

☐ When workers are required to change from street clothing into protective clothing, is a clean changeroom with separate storage facility for street and protective clothing provided?

☐ Are workers required to shower and wash their hair as soon as possible after a known contact has occurred with a carcinogen?

☐ When equipment, materials, or other items are taken into or removed from a carcinogen regulated area, is it done in a manner that will not contaminate non-regulated areas or the external environment?

TIRE INFLATION

☐ Where tires are mounted and/or inflated on drop center wheels, is a safe practice procedure posted and enforced?

☐ Where tires are mounted and/or inflated an wheels with split rims and/or retainer rings, is a safe practice procedure posted and enforced?

☐ Does each tire inflation hose have a clip-on chuck with at least 24 inches of hose between the chuck and an in-line hand valve and gauge?

☐ Does the tire inflation control valve automatically shut off the air flow when the valve is released?

☐ Is a tire restraining device such as a cage, rack or other effective means used while inflating tires mounted on split rims, or rims using retainer rings?

☐ Are workers strictly forbidden from taking a position directly over or in front of a tire while it's being inflated?

EMERGENCY ACTION PLAN

☐ Are you required to have an emergency action plan?

☐ Does the emergency action plan comply with requirements of T8CCR 3220(a)?

☐ Have emergency escape procedures and routes been developed and communicated to all employers?

☐ Do workers, who remain to operate critical plant operations before they evacuate, know the proper procedures?

☐ Is the worker alarm system that provides a warning for emergency action recognizable and perceptible above ambient conditions?

☐ Are alarm systems properly maintained and tested regularly?

☐ Is the emergency action plan reviewed and revised periodically?

☐ Do workers now their responsibilities:

☐ For reporting emergencies?

☐ During an emergency?

☐ For conducting rescue and medical duties?

INFECTION CONTROL

☐ Are workers potentially exposed to infectious agents in body fluids?

☐ Have occasions of potential occupational exposure been identified and documented?

☐ Has a training and information program been provided for workers exposed to or potentially exposed to blood and/or body fluids?

☐ Have infection control procedures been instituted where appropriate, such as ventilation, universal precautions, workplace practices, personal protective equipment?

☐ Are workers aware of specific workplace practices to follow when appropriate? (Hand washing, handling sharp instruments, handling of laundry, disposal of contaminated materials, reusable equipment.)

☐ Is personal protective equipment provided to workers, and in all appropriate locations?

☐ Is the necessary equipment li.e. mouth-pieces, resuscitation bags, other ventilation devices) provided for administering mouth-to-mouth resuscitation on potentially infected patients?

☐ Are facilities/equipment to comply with workplace practices available, such as hand-washing sinks, biohazard tags and labels, needle containers, detergents/disinfectants to clean up spills?

☐ Are all equipment and environmental and working surfaces cleaned and disinfected after contact with blood or potentially infectious materials?

☐ Is infectious waste placed in closable, leak proof containers, bags or puncture-resistant holders with proper labels?

☐ Has medical surveillance including HBV evaluation, antibody testing and vaccination been made available to potentially exposed workers?

☐ Training an universal precautions?

☐ Training on personal protective equipment?

☐ Training on workplace practices which should include blood drawing, room cleaning, laundry handling, clean-up of blood spills?

☐ Training an needlestick exposure/management?

☐ Hepatitis B vaccinations?

ERGONOMICS

☐ Can the work be performed without eye strain or glare to the workers?

☐ Does the task require prolonged raising of the arms?

☐ Do the neck and shoulders have to be stooped to view the task?

☐ Are there pressure points on any parts of the body (wrists, forearms, back of thighs)?

☐ Can the work be done using the larger muscles of the body?

☐ Can the work be done without twisting or overly bending the lower back?

☐ Are there sufficient rest breaks, in addition to the regular rest breaks, to relieve stress from repetitive-motion tasks?

☐ Are tools, instruments and machinery shaped, positioned and handled so that tasks can be Performed comfortably?

☐ Are all pieces of furniture adjusted, positioned and arranged to minimize strain on all parts of the body?

VENTILATION FOR INDOOR AIR QUALITY

☐ Does your HVAC system provide at least the quantity of outdoor air required by the State

☐ Building Standards Code, Title 24, Part 2 at the time the building was constructed?

☐ Is the HVAC system inspected at least annually, and problems corrected?

☐ Are inspection records retained for at least 5 years?

CRANE CHECKLIST

☐ Are the cranes visually inspected for defective components prior to the beginning of any work shift?

☐ Are all electrically operated cranes effectively grounded?

☐ Is a crane preventive maintenance program established?

☐ Is the load chart clearly visible to the operator?

☐ Are operating controls clearly identified?

☐ Is a fire extinguisher provided at the operator's station?

☐ Is the rated capacity visibly marked an each crane?

☐ Is an audible warning device mounted an each crane?

☐ Is sufficient illumination provided for the operator to perform the work safely?

☐ Are cranes of such design, that the boom could fall over backward, equipped with boomstops?

☐ Does each crane have a certificate indicating that required testing and examinations have been performed?

☐ Are crane inspection and maintenance records maintained and available for inspection?

HAZARD CONTROL

Introduction

The federal OSHA Safety and Health Program Management Guidelines state (in part):

(c)(3)(i) Hazard Prevention and Control. So that all current and potential hazards, however detected, are corrected or controlled in a timely manner, establish procedures for that purpose, using the following measures:

(A) Engineering techniques where feasible and appropriate;

(B) Procedures for safe work which are understood and followed by all affected parties, as a result of training, positive reinforcement, correction of unsafe performance, and if necessary, enforcement through a clearly communicated disciplinary system;

(C) Provision of personal protective equipment; and

(D) Administrative controls such as reducing the duration of exposure.

The State of California Cal/OSHA Regulations for Injury and Illness Prevention Program states (in part):

(2) Include a system for ensuring that employees comply with safe and healthy work practices.

(3) Include a system for communicating with employees in a form readily understandable by all affected employees on matters relating to occupational safety and health, including provisions designed to encourage employees to inform the employer of hazards at the worksite without fear of reprisal.

(6) Include methods and/or procedures for correcting unsafe or unhealthy conditions, work practices and work procedures in a timely manner based on the severity of the hazard:

(A) When observed or discovered: and

(B) When an imminent hazard exists which cannot be immediately abated without endangering employee(s) and/or property, remove all exposed personnel from the area except those necessary to correct the condition.

The material that follows is based on a paper that was published in 1976 by *Protection*, the official journal of the Institution of Industrial Safety Officers, United Kingdom (now the Institution for Occupational Safety and Health), and is printed with permission from Paramount Publishing Ltd.

The need for some type of a hazard control system is well understood by anyone concerned with the problem of occupational accidents. It has long been recognized that accidents can be prevented provided that the hazards that cause them can be identified and controlled.

From the author's experience, two major problems have to be overcome before any hazard control system can become effective. The first problem is that the people most concerned with the accident problem (i.e., the first-line supervisor and those at the work level) do not systematically look for hazards. The average supervisors, faced with the everyday problems of getting the job done, do not on their own initiative take time to search for unsafe acts or unsafe conditions on a daily basis. Hazard searching, at most, becomes a periodic ritual inspection or a look now and then. The second problem is that when hazards are discovered they are seldom processed in a systematic and logical manner. Management, possibly reacting to union pressure, may spend vast sums of money to correct a relatively low-risk hazard while other, high-risk hazard problems await action buried in some department's maintenance backlog.

The Hazard Control System

To help overcome the major problem areas mentioned previously, a three-part hazard control system is recommended. Part one is the Supervisor's Hazard Control Notebook, part two is the Department Hazard Control Log, and part three consists of Safety Recommendations. The following sections provide brief descriptions of these three parts.

Hazard Control Notebook

Each first-line supervisor or team leader is issued a Hazard Control (HC) Notebook. They are expected to carry this notebook on their person while at work and to evidence activity in the program by recording a minimum of one hazard control action for each shift worked. This approach has a double advantage. First, it motivates the supervisors to look for hazards more systematically. Second, it gets them more into the habit of writing down the hazards they discover so they are not forgotten in the rush. Employees are also encouraged to look for hazards and report them to their supervisors, who will then record the hazards in their notebooks.

The intent should be for supervisors to evidence systematic hazard control activity. This can be either an observation or report of a physically unsafe condition, or contact with a team member to discuss the safety aspects of a specific task. These observations should note not only unsafe work practices, but also positive recognition for the employee when he or she is seen following the proper procedure.

To help supervisors systematize their daily search for hazards, they can be trained to focus their surveys by hazard source–cause category. In Figure 6.7, a simple nine-point source–cause category guide is illustrated. In this way, a supervisor can pick one search category per day. This search can also be alternated with audits of job hazard analysis compliance.

After a hazard has been recognized and reported, the procedure requires that first-line supervisors take the following action steps: (1) write the hazard down in their notebooks; (2) risk-rate the hazard and discuss with the person reporting it; (3) take immediate temporary control action (ITC); (4) rerate after ITC has been taken and communicate to the person reporting it; and (5) if permanent control has not been completed within 24 hours, the hazard is transferred to the department hazard control log.

A typical format for a supervisor's HC notebook is shown in Figure 6.8.

After the hazard has been recorded, the supervisor has to evaluate the potential effect of the hazard in order to decide on the priority for its control. Measuring the risk value of a hazard is a difficult thing for anyone because feelings and emotions usually become involved. The following system has proved to be a helpful guide.

Risk Rating System

Consider each of the following criteria and assess points as appropriate:

- Severity potential for injury or damage — high (6 points); moderate (2 points); low (1 point)
- Frequency of employee exposure — high (3 points); moderate (2 points); low (1 point)
- Probability of accident occurrence — high (3 points); moderate (2 points); low (1 point)
- Violation of safety codes, standards, rules, etc. — yes (1 point); no (0 points)

Having established the risk factor on a points basis, the priority for control may be obtained from the following table:

Priority Rating	Risk Rating
E — Emergency	10–13 points
A — 1 week	8–10 points
B — 1 month	6–8 points
C — 3 months	4–6 points
D — 6 months	4 points

Besides providing a system to assist supervisors in increasing the frequency of their hazard surveys, the HC notebook is also a useful training tool. Experience with the program has shown that new supervisors usually find and record a majority of unsafe conditions. However, as they continue to develop their knowledge and skills in the principles and procedures of hazard control, the proportion of unsafe work practices increases.

A notebook can be assessed for the quantity and quality of the entries. A method used by some organizations is as follows:

Score:

$$0\text{--}100 = \frac{\text{total no. hazards recorded}}{\text{total no. shifts worked}} \times \text{quality}$$

Time Discovered:____ Date: _____   Time Discovered: ___ Date: _____
                                                Task Observation
                                                Job being performed:
                                                □ Safe Work Practice   □ Unsafe Work Practice

Describe: Describe:

                    Priority Rating   Priority Rating
Reported By:_____   Reported By: _____
      Hazard Source and Cause Control   **Hazard Source and Cause Control**

Describe: (Immediate Temporary Control) Action   Describe: (Immediate Temporary Control) Action

Describe: (Permanent Control) Action   Describe: (Permanent Control) Action

Time Completed: _____   Time Completed: _____
Completion Corrected within 24 hours □ or   Completion Corrected within 24 hours □ or
Entered in Hazard Control Log □ Date: _____   Entered in Hazard Control Log □ Date:_____

Figure 6.8   Sample of a hazard checklist simplified by grouping hazard categories into major
            source areas.

EVALUATION OF HAZARD CONTROL ACTIVITY

Description of Hazard	Max.	Control of Hazard	MAX.
Poor identification (location of unsafe condition or employee name not listed) Example: "Fork truck speeding" (No truck number, location, or identification	−5	Future control listed, but no scheduled date or follow-up	−5
Description vague, i.e., "Slipping hazard," instead of "Hydraulic oil on steps."	−10	Questionable or vague control	−10
Describes result instead of hazard, i.e., "Man receiving hand cuts from cable," instead of "working without gloves."	−15	Improper or inadequate control (instruction only when unsafe condition could have been eliminated).	−15
		No control listed	−20

The supervisor's HC score is derived by multiplying the quantity score by the quality score. Example:

Number of HC Entries	= 16
Number of HC shifts worked	= 20
Quantity score 16/20	= 80%
Average quality score of 5 HC	= 80

Supervisor's HC Score = .80 × .80 = 64

Figure 6.9   Example of a hazard notebook entry quality evaluation technique.

Example:

$$\frac{16 \text{ hazards recorded}}{20 \text{ shifts worked}} \times 80 \text{ quality} = 64 \text{ score}$$

Quality of the entries is obtained by randomly selecting five entries and assessing the average quality according to a criteria guideline. A simplified version of this system is as follows:

Hazard description	0–40 points
Risk rating	0–10 points
Hazard control	0–50 points

See Figure 6.9 for an illustration of another quality evaluation technique.

Using this approach enables department managers to identify nonperforming supervisors and assist them to improve their skills in recognizing, understanding, and controlling hazards.

Department Hazard Control Logbook

When permanent control action is not completed within 24 hours, the supervisor must enter the hazard in the Department Hazard Control Logbook (see Figure 6.10 for a typical format).

It is important to note that the hazard must be entered in the log where the hazard exists. This may result in the supervisor recording the hazard in two logs. For example, a supervisor in Department A finds a hazard in Department B. In this case, the supervisor must enter the hazard in the Department B logbook and also in the Department A logbook. The double entry of the hazard is necessary in order to inform the proprietary department of the hazard so that control action is taken and also to enable follow-up by the supervisor who discovered the hazard.

After entering the hazard in the log, the supervisor is expected to follow up periodically during the span of time that the job should be done (priority rating). When the hazard is not corrected within the priority rating period, the supervisor takes the problem to the next level for assistance and action.

The supervisor's risk ratings on a hazard should stand unless they participate in a rerating. If a hazard goes beyond the original scheduled date of correction, the hazard is rerated in the department with the consent of all concerned. If the originator disagrees with the rerating, then the hazard must be referred to a higher level for a final decision.

Experience with the logbook phase of this program has shown that follow-up must be maintained by management to ensure effectiveness. To provide department managers with a status of hazard control action, supervisors should provide monthly reports of outstanding hazards on an exception basis. In these reports, only those hazards that have overrun their priority ratings by 30 or more days are listed.

Copies of all reports should also be sent to the safety department. This ensures that plant-wide or major problems can be monitored by safety and top management.

A hazard control program, like any other management system, needs follow-up. Some supervisors still do not conscientiously look for hazards every day, and then at the end of the month they merely fill in their notebooks to meet the paper requirement. Second-line supervisors should be responsible for measuring the effectiveness of the system through random audits.

Hazard control is one of the five systematic action programs that are implemented by first-line supervisors. Of these five programs, hazard control is considered the most important and productive. In fact, it is fair to say that hazard control activity is the barometer of an organization's safety results. When this program is effectively implemented, the frequency of accidents will be steadily reduced.

Safety Recommendations

Employees should be informed that safety and health hazards must be reported to their supervisors for investigation and processing through the Hazard Control System. Despite this requirement, the reality is that for various reasons some employees

Hazard Control Log

Section _____ Location _____

Date Entered	Hazard	Control Action		Hazard Detected/Entered By	Priority*	Work Order Service Request	Date Completed and Signature
		ITC Taken	Recommended Permanent Control				

Reviewed by: _____ Date: _____

Reviewed by: _____ Date: _____

Priority

E Emergency C Three Months

A One Week D Six Months

B One Month Subject to review

* Priority rating must be after ITC action has been taken.

Figure 6.10 Sample format for hazard log.

may be reluctant to take certain safety problems to their supervisors. Recognizing this issue, OSHA requires employers to establish methods for employees to report hazards without fear of reprisal, even anonymously.

The Hazard Control System must, therefore, include a system of methods by which employees can report hazards. This serves as a relief valve and helps to point out problems that otherwise might go uncorrected. This need can be met by various means. Examples are

- A safety committee that has representatives from all departments
- An in-house safety department that is available to receive reports of problems
- A "hotline" on which employees can call in and report safety and health problems
- A formal safety recommendations (SR) program

One or more of these systems should be in place and known to all employees.

FURTHER READING
Books

Ridley, J. et al. *Safety at Work.* London: Butterworths, 142–153, 1977.
Manuele, F. A. *On the Practice of Safety.* New York: Van Nostrand Reinhold, 169–210, 1993.
Colvin, R. J. *The Guidebook to Successful Safety Programming.* Boca Raton, FL: Lewis Publishers, 77–93, 1992.
Gordon, H. L. et al. *A Management Approach to Hazard Control.* Bethesda, MD: Board of Certified Hazard Control Management, 1994.
Holt, A. St. J. and Andrews, H. *Principles of Health Safety at Work.* Leicester, U.K.: IOSH Publishing, 76–87, 1993.
Grimaldi, J. V. and Simonds, R. H. *Safety Management.* Boston, MA: Irwin, 1989.
Asfahl, C. R. *Industrial Safety and Health Management.* Englewood Cliffs, NJ: Prentice-Hall, 1990.
Brauer, R. L. *Safety and Health for Engineers.* New York: Van Nostrand Reinhold, 1990.

Other Publications

Lack, R. W. A hazard control system for effective results. *Protection* 13(5): 2–4, June 1976.
U.K. Health and Safety Executive. Successful Health and Safety Management 38-44 Performance Standards to the Control of Hazards and Risks, Ref. No. HSG65, 1992.
U.S. Department of Labor Occupational Safety and Health Administration, OSHA.
Safety and Health Program Management Guidelines: Issuance of Voluntary Guidelines. *Fed. Regist.* 54 (16), January 26, 1989.

JOB HAZARD ANALYSIS
Introduction

The federal OSHA Safety and Health Program Management Guidelines state (in part):

(c)(4)(i) Ensure that all employees understand the hazards to which they may exposed and how to prevent harm to themselves and others from exposure to these hazards, so that employees accept and follow established safety and health protections.

The State of California Cal/OSHA Regulations for Injury and Illness Prevention Programs state (in part):

(7) Provide training and instructions:

(B) To all new employees;

(C) To all employees given new job assignments for which training has not previously been received;

(D) Whenever new substances, processes, procedures or equipment are introduced to the workplace and represent a new hazard;

(E) Whenever the employer is made aware of a new or previously unrecognized hazard; and

(F) For supervisors to familiarize themselves with the safety and health hazards to which employees under their immediate direction and control may be exposed.

The material that follows is based on a paper published in 1980 by *Protection*, the official journal of the Institution of Industrial Safety Officers, United Kingdom (now the Institution for Occupational Safety and Health), and is printed with permission from Paramount Publishing Ltd.

Job hazard analyses (JHA) or job safety analyses (JSA) have been generally utilized by industry for many years as a key part of their accident prevention programs. Despite this, industry in general is faced with a continued accident problem, particularly in the human element area. It is toward this area of safety that the JHA program is aimed. Why do JHAs sometimes fail to get the message across? What should be done differently to achieve better results? How can the paper aspect be handled? These and many other questions are being encountered today by safety professionals and others concerned with improving the effectiveness of industrial safety programs.

Much has already been written on the basics of JHA or JSA systems, and most readers will be very familiar with this type of program. Therefore, the next section will be primarily concerned with identifying some of the problems with the program and outlining various solutions.

What Are the Problems?

The study of accidents where the key cause was found to be an unsafe work practice usually points out many, if not all, of these typical problems:

- No written JHA existed for the job being performed.
- The hazard involved was not included in the JHA.
- The JHA was written as a generality (hazards not clearly described and/or remedies not specific).
- The practice was not communicated to the employee.
- The employee did not follow the specific methods outlined in the JHA.

Conversely, it has also been found that accidents rarely occur when there exists a well-written JHA, containing a clear description of the hazards involved, that has been systematically reviewed with and followed by the employee concerned.

The solution to these problems appears at first glance to be deceptively simple. However, the author has found in 30 years work with these programs that the problems are far from easy to solve out in the real world at the industrial work level.

How to Improve the Effectiveness of Your Job Hazard Analysis Program

Suggested steps which have proven effective in overcoming the problems mentioned are described in the following sections.

Assessment of the Program

A common objection among many first-line supervisors to the JHA program is, "why all the paper?" The feeling is that, apart from a few important safety rules contained in the employee handbook, their employees are very experienced and, therefore, verbal safe practice instructions are sufficient. Unfortunately, those of us involved in accident prevention are only too well aware that when a JHA is verbal, one may as well call it useless. In fact, if a JHA is not written, the best position to adopt is to assume it does not exist because the probability of it being communicated and implemented correctly is very low.

Another aspect of this resistance to the paper is that some supervisors feel the program will expose their weaknesses. These supervisors, who are afraid to expose themselves to professional critics, will need help to understand the principles and procedures of the JHA program and how the system will benefit them personally.

A different problem from the above is the organization of the paper. A plant at which the author worked for a number of years had mostly young, well-educated supervisors. These people had no objections to writing JHAs. In fact, they turned them out by the hundreds! The problem was how to find them when they were needed for later reference. Needless to state, if they could not find their JHAs, they obviously were not being utilized in any systematic manner.

Many companies have established classification systems to organize their JHAs. A basic JHA classification system should consist of the following elements:

- Types of JHAs – To avoid excessive repetition, it is best to divide JHAs into three types:
  a. General or common JHAs will cover hazards associated with activities common to many jobs. Examples include manual lifting, using ladders, walking on slippery surfaces, etc.
  b. Specific JHAs will cover hazards related to specific tasks.
  c. On-the-spot JHAs are utilized whenever a job is being done which involves any high-risk hazard and for which there is no general or specific JHA. As the name implies, this JHA is usually a verbal discussion at the job location, and the supervisor should be encouraged to convert this JHA to writing for future reference.

- JHA Classification and Numbering system — The following is a plant-wide system that also adapts well to computerization:

Plant	one digit assigned
Major department	one digit assigned to each
Subdepartment	one digit assigned to each
Area or section	one digit assigned to each
Classification	one digit per classification table (see below)
Sequential number	two digits

The classification element is broken down into the nine (9) major hazard source areas:

1. Work area: includes ladders, platforms, and scaffolds
2. Machines: anything that "produces"; e.g., lathes presses, saws, and grinders
3. Equipment: anything that "conveys" e.g., pumps, pipes, valves, tanks, cranes, and conveyors.
4. Portable tools: all types of portable hand tools, excluding electrical
5. Materials — traumatic: includes acids, caustics, flammables, explosives
6. Materials — toxic: includes toxic and systemic poisons
7. Electrical: includes all types of electrical conductors and equipment
8. Vehicle: includes all types of power mobile equipment, railroads, and marine
9. Multicategory

Using this numbering system will provide a maximum of seven numbers. Here are a few examples:

1000601 would be a plant-wide general JHA related to materials — toxic.
0200601 would be a department-wide general JHA related to materials — toxic.
0220601 would be a subdepartment specific JHA related to materials — toxic.

Each first-line supervisor or section office should be provided with a JHA manual containing all the general and specific JHAs relating to their jobs. The manual should contain an index having a general layout as follows:

JHA INDEX

DEPARTMENT _____

SUBDEPARTMENT _____   CLASSIFICATION

AREA OR SECTION _____   _____

JHA NO.	JSP TITLE	JHA TYPE	DATE ORIGIN	DATE REVISED	DATE REVIEWED

- Determining the Need for New JHAs — To help combat the problem of JHA coverage, first- and second-line supervisors should periodically review their JHAs to determine whether new JHAs may be needed. Sources of information for this review include:

a. Listing all jobs supervised and noting which ones are not covered — Write JHAs for any that involve high-risk hazards.
b. Accident experiences — Every accident should cause an automatic JHA review and revision as needed.

    c. Task observations by the supervisor — This should be a systematic effort by
the supervisor to look for and anticipate acts or conditions that might result
in injury.

    d. Reports and recommendations from the supervisor's crew

    e. New or modified operations

    f. Experiences of other plants or organizations

Reviewing the Quality of Written Job Hazard Analyses

After looking at your JHA coverage, the next step in upgrading the quality of
your program should be to look at what your JHAs are saying. Start with the center
column (hazard). The major weak point of the JHA program begins with the description of the hazard. A format for a JHA, including guidelines for the evaluation of
its quality, is shown in Figure 6.11.

What do your JHAs tell your people? Do they provide a list of clearly described
hazards and specific control actions, or do they lapse into meaningless generalities?
Many standard procedures on developing JHAs include instructions for identifying
hazards such as the following examples:

- Is there a danger of striking against, being struck by, or otherwise making injurious
  contact with an object?
- Can a person be caught in or between objects?

It is suggested that a better approach toward identifying the hazard is to have
the supervisor and his or her crew define what a person might do wrong, which
could produce an accident. In other words, the hazard should clearly describe the
actual or potentially unsafe work practice.

The reason behind this approach is that studies of many accidents where an
unsafe work practice was involved have indicated that supervisors often do not know
what their people are doing wrong and what they should do differently to prevent
accidents. By leading their people to describe the hazard, they are at the same time
helping their people to recognize the risk of their actual or potentially unsafe actions.

To illustrate this approach, here are some examples:

Standard Approach:

Step	Hazard
Climbing ladder	Fall from ladder
	Ladder slipping
	Slip off ladder rung

Suggested Approach:

Step	Hazard (Unsafe Work Practice)
Climbing ladder	Climbing with tools or equipment in one hand
	Failure to secure ladder
	Placing toe of shoe on ladder rung

Job Hazard Analysis

JHA No.

Page ___ of ___ pages

Type: □ General □ Specific □ On-Spot

Job Classification of Involved Employees	
Task/Job Title	
Department	Personal Protective Equipment Required
Area	The following general JHA's must be read and understood before starting job.

Basic Job Steps	Potential Hazards What the person is likely to do wrong that could cause an accident.	Control Action What the person must do to control the hazard.

Basic Job Steps	Does it describe what is done (not how it is done) using brief action statements in sequence?	10
Potential Hazards	Does it describe what may be done that could cause an accident (not the injury or the accident).	40
Control Action	Does it describe specifically what the person must do to control the hazard?	50
Total		100

Reviewer's Comments

Prepared by:

Check One
□ New JHA written
□ Old JHA revised
□ Old JHA reviewed

Signature: _____ Date: _____

Distribution   Safety   Originator   Department

Figure 6.11 Example format for job hazard analysis which includes quality evaluation.

One of the advantages of this approach is that, once the supervisor has identified and clearly described what a person might do wrong, it is more likely that an equally specific hazard control action will also be stated. To illustrate this point, here is another example:

Standard Approach — Survey Assistant:

Step	Hazard	Control
Climbing steep rocky areas	Slip or fall	Use proper climbing technique

Suggested Approach:

Step	Hazard (Unsafe Act)	Control
Climbing steep rocky areas	Placing full weight on a hand hold before testing it	Always test the hand hold by pulling with your full strength before you test it with your full weight

In the author's experience, most supervisors using this method do have difficulty in describing the specific unsafe work practice. However, with constant practice plus regular training and guidance, their understanding of the principles and procedures improves and, as a result, the quality of their written efforts is steadily upgraded.

In fact, to be effective, a JHA program must be a constantly ongoing system. All JHAs should be reviewed by the supervisor and crew at least once every year, and this is an opportunity to improve the JHA and make it more specific. In this process, it is most important for the supervisor to let the employees become involved in making suggestions for changes, even to the point of challenging them to find something that needs to be changed.

What type of a hazard should go into a JHA? A suggested guideline would be as follows:

- The hazard should normally be an existing or potential high-risk, unsafe work practice, since unsafe conditions will be dealt with through plant hazard control procedures.
- It is important for supervisors to study the potential. In other words they should consider the job when conditions are not normal or under an emergency situation, as well as the routine. They should practice using their intuition and encourage their people to do the same so that they can anticipate mistakes the people might make which could produce severe injury.
- Many JHAs are multipage and list literally dozens of hazards, many of which are relatively minor in terms of their ultimate potential to produce injury. This tends to dilute the effectiveness of the total JHA. It is recommended that a JHA list only those hazards considered to be of a high-risk nature. Some suggested criteria for a high-risk hazard would be
  a. Any hazard that can produce permanent-partial disability or more
  b. Any hazard that can produce an injury resulting in days away from work
  c. Any hazard that can potentially produce severe injury or significant damage

Obtaining Better Utilization of Job Hazard Analyses

A JHA is merely a piece of paper until somebody converts it into action. How can supervisors get their people to abide by their JHAs rather than take short-cuts? As we said earlier, the first step is to look at the hazard description and control action. "What are we telling our people?" The second step is that supervisors must adopt a more systematic approach. This would include regular review of all their JHAs and daily task observations to ensure compliance with the safe practices outlined in each JHA.

Supervisors often fail to do an effective job of communicating JHAs. It would be beneficial for them to constantly remember the old adage, "If the student hasn't learned, the teacher hasn't taught!" They should frequently test to make sure their people understand their JHAs. One way is to show an employee a JHA and ask, "What does this paragraph say?" The employee's answer will indicate

- Whether the employee understands it
- Whether the JHA needs revising

Supervisors should frequently test for JHA understanding because only when their people can repeat their JHAs can they be sure the message got across. They should ask questions such as, "What are the key parts of your JHA?" or, "Tell me the hazards of your JHA. I expect you to know all five of them."

By adopting the systematic approach, supervisors will avoid the trap of tolerating JHA violations and thus permitting their people to slip into old habits. The systematic approach should also be targeted according to the risk potential of the hazard. The higher the risk, the more repetition and emphasis supervisors must provide to keep their people from making judgments. They should use this repetitive process in order to make the impression deep enough that it becomes a part of the subconscious.

Managing the Job Hazard Analysis Program

Management Level	Responsibilities
First-line supervisor	1. Minimum standard — write, revise, or review a minimum of one JHA per month. 2. Conduct on-the-spot JHAs as needed. 3. Audit for crew compliance with their JHAs.
Second-level supervision	1. Set up a development schedule for jobs involving high-risk hazards. 2. Set up department manual and index. 3. Review and approve JHAs. 4. Provide training and guidance. 5. Audit effectiveness.
Department manager	1. Provide a department procedure defining who does what to ensure that the plant JHA policy is implemented in the department. 2. Audit the effectiveness of the program.

3. Identify department-wide problems and set objectives for continuous improvement of the program.
4. Establish a department general JHA index and development schedule.
5. Provide advanced training and guidance.

Summary and Conclusions

In the writer's opinion, based on years of work with JHA programs, such a program is still the most effective and practical approach for occupational safety and health education and training relative to job-specific hazards and their control. Most of the problems experienced with this program are concerned with the implementation of the process rather than the process itself.

The key to solving most of the problems with this program is to establish a constantly improving system which is aimed at total involvement of all levels concerned. To achieve this objective, the program must require systematic activity by supervisors and work teams. In addition, the backup management line and safety professionals must provide the necessary training, guidance, and management leadership to ensure continuous improvement.

JHAs are vital because some of the worst catastrophes are caused by small mistakes that could easily have been predicted by the supervisor and people involved.

FURTHER READING

Books

Boley, J. W. *A Guide to Effective Industrial Safety.* Houston, TX: Gulf Publishing, 83–91, 1977.

Colvin. R. J. *The Guidebook to Successful Safety Programming.* Boca Raton, FL: Lewis Publishers, 81–84, 1992.

Kohn, J. P., Friend, M. A., and Winterberger, C. A. *Fundamentals of Occupational Safety and Health.* Rockville, MD: Government Institutes Inc., 214–227, 1996.

Petersen, D. *Techniques of Safety Management.* Goshen, NY: Aloray, 56–60, 1989.

Swartz, G. *Job Hazard Analysis: A Guide to Identifying Risks in the Workplace,* Rockville, MD: Government Institutes, 2001.

U.S. Department of Labor, Mine Safety and Health Administration. *Job Safety Analysis. Manual No. 5.* Washington, D.C.: National Mine Health and Safety Academy Safety, 1989.

U.S. Department of Labor, Occupational Safety and Health Administration. *Job Hazard Analysis.* Washington, D.C.: OSHA, 3071, 1987.

Other Publications

Lack, R. W. Job Safe Practices. *Protection* 17(2): 9–11, February 1980.

U.S. Department of Labor, Occupational Safety and Health Administration. OSHA Safety and Health Program Management Guidelines: Issuance of Voluntary Guidelines. *Fed. Regist.* 54 (16), January 26, 1989.

SAFETY MEETINGS

Introduction

The federal OSHA Safety and Health Program Management Guidelines state (in part):

(C)(1)(IV) Provide for and encourage employee involvement in the structure and operation of the program and in decisions that affect their safety and health, so that they will commit their insight and energy to achieving the safety and health program's goal and objectives.

Comment: Forms of participation which engage employees more fully in systematic prevention include (1) inspecting for hazards and recommending corrections and controls; (2) analyzing jobs to locate potential hazards and develop safe work procedures; (3) developing or revising general rules for safe work; (4) training newly hired employees in safe work procedures and rules, and/or training their co-workers in newly revised safety work procedures and rules, and/or training their co-workers in newly revised safety work procedures; (5) providing programs and presentations for safety meetings; and (6) assisting in accident investigations.

Note that these forms of employee participation embrace each one of the five safety action programs discussed in this section and underline the critical need for effective communications and employee involvement.

The state of California Cal/OSHA Regulations for Injury and Illness Prevention Programs state (in part):

The program shall at a minimum: (3) include a system for communicating with employees in a form readily understandable by all affected employees on matters relating to occupational safety and health ... substantial compliance with this provision includes meetings, training programs, posting written communications, a system of anonymous notification by employees about hazards, labor/management safety and health committees, or any other means that ensures communication with employees.

These requirements reinforce government concerns that safety and health programs cannot be truly effective unless and until those at the work level are involved in the accident prevention process. The Cal/OSHA Construction Safety Orders take this one step further by requiring employers to ensure that employees performing construction-related work attend a tailgate safety meeting once every 10 workdays.

Several later chapters in this book will discuss communication and training techniques, so this chapter will be confined to discussing safety meetings as an accident prevention system implemented at the work level by first-line supervisors or team leaders.

Some Problems with Safety Meetings

In conventional safety meetings, you may find a group of up to 20 people sitting in a meeting which is being conducted by a supervisor who does not know what to say or how to say it. In this case, it is not surprising that the meeting turns into a gripe session and ends up nowhere. Alternatively, the supervisor may have some relevant information, but lacks the communication skills to get the message across.

Over 30 years ago, Kaiser Aluminum and Chemical Corporation did a study of 10,000 safety meeting reports written by the company's supervisors. This study was lead by Homer K. Lambie, corporate safety director.

Each of the these reports was evaluated against the question, "Is there any statement indicating the group will do something differently after the meeting to solve a single safety problem?" The result was shocking to the management at that time, for the study found that in practically every case nothing positive happened. The group just discussed generalities with no specific action conclusions.

Following this study, Mr. Lambie and others at Kaiser devised a simple five-step, problem-solving approach that would help supervisors to plan and conduct more effective meetings. This system was named the K–5 by Kaiser and has since been adopted worldwide.

The reason that this technique has proven so effective is that, even without professional communications training or speaking skills, a supervisor has a simple tool that helps in selecting a topic and processing its solution with the group. The key is that the focus is on specific behavior rather than attitudes or generalities. A sample planning sheet report form and evaluation format are illustrated in Figures 6.12 and 6.13.

This method is a powerful tool to help meeting leaders once they understand the philosophy and the techniques. Training in how to use this system is most important because the concept is quite profound. The key point that must be stressed in this training is selection of a meeting topic (the problem). When leaders select a very specific problem, they will be able to process it to a positive action conclusion relatively easily. Conversely, if they select a problem that is vague or general, they will end up with an equally vague or general solution. Some guidelines for preparation and conducting this type of meeting are as follows:

- Problem (The Unsafe Work Practice)
  a. When the problem is of a general type (e.g., using ladders improperly), such a problem must be dissected into the specific practices that make up this general problem. (How many and which unsafe work practices were involved in improper use?) Then decide on which unsafe work practice is the most critical in terms of injury potential.
  b. When the problem is one of feelings (attitude or an undesirable personal characteristic, e.g., people are not safety-minded), such a problem must be converted to the action that led to this judgment. (What did the members of the team do that led to the belief that they are not safety-minded?). Address their actions, not their feelings!
  c. When the problem is an unsafe condition, convert it to the action that produced it and conduct a meeting with those people whose action is involved. Unsafe conditions are not usually the subject of safety meetings with a team and would most logically be handled relative to their sources, as follows:

Supervisors Problem Solving Safety Meeting

Department:	Hints on Holding This Type of Meeting	
Section:	A. Before scheduled meeting, identify clearly and the behavioral safety problem which you and your group can solve.	
Supervisor:		
Date of Meeting:	B. Jot down a list of expected discussion points.	
	C. Let your group know 2 or 3 days in advance the subject matter and time of meeting.	
Employees Attending:	1. Problem (The unsafe work practice) we are doing or are likely to do the following:	
1.		
2.		
3.		
4.		
5.		
6.	2. Discussion — Reasons why this problem exists.	Things we can do to overcome these obstacles.
7.	I.	I.
8.		
9.	II.	II.
10.		
11.	III.	III.
12.		
13.	IV.	IV.
14.		
15.	V.	V.
16.		
17.	3. Conclusion: What we as a group will do to solve our problems.	
18.		
19.		
20.		
Number of Employees in Crew:	4. Follow-Up: What we must do to ensure that our conclusion is put into action.	
Reviewed By:	5. Remarks: What else was said which must be acted upon after this meeting?	
Name:		
Date:		

"The value of any meeting is measured by what the group will do differently after the meeting to solve a specific safety problem"

Distribution: Safety. Originator. Department

Figure 6.12 Example layout for a five-step problem-solving safety meeting report.

Unsafe Condition Source	**Meeting with**
Employee unsafe work practice	Team whose action produced it
Wear and tear	Production/engineering and/or maintenance
Design deficiencies	Engineering

    d. The rule of thumb is to meet with those whose action can solve the problem.

    e. The problem must be a description of what the members of the group are doing or are likely to do wrong, and it must be one that will be controlled by a change in the actions of the members of the group.

Evaluation of Problem-Solving Safety Meeting		Department		Date		
		Supervisor				
		Possible Rating	**Actual Rating**			
			1	**2**	**3**	
1 **The Problem**	A. The problem states clearly the action that must be discontinued or avoided. (NOT: an unsafe condition, an attitude or feeling, or a 'general' act or action.)...............	0–20	...	...	...	
	B. Are most of the people doing or likely to do this?......	0–10	...	...	...	
	C. Can a change in their action control the problem?......	0–10	...	...	...	
2 **Discussion**	Reasons: A. Are the reasons listed those which contribute to the problem?...	0–5	... ...	... ...	... ...	
	B. Do the items address the question "What causes us to make these mistakes?"...	0–5	...	...	...	
	Controls: A. Is each reason adequately controlled or eliminated?....	0–5	... ...	... ...	... ...	
	B. Is the necessary follow-up on the controls noted below?	0–5	...	...	...	
3 **Conclusion**	A. Does it require a change in their behavior?.................	0–15	...	...	...	
	B. Did each employee agree to the action?.....................	0–5	...	...	...	
4 **Follow-Up**	Does this section specify WHO will do WHAT, WHEN, to see that the conclusion sticks?.....................................	0–10	...	...	...	
5 **Remarks**	Employees' significant comments about this or other hazards listed...	0–5	...	...	...	
	Form Preparation — General: The form is completely filled in, including date, department, leader's name, and names of all employees attending the meeting..	0–5	... ...	... ...	... ...	
Total						

Reviewer's Comments:		**Actual Rating By**	
	1	Supervisor	
		Date	
	2	Reviewer	
		Date	
	3	Name	
		Date	

RL-4

Figure 6.13 Sample layout for quality evaluation of a five-step problem-solving meeting.

- Discussion
  a. *"Their reasons for the above action."* — The leader must list beforehand all the significant reasons the members of the group may have to justify the undesirable action. During the meeting, the leader should list any additional reasons that may be offered by the group.
  b. *"How do we overcome these reasons?"* — Listed here, in advance, are the facts to be considered and the steps necessary to overcome the reasons for the action. During the meeting, other controls may be added by the employees.
- Conclusion — The conclusion should outline clearly the action of the members of the group necessary after they leave the meeting to control the problem. Individual commitment tends to ensure the desired action.
- Follow-up — Follow-up is other action needed to ensure that the conclusion is carried out. Outline and schedule actions of identified individuals that facilitate or ensure the carrying out of the conclusion:
  a. On commitments made by the leader during the meeting.

   b. On all other incomplete control measures, including checks to see that the unsafe work practice does not recur.
- Remarks — Remarks are related or unrelated comments or observations for action or review after the meeting. It is important to list in this section related or unrelated items, information, or observations for these reasons:
    a. To give the leader a means to avoid deviation from the problem being solved by postponing nonrelevant discussion.
    b. To ensure review of those items not dealt with in the meeting.

An objection frequently raised about the five-step process is, "We have so many problems — if we only take one per meeting, will we ever solve all of them?" To counter this argument, it must be pointed out that learning and using the five-step process will result in a transfer of problem-solving skills. With practice, leaders will find that they and their crews will automatically start defining and solving more and more problems of all types, not only safety issues.

In the chapter on training techniques (Chapter 47), there will be much more information on how to prepare for and conduct meetings. A brief checklist for the guidance of safety meeting leaders follows.

Checklist for Presenting Successful Safety Meetings

- Planning meetings
  - Regularly schedule meetings throughout the year.
  - Remind attendees of date, time, and place.
  - Preferably schedule meetings in the morning; if one must be held in the afternoon, it should not be immediately after lunch, and never after 3 p.m.
  - Occasionally invite a guest such as a person from safety, industrial hygiene, security, or fire protection, a department manager, etc.
  - Select a location where the group will be comfortable with minimal distractions.
- Preparing for meetings
  - Identify the meeting topic (problem).
  - Research the meeting subject and materials, including reasons why the meeting problem exists and methods to overcome the objections.
  - Check out visual aids such as a flip chart, VCR and monitor, or overhead projector (including a spare bulb).
  - Come early and set up the meeting room.
  - Be there to greet your attendees.
  - Where appropriate, provide refreshments.
  - Provide a sign-in sheet.
- Presenting the meeting
  - Start meetings on time!
  - Circulate a sign-in sheet.
  - Review the meeting agenda.
  - Set clear time limits. Set an ending time.
  - Review old business action items carried over from the previous meeting.
  - Introduce the subject.
  - Lead a discussion.
  - Reach an action conclusion.
  - Establish follow-up.

- Present new business items and suggestions.
- Close the meeting on a positive note.
- Set the next meeting date and time.
- Clean up the room and leave it in the condition in which you found it.
- Other presentation hints
  - Avoid interrupting a speaker. Wait until he or she is finished before you comment.
  - Avoid judgmental comments. Turn the question back to the group. ("What do the rest of you think?")
  - Avoid negative comments. Keep the meeting on a positive note.
- Documenting meetings
  - Record all action taken or to be taken — who will do what, when.
  - Record all other suggestions or issues raised and note follow-up action.
  - Write up and distribute meeting reports promptly.

Hints for Working with Difficult People in Meetings

Participant's Meeting Behavior	Techniques for Handling
Overly talkative	• Pass the person's comments out to the group. Let the group take care of the problem as much as possible.
Highly argumentative	• Look for the positive points in their comments. Thank them and then move on. • Ask the group for their views. Let the group reject the idea. • As a last resort, talk to the person after the meeting in private.
Side conversations	• Avoid embarrassing the side talkers. • It is best to ignore side conversations unless they continue to the point where they have become an obvious distraction for the entire group; then pause in your presentation. As the meeting becomes quiet, usually side talkers will stop their discussion.
Rambling or off-the-subject issues	• Thank the person for their point and then bring the meeting back to the subject.
Gripes and pet peeves	• Ask the group for their comments. • Point out that the meeting cannot change policy. • Ask the person bringing up the issues how he or she would solve them. • Record the points for later investigation and/or referral to the appropriate person. • Discuss the issues with the individual in private after the meeting.

Summary and Conclusions

The benefits of group discussion and brainstorming have already been demonstrated by the use of quality circles, and they can be equally effective in solving safety problems. What is often not appreciated is that the average first-line supervisor or work team leader needs help in improving his or her skills in meeting leadership. Providing systematic training and retraining will help these leaders become more effective as managers and specifically as meeting facilitators.

The safety meeting system works best when it is used to provide special emphasis on the control of unsafe work practices that involve high-risk potential for serious injury and/or damage. As such, they are a critical element in the total accident prevention system.

FURTHER READING

Books

Bird, F. E. and German, G. L. *Practical Loss Control Leadership.* Loganville, GA: Det Norske Veritas (USA) Inc. 195–226, 1996.

Doyle, M. and Strauss, D. *How to Make Meetings Work.* New York: Berkley Publishing Group, Jove Edition, 1982.

Fettig, A. *World's Greatest Safety Meeting Idea Book.* Battle Creek, MI: Growth Unlimited, 1990.

Corfield, A. *Safety Management.* London: International Institute of Risk and Safety Management, 185–200, 1988.

Mosvick, R. K. and Nelson, R. B. *We've Got to Stop Meeting Like This!* Indianapolis, IL: Park Avenue Productions, 1996.

Other Publications

U.S. Department of Labor, Occupational Safety and Health Administration. OSHA Safety and Health Program management guidelines: issuance of voluntary guidelines. *Fed. Regist.* 54(16), January 26, 1989.

ACCIDENT INVESTIGATION

Introduction

The federal OSHA Safety and Health Program Management Guidelines state (in part):

(c)(2)(IV) Provide for investigation of accidents and "near miss" incidents, so that their causes and means for preventing repetitions are identified.

Comment: Accidents and incidents in which employees narrowly escape injury, clearly expose hazards. Analysis to identify their causes permits development of measures to prevent future injury or illness. Although a first look may suggest that employee error is a major factor, it is rarely sufficient to stop there. Even when an employee has disobeyed a required work practice, it is critical to ask, "Why?" A thorough analysis will generally reveal a number of deeper factors, which permitted or even encouraged an employee's action. Such factors may include a supervisor's allowing or pressuring the employee to take short cuts in the interest of production, inadequate equipment, or a work practice which is difficult for the employee to carry out safely. An effective analysis will identify actions to address each of the causal factors for an accident or near miss incident.

These words speak for themselves and really stand alone as a fitting introduction to this subject.

The State of California Cal/OSHA Regulations for Injury and Illness Prevention Programs require all employers to "include a procedure to investigate occupational injury or occupational illness."

The draft Cal/OSHA Model Injury and Illness Prevention Program for High Hazard Employers includes the following guidelines:

Accident/Exposure Investigations — Procedures for investigating workplace accidents and hazardous substance exposures include:

1. Visiting the accident scene as soon as possible;
2. Interviewing injured workers and witnesses;
3. Examining the workplace for factors associated with the accident/exposure;
4. Determining the cause of the accident/exposure;
5. Taking corrective action to prevent the accident/exposure from recurring; and
6. Recording the findings and corrective action taken.

Definitions

Before we begin a discussion on the techniques used for accident investigation, it will be important to establish a few basic terms related to this process. They are defined in the following sections.

Accident

In *Practical Loss Control Leadership* by Bird and Germain, an accident is defined as "an undesired event that results in harm to people, damage to property, or loss to process." A simplified version that I prefer because it includes near miss-type events is "any unplanned event that may result in injury and/or damage."

The British Health and Safety Executives in their publication *Successful Health and Safety Management* define the term accident as follows:

Accident includes any undesired circumstances which give rise to ill health or injury; damage to property, plant, products, or the environment; production losses, or increased liabilities.

Accident Investigation

The American Society of Engineers *Dictionary of Terms Used in the Safety Profession* defines an accident investigation as follows:

A determination by one or more qualified persons of the significant facts and background information relating to an accident, based on statements taken from involved persons, and inspection of the site, vehicles, machinery, or equipment involved.

I am indebted to Homer Lambie for his outstanding training on this subject and for the next two definitions.

Accident Source

The accident source is the specific unsafe condition that was most closely related to or directly produced the accident.

Accident Cause

The accident cause is the specific error/unsafe work practice of a person, which either contributed to or directly caused the accident.

Hazard

Fred Manuele, in his book *On the Practice of Safety*, defines the term hazard as the potential for harm or damage to people, property, or the environment. Hazards include the characteristics of things and the actions or inactions of people.
The British Management of Health and Safety at Work Regulations (1992) defines it as follows:

A hazard is something with the potential to cause harm (this can include substances or machines, methods of work, and other aspects of work organization).

My simplified version of these definitions is: A hazard is either an unsafe condition or an unsafe work practice (act).

Near Miss

One final definition that must be considered is a significant incident, or near miss. This is any accident involving a high potential for serious injury and/or damage. The ratio of accidents and incidents has been the subject of numerous studies. Three well-known accident ratio studies are illustrated in Figure 6.14. The importance of investigating damage-only and near-miss events is emphasized by these statistics. A later chapter will discuss a program for this type of investigation.

Accident Investigation and Reporting

Obviously, it is vital for management to encourage employees to report all incidents as these are warning flags of weak links in the unit's accident prevention system and, if not addressed, eventually a more serious event will point out the need.

Accident Ratio Studies

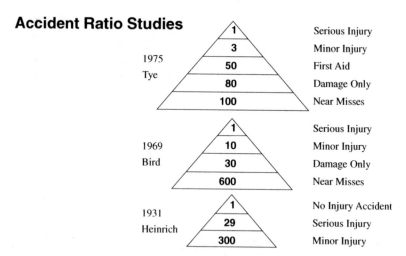

Figure 6.14 Three of the best-known accident ratio studies.

This section will provide an overview of a very broad and complex subject. Those readers desiring more in-depth information should consult the list of references at the end of the section.

Besides being required by law, the establishment of a written accident investigation and reporting policy and procedure is of vital importance. The policy-procedure as a minimum should include the elements outlined in the next five subsections.

Purpose and Definitions

- All accidents and incidents to be reported
- Accident types and their investigation
- Who is to be notified regarding occurrence of accidents, particularly serious events
- Reporting accidents to regulatory agencies

Apart from regulatory, insurance, and workers' compensation reporting requirements, an employer should establish an internal investigation report format, which preferably will be multipurpose (for any type of occupational accident). One exception might be traffic accidents on public highways. This type of accident involves reporting data that are somewhat unique, and frequently the form is provided by your insurance agency.

Principal accident types that should be considered are as follows:

- Employee occupational injuries and illnesses
- Damage-only accidents
- Significant incidents (near misses)
- Consultant/utility/telephone employee accidents on owner's property
- Vendor/contractor injuries and illnesses on owner's property
- Delivery vehicles and their occupants' accidents on owner's property
- Potential third-party liability accidents
- Visitor accidents on owner's property

In the event of third-party liability potential or a serious accident requiring a report to a regulatory agency, a team investigation is the recommended approach. Membership of an accident investigation team will vary with the circumstances. Membership may include:

- Immediate supervisor
- Next level of supervision/department head
- Safety and health professional
- Person(s) with technical expertise relevant to the accident situation
- Legal counsel
- Insurance representative

Reporting of accidents to regulatory agencies varies with the jurisdiction. In California, employers are required to report any work-related fatality or serious injury or illness. A serious injury is defined in the California Labor Code as:

... any injury or illness occurring in a place of employment or in connection with any employment which requires impatient hospitalization for a period in excess of 24 hours for other than medical observation or in which an employee suffers loss of any member of the body or any degree of permanent disfigurement.

Check with the appropriate agencies at your work locations to determine reporting requirements.

Procedure and Guidelines

The following are general guidelines for accident investigation and reporting.

Priorities Immediately after an Accident

- Get to the scene as quickly as possible.
- Size up the situation.
- Ensure that emergency services have been notified and render first aid care for the injured.
- Protect others from injury.
- Protect equipment and facilities.
- Secure the site.
- Notify the proper persons concerned.

Gathering Information

1. Speed is essential. As soon as you are aware of an accident, begin the investigation immediately. The longer you delay, the less likely you are to get the facts. People tend to forget details quickly and, if they have time to talk over the accident with others, they may include details that never happened.
2. Go to the scene of the accident and interview all witnesses at the scene.
   - Interview each witness separately.
   - Make it clear at the outset that the purpose of the interview is not to establish blame but to establish the facts of the accident in order to prevent a reoccurrence.

- Ask each witness to explain in his or her own words what he or she saw. Let the witness tell you the story without interruption; then ask questions for clarification.
- Record notes on all witnesses' reports or obtain signed statements. A hint on taking witnesses statements — have the witness describe in his or her own words what he or she saw. Questions to ask include:
  - What impressed you?
  - How close did you get?
  - What did you say?
  - What did you do?
  - At any moment did you notice the time?

Tell the witnesses to write down each point as they recall it occurred. Then as other things occur to them, they should write them down. The sequence is not important. Finally, ask each witness, "Is there anything else you remember that would help us understand the accident?"

- Do not complete the interview until you are satisfied that you have sufficient facts to establish the following:
- The position of all persons involved
- The job being done
- How the job was being done
- The unexpected happening
- The result of the unexpected happening

3. At the scene, note conditions carefully
   - Conditions of the working surface — wet or dry, oily, greasy, rough or smooth, obstructions, level or sloping, skid marks
   - Position of valves, switches, equipment, vehicles, tools, etc.
   - Lighting — general level of lighting, lights not working, etc.
   - Any unsafe conditions that may have produced or contributed to the accident; check work area, machines, equipment, tools, materials, electrical, vehicles
4. Make a sketch indicating layout and measurements.
5. Photograph the area. Take both a general view and close-ups of relevant items and equipment.
6. Keep site secured until at least this part of the investigation is complete. In cases of fatality or serious accidents, the site should remain secured until cleared by the law enforcement and regulatory agencies concerned.
7. Secure evidence — In the event of a potential insurance damage claim, third-party liability, or serious accident where there is legal liability potential, it may be prudent to secure certain equipment items and records. Check with legal counsel for guidance as to what should be secured and how it should be secured. Items to consider for securing include:
   - Operating and maintenance logs
   - Copies of agency reports
   - Complete accident investigation file with reports, notes, statements, sketches, pictures, etc.
8. Additional hints for accident investigations
   - Always bear in mind that people will not always tell you all the aspects of the accident due to their desire to protect themselves or protect their egos. Most accident victims will rationalize and tell the story differently whether they mean to or not.
   - The investigator should always be suspicious if the person changes his or her story.

- Example: On a construction project, two trucks collided. One was backing up and one was going forward. Both drivers said the other person was at fault. Investigation revealed that one truck was stationary and the other reversed into it without looking. Conclusion: It is very important to find out what motion took place prior to the accident.
- Example: A truck backing at a dump at night time backed so far it went over the side of the pile, overturned, and the driver was killed. The spotter at the dump said he signalled the driver to stop but the driver continued coming back. The investigator asked the following:
  Q: "How fast was he coming?"
  A: "Pretty fast."
  Q: "Did he stop before he went over?"
  A: "I think so."
  Q: "Any signals?"

At this point, the spotter became confused. On two occasions his story differed. Finally, he broke down and admitted he was not there at the time; he had gone to a nearby cafe for a cup of coffee.

Questions to ask include:
- What position were the people in?
- What did they do or not do?
- What made the person lose balance?
- Why did they fall?
- Where were they standing?
- How were they standing?
- Where was he or she supposed to have been at the time?

Analyzing the Facts

Having done the best possible job of obtaining the facts as to how the accident happened, the investigator is now ready to determine the cause and/or source of the accident. To do a thorough analysis, it is recommended that you use a format as illustrated in Figure 6.15.

Record all unsafe actions/work practices and unsafe conditions that in any way could have contributed toward producing the accident. Keep an open mind at this stage — do not prejudge the accident.

Next, determine the reasons why the actions occurred or the conditions existed. This may cause you to go back and reinterview witnesses or reexamine the equipment. This is useful toward ensuring that you have established all the facts of the accident, and it will also help to reveal underlying causes such as management system failures, inadequate training, improper maintenance, poor engineering design, etc.

Establishing Controls to Prevent Recurrence

- Review your analysis of the accident facts to determine the key causes and/or sources of the accident.
- Determine the action necessary to control those key unsafe actions and conditions.
- Take immediate temporary control measures as necessary to prevent further injury or damage.

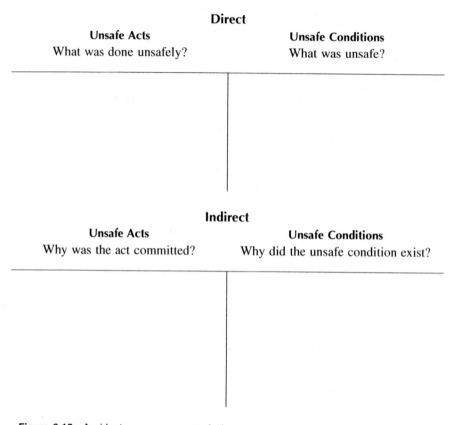

Figure 6.15   Accident cause-source analysis.

- Establish long-range or permanent actions to control the hazards in the future.
- Establish other controls, system improvements, etc., that are recommended. Identify which safety action programs may be involved in the correction process, such as job hazard analysis and hazard control.
- Follow through to ensure that all control actions have been taken.

Complete and Distribute an Accident Investigation Report

Figure 6.16 illustrates a sample format for this report.

Evaluation of Accident Reports

A sample quality evaluation format for an accident report is illustrated in Figure 6.17. The investigator's immediate supervisor should use this as a guide in reviewing accident reports. The evaluation will also assist the reviewer in pinpointing training needs and possible additional actions that may be needed in the investigation process.

Accident Investigation Report

Location	Plant or Facility		Location of Accident (the Name or Number of BLDG., Store, Dept., Floor, etc.)	
Employee	Employee Name	Badge Number	Date and Hour of Accident	
	Dept. or Section in which Employee Works		Employee's Job or Position	
	Describe the Injury and/or Damage:			Estimated Loss:
Description	Sequence of Events: (A) Employee's position in relation to surroundings; (B) how work was being done; (C) unexpected happening; (D) result of unexpected happening; (E) additional facts if available; (F) attach drawings and/or photographs whenever possible.			
Accident Cause & Source	Unsafe Act (What was Done Unsafely)		Unsafe Condition (What was Unsafe)	
	Why Act Was Committed		Why Condition Existed	
Hazard Control	**Hazard (Cause) Control** Immediate Control, Action		**Condition (Source) Control** Immediate Control/Action	
	Long Range		Permanent	
Follow-Up	Responsibility for Control Action/Est. Compl. Date		Responsibility for Control Action/Est. Compl. Date	
JHA	No. & Title Applicable Written Job Hazard Analysis		Hazard Identification ☐ Yes ☐ No	JHA Needs Revision ☐ Yes ☐ No Compl.Date
Control	Report Completed By: Date:		Reviewed By: Date:	Reviewed By: Date:

Figure 6.16 Typical accident investigation report format.

Accident Investigation Checklist

- During a regulatory agency "special investigation"
  - Cooperation — the basic rule
  - Legal counsel readily available
  - Escort at all times
  - Parallel photographs, samples, etc.
  - Representation at interviews

Evaluation of Accident Report			
Supervisor	**Department**	**Date**	
Report	Is the report accurate and complete?	5	
	Are all the sections completed, including date, loss estimate, employees involved, and JHA activity?		
Description	Is the description clear? Does the description include the sequence for events, employee's position in relation to surroundings, how was job being done, the unexpected happening and result of unexpected happening?	5	
Analysis of the Problem	Are all the cause and source factors contributing to the incident identified? Does the analysis include all direct people factors (unsafe acts) and indirect people factors (reasons why the employee acted as they did?) Does the analysis include all direct environment factors (unsafe conditions) and all indirect environment factors (reasons why the unsafe condition existed?) That is, equipment deterioration, improper or inadequate engineering, etc. NOTE: (a) Causes are the actions of people. (b) Sources are the conditions of things.	30	
Control Action	Does the control action protect the employee and prevent reoccurrence of the accident?	Immediate Remove the employee from the hazard, protect the employee form the hazard by temporary guarding, tell the employee how to do the job safely, etc.	20	
		Long Range Remove the hazard, guard the employee from the hazard, tell the employee how to control the hazard (Safety meeting or JHA).	30	
Follow-Up	Does this section specify WHO will do WHAT by WHEN?	10	
	Total	**100**	
Date:	Reviewed by:		

Figure 6.17 Example of quality evaluation for accident/incident investigation reports.

- Accident investigation kit
  - Chalk, pens, pencils, felt tip markers, China marker, flashlight, magnifying glass
  - Graph paper, lined paper, clipboard, tape, adhesive labels, tags with string hangers, ruler, tape measure
  - Sample containers — manila and plastic envelopes, containers with caps
  - Camera with flash film and photo log
  - Cassette recorder with tape
  - Barrier tape, warning signs
- Witness interviews
  DOs
  - Show concern.
  - Be friendly.
  - Listen carefully — do not interrupt.

- Ask broad, open-ended questions.
- Preferably, interview at the scene.
DON'Ts
- Sarcasm
- Pressuring witness
- Blaming anyone
- Asking leading questions
- Making assumptions
- Witness's statement tape-recorded
  - State the date, time, place, who you are, and with whom you are talking.
  - Record the witness's consent: "I am giving this statement of my own free will, and the facts and observations are true to the best of my knowledge and belief."
  - At the end of conversation, state the same information.
- Witness's statements written by witness
  - Ask witness to "write in your own words."
  - Describe as you saw it.
  - What impressed you?
  - How close did you get?
  - What did you say/see/do?
  - Did you notice the time?
  - Sign and date your statement and note time.
- Witness's statements written down for the witness
  - Write down the witness's statement in double-line spacing.
  - Date the statement and state the place in which it was signed.
  - Have the witness read statement, initial changes, and initial each page.
  - Put a clause at the end indicating the witness's consent.
- Witness's statement
  - The above statement was written down by according to what I have told him or her. I have read through the statement, and it accurately records my account of the facts.
  - Signature of witness and date
- Witness's interviews, investigators notes
  - Background information
  - Names of other parties at the scene
  - Where were the other witnesses/what were they doing?
  - Weather/lighting conditions
  - Physical limitations (hearing, eyesight)
  - Witness's location/observation/actions
- Photography
  - 35-mm camera is a must — Polaroid only for quick reference
  - Color film
  - Take pictures as soon as possible.
  - Take general shots of overall accident scene.
  - Take closeups.
- Photo log (*Note:* Much of this information can be covered in a photo ID board on the first shot of the film roll.)
  - Date/time/where picture taken
  - Name of photographer
  - What does the photograph show?
  - Camera, lens, and film used

- Sketches and diagrams
  - Scale drawing
  - Equipment or facilities involved
  - General topography
  - Signs
  - Witnesses
  - Key dimensions
- Securing evidence — guidelines:
  - Small items in envelopes; label each item, tag larger items
  - Store evidence in controlled area
  - Files indexed by injured person's name
  - Cross-reference the main file to bulky items
- Securing evidence — files should contain:
  - Witness/in-house statements and reports
  - Results of tests
  - Documents relating to machinery or equipment
  - Operating and staff schedules
  - Pictures, sketches, notes, logbook, photo log, list of evidence items, personnel records
  - Policies, procedures, regulations
  - Other relevant information

FURTHER READING

Books

American Society for Industrial Security (ASIS). *Basic Guidelines for Security Investigations.* Arlington, VA: ASIS, 1981.

Bird, F. E. and Germain, G. L. *Practical Loss Control Leadership.* Loganville, GA: Det Norske Veritas (USA) Inc. 69–121, 1996.

Boylston, R. P., Jr. *Managing Safety and Health Programs.* New York: Von Nostrand Reinhold, 95–108, 1990.

Colvin, R. J. *The Guidebook to Successful Safety Programming.* Boca Raton, FL: Lewis Publishers, 65–74, 1992.

Ferry, T. S. *Modern Accident Investigation and Analysis.* New York: John Wiley & Sons, 1988.

Ferry, T. S. *Accident Investigation for Supervisors.* Des Plaines, IL: American Society for Safety Engineers, 1988.

Hendrick, K. and Benner, L., Jr., *Investigating Accidents with STEP.* New York: Marcel Dekker, 1987.

Kohn, J. P., Friend, M. A., and Winterberger, C. A. *Fundamentals of Occupational Safety and Health.* Rockville, MD: Government Institutes Inc. 73–97, 1996.

Kuhlman, R. L. *Professional Accident Investigation.* Loganville, GA: Institute Press, 1977.

Lack, R. W. Ed. *The Dictionary of Terms used in the Safety Profession* 4th Edition. Des Plaines, IL. American Society of Safety Engineers, 2001.

National Safety Council (NSC). *Accident Investigation — A New Approach.* Washington, D.C.: NSC, 1983.

Ridley, J. et al. *Safety at Work.* London: Butterworths, 180–197, 1983.

Slote, L. et al. *Handbook of Occupational Safety and Health.* New York: John Wiley & Sons, 99–123, 1987.

Vincoli, J. W. *Basic Guide to Accident Investigation and Loss Control.* New York: Van Nostrand Reinhold, 1994.

Other Publications

Smith, S. L. Near misses: safety in the shadows. *Occup. Hazards* 56(9): 33–36, September 1994.

Tiedt, T. and Kindley, R. A lawyer's prospective on accident investigations. *Prof. Saf.* 32(8): 11–17, August 1987.

U.K. Health and Safety Commission Management of Health and Safety at Work. Management of Health and Safety at Work Regulations. Approved Code of Practice 1992.

U.S. Department of Labor, Occupational Safety and Health Administration. OSHA Safety and Health Program Management Guidelines: Issuance of Voluntary Guidelines. *Fed. Regist.* 54(16), January 26, 1989.

## CHAPTER 7

Training, Interest, and Motivation — Practical Applications That Work

Joan F. Woerner

CONTENTS

INTRODUCTION

Implementation and success with training, interest, and motivation programs for safety can be very challenging to the safety professional. Through many years of practical application, I have found several methods and techniques to be extremely beneficial. This chapter will give you specific examples that have worked over and over again and provide a positive direction in making a safety program successful.

EMPOWER EMPLOYEES TO CORRECT UNSAFE CONDITIONS

Employees generally can recognize hazards, but they do not have either the know-how or the authority to get unsafe conditions corrected. This lack of information creates problems for the safety professional, who gets targeted as the safety fix-it person all too frequently. It also creates an atmosphere of frustration and complaints

1-56670-370-0/02/$0.00+$1.50
© 2002 by CRC Press LLC

with employees who feel that safety issues are not being addressed or corrected in a timely manner.

To solve this problem, the safety professional must first be able to define the process or procedure to get unsafe conditions corrected. This can vary greatly from company to company, so it is important that you are well versed in safety maintenance prioritization and procedures before you take it out to the employees for implementation. To begin the process, the unsafe condition should be evaluated by severity of the hazard and cost to repair. Those conditions that have greater potential to cause severe injury should take precedence over low hazard potential conditions. Also, the cost of repair will make a tremendous difference in how quickly the condition can be corrected. A project requiring capital appropriations and a significant amount of engineering design will take much longer than correcting a low-cost maintenance item. We all have businesses to operate, and safety maintenance must be integrated appropriately into the day-to-day business.

Since every unsafe condition can not be corrected instantaneously, employees must be educated in the process of assessing and prioritizing the repair of unsafe conditions. Each employee should have access either to a system to input work orders for correction of unsafe conditions or to an individual in their immediate operating area who has the responsibility of performing this task. The use of a line or area safety representative to perform this job can also be effective if individual access is impractical.

Once a safety work order is input into the maintenance system, there must be a method to identify and handle unsafe conditions separately from other maintenance repair issues. It helps to put a safety identifier on the work order so it is easily recognized as a safety issue and given proper prioritization. There should also be a method for the person who issued the work order to check the status of the job. This provides feedback to the employee and helps alleviate the feeling that no one is working on the problem. Once completed, safety work orders can then be sent to the safety professional for tracking the progress of the system. Work orders also provide excellent documentation for audits and inspections.

UTILIZE YOUR IN-HOUSE RESOURCES AS INSTRUCTORS

We often fail to utilize our own in-house resources for safety training purposes. Many of us look to outside consultants to provide expertise. If this is the case in your operation, you need to stop and look at what you are missing. Every employee in your operation has some special skill or interest that could add value to your safety program. The best safety instructors are those with knowledge, operational experience, and an ability to communicate to others. Employees are much more receptive to a safety class taught by a knowledgeable peer than they would be to a class taught by an outside expert.

Many organizations provide train-the-trainer programs that last from three days to several weeks. Identify your own in-house experts and then provide them with teaching skills and an opportunity to write their own program. Once they are trained, they are there at your facility, not only as on-site instructors, but as experts to assist in accident investigations, problem solving, and advising. Some examples of areas in which operators make great instructors include: forklift operators and forklift maintenance mechanics are naturals as forklift instructors; employees who have experienced injuries and illnesses due to repetitive motion or have interest in competitive athletics make excellent exercise or ergonomic instructors; emergency responders make great first aid, CPR, confined space rescue, and Hazardous Waste Operations and Emergency Response (HAZWOPER) instructors.

Make sure that the employees who are chosen as trainers have respected operating skills, are knowledgeable in the area they are teaching, have a personal commitment or interest in the subject, and have an ability to communicate in front of a group. You do not want to set someone up for failure. If you do it right, you will find a commitment to safety that you have never experienced before.

MANAGEMENT COMMITMENT AND VISIBILITY = EMPLOYEE AWARENESS AND MOTIVATION

Without true management commitment and visibility throughout the operation, the safety program will never reach its greatest potential. There is nothing more noticeable to employees than a plant manager who regularly makes himself or herself visible and accessible by walking through the operation and randomly stopping and talking to employees about safe work practices. When the practice is also carried out by department managers and supervisors, it sends a very strong message to employees that management is committed to safety. Be careful, though, because this can backfire if not done correctly. Make sure your management staff and supervisors are well versed in safety issues and regulations before they actively participate in this type of activity. Sincerity and follow-up are also critical. The average employee is amazingly sensitive to management behavior. If managers do not set the appropriate example and provide reasonable responses to safety questions and comments, they will immediately lose credibility with the workforce. One additional way to get increased management visibility is to have them periodically stand at the facility entrance during the shift change and greet employees as they come and go to work with a safety hat, pin, pen, coffee cup, free lunch ticket, etc. The safety giveaways should tie in with a company safety theme or positive safety milestone that the employees have been working to achieve. This gives managers an opportunity to be visible and to personally thank employees for their participation in the safety program.

## CHAPTER 8

Investigating and Reporting Incidents

George E. Swartz

CONTENTS

INTRODUCTION

A vital part of every safety and health program should be provision for the identification, investigation, and reporting of incidents. Injuries in the workplace are apparent because the injured worker is required to seek medical treatment. Incidents, on the other hand, are many times overlooked when it comes to investigating and correcting these situations. Management should recognize that incidents occurring each day in the workplace are costing the company thousands of dollars in damaged product, late shipments, property damage, work interruption, as well as the increased potential for serious injury.

Incidents can involve fires of any size, near misses, discovery of an unsafe condition, observance of unsafe behavior, or any event that requires an employee to seek first aid treatment. During the course of a normal workday, numerous exposures occur, which can cause injury to any employee or plant visitor. When incidents are overlooked or neglected, they eventually catch up with the worker and cause bodily harm. Today's incident could become tomorrow's fatality.

SUPPORT FOR INCIDENT REPORTING

Organizations that embrace comprehensive safety and health programs incorporate incident reporting and investigations into these programs. In order to prevent the injury, one must focus on the incidents that led to the injury. DuPont Corporation is a world leader in safety having achieved a lost workday case rate that is 45 times lower than the case rates of all other industries. Within their operational units, every employee is required to report every safety incident, no matter how trivial it may seem at the time. DuPont feels that such reporting may help prevent a similar pattern of failure or laxity when the consequences may indeed be serious. Prevention must be the reaction to incidents, without waiting for the injuries to take place.

Corporations measure safety performance through the use of incidence rates, lost time injury rates, and lost work days. Efforts to prevent the injuries or to lessen the severity leads to problem solving and investigation of the causes. However, there is much to be gained in the control of losses by putting as much effort into the investigation of incidents that do not result in injury. If management does not put forth the effort in correcting incidents, the increasing exposure to daily incidents will lead to injuries. An injury could be moments or weeks away from occurring, but given enough time and opportunity, it will surely take place.

Another program that strongly advocates the use of incident reporting is the OSHA Voluntary Protection Program (VPP). Sites that desire entry into VPP must have a comprehensive safety and health program — one that exceeds OSHA standards. More than 6.2 million sites are covered by OSHA regulations. Since VPP began in the early 1980s, only 600 sites have achieved merit or star status. STAR is the highest form of recognition that OSHA bestows on a site. Merit status allows for a strengthening of specific safety program elements, but the merit designation represents a superior program.

An organization must be utilizing a comprehensive incident investigation program to gain acceptance into VPP. Supervisors as well as employees should have access to the use of incident reports accompanied by the training in how to use them. If a site is not using incident reporting in the scope of its program, OSHA will not approve it as a VPP site. Standard injury investigations are acceptable to OSHA, but the program must go beyond correcting those things that cause harm to workers. OSHA's request for depth in these investigations requires an organization to demonstrate that all workplace hazard exposures, which could cause worker injury, should be evaluated and corrected.

DEFINING TYPICAL INCIDENTS

Most safety professionals define incidents as warning signs. Warning signs in the workplace are those situations or behaviors that could lead to damaged property or product, or injury to employees; but, for whatever reason, these incidents did not occur. Near misses are warning signs of bigger things to come. Some believe that the investigation of near misses should receive higher priority in some cases than actual injury investigation.

It stands to reason that in the course of a normal workday there will be breakdowns of machines and unsafe conditions created by fellow workers — all of which pose a threat. Many unsafe conditions go unseen by workers, either because they fail to really notice the danger or they are unfamiliar with the hazards that workers know are definitely unsafe. They either work around them or fail to alert other workers of the potential perils facing them.

New employees are particularly vulnerable to injuries, especially injuries with high severity. Many times, it is a lack of training and specific job knowledge that leads to personal injury. New employees may take shortcuts or fail to inquire about the safe way to do a job because they are trying to make a good impression and meet production quotas. New workers may be working with other workers who have a lot of seniority but very little safety training.

It is not uncommon for workers to take shortcuts in their jobs. The shortcuts could easily lead to injury. Workers are convinced through past practice that they can operate a machine without a guard on it just to run a few extra parts. The workers have committed this unsafe behavior before and have been rewarded by not being injured. If the supervisor in the department fails to observe this dangerous behavior or fails to identify the missing guard during a walk-through of the department, the potential for a serious injury exists. One employee, who was confronted by a supervisor regarding a missing guard on a punch press, told the supervisor, "It's all a part of my job to run this machine this way." The worker was reaching his hand into the die area to retrieve parts. So much for safe working procedures and safety awareness!

A new employee was observed pouring a hazardous chemical into a funnel while not wearing any eye or face protection. The supervisor reprimanded him for this unsafe behavior. A union steward was summoned to be a part of the discussion. The disciplined worker said, "I've done this many times before, and you have observed me doing the job this way but never said anything about it." The worker was correct in reminding the supervisor that management was delinquent in being observant regarding safety rules, but the worker was wrong in performing a job while endangering his face and eyes.

A common example of a workplace exposure is oil on a walking or operating surface. In one case, a worker slips on a spot of oil and nearly wrenches his back. Another worker slips on a similar spot of oil and falls; the only damage is to his dignity and his dirty pants. A supervisor steps on a similar oil spot, falls, and breaks his wrist. Any one of these three slips on the oil could have resulted in just a casual slip or a fall with the potential of a fractured skull. The means of preventing the falls should have focused on the origins of the oil spill. In this case, a forklift had a leaking hydraulic hose, and it went unrepaired for several weeks until the hose finally failed. In the meantime, the forklift was not undergoing a daily inspection at the beginning of the shift because management did not require inspections. As the forklift traveled throughout the plant, the puddles of oil were everywhere. Early intervention in the correction of the leak would have prevented a housekeeping problem as well as the injuries.

EXAMPLES OF INCIDENTS

When incidents occur, the close call or near miss indicates that:

- Someone could have been hurt
- Something could have been damaged
- Production could have been interrupted
- Profits could be lost

Many times, incidents are not reported because they are not taken seriously. If a close call occurs, most times it is forgotten because no one was injured.

Two forklift operators working in a brickyard were observed racing their forklifts a number of times by fellow employees. One employee told another, "They're going to get hurt one of these days doing that." The day arrived unexpectedly when one forklift bumped into the other and tipped over. The operator was not wearing a seat belt and attempted to jump clear of the tipping lift truck. His right leg was crushed by the overhead guard. No one bothered intervening in an attempt to prevent the unsafe behavior before the serious injury occurred. Neither of the operators was ever given formal operator training.

An employee was operating his electric-powered pallet jack in a warehouse and narrowly escaped a serious hand injury. He turned the unit too sharply, and the portion of the operating handle was jammed against the racking. The shock was hard enough to shatter the hand protector. The employee was not injured, but the repair to the handle cost $327. Four hours of downtime was also associated with the incident.

At a conveyor in a large warehouse, an employee was unjured when a stack of cartons toppled over on him. The electronic sensor that would prevent a build-up of boxes on the conveyor was not functioning. The employee did not report the malfunction and worked around it. On the day of the injury, the worker allowed too many boxes to accumulate at the drop-off end of the conveyor. The boxes jammed and when he went to pry them loose, a large group of boxes fell on him.

At a steel processing plant, a worker narrowly missed being killed when the crane block, hook, and steel coil being handled fell. The main cable on the crane parted, and the weight of that particular coil caused the system to fail. The worker recognized the signs of cable wear and failed to notify his supervisor. He was not performing the required daily crane inspections and obviously failed to identify the defective cable that he was trained to detect. The incident shut the department down for one full day. The real cause of the cable failure as discovered by a maintenance review of the trolly assembly, involved problems with the take-up reel.

An automotive mechanic stabbed himself in the forehead with a pair of needlenose pliers while repairing the brakes on a car. He suffered only a minor injury. Things could have been worse. He was not wearing the required safety glasses and was not using the correct tool for the job. He could have easily lost an eye, which could have cost the employer several hundred thousand dollars. The employer failed to enforce the safety rule requiring safety glasses and failed to remind workers to always use the correct tool.

PROGRAMS THAT WORK

Job hazard analysis (JHA) has been used by industry for more than sixty years. As a result, the JHA process focuses on the exact method in which to perform a task as well as how to perform it safely. When workers receive thorough safety instructions in all the tasks they perform, the likelihood of injury is reduced. The JHA's can be learning tools for all workers, and safe working procedures will aid in injury prevention. Many times the employees have the ability to correct hazards, and they have control over a safer worksite. The most effective programs allow for employee participation. Incident reporting, added to programs like JHA, are ideal for employee participation.

Worker training is essential to the prevention of injuries and compliance with OSHA regulations. Review and discussions on incident reports can help get the safety message across to the workers. Reports that are generated during the month can be used for safety meetings and discussions. Once employees become comfortable with the concept of indentifying incidents and reporting them, greater safety awareness is created. When the facility employees are taking it upon themselves to identify incidents and complete the appropriate report, management has to be complimented for taking the program so far.

HEINRICH'S THEORY AND BEYOND

During the 1930s, William Heinrich, a safety engineer, studied hundreds of injury investigation reports and came to the conclusion that over 80% of claims were a result of unsafe behavior on the part of the worker and that a chain of workplace events led up to each injury. Heinrich was a pioneer in this area and, if anything, his work has caused many to think about unsafe behavior and unsafe conditions.

Figure 8.1 identifies the classic Heinrich triangle with the appropriate designations for worker injuries. Many safety professionals do not agree with the Heinrich theory and challenge the percentages attributed to his injury totals. Some even question the methods used by Heinrich to arrive at the conclusions.

One premise that Heinrich did advocate was that if the cause of the injury were removed from this chain of events in advance, the injury would not take place. This belief is still popular today and is one of the basics in injury prevention. If a machine needs a guard, if a handrail is missing, if the worker lacks an appropriate piece of protective safety equipment or if local ventilation is needed for a particular job, the use of any of these safeguards will protect the worker. If any of these items are missing in the worker's environment, he or she could be exposed to a hazardous situation. Heinrich was correct on this point.

As identified in Figure 8.1, Heinrich believed that for each serious injury a number of less serious injuries and minor injuries also occur. The large, bottom portion of the triangle was added later to represent opportunities for incidents or exposures. The focus of a comprehensive safety and health program should be for the correction of day-to-day incidents so that injuries do not occur.

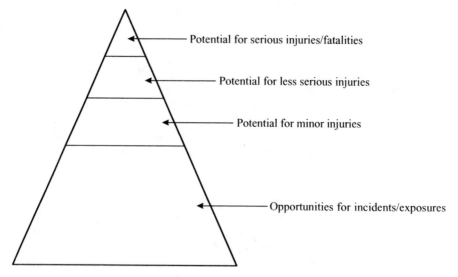

Figure 8.1   The injury pyramid, or Heinrich triangle, which illustrates that for each serious injury, a number of less serious and minor injuries also occur.

USE OF A FORM

Figure 8.2 illustrates an example of a form that could be used for reporting incidents. The form is on a single page and only takes a few moments to complete. It has been kept simple to allow for easy completion and less interruption to the working schedule in the department. The information required at the top of the form asks if the incident was a near miss, noninjury, damage to property, a discovery of an unsafe condition, unsafe behavior, or a fire loss. Item one asks for a description of what occured, along with additional details involving the incident.

Item 10 asks what was done to prevent reoccurrence. This is an important part of the form because corrective action is needed. If the condition is not corrected, the injury cannot be prevented. Perhaps the maintenance department can repair the problem. Item 11 asks for the steps taken to prevent reoccurrence by others. If a supervisor needs the maintenance department to repair something, a copy of the report must be passed on to that department for corrective action.

In addition to the main points identified in numbers 1, 3, 10, and 11, when the form is completed the following important additional information should be sought in order to be accurate:

- Name of the machine or equipment involved
- Amount of down time involved
- Number of employees involved in the incident
- Identify first aid administered
- Identify if a health hazard or exposure was involved
- Possibility of the incident taking place again
- Attach a sketch or plots on the back of the form identifying the details of the incident

Non-Injury Incident Report **Property Damage Report** **and/or** **Discovery of Potential Hazard**

Check all those that apply)

☐ Near Miss   ☐ Non-Injury   ☐ Property Damage   ☐ Incident/Event   ☐ Fire Loss

Market:_____ Shop Location: _____ Date Incident Occurred:_____

1) Describe the incident or what occurred:_____

_____

_____

_____

2) Machine or equipment involved: _____
3) Extend or damage/costs (describe): _____

_____

_____

_____

4) How much lost time or down time involved? _____
5) How many employees wee involved in the incident:_____
6) Was anyone injured?   Yes ☐   No ☐
7) Was first aid administered?   Yes ☐   No ☐
8) Was there a fire?   Yes ☐   No ☐
9) Was employee wearing required personal protective equipment?   Yes ☐   No ☐
10) What steps were taken **by you** to prevent recurrence: _____

_____

_____

11) What steps were taken by others (maintenance, management, outside sources, etc.) to prevent recurrence:

_____

_____

_____

12) Was a health hazard or exposure involved: Yes ☐   No ☐
13) Can this incident or event take place again: Explain on back side.   Yes ☐   No ☐
14) Identify additional factors or comments on back of this report.
15) Have you drawn a sketch of the details on the back of this report or attached a photo: Yes ☐   No ☐

Signature: _____   Date: _____

Forward Top Copy to:	Corporate Safety
Middle Copy:	Shop Files
Bottom Copy:	Market/Administrator's File

3400-64   Revised 11/91

Figure 8.2   Sample form for reporting and correcting hazards.

Management should insist on the immediate processing of incident reports so corrective action can be taken to prevent further losses. Every safety program, regardless of the size of the company, should include incident reporting. When management places a focus on these day-to-day events, injuries can be reduced along with the associated economic and human losses.

FURTHER READING

Books

National Safety Council, *Accident Prevention Manual for Industrial Operations*. 11th edition, Itasca, IL; National Safety Council, 1997.

Manuele, F. H., *On the Practice of Safety*. New York: Van Nostrand Reinhold, 1993.

Colvin, R. J. *The Guidebook to Successful Safety Programming*. Boca Raton, FL: Lewis Publishers, 1992.

Mottel, W., Long, J., and Morrision, D. *Industrial Safety is Good Business — The DuPont Story*. New York, Van Nostrand Reinhold, 1995.

Other Publications

Smith, S. L. Near misses: safety in the shadows, *Occupational Hazards*. September 1994, 33–36.

Bureau of Business Practice, Inc., Supervisors Safety Clinic, No. 249, (7A), 1982.

National Safety Council — Rubber and Plastics Section Newsletter, Prevent accidents by treating incidents as warnings, March–April, 1987.

Swartz, G. Incident Reporting — a vital point of quality safety programs, *Professional Safety*. December 1998, 32–34.

## CHAPTER 9

Principles, Objectives, and Techniques for Effective Health and Safety Management

Herman Woessner

CONTENTS

INTRODUCTION

Seven management objectives must be achieved by and incorporated in a safety and health system for it to be effective. This chapter will analyze the objectives and the principles upon which they are based and will describe proven techniques for implementing them.

1-56670-370-0/02/$0.00+$1.50

The techniques illustrated in this chapter by no means exhaust the universe of possibilities. They are simply measures that have proven to be effective over a wide range of public and private organizations with which I have been associated in a professional career spanning more than 25 years. Other safety practitioners, working with other types of organizations, at other times, undoubtedly have developed and will continue to develop different, equally effective methods for accomplishing safety objectives.

Although many different methods can be employed to accomplish the seven safety management objectives, the objectives themselves and the principles upon which they are based are universal. They define and determine the effectiveness of all health and safety programs.

ROLES AND RESPONSIBILITIES

The fundamental responsibility of a safety and health professional is to facilitate and coordinate the development and implementation of the organization's injury/illness prevention and loss control system. The key role to be played in fulfilling this responsibility is that of system leader, not hazard controller, which is often the role many practitioners assume, and some even relish, particularly the unmentored.

Controlling hazards (i.e., unsafe conditions and unsafe work practices) is primarily the responsibility of employees and supervisors. They are closest to the hazards and have the greatest capability for recognizing and controlling them. There can never be enough safety and health professionals present in the workplace, with enough authority, to effectively control the myriad of hazardous employee exposures that can occur on any given workday. Thus, most of the organizational time and energy of safety and health professionals should be spent on influencing and assisting managers, supervisors, and employees to successfully meet their responsibilities.

To influence the behavior of others, one must be able to effectively communicate to them how the desired behavior supports or enhances their valued interests and why it is more beneficial than competing behaviors. While everyone is interested in safety, the degree of interest among and between managers, supervisors, and employees varies, for each group and each individual holds values and interests that compete for time, energy, and resources with their safety interests.

To be a successful safety leader, the safety and health professional must first identify and understand the individual and organizational interests and competing values of the person or persons whose behavior he or she is trying to influence, and then address those interests by demonstrating how the desired safety behavior will positively affect them.

Assume, for example, that a safety and health practitioner for a private company discerns through interactions with a plant manager and his or her peers that cost control is one of the manager's major organizational concerns, and receiving a high performance appraisal from his or her boss is of significant personal interest. Also, assume that the safety and health professional wishes to have the manager demonstrate greater safety leadership to improve the quality of housekeeping and employee training in the plant. How should he or she proceed to convince the manager to lead the recommended safety improvement effort?

The answer lies in providing the manager with valued information that supports his or her organizational and personal interests. In this example, the valued information must relate to cost control and management effectiveness. The safety and health professional should, therefore, obtain and analyze all available records to determine the historical and potential loss experience, as well as the affect it has had and likely will have on the manager's budget and cost containment goals. In addition, the records should be analyzed to identify the significant causal factors, which, presumably, would include poor housekeeping and inadequate employee training.

If available, the following loss/cost experience and estimates should be analyzed, organized, and explained to the manager to help him or her understand and appreciate the need to take the desired action:

Loss Experience	Calculated Costs	Potential Losses
Injury and illness cases	Workers' compensation costs Increased training costs	Reduced productivity
Property damage incidents	Property damage cost	Impaired operations
Environmental impairment incidents	Environmental remediation cost	Impaired public relations
Hazardous materials exposures	Medical cost	Impaired labor relations
Business interruption incidents	Production loss	Impaired shareholder relations
Regulatory citations	Penalty cost	Impaired government relations
Legal cases	Legal cost	Impaired operations

The previous illustration presents a win-win approach. It gives positively valued information for valued actions, which is much more effective than threats and predictions of dire consequences. Too many safety and health practitioners rely on this "chicken little" approach to influence their organizations. They do so because scare tactics require little effort to develop and sometimes produce the desired results faster than would the systematic process of analyzing and marshalling positively valued information. However, the continual use of such tactics can eventually lead to a loss of creditability and effectiveness for the safety and health practitioner. No one appreciates, over the long run, the bearer of only negative news. Eventually, the messenger will become associated with the message.

The positive approach also is not limited by rank or organizational type, as are the negative tactics. Employees and public agencies, for example, are not influenced by the threat of an OSHA citation, for they are usually exempt from any penalties from failure to comply. On the other hand, they will respond to information that helps them to realize their personal and organizational goals.

The interests of persons and organizations, and the information required to influence them, will vary, but the approach need not. The challenge to the safety and health practitioner is to identify the organizational and personal interests that motivate individuals in his or her organization, and provide them with accurate and credible data in a manner that complements those interests. This is easier said than done, for it requires the safety and health professional to have good analytical and communication skills, organizational credibility, and a nonthreatening personality.

SAFETY AND HEALTH SYSTEM

A safety and health system is a network of interrelated and dynamic policies, goals, responsibilities, programs, plans, procedures, and standards designed to guide an organization to proactively and systematically identify and control workplace hazards and risks. The wide scope and complexity of such a system requires that it be written down to ensure its completeness, clarity, and consistency.

The development of a safety and health system should be planned and designed to accomplish the seven universal management objectives. It should not just evolve in response to promulgated OSHA regulations or as a reaction to injury/illness incidents experienced by the organization. This reactive approach invariably produces incongruent or duplicative program elements.

A quality control process needs to be an integral part of every safety and health system, for it is critical to the program's long-term effectiveness. Without a quality control process, any system will become inconsistent and outmoded over time. To prevent this from happening to the safety and health system, the responsibilities and operations comprising the quality control process should be written into the program, either as a separate element or as the quality control section in each safety procedure.

The quality control process should ensure that the performance and content of the entire health and safety system undergoes continuous evaluation and improvement. The process must provide for: (1) periodic, random observation and focused analysis of required system activities; (2) evaluation of written system elements in light of the performance observations and analyses; (3) feedback of the analytical and evaluation findings to persons with the authority and responsibility for correcting system deficiencies; (4) observation and analysis of the improvement actions taken and evaluation of the results.

OBJECTIVES AND TECHNIQUES

Safety Leadership

The primary safety management objective that must be achieved by and incorporated into an organization's safety and health system is to have top and middle management actively lead employees toward achievement of the organization's safety goals. Achievement of the other safety management objectives is largely dependent upon the quality and consistency of leadership demonstrated by management.

Leaders convey vision and values to their organizations through interactions and communications with subordinates. Subordinates judge their leaders' commitment to the vision and values by the frequency, consistency, and sincerity of their written and verbal statements, and even by their body language.

The significant measures of a leader's commitment to stated policies and objectives are the quantity and quality of the time and resources (human, financial, physical) he or she devotes to achieve them. Thus, managers must devote their time and the organizational resources they control in support of the safety and health system if they expect employees to fully appreciate its value and to enthusiastically work to achieve its goals. Moreover, managers must be consistent, positive, and sincere in their safety interactions/communications with employees if they expect to be believed and consistently followed.

Including a safety objective in the organization's mission statement is an effective, first step toward establishing safety as a valued organizational function. It places safety on the same level as the other critical mission objectives, such as profit and quality, and conveys that the safety and health system is integral to their achievement.

To illustrate how and why a safety objective can be incorporated in an organization's mission statement, consider the examples of ABC Company and The Public Agency. The stated mission of ABC Company is "to manufacture the highest quality widget at the lowest cost," while The Public Agency's mission is "to efficiently provide reliable public service."

ABC Company's mission statement could be revised to read: "The mission of ABC company is to safely manufacture the highest quality widget at the lowest cost." By inserting the word safely in the statement, management is committing the organization to the goal of injury/illness prevention. It implies that, even if the company succeeds in becoming the high quality, low cost leader of the widget manufacturing industry, its mission will not be fully achieved unless the risk and experience of employee injury is low.

A mission statement for The Public Agency could be rewritten to read: "The mission of The Public Agency is to provide public services in a courteous, expeditious, and efficient manner while protecting the health and safety of employees, the public, and the environment." Such a statement defines occupational and environmental protection as important operational objectives, and serves as a visual reminder to the organization of management's commitment to employee and public safety and to environmental protection.

Developing a safety policy statement that defines and reinforces the safety objective expressed in the mission statement is an essential next step to institutionalizing the organization's safety management commitment. The safety policy statement should describe the organization's core beliefs, commitments, and responsibilities regarding safety, and should connect the success of the safety and health system to the success of the organization's overall mission.

An example of a connective statement would be: "We believe that XYZ organization has a fundamental responsibility to protect the health and safety of our employees and that fulfillment of that responsibility is critical to the organization's overall mission. We further believe that establishment of an effective safety and health system is essential for meeting our safety responsibilities and for achieving our mission objectives."

The owner and/or top manager(s) in the organization should sign the finished policy statement and communicate it to the entire organization along with the mission statement.

The process of developing and/or revising the organization's mission and policy statement will undoubtedly serve to broaden management's understanding of its safety roles and responsibilities, and the statements themselves will serve as visual reminders to the organization's commitment to safety.

Another technique that management can use to lead the organization to improved safety performance is to develop and communicate process-directed and results-directed safety goals and measures. Establishing the objective of a daily, supervisory-conducted toolbox safety meeting with employees would be an example of a process-directed safety goal, while a 10% reduction in a facility's annual injury/illness rate as an objective is an example of a results-directed goal.

By establishing and communicating such goals, and visibly monitoring performance toward their achievement, management conveys that prescribed safety activity and desired injury/illness results are of value to the leadership and the organization.

Management should conduct regularly scheduled reviews of the facility's/company's safety performance versus the established goals to demonstrate its safety vision and interests, as well as to identify and correct any program deficiencies.

Safety and health professionals should seek to influence and assist management's safety leadership efforts. This can be done by using the valued information approach discussed earlier to convince them to adopt the system objectives covered in this chapter.

Employee Participation

A closely related safety management objective to that of having top and middle managers actively and visibly lead the organization's safety efforts is to achieve widespread employee participation in safety process activities. The more employees are involved in the program and the greater the involvement, the more likely they will feel responsible for ensuring its success.

Employees are much more likely to develop a sense of ownership for the safety and health system and to understand, appreciate, and support its requirements, if they help to develop and maintain it as well as be expected to implement it.

Management should encourage employees from every department and level of the organization to participate in the development and monitoring of the organization's safety and health policies and procedures, including the mission and safety policy statements. Establishing a steering task force for each policy/procedure and an implementation monitoring team in each department can do this.

The task force and team members must be trained to perform their assignments. Those involved in policy/procedure development should be informed by management about the purpose, objectives, and format of the proposed policy/procedure, and about the process for developing and writing it. Those team members involved in monitoring policy/procedure implementation activities should be trained in the requirements of the policy/procedure and in methods for behavior monitoring and reporting.

Management should recruit qualified volunteers from each department to serve as safety communicators, problem solvers, and/or trainers. These employees would be trained to assist supervisory personnel to communicate hazard awareness and hazard control information to their workers and to facilitate solutions for system implementation conflicts. Bilingual employees could be chosen as communicators and problem solvers for issues affecting individuals and ethnic groups for which English is a second language.

Employees can participate in an organization's safety and health system in many ways. The more participation, the better — as long as the participation is planned, the participants are adequately trained, and their contributions are recognized by management. To ensure that these requirements are met, they should be covered in a written employee safety participation plan and the plan should be incorporated as an element of the organization's safety and health system.

Knowledgeable Program Participants

Developing knowledgeable participants is the third safety management objective that is essential to establishing and maintaining an effective safety and health system. Managers, supervisors, workers, safety team members, program auditors, trainers, safety communicators, problem solvers, and safety department personnel are all system participants. The more they know about how to carry out their program responsibilities, the more successful they are likely to be.

A safety training program should be included in every organization's safety and health system to ensure that a sufficient amount of safety training occurs each year. The program should specify: (1) the frequency, duration, and type of training and retraining required by OSHA and by other elements of the safety and health system; (2) a minimum amount of safety leadership training for managers and supervisors; (3) training documentation requirements; (4) the persons or departments responsible for conducting or arranging the required training and for maintaining the training records. The program should also include a provision for conducting safety training for employees who, for whatever reason, miss a scheduled training session.

The safety and health professional should recommend that the top managers of the organization attend an in-house or external seminar on the principles and techniques of safety leadership and management every year, and should seek their approval to include a commitment for such training in the safety training program.

Many organizations hold annual planning or operational review meetings at which either staff or outside consultants make presentations on various topics applicable to the objectives of the organization. This type of setting would be ideal for a session on safety leadership/management. It could serve as the forum for annually refreshing managers about their safety leadership roles and responsibilities. To convince top managers of the value of including safety on the agenda of such meetings, the safety and health professional can point out that the safety concept supports and is consistent with their expressed commitments in the organization's mission and safety policy statements.

The safety training program, like all elements of the safety and health system, must be regularly monitored to determine if the frequency and quality of required activities are being met. The Program should outline reporting and follow-up actions in response to any training and record keeping deficiencies found by the monitors.

Hazard Identification, Evaluation, and Control

Establishing a process for identifying, evaluating, and controlling unsafe conditions and unsafe work practices is the fourth essential safety management objective.

The occurrence of an injury or illness is largely dependent upon the frequency and duration of exposure to hazards. The longer and/or more frequently an employee is exposed to an unsafe condition or commits an unsafe work practice, the greater the likelihood that a harmful encounter will occur. The severity of that encounter depends on the amount of energy or toxicity that can be imparted to the exposed employee by the hazardous condition or unsafe work practice.

Employees and their supervisors should be trained and authorized to conduct regular workplace inspections and task observations to identify inherent hazards, to assess the risks presented by the hazards, and to implement needed hazard/risk control measures in a timely manner. The frequency of operational inspections and job observations should be no less than monthly, and the time requirement for controlling an identified hazard should be based on an assessment of the risk.

Risk assessment is an evaluation of the probability of an injury/illness occurrence based upon the frequency and duration of employee exposure to a hazard, and an evaluation of the likely magnitude of the harm from such an occurrence based on an assessment of the amount of toxicity or energy that would likely be imparted from the hazard to an affected employee. The highest risk hazards have high energy/toxicity potential and frequent employee exposure. Control of these risks should have the highest priority.

The process of investigating, reporting, analyzing, and following up accident and near-accident events is a reactive process for identifying, evaluating, and controlling hazardous causal factors. Accidents and near-accident events are the probable consequence of the frequency and duration of exposure to hazards. Careful analysis of the accident investigation findings will lead to recognition of the hazardous exposures and to the development of effective prevention measures.

Both the proactive hazard identification, evaluation, and control process and the reactive accident and incident investigation process should be described in writing and established as elements of the safety and health system. Their ultimate objectives should be to discover and permanently correct the underlying management and organizational factors that either produced or failed to prevent the recognized hazards and risks.

The following Risk Assessment Guide can be used to assess the risk presented by identified hazards and to prioritize hazard control actions:

Average Number of Employees Exposed to Hazard Each Day A	Total Duration of Daily Exposure to Hazard B	Average Daily Frequency of Exposure to Hazard C	Total Daily and Yearly Exposure to Hazards in Hours $D = A \times B \times C$ $Y = D \times 256$ days	Assessment of the Likelihood of an Injury or Illness Occurring Based on the Total Daily and Yearly Exposure to the Hazard
Example: 10 exposed workers	Example: 1.5 hours of exposure	Example: 2 exposures per day	$D = 10$ employees $\times$ 1.5 exposure hours $\times 2$ exposures per day $Y = 30$ hours of worker exposure per day $\times$ 256 days in year $Y = 7{,}680$ annual exposure hours	(a) Very Likely (b) Likely (c) Somewhat Likely (d) Unlikely (e) Very Unlikely $\downarrow$

Assessment of the Maximum Amount of Energy and/or Toxicity Likely to be Transferred from/by the Hazard to the Exposed Employee(s)	Assessment of the Likely Harm to the Most Exposed Employee(s) from the Maximum Amount of Energy /Toxicity Transferrable from/by the Hazard	Risk Assessment matrix		(a)	(b)	(c)	(d)	(e)
			(1)	Class 1 Risk	Class 1 Risk	Class 1 Risk	Class 1 Risk	Class 2 Risk
			(2)	Class 1 Risk	Class 1 Risk	Class 2 Risk	Class 2 Risk	Class 3 Risk
Very Large Large Moderate Low Negligible	(1) Death (2) Permanent impairment (3) Temporary impairment (4) Injury but no impairment (5) Negligible injury		(3)	Class 1 Risk	Class 2 Risk	Class 3 Risk	Class 3 Risk	Class 4 Risk
			(4)	Class 2 Risk	Class 3 Risk	Class 4 Risk	Class 4 Risk	Class 4 Risk
Note: Harmful energy may be in the form of radiation, heat, electrical current, mechanical movement, chemical reaction, gravitation, combustion, corrosion, repetitive motion, etc. Toxicity may be in the form of airborne gases, vapors, particulates, fungal spores, bacteria etc., as well as toxic and reactive liquids/solids.	Note: Impairment is dismemberment, disfigurement, or functional disability.		(5)	Class 4 Risk	Class 4 Risk	Class 4 Risk	Class 5 Risk	Class 5 Risk

Risk Classes:

Class 1: Great Risk
Class 2: High Risk
Class 3: Moderate Risk
Class 4: Low Risk
Class 5: Negligible Risk

Hazard Control Priorities:

Priority 1: Control Class 1 Risks immediately
Priority 2: Control Class 2 Risks as soon as possible within 30 days
Priority 3: Control Class 3 Risks as soon as possible within 90 days
Priority 4: Control Class 4 Risks as soon as possible within 1 year
Priority 5: Control Class 5 Risks when practical

Performance Goals, Standards, and Measures

The fifth safety management objective that the safety and health system must achieve is the creation of system-directed and results-directed performance goals, standards, and measures.

System-directed goals seek to focus the organization's attention and efforts on implementation of the requirements set forth in the health and safety system, rather than on the expected consequence of those requirements. Results-directed goals, on the other hand, focus attention and efforts on the expected consequence in order to encourage the implementation of system requirements.

Thus, a goal to have supervisors and employees conduct a weekly safety inspection of their work areas in conformance with the provisions of the hazard identification and control process would be an example of a system-directed goal. It directly encourages specific supervisory and employee behavior known to be effective in the identification and control of hazards. A goal to have supervisors and employees reduce the number of injuries and illnesses in their work units by 10% is an example of a results-directed goal. It exhorts supervisors and employees to take unspecified actions to prevent and control the causes of injuries and illnesses.

Performance standards establish the preferred ways and means for achieving performance goals, and together with the goals serve as the measures for performance effectiveness. A performance standard for the previous example of a system-directed goal could be to have a team consisting of the unit supervisor and at least two unit employees conduct the weekly safety inspection using a checklist of safe work practices and good housekeeping requirements. The same performance standard could be used in support of the results-directed goal.

By comparing and evaluating system implementation activities and their results to performance goals and standards, organizations can measure the effectiveness of their safety and health system. The frequency, timeliness, and quality of required actions are the principle measures of performance relative to system-directed goals and standards, while injury/illness experience rates and associated costs are the key measures of results-directed goals.

The following performance indicators are some of the important measures that can be used to evaluate the overall effectiveness of the safety and health system and to point out areas in need of improvement:

Results Measures	System Measures
Daily Hazard Rate: daily average of unsafe conditions and unsafe work practices observed each month.	Frequency and quality of system-required workplace inspections
Monthly Repetitive Hazard Rate: number of repetitive unsafe conditions and unsafe work practices identified per month.	Frequency and quality of system-required job safety analyses
Total Case Rate: number of injuries and illnesses x 200,000 hours ÷ number of hours worked.	Frequency and quality of system-required safety meetings
Lost Time Case Rate: number of lost time cases x 200,000 hours ÷ number of hours worked.	Frequency and quality of system-required training sessions

Lost and Restricted Case Rate: number of lost time and restricted work cases x 200,000 hours ÷ number of hours worked.

Lost and Restricted Day Rate: number of days lost and restricted x 200,000 hours ÷ number of hours worked.

Total Workers' Compensation Reserves: the amount reserved by the insurance company to pay for current year's injuries/illnesses.

Experience Modification (Ex-Mod) Factor: the multiplier based on injury/illness experience used by insurance companies to determine workers compensation premiums. An ex-mod factor greater than one is high.

Property Damage and Business Interruption Costs: the direct and indirect costs resulting from accidents involving property damage and/or business interruption.

Vehicle Accident Incidence Rate: the number of vehicle accidents x 25,000, 100,000, or 1,000,000 miles ÷ number of miles driven in a year.

Total Vehicle Accident Costs: the direct and indirect costs to pay for accident and to restore vehicles.

Annual Number of OSHA Citations: the total number of citations received in a year.

Cost of OSHA Penalties: annual cost of penalties for OSHA citations

Frequency and quality of system-required program audits

Frequency and quality of system-required job safety observations

Frequency and quality of system-required industrial hygiene monitoring

Frequency and quality of system-required safety documentation

Frequency and quality of system-required safety performance appraisals

Number and frequency of system-encouraged employee safety suggestion

Timeliness of system-required responses to employee safety suggestions

Timeliness and quality of system-required accident/incident investigations

Number of employees properly wearing system-required personal protection equipment

Safety Performance Accountability

Accountability for performing up to standards is the sixth safety management principle and objective that must be achieved by and incorporated in a safety and health system. It provides system participants with the necessary impetus and reinforcement to consistently strive to achieve performance goals and objectives.

Accountability involves a supervisor's evaluation of the work performance of a subordinate over a specific period of time as measured against established performance goals and standards. The supervisor reviews his or her findings with the subordinate and takes steps to reinforce acceptable behaviors/results and to modify unacceptable behaviors/results. The supervisor has to acknowledge with praise, or some other form of recognition, subordinate performance that has met or exceeded standards, and must provide encouragement and support to improve performance that is not up to standard. This support can take many forms, ranging from verbal encouragement, to support for additional training, to providing needed tools and equipment, to discipline.

Discipline should only be used when it is necessary to correct persistently substandard performance. It should be employed as a last resort, only after positive accountability methods have failed to achieve acceptable performance.

Supervisors communicate value to their subordinates through the accountability process. The responsibilities for which supervisors hold their subordinates accountable will be a priority for subordinates because of their obvious importance. Supervisors would not spend the time and effort to hold subordinates accountable for achieving them if they were not important. Conversely, responsibilities for which subordinates are not held accountable by their superiors or for which accountability is sporadically or inconsistently administered, will be recognized by subordinates as being of less value than the responsibilities for which they are held accountable.

Accountability can not be an effective guarantor of performance, however, unless the means for achieving the desired goals and standards are adequate and available to the responsible persons. Employees cannot be expected to work safely and avoid injury, for example, unless they are regularly trained to recognize workplace hazards and are provided with the means to control the hazards. It is also not effective safety management to hold supervisors accountable for failing to prevent an employee's injury, unless the supervisor had, or should have had, the knowledge and means to prevent it. The key responsibility for which all superiors in an organization should be held accountable is the responsibility to provide their subordinates with sufficient information, resources, and oversight to meet their safety responsibilities.

The process for holding safety program participants accountable for meeting performance goals should be an integral yet distinct element of the written safety and health system. The Safety Performance Accountability process element should describe the actions to be taken by supervisors and subordinates in order to determine the appropriate system-directed and results-directed goals and standards against which the subordinate's performance will be measured, and should set the frequency, format, and documentation requirements for conducting performance evaluations. The process also should describe the actions to be taken by supervisors and subordinates in response to the evaluation findings.

One action that must be taken to ensure the effectiveness of the accountability process is to tie the evaluation findings to the organization's pay-for-performance system. While the prospect of gaining or losing a monetary reward alone can not guarantee desired performance results, it does convey value to the affected individuals in the organization and should serve as a powerful performance motivator.

Continuous Safety Improvement Program

The establishment of a program to ensure continuous safety improvement is the seventh safety management principle and objective that must be incorporated into every safety and health system. The Continuous Safety Improvement Program should provide for an assessment of the overall performance of the safety and health system over the past year by senior managers for the purpose of establishing an improvement plan and budget. An action plan unsupported by budgeted funds is only a wish list.

The Continuous Safety Improvement Program should be included as a distinct element in the written safety and health system. This element would guide the organization to:

- Collect performance and costs information for the purpose of identifying system strengths and weaknesses.
- Obtain input from affected and knowledgeable employees, supervisors, and managers on existing safety problems.
- Formulate an improvement plan based on the data collected and input received.
- Identify the financial, human, and material resources needed to accomplish the plan.
- Create organizational and department budgets to pay for planned activities.
- Establish plan responsibilities, completion dates, and an evaluation process.

CONCLUSION

The effectiveness of an organization's safety and health system is defined and determined by the degree to which it achieves the seven safety management objectives discussed in this chapter. Achievement of these objectives is greatly facilitated by including them as elements of the organization's written safety and health system. The principal responsibility of safety and health professionals is to facilitate, coordinate, and manage the development and implementation of the system.

REFERENCES

Adams, E. *Total Quality Safety Management*. Des Plaines, IL: American Society of Safety Engineers, 1995.

Bird, F. E. and Davies, R. J. *Safety and the Bottom Line*. Loganville, GA: Institute Publiching, Inc., 1996..

Bird, F. E. and Germain, G. L. *Practical Loss Control Leadership*. Loganville, GA: International Loss Control Institute, 1990.

Drucker, P. *The Practice of Management*. New York: Harper & Row, 1986.

Grimaldi, J. V. and Simonds, R. H. *Safety Management*, 5th edition, Homewood, IL: Richard D. Irwin, Inc.

Krause, T. R., Hidley J. H., and Hodson S. J. *The Behavior-Based Safety Process*, New York: Van Nostrand Reinhold, 1990.

Peterson, D. *Analyzing Safety System Effectiveness*. New York: John Wiley & Sons, 1996.

Peterson, D. *Techniques of Safety Management, A Systems Approach*, New York: Aloray, 1989.

Roland, H. E. and Moriarty, B. *System Safety Engineering and Management*, New York: John Wiley & Sons, 1990.

Swartz, G. *Safety Culture and Effective Safety Management*, Ibassa, IL: National Safety Council, 2000.

## CHAPTER 10

Process Safety Management — A Global Impact

Mark D. Hansen

CONTENTS

INTRODUCTION

As the Occupational Safety & Health Administration (OSHA) 29 CFR 1910.119 "Process Safety of Highly Hazardous Materials" was phased in during the mid-1990s, many companies needed to come to grips with these new requirements. In fact, many companies across the globe implemented much of the intent of 1910.119 because it made good business sense. OSHA took the next step in safety by making 1910.119 a performance-based regulation. What it meant to the many companies that had to comply with 1910.119 was that they had to implement the program elements identified in Process Safety Management (PSM) by including safety considerations in the design of systems. In other words, they were required to engineer the hazard out *first*, by design. If it could not be designed out, it had to be controlled using safety devices supported with warning devices. Lastly, procedures and training had to be implemented. The PSM program elements will be discussed in detail in this chapter.

The systems approach to PSM revolves around identifying potential hazard severity and probability to determine a risk assessment. The methods for tailoring this approach to particular applications will be discussed in detail. Several example risk assessment matrices will be provided to illustrate the different approaches to eliminating and controlling hazards.

The benefits of implementing PSM's 14 program elements will be discussed and some inherent requirements for success will be identified. This will include what the standard (29 CFR 1910.119) requires employers to do to successfully implement each element.

Methods of implementing rigid structures, to guarantee the process safety function is included in all areas required by the standard, will be discussed. Also, methods for ensuring that these methods are flexible and responsive to change via the management of change functions will also be discussed.

HISTORICAL BACKGROUND AND PHILOSOPHY

A number of serious incidents have occurred over the last 15 years, resulting in increased attention to companies that use, manufacture, store, or handle highly hazardous materials. These incidents have caused considerable human casualties and property losses, prompting regulators worldwide to enact legislation mandating process safety programs. Figure 10.1 gives a global perspective of recent disasters that have had a significant impact on the process and refining industries.

The Flixborough, Seveso, and Mexico City disasters resulted in new regulations in the United Kingdom, Italy, and Mexico, respectively. The Bhopal, India, incident in 1984 involved a release of toxic methyl isocyanate, causing the deaths of over 2500 nearby residents. The Bhopal incident prompted the U.S. Environmental Protection Agency (EPA) and OSHA to develop new regulations relative to chemical releases, spills, pollution controls, and worker/community "right-to-know." Some of the serious industrial losses are summarized in Tables 10.1 through 10.7.

OSHA 29 CFR 1910.119 and the Clean Air Act Amendment (CAAA) paragraph 301(r)/304 are the result of industry incidents involving fires, explosions, and/or toxic releases since 1988. The regulations demand more stringent management controls, including more thorough and effective training, review, documentation, and auditing requirements to ensure that all personnel involved in an operation understand and accept their roles in process safety.

These regulations require employers to initiate considerable management changes and improvements to their industrial process operations. Implementation of a sound PSM system will ensure that process hazards are properly managed, and that formalized engineering and administrative controls become an integral part of the design, construction, operation, and maintenance, including startups, shutdowns, turnarounds, and other phases of a facility's life.

A Global Perspective of Recent Disasters

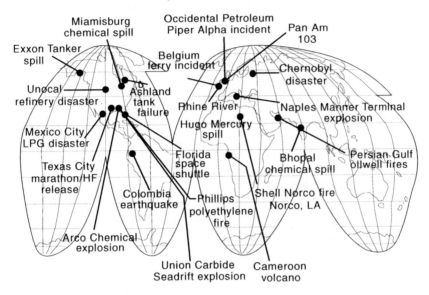

Figure 10.1 A global perspective of recent disasters.

Table 10.1 A Review of Serious Industrial Losses

Incident Location and Description	Damage Estimate and Description
Baton Rouge, LA (December 1989) Refinery pipeline rupture	$43 million; damage to houses off-site
Pasadena, TX (October 1989) Petrochemical — vapor cloud explosion	$500 to $725 million; severe business interruption
St. Croix, Virgin Island (September 1989) Refinery — Hurricane Hugo	$60 million; extensive damage to storage tanks
Martinez, CA (September 1989) Refinery — H_2 and hydrocarbon release	$50 million; massive fire fighting effort
Morris, IL (June 1989) Petrochemical — vapor cloud explosion	$41 million; 40 acres surrounding plant damaged
Richmond, CA (April 1989) Refinery — H_2 gas fire	$90 million; two years to repair plant
Antwerp, Belgium (March 1989) Petrochemical — column exploded	$77 million; business interruption of $267 million

Table 10.2 Loss Trends (Five-Year Intervals)

Period	Number of Losses	Losses (Billions of Dollars)
1959–63	6	$0.15
1964–68	13	0.48
1969–73	24	0.51
1974–78	30	1.18
1979–83	45	1.29
1984–88	32	1.54

Table 10.3 Loss Distribution by Type of Complex

Operation	Percent	Losses (Millions of Dollars)
Refineries	40	$40.1
Petrochemical	17	34.4
Terminals	13	25.8
Plastics/rubber	9	23.8
Chemical	8	18.5
Natural gas	7	52.2
Pipelines	2	45.6
Miscellaneous	4	24.6

Table 10.4 Cause of Losses

Cause	Percent	Losses (Millions of Dollars)
Mechanical failure	41	$36.0
Operational error	19	38.6
Unknown	17	25.9
Process upset	10	40.7
Natural hazard	5	43.2
Design error	4	60.5
Sabotage/arson	4	19.0

Table 10.5 Equipment Involved in Cause

Equipment	Percent	Losses (Millions of Dollars)
Piping	31	$41.9
Tanks	17	40.5
Reactors	13	28.5
Miscellaneous	9	34.7
Process drums	7	25.5
Marine vessels	6	32.0
Unknown	5	25.0
Pumps	5	19.2
Heat exchanger	3	24.0
Process towers	3	53.8
Heaters/boilers	1	28.8

In essence, CAAA Paragraphs 301(r)/304 require hazard assessments for certain chemicals to determine the potential effects from accidental releases of both toxic and/or flammable chemicals, including downwind effects and potential exposures affecting populations. Risk management plans are also required and are intended to help prevent or minimize accidental releases, and provide prompt emergency response following such releases.

Table 10.6 Equipment Involved in Cause

Type	Percent	Average Losses (Millions of Dollars)
Fire	38	$30.8
Vapor cloud explosion	36	45.5
Other explosions	24	24.6
Miscellaneous	2	16.0
Normal operation	76	

Note: 24% of the total losses occurred when plants were not in the "normal operating mode." The normal operation category includes losses during start-up, shutdown, on-line maintenance, or turnaround.

Table 10.7 Source of Ignition

Source	Percent	Losses (Millions of Dollars)
Unknown	53	$38.5
Open flame	12	32.0
Chemical reaction	9	26.0
Electrical equipment	5	16.5
Internal combustion engine	6	49.0
Auto-ignition	4	34.1
Lightning	3	39.1
Hot surface	3	39.4
No ignition	3	18.7
Sabotage/arson	2	43.5
Static	1	14.5
Cutting/welding	1	16.0

CAAA regulations were promulgated in November 1993 and provide guidance to prevent or at least detect accidental releases. The promulgation of 29 CFR 1910.119, which was published August 26, 1992, has been promulgated after resolving several contractor safety issues.

Similar regulations were enacted in New Jersey, Delaware and California and have been proposed in several other states. Process safety legislation has also been promulgated in the United Kingdom, Canada, Australia, Netherlands, Mexico, and Italy. Similarly, industry associations in the U.S. developed their own process safety standards by publishing the following programs:

1. American Petroleum Institute (API), Recommended Practice RP-750, "Management of Process Hazards"
2. American Institute of Chemical Engineers (AIChE), Center for Chemical Process Safety (CCPS), Guidelines for Technical Management of Chemical Process Safety
3. Chemical Manufacturer's Association (CMA), Responsible Care Program

These laws and standards have had an unprecedented impact on industry's ability to more safely manage its operations.

PROCESS SAFETY MANAGEMENT (PSM) DEFINITION

PSM is the application of management systems to policies, procedures, and practices for the identification, understanding, and control of process hazards. The goal is to prevent episodic process-related injuries and incidents.

Process safety, as defined by OSHA, involves the operation of facilities that handle, use, process, or store hazardous materials/chemicals in a manner free from episodic or catastrophic incidents. Process safety applies to chemical and refinery processes, and to other processes, such as mechanical and electrical processes, in the manufacturing, aerospace, nuclear, transportation, and distribution industries.

PROCESS SAFETY, SYSTEM SAFETY, AND INDUSTRIAL SAFETY

System safety engineering programs that began in the United States military in the 1960s were the first successful efforts that addressed process safety management. These long-standing programs have matured over the past decade and have become highly successful in fulfilling integrated process safety requirements in the aircraft, missile, spacecraft, and nuclear industries. Although these programs concentrated on major military hardware systems and their design, operation, and maintenance, the principles they use for identifying, evaluating, and reducing process-related hazards also apply to the chemical and petroleum industries.

Furthermore, process safety differs from traditional personnel or industrial safety activities, which have been in place in industry for many years. Traditional industrial losses have resulted in personnel injuries from falls, slips and trips, and vehicle accidents as well as property damage from material handling mishaps, utility failures, etc. Process-related losses involve operating equipment and processes, whether chemical, mechanical or electrical systems, and/or subsystems. Process safety is primarily concerned with the use, manufacture, handling, and storage of hazardous materials/chemicals, and with the design, construction, operation, and maintenance of these facilities.

FACILITY MANAGEMENT COMMITMENT

Facility management plays a vital role in the success of a PSM system. Top management commitment and support combined with similar commitments by a facility's line management and staff personnel helps ensure that PSM programs can be successfully implemented. Management's role should be to:

- **Demonstrate commitment.** Promote interest and instill commitment among employees to develop effective PSM controls to reduce and/or minimize catastrophic failures; establish corporate policy to ensure fulfillment of commitment.
- **Allocate resources.** Delegate personnel to implement PSM system elements through assigned duties and responsibilities; assign accountability to meet overall objectives.

PROCESS SAFETY MANAGEMENT ELEMENTS

1. Process safety information

2. Process hazard/risk analysis

3. Management of change

4. Mechanical integrity

5. Operating procedures

6. Safe work practices

7. Process safety reviews

8. Training

9. Contractors

10. Emergency response

11. Incident investigation

12. Audits

13. Employee participation

14. Trade secrets

Figure 10.2   Process safety management elements.

- **Set standards.** Ensure that adequate procedures, drawings, and documentation control systems are established and maintained to implement the program; ensure that training programs are developed and updated as needed; perform audits periodically.
- **Define priorities.** Establish priorities to meet specific objectives relative to personnel, process technology, facility, and equipment integrity for the reduction of process hazards.
- **Measure and assess performance.** Develop benchmarks for performance and assess needs for continuing improvement of program effectiveness.

PROCESS SAFETY MANAGEMENT ELEMENTS

An effective PSM program consists of various integrated management elements that can result in a systematic and formalized program to reduce and/or eliminate workplace process hazards, thus meeting industry standards and regulatory requirements.

A list of the conventional 14 PSM elements is shown in Figure 10.2. Collectively, these elements, when implemented satisfactorily, can achieve an optimum level of process safety. They have been developed from various sources — primarily OSHA, EPA, API, CCPS and CMA — and also meet the basic requirements of the United Kingdom's *The Control of Industrial Major Accident Hazards (CIMAH) Regulations of 1990.*

Each of these elements is summarized below.

ELEMENT 1: Process safety information. A compilation of written safety records and documentation that provides the foundation of process safety information. It establishes the means by which the hazards of a chemical process or mechanical and electrical operation are identified and understood. The element consists of hazards of materials used, process and mechanical design of equipment/systems, and documentation (i.e., engineering drawings, equipment specifications, process materials, and relief system design).

ELEMENT 2: Process hazard analysis. The heart of the process safety management system consists of the systematic identification, evaluation, and elimination and/or control of process-related hazards to prevent catastrophic incidents. Various hazard analysis methodologies are suggested to evaluate and assess process hazards, and determine the consequences of catastrophic failures.

ELEMENT 3: Management of change. Written procedures to manage changes in technology, facilities, and equipment including personnel. Changes, although inevitable, must be properly managed to avoid serious process incidents. Basis for change, safety and process operating impacts, consequences of deviations, engineering, and administrative control of facility changes must be considered.

ELEMENT 4: Mechanical integrity. An on-going program of written documentation and activities to maintain the integrity of process equipment in an as-built condition. Effective preventive and predictive maintenance programs, testing, inspection, and quality control systems for process equipment, including mechanical and electrical systems and subsystems, are necessary to ensure safe operation of the facility and prevent premature failures or emergencies (i.e., reliability engineering, quality assurance, preventive maintenance, materials of construction, and fabrication and inspection procedures).

ELEMENT 5: Operating procedures. Written procedures providing clear instructions for safely conducting activities in a facility for each operating phase and procedure, operating limits, and safety/health considerations including steps taken to correct and/or avoid deviations.

ELEMENT 6: Safe work practices/hot work permits. Written procedures to ensure the safe conduct of process equipment operations, maintenance, and modification activities involving lockout/tagout of energy sources, confined space entry, line breaking and entering, hot work/welding operations, and use of heavy equipment (e.g., cranes).

ELEMENT 7: Process safety reviews. Formal programs to identify, evaluate, and reduce hazards by intensive reviews at various stages of a project such as design, construction, installation/assembly, commissioning, start-up, operation, maintenance, and demolition. These reviews are divided into capital project, pre-start-up, and process hazard analysis reviews. Process hazard analysis reviews involve a review team that uses one or more of the analytical techniques in Element 2 to evaluate and control process hazards.

ELEMENT 8: Training. An essential ingredient to effectively keep operating personnel knowledgeable in maintaining and safely operating process equipment and machinery. Personnel must also be kept informed regarding the hazards of the processes and chemicals they are using. Specific hazards and safe practices applicable to job tasks should be stressed.

ELEMENT 9: Contractors. Provisions for instructions, including safety rules to contractors working on or near company process equipment and/or machinery, to ensure they can safely perform their tasks. This element includes informing contractors regarding potential emergency actions.

ELEMENT 10: Emergency response. The establishment and maintenance of a comprehensive emergency response program involving emergency plans, defined accountabilities, exercises/drills, and periodic updates. Emergency equipment and procedures should reflect credible catastrophic scenarios that might occur in the facility.

ELEMENT 11: Incident investigation. The formation of an investigative team of knowledgeable persons to evaluate incidents that could have or did result in a major accident in the workplace.

ELEMENT 12: Audits. Periodic, detailed examinations of process facilities by a qualified team to determine compliance with process safety elements, which provide corrective feedback and recommendations where improvements are needed.

ELEMENT 13: Employee participation. Employers must have a written plan outlining employee participation in the conduct and development of Process Hazard Analyses (PHAs) and in the development of other elements of PSM. Employees and their representatives must have access to the information developed for PSM activities.

ELEMENT 14: Trade secrets. Employers must make information necessary for compliance with the regulations available to those persons (1) compiling process safety information; (2) developing PHAs and operating procedures; and (3) involved in accident investigations, emergency planning and response, and compliance audits, without regard to the possible trade secret status of such information. Employers may require confidentiality agreements with personnel receiving trade secret information. Employees and their designated representatives shall have access to trade secret information contained in PHAs and other required documents.

Figure 10.3 provides a comparison of PSM elements as they apply to each standard and U.S. regulatory requirements for OSHA and EPA.

THE SYSTEMS APPROACH TO PROCESS SAFETY MANAGEMENT

Few programs or initiatives have had the potential of affecting so much of a facility's operation as PSM. Initial evaluations by many companies have determined that the requirements set forth by OSHA and EPA will have an enormous impact, and that it will take three to four years to fully implement the fourteen (14) PSM elements.

Several major oil companies have conservatively estimated that, for an average refinery producing 100,000 barrels of gasoline per day, compliance will cost between $10-15 million to implement and will require an additional 25 to 40 more employees and/or contractors to initially implement a PSM program for compliance. These are conservative estimates, but they indicate the significant impact the regulations will have on the industry. Similar costs are expected to be incurred by other industries affected (i.e., chemicals plant processes, explosive and pyrotechnic manufacturers, and gas plant operations).

Title	Guidelines for Management of Process Hazards API RP750 Section	Process Safety Management of Highly Hazardous Chemicals OSHA 1910.119 Paragraph	Guidelines for Technology Management of Chemical Process Safety AICHE, CCPS Chapter	Process Safety Code of Management Practices CMA Code for Responsible Care Tabs	1990 Clear Air Act Amendment OSHA-CAAA S 1630 Air Toxics Section
General	1	a, b, c	1, 2, 3	1, 2, 3	301, 304
Process safety information	2	d	4	5.5	301 Regs, 304 std
Process hazard analysis	3	e	4, 6	5.7	301 Regs, 304 std
Management of change	4	l	7	5.10.8	304 Standard
Operating procedures	5	f	4		301 Regs, 304 std
Safe work practices	6	k	4	7.4	301 Regs, 304 std
Training	7	g	10	7.7	301 Regs, 304 std
Assuring quality & mechanical integrity of critical equipment	8	j	8	6.10	301 Regs, 304 std
Pre-start-up safety review	9	i	5		304 Standard
Emergency response & control	10	n		6.16	301 Regs, 304 std
Investigation of process related incidents	11	m	11	4.10	301 Regs, 304 std
Audit of process hazards management systems	12	o	13	6.7	301 Regs, 304 std
Contractors		h		7.13	304 Standard
Capital projects			5		
Human factors			9	7.9	
Standards, codes, & regulations			12	6.4	304 Standard
Waste management				2.5–2.7	
References	13	Appendices A, B, D	14	9	

Figure 10.3 Process safety management comparison of basic program elements.

Key Elements

A large proportion of the cost will occur in technical, operation, and maintenance activities to formalize necessary records, documentation, drawings, equipment specifications, operating procedures, and inspection and testing programs. Over the years, engineering drawings (e.g., piping and instrumentation drawings, process flow diagrams, and process equipment specifications and descriptions) have typically become outdated. Additionally, operating procedures, including relief system designs, materials of construction, and quality assurance programs may need updating to reflect current design after changes to operations or facilities.

In addition, the consequences of process deviations to equipment design and operations, as well as their relative safety impact must be assessed. Safe operating limits must be defined and procedures developed to address how to correct and/or avoid dangerous process deviations. Two important elements in the proposed regulations are directed at management of change and Process Hazards Analysis (PHA).

Management of change will affect all aspects of the facility's operation — technology, instrumentation, equipment, and personnel. Process changes often become necessary to improve a unit's design or operation and, in some operations, may be as frequent as 50 to 100 changes per year. Moreover, changes may adversely impact a facility's design, operation, and/or maintainability. Comprehensive, yet flexible, management of change procedures must be developed and implemented to meet these concerns.

PHA is the analytical heart of the PSM system. When properly performed, PHA can result in improvements affecting design, operation, maintenance, and training.

PHAs can be accomplished by using different hazard identification and analytical techniques. Common examples are:

- Preliminary Hazard Analysis
- What If or What If/Checklist
- Hazard and Operability Study (HAZOP)
- Failure Mode and Effects Analysis (FMEA)
- Fault Tree Analysis (FTA)

Each technique has specific advantages and disadvantages in hazard identification, evaluation, and control.

Once process hazards have been identified and their potential causes evaluated, the risk (i.e., consequences and frequency) can be evaluated. In simpler terms, the following questions can be answered:

- What can go wrong?
- What are the causes?
- What are the consequences?
- How likely is it?
- What are the risks?
- Is the risk acceptable?

Risks can be evaluated qualitatively and/or quantitatively. Qualitative risk analysis is the most widely used method of evaluating episodic risk. A common approach to qualitatively analyzing hazards to determine potential risks is the use of a risk matrix method as illustrated in Figures 10.4 and 10.5. These techniques, although based on intuitive reasoning and the analyst's experience, can be very effective in determining the relative risk of each incident scenario and in identifying risk reduction methods. The objective of the risk matrix is to determine how the severity and/or frequency of the identified hazards can be reduced through some mitigating action. For example, water spray system installation, reduction of hazardous material inventory, or improvement in the flare system design can be used to reduce risk from high to moderate or moderate to low.

Quantitative risk analyses are performed using engineering techniques and computer models to analyze incident scenarios to determine their impacts — more specifically, what damage may occur from a fire, explosion, and/or toxic release and how frequently will it occur. The potential impacts (e.g., fatality, injury, equipment loss, and/or environmental damage) are systematically estimated to determine the magnitude and severity of the incident.

A Management Function

PSM is comprehensive and addresses all aspects of a line management function responsibility. Organizationally, many companies form process safety groups that report directly to the plant manager or are once removed from the plant manager via the environmental, safety, and health (ESH) manager. The process safety group is typically staffed with technical personnel who have operations experience, and it is charged with the responsibility to coordinate and facilitate the PSM program implementation. This requires direct interaction among employees and departments responsible for implementation of PSM elements (e.g., technical, operations, maintenance, ESH, and training.)

Computer Programs

The required documentation and maintenance of procedures, records, and drawings have resulted in some companies using new and/or modified computer systems to help ensure proper documentation and revision. Specific programs have been developed to record and document test and inspection records, preventive maintenance, and equipment specifications. Other programs are used to monitor real-time process equipment parameters such as pressures, temperatures, flow-rates, and maintenance of daily logs for future reference and planning. Computer Aided Drafting and Designing (CADD) systems are being used to update engineering drawings. More sophisticated systems include detailed schematics and assembly drawings, Material Safety Data Sheets (MSDS), training programs, and operating procedures.

CONCLUSIONS

Recent worldwide experience has shown that a majority of the 100 largest property losses have occurred since 1957 as shown in Figure 10.5. As previously

Severity Category	A - Frequent Will occur twice or more in system lifecycle	B - Probable Will occur at least once in system lifecycle	C - Remote Possible, but unlikely to occur in system lifecycle	D- Improbable A credible accident event cannot be established

I. Catastrophic
Personnel:
  Death
Facilities/Equipment/Vehicles:
  System loss, repair impractical, requires salvage or replacement
	IA	IB	IC	ID

II. Critical
Personnel:
  Severe injury/occupational illness. Requires admission to a health care facility.
Facilities/Equipment/Vehicles:
  Major system damage. Damage greater than $1,000,000 or which causes loss of primary mission capability
	IIA	IIB	IIC	IID

III. Marginal
Personnel:
  Minor injury/occupational illness. Lost time accident of more than one day that does not require admission to a heath care facility.
Facilities/Equipment/Vehicles:
  Loss of any non-primary mission capability, or damage more than $200,000 but less than $1,000,000 with no direct impact on Mission capability
	IIIA	IIIB	IIIC	IIID

IV. Negligible
Personnel:
  Less than minor injury/occupational illness. May or may not require first aid but lost time is less than one day.
Facilities/Equipment/Vehicles:
  Damage equal to or less than $200,000
	IVA	IVB	IVC	IVD

Figure 10.4 The risk matrix method is a common approach to qualitatively analyzing hazards to determine potential risks.

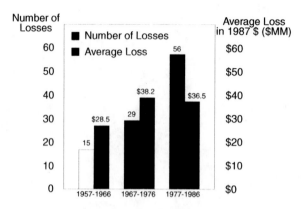

Figure 10.5 The 100 largest property damage losses in the hydrocarbon-chemical industries (1957–1986).

discussed, the federal goverment and some state governments have implemented extensive legislation for industry process safety and risk management programs. Companies affected by these regulations must accept the challenge to continue improving the safety of their operations.

Companies are continually striving to be better corporate citizens because it makes good business sense and because "it is the right thing to do." A safe and healthful workplace is essential for employees, contractors, communities, and the environment. In addition, corporate management recognizes the importance of PSM programs in a global marketplace.

Recently, companies have been working toward improving process safety. Initial audits and reviews are now being conducted to evaluate the impact of OSHA and EPA PSM regulations, and to identify areas where improvements are required. Process Hazard Analyses are being conducted by experienced teams to provide preliminary screening of process hazards in their facilities. These plans assess what is required to attain full compliance with the regulations.

Sufficient time is needed, however, to evaluate older facilities, develop plans, and allocate resources to address needed improvements. A cooperative, remedial action plan, working with federal and state agencies, should be pursued with appropriate authorities to help ensure program compliance and success.

Realistically, risks are an unavoidable, but controllable, aspect of petroleum and petrochemical operations. Moreover, risk is an unavoidable part of our day-to-day activities. As individuals and as a society, people often engage in risky activities because they believe that the rewards and potential benefits outweigh any potential harm. Likewise, risk management decisions for "industry-imposed" risks involve value judgments that integrate technical, social, economic, and political concerns with scientific risk assessments.

Thus, the management of process hazards and the prevention of catastrophic incidents in the workplace requires a recommitment of manpower and financial resources. Assuming this PSM commitment is made, many positive benefits beyond ESH performance will result, including improved quality and reliability, and enhanced worker and community safety.

The complexity, magnitude, and impact of the 14 PSM program elements and the interrelationships among facility operating departments require the development of a formalized, comprehensive, and organized network to implement an effective program. Most important, management and employees must learn to apply these systems to ensure that process safety becomes integrated into day-to-day operations.

FURTHER READING

Addamo, F. Imaging: A critical component of process safety management (PSM). *Document Imaging & Windows Imaging*, May/June 1993, 21–28.

Alderman, J. A. PHA-A Screening Tool. Presented at 1990 Health & Safety Symposium, New Orleans, LA, March, 1990.

American Petroleum Association, *Management of Process Hazards*, API Recommended Practice (RP) 750, 1st edition, Washington, D.C.: American Petroleum Association, 1990.

Baum, D. Client/server development tools for Windows, *Computerworld*, April 26, 1993, 73–75.

Bird, F. E. and Germain, G. E. *Practical Loss Control Leadership*, Atlanta, GA: Institute of Publishing, 1990.

Center for Chemical Process Safety (CCPS). *Guidelines for Technical Management of Chemical Process Safety*, New York: American Institute of Chemical Engineers, 1989.

DeHart, R. E. and Gremillion, E. J. Process Safety — A New Culture. Presented at Occupational Safety & Health Summit, Washington, D.C., February, 1992.

Department of Defense Military Standard, System Safety Program Requirements, MIL-STD 882C. Washington, D.C., February, 1993.

Environmental Protection Agency (EPA). Chemical Process Safety Management, Section 304. EPA, 1990 Clean Air Act Amendment, promulgated November, 1990, *Federal Register*.

Environmental Protection Agency, Prevention of Accidental Releases, Section 301, Paragraph 1, 1990 EPA Clean Air Act Amendment, promulgated November, 1990, *Federal Register*.

Guidelines for Hazardous Evaluation Procedures, 2nd edition. New York: AICE, 1992.

Hawks, J. L. and Mirian, J. L. Create a good PSM system. *Hydrocarbon Processing,* August 1991.

Krembs, J. A. and Connolly, J. M. Analysis of Large Property Losses in the Hydrocarbon & Chemical Industries. Presented at 1990 NPRA Refinery and Petrochemical Plant Maintenance Conference, May 23–25, San Antonio, Texas.

Lees, F. P. *Loss Prevention in the Process Industries*. London: Butterworths, 1980.

Large Property Damage Losses in Hydrocarbon-Chemical Industries — A Thirty Year Review, 13th ed. M&M Consultants, 1996.

O'Lone, E. J. Datapro, Document Imaging Systems, Management Issues, *Client/Server Computing,* June 1993, 1–8.

OSHA's Process Safety Management of Highly Hazardous Chemicals (29 CFR 1910.119) in Plain English, NUS Training Corporation, Houston, TX, undated.

Process Risk Reviews Reference Manual, Wilmington, DE: E. I. du Pont deNemours and Company, Safety and Environmental Resources Chemicals and Pigments Department, 1990.

Occupational Safety & Health Administration (OSHA) 1910.119, Process Safety Management of Highly Hazardous Chemicals, *Federal Register,* August 26, 1992.

Renshaw, F. M. A major accident prevention program, *Plant/Operations Progress,* July, 1990.

Sherrod, R. M. and Sawyer, M. E. OSHA 1910.119 Process Safety Management: Its Effect on Engineering Design. Presented at System Safety Conference, Houston, TX, July, 1991.

## CHAPTER 11

Modern Management Principles and Their Application to Safety and Health

Richard W. Lack

CONTENTS

INTRODUCTION

It is beyond the scope of this chapter to present complete in-depth review of the art and science of management. For this reason, the object will be to provide the reader with more of an overview of the management process and its application in the safety and health setting.

The chapter will first deal with definitions and the basic functions of professional management. The following sections will address key points for special emphasis and problem solving, as well as setting safety goals, standards, and measurements before closing with some new and emerging trends.

The importance of management is emphasized by the fact that of the 12 special interest groups among the 35,000-plus membership of the American Society of Safety Engineers (ASSE), one of the largest is the Management Practice Specialty, with over 3,500 members. The next largest practice specialty is the Risk Management and Insurance Practice Specialty, which is of course a management-related function.

Regulatory agencies are also increasingly emphasizing the need for effective management of the safety and health process. The federal Occupational Safety and Health Administration (OSHA) published its *Safety and Health Program Management Guidelines* in 1989. In 1991, the Health and Safety Executive in the United Kingdom published its excellent booklet, *Successful Health and Safety Management*. This was followed a year later by the *Management of Health and Safety at Work Regulations 1992 Approved Code of Practice*.

DEFINITIONS

The following are some common definitions of terms used in the profession of management. The source of each is indicated in parentheses.

Management — A social process of getting things done effectively through people, organizing the elements of productive enterprise money, materials, equipment, and peoples in the interest of economic ends. The functions of management are a series of activities often described as planning, organizing, staffing, directing, coordinating, and evaluating. (American Society of Safety Engineers (ASSE))

Planning function — The work of manager performs to identify and group the work to be done so it can be accomplished effectively by people. (Louis Allen)

Leading function — The work a manager performs to cause people to take effective action. (Louis Allen)

Controlling function — The work a manager does to assess and regulate work in progress and to assess results secured. (Louis Allen)

Accountability management — The obligation to perform responsibly and exercise authority in terms of established performance standards. (Louis Allen)

Authority — The powers and rights of a person or a position. The more authority people can be given to make decisions concerning the work assigned to them, the more completely they can do that work without going to a higher level and the more interest and initiative they will tend to show in the work. (Louis Allen)

Responsibility — Work which is a continuing obligation of a position. Responsibility is the physical and mental effort or work a person is obligated to perform in order to fulfill the requirements of the position. (Louis Allen)

Line relationship — The command relationship between those persons and positions directly accountable for accomplishing key objectives and therefore vested with the authority necessary to achieve those objectives. (Louis Allen)

Staff relationship — The relationship of those positions and components that provide advice and service to others which are accountable for direct accomplishment of key objectives. (Louis Allen)

Line function — Management activity directly related to the production operations of an organization. Persons performing line functions have responsibility and authority for accomplishing the mission or objectives of the organization and are held accountable for its effective operation. (ASSE)

Staff function — An auxiliary service which operates in support of production or line management. Rather than performing auxiliary functions themselves, line management personnel depend upon the advice, guidance, or service of specialized staff departments while retaining responsibility, authority, and ultimate accountability for production success. (ASSE)

THE WORK OF MANAGEMENT

According to Peter Drucker in his classic book on management, *The Practice of Management,* the work of the manager incorporates five basic operations.

Step 1: The manager *sets objectives* and decided what has to be done to reach those objectives.

Step 2: The manager *organizes.* This is accomplished through analyzing the activities, decisions, and relations needed. These activities are then divided into manageable jobs which are grouped into a unit organizations structure. People are then selected for the management of these units and for the jobs to be done.

Step 3: The manager *motivates and communicates.* Through constant communication and good leadership practices, the manager builds a team that will enthusiastically and effectively accomplish their objectives.

Step 4: The manager performs *job measurement.* To accomplish this, the manager establishes performance standards which are agreed upon and accepted by the team members. Once standards have been established, the manager can analyze the performance, appraise it, interpret it, and communicate the findings to the team and individual members as well as to higher levels of management.

Step 5: In perhaps the most crucial step, the manager *develops people.* The effective manager is constantly seeking ways to help the team members develop their professional knowledge and skills by means of training, both on the job and through seminars, conferences, study courses, etc.

THE FUNCTIONS OF PROFESSIONAL MANAGEMENT

A checklist of management functions based on the Louis Allen system is as follows:*

1. Management planning
   - Forecasting
   - Establishing objectives
   - Programming (resources and people)
   - Scheduling
   - Budgeting
   - Developing policies
   - Establishing procedures
2. Management organizing
   - Establishing organizational structure
   - Delegating
   - Establishing relationships (between units)
3. Management leading
   - Decision making
   - Communicating
   - Motivating people
   - Selecting people
   - Developing people
4. Management controlling
   - Establishing performance standards
   - Measuring/evaluating performance standards
   - Correcting performance deviations

For example, the duties involved with *planning* may consist more of telling what and less of explaining how. *Organizing* duties may concentrate more on providing the resources. *Leading* may be focussed more on developing people, while *controlling* may be more concerned with checking results. Those readers not actually practicing these functions are strongly encouraged to increase your knowledge by reading (see Further Reading at the end of this chapter), attending seminars, or enrolling in courses offered by local colleges and universities.

KEY POINTS FOR SPECIAL EMPHASIS

In other chapters, and in earlier sections of this chapter, we have discussed safety and health organization, the functions of the safety and health professional, line and staff responsibilities, and the various elements of effective safety and health systems. The purpose of this section, therefore, is to highlight the distinctions between some important terms and to provide a central focus on priorities, since the theme of this entire book is the *essentials* of safety and health management.

* The management duties outlined above should be considered as a general guideline. In organizations which are managed with greater emphasis on employee participation in the management process, some of these duties may shift or receive greater emphasis.

I am indebted to Dr. John Grimaldi for drawing my attention to the distinction between *administration* and *management* of the safety and health process. Readers are referred to Dr. Grimaldi's book *Safety Management* (5th edition) for further information.

The issue does appear somewhat divided. For example, Louis Allen defines administration as the "total activity of a manager". The ASSE *Dictionary of Terms Used in the Safety Profession* defines management and administration similarly. Readers may reach their own view, *administration* of the safety and health process refers to the development of safety and health policy, procedures, and systems, their *assessment,* and their documentation.

Further to this discussion, I am also indebted to Dr. Michal Settles for her research on this subject, which is outlined to Chapter 38 of this book. Dr. Settles points out that an administrator is more focussed on his/her specialty area than a manager, who usually has a more across the board application.

Readers are also referred to Dr. Settle's chapter for further information on leadership skills. Stephen Covey, Warren Bennis, and others have conducted considerable research on the traits and competencies of managers and leaders. Managers tend to be more analytical problem solvers and leaders more visionary, inspiring, and risk taking. A final comment on leadership comes from Anita Roddick's book entitled *Body and Soul,* from which I quote in part her comments on this subject, which she calls "leading from behind." Ms. Roddick states.

Leadership is not commandership — it is about managing the future. The principal forces that motivate a leader are an incredibly high need for personal achievement and a different vision of the world — one who marches to a different drumbeat, who does not see himself as part of the mainstream, its essentially an outsider. You don't necessarily have to be charismatic, you just have to believe in what you are doing so strongly that it becomes a reality. ...

Ending on those stirring words seems to be a fitting point to close this discussion and allow the reader to ponder these critical concepts.

Management of the safety and health systems refers to their *implementation* by management *action* utilizing basic professional management techniques (planning, leading, organizing, and controlling). In short, management is the control of resources with the objective of maximizing their value and contribution for the good of the enterprise.

The practice of effective safety management thus presents some unique challenges to its practitioners. In addition to their technical knowledge, they not only must develop administrative skills in organizing and assessing the system, they also must practice broad professional management skills so they are able to communicate, sell, and persuade others to accept and implement the process.

Another aspect that requires special emphasis is the matter of *risk*. The ASSE *Dictionary of Terms Used in the Safety Profession* defines risk as:

A measure of the combined probability and severity of potential harm to one or more resources as a consequence of exposure to one or more hazards. Mathematically, risk is the simple product of severity and probability that one or more hazards will impose upon one or more resources. In insurance, a person or thing insured.

In Grimaldi and Simond's *Safety Management,* risk is defined as:

The assumed effect of an uncontrolled hazard, appraised in terms of probability it will happen, the maximum severity of any injuries or damages, and the public's sensitivity to the occurrence.

Fred Manuele has discussed this issue in several articles and in his book *On the Practice of Safety*:

The important point for safety and health professionals to keep in mind is that it is vitally important to help their management stay focused on the identification and control of severe risks or, more accurately, high-risk hazards. This approach is incorporated in the process safety management approach which is now mandated by OSHA in the U.S. for chemical process and related industries. Nevertheless, other industries and organizations should adapt this concept in order to minimize their exposure to disastrous or very severe events.

A simplified risk evaluation and control process is as follows:

- **Investigation** to identify risks. What risks do we have? An aggressive inquiry by a top management committee is conducted to determine high-risk hazards and their loss possibilities. This stage is not left until every likelihood of something going seriously wrong is investigated.
- **Elimination** of very severe risks, if feasible
- **Assurance** that all severe risks are controlled, preferably through engineering or, if not feasible, through appropriate management hazard control systems
- **Correction** of uncontrolled severe risks
- **Follow-up** periodically to assure maintenance of controls and to ascertain that no new severe risks have been introduced

Safety and health professionals should systematically work with their management to be sure that recognized severe risks are under control. For example, if your employer is in the retail or lodging business, do you have adequate security measures and systems to protect employees and the public from acts of criminal violence? How about the dangers of heavy mobile equipment, aircraft jet blast, hazardous atmospheres in confined spaces, excavations in unstable soils, natural gas- or propane-fired systems located where leaking gas might accumulate, crane operations — the list could go on. Suffice it to say, be sure you are aware of the most severe recognized risks in your workplaces and that systems are in place for their control.

PROBLEM SOLVING AS IT RELATES TO SAFETY
AND HEALTH MANAGEMENT

Introduction

The word problem seems to be falling into disfavor in modern professional management parlance because of its negative connotations. Some readers may, therefore, choose to substitute another, more positive or opportunistic word such as situation, issue, question, or challenge. For the purpose of this discussion, however, and to avoid any confusion, we will continue to use the term problem.

Surveys conducted among persons in a management position, particularly those at the first line or team leader level, indicate that identifying and solving problems is a major part of their daily activities. Indeed problem solving is the essence of prevention. Obviously, it is better to try and find the problems in a proactive sense rather than wait for the problems to be actualized and then have to react to them. Among all the qualities of effective professional management, problem solving, and problem prevention are some of the most essential skills.

The following summary of the principal elements in this process is based on the teachings of Homer K. Lambie.

Definition of a Problem

For the purpose of the safety and health aspects of professional management, a problem may be defined in the following manner:

A problem is any action or condition, actual or potential, that may produce adversity in terms of loss, injury, or poor human relationship.

Types of Problems

All problems can be classified into one of the following categories:

- Engineering/environmental — physical/chemical/biotechnical
- Human element — behavioral
- Human element — mental/emotional

Methods of Identifying/Recognizing Problems

Problems may be all around you, but you may be unaware of them until you deliberately and systematically take the time to search and discover them. The key is to continuously practice and develop your skills in problem recognition. Here are areas in which to begin the process:

- **Deviations from Standards.** Please refer to the section in this chapter on safety standards and measurements. Standards may be related to safety engineering, safety systems, or safety performance standards. Using an audit approach will lead you to deviations in all these categories.
- **Abnormal Situations (Human Behavior).** Finding this type of problem taxes your powers of observation. For example, if your observe a number of improper or incorrect actions by people following your unit's lock and tag procedure, you should anticipate that if this problem is not addressed the potential exists for a serious event to occur.
- **Loss/Injury Experience — In-House and Others.** Another excellent way to identify problems is to review your organization's past experience. Scan your workplace injury/illness records (OSHA log in the U.S.). Also look at your accident investigation reports for accidents involving damage and/or a significant

incident (near miss) for more clues. Another very helpful area to find problems is by looking at the experience of others. Regulatory agencies publish injury experience statistics and reports. National and trade associations publish reports. The National Safety Council (U.S.) also publishes statistics and reports. Finally, your own professional network of fellow professionals is a great resource for accident experience information.

- **Communication/Dialog with Employees.** How many times has it happened after an accident that someone comes forward and says, "We knew something like this would happen; we have reported this problem many times and nobody paid any attention?" To avoid surprises like this, it is vitally important for managers to cultivate a climate that encourages employees to communicate problems. You can do this a number of ways, but always remember that perception is half the battle. You must demonstrate by your actions that you want to hear about problems. Here are some things to do to build up positive communications:
  - Ask people about problems.
  - Tell them frequently that you want to hear about problems.
  - Act promptly on problems when your people bring them to you.
  - Follow through systematically on the resolution of all pending problems.
  - Provide recognition for those who identify significant problems.
- **Looking/Searching Intensively.** The old saying goes, "You can look and not find, but you will never find if you don't look!" This point has been discussed already; however, it is so important that it deserves a special category to ensure proper emphasis. Many of us become so wrapped up with our daily tasks that soon we are in danger of developing blind spots to problems — problems of a magnitude and so glaring that we almost stumble over them, and yet we still do not recognize them for what they really are. It takes discipline to systematically set aside time to walk around and concentrate on looking with your full mind to see if you can uncover problems. To aid your search, it is best to program your surveys, using the 10-point approach as a guide. These 10 points are:
  1. Work area
  2. Portable tools
  3. Machines
  4. Equipment
  5. Materials — health
  6. Materials — traumatic
  7. Electrical
  8. Vehicle
  9. Multicategory
     - Lockout-tagout
     - Confined space entry
     - Excavations
     - Ergonomics
  10. Related
     - Medical health
     - Security
     - Fire protection
     - Emergency procedures

Refer also to the sections on hazard control and the safety manual in this chapter plus the chapters on health for more information on hazard sources and causes.

Evaluation of Problems

As a problem becomes evident to you, the most important first step is to write it down. There is an old Chinese saying, "The finest memory is worse than faded ink." Cultivate the habit of carrying a notebook with you at all times so you can jot down problems as they flash into your mind. Many a serious accident has occurred where the potential problem had been previously recognized, but the manager concerned forgot to write it down. Homer Lambie always says, "The beginning of the solution of any problem is when it is defined in writing." He also cautions that a general problem is not solvable until it is dissected down to the specific problems that collectively produce the general situation.

The hardest step is Step 1 in the problem-solving process, which is defining the problem in simple and specific terms. To help with this process, go to Step 2, which is to define the source and/or cause of the problem.

- **Source** always relates to a physical condition.
- **Cause** relates to the human action or lack of action that produces the problem.

Once the source and/or cause is/are identified, go to Step 3, which is to take any immediate temporary control (ITC) measures that may be appropriate. Step 4 is to measure the problem by the severity of its ultimate potential to produce adversity. In hazard control this is known as risk rating or risk assessment. There are numerous techniques for risk assessment, some of which are discussed in the section on hazard control in this chapter. A more technical system is also discussed in Chapter 10, *Process Safety Management*.

A simple approach is to consider the criteria, frequency, severity, and probability. If any of these factors are high, some measure of control action is necessary.

Controlling Problems

The first consideration in problem solution, which was already mentioned in the evaluation stage, is to always consider what ITC measures must be taken to avoid loss and/or injury. After this you must decide on a plan of action by identifying the number of steps and methods of controlling the problem. Considerations will include effort, money, and risk. The priority for control will be determined by the risk rating after temporary control measures are in place.

Processing any problem involves three options, known as CS or P.

- **Correct** the problem.
- **Schedule** the problem for correction.
- **Pass** the problem on to the person having the authority and/or the ability to solve it.

Summary and Final Hints on Problem Defining and Solving

One of the key qualities of effective leaders is their skill at problem solving and problem prevention. There is nothing wrong with coming up with a general problem. In fact, it is often a good place to start. However, it is just the start; after this you must go on to dissect or expand the problem into its specific components and then solve them one by one.

In the realm of problems involving the human element, be sure you have the facts and concentrate on what the person is doing wrong or is likely to do wrong. This is necessary to be sure you are dealing with a realistic situation versus an inaccurate one. In other words, separate the facts from fantasy.

Convert problems of a mental or emotional nature to the underlying source(s) or cause(s). Look for what is really triggering the emotional response. Then convert the "I don't like...," "I feel...," etc., to the specific hazard(s) and solve them through engineering or training.

Another very important principle is to always give priority to high-risk issues. Remember the old saying, "Don't worry about ants when tigers are in your path."

To recap, then, here are the steps:

- Problem — must be specific
- Source/cause
- Risk value
- Priority
- ITC
- Control — CS or P
- Follow-up

SETTING SAFETY GOALS, STANDARDS, AND MEASUREMENTS

The material in this section is based on research conducted by the author and contained in a presentation on this topic at the American Society of Safety Engineers Annual Professional Development Conference, Las Vegas, Nevada, June 1994.

With all the trends that have been discussed earlier in this chapter, it will be vital for safety and health professionals to continuously develop their skills in the long-range strategic planning process and, in particular, goal setting, standards, and measurements.

The next section includes definitions for the terms used in this process and examples of their applications.

Definitions

Ethics — A code of moral principles that guide people in the organization. Test questions:
- Is it legal?
- Is it balanced? Is it fair?
- Does it promote win-win relationships?
- How would it make us feel if it were published in the newspapers or known in the community?

Goal/Objective — The end result toward which effort is directed: a desired outcome.

Mission — A clear, definite, and motivational point of focus — an achievable goal; a finish line to work toward. Mission is purpose-based.

Mission Statement — States what business the organization is in and may include ranking. It provides the overall direction and scope of the organization. Key words are accomplishments, contributions, values, and principles.

Purpose — The board ongoing aim or intention of an organization. This may have two components:

- External — what the people in the organization want to do
- Internal — what the people in the organization want it to be

Strategic Plan — The process that must be completed to achieve a goal. This includes assignment of responsibility and completion dates.

Values — The core qualities that people in the organization consider most important and essential for success of their mission. Values form the basis for beliefs.

Vision — Seeing a desired state as though it exists today: Seeing the potential in or necessity of opportunities right in front of you. Visualization is a way of seeing the "what might be done". Vision is values-based.

Definition Examples

Purpose — External — Giro Sport Design: "Design of great products that change the face of the industry, improve safety, and make people happy."

Purpose — Internal — Circuit Technology Group, Hewlett-Packard: "…To continually work towards building a work environment that helps employees attain a happy meaningful existence — to attain self-actualization."

Mission Statement — The mission of the American Society of Safety Engineers is to foster the technical, scientific, managerial, and ethical knowledge, skills, and competency of safety, health, and environmental professionals for the protection of people, property, and the environment and to enhance the status and promote the advancement of the safety profession. The American Society for Industrial Security will establish, develop, and promote excellence in the security profession by assuring:

- High-quality educational programs.
- Responsiveness to members' needs.
- Standards for professional and ethical conduct.
- A forum for the debate and exchange of ideas.
- Promotion of the organization and the profession.
- Strategic alliances with related organizations.

Values — Examples are family, health, career, religion, honesty, integrity, credibility, justice, liberty, freedom, and respect for the rule of law.

Vision — The American Society for Industrial Security will be the foremost organization advancing the security profession worldwide.

Safety System Standard

Example Title: Lockout-Tagout

Key elements:

- Written procedure
- Standardized devices
- Step-by-step procedure
- Group lockout-tagout
- Contractors

Training requirements:

- All affected employees trained
- Authorized employees receive specific training
- Entry supervisors trained

Management controls/documentation:

- Documentation of energy sources and isolation points for each piece of equipment
- Training records
- Annual audits

Safety Performance Standard

Example:

Title: Safety inspections

Key elements:

- Written procedure
- Inspection responsibility areas assigned

Minimum performance standard:

- Each supervisor inspects assigned areas each month
- Report findings and action

Management controls/documentation:

- Immediate supervisor evaluates and rates
- Immediate supervisor trains as needed
- Upper management conducts periodic audits

Goal Setting

Many different techniques are available for defining goals. A simple system that works well is the so-called "SMART" technique. The criteria for an effective goal using this system are that each goal must be

Specific — Generalities mean zero! Dissect a general goal into its *specific* elements which must be the *exact outcomes* to be accomplished.
Measurable — The goal must be measurable by quantity and by time (how much and by when).
Achievable — The goal must be believable and conceivable. (Is the goal within your group's reach?)
Realistic — The goal must be results oriented, practical, and controllable. (Can it be accomplished in the "real world?")
Time Bounded — There must be a target or deadline for accomplishment of the goal.

The following are some other tips on goal setting based on the teachings of Brian Tracy:

- There must be a strong degree of *desire* to accomplish the goal. The opposite of desire is fear, which is a powerful demotivator. Dwelling on desire will help override the natural inertia which prevents people from getting started.
- The people must *believe* it is possible to achieve the goal.
- It is said that a wish is a goal without any energy in it. *Writing* the goal intensifies the desire and strengthens resolve to achieve the goal.
- List all the *benefits* of achieving the goal. The more "reasons why" that are identified, the more people will be motivated to achieve the goal.
- Be sure to define the *starting point* because this gives a baseline from which to measure progress.
- It is said there are no unrealistic goals, merely unrealistic deadlines. Set *deadlines* on all tangible (measurable) goals. Do not worry if the deadline is missed; set another one.
- *Planning backward* is a powerful exercise; start from the goal and work backward, identifying the steps involved.
- Identify *obstacles* that stand in the way of the goal. Focus attention on removing the biggest major obstacle first.
- Identify *limiting steps*. Is there some additional knowledge or information needed to achieve the goal? Can the training be done in-house or must it be outsourced?
- Make a list of all the people needed to *help* achieve the goal.
- Having done all of the above, next make a *plan* and write it out in detail. List the actions by time and priority.
- As Stephen Covey says, "Begin with the end in mind."
- The next step is to organize the list of all the things that will have to be done to accomplish the goal — Brian Tracy calls this project management. Arrange these things to be done either in *sequential* order (dependent on each other being completed) or as *parallel* activities that can be worked on at the same time.
- Next, *identify the team* needed to accomplish the goal and *delegate* to them the various parts of the project.
- Continually monitor and measure to ensure that every part of the project is on track.
- Encourage all involved to *visualize* the goal so their subconscious minds go to work to program them toward the goal — remember the saying, "What you see is what you get."
- Final words of wisdom: *never, never give up!* Encourage the group to discipline themselves to systematically keep doing something that moves them toward the goal. Nothing succeeds like success, and this helps to keep everyone positive and motivated.

Safety Standards

Safety standards may be divided into three broad categories. First are *safety engineering standards*. These standards relate to the physical environment of the people. Standards of this type include those established by the various regulatory agencies. Many others are established by national organizations such as the American National Standards Institute, engineering societies, trade associations, etc. Other standards are found in industrial hygiene standards for employee exposure, environmental standards, fire, security, and building codes, and the list goes on and on.

The second category includes *system safety standards*. System safety standards are those safety systems established by organizations as necessary for the implementation of an effective safety plan. Examples might be safety inspections, hazard/risk assessment and control, job hazard analysis, or accident investigation. Each of these standards must outline the procedures for implementation, training requirements, documentation, and management controls.

The third category of safety standards is the *safety performance standard*. A performance standard tells the performers what must happen for them to accomplish the goal successfully.

Taking safety inspections as an example to illustrate the practical application of this process, the goal might be to establish a process that will ensure that all workplaces are systematically inspected and maintained to establish safety standards and procedures. The applicable engineering standards such as OSHA regulations would be identified and may be encapsulated into simple, easy-to-use checklists.

Next a safety system standard must be developed that will provide a guideline for how this system will be implemented. Elements that this procedure must address will be assignment of responsibility areas, as self-survey system, and a follow-up control system to ensure that the inspection findings are corrected in a timely fashion.

Finally, an integral part of this system will be safety performance standards. These standards must describe who does what, when. Figure 11.1 illustrates typical performance standards for each level in the basic safety action system.

In the example we are discussing, the standard will state who inspects and at what frequency. Also, how do the inspectors report and follow through on their findings?

The training requirements must be identified for each level involved in that system. Finally, the standard must define what management does to monitor the activity and results.

Measurements

The final phase of the goal-setting process is measurement. The Deming total quality management (TQM) philosophy points out that measurement is a vital step in that it provides data that can serve as a basis for system improvement. Without measurement, how can we tell if the goal has been achieved?

Tom Peters says, "What gets measured gets done!" This is a succinct implication that without measurement our system will be hit or miss at best.

Using the safety inspections example mentioned above, if the standard requires every first-line supervisor or work team to do a self-inspection of their workplace each month, what are the chances that this will consistently happen in all areas unless there is measurement?

Many techniques can be used for the measurement of safety systems. The principal ones involve measurement of system result such as injury experience, accident costs, and trends.

Measuring results is most commonly used, but those in the profession have long since recognized that this is a very crude measurement technique because it does not address the performance failures that produced these results.

Measurement of safety system activity can be on a quantitative or a qualitative basis. Using the safety inspection report example (Figure 11.2), the quantitative

MANAGMENT LEVEL	RESPONSIBILITIES	METHODS BY WHICH ACCOMPLISHED
Department manager	1. Provide policy procedures and ensure they are reviewed and accepted by all supervisors. 2. Provide necessary dept. procedures defining who does what to ensure unit policy is implemented in the dept. 3. Set up management controls to ensure the programs are effective. 4. Sample the 2nd line's work and provide advanced training and guidance as needed. 5. Identify department-wide problems and set up objectives for continued improvement of safety results.	1. Delegate to dept. safety coord. the responsibility to assist with administration of dept. safety program (Reports-problems-objectives-audits-controls). 2. Conduct dept. meeting to discuss problems and review progress. 3. Random inspection to audit effectiveness.
Second line supervisor	1. Provide training in the basic safety action programs. 2. Evaluate safety activity and provide necessary training and guidance. 3. Assist supervisors in solving problems beyond their ability or authority to control. 4. Set up management controls to ensure the supervisors safety action programs are effective. 5. Identify problems with the safety action programs and determine methods for their solution.	1. Review HC logs weekly. 2. Evaluate all supervisors safety activity reports and measure progress. 3. Review all accident reports and investigate all those considered significant. 4. Conduct spot-audits of self-surveys and HC notebooks. 5. Conduct spot-audits of critical procedures (lock and tag, eye protection). 6. Inspect area for major housekeeping and hazard problems. 7. Random audit of supervisor safety meetings. 8. Random audit JHA/JSA compliance. 9. Plan and schedule area JHA/JSA manuals program.
First line supervisor or team leader	1. Systematically look for problems (hazards), identify risk rate and process their control or correction. 2. Guide and assist the employee through encouraging joint discussion to solve mutual problems. 3. Provide repetition and emphasis on recognized high-risk hazards.	1. Use of HC notebook and HC log. 2. Monthly safety self-survey. 3. Monthly JHA/JSA meeting. 4. Monthly crew safety meeting. 5. Accident investigation.
EMPLOYEE	Rcognize, understand, and control job HAZARDS (Unsafe acts and conditions)	

Figure 11.1   This diagram illustrates the typical safety performance standards in the basic safety action system for various levels in the organization.

measurement would be the receipt of the report each month as evidence of system implementation. Hence, quantitative measurement is satisfied as 100% compliance with the standard. The quality of the report in terms of hazards identified and controlled, however, may be 100 or it may be zero! A qualitative measurement as shown on the form enables the inspector and/or a reviewer to better assess the effectiveness of their inspection.

Other measurement techniques involve going beyond the paper and evaluating what is actually happening in the workplace. This type of measurement compares the paper to reality in the workplace and is, of course, the audit.

Figure 11.3 illustrates an example audit format for the self-audit of a typical safety management system. On page 3 of this audit, paragraph B, Safety System

Safety Inspection

Department	Inspection Responsibility Area/Equipment Assigned	Inspected By		Date:

Problem			Control Action	Follow-Up
Describe WHAT is wrong and where it is located	Give reasons WHY it exists (source/cause)	Priority Code	Describe actions taken or planned to control the problem and its source/cause	Code · Name/Date

		Weight	Reviewer's Comments	Problem Priority Code
Problem	Is the problem clearly described and located?	10		**E** Emergency
	Are the reasons why the problem exists identified?	30		**A** 1 Week
Priority	Do the priorities reflect an understanding of the potential effects of the problem listed?	10		**B** 1 Month
Control	Did the action control the problems and the reasons for their existence?	40		**C** 3 Months
				D 6 Months
Follow-Up	Does the follow-up specify who will do what, when, to ensure action completed?	10		**Follow-Up Code**
				C Corrected (Date)
	Total	100		**S** Scheduled (Date)
				P Passed to (Name)

Signature: _____ Date: _____

Distribution: Safety   Originator   Department

Figure 11.2   Format for a safety inspection report with a qualitative evaluation.

Activity Questions 1 through 5 are aimed at assessing the effectiveness of the safety inspection system.

So far in measurement we have looked at the quantity, quality, and actual results of the system. How about the perception of the people implementing the system? Also, how about underlying issues, such as breakdown in human communications or such indirect factors as lack of knowledge and/or understanding due to inadequate training?

To assess these root cause aspects, a number of techniques have been tried. Two that are commonly utilized are attitude and perception surveys and interviews. These surveys can often provide valuable clues to weak links that may be lurking in your safety systems. A word of caution — these surveys can have a negative impact if they are not professionally administered and interpreted. These aspects of program measurement will be discussed in other chapters of this book.

Safety System Performance Assessment

Figure 11.4 illustrates a comparison between the more traditional management control systems and a performance management approach as described by Daniels and Rosen in their book of the same title. This type of management system is commonly used in production operations, but its adaptation to safety performance management has been slow. This will be a trend to watch in the future.

Conclusion

Safety goals, standards, and measurements are literally the key to the future of safety management. It must be the goal of all safety professionals to continuously develop their skills in this aspect of professional management.

If safety and health professionals wish to remain on the team in their respective organizations, it is absolutely vital to serve by helping management set up and maintain comprehensive and effective safety systems which will ensure that the enterprise is successful in an increasingly competitive world.

IMPLEMENTATION OF SAFETY MANAGEMENT SYSTEMS

One of the frustrating difficulties most organizations have to deal with is that, having adopted a particular type of safety system, they still have to get the system implemented and producing effective results. To help ensure that this process takes place with a minimum of delays and misunderstandings, a step-by-step method follows:

Step	Description
Procedure	The system procedure must be reviewed with the various management levels and accepted by them.
Basic training	All levels must be trained in the basics of the system, including their responsibilities and the minimum action required. The training will also cover paper flow of reports or other documentation.
Advanced training	Advanced training must be provided for management levels above the first line level to explain their responsibilities and the minimum action required. This will include evaluation of reports and feedback to their first-line supervision.
Management controls	The procedure must include systems for auditing the effectiveness of the system.

SELF AUDIT

Safety Managment System

Organization _____ Date _____
Branch, Division, or Subsidiary _____

Question	Answer		Remarks
	Yes	No	
Basic Safety Management System Audit			
A. Administration			
1. Are the unit safety objectives defined in writing?			
2. Has one person been designated as safety and health coordinator?			
3. Is there a unit executive safety committee or group, and do they meet on a regular basis?			
4. Does the executive safety committee review all new or revised policies?			
5. Is there a unit safety manual?			
6. Is the manual available in each department?			
7. Are supervisors familiar with the manual?			
8. Is there a formal safety orientation program for new employees?			

SELF AUDIT

Safety Managment System

Organization _____ Date _____
Branch, Division, or Subsidiary _____

Question	Answer		Remarks
	Yes	No	
9. Have supervisors received formal training in the unit safety program responsibilities?			
10. Is there a formal system which includes safety performance in salary reviews?			
• Is the program in writing?			
• Does each supervisor have objectives?			
• Does the program ask for completion of:			
G safety inspection?			
G hazard control action?			
G accident & incident investigation reports?			
G job safe practices?			
G safety meetings?			
• Does each program have an established minimum program activity standard?			

Figure 11.3 Example format for self-audit of safety management systems.

SELF AUDIT

Safety Managment System

Organization _____ Date _____
Branch, Division, or Subsidiary _____

Question	Answer		Remarks
	Yes	No	
B. Safety System Activity			
Basic Safety Management System Audit			
1. Is there a written safety inspection program?			
2. Is the unit divided into departmental safety inspection responsibility areas and these in turn divided into specific areas for each supervisor?			
3. Are there regular safety inspections?			
4. Are safety inspection reports written?			
5. Is there a follow-up and control system for identified safety problems?			
6. Is there a written hazard control system with a standard for supervisor action?			
7. Are supervisory hazard surveys documented?			
8. Is there a system with priorities assigned for the follow-up and control of all identified hazards?			

SELF AUDIT

Safety Managment System

Organization _____ Date _____
Branch, Division, or Subsidiary _____

Question	Answer		Remarks
	Yes	No	
9. Is there a procedure for handling employee suggestions or complaints regarding safety and health?			
10. Do members of top management conduct regular safety inspections?			
11. Are there written procedures, such as: job safe practices, job safety analysis or similar to cover all jobs involving high risk potential hazards?			
12. Are there procedures available for systematic re-training and review with new employees?			
13. Do all employees attend a safety meeting a minimum of once per month for field personnel or quarterly for officer personnel?			
14. Are minutes kept of all safety meetings?			

Figure 11.3 (Continued)   Example format for self-audit of safety management systems.

SELF AUDIT

Safety Management System

Organization _____ Date _____
Branch, Division, or Subsidiary _____

Question	Answer		Remarks
	Yes	No	
• Forklifts			
• Cranes, hoists and slings			
• Wire rope			
• Compressors and air receivers			
• Elevators			
2. Are the following procedures established, reviewed, and audited for compliance. Are employees authorized and trained?			
• Lockout/Tagout			
• Permit Required Confined Space Entry			
• Hot Work Permit			
• High Voltage Electrical Procedures			
• Pipe Blinding and Blanking			
• Excavations			

SELF AUDIT

Safety Management System

Organization _____ Date _____
Branch, Division, or Subsidiary _____

Question	Answer		Remarks
	Yes	No	
Basic Safety Management System Audit			
15. Is there a written procedure to follow when an accident occurs?			
16. Are no-injury (damage and/or significant incident) accidents investigated?			
17. Are all accident reports reviewed and signed by at least one level above the originator?			
18. Do top management review at least all lost workday injury accident reports?			
C. Critical Safety Procedures			
1. Are inspections, tests and preventive maintenance conducted on the following:			
Are employees authorized and trained?			
• Scaffolds			
• Man-lifts/powered platforms			

Figure 11.3 (Continued) Example format for self-audit of safety management systems.

ACCIDENT PREVENTION PERFORMANCE ASSESSMENT

Employee Performance Action Step	Antecedents	Employee Perceived Consequences	Management Control System		
			Data System	Feedback System	Review System
1. Not looking for hazards.	• Job description • Job instruction • Tools and materials • Work environment	• More time to do the job • Avoids paperwork • Makes sense to them • Risk of supervisory reprimand • Accidents may occur • Unfavorable reaction when report another person's unsafe actions			
2. Looking for hazards.	• Supervisor observing and providing guidance • Safety Manual • Procedures • Training	• "Fault finding" after accidents occur • More hassle follow up, etc. • Jobs take longer • Promotion/pay increase may be affected if do not comply • More paperwork • May be less accidents	• Safety Inspection Reports • Accident Reports • JHA JSA/JSP • Safety Meeting Reports • Hazard Control System	• Periodic discussion	• Quarterly MBO • Annual Performance Appraisal
PERFORMANCE REINFORMCENT PLAN	• Set goal • Meet w/employees, explain measures, goal and baseline • Ask employees for suggestions	• Supervisor personally says something positive to each employee whose performance is above baseline* • Set up graph • Set up incentives • Set up awards • During the week surprise those groups with coffee and donuts who have a higher performance than previous week* * Reinforcements that produce Positive Immediate Certain Response	• Each employee keeps daily tally of hazards recognized • Manager and supervisor randomly check reports and audit for accuracy	• Group maintains graphs of safe hours worked and hazards recognized • Individual graphs for each supervisor • Manager maintains graphs and writes reinforcing statements on them • Manager explains baseline and graph to all when posting • Quarterly performance appraisals	• Manager conducts review meetings with each supervisor each month • Supervisor holds monthly meeting with crew, asks them about reinforcers and ideas

Figure 11.4   Illustration of traditional safety performance controls vs. over-performance management techniques which increase employee involvement.

SAFETY MANAGEMENT SYSTEM ACTION PLAN

Dept. _____ Program _____ Plan originated_____

reviewed _____

Dept. Manager approved _____ revised_____

BASIC STEP	NO.	STEP DESCRIPTION	RESPONSIBILITY	COMPLETION DATE SCHED.	ACTUAL
Procedure	1	Procedure reviewed and accepted by all team leaders first line supervisors.			
	2	Dept. manual accessible to all team leaders/supervisors.			
Implementation training	3	Basic training program developed in writing stating that the team leader/first line supervisor must know.			
	4	All second line supervisors trained in basics.			
	5	All team leaders/supervisors trained in basics.			
Minimum action required	6	Dept. procedure issued to state the specific minimum action required by each team leader/first line supervisor.			
Follow-up	7	Dept. procedure issued to state who does what by when the paper.			
Advanced training for reviewer	8	Advanced training program developed in writing stating what the reviewer does.			
Review of the system action	9	Dept. procedure issued to state what the second line supervisors do with the paper and by when.			
Sampling of the reviewer's work	10	Dept. procedure to state how second line supervisors set up management controls to ensure that the system is effective.			
Additional management controls	11	Additional systems steps, e.g., index and review schedule and line employee training schedule.			

Figure 11.5   Sample format for the implementation of safety systems by the various levels of department management.

Figure 11.5 illustrates a typical format for implementing safety systems within individual departments. Refer also to Figure 11.1 for a sample illustrating the specific actions in the basic safety systems by various management levels.

NEW AND EMERGING TRENDS IN PROFESSIONAL MANAGEMENT

Management is a dynamic evolving process, and in today's highly competitive business climate there has been an almost constant stream of new techniques and systems, each espoused by the latest guru.

The 1980s saw an increasing interest in the total quality management (TQM) movement based on the teachings of Deming, Juran, and others. Many companies and organizations went through a process of reorganizing or restructuring in order to reduce costs and improve profitability. These efforts usually involved reducing layers of management and/or combining job functions.

The 1990s have brought some refinements to the scene through the work and teachings of such proponents as Stephen Covey (*Seven Habits of Highly Effective People*), Tom Peters (*In Search of Excellence*), Peter Senge (*The Fifth Discipline*), and Peter Block (*Stewardship and the Empowered Manager*), to name but a few. Nevertheless, to quote John Parkington of the Wyatt Company, "Downsizing is still the diet of corporate choice."

The latest technique is termed reengineering, as championed by authors Michael Hammer and James Champy in their best-selling book entitled *Reengineering the Corporation.*

The most positive outcome of all this turmoil in the management scene is that organizations are now more focussed on employee involvement and empowerment, and the old bureaucratic top-down management culture is being replaced by a much more open collaborative teamwork approach.

SOME CONCLUSIONS FOR THE SAFETY AND HEALTH PROFESSIONAL OF THE FUTURE

Flexibility and continuous learning will have to be the watchwords of tomorrow's safety and health professional. It is highly likely that you will find yourself with new partners, both internal and external. Furthermore, you will need to be constantly looking at ways you can outsource and/or offload some of your responsibilities in order to concentrate on the functions that will provide you with maximum leverage and results.

To be successful, the safety and health professional should concentrate on being recognized as a policy setter, auditor, and internal adviser. Communication skills will be vital as you work with your management to define the organization's needs and then go out and sell your solutions in terms of improvements to their management systems.

The goal should be to become recognized as a valued resource. Do this by looking for ways in which you can contribute to the bottom line. Always keep in mind that your management's agenda is your agenda. They are your customer. So be sure to blend your agenda to theirs.

As we said, be flexible and creative, try different ways, be prepared to compromise, and take it step by step. To quote Peter Drucker. "It is more important to do the right things than to do things right." If you want to change results, you may need to change your strategy.

Here are some final hints based on the teachings of security management consultant Dennis Dalton:

- Identify your agenda; make sure it compliments rather than competes with management's agenda.
- Understand your business.
- Speak in business managers' terms.

- Meet regularly with managers.
- Walk around and observe — is Safety and Health helping?
- Solicit customer feedback.
- Listen to what your employees are saying.
- Be aware of new approaches and ideas — find out what others are doing.
- Show you care — the human touch.
- Really *listen.*
- Focus on the *positive.*
- Maintain *focus on what you do best.*

Finally, in terms of safety and health systems, be looking regularly at how these systems can be improved by using either state-of-the-art computer technology or new management tools such as flow diagrams, fishbone diagrams, or new computerized training and simulation systems.

FURTHER READING

Books

Adams, J. D. et al. *Transforming Leadership from Vision to Results.* Alexandria, VA: Miles River Press, 1986.

Ailes, R. *You Are the Message.* New York: Doubleday, 1988.

Albrecht, K. *Successful Management by Objectives.* Englewood Cliffs, NJ: Prentice-Hall, 1978.

Allen, L. A. *Common Vocabulary of Professional Management.* Palo Alto, CA: Lois A. Allen Associates, 1969.

Allen, L. A. *The Management Profession.* New York: McGraw-Hill, 1964.

Allesandra, A. J. and Hulnsaker, P. L. *Communicating at Work.* New York: Simon & Schuster, 1993.

American Society of Safety Engineers (ASSE). *The Dictionary of Terms Used in the Safety Profession.* 4th ed. Des Plaines, IL: ASSE, 2001.

Balge, M. Z. and Krieger, G. R. *Occupational Health and Safety.* 3rd ed. Itasca, IL: National Safety Council, 2000.

Bellman, G. M. *Getting Things Done When You Are Not in Charge.* San Francisco, CA: Berrett-Koehler, 1992.

Bennis, W. *On Becoming a Leader.* Reading, MA: Addison-Wesley, 1989.

Bennis, W. and Nanus, B. *Leaders — The Strategies for Taking Charge.* New York: Harper & Row, 1985.

Bird, F. E. and Davies, R. J. *Safety and the Bottom Line.* Loganville, GA: Institute Publishing Inc. 1996.

Bird, F. E. and Germain, G. L. *Practical Loss Control Leadership.* Loganville, GA: Det Norske Veritas (USA) Inc. 1996.

Block, P. *Stewardship.* San Francisco, CA: Jossey-Bass, 1987.

Blanchard, K. and Johnson, S. *The One-Minute Manager.* New York: Berkley Publishing Group, 1984.

Blanchard, K. and Peale, N. V. *The Power of Ethical Management.* New York: Fawcett Crest, 1988.

Byham, W. C. and Cox, J. *Zapp! The Lightning of Empowerment.* New York: Ballantine, 1988.

Cohen, A. R. and Bradford, D. L. *Influence without Authority.* New York: John Wiley & Sons, 1990.

Conger, J. A. *Winning 'em Over — A New Model for Management in the Age of Persuasion.* New York: Simon & Schuster, 1998.

Connellan, T. K. and Zemke, R. *Sustaining Knock-Your-Socks-Off Service,* New York: American Management Association, 1993.

Covey, S. R. *The Seven Habits of Highly Effective People.* New York: Simon & Schuster, 1989.

Covey, S. R., Merrill, R. A., and Merrill, R. R. *First Things First.* New York: Simon & Schuster, 1994.

Dalton, D. R. *Security Management: Business Strategies for Success.* Woburn, MA: Butterworth–Heinemann, 1994.

Daniels, A. C. and Rose, T. A. *Performance Management — Improving Quality and Productivity through Positive Reinforcement.* Tucker, GA: Performance Management Publications, 1984.

Dawson. R. *Secrets of Power Persuasion.* Englewood Cliffs, NJ: Prentice-Hall, 1992.

Decker. B. *The Art of Communicating.* Los Altos, CA: Crisp Publications, 1988.

Drucker, P. F. *The Practice of Management.* London: Heinemann, 1963.

Drucker, P. F. *The Frontiers of Management.* New York: Harper & Row, 1986.

Drucker, P. F. *The Effective Executive.* London: Heinemann, 1967.

Findlay. J. V. and Kuhlman, R. L. *Leadership in Safety.* Loganville, GA: Institute Press, 1980.

Fulton, R.V. *Common Sense Supervision.* Berkeley, CA: Ten Speed Press, 1988.

Gellerman, S. W. *Motivation in the Real World.* New York: Penguin Books, 1993.

Gordon. H. L. et al. *A Management Approach to Hazard Control.* Bethesda, MD: Board of Certified Hazard Control Management, 1994.

Grimaldi. J. V. and Simonds, R. H. *Safety Management.* Boston, MA: Irwin, 1989.

Grose, V. L. *Managing Risk — Systematic Loss Prevention for Executives.* Englewood Cliffs, NJ: Prentice-Hall, 1987.

Hammer, M. S. and Champy, J. *Re-engineering the Corporation.* New York: HarperCollins, 1993.

Health and Safety Executive. *Successful Health and Safety Management.* London: Her Majesty's Stationery Office, 1992.

Hopkins, T. *How to Master the Art of Selling.* Scottsdale, AZ: Warner, 1982.

Hughes, C. L. *Goal Setting — Key to Individual and Organizational Effectiveness.* New York: American Management Association, 1965.

Jaffe, D. T., Scott, C. D., and Tobe, G. R. *Rekindling Commitment.* San Francisco, CA: Jossey-Bass, 1994.

Janov, J. *The Inventive Organization.* San Francisco, CA: Jossey-Bass, 1994.

Kazmier, L. J. *Principles of Management.* New York: McGraw-Hill, 1974.

Knowdell. R. L., Branstead, E., and Moravec, M. *From Downsizing to Recovery: Strategic Transition Options for Organizations and Individuals.* Palo Alto, CA: CPP Books, 1994.

Kohn, J. P. and Ferry, T. S. *Safety and Health Management Planning.* Rockville, MD: Government Institutes, 1999.

Kouzes, J. M. and Posner, B. Z. *The Leadership Challenge: How to Get Extraordinary Things Done in Organizations.* San Francisco, CA: Jossey-Bass, 1987.

Krause, J. R., Hidley, J. H., and Hodson, S. J. *The Behavior-Based Safety Process.* New York: Van Nostrand Reinhold, 1990.

Lack, R. W. ed. *The Dictionary of Terms used in the Safety Profession.* 4th Edition, Des Plaines, IL, American Society of Safety Engineers, 2001.

Lund, D. R. and Finch, L. W. *Lessons in Leadership.* Staples, MN: Nordell Graphic Communications, 1987.

Manuele, F. A. *On the Practice of Safety.* Second edition. New York: Van Nostrand Reinhold, 1997.

Nelson, R. 1001 *Ways to Reward Employees.* New York, NY: Workman Publishing Co. Inc, 1994.

Nirenberg, J. *The Living Organization — Transforming Teams into Workplace Communities.* New York: Irwin, 1993.

Pater, R. *Making Successful Safety Presentations.* Portland, OR: Fallsafe, 1987.

Pater, R. *The Black-Belt Manager.* Rochester, VT: Park Street Press, 1988.

Peters, T. *Thriving on Chaos.* New York: Alfred A. Knopf, 1987.

Peters, T. and Austin, N. *A Passion for Excellence.* New York: Random House, 1985.

Petersen, D. *Analyzing Safety System Effectiveness.* New York, NY: John Wiley & Sons, 1996.

Petersen, D. *Safety by Objectives.* Goshen, NY: Aloray, 1978.

Petersen, D. *Safety Management.* Goshen, NY: Aloray, 1988.

Petersen, D. *Techniques of Safety Management.* Goshen, NY: Aloray, 1989.

Pinchot, G. and Pinchot, E. *The End of Bureaucracy and the Rise of the Intelligent Organization.* San Francisco, CA: Berrett-Koehler, 1994.

Pope, W. C. *Managing for Performance Perfection: The Changing Emphasis.* Weaverville, NC: Bonnie Brae Publications, 1990.

Ray, M., Rinzler, A. et al. *The New Paradigm in Business.* New York: Tarcher/Perigee, 1993.

Rose, C. *Accelerated Learning.* New York: Dell Publishing, 1985.

Roddick, A. *Body and Soul.* New York: Crown Publishers, 1991.

Rosenbluth, H. F. and McFerrin, Peters, D. *The Customer Comes Second.* New York: William Morris & Co., Inc., 1992.

Sashkin, M. and Kiser, K. *Putting Total Quality Management to Work.* San Francisco, CA: Berrett-Koehler, 1993.

Schein, E. H. *The Corporate Culture Survival Guide.* San Francisco, CA: Jossey-Bass Publishers, 1999.

Senge, P. M. *The Fifth Discipline.* New York: Doubleday, 1990.

Senge, P. M. et al. *The Fifth Discipline Fieldbook — Strategies and Tools for Building a Learning Organization.* New York: Doubleday, 1994.

Taylor, H. L. *Delegate — The Key to Successful Management.* New York: Warner Books, 1989.

Thomas, H. G. *Safety, Work, and Life — An International View.* Des Plaines, IL: American Society of Safety Engineers, 161–185, 1991.

Tracey, W. R., Ed. *Human Resources Management and Development Handbook.* New York: Amacom, 1994.

Tracy, B. *Action Strategies for Personal Achievement.* Series of Tapes. Niles, IL: Nightingale Conant, 1993.

Tracy, B. *The Effective Manager Seminar.* Series of Tapes. Niles, IL: Nightingale Conant, 1994.

Tracy, B. *Advanced Selling Strategies.* New York, NY: Simon & Schuster Fireside Books, 1996.

Tracy, B. *The 100 Absolutely Unbreakable Laws of Business Success.* San Francisco, CA: Benett-Koehler, 2000.

Walsh, T. J. et al. *Protection of Assets Manual.* Santa Monica, CA: The Merritt Company, latest printing updated monthly.

Walton, D. *Are You Communicating?* New York: McGraw-Hill, 1989.

Walton, M. *The Deming Management Method.* New York: Putnam Publishing, 1986.

Wheatley, M. J. *Leadership and the New Science.* San Francisco, CA: Berrett-Koehler, 1992.

Other Publications

Barenklau, K. E. Developing standards for safety work activities. *Occup. Hazards Natl. Saf. Manage. Soc. Focus,* 145–148, October 1989.

Barenklau, K. E. Effectively measuring safety involves consequences and cause. *Occup. Saf. Health,* 41–49, March 1986.

Blair, E. H. Which competencies are most important for safety managers? *Prof. Saf.* 44(1), 28–32, January 1999.

Carder, B. Quality theory and the measurement of safety systems. *Prof. Saf.,* 39(2), 23–28, February 1994.

Carr, C. Ingredients of good performance. *Training* 30(80), 51–54, August 1993.

Cooper, D. Goal setting for safety. *Saf. Health Pract. (U. K.)* 11(11), 32–37, November 1993.

Creswell, T. J. Safety and the management function. *Occup. Hazards,* 31–34, December 1988.

Crutchfield, N. and Waite, M. A vision and mission for safety. *Occup. Hazards Natl. Saf. Manage. Soc. Focus,* 65–68, September 1990.

Deacon, A. The role of safety in total quality management. *Saf. Health Pract. (U.K.)* 12(1), 18–21, January 1994.

Earnest, R. E. What counts in safety? A non-injury based measurement system. *Natl. Saf. Manage. Soc. Insights Manage.,* 2–6, 2nd quarter 1994.

Esposito, P. Applying statistical process control to safety, *Prof. Saf.* 38(12), 18–23, December 1993.

Frederick, L. J, Winn, G. L. and Hungate, A. C. Characteristics employers are seeking in today's safety professionals. *Prof. Safety* 44(2), 27–31, February 1999.

Lack, R. W. Safety management — accountability or sideshow? *Saf. Pract. (U. K.)* 3(8), 4–7, August 1985.

Le Clerg, R. E. Why the "safety" function? *Occup. Hazards Natl. Saf. Manage. Soc. Focus,* May 1989.

Northage, E. I. Auditing: a closed control loop. *Occup. Hazards Natl. Saf. Manage. Soc. Focus,* 87–90, May 1992.

Petersen, D. Integrating safety into total quality management. *Prof. Saf.* 39(6), 28–30, June 1994.

Ragan, P. and Carder, B. Systems theory and safety. *Prof. Saf.* 39(6), 22–27, June 1994.

Petersen, D. The barriers to safety excellence. *Occup. Hazards* 37–42, December 2000.

Swartz, G. Consider a safety audit program. *Prof. Saf.* 28(4), 47–49, April 1983.

Top, W. N. What makes safety management flourish? *Occup. Hazards Natl. Saf. Manage. Soc. Focus,* 67–70, June 1992.

## CHAPTER 12

Addressing the Human Dynamics of Occupational Safety and Health

E. Scott Geller

CONTENTS

INTRODUCTION

Along with the design of safer industrial operations and equipment has come increased realization that the human dynamics in a safety achievement system need further attention. In other words, it seems safety professionals are becoming increasingly aware that the next successive approximation to achieving an injury-free workplace requires concerted efforts in the realm of psychology. However, the theory, research, and tools in psychology are so vast and often so complex that it can be an overwhelming task to choose a particular approach or strategy to apply. Indeed, there is an apparent endless market of self-help books, audiotapes, and videotapes addressing concepts seemingly relevant to the human element in industrial health and safety. How can one make an informed decision about which psychological perspective to take without earning an advanced degree in psychology? Can one trust the opinions of a psychologist-consultant who is selling a training or intervention program that represents his or her own perspective?

I do not have a quick-fix answer to the dilemma of choosing a particular psychological paradigm for industrial safety, but I offer some recommendations and guidelines in this chapter. Obviously, my particular training, research, and scholarship in psychology over the past 40 years has influenced my perspective, and this chapter reflects my professional experience and biases. I have attempted, however, to role-play professor rather than psychologist-consultant in preparing these recommendations. First, let us consider two divergent approaches to understanding and managing the human dynamics of safety and health.

BEHAVIOR-BASED VERSUS PERSON-BASED APPROACHES

Most of the myriad of opinions and recommendations given to address the human dynamics of occupational safety and health can be classified into one of two basic approaches to producing beneficial change in people — a person-based approach and a behavior-based approach. Indeed, most of the numerous psychotherapies available to treat developmental disabilities and psychological disorders (from neuroticism to psychosis) can be classified as essentially person-based or behavior-based. That is, most psychotherapies focus on changing people from the inside out (i.e., thinking people into acting differently) or from the outside in (i.e., acting people into thinking differently). In other words, person-based approaches attack individual attitudes or thinking processes directly by teaching clients new thinking strategies or giving them insight into the origins of their abnormal or unhealthy thoughts, attitudes, or feelings. In contrast, behavior-based approaches attack the clients' behaviors directly by changing relationships between behaviors and their consequences.

Many clinical psychologists use both person-based and behavior-based approaches with their clients, depending upon the nature of the problem. Sometimes the same client is treated with both person-based and behavior-based intervention strategies. I am convinced both of these approaches are relevant in certain ways for improving health and safety. This chapter provides a rationale for using both approaches to manage safety and health in the workplace, and offers a framework and guidelines for doing so.

The Person-Based Approach

Imagine observing two employees pushing each other in a parking lot, as a crowd gathers around to watch. Is this aggressive behavior, horseplay, or mutual instruction for self-defense? Are the employees physically attacking each other to inflict harm, or does their physical contact indicate a special friendship and mutual understanding of the line between aggression and play. Perhaps longer observation of this interaction, with attention to verbal behavior, will help determine whether this physical contact between individuals is aggression, horseplay, or a teaching/learning demonstration. However, a truly accurate account of this event might require an assessment of each individual's intentions or feelings. It is possible, in fact, that one person was aggressing while the other was having fun, or that the interaction started as horseplay and progressed to aggression (from the perspective of personal feelings, attitudes, or intentions of the two individuals).

This scenario illustrates a basic premise of the person-based paradigm — namely that focusing only on observable behavior does not explain enough. People are much more than their behaviors. Concepts like intention, creativity, intrinsic motivation, subjective interpretation, self-esteem, and mental attitude are essential to understanding and appreciating the human dynamics of a problem. Thus, a person-based approach applies surveys, personal interviews and focus-group discussions to find out how individuals feel about certain situations, conditions, behaviors, or personal interactions.

A wide range of therapeutic approaches fall within the general framework of person-based, from the psychoanalytic techniques of Sigmund Freud, Alfred Alder, and Carl Jung to the client-centered humanism developed and practiced by Carl Rogers, Abraham Maslow, and Viktor Frankl (cf. Wandersman, Poppen, and Ricks, 1976). Humanism is the most popular person-based approach today, as evidenced by the current market of pop psychology videotapes, audiotapes, and self-help books; although some popular industrial psychology tools (such as the Myers–Briggs Type Indicator and other trait measures of personality, motivation, or risk-taking propensity) stem from psychoanalytic theory and practice.

The key perspectives of humanism included in most pop psychology approaches to increasing personal achievement are:

- Everyone is unique in numerous ways and the special characteristics of individuals cannot be understood or appreciated by applying general principles or concepts (e.g., the behavior-based principles of performance management, or the permanent personality trait perspective of psychoanalysis).
- Individuals have far more potential to achieve than they typically realize and should not feel hampered by past experiences or present liabilities.
- The present state of an individual in terms of feeling, thinking, and believing is a critical determinant of personal success.
- One's self-concept influences mental and physical health, as well as personal effectiveness and achievement.
- Ineffectiveness and abnormal thinking and behavior result from large discrepancies between one's real self ("who I am") and ideal self ("who I would like to be").
- Individual motives vary widely and come from within a person.

Readers familiar with the writings of W. Edwards Deming (1982, 1993) and Stephen R. Covey (1989, 1990) will recognize that these renowned industrial consultants would be classified as humanists (or person-based).

The Behavior-Based Approach

The behavior-based approach to applied psychology is founded on behavioral science as conceptualized and researched by B. F. Skinner (e.g., 1938, 1971, 1974). In his experimental analysis of behavior Skinner rejected, for scientific study, unobservable inferred constructs such as self-esteem, cognition strategies, intentions, and attitudes. He researched only overt behavior and its environmental, social, and physiological determinants. Therefore, the behavior-based approach starts with an identification of overt behaviors to change and environmental conditions or contingencies (i.e., relationships between designated target behaviors and their supportive consequences) that can be manipulated to influence the target behavior(s) in desired directions.

This approach has been used effectively to solve environmental, safety, and health problems in organizations and throughout entire communities by first defining the problem in terms of relevant observable behavior, and then designing and implementing intervention programs to decrease behaviors causing the problem and/or increase behaviors that can alleviate the problem (e.g., Elder, Geller, Hovell, and Mayer, 1994; Geller, Winett, and Everett, 1982; Goldstein and Krasner, 1987; Greene, Winett, Van Houten, Geller, and Iwata, 1987). The behavior-based approach to occupational safety is reflected in the research and scholarship of several safety consultants (e.g., Geller, 1996, 2001; Krause, Hidley, and Hodson, 1996; McSween, 1995; Petersen, 1989) and is becoming increasingly popular for industrial applications.

Considering Cost Effectiveness

When people act in certain ways they usually adjust their mental attitude and self-talk to be consistent with their actions. And when people change their attitudes, values, or thinking strategies, certain behaviors change as a result. Thus, person-based and behavior-based approaches to changing people can influence both attitudes and behaviors, either directly or indirectly. Furthermore, most parents, teachers, first-line supervisors, and safety captains have used both of these approaches in their attempts to change other persons' knowledge, skills, attitudes, or behaviors. Thus, when we lecture, counsel, or coach others in a one-on-one or group situation, we are essentially using a person-based approach; and when we recognize, correct or punish others for what they have done, we are operating from a behavior-based perspective. Unfortunately, we are not always effective with our person-based or behavior-based change techniques, and often we do not know whether our intervention techniques worked as intended.

In order to apply person-based approaches to psychotherapy, clinical psychologists receive specialized therapy or counseling training for four years or more, followed by an internship of at least a year. Such intensive experiential training is necessary because tapping into an individual's perceptions, attitudes, and thinking styles is a demanding and complex process. Also, these inside dimensions of people are extremely difficult to measure reliably, making it cumbersome to assess therapeutic progress and obtain straightforward feedback regarding one's therapy skills. Consequently, the person-based therapy process can be very time consuming, involving numerous one-on-one sessions between professional therapist and client.

In contrast, the behavior-based approach to psychotherapy was designed for administration by individuals with minimal professional training. From the start, the idea behind the behavior-based approach was to reach people in the settings where their problems occur (e.g., the home, school, rehabilitation institute, workplace) and teach the managers or leaders in these settings (e.g., parents, teachers, supervisors, friends, or co-workers) the behavior-change techniques most likely to work under the circumstances (cf. Ullman and Krasner, 1965).

Forty years of research has shown convincingly that this on-site approach is cost effective, primarily because behavior-change techniques are straightforward and relatively easy to administer, and because intervention progress can be readily monitored by ongoing observation of target behaviors. Thus, intervention agents can obtain objective feedback regarding the impact of their intervention techniques, and accordingly refine or alter components of a behavior-based process.

A Need for Integration

A common perspective, even among psychologists, is that humanists (as in person-based) and behaviorists (as in behavior-based) represent opposite poles of an intervention continuum (Wandersman et al., 1976). Behaviorists are considered cold, objective, and mechanistic, operating with minimal concern for people's feelings; whereas humanists are thought of as warm, subjective and caring, with limited concern for directly changing another person's behaviors or attitudes. In fact, the basic humanistic approach to therapeutic intervention is termed nondirective or client-centered, referring to the paradigm that therapists, counselors, or coaches do not directly change their clients but rather provide empathy and a caring and supportive environment that enables clients to change themselves (i.e., from the inside out).

Given the conceptual foundations of humanism and behaviorism, it is easy to build barriers between person-based and behavior-based perspectives, and assume one must follow either one or the other approach when designing an intervention process. In fact, many consultants in the safety management field market themselves as using one or the other approach, but not both. I believe an integration of these approaches is not only possible but also necessary for optimal safety and health management. In other words, all factors contributing to the safety of an organizational culture can be classified as environmental (e.g., equipment, tools, machines, engineering, housekeeping), personal (e.g., attitudes, motives, thinking strategies, personality), or behavioral (e.g., working safely or unsafely, acting on behalf of the safety of others). And aspects of both person-based and behavior-based psychology are relevant for addressing the two human dimensions of this *Safety Triad* (Geller, Lehman, and Kalsher, 1989).

Building Bridges between Person-Based and Behavior-Based

The founder of contemporary behaviorism (B. F. Skinner) is probably among the most frequently misunderstood researchers and scholars of this century. Most people, from psychologists to the general public, have presumed that Skinner's behavior-based approach to psychological research and societal application has no regard for feelings, attitudes, or thinking styles (i.e., human factors inside the person). This is just not true (cf. Skinner, 1974). Skinner's paradigm was that only observable behavior should be the research domain of psychology, but he certainly did not deny inside reactions or interpretations to outside (observable) events. In fact, Skinner (1971) claimed that perceptions of freedom (today we might substitute empowerment for freedom) are determined by the nature of the response-consequence contingencies controlling one's behavior.

When we are performing to achieve rewarding consequences (i.e., controlled by a positive reinforcement contingency) we feel free (or empowered); but when we are working to avoid unpleasant or punishing consequences (i.e., technically referred to as control by negative reinforcement), we feel controlled (i.e., our empowerment is sapped). This is the main reason proponents of the behavior-based approach advocate the use of positive consequence to motivate behavior change.

The behavior-based approach also advocates that reinforcing consequences be soon and certain. However, behaviorists also recognize that delayed consequences can influence behavior if their occurrence is certain. For example, employees work for weekly wages; students study for end-of-the-term grades; professors write research reports for delayed professional recognition, tenure, and promotion; and consultants initiate work on a contract for eventual reimbursement for expenses and fees. To explain the control of behavior with delayed consequences, behaviorists again consider person or inside factors. More specifically, it is presumed that we use rules (e.g., by talking to ourselves) to connect behaviors with delayed consequences, and then through self-talk (to remind ourselves of the rules) we can maintain our personal motivation for earning eventual rewards (cf. Malott, 1992).

Probably the most important characteristic of consequences is certainty, a prime reason why the safety professional's job is so challenging. If a response-consequence is certain, we are motivated to perform, even if the other consequence characteristics are negative and delayed. Consider, for example, the great amount of societal control achieved with the threat of delayed penalties for various undesirable behaviors (e.g., from drunk driving to tax evasion). Such threat approaches are effective as long as we believe the delayed and negative consequences have a high probability of occurring. Of course, under the control of delayed and certain negative consequences, we feel controlled, and as a result we might procrastinate, accomplish inferior work, or attempt to beat the system.

Negative consequences for at-risk behaviors are quite rare. In fact, individuals can work at-risk for years and never receive an injury. As a result most people internalize the rule or response-consequence contingency as, "this unsafe behavior gives me convenience, comfort, or a faster job and never gives me a negative consequence." Thus, it is natural to develop the attitude, "It will never happen to me."

If people look beyond their individual contingencies to the plant population as a whole, the undesirable consequences (i.e., injuries) for at-risk behavior become much more certain. In other words, although the probability of an injury to a certain individual is minuscule, the probability that someone at the industrial site will get hurt in the next six months as a result of at-risk behavior is probably near certainty at most large plants. When employees truly care about this group contingency and believe their own actions can reduce the probability a co-worker will get injured on the job, new levels of health and safety excellence are achievable.

A caring attitude for others and a belief that one can make a difference relate directly to certain person factors from humanistic psychology, including self-esteem ("I am valuable"), self-efficacy ("I can do it"), optimism ("I expect the best"), and a sense of personal control and belonging. Research psychologists have shown that these person characteristics (or belief states) can be increased or decreased by environmental and interpersonal conditions, and promoting these states increases the probability that one individual will help another (Schroeder, Penner, Dovidio, and Pilavin, 1995). Consequently actively caring for the safety of others can be enhanced by establishing conditions, situations, or contingencies that cultivate key person factors of humanistic psychology.

The highest level on Maslow's need hierarchy, for example, is self-transcendence, a need state quite analogous to the actively caring concept (Maslow, 1971). Thus, optimal safety and health management can be achieved by integrating person-based and behavior-based approaches in special ways. Indeed, B. F. Skinner (1978) himself said, "Behaviorism makes it possible to achieve the goals of humanism more effectively." Let us get more specific with regard to processes derived from behavior-based and person-based psychology that need to be implemented in corporate cultures to improve safety and health.

THE "DO IT" PROCESS

The objective of a behavior-based and person-based approach to safety management is to change behavior directly in such a way that the resultant attitude is positive; to achieve this objective, a continuous process is followed, as typified by the acronym "DO IT," which my associates and I use to teach this process. The behavior change process involves the following four steps:

D = Define the target behavior to increase or decrease.
O = Observe the target behavior during a pre-intervention baseline period to set behavior-change goals, and perhaps to understand natural environmental or social determinants of the target behavior.
I = Intervene to change the target behavior in desired directions.
T = Test the impact of the intervention procedure by continuing to observe and record the target behavior during the intervention program.

The cost-effectiveness of the intervention is evaluated and a decision is made whether to continue the program, implement another intervention strategy, or define another behavior for the DO IT process.

The DO IT process for safety and health achievement is easier said than done. It begins with the clear and concise definition of a behavior (e.g., using equipment safely, lifting correctly, locking out power appropriately, looking out for the safety of others) or the outcome of behaviors (e.g., wearing personal protective equipment, working in a clean and organized environment, using a safety belt). If two or more people obtain the same frequency recordings when observing the defined behavior or behavioral outcome during the same time period, the definition is sufficient for an effective DO IT process. Baseline observations of the target behavior should be made and recorded before implementing an intervention program.

Designing Interventions

When designing interventions to change behavior, behavior managers follow a simple ABC model (**A** for Activator, **B** for Behavior, and **C** for Consequence). Activators direct behavior (as when the ringing of a telephone or doorbell signals certain behaviors from residents), and consequences motivate behavior (as when residents answer or do not answer the telephone or door depending on current motives or expectations developed from prior experience at telephone or door answering).

As discussed above, the most motivating consequences are soon, sizable, and certain. Safe behaviors are not usually reinforced by soon, sizable, and certain consequences. In fact, safe behaviors are often punished by soon and certain negative consequences (e.g., inconvenience, discomfort, slower goal attainment).

The consequences that motivate safety professionals to promote safe work practices (i.e., the reduction of injuries and associated costs) are delayed, negative, and uncertain (actually improbable) from an individual's perspective. Moreover, at-risk behaviors avoid the discomfort and inconvenience of most safe behaviors, and usually allow people to achieve their production and quality goals faster and easier. Indeed, sometimes production supervisors inadvertently activate and reward at-risk behaviors (or short cuts) in their attempts to achieve more production.

Because activators and consequences are naturally available throughout our everyday existence to support at-risk behaviors in lieu of safe behaviors, safety management can be considered a continuous fight with human nature. Hence, the development and maintenance of safe work practices often requires the implementation of intervention strategies to keep people safe. These intervention strategies serve as activators or consequences, or they combine both approaches; and they focus on increasing safe behaviors and/or decreasing at-risk behaviors. Let us consider some actual intervention techniques that were quite effective at increasing safe work practices. The interventions that included consequences were more effective (although usually more costly). And those that involved employees in designing, implementing, and refining the behavior change procedures resulted in larger and longer-term change.

Worker-Designed Safety Slogans

Signs and slogans are activators with only limited behavioral influence, unless they signal the occurrence and availability of motivating consequences. Furthermore, the same signs (e.g., "Caution: Hazardous Area," "Please Hold Handrail," "Buckle Up") eventually blend into the woodwork and are not noticed. However, changing signs periodically can increase attention and awareness; and if employees design the messages or slogans (perhaps as a result of a company contest), awareness and attention can be increased before and after sign display.

In 1985, employees and visitors driving into the main parking lot for Ford World Headquarters in Dearborn, Michigan passed a series of four signs arranged with sequential messages. The messages were changed periodically as selected from a pool of 55 employee entries in a limerick contest for safety-belt promotion. Two creative examples were:

Sign 1: Mary, Mary Quite Contrary
Sign 2: Just Wouldn't Buckle, You Know
Sign 3: She Had a Fuss with a Greyhound Bus
Sign 4: Now She's Planted All in a Row.

Sign 1: Think of Those Who Died in Cars
Sign 2: And How Their Family Felt
Sign 3: How Much Trouble Can It Be
Sign 4: To Tug and Click a Belt.

Near-Miss and Corrective Action Reporting

A proactive stance to occupational safety requires discussion and problem solving regarding near misses (i.e., workplace incidents that could have resulted in an injury had the timing or body positioning been slightly different). Motivating near-miss reporting and discussion is not easy, especially if the reporting process is inconvenient and the instigators or victims of an injury or near miss are viewed by others as careless, dumb, accident prone, or thoughtless.

In the mid-1980s, Air Products and Chemicals in Allentown, Pennsylvania attempted to encourage near-miss reporting by recognizing near-miss reporters in the company newspaper. The entire Body and Assembly Division of Ford Motor Company also promoted attention to near-miss reporting by having reporters reenact their near misses on videotape for later analysis and corrective action. It is noteworthy that these near-miss reports considered factors in each dimension of the *Safety Triad* (i.e., environment, person, and behavior) when searching for factors contributing to the incidents and when deriving recommendations for corrective action.

Group Safety Share

At the start of group meetings at the Hercules plant in Portland, Oregon, facilitators ask participants to report something they have done for safety since the last meeting. Because this safety share is used to open all kinds of meetings, safety is given a priority status and integrated into the various other functions of the organization (from marketing to quality control). Employees expect to be asked about their safety achievements, and thus they prepare accordingly, sometimes going out of their way for safety in order to report an impressive safety share.

This awareness booster focuses on achieving safety (i.e., "What have you done for safety?"), and contributes to teaching the culture that safety is not only loss control (i.e., an attempt to avoid failure) but belongs in the same achievement system as productivity, quality, and profits.

Private and Public Commitment

Intentions are verbal statements to ourselves (i.e., private) or to others (i.e., public) that we plan on doing something. Whether private or public, such statements activate personal commitment to emit particular behaviors. However, when intentions are given publicly, the commitment is strengthened through peer influence. Public intentions can activate peer support as reminders (e.g., "Didn't you say you were going to use your ear protectors?") and as consequences (e.g., "I see you're reaching your personal safety goal, congratulations.").

Years ago when the nationwide use of safety belts was below 15% (i.e., before state belt-use laws), many companies more than doubled the use of safety belts in company and private vehicles by distributing "Buckle-Up Promise Cards" and encouraging their employees to sign them (e.g., General Motors and Ford Motor Company in Detroit, Michigan; Corning Glass in Blacksburg, Virginia; Burroughs

Welcome Company in Greenville, North Carolina; and the Reeves Brothers Curon Plant in Cornelius, North Carolina). Logan Aluminum in Russellville, Kentucky instituted a "Public Safety Declaration" whereby employees signed a poster at the plant entrance that specified a safe-behavior commitment for the day (e.g., "We wear safety glasses in all designated areas.") Delta Airlines in Atlanta, GA recently asked maintenance employees at a "Safety Awareness Day" to sign a "Declaration of Interdependence" as a group commitment to look out for each other's safety and health.

Actively Caring Thank-You Cards

Kal Kan of Columbus, Ohio, the Hercules plant in Portland, Oregon, and two Hoechst Celanese plants (Celco in Narrows, Virginia, and Celriver in Rock Hill, South Carolina) have provided employees with thank-you cards for distribution to co-workers when observing them actively caring or going beyond the call of duty for another person's safety. For some variations of this process, the observers described the actively caring behaviors they saw on the thank-you cards, and thereby increased their operational understanding of actively caring behavior.

In one program, the actively caring thank-you cards could be exchanged for a beverage in the company cafeteria. For another program, the actively caring thank-you cards contained a removable recognition sticker that could be attached to a variety of objects, the most popular being a hard hat.

Two of these intervention programs offered a unique consequence that extended the actively caring concept beyond the workplace. Specifically, when deposited in a special collection container, each thank-you card was worth $1.00 toward corporate contributions to a local charity or to needy families in the community. The employees who designed one of these programs called their actively caring cards "START" for Safety Through Awareness, Recognition, Teamwork. In 1996, the 64 employees at the Hercules plant collected enough actively caring thank-you cards to spend $1,750 on toys for needy children on Christmas Day.

This intervention process was an incentive/reward program, since the employees knew they could receive a thank-you card for emitting certain behaviors (an incentive), and they actually received this reward when meeting an actively caring criterion. However, this intervention was very different from the typical incentive/reward program used in industry to manage safety and health. Let us turn now to a discussion of guidelines for designing an effective incentive/reward intervention to manage safety and health.

INCENTIVES AND REWARDS

Referring to the three-term contingency of behavior-based psychology (i.e., Activator-Behavior-Consequence), an incentive is an activator that announces the availability of a particular pleasant consequence (i.e., a reward) following the occurrence of certain desired (e.g., safe) behavior. Activators that announce the availability of unpleasant consequences (i.e., a penalty) following the occurrence of certain undesired (e.g., at-risk) behaviors are termed disincentives.

The power or motivational influence of incentives and disincentives are determined by consequences. Rules or policies (i.e., disincentives) that are not consistently enforced (e.g., with penalties for noncompliance) are often disregarded; and if promises of rewards (i.e., incentives) are not fulfilled when the correct behavior is emitted, subsequent incentives might be ignored.

Doing It Wrong

The definitions of incentives and rewards imply that specific target behavior(s) must be specified to complete the three-term contingency of activator-behavior-consequence. However, most incentive programs for occupational safety do not specify behavior. That is, the most common application of incentives for safety management specifies a certain reward that employees can receive by avoiding a work injury (or by achieving a certain number of safe work days). Many of these nonbehavioral, outcome-based incentive programs implicate substantial peer pressure because they use a group-based contingency (i.e., if anyone in the company or work group is injured, everyone loses their reward).

So what behavior does such an outcome-based, group-contingency incentive/reward program motivate? Obviously, if workers link certain safe behaviors directly with a high probability of avoiding an injury, then an outcome-based incentive program can have a beneficial impact. However, as discussed above, most safe and at-risk behaviors are rarely followed by supportive consequences (i.e., injury avoidance following safe behaviors and injury following at-risk behaviors). Thus, the most likely behavior to be influenced by an outcome-based incentive/reward program is injury reporting.

If having an injury loses one's reward (or worse, the reward for an entire work group), there is pressure to avoid reporting an injury if possible. For example, I have observed coworkers cover for an injured employee in order to keep accumulating safe days and not lose their reward possibility. Hence, these incentive programs do decrease the numbers (i.e., the reported injury rate), at least over the short term, but corporate safety is obviously not improved. In fact, such outcome-based incentive programs often lead to a detrimental attitude of apathy or helplessness regarding safety achievement. In other words, employees develop the perspective that they cannot really control their injury records, but they must cheat or beat the system to celebrate the achievement of an injury reduction goal.

Doing It Right

This discussion leads logically to the following seven basic guidelines for establishing an effective incentive/reward program for managing the human dynamics of industrial safety and health.

- The behaviors required to achieve a safety reward should be specified and perceived as achievable by the participants.
- Everyone who meets the behavioral criterion should be rewarded.
- Groups should not be penalized (e.g., lose their rewards) for failure by an individual *unless* the group can control the individual's performance.

- It is better for many participants to receive small rewards than for one person to receive a big reward.
- Contests should not reward one group at the expense of another.
- Progress toward achieving a safety reward should be systematically monitored and publicly posted for all participants.
- The rewards should be displayable and represent safety achievement (e.g., coffee mugs, hats, shirts, sweaters, blankets, or jackets with a safety logo or message).

An Exemplary Incentive/Reward Program

In 1992, I consulted with the safety incentives committee of a Narrows, VA Hoechst Celanese company of about 2000 employees in the development of an exemplary plant-wide incentive program that followed each of the guidelines given above. The incentives committee, including four hourly and four salary employees, met several times to identify specific behavior-consequence contingencies (i.e., what behaviors would earn what rewards). The resultant program plan was essentially a credit economy whereby certain desirable safe behaviors achievable by all employees earned a certain numbers of credits. At the end of the year, the participants could exchange their credits for their choices among a number of different prizes (all containing a special safety logo).

The variety of behaviors earning credits included: attendance of monthly safety meetings; special participation in safety meeting; leading a safety meeting; writing, reviewing, and revising a job safety analysis; and conducting periodic audits of environmental and equipment conditions, and certain work practices. For a work group to receive their credits, the results of environmental and PPE audits had to be posted in the relevant work area. Only one behavior was penalized by a loss of credits — the late reporting of an injury.

At the start of a new year, each participant would receive a safety credit card on which individual credit earnings could be tallied per month. Some individual behaviors earned credits for the person's entire work group, thus promoting teamwork and group cohesion. It is noteworthy that this kind of incentive/reward program exemplifies a basic behavior-based principle for safety and health management — observation and feedback. In other words, for this intervention, employees were systematically observed with regard to their performance of certain desirable behaviors, and they received soon, certain, and positive feedback (i.e., a reward) after emitting a target behavior.

An incentive/reward program is only one of several methods to increase safe work practices with observation and feedback. Let us turn now to a more generic discussion of behavioral observation and feedback as a key principle for managing the human aspects of occupational safety and health.

Behavioral Observation and Feedback

Ask any safety manager, industrial consultant, or applied psychologist whether he or she has heard of the "Hawthorne Effect," and, it is likely, the answer will be yes. They might not be able to describe any details of the studies that occurred between 1927 and 1932 at the Western Electric plant in the Hawthorne community (near Chicago) that led to the classic Hawthorne Effect. However, most will be able to paraphrase the well-known finding from these studies that the hourly output rates of the employees studied increased whenever an obvious environmental change occurred in the work setting.

The explanation of the Hawthorne results is also well known, and recited as a potential confounding factor in numerous field studies of human behavior. Specifically, it is commonly believed that the Hawthorne studies showed that people will change their behavior in desired directions when they know their behavior is being observed. The primary Hawthorne sources (Mayo, 1933; Roethlisberger and Dickson, 1939; Whitehead, 1938) leave us with this impression, and in fact this interpretation seems intuitive. The fact is, however, this interpretation of the Hawthorne studies is not accurate — it is nothing but a widely disseminated myth.

Hawthorne Workers Got Feedback

Dr. H. McIlvaine ("Mac") Parsons, a good friend of mine and world-renowned industrial and behavioral psychologist, conducted a careful reexamination of the Hawthorne data and interviewed eyewitness observers, including one of the five female relay assemblers who were the primary targets of the Hawthorne studies. Parsons' findings were published in a seminal article titled "What happened at Hawthorne?"

What happened was the five women observed systematically in the Relay Assembly Test Room received regular feedback about the number of relays each had assembled. "They were told daily about their output, and they found out during the working day how they were doing simply by getting up and walking a few steps to where a record of each output was being accumulated" (Parsons, 1980, p. 58).

From his scientific detective work, Parsons concluded that performance feedback was the principal extraneous, confounding variable that accounted for the Hawthorne effect. It is not insignificant, however, that the performance feedback was important to the workers (i.e., they were apt to attend to it) because their salaries were influenced by an individual piecework schedule — the more relays each employee assembled, the more money each earned.

Feedback for Safety

For anyone who has studied the behavior-based approach to performance management, the only surprise in Parsons' research is that the critical role of performance feedback had not been documented in the original reports of the Hawthorne studies. To these folks the finding that feedback was the critical change variable is common sense. For example, numerous research studies have shown that posting the results of behavioral observations related to safety, production, or quality yields beneficial change in the targeted work behaviors. If desired work behaviors are targeted, they increase in frequency of occurrence; when undesired behaviors are targeted for observation and feedback, they decrease in probability (e.g., Geller, Eason, Phillips, and Pierson, 1980; Kim and Hamner, 1982; Komaki, Heinzmann, and Lawson, 1980; Krause et al., 1996; Sulzer-Azaroff, 1982; Williams and Geller, 2000).

It is noteworthy that when asked about their preference for working in the test room rather than the regular department, the five Hawthorne employees did not mention the feedback intervention, but gave a variety of other reasons, including smaller groups, no bosses, less supervision, freedom, and the way we are treated (Roethlisberger and Dickson, 1939, pp. 66–67). This suggests it is inadvisable to rely on only verbal report to discover the factors influencing work performance. Sometimes people are not aware of the basic contingencies controlling their behaviors. Through systematic and objective behavioral observation these factors can be uncovered and corrective feedback given.

Most of the published research that showed significant benefits of observing work practice and then posting feedback charts used outside observers (i.e., research assistants) to do the behavioral observing and data posting. Hiring extra staff for such tasks is certainly not practical for most organizations, and actually is not an optimal approach to safety management. In fact, using outsiders as observers instead of peers can make the process seem threatening ("Will the data be used against me?") and can decrease group cohesion or teamwork.

Therefore, the employees themselves should:

- Decide what behaviors to observe.
- Conduct regular behavioral observations.
- Deliver one-on-one feedback sessions with fellow employees.
- Calculate daily percentages of safe behaviors in circumscribed work areas.
- Post the group data in conspicuous locations.
- Design and implement strategies (e.g., training sessions, goal setting, incentives, recognition celebrations) to motivate participation in the observation feedback process.
- Refine components of the process with a continuous improvement perspective.

Employee Involvement Is Key

The idea of watching fellow employees for their safe and at-risk work practices (and being watched by fellow employees for the same behaviors) is a novel concept to many individuals and can appear threatening at first. Accomplishing such a paradigm shift can be quite challenging in some work cultures, and can only be accomplished by establishing conditions and contingencies that build and nourish interpersonal trust throughout the work culture (Geller, 1999).

The process cannot include any punishment contingency (e.g., no names recorded for at-risk behavior) and participation should be strictly voluntary (although reward or recognition contingencies can be implemented to motivate involvement). In addition, the employees (e.g., a representative design team) should have as much choice as possible in the operational details of their observation/feedback process. Providing opportunities for choice is motivating, and helps to develop interpersonal trust.

The Process Is Critical

Obviously, there is no best protocol for an observation/feedback process for safety management. The process must fit the culture, and this can only happen if the employees themselves decide on the procedural details consistent with certain principles and guidelines. It is possible, for example, that a full-blown observation/feedback process will not be feasible at first, but successive approximations to a complete, plant-wide process are possible. Thus, a plant might start small (e.g., with brief and infrequent observation/feedback of a select number of behaviors in one work area), and then expand the process as confidence and interpersonal trust develop through experience.

It is easy to get caught up in the particulars of an observation/feedback process (e.g., duration, frequency, and sampling of observations; timing, location, and length of feedback sessions), but it is more important to get the process going. The true value of the process is not in the behavioral data themselves (which are no doubt biased by numerous confounding variables), but in the behavior-focused interactions between employees.

I have seen an employee safety observation/feedback process work well at numerous companies, and the implementation procedures have varied widely. The employees at an ExxonMobil plant, for example, designed an innovative observation/feedback process whereby each month every employee schedules three to five observation/feedback sessions with any two other employees. Thus, on days and at times selected by the observee, two observers show up at the observee's worksite and use a standard behavioral checklist to conduct a systematic 30-minute observation session, followed by interactive feedback.

Those concerned with obtaining a representative and objective sample of work practice will readily criticize this procedure of allowing employees to choose their own observers and observation times. However, the critical ingredient here is the process of people willingly observing and coaching others in order to increase safe work practices. Although those observed are on their best behavior during assigned observation sessions, at-risk acts are observed and often an observee's demonstration of safe behavior becomes a valuable learning experience for the observers.

Employees at a Hoechst Celanese plant in Rockville, SC developed a very different observation/feedback process. Termed the "planned 60-second actively caring review," all employees attempt to complete one feedback card and brief coaching session every day. The feedback card is completed after a one-minute observation of an employee's safe and at-risk work practices in five general categories: body position, personal apparel, housekeeping, tools/equipment, and operating procedures. The feedback cards are collected each day and the results tallied and posted. The number of feedback cards collected per department is exhibited in a large display case at the plant entrance, along with weekly department goals.

These one-minute observation sessions are briefer than I had advised, but the employees felt that plant-wide acceptance of longer sessions would be unlikely at the time. Again, the process of employee involvement in behavior-focused observation and feedback is the most important feature of this intervention; perhaps longer sessions will come later. The members of the design team who developed their process reminded me that one minute was sufficient for the one-minute manager's goal setting, praising, and reprimanding (Blanchard and Johnson, 1982), so why not one-minute safety observations?

After one minute of behavior-based observing, employees are urged to review the feedback card with the observee. This one-on-one interaction is termed coaching, and is based on principles from both behavior-based and person-based psychology. The next section reviews these principles, and presents guidelines for effective safety coaching. When large numbers of employees practice effective safety coaching, interpersonal trust and teamwork increase naturally.

COACHING FOR SAFETY AND HEALTH MANAGEMENT

The coaches of winning athletic teams practice the basic observation and feedback processes necessary for effective safety management, and follow most of the guidelines reviewed here. In other words, the best team coaches observe the behaviors of their players, record their observations in systematic fashion (e.g., on a team roster, behavioral checklist, or on videotape), and then offer specific behavioral feedback (both correcting and supporting) to the team members. Often this behavioral feedback is given both individually (in one-on-one conversation) and in group sessions (e.g., by reviewing videotapes of the team competition). And the most effective coaches give this feedback so the team members learn from the exchange *and* are motivated to continuously improve.

The five letters of the word "COACH" can be used as a mnemonic to remember the basic characteristics of the most effective coaches, whether coaching for a winning athletic team or for a work group seeking improved safety performance.

C for Care

Safety coaches truly care about the health and safety of their co-workers, and they act on such caring. In other words, they actively care. In the next section, I discuss ways to increase actively caring. For now, just consider the value of initiating a one-on-one coaching process with an actively caring attitude. When people realize the words and body language of a coach that he or she cares, they will listen to the coach's advice. In other words, when people know you care, they will care what you know.

Covey (1989) explains the value of caring and interdependence (exemplified by appropriate safety coaching) with the metaphor of an emotional bank account. The idea is that individuals establish an emotional bank account with others through personal interaction. Deposits are made when the holder of the account perceives a particular interaction to be positive (e.g., they feel recognized, appreciated, or listened to). Withdrawals from individual's emotional bank account occur whenever that individual feels criticized, humiliated, or less appreciated (e.g., from personal interaction).

It is sometimes necessary, of course, to offer constructive criticism to others or even to state extreme displeasure with another person's behavior. But if such negative discourse occurs on an overdrawn or bankrupt account, corrective feedback will have limited impact. In fact, continued withdrawals from an overdrawn account can result in defensive or countercontrol reactions (from simply ignoring the communication, to emitting overt behavior that discredit the source or undermine the process or system implicated in the communication). Thus, safety coaches need to demonstrate a caring attitude through their personal interactions with others, thereby maintaining healthy emotional bank accounts (i.e., operating in the black).

O for Observe

Safety coaches observe the behavior of others objectively and systematically with an eye for supporting safe behavior and correcting at-risk behavior. Behavior that illustrates going beyond the call of duty for the safety of another person should be especially noted for supportive coaching. This is the sort of behavior that contributes significantly to safety achievement and improvement and can be increased through rewarding feedback.

Observing behavior for supportive or corrective feedback is easy if the coach knows exactly what behavior is desired and undesired (an obvious requirement for athletic coaching) and takes the time to observe occurrences of these behaviors in the work setting. It is often advantageous (usually essential) to develop a checklist of safe and at-risk behaviors in the various work settings of an industrial complex and rank these in terms of risk. As discussed above, the workers themselves should develop their behavioral checklists. Risk rankings can be derived from careful study of injury records (from first-aid cases to lost-time injuries) and reports of near misses. Promoting the reporting of near misses to learn ways of preventing injury is another important challenge for the effective safety coach.

A for Analyze

Safety coaches appreciate the ABC model (**A**ctivator, **B**ehavior, **C**onsequence) in interpreting their observations. In other words, they understand there are usually observable reasons for behaviors. As discussed in more detail above, activators (e.g., signs, memos, instructions, policies, mission statements) direct behavior, whereas consequences (e.g., praise, feedback, reprimands, recognition ceremonies) motivate behavior.

Coaches realize, for example, that certain at-risk behaviors occur because they are directed by such activators as work demands, at-risk example setting by peers, and inconsistent or mixed messages from management. And at-risk behaviors are often motivated by one or more positive consequences (e.g., comfort, convenience, work breaks, and approval from peers or supervisors). See Geller (2000b) for more details on behavioral safety analysis.

Of course, safety coaches also realize their behavioral observation, analysis, and communication can activate the occurrence of safe behaviors (e.g., through instructions, reminders, individual and group discussions) and motivate their reoccurrence (e.g., through verbal feedback, recognition, and group celebrations). This brings us to the next letter of COACH.

C for Communicate

Effective coaching requires basic communication skills, including strategies for active listening and persuasive speaking. Indeed, communication training sessions for employees that incorporate role-play exercises can be invaluable in developing each individual's confidence and competence in sending and receiving behavioral feedback.

Communication training should illustrate the need to separate behavior (i.e., actions) from person factors (i.e., attitudes and feelings), thus enabling corrective feedback to occur at the behavioral level without stepping on feelings. For example, people need to understand that anyone can act at-risk without even realizing it (as in unconscious incompetence) and performance can only improve with behavior-specific feedback. Thus, even corrective feedback for at-risk behavior is appreciated when it is given appropriately, regardless of the work status of the feedback sender.

Whether supportive or corrective, feedback should be specific (with regard to a particular behavior) and timely (occurring soon after the target behavior); and it should be given in a private, one-on-one situation to avoid potential interference or embarrassment from others. In addition, corrective feedback is most effective if the alternative safe behavior is specified and potential solutions for eliminating barriers to performing the safe behavior are discussed.

Whether giving supportive or corrective feedback, the communicator of behavioral feedback must actively listen with appropriate body language and verbal responses when the feedback recipient reacts to support or correction. Through active (or empathic) listening, the safety coach shows sincere concern for the feelings and commitment of the feedback recipient. The best listeners give empathic attention with facial cues and posture, paraphrase to check understanding, prompt for more details (e.g., "Tell me more."), accept feelings as stated without interpretation, and avoid autobiographical statements that can stifle self-disclosure (e.g., "When I worked in your department, I always worked safely.").

H for Help

The word help summarizes the essential mission of a safety coach. In other words, the goal of safety coaching is to help an individual prevent an unintentional injury by supporting safe work practices and correcting at-risk work practices. And it is critical, of course, that the feedback recipient accepts a coach's helping communication. The four letters of "HELP" offer a mnemonic for remembering four words that suggest ways to obtain acceptance of a coach's advice, directions, or feedback.

H for Humor

Safety is certainly a serious matter, but sometimes a little humor can be inserted in our safety communications as a way of increasing interest and acceptance. In fact, some research has indicated that laughter can reduce stress and even benefit an individual's immune system (*National Safety News*, 1985); and we know from personal experience that humor can enhance the constructive gains from a communication by increasing our attention and decreasing our resistance to an appeal for change.

E for Esteem

Esteem or, more specifically, self-esteem is the personal perception of self-worth ("I am valuable."), and is critical to increasing safe work practices. People who feel inadequate, unappreciated, or unimportant in a particular work setting are not as likely to go beyond the call of duty to benefit the safety of themselves or others, as are people who feel capable and valuable. Thus, the most effective coaches choose their words carefully (i.e., emphasizing the positive over the negative) in an attempt to build or avoid lessening another person's self-esteem. In other words, effective coaches make many more deposits than withdrawals in people's emotional bank accounts (Covey, 1989).

L for Listen

One of the most powerful and convenient ways to build self-esteem is to listen attentively to another person. This sends the signal that you care about the person and his or her situation. The person can in turn interpret this active listening as a reflection of self-worth (e.g., "I must be valuable to the organization because my opinion is considered and appreciated."). And, after a coach actively listens, his or her message is more likely to be heard and accepted.

P for Praise

Praising another person for specific accomplishments is another powerful way to build self-esteem. And if the praise targets a particular behavior, the probability of the behavior occurring again increases. This reflects the basic behavioral science principle of positive reinforcement, and motivates people to continue their safe work behaviors and to look out for the safety of their associates. In other words, behavior-focused praising is a powerful rewarding consequence which not only increases the behavior it follows, but also increases a person state (i.e., self-esteem) which in turn increases the individual's willingness to act for the safety of others.

Summary

Safety coaching is a key process for managing industrial safety and health. Obviously, training and practice are essential to the development of safety coaching skills, but also critical is a work culture that enables and supports interpersonal coaching for safety improvement. For some cultures, this requires some dramatic paradigm shifting (e.g., from fault finding to fact finding, from a focus on discipline for safety violations to a focus on recognition and support for safety achievements, and from a fatalistic perspective to a make-a-difference attitude).

Thus, a large-scale safety coaching process may take significant time and resources to achieve, but the long-range outcome will be well worth the effort — an organization of people who coach each other consistently and effectively with the purpose of increasing safe work practices. This coaching process increases peoples' senses of self-esteem, empowerment, and belonging. The next section discusses these three person states, and explains the value of building their states among employees at all levels of an organization.

AN ACTIVELY CARING MODEL

In order to facilitate the design of the most cost-effective intervention process for a particular target behavior, setting, and audience, Geller et al. (1990) proposed a systematic scheme for classifying and evaluating behavior change techniques. This system classifies intervention programs, which usually combine several behavior change techniques, into multiple tiers or levels, and each tier is defined by its overall cost-effectiveness and personal intrusiveness.

At the top of the multiple intervention hierarchy (i.e., Level 1), the interventions are least expensive per person. At this level, intervention techniques (e.g., activation through signs, billboards, and public service announcements) are designed to have broad-based appeal with minimal personal contact between target individuals and intervention agents. Geller et al. hypothesized that people normally uninfluenced by initial exposure to these types of interventions (i.e., Level 1) will also be uninfluenced by repeated exposures to interventions at the same level. These people require a more intrusive and costly (i.e., higher level) intervention program.

A key proposition in the multiple intervention level model proposed by Geller et al. (1990) and refined by Geller (1998) is that people influenced by an intervention program (at a particular level of cost-effectiveness and intrusiveness) should not be targeted for further intervention, but instead they should be enrolled as intervention agents for the next (i.e., higher) level of behavior change intervention. In other words, preaching to the choir is not as beneficial as enlisting the "choir" to preach to others (cf. Katz and Lazarfeld, 1955). This section explores ways of identifying those individuals most likely to become intervention agents for industrial safety and health, as well as ways to increase participation as an intervention agent or intervention target.

Actively Caring

Intervention agents are individuals who care enough about a particular problem or about other people to implement an intervention strategy that could make a beneficial difference. In other words, intervention agents are individuals who actively care. Actively caring behaviors can be classified into three categories, depending upon the target of the intervention — environment, person, or behavior (Geller, 1996). Thus, when people intervene to reorganize or redistribute resources in an attempt to benefit the safety of others (e.g., clean up another's work area, report an environmental hazard, practice appropriate energy control and power lock-out procedures), they are actively caring from an environmental focus.

Actively caring from a person focus is behaving in an attempt to make another person feel better (e.g., intervening in a crisis situation, actively listening in one-to-one communication, verbalizing unconditional positive regard for someone, sending a get-well or thank-you card).

Finally, behavior-focused actively caring is doing something to influence another individual's behavior in desired directions (e.g., giving rewarding or correcting feedback, demonstrating or teaching desirable behavior, designing or implementing a behavior-change intervention program).

Actively Caring States

With nonhumans as experimental subjects, behaviorists have influenced marked changes in performance when altering certain physiological states of their subjects (e.g., through food, sleep, or activity deprivation). Similarly, behaviorists have demonstrated significant behavior change in both normal and developmentally disabled children as a function of simple manipulations of the social context (e.g., Gewirtz and Baer, 1958a,b) or the temporal proximity of lunch and response-consequence contingencies (Vollmer and Iwata, 1991).

Although behaviorists typically refer to these manipulations of physiological (e.g., food deprivation) or psychological (e.g., social deprivation) conditions as establishing operations (Michael, 1982; Vollmer and Iwata, 1991), these independent variables are certainly analogous to the person-based concepts of expectancies, personality states, and intrinsic motivation. In other words, certain operations or environmental conditions (past or present) can influence (or establish) physiological or psychological states within individuals, which in turn can affect their behavior. From the behavior-based approach discussed above, the basic mechanism of this impact is through enhancing the quality and quantity of positive consequences achievable by designated target behaviors.

The author has proposed that certain psychological states or expectancies affect the propensity for individuals to actively care for the safety or health of others; and furthermore, that certain conditions or establishing operations (including activators and consequences) can influence these psychological states (Geller, 1991, 1996, 2000a, 2001). These states are illustrated in Figure 12.1, a model the author and his associates have used numerous times to stimulate discussions among industry employees of specific situations, operations, or incidents that influence their willingness to actively care (e.g., to participate actively in safety achievement efforts).

Factors consistently listed as determinants of self-esteem include communication strategies, reinforcement and punishment contingencies, and leadership styles. Participants have suggested a number of ways to build self-esteem, including:

- Providing opportunities for personal learning and peer mentoring.
- Increasing recognition for desirable behaviors and personal accomplishments.
- Soliciting and following up a person's suggestions.

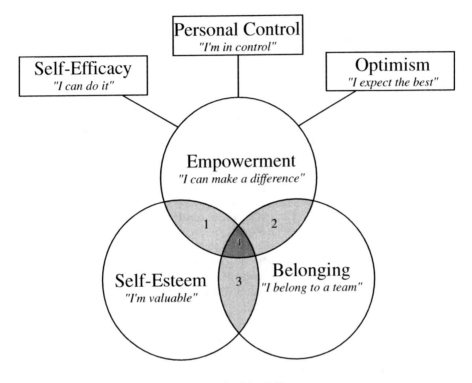

1. I can make valuable differences.
2. We can make a difference.
3. I'm a valuable team member.
4. We can make valuable differences.

Figure 12.1   The actively caring model for occupational safety and health. (Adapted from Geller, 1996).

Common proposals for increasing an atmosphere of belonging among employees have included decreasing the frequency of top-down directives and quick-fix programs, and increasing team-building discussions, group goal setting and feedback, group celebrations for both process and outcome achievements, and the use of self-managed (or self-directed) work teams.

In the management literature, empowerment typically refers to delegating authority or responsibility, or sharing decision making (Conger and Kanungo, 1988). In contrast, the person-based perspective of empowerment focuses on the reaction of the recipient to increased power or responsibility. In other words, this view of empowerment requires the personal belief "I can make a difference," and this belief is strengthened with perceptions of personal control (Rotter, 1966), self-efficacy (Bandura, 1977) and optimism (Scheir and Carver, 1985; Seligman, 1991). Such an empowerment state is presumed to increase motivation (or effort) to "make a difference" (e.g., to go beyond the call of duty), and there is empirical support for this intuitive hypothesis (e.g., Bandura, 1986; Barling and Beattie, 1983; Ozer and Bandura, 1990; Phares, 1976).

Employees at my actively caring training sessions have listed a number of ways to increase empowerment, including:

- Setting short-term goals and tracking achievements.
- Offering frequent rewarding and correcting feedback for process activities (e.g., behavior that reflects actively caring for health or safety) rather than only for outcomes (e.g., total recordable injury rate).
- Providing opportunities to set personal goals, teach peers, and chart "small wins" (Weick, 1984).
- Teaching employees basic behavior-change intervention strategies (e.g., response feedback and recognition procedures), and providing time and resources for them to implement and evaluate intervention programs.
- Showing employees how to graph daily records of baseline, intervention, and follow-up data.
- Posting response feedback graphs of group performance.

Research Support for the Actively Caring Model

A number of empirical studies, mostly in the social psychology literature, support the individual components of the actively caring model depicted in Figure 12.1. The bystander intervention paradigm (Darley and Latane, 1968) has been the most common (and rigorous) laboratory technique used to study variables related to actively caring behaviors. With this approach, one or more of the person states presumed to affect actively caring behavior (i.e., self-esteem, empowerment, and/or belonging) were measured or manipulated among subjects, and subsequently these individuals were placed in a situation where they had an opportunity to help another individual who presumably encountered a personal crisis (e.g., falling off a ladder, dropping personal belongings, or experiencing an illness or heart attack). The latency in attempting to help the other person was the primary dependent variable, studied as a function of a subject's social situation or personality state.

All actively caring behaviors studied in these experiments were person-focused (i.e., helping a person feel better) or environment-focused (i.e., redistributing resources to benefit someone else). The actively caring behaviors were never behavior-focused (i.e., attempting to change another individual's behavior in beneficial directions).

Self-Esteem

Michelini, Wilson, and Messe (1975) and Wilson (1976) measured subjects' self-esteem with a sentence completion test and then observed whether they helped another individual in a bystander intervention situation. Subjects with high self-esteem were significantly more likely than subjects with low self-esteem to help another individual pick up dropped books (Michelini et al., 1975), and to exit an experimental room to help someone in another room who screamed he had broken his foot following a mock explosion (Wilson, 1976). Analogously, subjects with higher self-esteem scores were more likely to help a stranger by taking his place in an experiment that would presumably give them electric shocks (Batson, Bolen, Cross, and Newinger-Benfiel, 1986).

Belonging

By systematically manipulating a person's sense of belonging in groups of two and four, Rutkowski, Gruder, and Romer (1983) found group cohesion to reverse the usual bystander intervention effect (i.e., the finding of more individual intervention in smaller groups). Cohesiveness was manipulated by having groups discuss topics and feelings related to college life. Both frequency and speed of helping a "victim" who had ostensibly fallen off a ladder was greater for the cohesive groups. Indeed, the most actively caring behavior was found among subjects in the high-cohesion/four-person group condition.

In a retrospective real-world study, Blake (1978) studied real-world relationships between group cohesion and the ultimate in actively caring behavior — altruistic suicide. He collected data from official records of Medal of Honor awards given during World War II and Vietnam. The independent variable was the cohesiveness of combat units (estimated by group training and size), and the dependent variable was percentage of grenade acts — voluntarily using one's body to shield others from exploding devices. The smaller, more elite, specially trained combat units (e.g., the Marine Corps and Army airborne units) accounted for a significantly larger percentage of grenade acts than larger, less specialized units (e.g., Army non-airborne units).

Personal Control

The personal control factor of the actively caring model represents one of the most extensively researched individual difference variables in psychology, and refers to a general expectancy regarding the location of forces controlling an individual's life (i.e., internal verses external factors).

Persons with an internal locus of control believe they normally have personal control over important life events as a result of knowledge, skills, and abilities. In contrast, individuals with an external locus of control believe factors like luck, chance, or fate have significant influence in their lives (Rotter, 1966; Rushton, 1980). From a behavioral perspective, externals generally expect to have less personal control over the pleasant and unpleasant consequences in their lives than do internals.

Those high-esteem subjects who showed more actively caring behavior than low-esteem behavior in Wilson's (1976) bystander intervention study (discussed above) were also characterized as internals, in contrast to the lower self-esteem externals who were less apt to display actively caring behavior. Similarly, Midlarsky (1971) observed more internals than externals willing to help another person perform a motor coordination task that included receiving electric shocks. In addition, those who helped at an accident scene scored significantly higher on personal control (i.e., internals) and self-esteem than those who only stopped and watched (Bierhoff, Klein, and Kramp, 1991).

Sherrod and Downs (1974) asked subjects to perform a task while hearing loud, distracting noise. They manipulated subjects' perception of personal control by telling half they could terminate the noise (if necessary) by notifying them through an intercom. The subjects who could have terminated the noise (but did not) were significantly more likely to comply with a later request by another individual to take some extra time (with no extrinsic benefits) to solve math problems.

Self-Efficacy

Self-efficacy refers to people's beliefs that they have the personal skills and resources to complete a task successfully (Bandura, 1977). Other individual difference factors relate significantly to this construct, including self-esteem (Rosenberg, 1965), locus of control (Rotter, 1966), and learned hopefulness (Zimmerman, 1990). Thus, research that showed more actively caring behavior from internals with high self-esteem (e.g., Bierhoff et al., 1991; Midlarsky, 1971) indirectly supported this factor as a potential determinant of actively caring behavior.

Zimmerman (1990) defined empowering experiences as experiences providing opportunities to learn skills and develop a sense of personal control. He proposed empowerment to be a product of learned hopefulness. In other words, people become empowered as they gain control and mastery over their lives and learn to use their skills to affect life events.

Optimism

Optimism is the learned expectation that life events, including personal actions, will turn out well (Scheir and Carver, 1985; Seligman, 1991). Researchers have manipulated optimistic states (or moods) among individuals by giving them unexpected rewards or positive feedback and then observing frequency of actively caring behaviors.

Isen and Levin (1972), for example, showed that individuals who found a dime in the coin return slot of a public phone (placed there by researchers) were more likely to help a stranger who dropped a folder of papers than were individuals who did not find a dime. Similarly, students given a cookie while studying at the university library were more likely than those not given a cookie to agree to help another by participating in a psychology experiment. Carlson, Charlin, and Miller (1998) reviewed these and other studies that showed direct relationships between an optimistic mood state and actively caring behavior.

Direct Tests of the Actively Caring Model

My colleagues and I conducted a series of studies to test the actively caring model in field settings, and the results were quite promising. We developed a safety culture survey (SCS) for industrial application, which includes a measure of each person factor hypothesized to influence actively caring (see Figure 12.1). The SCS also assesses the respondent's willingness to actively care in various ways (i.e., from a person, environment, and behavior focus). To date, we have administered this survey at over 100 industrial sites, and have shown rather consistent support for the actively caring model.

The step-wise regression analyses at our first nine sites resulted in high regression coefficients (i.e., .54 (n = 262), .57 (n = 307), and .71 (n = 207)) at the three plants studied by Geller and Roberts (1993); and .52 (n = 328), and .68 (n = 202) at the two plants studied by Geller, Roberts, and Gilmore (1996). The personal control factor was usually the most influential in predicting willingness to actively care, with group belonging predicting significant independent variance in actively caring propensity at all but one of the plants. Self-esteem, optimism, and self-efficacy have always correlated highly with each other and willingness to actively care, but usually only one of these factors predicted independent variance in actively caring (i.e., above that predicted by personal control and belonging).

In one test of our actively caring model (Roberts and Geller, 1995), we studied relationships between workers' on-the-job actively caring behaviors and prior measures of their self-esteem, optimism, and group cohesion. More specifically, employees (n = 65) were instructed to give their co-workers special actively caring thank-you cards (redeemable for a beverage in the cafeteria) whenever they observed a co-worker going beyond the call of duty (i.e., actively caring) for another person's safety. Those employees who gave or received an actively caring thank-you card scored significantly higher on measures of self-esteem, and group cohesion than those who did not give nor receive an actively caring thank-you card.

In another field study, my students asked individuals (n = 156) who had just donated blood at a campus location to complete a 60-item survey that measured each of the five person factors in Figure 12.1. The high return rate of 92% was consistent with an actively caring profile, but most remarkable was that this group scored significantly higher ($p < .001$) on each of the five subscales than did a group of students (n = 292) from the same university population (Buermeyer, Rasmussen, Roberts, Martin, and Gershenoff, 1994).

SUMMARY

This chapter illustrated the value of integrating behavior-based and person-based psychology to address the human dynamics of occupational safety and health. The purpose of this integration is to develop a culture in which everyone shares the responsibility of organizational safety by looking out for conditions and behaviors that are risky and then intervening to make them safe.

Techniques for increasing the occurrences of safe behavior and decreasing the frequency of at-risk behaviors have been developed and tested by behavioral scientists. However, these techniques are only useful if people use them on a regular basis, and this requires consideration of issues and concepts in person-based psychology.

For example, certain person states or individual conditions increase one's willingness to apply behavior-based techniques (e.g., observation, feedback, and rewards) for the benefit of another person's safety. In other words, these person states influence actively caring. Understanding these person states (i.e., self-esteem, belonging, self-efficacy, personal control, and optimism) can lead to the development of action plans to increase these states among members of a work force. Incorporating behavior-based strategies in these action plans can make them more effective.

Interpersonal coaching is a key strategy for increasing both safe behaviors and the person states that facilitate actively caring. In other words, when people interact effectively with others to support safe behaviors and correct at-risk behaviors, they not only activate beneficial behavior change, they also send a message to others that they are valuable team members (thereby building self-esteem, and belonging) and that people can make beneficial differences in the safety of an organization (thereby building self-efficacy, personal control, and optimism).

The five letters of coach represent a useful mnemonic for remembering an invaluable sequence of events that integrate behavior-based and person-based psychology to effectively manage the human element of occupational health and safety (i.e., **C** = care, **O** = observe, **A** = analyze, **C** = communicate, and **H** = help).

The first word is care because of the person-based principle that people will care what you know after they know you care. When people care enough to give supportive or corrective feedback on behalf of another person's safety, and such caring is recognized among individuals in a work group or throughout an entire plant, the next step (observation) is accepted and appreciated.

Careful and systematic behavioral observations are followed by an ABC analysis. In other words, the directive activators and motivational consequences of the situation are considered to understand causes of the observed behaviors and to develop interventions for increasing safe behaviors and decreasing at-risk behaviors.

After observation and analysis, the useful information is communicated effectively and objectively to the person observed. If the communication is constructive and accomplished in an atmosphere of win/win collaboration and mutual trust, it will be accepted and appreciated. Such actively caring communication will truly help the individual and the ongoing management of occupational safety and health.

REFERENCES

Bandura, A. *Social Foundations of Thought and Action.* Englewood Cliffs, NJ: Prentice-Hall, 1986.

Bandura, A. Self-efficacy: toward a unifying theory of behavioral change. *Psychological Review* 84: 191–215, 1977.

Barling, J. and Beattie, R. Self-efficacy beliefs and sales performance. *Journal of Organization Behavior Management* 5: 41–51, 1983.

Batson, C. D., Bolen, M. H., Cross, J. A., and Neuringer-Benefiel, H. E. Where is altruism in the altruistic personality? *Journal of Personality and Social Psychology* 1: 212–220, 1986.

Bierhoff, H. W., Klein, R., and Kramp, P. Evidence for altruistic personality from data on accident research. *Journal of Personality* 59(2): 263–279, 1991.

Blake, J. A. Death by hand grenade: altruistic suicide in combat. *Suicide and Life-Threatening Behavior* 8: 46–59, 1978.

Blanchard, K. H. and Johnson, S. *The One-Minute Manager.* West Caldwell, NJ: William Morrow & Co, Inc., 1982.

Buermeyer, C. M., Rasmussen, D., Roberts, D. S., Martin, C., and Gershenoff, A. B. Red Cross blood donors vs. a sample of students: an assessment of differences between groups on "actively caring" person factors. Paper presented at the Virginia Academy of Science, Harrisonburg, VA, May 1994.

Carlson, M., Charlin, V., and Miller, N. Positive mood and helping behavior: a test of six hypotheses. *Journal of Personality and Social Psychology* 55: 211–229, 1998.

Conger, J. A. and Kanungo, R. N. The empowerment process: integrating theory and practice. *Academy of Management Review* 13: 471–482, 1988.

Covey, S. R. *The Seven Habits of Highly Effective People.* New York: Simon and Schuster, 1989.

Covey, S. R. *Principle-Centered Leadership.* New York: Simon and Schuster, 1990.

Darley, J. M. and Latane, B. Bystander intervention in emergencies: diffusion of responsibility. *Journal of Personality and Social Psychology* 8, 377–383, 1968.

Deming, W. E. *Out of the Crisis.* Cambridge, MA: Massachusetts Institute of Technology, Center for Advanced Engineering Study, 1982.

Deming, W. E. *The New Economics.* Boston, MA: MIT Press, 1993.

Elder, J. P., Geller, E. S., Hovell, M. F., and Mayer, J. A. *Motivating Health Behavior.* New York: Delmar Publishers, Inc., 1994.

Geller, E. S. *Corporate Safety-Belt Programs.* Blacksburg, VA: Virginia Polytechnic Institute and State University, 1985.

Geller, E. S. Prevention of environmental problems. In L. Michelson and B. Edelstein, Eds. *Handbook of Prevention* New York: Plenum Press, 361–383, 1986.

Geller, E. S. If only more would actively care. *Journal of Applied Behavior Analysis* 24: 607–612, 1991.

Geller, E. S. *The Psychology of Safety: How to Improve Behaviors and Attitudes on the Job.* Radnor, PA: Chilton Book Company, 1996.

Geller, E. S. *Application of Behavior Analysis to Prevent Injury from Vehicle Crashes,* 2nd ed., Monograph No. 2, Cambridge, MA, Cambridge Center for Behavioral Studies Monograph Series: Progress in Behavioral Studies, 1998.

Geller, E. S. Building interpersonal trust: key to getting the best from behavior-based safety coaching. *Professional Safety* 44(4): 16–19, 1999.

Geller, E. S. Actively caring for occupational safety: extending the performance management paradigm. In C. M. Johnson, W. K. Redmon, and T. C. Mawhinney, Eds. *Organizational Performance: Behavior Analysis and Management.* New York: Springer-Verlag, 2000a.

Geller, E. S. Behavioral safety analysis: a necessary precursor to corrective action. *Professional Safety* 45(3): 29–32, 2000b.

Geller, E. S. *The Psychology of Safety Handbook.* Boca Raton, FL: CRC Press, 2001.

Geller, E. S., Berry, T. D., Ludwig, T. D., Evans, R. E., Gilmore, M. R., and Clarke, S. W. A conceptual framework for developing and evaluating behavior change interventions for injury control. *Health Education Research: Theory and Practice* 5: 125–137, 1990.

Geller, E. S., Eason, S. L., Phillips, J. A., and Pierson, M. D. Interventions to improve sanitation during food preparation. *Journal of Organizational Behavior Management* 2: 229–240, 1980.

Geller, E. S. and Lehman, G. R. The "buckle-up promise card": a versatile intervention for large-scale behavior change. *Journal of Applied Behavior Analysis* 24: 91–94, 1991.

Geller, E. S., Lehman, G. R., and Kalsher, M. R. *Behavior Analysis Training for Occupational Safety.* Newport, VA: Make-A-Difference, Inc., 1989.

Geller, E. S. and Roberts, D. S. Beyond behavior modification for continuous improvement in occupational safety. Paper presented at the FABA/OBM Network Conference, St. Petersburg, FL, January 1993.

Geller, E. S., Roberts, D. S., and Gilmore, M. R. Predicting propensity to actively care for occupational safety. *Journal of Safety Research* 27: 1–8, 1996.

Geller, E. S., Winett, R. A., and Everett, P. B. *Preserving the Environment: New Strategies for Behavior Change.* Elmsford, NY: Pergamon Press, 1982.

Gewirtz, J. L. and Baer, D. M. Deprivation and satiation of social reinforcers as drive conditions. *Journal of Abnormal and Social Psychology* 57: 165-172, 1958a.

Gewirtz, J. L. and Baer, D. M. The effects of brief social deprivation on behaviors for a social reinforcer. *Journal of Abnormal and Social Psychology* 56: 49–56, 1958b.

Goldstein, A. P. and Krasner, L. *Modern Applied Psychology.* New York: Pergamon Press, 1987.

Greene, B. F., Winett, R. A., Van Houten, R., Geller, E. S., and Iwata, B. A. *Behavior Analysis in the Community: Readings from the Journal of Applied Behavior Analysis.* Lawrence, KS: University of Kansas, 1987.

Isen, A. M. and Levin, P. F. Effect of feeling good on helping: cookies and kindness. *Journal of Personality and Social Psychology* 21: 384–388, 1972.

Katz, E. and Lazarfeld, P. E. *Personal Influence: The Part Played by People in the Flow of Mass Communication.* Glencoe, IL: Free Press, 1955.

Kim, J. and Hamner, C. Effect of performance feedback and goal setting on productivity and satisfaction in an organizational setting. *Journal of Applied Psychology* 61: 48–57, 1982.

Komaki, J. A., Heinzmann, T., and Lawson, L. Effect of training and feedback: components analysis of a behavioral safety program. *Journal of Applied Psychology* 67: 334–340, 1980.

Krause, T. R., Hidley, J. H., and Hodson, S. J. *The Behavior-Based Safety Process* (Second Edition). New York: Van Nostrand Reinhold, 1996.

Malott, R. W. A theory of rule-governed behavior and organizational behavior management. *Journal of Organizational Behavior Management* 12(2): 45–65, 1992.

Maslow, A. *Motivation and Personality.* New York: Harper & Row, 1970.

Maslow, A. *The Farther Reaches of Human Nature.* New York: Viking, 1971.

Mayo, E. *The Human Problems of an Industrialized Civilization.* Boston: Harvard University Graduate School of Business Administration, 1933.

McSween, T. E. *The Values-Based Safety Process: Improving Your Safety Culture with a Behavioral Approach.* New York: Van Nostrand Reinhold, 1995.

Michael, J. Distinguishing between discriminative and motivational functions of stimuli. *Journal of the Experimental Analysis of Behavior* 37: 149–155, 1982.

Michelini, R. L., Wilson, J. P., and Messe, L. A. The influence of psychological needs on helping behavior. *The Journal of Psychology* 91: 253–258, 1975.

Midlarsky, E. Aiding under stress: the effects of competence, dependency, visibility, and fatalism. *Journal of Personality* 39: 132–149, 1971.

Laughter could really be the best medicine. *National Safety News* April, 1985.

Ozer, E. M. and Bandura, A. Mechanisms governing empowerment effects: a self-efficacy analysis. *Journal of Personality and Social Psychology* 58: 472–486, 1990.

Parsons, H. M. Lessons for Productivity from the Hawthorne studies. *Proceedings of 1980 Human Factors Symposium* sponsored by the Metropolitan Chapter of the Human Factors Society. New York: Columbia University.

Parsons, H. M. What happened at Hawthorne? *Science* 183: 922–932, 1974.

Petersen, D. *Safe Behavior Reinforcement.* Goshen, NY: Aloray, Inc., 1989.

Phares, E. J. *Locus of Control in Personality.* Morristown, NJ: General Learning Press, 1976.

Roberts, D. S. and Geller, E. S. An actively caring model for occupational safety: a field test. *Applied and Preventive Psychology* 4: 53–59, 1995.

Roethlisberger, F. J. and Dickson, W. J. *Management and the Worker.* Cambridge, MA: Harvard University Press, 1939.

Rosenberg, M. *Society and the Adolescent Self-Image.* Princeton, NJ: Princeton University Press, 1965.

Rotter, J. B. Generalized expectancies for internal versus external control of reinforcement. *Psychological Monographs* 80(1), 1966.

Rushton, J. P. *Altruism, Socialization, and Society.* Englewood Cliffs, NJ: Prentice-Hall, 1980.

Rutkowski, G. K., Gruder, C. L., and Romer, D. Group cohesiveness, social norms, and bystander intervention. *Journal of Personality and Social Psychology* 44: 545–552, 1983.

Scheier, M. F. and Carver, C. S. Optimism, coping, and health: assessment and implications of generalized outcome expectancies. *Health Psychology* 4: 219–247, 1985.

Schroeder, D. A., Penner, L. A., Dovidio, J. F., and Piliavin, J. A. *The Psychology of Helping and Altruism.* New York: McGraw-Hill, Inc., 1995.

Seligman, M. E. P. *Learned Optimism.* New York: Alfred A. Knopf, 1991.

Sherrod, D. R. and Downs, R. Environmental determinants of altruism: the effects of stimulus overload and perceived control on helping. *Journal of Experimental Social Psychology* 10: 468–479, 1974.

Skinner, B. F. *The Behavior of Organisms.* Acton, MA: Copley Publishing Group, 1938.

Skinner, B. F. *Beyond Freedom and Dignity.* New York: Alfred A. Knopf, 1971.

Skinner, B. F. *About Behaviorism.* New York: Alfred A. Knopf, 1974.

Skinner, B. F. *Reflections on Behaviorism and Society.* Englewood, NJ: Prentice Hall, 1978.

Sulzer-Azaroff, B. Behavioral approaches to occupational health and safety. In L. W. Frederiksen, Ed. *Handbook of organizational behavior management.* New York: John Wiley & Sons, 505–538, 1982.

Ullman, L. P. and Krasner, L., Eds. *Case studies in behavior modification.* New York: Holt, Rinehart & Winston, 1965.

Vollmer, T. R. and Iwata, B. A. Establishing operations and reinforcement effects. *Journal of Applied Behavior Analysis* 24: 279–291, 1991.

Wandersman, A., Popper, P., and Ricks, D. *Humanism and Behaviorism: Dialogue and Growth.* New York: Pergamon Press, 1976.

Weick, K. E. Small wins: redefining the scale of social problems. *American Psychologist* 39: 40–49, 1984.

Whitehead, T. N. *The Industrial Worker.* Cambridge, MA: Harvard University Press, 1938.

Williams, J. H. and Geller, E. S. Behavior-based intervention for occupational safety: critical impact of social comparison feedback. *Journal of Safety Research* 31: 135–142, 2000.

Wilson, J. P. Motivation, modeling, and altruism: a person X situation analysis. *Journal of Personality and Social Psychology* 34: 1078–1086, 1976.

Zimmerman, M. A. Toward a theory of learned hopefulness: a structural model analysis of participation and empowerment. *Journal of Research in Personality* 24: 71–86, 1990.

## CHAPTER 13

Improving Safety Performance Through Cultural Interventions

Steven I. Simon and Rosa Antonia Carillo

CONTENTS

INTRODUCTION

Safety and health managers face increasing pressures from government and from consumer and community groups to meet tougher environment, safety, and health demands. This, coupled with downsizing, streamlining, and trends toward employee empowerment, have created the need for safety professionals to develop new frameworks. They must redefine their roles in flatter organizations and find new models to diagnose complex problems, design solutions, and do more with less.

Traditional reliance on safety engineering and identification of human error is no longer sufficient to meet these demands. In response, tremendous cultural changes are taking place right now in our organizations. Some companies are responding by moving tasks that were formerly performed by safety staff over to line managers. Others are moving responsibilities like accident investigations, audits, and communication to safety teams. In both these cases, the safety professional serves in a new role of facilitator, advisor, or team member.

1-56670-370-0/02/$0.00+$1.50

The cultural interventions described in this chapter have been used by companies throughout the United States to reduce accidents. This approach focuses on culture for two reasons: first, because reengineering and implementing a self-directed workforce require cultural change in most organizations, and second, because more and more research indicates that there is a correlation between safety performance and how well an organization manages its social aspects such as communication and cooperation. This makes it imperative for the safety manager to understand how culture affects organizations and to understand its impact on safety performance in an organization.

RESISTANCE TO CHANGE

One of the most difficult issues to manage in a change effort is the resistance to change. One obstacle is the rule-bound tradition inherited from a time when tight control over policies and procedures through enforcement and training was thought to be the ultimate answer to accident reduction. This tradition keeps organizations from making many of the changes needed to adapt to evolving conditions. Another obstacle is that many of the changes needed are viewed negatively by one or more levels of the organization. Some examples of the comments taken from interviews with employees, supervisors, plant managers, and corporate executives may make the dilemma more concrete (Simon, 1991, 1992, 1993).

Employees say that management is the problem. "Safety departments aren't providing the kind of support they used to provide. They (managers) cut back on safety staff and gave safety duties to line managers, but managers and supervisors aren't adequately trained to spot safety problems." Furthermore, "they've instituted cross-functional teams which means sending employees out to do jobs they're not properly trained to do safely. This sets up accidents waiting to happen."

Supervisors say that safety concerns are out of balance. "Sure, safety is important, but we have to keep our jobs, too. We don't want anyone to get hurt, but it's impossible to eliminate all risk. Now they (managers) want us to take work time out for behavior observation and positive reinforcement. Safety is its own reward. That ought to be enough."

Plant managers say one problem is corporate headquarters and another is the workforce. "Corporate sets tough production and cost-cutting goals that don't always support safety. Staff and budget cutbacks make it harder and harder to cover all bases even though the staff is working 60-hour weeks or more." Regarding the workforce, "Studies show that accidents are avoidable for the most part. It's just that people don't take personal responsibility for their own safety. With all the time we spend on safety, I don't understand why accidents still happen."

Corporate executives say the same pressures exist in all plants, but some plants have more accidents. "Since the same programs are available to all facilities, the problem must be in implementation. If we could just get all the plants to do what our best plants do, we wouldn't have a safety problem." Yet, programs that are successful in one plant too often fail when they are transplanted to another.

Strong delineation between "us" and "them" (i.e., union vs. management, plant management vs. corporate) is common across companies and blocks the ability of all parties to work together to find solutions to common problems. Bringing people together requires understanding how these attitudes or beliefs come into existence and how they are maintained. It is necessary to learn under what circumstances it is possible for an entire group to reexamine its attitudes and change them. Many of the conflicts described have their roots in the cultural norms and assumptions of organizational members. Thus, we need a framework to understand how culture forms and how it can change, and we need the skills to facilitize the change.

Some safety behavior consultants reject the notion that you should manage attitudes, and yet they speak about the importance of culture. They say, "Manage behaviors, not attitudes." "You can't see attitudes, but you can observe behavior. Get people to behave the way you want them to behave and their attitudes will change." These beliefs are based on Pavlov's and Skinner's behaviorism theories, as well as a fear of invading people's privacy.

This concern has merit, but it does not represent a complete picture of the cultural approach. It is not the intention of the authors to suggest that management involve itself in changing individual attitudes. Yet research indicates that group attitudes are formed by culture and that culture plays a crucial role in the productivity, quality, and safety performance of our organizations.

The problem with many behavioral safety programs is that they fail to take into account the influence of culture on individual behavior and the role of free will or consciousness on individual decision-making. This may be the reason that many behavior-based programs end up being perceived as "I Spy" programs where co-workers are expected to tell on each other (Simon, 1994).

The cultural approach, on the other hand, helps people realize that individual actions spring from group norms and traditions, as well as individual traits. Once people are aware of these influences, they can make better choices. Only then can they consciously create safer organizations without the use of manipulation or coercion.

In the following section, Changing the Safety Culture, the cultural approach provides safety managers with a new lens to get at cultural problems and design comprehensive solutions. Next, Changing the Organization to Support the New Culture, outlines the key areas that must be evaluated and aligned to support cultural changes.

CHANGING THE SAFETY CULTURE

Instead of pouring knowledge into people's heads, you need to help them grind a new set of eyeglasses so they can see the world in a new way. That involves challenging the implicit assumptions that have shaped the way people have historically looked at things.

John Seeley Brown
Vice President, Zerox Parc

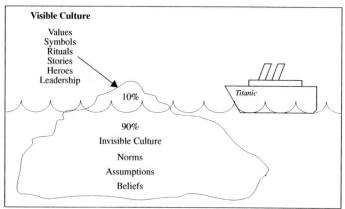

90% of a company's culture is invisible. It is the unspoken rules and beliefs that determine how the work gets done.

Figure 13.1   The culture iceberg.

The Culture Iceberg (Figure 13.1) illustrates the components of culture and shows why changing the culture is a great challenge. The visible part (values, symbols, rituals, leaders, stories, language), which translates into policies, procedures, and accident records, is only a small percentage of the whole. The much larger invisible part, norms and assumptions, is the real driver in organization behavior. The unaware manager, like the captain of the *Titanic,* who focuses only on the visible and ignores the invisible culture, will crash.

Starting with the visible culture, values are guiding principles, spoken or written, of a culture. Symbols, such as posted achievements, or rituals, such as safety meetings, would alert the newcomer to the idea that safety is important here. Stories might be told that represent positive or negative experiences with safety. These stories usually involve either heroes and heroines or villains.

Leadership is highly visible and is the most powerful component of culture. People look to leaders to set the example and set the tone for trust and open communication. Depending on the size of the company, several levels of leadership (including hourly workers) are needed to create a high-performance culture.

Below the iceberg, people experience their beliefs or attitudes about safety as personal and originating from within. However, there is substantial research showing that people's beliefs about how things work arise from the norms or expectation of the groups in which they hold membership.

Research on the importance of norms (defined as any uniformity of attitude, opinion, feeling, or action shared by two or more people) in determining attitudes toward production is well documented by Robert Black and Jane S. Mouton in *Productivity: The Human Side.* It was their opinion that the same phenomena, convergence and cohesion, which shape attitudes and/or norms toward productivity shape attitudes toward safety.

Briefly, convergence is the tendency of human beings to adopt the point of view of people whom they perceive as part of their group. Studies show that even if a group member initially disagrees with the group, given enough time he or she will adopt the group's point of view (or leave the group). The second force, cohesion, is our tendency to form groups with people we like and who hold opinions, attitudes, and ideas similar to our own.

This research on norms has deep implications for understanding how safety attitudes form and what one has to do to change them. If you train individuals in new behaviors or offer individual rewards for changing, but leave the group norms unchanged, the individual will most likely revert to the old behavior. Thus, to change organizational behavior, focus on changing the group norms, not the individual.

Culture as a management concept is fully explained in Edgar Schein's *Organizational Culture and Leadership* (1991). Schein goes beyond group norms to explain the fabric of social dynamics that shape and control the behaviors of the entire organization. Schein defines culture as the assumptions that a given group has invented, discovered, or developed for survival. Cultural assumptions can be thought of as explanations of how the world works, which is why cultural change is frequently extremely difficult.

For example, it was assumed at one time that the sun rotated around the earth. This assumption made people feel secure because it provided an explanation for something that was unverifiable at that time. When Copernicus tried to challenge that assumption, he was ostracized and ignored. The reaction to Copernicus is typical of our resistance to changing many assumptions. Our assumptions stabilize the world even if they are wrong. Giving them up often means dealing with anxiety until new ones take root. This anxiety is disruptive because people have difficulty concentrating on primary tasks while key assumptions are in transition. It is natural, then, for people to resist changes that produce anxiety.

How this dynamic applies to safety can be seen in the following example. Printers worked with certain inks and solvents that for years were thought to be safe. Respiratory illnesses began to show up in retired printers and were traced back to these chemicals, so respirators became required equipment for printers when they used those chemicals. Years after the medical information surfaced, printers at a major newspaper were refusing to wear the respirators. When interviewed, they stated that the equipment was too cumbersome to wear. When pressed about the long-term effects, they said, "We know how to take care of ourselves. We've been doing this for 30 years and nothing has happened."

In other words, work groups form assumptions about the way to stay safe based on their experience. These discoveries are passed on to new members as the correct way to do things. Once an assumption becomes a part of the culture, it drops into the unconscious, and rejecting it would require invalidating years of experience. This is unfortunate when it is based on faulty experience, as was the case in our printer example.

If it is difficult to change an assumption where medical evidence is available, imagine the difficulty in changing an assumption that is counterintuitive. One such example is changing the belief that safety is the safety department's responsibility. Because safety professionals are the experts, they are commonly viewed as the people most likely to enforce safety and implement changes. In reality, line management holds the authority and resources to enforce safety and line workers are the final implementers, but the old belief is held in place by years of experience and tradition.

Until now, we have discussed the importance of norms and assumptions, but not how to change them. It is extremely difficult to change assumptions because they are so deep within us that changing them feels like we are giving up our freedom or a part of our inner selves. So forget the notion of getting the behavior you want by establishing a written policy. Policies, like laws, do not determine behavior unless, of course, the policy is already a reflection of a cultural norm. Prohibition is a good example of a failed attempt to change behavior through laws. Seat belt usage, on the other hand, increased after laws were passed because many, many years of concerted effort had won significant support for seat belt use before the laws were enacted. If the culture is to change, a specific set of circumstances must exist.

First, there must be an urgent need to change and it must be communicated to people. Second, the resources and capabilities to change must exist. Third, an action plan to guide the transition from the old to the new culture must be developed. Leadership is responsible for detecting the need for change, communicating it to the organization, gaining consensus on solutions, and guiding implementation.

The safety manager's role in cultural change is to be a catalyst for change, a collaborator, advisor, and facilitator, but he or she cannot bring about change alone because the authority and resources for change lie in the hands of line management. Nevertheless, his or her specialized knowledge and experience play a critical role in designing and implementing the strategies to improve the safety culture. Two proven analysis and action planning models are described next — the norms-changing workshop and the organizational influences model.

Norms-Changing Workshop

The cultural change technology that works to change norms and assumptions beneath the iceberg consists of six steps:

1. Help people identify unsafe behaviors and articulate the unstated norms and assumptions behind those behaviors, why they came into existence, and why they need to change.
2. Get agreement on desired norms and assumptions from key players.
3. Create an action plan to change the norms and assumptions.
4. Acquire commitment and involvement of all levels of the organization for a specific action plan to change the norms and assumptions.
5. Set up the systems to support the new norms and assumptions.
6. Implement and follow up.

The norms-changing worksheet (Figure 13.2) can be used for steps 1 to 4. It is best to gather a group of 8 to 10 people who work in the same area. Keep managers and employees in separate groups because each group has different norms. Use the examples in Figure 13.2 to get people started or use your own, while following these instructions:

1. Write out a specific goal to improve a specific area of safety.
2. Identify unsafe behaviors/actions.
3. Identify the unstated norm(s) that drives the undesired behavior.

Unsafe Behaviors/Actions	Unstated Norms	Why Norms Exist	Desired Norms/Behaviors
Not locking out	I've always done it this way 25 yrs	We have actually been trained to take shortcuts	Lockout-Tagout
Reaching into moving equipment	I don't get done with my work as soon	Get longer breaks	Train the safe way
Bypassing the guards	Production is the first priority (Get the numbers)	Get more production, get more leeway	Production, quality, safety & dedication are all equal
Taping blades open		Design of some equipment goes against the job	
Climbing on equipment	Keep the line going	Belief it's not going to happen to me	Everyone work together
Not wearing gloves	Must be OK if no one says anything		Look out and speak up for each other
Using wrong tools	I've done it so long and nothing has happened	People pick up shortcuts	Use proper tool
		Negative Peer pressure — "Don't go too fast"	Pride in job

* This worksheet is an unaltered document that was prepared by hourly employees.

Figure 13.2   Norms-changing worksheet. The goal is to eliminate shortcuts and assume a safe attitude about hand safety.

4. State why the undesired norm exists.
   - What are the rewards for following the undesired norm?
   - Are there negative consequences for following the norm?
   - Are there negative consequences for going against the norm?
   - Are there problems in the design of the guards, procedures, or equipment that push people toward unsafe actions?
5. Specify desired norms (which is also the desired culture).

CHANGING THE ORGANIZATION TO SUPPORT THE NEW CULTURE

In this section, step 5 of the norms-changing process will be expanded to set up the systems to support the new norms and assumptions. For the new culture to take root, the formal organization must be examined and changed to support the new culture. This task requires a comprehensive model or map of the various components of an organization that influence safety performance because attempting to change the safety culture without getting support from the rest of the organization is not possible. Figure 13.3 is one model that breaks out the organizational factors that must be considered when reshaping a culture. Culture would be at the heart, while the supporting elements that produce the company's product or service surround the culture.

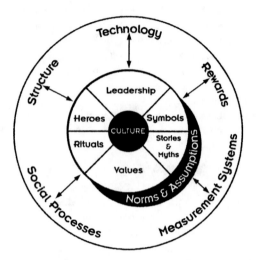

Figure 13.3   Model of organizational factors affecting cultural change.

Five organizational components surround the culture:

- Structure
- Technology
- Rewards
- Measurement systems
- Social processes

The arrows indicate that these components are a set of closely connected systems. Therefore, each component must support the others, much like the parts of a motor-cycle must work together. It is not enough to have the best engine, transmission, wheels, and so on. The various components of the motorcycle must fit with each other and complement each other. Likewise, the various parts and design elements of organizations must fit with each other to achieve the desired results.

Although all five organizational components are important, a recent research study conducted by the authors indicates an especially strong correlation between the perception of managers and employees of the leadership, social process, and technology organizational components and the actual safety performance. Culture ratings were obtained from nine sites of a national organization and then compared to workers' compensation costs at each site. The perception of employees and managers of the role of leadership, the degree of open communication, and the positive fit between work flow/work assignments and safety all correlated strongly to positive safety performance as measured by low workers' compensation costs. This research suggests that companies with high scores in these three areas tend to have excellent safety records.

A brief description of the function and content of each component follows with examples of what types of organizational changes might be required to develop safety excellence through a cultural change process:

- **Structure** — These are the policies, procedures, and formal communication structures (such as safety departments, committees, safety meetings, and training) designed to get individuals to perform tasks in compliance with organizational objectives. An example of structural changes to support a new culture is recognizing that safety is a line responsibility or empowering employee-led safety teams to handle safety functions formerly done by safety personnel.
- **Technology** — This represents the way the work gets done, the work flow, the type of equipment used, and how jobs are designed. An organization's safety technology focuses on safety equipment, job planning, and hazards correction. Larger companies are additionally focused on information systems. An example of technological change is using computers and videos to communicate in place of the training and administrative tasks usually performed by the safety function.
- **Rewards** — National surveys indicate that people do not feel rewarded or recognized for safety performance. The rewards system provides a great deal of leverage for change because people perform in accordance with their expectation for rewards which may be extrinsic (bonuses, incentives) or intrinsic (job satisfaction, one's own health). For example, when rewards for production outweigh rewards for safety or quality, production will be the higher priority. Care must be taken in changing reward structures to reward desired outcomes. Improper incentive designs have been known to encourage employees to hide minor accidents and suppress information.
- **Measurement Systems** — A recent survey at a major chemical company indicated that safety managers felt that using the injury incidence rate as the sole measure of safety performance was antiquated and a morale detractor. They felt workers' compensation costs and severity rates were good after-the-fact measures, but did not contribute to accident prevention. Changing these after-the-fact measures to process measurements such as perception surveys and statistical process control can help prevent accidents by pointing out weaknesses in the system. These systems can also be used to measure progress in culture change efforts.
- **Social Processes** — These include the informal communication systems, the degree of employee empowerment and participation in safety, cooperation, conflict, politics, openness, and trust. The social processes of an organization deeply affect morale, productivity, and safety performance. Any attempt to improve performance requires examination of these processes and change as necessary. An example of changing a social process to improve safety is empowering employees to issue work permits, a task normally done by supervisors.

Figure 13.4 applies the model to safety performance. It lists five analytical questions to help determine if the organizational factors are supporting the desired safety culture. Going through the model asking these questions will indicate which systems are out of balance. The next step is to determine how each factor needs to be changed.

Use of this model to change the culture of an organization will be demonstrated by showing the results from a petrochemical division meeting where safety supervisors met to discuss how to successfully implement an accident investigation process to conduct accident root cause analysis called "TapRoot."

The TapRoot objectives are:

- Discovering the root cause of accidents
- Fact finding, not fault finding

Do we have the tools, equipment
and knowledge that we need to
support the desired culture?

Are the current organi-
zations for safety,
including policies and
procedures, set up to
support the desired
culture?

Is line management
responsible for safety?

Are desired cultural
outcomes rewarded?

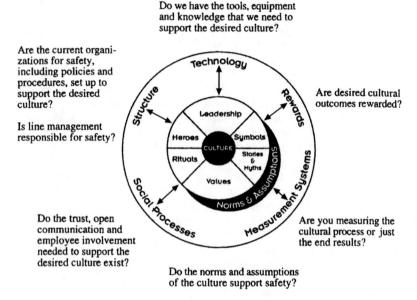

Do the trust, open
communication and
employee involvement
needed to support the
desired culture exist?

Are you measuring the
cultural process or just
the end results?

Do the norms and assumptions
of the culture support safety?

Figure 13.4   Organizational factors influencing the safety culture.

- Increasing near-miss reporting
- Employee and supervisor involvement.

The safety supervisors anticipated cultural as well as organizational resistance, since the new process would require employee involvement and plant resources.

As Figure 13.5 shows, they examined each aspect of the model and asked what would have to change to support TapRoot to achieve the desired outcomes from the TapRoot process. By way of illustrating the next step, here are their answers to the Social Process question and the steps the safety managers felt would have to be taken in order to align the social process to support TapRoot:

Does the trust, open communication, and employee involvement needed to support TapRoot exist?

1. Current state of social processes
   a. There is a climate of fear (vs. trust).
   b. There is a lot of communication of inaccurate information.
   c. Our personnel subscribe to the sooner or later myth (sooner or later there will be an accident).
   d. Our employees view their relationship to management as Us vs. Them.
   e. Our safety meetings have the tone of a Spanish inquisition.
   f. Injured people are viewed as heroes or martyrs.
   g. Leadership needs to establish safety as a core value, not just a compliance requirement.

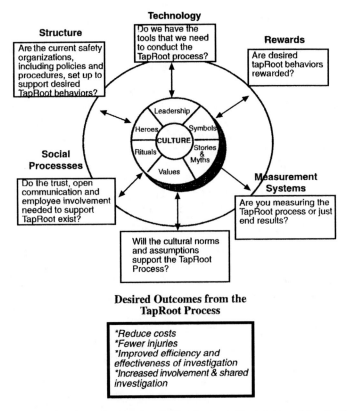

Technology
Do we have the tools that we need to conduct the TapRoot process?

Structure
Are the current safety organizations, including policies and procedures, set up to support desired TapRoot behaviors?

Rewards
Are desired tapRoot behaviors rewarded?

Leadership
Heroes   Symbols
CULTURE
Rituals   Stories & Myths
Values

Social Processses
Do the trust, open communication and employee involvement needed to support TapRoot exist?

Measurement Systems
Are you measuring the TapRoot process or just end results?

Will the cultural norms and assumptions support the TapRoot Process?

Desired Outcomes from the TapRoot Process

*Reduce costs
*Fewer injuries
*Improved efficiency and effectiveness of investigation
*Increased involvement & shared investigation

Figure 13.5   Analyzing the culture and systems to support installation of the TapRoot investigation process to achieve desired outcomes.

2. Alignment of social processes needed
   a. Break down the fear and begin to build trust, communicate history of incident occurrences, and show lack of recriminations.
   b. Get top management to enforce value of accuracy and fact-finding during incident investigation; communicate their endorsement to mid-level managers/supervisors with history of management by fear climate.
   c. Communicate success stories from last year's investigations to create new stories.

MAKING THE CHANGE HAPPEN

- **Involvement is the key to cultural change.** Involve members in identifying current norms and assumptions and in exploring alternative solutions. Include the people affected by the change in discovering why the change is needed, what the obstacles are to changing, and how they might be overcome.
- **Change leadership is in the hands of those who control the resources.** A cultural change effort must include all the leaders of the existing culture. That usually means top management, union leaders, and key opinion leaders from the ranks. Without strong support from leadership, the culture will not change.

- **There has to be a convincing reason for the change.** Leaders must answer the question, "Why change?" The only acceptable answers will arise out of business necessity. As much as we would like to think that doing the right thing as a reason for change should work, it does not always seem to. This is because change is hard work; it is disruptive and it is expensive.
  - All successful change efforts are based on a business need such as lowering workers' compensation costs, reducing Occupational Safety and Health Administration (OSHA) and Environmental Protection Agency (EPA) fines, fear of damage to reputation in the community, or loss of business to competitors.
  - One of the key roles for a safety professional is making management aware of the need to change. Tools to accomplish this task include educating managers on the effect of their behavior on the safety culture, the cost of accidents to overall morale and productivity, and the benefits to the bottom line of accident prevention. If management will not listen, bring in an outside consultant to support you, use perception surveys to gather cultural data, and educate yourself with case histories of the successes and failures of other companies facing similar issues.
- **Have a plan to get there.** You need a vision of where you want to go. This is the end result. Then, collect objective data to determine where you are now. Next, make a step-by-step plan to get where you want to go. One valuable tool for determining where you are now is a perception survey that measures norms and assumptions as well as management systems.
- **No one program fits all.** There are many factors to consider, such as where the organization is starting from, before choosing an approach. If the organization has not allowed much participation, designing campaigns may be the place to start. There is a continuum that must be followed before attempting employee empowerment.
- **Work on the norms, beliefs, and assumptions** of the invisible culture as well as the technical and administrative aspects of the organization.

SUMMARY

The cultural view of the organization described in this chapter presents a number of implications for improving safety performance. First, it implies a challenge to safety professionals to become agents of behavioral and cultural change. This is based on the recognition that competence in managing the social forces in organizations is equally important to technical competence in achieving safety excellence.

Second, this is a contingency view — that is, the right design for one company, division, or plant may not necessarily be the right one for another. What techniques or methods you choose depends on the technology, the people and assumptions of your organization, and challenges your company is facing.

Third, it implies that safety managers must utilize a comprehensive organizational framework to analyze safety performance and determine the root causes of problem. No longer can they rely on single-factor methods or theories to prevent accidents. Only from a comprehensive systems perspective is it possible to understand how to affect long-term safety improvements. This perspective will help managers realize why simplistic approaches such as changing a job description or changing a policy are ineffective.

Finally, this chapter provides a conceptual framework for safety leadership that extends beyond the traditional realm of management. It stretches the manager from policy maker to culture shaper. Yet, no leader stands alone. Without followers there can be no leadership. So, while it is true that a company would be unable to change its culture without a leader, a leader cannot change a culture without the support of the people. It takes everyone working together to make the change. The challenge to the leader is to involve, coach, and empower the employees to understand, accept, and act on the need to change.

FURTHER READING

Books

Beckhard, R. and Harris, R. T. *Organizational Transitions,* 2nd ed., Reading, MA: Addison-Wesley, 1978.

Beckhard, R. and Pritchard, W. *Changing the Essence.* San Francisco: Jossey-Bass, 1991.

Blake, R. and Mouton, J. S. *Productivity: The Human Side.* Self-published, 1981.

Bryson, J. M. *Strategic Planning for Public Nonprofit Organizations.* San Francisco: Jossey-Bass, 1988.

Cummings, T. G. and Huse, E. *Organization Development and Change.* St. Paul, MN: West Publishing, 1989.

Nadler, D. A., Gerstein, M. S., Shaw, R. B. et al. Eds. *Organizational Architecture: Designs for Changing Organizations.* San Francisco: Jossey-Bass, 1992.

Nadler, D. A. and Tushman, M. L. *Strategic Organization Design.* Glenview, IL: Scott, Foresman, 1988.

Nadler, D. A., Gerstein, M. S. Shaw, R. B. et al. Eds. *Organizational Architecture: Designs for Changing Organizations,* San Francisco: Jossey-Bass, 1992.

Perrow, C. *Normal Accidents: Living with High-Risk Technologies.* New York: Basic Books, 1984.

Schein, E. H. *Organizational Culture and Leadership.* San Francisco: Jossey-Bass, 1991.

Simon, R. A. and Simon, S. I. *Grassroots Safety Leadership: A Handbook for Designing and Implementing Culture-Based Safety Improvement Strategies.* Seal Beach, CA: Culture, Change Consultants, 1993.

Thompson, A. A., Jr. and Strickland, A. J., III. *Strategic Management Concepts and Cases,* 5th ed., Homewood, IL: Richard D. Irwin, 1990.

Other Publications

Asch, S. E. Studies of independence and conformity: a minority of one against a unanimous majority, *Psychol. Monogr.,* 70(9): No. 416, 1956.

Sherif, M. A study of some social factors in perception. *Arch. Psychol.,* 187, 1935.

Simon, R. A. and Simon, S. I. The Relationship of Culture Perception to Worker's Compensation Costs at Nine Sites. Culture Change Consultants, Seal Beach, CA, 1994.

Simon, R. A. An Open Systems Approach to Improving Safety Performance. Master's thesis, Pepperdine University, Malibu, CA, 1993.

Simon, R. A. Unpublished Safety Culture Survey Reports, 1991, 1992, 1993, 1994.

Tarrants, W. E. Emerging trends in safety and health, in *Accident Prevention Manual for Business and Industry,* 10th ed. Itasca, IL: National Safety Council, 1992.

## CHAPTER 14

A Holistic, Integrated Approach to Improving Unsafe Attitudes and Behaviors Can Produce Lasting Change

Michael D. Topf

CONTENTS

INTRODUCTION

Safety, health, and environmental improvement processes that focus on changing employee attitudes, thinking, and behaviors are effective in preventing accidents, injuries, health issues, and environmental incidents.

Safety, health, and environmental professionals and business leaders continue to search for methods to reduce the impact that work-related injuries and incidents have both on the physical well-being of employees and the bottom line of companies. Managing safety, health, and environmental performance through the use of holistic, integrated approach is an idea whose time has come. It is no longer a question of whether to address safety, health, and environmental behaviors, but what is the best methodology to use.

Implementing effective employee safety programs that address behaviors is one way to improve safety performance, and many such programs have had an impact. The emphasis in the past has been primarily on external regulations and internal policies and procedures — what to do and how to do it. In most companies today, all levels of employees generally know what to do and what not to do regarding personal health and safety, as well as environmental performance. The problem is they do not always accept and comply with what is required. Other priorities from within and outside oneself also offer many opportunities to make unsafe choices on and off the job. Awareness and attention factors also play a large role in causing injuries and health or environmental incidents. Most companies are demanding improvements beyond those that current programs have achieved. Mere incremental and short-lived improvements, often achieved through traditional behavior-based improvement approaches, are no longer adequate for many companies, which now seek what amounts to quantum changes in safety, health, and environmental performance. Another issue is that in their zeal to lower their incident rates, many leaders do not promote the kind of environment where employees feel open to accurately report incidents and near misses. They can work to the overall disadvantage of all employees when fear of consequences causes people to suppress their communication, participation and involvement in improvement efforts.

Traditional behavior-based strategies, which emphasize third-party observation and correction of others' behaviors, have their limitations. Although there may be a brief period during which behaviors change and safety performance improves, these results often diminish when direct supervision or constant reinforcement are no longer present. They are also ineffective for generalizing safe performance to other job situations than those for which the person was trained as well as off-the-job behaviors. They are a fine tool to use to reinforce safe behaviors, but are most effective when used after a foundation of constructive safety attitudes and beliefs has been established not as a means to an end. A strict behavioral-based approach does not deal with our cognitive powers and the human ability to rationalize, explain, and excuse their behaviors for a multitude of motivating factors.

Due to the trend in industry today to place greater levels of responsibility on individual line employees, along with a high degree of unsupervised, independent working situations and off-the-job accidents and injuries, companies need an innovative approach to the self-observation and self-management of risk behaviors as well as the management and support of others, whether line or management employees.

There is controversy today over whether the quest for behavioral safety, health, and environmental improvements should begin with a cognitive approach to changing people's attitudes, beliefs, and values to produce a positive change in behavior, or whether it should begin with a behaviorist approach to changing specific safety-related behaviors. With the latter the hope is that these will turn into new habits that will generalize to other work situations, and in the long run, attitudes, beliefs, and values will also change for the better. Can increased awareness and attitude changes create behavior change? Can cognitive methods ever be considered to be effective behavior improvement methods? The answer is yes. Whatever will work to change unsafe attitudes and behaviors is valid. Methods to increase awareness as a starting point, coupled with a cognitive approach to educate people and provide them with insights, ideas, and knowledge that addresses the attitudes, beliefs, and thinking that underlie their behaviors have been used successfully for many years.

People can be taken through an introspective process that allows them to rise above prior conditioning by recognizing which attitudes, beliefs, and behaviors work and don't work for their own and others safety, health, and well-being as well as environmental excellence. They then can change their own behaviors based on their value systems and their key motivating factors. Behavior-based observations are to few and far between to provide the constant, conscious level of awareness that is necessary to be most effective in preventing incidents. People need methods and skills to recognize their own unsafe thinking and behaviors any time, any place. This includes management of line employees!

A holistic, integrated behavior improvement approach is required for lasting improvements. This involves a seamless process, which includes five types of behavior-improvement strategies. These strategies interact to uncover the attitudes and beliefs that drive unsafe behaviors, and then increase individual responsibility for the self-management of safe behaviors. The process then continues to shape a reinforcing culture in which achieved changes can be sustained.

This awareness approach to attitude and behavior improvement includes:

- **Cognitive behavior management** — Understanding how attitudes, beliefs, values, thoughts, and other aspects of people's thinking process contribute to both automatic and calculated behaviors that are unsafe; self-observation and management are key
- **Affective behavior management** — Discovering how our emotions and feelings about ourselves and others and our situations and circumstances inform and influence our choices and behaviors
- **Reality behavior management** — Using peer, leadership, and management strategies along with coaching and counseling techniques in dealing with current behaviors and their consequences help achieve a shift to personal responsibility and self-management leading to realistic corrective action plans
- **Management by commitment** — Once individuals realize that they are inherently committed to safety, health, and the environment, and their labor and management leadership shares that commitment, they are motivated by the mutual benefit to themselves, the company, and their community to act more safely in greater cooperation with company requirements.
- **Traditional behavior management** — Once a foundation is laid consisting of essential values, beliefs, attitudes, and behaviors, observation and feedback methods are a valuable means to support and sustain what has been implemented on an on-going basis.

The interplay of these three elements — thought, action, and commitment — can help all levels of individuals in organizations become aware of the thought processes and interpersonal influences that cause people to place themselves and others at risk, either by acting automatically or in a calculated manner. The process continues to shape actions consistent with this awareness and to provide an environment in which safe behavior is supported and reinforced.

Improvement processes that focus on employee awareness, attitudes, and thinking are effective tools for significantly preventing injuries and incidents on and off the job. These processes provide a substantial return on investment in terms of employee productivity and improve an organization's competitiveness.

An awareness and cognitive approach to attitude and behavior improvement, unlike strict behavior-based techniques, makes people think about what they are going to do before they do it. People feel more comfortable about their actions.

Safety programs focusing only on changing behavior without addressing awareness and attitudes often provide short-term results and little return on investment if safe behaviors don't last.

Many programs have a positive impact on safety and then they begin to fade. The strictly observation-based, behavior-change programs tend to distract employees because they are invasive and may also develop negative feelings toward being observed which can create resistance to the process.

EMPLOYEE SAFEGUARDS

Enhancing all employees' abilities to be safe can be accomplished by providing line employees, supervisors, and managers with the necessary skills to manage themselves and others to maintain and improve safety, health, and environmental performance. It is important to understand that most accidents, injuries, and health and environmental incidents do not exist from exotic causes. People are not scientific in their behaviors. We need simple, yet effective, practical awarenesses and skills to observe and manage within ourselves and others the kinds of attitudes and behaviors that cause injuries.

Behavior improvement and management can be accelerated greatly by focusing on the following:

- **Safety and awareness** — Learning techniques to observe one's own safety attitudes, thinking, and behaviors will help people learn how to identify and heed early warning signs that can lead to an incident.
- **Managing risk behaviors** — Heightening awareness of attitudes, beliefs, and values, can lead to behavior change. Attitude change can precede behavior change. People become motivated to make safe choices anywhere.
- **Safety and the thinking process** — Learning to control and manage automatic responses for appropriate behaviors can improve safety health and environmental performance.
- **Personal responsibility** — Increasing the level of ownership or responsibility for safety, health, and the environment is the key to effective individual actions.
- **Leadership's role** — Enabling others to manage risk behaviors constructively can be accomplished by providing essential skills for empowerment and behavior management. Management and labor leadership need to join forces to lead, guide, and direct all levels of employees to take responsibility for safety, health and the environment.
- **Safety and personal commitment** — People can be helped to recognize what is at stake for themselves, their families, their co-workers, and the company, as well as how to make appropriate choices. No one wants to get hurt, we just have attitudes and beliefs that lead us to think we can take risks and get away with it.

THE ROLE OF RESPONSIBILITY

Preventing incidents requires awareness and training that emphasizes safety as a function of personal responsibility. It is necessary to address the common understanding of what responsibility is along with people's general resistance to accepting it regarding safety, health, and the environment. The general perception relates responsibility to blame rather than a recognition of one's own role and power in ensuring our own and other people's safety, health, and well-being. Individuals can learn to take responsibility for their safety and health, as well as the safety and health of others. When people are truly empowered, they understand their ability to protect themselves, their co-workers, and families and take the initiative to take whatever actions necessary to accomplish this. This requires an attitude change for many people, which is accomplished by providing an understanding of the beliefs that create a negative perception of what responsibility is. Structures and systems need to be set up that will allow employees to take constructive actions and make a difference in their safety improvement process.

HUMAN MECHANISMS

Researchers, production, maintenance, administration and sales employees (both line and management) are making decisions every day that can allow risky behaviors to cause injury to themselves and others. Whether the decision is to stand on a chair in order to get something off a shelf or to not lockout equipment, the issues of inconvenience, the fear of losing time, and the belief that "it will never happen to me" all contribute to unsafe behaviors.

One key to implementing successful safety, health, and environmental improvement programs is to train people to manage themselves by observing their own behaviors. This is not as complex as it may appear. Most people are very poor observers of their own "play" because they are key characters in it. An awareness approach to behavior management addresses the basic human mechanisms or attitudes, beliefs, and behaviors that lead individuals to place themselves and their co-workers at risk, and enables them to observe their thought processes, actions, and consequences.

The following human mechanisms can effect a person's ability to stay safe:

- **Inattention** — daydreaming, distractions, or thinking about things other than the job, not listening to instructions, and losing focus while working on repetitive or familiar activities
- **Decreased awareness** — not remaining alert to danger at all times and not being on guard for unexpected dangers because of complacency and inattention from the above noted issues
- **Lack of focus** — difficulty concentrating on tasks increases the possibility of accidents and injuries
- **Making distinctions** — unable to see, hear, and smell potential safety hazards as a result of thinking, attitudes, and behaviors and thus being unable to take appropriate actions to remain safe

- **Taking responsibility** — not responding effectively when safety demands immediate action, due to an inability to understand how to change attitudes that may
  encourage or block the appropriate response for safety
- **Resisting change** — resisting certain safety behaviors and feedback from others
  due to past negative experiences, attitudes, and beliefs regarding safety; not changing habits as needed for safety because of a lack of awareness of what is really
  at stake

Belief systems can cloud a person's ability to deal with current reality and recognize
the true level of risk or one's ability to protect themselves regardless of the circumstances. Macho attitudes or the "I'll do it my way" syndrome can interfere with a
common sense approach to safe behavior.

- **Heeding early warning signs** — ignoring the mental alarm system, or the inner
  voice, that can alert people to danger and necessary actions to avoid accidents,
  injuries, and health and environmental incidents
- **Keeping safety agreements** — ignoring and/or not communicating and discussing
  safety rules and regulations that people don't agree with or think are necessary
  put them at risk

Once again, attitudes and beliefs that cause people to resist rules or do it their own
way for a variety of reasons can contribute to noncompliance and incidents.

INCREASING AWARENESS

A responsibility-based, awareness approach is a viable alternative that enhances
existing safety programs, activities, and strategies. It focuses on first assessing and
redesigning the existing culture and environment of an organization so that safety,
health, and the environment become essential elements in productivity.

It is essential to consider safety, health, and the environment as key aspects of
an overall quality management system. As a result of increased safety awareness,
one can expect to see increased productivity and quality.

Approaching behavior improvement from this perspective places value on previously established safety requirements, while also emphasizing personal well-being,
which puts safety programs into proper perspective for all employees. Wearing safety
goggles, for example, is not meant to please the bosses, but to protect one's eyes
from being injured. Rather than cut corners, management must communicate that
the company wants all employees to take the time to do a job safely.

EMPHASIZING ATTITUDE

In the ideal world, implementing only safety tactics, such as rules, regulations,
and procedures, can be successful in reducing and preventing injuries, if people
adhere to these tactics consistently. Unfortunately, this ideal world does not exist.

Changing attitude as well as behavior, by emphasizing awareness, has proved
successful in increasing safe behavior and productivity on an ongoing basis. The

result is that people have a higher regard for personal health and safety as well as environmental excellence.

One maintenance worker reported a major shift in his attitude after participating in this awareness approach to safety training that we conducted. Prior to the training, he often would perform repair jobs without protective equipment, because obtaining and using the equipment was inconvenient. During the training, he became aware of his belief that nothing would happen to him because of his years of experience. He also had the attitude that safety was just for the company to protect them from any liability. He reported that during the process of the training he had an insight into his existing beliefs and recognized that safety was really for him, not just for the company. "It's my eyes and body that are at stake." "Safety is for me." As a result of that awareness, he now determines which protective equipment he will need before the repair job and brings it with him to the site. "Even if it's a one-in-a-thousand chance of getting hurt, it's not worth it to me. I value my health too much," he said in a follow-up interview.

MEASURING SAFETY

Safety can be measured by qualitative as well as quantitative measures. Traditional quantitative measures such as lost times, recordable rates, and minor injuries, are not necessarily an accurate measure of safety performance. Behavior-based programs that focus only on observing the behavior of others must, by definition, rely on quantitative results to measure safety and productivity. Results are charted by observing the frequency or occurrence of safe or unsafe behaviors. Although these methods are very important, relying on these traditional measurement techniques for after-the-fact Occupational Safety and Health Administration (OSHA) incident reporting, as well as, downsteam behavioral tracking does not provide a complete view. A low incident rate does not mean people are behaving safely, nor does the observation of compliance to requirements mean that the person has developed the attitudes and beliefs that will stay with that person wherever he or she goes, influencing their behavior in a positive way when working independently at work or at home, nor does it translate into other domains of quality and production.

Along with the reduction in recordable injuries and first aid cases and observation of safe behaviors positive changes can be observed in people's attitudes and willingness to work with the safety process and contribute to on-going continual improvement efforts out of their personal commitments to themselves and others. An indication of this is the observation of greater activity, participation, and cooperation from all levels of employees working together on corrective-action teams that deal with safety problems and issues.

Analyzing qualitative results is an essential method of measuring the effectiveness of safety programs, especially those that focus on attitude change. Many behavior-based methods do not account for qualitative results because measurable statistics for comparison are not available. Qualitative results, which indicate a change in attitude, can be seen by observing the following types of behaviors:

- Increased participation at safety meetings; voluntary discussion of on- and off-the-job hazards, behaviors, and improvements
- Increased natural caring and self-motivation for safety and health through greater use of personal protective equipment and compliance with company requirements
- Increased discussions on the importance of safety, health, and the environment to the individual and company
- Increased acceptance of feedback from management, safety representatives, and other workers regarding safety, health, and environmental procedures
- Increased willingness of employees at all levels to be proactive regarding safety, health, and environmental issues
- Greater solicited and unsolicited communication of safety issues between line employees and management, and among departments during normal work routines
- Increased solicited and unsolicited reporting of observed hazards to maintenance and engineering departments, along with potential solutions

Behavioral changes often are noticed in comments from employees. For example, many employees state that they wear safety glasses at home while cutting the grass, which they never did prior to the emphasis on awareness and personal responsibility as a part of the safety process.

Although attitude is difficult to measure directly, it can be more accurately observed through the employees' cooperation, moods, communication and behaviors, and it is important in evaluating the effectiveness of safety programs.

ENVIRONMENT AND CULTURE

Organizations, like individuals, possess certain traits and prevailing attitudes that are a result of historical backgrounds, morals, and philosophies. Individuals also are influenced by company values, attitudes, and priorities regarding quality, safety, and production. It is essential to assess the existing culture and environment for the various perspectives on norms, values, attitudes, and beliefs regarding safety in order to determine influences on employee thinking and behavior.

For example, the culture at some organizations may be so production-oriented that risk taking is considered acceptable to get the job done. Supervisors often stress the end result without mentioning that safe procedures are essential to getting the job done well or in a quality manner. Workers often cut corners to remain productive, rather than take the time to do the job safely. When left unaddressed by management or supervision, unsafe behavior can be viewed as being encouraged and considered acceptable. Behavior change includes managers and supervisors.

Eliminating an environment that condones risk behavior requires an organization to focus on changing the prevailing attitudes that influence behaviors. Rather than emphasizing statistics alone, organizations from the CEO down, must communicate that the primary concern is the human being behind the statistic.

Safety processes must explore and emphasize the underlying causes of the behaviors that lead to injuries. This is best accomplished by allowing employees to share experiences with other employees from an educational point of view, not a punitive one. They must feel safe to do this. Safety professionals, managers, and supervisors must change their approaches in order to involve a nonpunitive, educational discussion in which an injured employee can explain to others how the injury occurred. When open communication can occur, incidents or near incidents can be valuable learning tools when the informantion shared can be used to prevent others from being injured. As a result of this type of approach, individuals should learn to prevent future accidents and injuries and continue to do so in an environment that provides reinforcement for safety, health, and environmental learning.

COMMITMENT AND COMMUNICATION

To prevent incidents in an organization, there must be a genuine concern for the safety, health, and well-being of every employee. Increasing productivity, while emphasizing safety, health, and the environment requires everyone in an organization to remain committed to safety by proactively refining and improving the safety process.

To encourage commitment, it is important for top-level management to clearly define and communicate the importance of safety and what is required so that individuals take responsibility for their safety, health, and well-being and the safety, health, and well-being of their co-workers. In addition, managers and supervisors as well as labor or line safety representatives must be skilled in leading others for safety, health, and environmental excellence. Finally, all employees must be aware that they have a direct impact on the overall safety of the organization.

LEADERSHIP FOR SAFETY

To create the kind of safety, health, and environmental culture where people are willing to observe and give and receive feedback from anyone regardless of his or her position, requires an attitude shift. This attitude shift gives every man and woman the perspective that he or she is a leader when it comes to safety, and each individual has a responsibility for every person's safety.

It is essential to provide leadership training and skills for line personnel, management, and labor safety leadership as part of an overall safety program. Unlike many leadership courses, a holistic, integrated behavior improvement process also focuses on the underlying human mechanisms that cause leadership to resist taking actions or avoid following through on managing safety on a moment to moment basis.

Company safety leaders are encouraged to intercede immediately upon discovering an unsafe behavior or condition, based on an understanding of the potential consequences of delaying their involvement. They must learn to provide constructive feedback in a coaching manner so that employees will take responsibility for correcting their behaviors. If done in a noncritical manner, employees tend to be less defensive and learn to take responsibility for their actions and behave in accordance with the requirements.

Once again, management must support this process to demonstrate its effectiveness and credibility. In addition, it is important to ensure that leadership applies the same requirements for everyone, not just the line staff.

Leadership's accountability for safety can be approached from two perspectives:

- **Preventive** — properly identifying potential hazards due to unsafe conditions, attitudes, and behaviors, reinforce existing safety requirements, resolve problem areas, and change unsafe behaviors.
- **Post-incident/injury assessment and correction** — determining the accountability and responsibility of the individuals involved in an incident and implementing strategies to prevent further unsafe acts. (It is essential that this aspect be handled from an educational perspective to ensure employee cooperation and involvement.)

Leadership's role in safety, health, and environmental awareness and improvement includes encouraging participation, and sharing and developing ideas and concepts with everyone who is at risk within an organization. This provides purpose and meaning to participating in the process of increasing safety, health, and environmental excellence.

A holistic, integrated process that combines a variety of behavioral and cognitive methods and strategies will stand the best chance of working effectively with all the individual differences in an employee population. This process will assist greatly in the development of the kinds of positive safety, health, and environmental attitudes and behaviors that open the possibility for breakthroughs in performance to occur.

ACKNOWLEDGMENTS

My thanks, appreciation, and acknowledgment to the many men and women who, through their commitments to themselves and each other, have made significant differences in their safety, health, and well-being.

FURTHER READING

Berger, L. and Sikor, J. *The Change Management Handbook: A Road Map to Corporate Transformation.* New York: Irwin, 1994.

Fournies, F. and Plunkett, C. *Participative Management: Implementing Empowerment.* New York: John Wiley & Sons, 1991.

Fournies, F. *Why Employees Don't Do What They're Supposed to Do.* City: Liberty Hall Press, 1988.

Rokeach, M. *The Nature of Human Values.* New York: The Free Press, 1973.

William, R. M. Values. In *International Encyclopedia of the Social Sciences,* B. Sills, Ed. New York: Macmillan, 1968.

## CHAPTER 15

The Basic Essential of Safety Management: Supportive Organizational Culture

Judith A. Erickson

CONTENTS

1-56670-370-0/02/$0.00+$1.50
© 2002 by CRC Press LLC

INTRODUCTION

An organization is a complex institution with many different people who have diverse education, skills, and experience. These people perform many types of tasks to ensure the success of a company's operation. Safety is only one element that functions in an organization. Unlike most of the other operations in a company, this safety effort is part of nearly every program, activity, and department at one time or another.

There is overwhelming scientific evidence demonstrating that the organization, through its culture, determines the level of safety performance that will be achieved.[1,2] Therefore, safety cannot be addressed and evaluated in isolation if one expects to achieve long-term results. Dependence upon piecemeal techniques that emphasize the safety function to the exclusion of other organizational concerns yields short-term solutions at best.

Treating safety as an important and integrated part of the company, considering both safety and organizational issues, is the only way to optimize safety performance.

ORGANIZATIONAL CULTURE

Organizational culture is also known as corporate culture, workplace culture, or management philosophy. They all address the same concepts: the worldview of upper management as embodied in their assumptions, values, and behavior.

Schein[3] defines culture as "a pattern of basic assumptions — invented, discovered, or developed by a given group as it learns to cope with its problems of external adaptation and internal integration — that has worked well enough to be considered valid and, therefore, to be taught to new members as the correct way to perceive, think, and feel in relation to those problems." Culture implies a way of management thinking and feeling that is transmitted to all employees to define the beliefs and the reality of their company. The end result is that employees become aware of the organization's meanings, learn what is expected of them, and learn how to behave. Generally, the philosophies of the founder(s) or the current top management form the basis for a company's culture.

Organizational culture can be understood by examining the assumptions, values, and behavior of top management.

Assumptions

Assumptions are unconscious and taken for granted. They represent the basic philosophy, or worldview, of management. Within the organization, the shared assumptions of management underlie the corporate culture. They guide and shape values and behavior as they tell people how to perceive, think about, and feel about things.[4] They are fundamental in determining the way the world is viewed by individuals[5] and are basic to the way management views, treats, and communicates with the employees.

Two examples of assumptions concern the human nature of people and the nature of human relationships.

Human Nature

Human nature may be viewed as good or evil or neither good nor evil. In essence, the view of human nature concerns what is meant by being human.[6] Management may put people into categories such as right/wrong or either/or.[7]

If people are viewed as bad, they may be thought of as being lazy or selfish. That might mean to management that employees have to be closely monitored to make sure they are doing their jobs properly, and they are not trying to get away with anything, or cut corners. Management may believe that employees need to be told what to do and need all decisions made for them.

When human nature is viewed negatively, a positive safety message is not communicated. Management shows, by its actions and behavior, that concerns for employee safety and welfare are minimal or do not exist at all. Safety, if mentioned at all, is paid only lip service. However, employees are generally able to recognize the difference between what management says and what it does.

When people are viewed as good, they are thought of as hard working, committed, and conscientious. If they are given latitude to do their jobs, they are expected to perform well and be an asset to the company. Management may consider employees trustworthy and confident to work on their own and who are independent thinkers capable of sound judgments.

When human nature is viewed in a more positive light, the importance of safety is readily communicated. Management actively demonstrates its support and commitment to the safety and health professionals and their effort and to the employees.

Management beliefs concerning injury and accident causation influence how the safety message is communicated. Erickson found that in low safety-performing organizations, employees were blamed for causing their injuries and accidents, whereas employees in high safety-performing organizations were not blamed for these events.[1] Because of their negative assumptions about human nature, some managers may believe that employees cause their injuries and accidents.[8,9] They may believe that the injuries and accidents result from carelessness, inattention, bad attitudes, or other negative employee attributes.

Human Relationships

Another group of assumptions deals with the nature of human relationships. These assumptions determine the accepted way for people to relate to each other and how power and love are distributed in the organization.[6,10] They include elements such as power distribution, the structure of work, control, influence, involvement, and individuality. An example of these assumptions is the organization chart of a company, that is, who reports to whom. Another example is organizational communication, or who speaks to whom. For example, are employees encouraged to speak freely to management in order to offer suggestions and new ideas? Or is there only top-down communication where employees are told what to do, and no feedback is expected from them?

The nature of human relationships determines the right way for people to act with each other. These relationships may be cooperative or competitive, equal or hierarchical, or group-oriented or individualistic. How human relationships are viewed by management influences how management treats employees, who makes decisions and suggestions, and how communication takes place in the organization.

Based on their assumptions, upper levels of management assume they have the right attitude and their choices reflect their view of reality. Many management practices are based on their assumptions about human nature and reality.[11] Therefore, depending upon the assumptions regarding people, employees may be treated as a cost of doing business or as a valuable resource. These assumptions are expressed as management's values.

Values

The values held by upper management serve as a guide. They represent the organization's standards that influence nearly every aspect of the working environment, including the safety and health function. They represent preferences, including the right way to conduct organizational activities.

Examples include organizational standards of desired ends[12] and preferred actions to attain these end points.[13] For example, maybe profit is the primary goal and to achieve this end point because production is stressed over safety. These values are actually references that indicate the correctness of certain beliefs and practices over others,[14] which act as an informal control system to let people know what is expected of them.[15]

When people begin working for a company, they start to learn what the company values as they begin to learn its culture. In other words, they learn what is important and how to behave. Organizational values, which are derived from the assumptions, are reflected in management's opinions, actions, and behavior.

Management Actions and Behavior

Behaviors are what people do. They are a result or an extension of their values, beliefs, and attitudes. The behavior patterns in an organization reflect the organizational culture or the corporate values held by management. Management's actions are symbolic; they provide a living model of its vision.[16] For example, when top management attends safety meetings, it gives the message to employees that safety is important to the organization.

The organization's values are transmitted to the employees through management's opinions, actions, and behavior. By using these criteria of management actions and behavior, employees and safety and health professionals can determine how management values the organizational safety function through the following:

- What management does[17,18,19] — Does management wear personal protective equipment in designated areas? Does it give only lip service to the safety effort? Is management generally supportive of the safety and health function? Does it actively, continuously, and visibly demonstrate this support? Is safety discussed during regular company meetings?
- What management pays attention to[10,20,21] — Does management give feedback to employees for their suggestions? Are the work environment and tasks designed for employee safety and well-being? Are employees properly selected and trained for their jobs?
- What management ignores[22] — Does management ignore poor housekeeping? Is safety mentioned in organizational meetings? Does management ask employees for work or safety suggestions or do only top level people in the organization

make decisions? Does management consult with safety and health professionals before making safety-related purchases or introducing new processes or chemicals?

- Measures and controls management uses[10,21] — Is production more important than employee safety? Does management encourage the reporting of near misses? Are injuries and illnesses honestly reported or are they under-reported? What is the relationship between employee welfare and productivity?

It is through management's actions and behavior that employees become aware of the organization's meanings and learn what is expected of them and how to behave. Through a reward and punishment system, among other means, the culture in an organization is maintained thereby selecting certain beliefs and values.

TRANSMISSION AND MAINTENANCE OF CULTURE

Management has a great deal of discretion in how it orients its behavior and attention,[19] toward safety and health.

Reward and Punishment System

The establishment of an organizational reality occurs through a reward and punishment system, thereby maintaining management's assumptions, values, and beliefs. In this manner, certain beliefs and values may be selectively reinforced and thereby become the dominant patterns of behavior in the organization.[23] Rewarded behaviors will be repeated and, as they are rewarded and repeated, they become unconscious.

People give meanings to their environments by interpreting their surroundings, and behaving accordingly.[24] Through paying attention to their experiences, individuals develop perceptions of what their company truly values.[25] Appropriate behavior is also discovered when people mistakenly violate organizational assumptions or voice dissenting views within a group.[26]

Types of Communication

Communication within an organization also helps to define its reality. Communication takes place in many different ways. Many symbols with organizational meaning are used in the communication process. Some examples of symbols include events, behavior, objects, jargon, dress codes, and physical layout. For example, an office or workstation layout indicates how much privacy or mobility employees have,[27] thereby either limiting or encouraging employee socialization. In other words, the physical layout and design of an organization might be someone's idea of how relationships should be conducted.[28]

Miscommunication is common in organizations.[25] If there are discrepancies between what management says and what it does, this is known as espoused theories and theories-in-use,[29] or as lip service. There can be a loss of management credibility when employees discover that management does not believe or value what it says (e.g., the importance of safety and health). In such instances, there is also lower safety performance.[1]

Young and Post studied 10 leading companies that had excellent internal communications.[30] From this study, they found eight aspects related to effective communication. The most important of these aspects was top management's leadership and belief in employee commitment as necessary in achieving the organization's goals. Management spent time listening to employees and their concerns and suggestions. Second, management followed up its words with action; there was no lip service. Third, two-way communication, downward and upward, was stressed. Fourth, communication between top management and employees was face-to-face, not just with memos. Fifth, all employees received information about company changes. Management felt that it had a responsibility to inform employees before they heard any rumors from the grapevine. Sixth, any bad news, such as product failures, was honestly reported to employees. When bad news was candidly reported, management gained increased credibility among employees. Seventh, communication was tailored to the audience receiving it, according to their different information needs (i.e., safety and health, accounting, research and development). Eighth, communication was an ongoing process. Managers communicated facts as soon as they were known and informed employees why certain decisions and changes were made. Communication, especially during times of crisis or change, was continual.

Based on current research and literature, there is a significant correlation between these communication elements and higher safety performance.

So, what is organizational culture? Fundamentally, it is the philosophy or basic value system, derived from assumptions, that is held by top management. This philosophy is transmitted to employees and maintained in the organization through management's actions and behavior. The end result is that employees become aware of what is meaningful in the organization. They learn what is expected of them and how to behave.

IMPORTANCE OF UNDERSTANDING CORPORATE CULTURE

An understanding of culture is essential to a basic understanding of what goes on in organizations, how to run them, or how to improve them.[3] A system of informal rules or cultures spell out how people are to behave.[4] They form the basis for everything that is valued in the company and have an impact on everything that goes on in the organization, including safety. When safety professionals are aware of their company's culture or philosophy, they can tell where and how it hurts or helps their safety program.

Management is concerned with the overall success of the organization's mission. Management's goals are to increase profits, have a competitive edge, be responsive to customer needs, lower operating costs, and increase quality. In many companies, the safety and health function is handled differently than production, quality, and other company concerns.

Management may view the safety effort in many different ways: as important and necessary, as a cost of doing business, as an economic restraint, or as a necessary evil. Some managers may be actively supportive and committed to safety, while others may totally ignore the safety concerns of the workforce.

Safety and health professionals attend many conferences and seminars to learn new and innovative ways to make their workplaces safer. However, having this new knowledge is one thing and being able to implement it may be another. Management's support in implementing the safety professionals' suggestions depends primarily on the company's culture — in other words, how it values or views safety.

What this means to safety and health professionals is that with all of their knowledge, expertise, and commitment, they cannot, by themselves, effectively provide a safe and healthful working environment. They need management support. If management values employee safety, the safety program will be successful. However, if management does not value employee safety, the safety professional will not be able to provide an optimum safety program. These assertions are supported by scientific research.[1,2]

High- Versus Low-Safety-Performing Companies

Management characteristics associated with high and low-safety-performing organizations have been identified and quantified.

High-Safety-Performing Companies

In high-safety-performing companies, the safety and health function was visible and in a top management position. Safety and health issues were integrated throughout the organization. They were equal to other organizational concerns and were discussed at company meetings.

Management understood and appreciated the role and effort of the safety and health professionals. It recognized the knowledge and expertise of safety and health professionals and sought their advice for new purchases or processes that could affect the safety, health, or well-being of employees. Occupational safety and health professionals were encouraged to attend conferences, seminars, and continuing education. Enlightened management seemed aware of the importance of professional currency and its relevance to the organization's mission and goals.

An adequate allocation of resources was provided for the safety and health effort. Management demonstrated its commitment by going beyond occupational safety and health legislation and regulations to ensure a safe and healthful workplace. Production was not stressed over employee welfare and employees participated in decision making concerning their jobs and tasks. All injuries and accidents were thoroughly investigated.

The environment was clean, well designed, and free of hazards. The work environment and performance tasks were designed for the employees. Management actively and visibly supported the safety and health effort. Management and employees seemed to share similar value systems. Communication was honest, open, and understandable. There were ethical and moral concerns for employee welfare. Employees were properly selected and trained for their jobs. Employees were also trained whenever new equipment, processes, or chemicals were introduced. Employees were treated as valuable resources. They were encouraged to be innovative and make suggestions. Management provided positive feedback to these suggestions. Employees' morale was high and they were committed to the organization.

Low-Safety-Performing Companies

In low-safety-performing organizations, the safety and health function occupied a low position on the organization chart. Safety and health was not a priority in company meetings, and it was not integrated in the organization. It appeared that management had abdicated its responsibility for the employees' safety and health. The responsibility had been relegated to the occupational safety and health professionals who usually occupied a staff position, without authority, in the company.

Management paid only lip service to the safety and health effort. Production was stressed over employee concerns, and employees were blamed for causing injuries and accidents. Management seemed to treat employees as a means to an end. There was no commitment from the employees, only compliance. Employee morale was low and the value system of management seemed incongruent with other organizational members.

Overall, management did not seem to understand the occupational safety and health function and showed little or no overt interest in the safety and health effort.

ORGANIZATIONAL FACTORS

Organizational factors, determined by the management philosophy, are reflections of the company's culture. They include organizational structure, organizational importance of safety, safety responsibility and accountability, communication, management behavior, employee involvement, and employee responses and behavior. They represent a dynamic, ongoing system within the company. For convenience and identification purposes, these factors have been separated into seven categories. However, they must be considered together when analyzing a safety program because, in reality, these organizational factors are interrelated and interactive.

Organizational Structure

Top management controls the organizational structure. Elements of this structure directly relevant to the level of safety performance include placement of safety on the organizational chart; the number and type of rules and regulations; who does the decision making and when; and job, tool, and equipment design.

The Organizational Chart

The organizational chart depicts the positions, in a hierarchical order, within the organization — who reports to whom. The placement of safety professionals on the organizational chart is of utmost importance because when safety professionals are at a higher level, such as in top management or corporate positions, higher safety performance results. Safety professionals must hold a visible management position in order to be effective.

Rules and Regulations

The number of rules and regulations, or how loose or restrictive they are, has an effect on safety performance. When companies have a lot of rules and regulations, or if the rules and regulations are restrictive, there is lower safety performance.

Decision Making

If management makes all or most of the decisions in a company, safety performance suffers. Safety performance is higher when management does not make all decisions but, rather, encourages employees to participate in decision making. Enlightened management is aware of the importance of employees being able to make decisions about their work environments. Employees then have some degree of control over the way in which they perform their task assignments. There are definite advantages to a company when employees are allowed to make decisions. For example, problems can be solved more quickly, more people provide input into decisions, and employees are less likely to feel alienated from those who make decisions that affect their lives. The evidence demonstrates that when employees are allowed to make decisions that affect them, in addition to higher safety performance, there are increases in productivity, commitment to work, motivation, and job satisfaction.

Job Design

Job design involves the structure of jobs. When employees have discretion in what, when, and how they do their jobs, safety performance is higher. There is evidence that increased work-related stress may occur when employees do not have any discretion in the way that they perform their jobs.[13]

Job design also includes the type of tools and equipment that will be used to do the work of the company. There is higher safety performance in those companies that design safety into the performance tasks and the work environment. Regular preventive machinery maintenance is also related to higher safety performance.

Organizational Importance of Safety

The organizational importance of safety is manifested by how well safety and health concerns correspond with overall business matters. The perception of safety versus other organizational concerns is a major indication of its importance to top management. Examples include the formal policy statement, integration of safety, safety versus production, equality of importance, allocation of resources, and priority setting. The amount and quality of management attention toward the workplace environment and the selection and training of employees are also indications of the relative importance of safety.

The Written Policy

All companies should have a formal written policy that includes direct reference to safety concerns. However, such a policy is meaningless unless management's actions back up the written words. The study demonstrated that a written policy, by itself, was not related to high safety performance.[1]

Integration of Safety

Safety and health cannot operate in isolation in an organization. Safety should be integrated throughout the company since the safety effort is involved with nearly every part of the organization at one time or another.

Safety Versus Production

When production is considered more important than safety, lower safety performance results. Employees should not be expected to take safety and health risks because of production demands.

Equality of Importance

Safety must be viewed or treated as being equal in importance to other organizational concerns. Otherwise, safety performance is lower.

Resource Allocation

A main indicator of power in a company is the ability to control and allocate resources. Therefore, the amount of resources allocated to the safety effort is an indicator of management's perception of the importance of safety. High-safety-performing companies allocate sufficient resources to the safety effort.

Safety as a Priority

Safety should be a priority in all company meetings, sharing the same importance as other organizational matters. When safety is not considered as important as other company matters, other goals are emphasized, and safety is devalued or abandoned.

Workplace Environment

The work environment should be clean and well designed; these considerations result in higher safety performance in the organization.

Employee Selection and Training

Employees should be selected and trained for their positions. They should be physically and mentally capable of performing their tasks. In addition, periodic training to update skills and knowledge, should be provided to all employees. Training should also be provided whenever new processes, equipment, chemicals, etc., are introduced and whenever employees are transferred to other departments. Of interest is the fact that the study did not indicate than an initial safety orientation was related to higher safety performance.[1]

Safety Responsibility and Accountability

Personal Responsibility

Managers, supervisors, and employees should feel personally responsible for safety. When everyone in the company takes responsibility for safety, there are fewer injuries and accidents.

Performance Appraisals

Safety performance of individual employees should be part of their performance appraisals. Operations managers and supervisors should also be held accountable for the safety performance of their work teams. However, care must be taken on the part of management to use this information to increase safety performance, not as a punitive device.

Injury and Accident Investigation

Thorough and objective investigations (root cause analysis) should be carried out as soon as possible after the occurrence and should not stop at just the proximal cause.

Communication

Communication is one of the most important factors in any organization and is a critical component for an effective safety program. Elements of communication include honesty and direction.

Honesty

Companies that have honest, open, and understandable workplace communication have higher safety performance. This type of communication eliminates misunderstandings and differences in interpretation, which are necessary for employee trust and commitment. However, when management gives only lip service to safety and health there is lower safety performance. When there is little or no management support to follow the spoken words, employees become aware of the differences between what management says and what it does.

Direction

For lower injuries and accidents, information should flow up, down, and across the organization. Employees should be encouraged to provide information, feedback, and suggestions to management. The traditional management model, with its command and control philosophy, relies on top-down communication. The organization with this type of management has lower safety performance.

Near Misses

The reporting of near misses by employees should be encouraged. All such instances should be discussed and evaluated by managers, supervisors, and employees to help prevent future events. The importance of this cannot be overly stressed since it is common knowledge that the difference between a near miss and an injury could be just a millimeter or a millisecond.

Management Behavior

Management actions and behavior are expressions of its values and beliefs. If management values and believes in safety, its actions are going to demonstrate that.

Beyond Legislation

The company that does not rely solely on safety and health legislation and regulations experiences higher safety performance. Management of these organizations realizes that the legislation and regulations are but a framework for an effective and efficient safety program.

Treatment of Employees

When employees are treated as valuable resources rather than as costs of doing business, there are fewer injuries. When employees are treated as individuals with caring and respect, they respond accordingly. Supportive management encourages them to take initiative and allows them autonomy and control over their work. Other positive management actions toward employees include offering feedback, encouraging suggestions, face-to-face communication, and demonstrating an ethical concern for employee welfare.

Beliefs About Injury Causation

When employees are blamed for their injuries, safety performance is lower. Blaming provokes defensive behaviors that interfere with objectivity and cooperation. Employees may be blamed for their mishaps because of the assumptions of human nature that are held by management, such as viewing employees as lazy, not committed to the company, not caring, and so forth. This perception may also be based on Heinrich's virtually unsubstantiated claim that 85% of workplace injuries are attributable to unsafe acts by employees.[31] There has never been any scientific research to back up this claim.

Conversely, safety performance is higher in those companies where employees are not blamed for their injuries and accidents. In these companies, concerned management goes beyond immediate causes in injury investigations. It looks for root causes and long-term solutions.

Safety management and accident investigation texts stress the importance of investigating mishaps beyond the apparent proximal cause of injuries and accidents to determine what is actually going on in the organization. Two classic texts in these fields, by Ferry[32] and Grimaldi and Simonds[33] instruct the reader to look at the organizational systems that allow injuries and accidents to occur. Examples of system failures include inefficient operations, negative management practices, poor operating procedures, and so forth.

Emphasizing the behavior of employees as the apparent cause of injuries and accidents is fairly common in injury and accident investigation. Behavior is defined as the manner in which one conducts oneself as a response to one's environment. The way a person conducts him- or herself at work can be the result of many complex and intervening variables. Therefore, attributing injury and accident causation to solely employee behavior is, in essence, ignoring any and all elements that may be influencing that behavior. Examples of some of these elements include:

- The environment — temperature, air quality, humidity, etc.
- Fatigue
- Job complexity
- Adequacy of training
- Ineffective communication
- Production pressures
- Maintenance errors
- Psychological stress: work-related, financial, marital, etc.
- Mental and physical capability to perform task(s)
- Injury and accident investigations that provide objective, thorough root cause analyses
- Working overtime
- Organizational culture

Therefore, before stating that employee behavior is the primary cause of injuries and accidents, one must first consider all of these variables since they are assumed to cause, affect, or influence the outcome. They must be taken into account and explained before one looks at employee behavior as the fundamental cause of injuries and accidents.

Ethics

When management expresses an ethical concern for employees, safety performance is higher. Management has a moral and ethical responsibility to ensure that employees and others do not experience harm or are exposed to risks as a result of organizational activities. Inattention to moral questions is not good for the conduction of business. The safety, health, and welfare of the workforce is such a moral question.

Ultimately, the safety and health of the employee is dependent upon the value system, and thereby, the assumptions of the management of the organization.

This category, the way employees are treated, and the next category, employee involvement, are most predictive of whether or not safety performance will be high or low. There are a couple of reasons to take a second look at these data because these findings are significant.

First, the management characteristics related to high safety performance are the same characteristics reported in the management literature as being associated with higher job satisfaction, productivity, and quality. This is what safety professionals have been saying for years.

Second, how much employees are valued by a company seems to be the primary determinant of the level of safety performance. What this seems to mean is that, no matter how well safety professionals may speak management's language, unless employees are truly cared for and respected, a safety program will never be the best that it could be.

In high-safety-performing companies, management views employees as a valuable resource with abilities and limitations that require motivation, participation, satisfaction, leadership, incentives, rewards, and status. For it is management, and only management, that can satisfy these requirements.[34]

Employee Involvement

When employees are aware of management's genuine interest in them, that what they say and do is valued by management, they will respond in kind. In this type of an environment, employee innovative thinking, suggestions, and decision making evolve, to the benefit of the employee and the organization alike. One benefit is fewer injuries. This involvement can take a number of forms, such as decision making, suggestions, and innovative thinking.

Decision Making

When employees are encouraged to make decisions about their work practices and their safety concerns, safety performance is higher. When employees make decisions about their work, they have some degree of control over the way they perform their tasks. By being allowed this freedom, employees will generally have higher morale and job satisfaction, as well as fewer injuries.

Suggestions and Innovative Thinking

When employees are encouraged to make suggestions and to think creatively, they have higher safety performance. The organization that encourages suggestions and innovative thinking implies openness and trust between management and employees. In these organizations, employees are not hesitant to offer suggestions because they know management will provide encouragement and welcome new ideas. Also, employees then feel they have more of a buy-in into the organization.

Employee Responses and Behavior

People respond to their environments and behave accordingly. This includes the way employees behave in their companies. Organizational commitment and the morale status of employees are related to the way they are treated within the company. A management philosophy based on respect, honesty, and cooperation results in higher employee morale and commitment as well as higher safety performance. In addition, decreased absenteeism and turnover are associated with higher safety performance.

SUBCULTURES

Within the structure of an organization, with its different departments and work groups, there sometimes are individuals or groups of employees who do not accept the prevalent organizational values. They have a value system that is different from that of management that serves as guides to their intentions and behavior. The organizational value system provides guides for organizational accomplishments. Therefore, if the organizational goals are independent of or antagonistic to those of the individual, there will be basic incongruities between the employees and management.[35] When there is an incongruency between the two values systems, there may not be a proper fit between the employees and management.

Some employees have different backgrounds, education, and experiences than management. These employees may also have different values than those persons within the dominant culture. Often these employees are members of the same occupational group and share standards for appropriate behavior, work codes, common skills, and a knowledge base.[36] They also share similar identities and values.[37,38] Safety and health professionals are such an occupational group.

Members of the same occupational group are more likely to find shared values and understandings with each other.[39] Therefore, they often rely upon one another for understanding and validation of their experiences and perceptions. Through professional meetings, symposia, and conferences, the bonds of occupational identification are strengthened. These activities emphasize the common issues and concerns of the professionals.[40]

Sometimes occupational groups have limited acceptance and participation in the prevailing culture. These groups may find that their not totally accepted or conflicting values complicate attempts at interaction and coordination of their function with the rest of the organization. This may be a reason why some safety and health professionals experience difficulty getting management's attention concerning safety matters. If this is so, just speaking management's language concerning injury- and illness-related costs may be insufficient. Rather, the different value systems between these two groups may be what are interferring with communication. Management may not be able to truly hear what the safety and health professionals are saying. As a result, these professionals may experience frustration in trying to do their jobs — providing a safe and healthful workplace. If their values are not congruent with those of the dominant culture, safety and health professionals may be considered a subculture within their organizations.

The relationship of subcultures, or occupational communities to the dominant organizational culture may be competitive or cooperative or it may be one of peaceful coexistence.

When employees have core values that are in conflict with the core values of the prevailing culture, countercultures may develop. For both the organization and employees, these value issues are of critical importance. An example of a counterculture is a union that is in an adversarial relationship with management.

In high safety-performing organizations there was a good fit between management and employee value systems. Conversely, there was not a good fit of values in companies with low safety performance.

CHANGING ORGANIZATIONAL CULTURE

During the last 20 years there have been many major developments in the business community. There are many reasons why companies are changing: increasing global competition, development of new technologies, increasing knowledge base of employees, new regulations and legislation, increasing numbers of mergers and acquisitions, and increasing size and complexity of organizations, among others. In order to be able to respond effectively and meet competitive challenges, businesses have been undergoing a number of changes. As a result, many American businesses are currently in transition. The way business has been conducted is changing. This transition has already begun in a number of companies and is gaining momentum.

Paradigm Shift

The traditional management style has been seen as order giving and having all the answers. It has been described as controlling, rigid, and not welcome to change. This type of management style sees employee safety and health as a necessary evil, and regards injuries and accidents as a cost of doing business.

This old hierarchical command and control management style is being replaced by one that is visionary, flexible, innovative and responsive. This shift in management philosophy means that organizations will be able to respond more quickly to change in order to be competitive and that employees, instead of being viewed as a means to an end, will be viewed as valuable resources, parties to attaining the organization's goals and objectives.

The new management style is visionary, empowering, and flexible. It values creativity and views employees as valuable resources — and this includes employee safety considerations. These managers realize that they have to become the role models in the company that communicate the importance of safety throughout the organization. They are aware that what they value and how they behave are the primary methods for getting a positive safety message across to the employees.

However, changing a company's culture is not an easy venture.

How Cultures Change

Organizational cultures run very deep. They represent top management's personal identity and involvement in a certain belief system. These basic beliefs and values are resistant to change.

There is no standard approach to for changing corporate culture. However, for new systems of meaning to develop, or for a philosophical change to occur, it must happen throughout the organization in order to be successful. Changes must involve all of the elements of a company's culture: the underlying management assumptions, management's values, actions and behavior, and the reward/punishment system. Management must be willing to examine its basic philosophy toward work and the employees. In other words, this transition requires a paradigm shift (a change in the organizational culture).

Some organizational cultures are sufficiently adaptable, which enables them to foresee changing events and to respond accordingly. However, the impetus for most cultural change seems to come from the external environment.[3,41] There may be crises or changes in the political or social environments.

Corporate cultures do change and these changes usually occur because:

- A major crisis has occurred such as an explosion or there has been a multiple fatality accident.
- The company is involved with survival issues. It either changes the way it does things or it is not successful.
- There have been mergers, takeovers, and/or new top management.

Also, more managers have come to realize that its way of doing business is not the best way and it wants to change. This type of management philosophy is an example of the paradigm shift that is occurring in companies. These are also the companies that consider their employees valuable resources. They ensure that their employees have safe and healthful working conditions.

THE RESEARCH

Throughout the safety and health management literature, the necessity of obtaining visible, explicit, and continual management support has been repeatedly stressed as the single most important factor between a successful and unsuccessful safety and health program.[33,42,43,44,45,46] It is generally believed that if management holds erroneous assumptions about the causes of problems that the safety and health program seeks to resolve, or if there is a mismatch of values, beliefs, and assumptions held by the safety and health professionals and management, the program will most likely fail.

Practicing safety and health professionals have been saying for years that management must play the major role to ensure the success of safety programs. Specifically, they have been saying that management should, among other elements:

- Be committed to safety
- Manage safety like other organizational concerns

- Integrate safety and health into the entire organization
- Become personally involved in the safety effort
- Assume accountability for safety

Historically, these statements have been based on hunches and/or anecdotal experiences. Until fairly recently there had been no research to back up these claims. Now there is. There is compelling scientific research demonstrating that the management philosophy, or organizational culture, of an organization is the key to high safety performance.

To explore the effect of organizational culture on safety performance, a nationwide study was conducted. This study, *The Effect of Corporate Culture on Injury and Illness Rates within the Organization*, showed that the safety performance of an organization is directly related to the management philosophy of that organization.[1] Those elements most predictive of high safety performance include a positive management commitment to safety and to employees, open communication, encouragement of employee innovation and suggestions, and management feedback to employees, among other elements.[1,47,48,49]

Other studies also demonstrate the definite relationship between management philosophy and safety performance. For example, the State of Maine conducted an in-depth study of that state's companies with the highest number of worker injury claims. The Maine Top 200 Program, representing about 1% of the state's employers, accounted for 45% of the workplace injuries, illnesses, and fatalities. The research demonstrated that those companies with the lowest lost time injury rates had the highest level of management commitment and employee involvement.[2]

The Maine Top 200 Program findings have been incorporated into the Occupational Safety and Health Administration's (OSHA) standards. As an example, the first core element in OSHA's Proposed Safety and Health Program Rule, which requires employers to establish safety and health programs, is management leadership and employee participation.[50] OSHA, realizing the paramount importance of organizational culture to safety performance, has included corporate culture as part of its Strategic Plan for Fiscal Years 1997–2002. One of its three strategic goals is: "Change workplace culture to increase employer and worker awareness of, commitment to, and involvement in, safety and health."[51]

Practical Considerations of the Research

Organizational Issues Are the Key

The current research clearly demonstrates that organizational issues are the primary determinant of the level of safety performance in an organization.

Positive Treatment Means Positive Results

It has been demonstrated that the management characteristic most predictive of high safety performance is a positive employee environment $(P < .01)$.[1] Specific characteristics within this category included respecting and caring for employees, open communication, and employee participation and involvement. This is an extremely important and significant finding for a couple of reasons. First, it has nothing to do with safety, as it is generally defined. What this seems to mean is that safety professionals are essentially out of the loop when it comes to influencing the company to have an optimal safety program. Only management can do that.

The second significant aspect to this finding is that the management characteristics of respecting, caring for, and empowering employees that increase safety performance, are identical to the management characteristics reported in management books as associated with increased quality and productivity, and in the occupational psychology literature as being positively related to increased worker job satisfaction and decreased occupational stress.

In current management books, there is an increasing emphasis on the importance of soft skills such as trustworthiness, communication, interpersonal skills, and integrity. These soft skills are linked with increases in productivity, job satisfaction, commitment, and quality. These elements create more successful employees and more successful companies. The messages in these books revolve around similar themes. Examples include:

- Paying attention to employees as individuals keeps them committed and productive.[52]
- A vital component of corporate success is employee fulfillment.[53]
- Employees' relationships with their superiors determine, to a great extent, employee loyalty, job satisfaction, and productivity.[54]
- A company's most important source of competitive advantage is a management philosophy that focuses on developing and leveraging the individual's unique talents and skills.[55]
- Business performance improves by toppling outdated hierarchies, encouraging the flow of ideas, and sweeping away artificial boundaries.[56]
- By empowering employees, by giving them input into things that affect them, and by getting out of their way, management can get the best out of people.[57]
- Job strain lessens and productivity increases when workers are given more control over what they do.[58]

Maybe Money Does Not Talk

When referring to safety programs and their financial impact on the bottom line, safety professionals have traditionally been taught to speak management's language. However, if management does not really understand what the safety effort is all about, or what safety and health professionals actually do, it will not necessarily hear this message. Management may consider only at the salary of the safety professional to be an unnecessary cost to the company. This seems to be supported by the fact that many safety and health professionals are being downsized and demoted across the country. What seems to be happening here, based on the research, is that if management does not genuinely care for and respect employees, or share the same values as the safety professionals, this financial approach may not be as meaningful or influential as previously thought.

Safety Programs Alone Do Not Work

The traditional safety management techniques such as regulatory compliance, safety audits, and hazard analysis, are not enough to realize an optimal safety and health program. Safety and health professionals have been conscientiously performing these duties for years and have come to realize that they alone cannot provide the best safety program by themselves.

Safety Programs Should Be Evaluated by Organizational Factors

As stated earlier, the organizational factors are expressions of the company's culture. Therefore, the compelling evidence linking corporate culture, or management philosophy, to safety performance strongly recommends that safety programs be evaluated by the organizational factors that affect their effectiveness and efficiency. All of these factors, including employee participation and management leadership, can be measured because they have been scientifically studied and quantified. Therefore, the measurements exist.

These research findings have direct significant practical value for companies interested in increasing their safety performance and, as a result, their bottom line. Interestingly, many of the implications of these findings seem to be outside the realm of traditional safety management thinking. Therefore, it may be appropriate for some safety professionals to step outside of their usual way of looking at what they do and adopt a different perspective. The scientific research overwhelmingly supports this position.

What Safety Managers Can Do

An optimal safety program can be achieved through an integrated approach, recognizing the interactive and interrelated aspects of this function, and by not treating it in a piecemeal fashion. In an organization, safety does not exist in isolation. Therefore, it only makes sense that it should not be treated that way, as if it has nothing to do with anything else that is going on in the company.

Increasing safety performance can occur by looking at all those organizational factors that impact safety. What is needed is an integrated approach, not just zeroing in on injuries and accidents. One should be able to view in one place all of the organizational factors that affect safety. A method for exploring and communicating organizational factors, the organizational issues impacting safety performance, should be identified and established.

Communicating the safety message is related to all of the other messages management sends. Therefore, how management communicates about all company matters, including the importance of safety and health determines, in part, how the employees will accept the messages. Of paramount importance is that everyone in the company should be perceiving safety in the same way. This means that in addition to the influence of communication, individual perceptions also play a major role.

PERCEPTIONS

Perceptions are defined as the way people organize and interpret their sensory input to give meaning to their environment, which they then call reality. This reality is personally defined. It is different for each of us. People need to make sense of their environments. Many different groups of people, with various perspectives, make up an organization. These different perspectives arise because of individuals' positions in the company, assumptions, education, experiences, and so forth. There is a problem for safety professionals when these individual perceptions interfere with the safety effort. However, once the differences in perceptions are identified, a common safety- and health-related language can be developed for all employees, including management. The importance of a common safety language cannot be overemphasized because once everyone is speaking the same language, with an awareness of others' perceptions, true safety communication and dialogue can begin. Becoming aware of others' perceptions can be achieved through administration of scientifically valid perception surveys.

Perception survey use is gaining in popularity. Management may want to get feedback on past performance or to get information for future direction. It may want to determine employee job satisfaction or gain an understanding of safety and health. Once management has the knowledge of what the employees perceive to be true, it has the influence and power to address and correct any problematic issues.

However, not all perception surveys are equal. A perception survey must be scientifically valid or, otherwise, it reveals nothing of value to the company. The survey must have been rigidly tested to ensure that the right questions are being asked in order to guarantee meaningful answers. The basic challenge when using a perception survey is asking the right questions and obtaining the right answers in order to have something to work with. When a scientifically tested perception survey is used, policies and programs can be successfully designed. Anonymity and confidentiality is another crucial aspect to survey administration. Employees need to know that their identities will be protected in order to respond with honest answers.

CONCLUSION

An organization is a complex institution with many different people who have diverse education, experience, and skills. These people perform various tasks and duties to ensure the success of the company's operation. Safety is but one element that functions in a company. However, unlike most other organizational operations, the safety effort is part of nearly every program, department, and activity in the company. Its ultimate success is directly influenced by the corporate culture of the organization.

Each organization has its own culture and this culture directly influences the success of the safety and health program. The assumptions, values, and actions and behavior of management determine how employees will be treated in the organization. If employees are regarded by management as valuable resources and are treated with respect, safety performance is higher.

Given these research findings, how should safety be done? How should safety be managed? Safety used to be approached linearly. Its various elements were treated in a piecemeal fashion. Sometimes they still are. This may have worked in the past, but it is not sufficient for the future. One of the reasons the old system does not work is because safety is an integrated and interrelated organizational function. It does not operate in isolation in a company.

To truly assess and improve a safety and health program, a more holistic and integrated approach is necessary, not just zeroing in on injuries and accidents. There is a need to move beyond, yet still include the traditional safety management approaches of regulatory compliance, engineering, job safety analyses, hazard analysis, workers' compensation data, and so forth. Historically, safety and health interventions such as incentive, behavior-based safety, and stress-reduction programs have been aimed at the employees. Therefore, they have perpetuated the management perception that safety performance is an individual, not an organizational, problem.

The research evidence compels safety professionals to rethink the way they do safety. Traditional safety management methods are necessary, but they are not sufficient for an optimal safety program. Something else is needed. It only makes sense to use a proven integrated systems approach that emphasizes organizational factors, direct expressions of the corporate culture, in order to improve safety program effectiveness.

Forward-thinking companies are changing their corporate cultures from the old to the new management style. These companies are going to be the most successful as they go into the millennium. These are the companies that will incorporate safety into all of their operations because they know that safety is as an important and integral part of the company and that companies with lower injury and accident rates have the competitive edge.

Organizational factors have been scientifically demonstrated to have a direct impact on safety performance and to provide definite non-safety-related benefits to companies. Through the evaluation of organizational factors, not only does safety performance improve, but there are also positive effects on productivity, employee commitment, communication, and profitability. These more encompassing company-wide effects should be the goals of professional safety managers, not just narrowly focusing on reducing injuries and accidents.

This is the sort of vision safety and health managers need to demonstrate in order to elevate their importance to the overall corporate scheme and to take on the added level of respect it deserves.

REFERENCES

1. Erickson, J. A. The effect of corporate culture on injury and illness rates within the organization, Ph.D. thesis, University of Southern California, Los Angeles, 1994.
2. Most, I. G. The quality of the workplace organization and its relationship to employee health, in *Abstracts of Work Stress & Health '99: Organization of Work in a Global Economy American Psychological Association/National Institute for Occupational Safety and Health Joint Conference*, Baltimore, MD, 1999, 179.

3. Shein, E. H. *Organizational Culture and Leadership*. San Francisco: Jossey-Bass, 1988.
4. Argyris, C. *Increasing Leadership Effectiveness*. New York: Wiley-Interscience, 1976.
5. Burrell, G. and Morgan, G. *Sociological Paradigms and Organizational Analysis: Elements of the Sociology of Corporate Life*. Portsmouth, NH: Heinemann, 1985.
6. Schein, E. H. Coming to a new awareness of organizational culture. *Sloan Management Review* 25: 3–16, 1984.
7. Reilly, B. J. and DiAngelo, J. A. Communication: a cultural system of meaning and value. *Human Relations* 43: 129–140, 1990.
8. De Joy, D. M. Spontaneous attributional thinking following near-miss and loss-producing traffic accidents. *Journal of Safety Research* 21: 115–124, 1990.
9. Petersen, D. Safety's paradigm shift, *Professional Safety*. 36: 47–49, 1991.
10. Schein, E. H. The role of the founder in creating organizational culture. *Organizational Dynamics* 12: 13–28, 1983.
11. Koprowski, E. J. Cultural myths: clues to effective management. *Organizational Dynamics* 12: 39–51, 1983.
12. Ranson, S., Hinings, B., and Greenwood, R. The structuring of organizational structures. *Administrative Science Quarterly* 25: 1–17, 1980.
13. Stubbart, C. I. Review of power and shared values in the corporate culture. *Administrative Science Quarterly* 33: 333–336, 1988.
14. Trice, H. M. and Beyer, J. Studying organizational cultures through rites and ceremonies. *Academy of Management Review* 9: 653–669, 1984.
15. Deal, T. D. and Kennedy, A. A. *Corporate Cultures: The Rites and Rituals of Corporate Life*. Reading, MA: Addison-Wesley, 1990.
16. Peters, T. *Thriving on Chaos*. New York: Harper & Row, 1988.
17. Cohen, H. H. and Cleveland, B. J. Safety program practices in record-holding plants. *Professional Safety* 28: 26–33, 1983.
18. Lorsch, J. W. Strategic myopia: culture as an invisible barrier to change, in *Gaining Control of the Corporate Culture*. Kilmann, H. C., Saxton, M. J. and Serpa, R., Eds. San Francisco: Jossey-Bass, 1985, 84–102.
19. Peters, T. J. Management systems: the language of organizational character and competence. *Organizational Dynamics* 9: 3–26, 1980.
20. Perrow, C. The organizational context of human factors engineering. *Administrative Science Quarterly* 28: 521–541, 1983.
21. Schein, E. H. Organizational culture. *American Psychologist* 45: 109–119, 1990.
22. Kilmann, R. H., Saxton, M. J., and Serpa, R. Issues in understanding and changing culture. *California Management Review* 28: 87–94, 1986.
23. Sethia N. K. and Von Glinow, M. A. Arriving at four cultures by managing their reward system, in *Gaining Control of the Corporate Culture*. Kilmann, R. H., Saxton, M. J., and Serpa, R., Eds. San Francisco: Jossey-Bass, 1985, 400–420.
24. Akin, G. and Hopelain, D. Finding the culture of productivity. *Organizational Dynamics* 14: 19–32, 1986.
25. Sathe, V. Implications of corporate culture: a manager's guide to action. *Organizational Dynamics* 12: 5–23, 1983.
26. Janis, I. L. *Victims of Groupthink*. Cambridge: Belknap Press of Harvard University Press, 1972.
27. Ashforth, B. E. Climate formation: issues and extensions. *Academy of Management Review* 10: 837–847, 1985.
28. Gorman, L. Corporate culture: why managers should be interested. *Leadership and Organization Journal* 8: 3–9, 1987.

29. Argyris, C., Putnam, R., and Smith, D. M. *Action Science*. San Francisco: Jossey-Bass, 1985, 81–82.

30. Young, M. and Post, J. J. Managing to communicate, communicating to manage: How leading companies communicate with employees. *Organizational Dynamics* 12: 31–43, 1993.

31. Heinrich, H. W. *Industrial Accident Prevention*. New York: McGraw-Hill, 1931.

32. Ferry, T. S. *Modern Accident Investigation and Analysis: An Executive Guide*. New York: John Wiley & Sons, 1981.

33. Grimaldi, J. and Simonds, R. *Safety Management*. Homewood, IL: 1989.

34. Drucker, P. F. *The Practice of Management*. New York: Harper Business, 1993.

35. Argyris, C. Personality versus organization. *Organizational Dynamics* 3: 3–17, 1974.

36. Van Maanan, J. and Barley, S. R. Occupational communities: culture and control in organizations, in *Research in Organizational Behavior*. Staw, B. M. and Cummings, L. L., Eds. Greenwich, CT: JAI Press, 1984, 287–365.

37. Gregory, K. L. Native-view paradigms: multiple cultures and culture conflicts in organizations. *Administrative Science Quarterly*. 28: 359–376, 1983.

38. Trice, H. M. and Beyer, J. Studying organizational cultures through rites and ceremonies. *Academy of Management Review* 9: 653–669, 1984.

39. Ouchi, W. G. and Wilkins, A. L. Organizational culture. *Annual Review of Sociology* 11: 457–483, 1985.

40. Beyer, J. M. and Trice, H. M. How an organization's rites reveal its culture. *Organizational Dynamics* 15: 5–24, 1987.

41. Jelinek, M., Smircich, L., and Hirsch, P. Introduction: a code of many colors. *Administrative Science Quarterly* 28: 331–338, 1983.

42. Ferry, T. S. *Safety and Health Management Planning*. New York: Van Nostrand Reinhold, 1990.

43. Geyer, S. Safety investment yields future dividends. *Safety and Health* 14: 23–25, 1991.

44. Hammer, W. *Occupational Safety Management and Engineering*, 3rd ed. Englewood Cliffs, NJ: Prentice-Hall, Inc., 1985.

45. Manuele, F. A. *On the Practice of Safety*. New York: Von Nostrand Reinhold, 1991.

46. Smith, M. J., Cohen, H. H., Cohen, A., and Cleveland, R. J. Characteristics of successful safety programs. *Journal of Safety Research* 10: 5–15, 1978.

47. Erickson, J. A. The relationship between corporate culture and safety performance. *Professional Safety* 42: 29–33, 1997.

48. Erickson, J. A. The relationship between corporate culture and safety culture, in *Safety Culture and Effective Safety Management*. Swartz, G., Ed. Chicago, IL: National Safety Council, 1999, 73–106.

49. Erickson, J. A. Corporate culture: the key to safety performance. *Occupational Hazards* 62: 45–50, 2000.

50. Occupational Safety and Health Administration. *Draft proposed safety and health program rule*. 29 CFR 1900.1, Docket No. S & H-0027. http://www.oshaslc.gov/SLTC/Safetyhealth/nshp.htm/. 1999.

51. Occupational Safety and Health Administration, *OSHA Strategic Plan FY 1997–FY 2002*. http://www.osha.gov/oshinfo/strategic/, 1997.

52. Ulrich, D., Zenger, J., and Smallwood, N. *Results-Based Leadership*. Boston, MA: Harvard Business School Press, 1999.

53. Katzenbach, J. R. *Peak Performance: Aligning the Hearts and Minds of Your Employees*, Boston, MA: Harvard Business School Press, 2000.

54. Buckingham, M. and Coffman, C. *First, Break All the Rules: What the World's Greatest Managers Do Differently*. New York: Simon & Schuster, 1999.

55. Ghosal, S. and Bartlett, C. A. *The Individualized Corporation: A Fundamentally New Approach to Management.* New York: Harper Business, 1999.
56. Ashkenas, R. N., Ulrich, D., Prahalad, C. K., and Jick, T. *The Boundaryless Organization: Breaking the Chains of Organizational Structure.* San Francisco: Jossey-Bass, 1995.
57. Bennis, W. G. *On Becoming a Leader.* Cambridge, MA: Perseus, 1994.
58. Karasek, R. and Theorell, T. *Healthy Work.* New York: Basic Books, 1990.

## CHAPTER 16

Safety Performance Measurement: New Approaches

Jerry L. Williams

CONTENTS

INTRODUCTION

Industries have been using qualitative and quantitative measures for centuries to determine the effectiveness of their business decisions. The measures have changed, as industrial processes became more automated, technical and global in nature. The measures used, identify the inputs and outputs in financial, or quality terms. Measures are used to allow the management team to modify their systems or processes as their trends indicate a negative return from the present methods. Industries have been using uniformly accepted business measures to develop benchmarks from other companies with world class operations. Industries use "best practices" to set up their industrial line operations. Best practices provide employees with defined steps to create each product and provide drawings and pictures to show the employees how the final product should look. These practices ensure consistent quality production with very few variables. Measures are taken from these best practices, and quality or numerical deficiencies usually result from a failure to follow the best practices.

Safety performance has primarily been measured in output terms only, using the Occupational Safety and Health Administration (OSHA) records and rates. These measures have been used since the inception of OSHA in 1970, and similar measures (American National Safety Institute (ANSI) method of tracking outputs, injury rates) were used prior to OSHA.

Measures are only useful if they are accurate, consistent, and can be compared against a norm. OSHA rates have a minimum of thirteen variables for each entry and do not permit consistency or reliability when comparing the rates against other companies. OSHA rates do not accurately identify the effectiveness of any safety process. As an example, if a company has a 16.0 LWDI compared against the national average for the industry standard industrial classification (SIC) code of 24.0, does this mean they have a quality lockout safety process? Unless the rates are completed only for injuries involved in the lockout processes, you would never be able to measure the lockout performance, nor compare that performance to other companies or plants. Not having a work related injury is not necessarily an indicator of a quality safety and health process.

Safety performance measures are used to identify:

- The areas where changes or modifications are needed, and where time, energy, and resources are best used.
- Areas of success, which can be used as a basis for best practices.
- Possible liabilities to the company.
- The value added to the profit and loss of a company.

Safety performance measures must meet specific criteria to quantify and demonstrate value:

- Measure against a defined standard (standards define what is acceptable or not).
- Measure performance, not just written documents. Performance measures validate what is written and that the safety process is actually performed.
- Measure to a specific safety process. To identify areas of success or needs for improvement, each safety process must be evaluated separately. This is an ISO QS 9000 principle where you measure against what the company says it does.
- Measures must be repeatable (the measures cannot be subjective). Measures must be based on fact and must be reliable.
- Measures must be understandable and easy to relate to a business value.
- Measures must directly relate to the bottom line of a company.
- Measures must be compatible with other business measures of the company.

We use a process of measuring, similar to job safety analysis. We break down all safety processes into small segments (safety areas), then measure each segment.

The total measures for each segment gives you the total measure for each safety process. The total measures for each safety process equals the total safety performance for the plant, division or company. We will explain this concept throughout this chapter.

MEASURE AGAINST A DEFINED STANDARD

Safety standards define the hazards, risks involved, the action, and (design controls) required to eliminate those hazards. There are two types of safety standards. The first defines design requirements of equipment, machines or processes. The second involves safety standards that establish controls in the form of safe work practices and safety devices to eliminate the hazards. Safety standards must be in written form, so that everyone is working from the same blueprint. A simple QS ISO 9000 philosophy is that if standards are not written, they do not exist.

The key components and elements for effective, measurable safety performance standards are:

- Safety standards *must be* written in a language that is simple, practicable, and understandable. Writing standards similar to OSHA regulations will only confuse the end users, (the company). OSHA regulations average 52 words per sentence and have a reading ease of 12th grade 5%. This means that only 5% of high school graduates could understand the OSHA regulations as written.
- Safety standards must be written so that all the affected employees can relate the standard in their actual work environment.
- Safety standards must be written in a language that provides a clear picture of the requirements and what is acceptable.
- All safety standards must be written in a manner that the actions required can visibly be seen and thus measured. Stay away from terms such as be safe, use caution, should, may or other terms that cannot be measured. For example, flight attendants announce, during the landing of an aircraft, that passengers should use caution when opening the overhead bins. If we were to measure the performance of the passengers in opening the bins, what would we use as a basis? If the standard was changed to open the overhead bins slowly because the baggage may have shifted in flight, this standard could be measured.
- Safety standards must be developed that will identify specific safe work practices or actions that employees must perform in order to prevent an injury. For example, lockout locks must be used for every lockout procedure, and one lock per person per affected energy source is required.
- Safety standards must be written in a manner that will allow employees to understand the hazards, the risk involved and know the required controls. For example, using the same lockout lock issue, one lockout lock per person per affected energy source is required to ensure that each authorized employee is protected from the unexpected release of the power or energy sources that could result in serious injury or fatality.

Safety professionals need to help develop the company-wide safety and health standards through a team process. A minimum basis for safety and health standards would be OSHA regulations and standards. Safety professionals also need to look at applicable ANSI, American Society of Mechanical Engineers (ASME), National Fire Prevention Association (NFPA), state codes and related industrial association models. Remember that most OSHA regulations were adopted from consensus standards such as ANSI and other, however, they were dated 1970 and before. All ANSI standards are modified and updated at least every five years. We suggest that companies develop their safety and health standards using teams of hourly and salaried employees that may be affected by the standard. For each standard, the team would be responsible for identifying the intent, the actions to be taken and the measures to be used. The intent is a simple statement of the purpose and scope of the safety and health standard. Why do we need it, what does it do for us, and who does it involve? The actions describe the safe work practice areas or need for safe work practices. For example, in the lockout lock safety process, the standard would read: "All lockout locks must be used for lockout only, be identified to the employee using the lock, and requires one lock per employee per affected energy source." The measures describe how we will identify if the safe work practices are being followed. The measures define what will be measured, how often, and in what manner. For example (lockout locks as stated above): "The measures would be to observe every lockout lock issued on a quarterly basis and ensure each lock is identified in some manner to the employee the lock was issued to, and ensure the lock was a designated lockout lock. In addition look for these locks on equipment or machines or toolboxes not involved with lockout. Then observe 10% of all lockout processes during the month and ensure that for each person working the lockout has a lockout lock for each person per energy source, or a group lockout box is used." Listing the intent, action and measures will help the team develop an understandable, measurable effective safety standard. The safety professional may want to identify all the safety issues to be addressed per safety process and provide the team with ideas and ranges of acceptable action that can be taken.

Safety and health standards need to be integrated into the day-to-day operations of a company. To do this, we have found that safety and health standards are best when they are modified into the form of a question. For example, one standard for lockout locks would be all locks must be assigned to the person using the locks and the locks must identify to that person in some way. The question would be "How do we identify the assigned locks to each authorized employee?" When the question is answered, the answers become the company's safe work practice for that area. We have learned that regardless of how well safety and health standards are written, some managers and employees will still not understand them. However, we learned that almost everyone is comfortable answering questions. We suggest that a team of hourly and salaried employees affected by each safety process be developed to create their own site-specific safety process. The team will use the standards and questions to complete their safety process. The team would also be responsible for developing the following:

- Safe work practices for each safety element of each safety process. (Answer the questions.)
- Develop a plan of implementation for the process. (What equipment is needed, what training is required, who will do the training, how will the training be

documented, how will the equipment be issued, what is the time table, what comes first, and the completion date?)

- Develop a budget for first year and thereafter. (What will it cost the company for all aspects of this safety process?)
- Develop defined measures to ensure the safety process is effective. (What will the measures be, who will do the measures, how often will the measures be taken, what will happen to the results, who is responsible for making changes if the measures indicate problems?)
- Identify new equipment needs. (List the specific equipment needed and the cost. What is the purpose of the equipment and how does it affect the safety process?)
- Define how the process will be audited and when.

We have learned that safety performance measures can only be taken when effective standards are developed and integrated into the day-to-day operations by creating safety processes for each standard.

Safety processes are often called safety programs. We use the term process because process indicate how the practices and procedures will be completed. A program indicates what is needed. Examples of safety processes would be: lockout, permit required, confined space, hazard communications, personal protective equipment (PPE), hearing conservation, machine guarding, and others. These processes can be measured, as explained in the next section.

MEASURE PERFORMANCE, NOT JUST WRITTEN DOCUMENTS

Safety performance measures often involve reviews of written safety processes. The purpose of a written safety process is to document the actual practices that are supposed to be followed. Reviews of written safety processes tell you that the company has documented safety procedures for a specific safety area, and if the process is in conformance with the company safety and health standards.

Safety processes need to be written as actual practices rather than a compliance document. Compliance documents are difficult to measure since most of the time they only state what issues need to be addressed rather than how the company has addressed the standard issues. For example, a safety process sentence, written as a compliance document, might read "Lockout requires specific identified locks issued to each authorized employee." How do you measure this? On the other hand, a safety process written as an actual practice might read "Lockout locks will be Master brand with a picture of the authorized employee on the front of the lock." You measure this by looking at each authorized employee's lockout lock to ensure they have the employee's picture on the lock. If you observe 100% of all authorized employees and find all the locks issued are Master locks and have a picture of that employee on the lock, then the measures show a 100% conformance to the safety standard, and the practice is effective. We have set a measurement standard using percentages of total reviewed, observed or interviewed, and ratings that correlate with these percentages. They are as follows:

- 0 ratings are less than 50% of the standards were effectively addressed in the records review (less than 50% of the employees interviewed knew the correct safe work practices for the safety process, and less than 50% of the observation made met the safe work practices defined in the written safety process)
- A 1 rating equals 51 to 99% of the standards effectively addressed in the records review, 51 to 99% of the employees interviewed knew the correct safe work practices for the safety process, and 51 to 99% of the observations made met the required safe work practices)
- A 2 rating means that 100% of all standards were effectively addressed (100% of all employees interviewed knew the correct safe work practices, and 100% of the observations made met the safe work practices for the specific safety process)

Ratings and percentages are given to each safety process and each element within that safety process.

Example 1

Using lockout, the safety elements would be:

- Equipment evaluation
- Lockout procedures for lockout set
- Lockout procedures for release of lockout set
- Lockout procedures for group lockout set
- Procedures for lock removal set
- Procedures set for lockout equipment
- Procedures for periodic inspections set
- Procedures for temporary test set
- Procedures for training on lockout set
- Energy source labeled
- Contractor control procedures set
- New equipment EID (Energy Isolating Devices) procedures set
- Shift change procedures set
- Certifications completed quarterly.

Each of these areas will be identified in the written safety process with defined actions, controls, or safe work practices assigned to each. The written safety process tells how a company will meet the standards without restating the standards. Each of the previously mentioned safety issues for lockout will be measured by records review, observation, or employee interviews or a combination of all three.

The idea behind measures is to validate that each area has been addressed. The records review evaluates the written lockout process to determine if each of the above areas were effectively addressed and met the safety and health standards developed. Each area will be evaluated and rated. In this example, a maximum of 56 points of measure can be obtained (14 areas × the maximum value of 4 × 14 safety issues = 56). The measures will show which areas were successful and which areas need to be improved. The measures will also give you an overall measure for the safety process.

Example 2

Possible points are developed from each safety area. Using equipment evaluations for lockout, we have 145 machines that must be evaluated. One point is given for each machine. In addition if we use observations to validate each machine was evaluated this equals another 145 possible points. The total for equipment evaluation would be 290. One point is given for each machine that has been evaluated and validated by observation. Then we take the total number evaluated and validated and divide by the total possible. This provides a percentage of efficiency.

The percentage level is then correlated to a 0, 1, or 2 rating level as stated above, and the equipment evaluation safety area would be measured. This area would be given a percentage value and a rating. When adding this area to all others, we simply add the total possible ratings used. This area gets two maximum points for records review and two maximum points for observation. This gives the safety area a total of four points possible.

The following are specific examples to explain the measurement process:

- If 45 out of 56 rating points were scored, the overall percentage would be 80% or a 1 rating (45 divided by 56). After the measures are completed, management only has to improve those areas that score less than 100% or any ratings of 0 or 1. This type of measurement system evaluates the performance of the written safety process against the company standard rather than just ensuring that all the paperwork is in place.
- An example, still using lockout, would be:
  1. Equipment evaluation: The company has 145 specific equipment that needs to be evaluated. Of that 145 only 100 evaluations were completed. This results in 68% (100/145) or a 1 rating.
  2. Lockout procedures set, seven steps for lockout, of the seven, seven were identified in the written process. This equals 100% conformance to the standard if you use just the written review. If you use interviews, and out of 75 authorized employees interviewed only 65 employees were able to state the seven steps, the percentage would be 86% or a 1 rating.
  3. Lockout release procedures set — The records review show all five steps for release were identified but out of the 75 interviewed only 65 knew the procedures. This provides an 86% efficiency level or a 1 rating.
  4. Group lockout procedures set — Five steps were set up and this meant 100% conformance to the standard. The interviews showed that 55 out of 75 interviewed knew the procedures. This provides a 73% efficiency level or a 1 rating.
  5. Lock removal procedures — All procedures were set and all employees interviewed knew the procedures. This provides 100% efficiency or a 2 rating.
  6. Lockout equipment — All required equipment was identified and all interviews and observation showed the equipment was available and used. This equals 100% conformance to the standard.
  7. Procedures for periodic inspections were set and 100% of the employees interviewed knew the procedures and were part of that process this year. This equals 100% conformance and a 2 rating.
  8. Procedures for temporary test set — The procedures were set and met the company standards however less than 35 employees interviewed knew the correct procedures. This equals a 46 % efficiency level or a 0 rating.

9. Procedures for training on lockout set — All authorized and affected employ-
    ees were trained, however out of 75 interviewed only 58 knew the procedures
    for all aspect. This equals a 77% efficient level or a 1 rating.
10. Energy source labeled — Only 100 out of 145 machines were labeled as
    required. This equals an efficiency level of 68% or a 1 rating.
11. Contractor control procedures set — All procedures were set. Interviews of
    all contractors showed only 15 out of 35 contractor employees knew the
    procedures. This provides a 42% efficiency level or a 0 rating.
12. New equipment EID procedures — All knew equipment was provided with
    EID. This equals 100% efficiency and a 2 rating set.
13. Shift change procedures set — All procedures were set but only 45 out of 75
    employees interviewed knew the procedures. This provides a 60% efficiency
    level or a 1 rating.
14. Certifications completed quarterly — All were completed quarterly giving an
    efficiency level of 100% and a 2 rating.

Results

- Equipment evaluation: records review 2, observation 1
- Lockout procedures for lockout set: 2 records, 1 interview
- Lockout procedures for release of lockout set: 2 records, 1 interviews
- Lockout procedures for group lockout set: 2 records, 1 interview
- Procedures for lock removal set: 2 records, 2 interviews
- Procedures set for lockout equipment: 2 records, 2 observations
- Procedures for periodic inspections set: 2 records, 2 interviews
- Procedures for temporary test set: 2 records, 0 interviews
- Procedures for training on lockout set: 2 records, 1 interview
- Energy source labeled: 1 record
- Contractor control procedures set: 1 record, 0 interviews
- New equipment EID (Energy Isolating Devices) procedures set: 2 records,
  2 observations
- Shift change procedures set: 2 record, 1 interview
- Certifications completed quarterly: 2 records, 1 interview

This gives a total of 28 possible ratings for records review. The recorded example
gave a total of 25. The percentage for records review is 89% efficiency level or a
1 rating. The total for interviews equals 28 points and the recorded level was 16.
The efficiency level was 16 divided by 28, or 57% or a 1 rating. The total for the
safety process was 56 with a recorded value of 39 or 39 divided by 58 or 67% and
a 1 rating. Management can now look at the safety issues that did not record a
2 rating as a target for improvement.

MEASURE SPECIFIC SAFETY PROCESSES

We have organized our Safety and Health Standards into categories. This helps
to keep safety processes with similar administrative requirements in the same area.
We suggest that each company define their own method. As an example, we broke
down all standards into the following areas:

- Records and OSHA/EPA management — all records required to be maintained, inspection reports, citations, abatement plans, postings and other related documents
- Required written safety and health processes — all safety processes that require administrative controls, employee training, or defined controls and should be in written form (lockout, permit required confined space, hazard communication, etc.) (While OSHA does not require safety process to be in written form, (yet) it is difficult from an ISO QS 9000 or measurement basis to effectively evaluate any safety process that is not in written form)
- Physical safety controls — requirements for physical controls for equipment, buildings, and processes (e.g., ladders, welding, steps, and fall protection)
- Safety team processes — the how-to aspects for injury investigations, safety committees, safety training, job analysis, medical case management, and other safety and health management areas
- Safety recognition and incentive awards systems

Each safety process would require a member of the management team to act as the champion or leader for the process. The leaders need to be selected by the management team and need to be directly affected by the safety process they will lead. For example, lockout and permit-required confined space should belong to the maintenance department, press operations belongs to the press manager, while hearing conservation belongs to the nurse, and so on. The best way to select a safety process champion or leader is to use a simple philosophy that the people that do it, own it.

So why do safety processes need a champion or leader and what do safety professionals do? The leaders are needed to ensure that the safety standards set by the company are specifically addressed in their safety process and that hourly and salaried employees are involved with the development of that safety process. The leaders are responsible to ensure that safe work practices are set up, employees are effectively trained, supervisors lead the process and all aspects of the process are integrated into the day-to-day operations of the company. This method allows for greater involvement of employees at all levels. The people that are affected by the safe work practices, then own the safety and health issues. They are responsible for the safety process success or failure. Safety professionals help to facilitate the safety process development and ensure that all answers to the questions meet the company standards. Safety professionals also help the teams to understand the consequences of their decisions and to know range of choices they have.

The safety process leaders can also help when completing the measures for each safety process. They can identify all affected people, information on the training available, changes made since implementation, and identify the measures they have taken themselves. When the safety performance measures are completed and they indicate a need for improvement, the safety process leaders are responsible for making changes to the process to ensure success. When safety process or areas within that process record a 0 or 1 rating level, then revisions are needed. The use of the safety process leaders and their teams will greatly expand the resources needed for improvement. The safety professional will be responsible for monitoring the activities of each team, rather than completing all the changes themselves. The results will come faster and with greater return than from placing all the responsibility for modifications on the safety professional.

As we described earlier for the lockout safety process, each safety process will have specific areas that need to be addressed. Each will be measured using records reviews, interviews and or observations. Each area will record a percentage of effectiveness (the outcome divided by the total possible outcome) and then that percentage will correlate with the rating number. Using individual safety processes as a focus for measurements will provide the company with specific detail for evaluating each area of a safety process, then the safety process as a whole, then the company as a whole (adding the results of all safety processes). The total measures for each safety process and the measures for each area within that process will identify the areas that need additional improvements and modifications. Management will then only have to focus on the areas that require improvements (those areas that recorded measures of less than 100% and a rating of 0 or 1).

MEASURES MUST BE REPEATABLE AND BASED ON FACTS

The system used for all safety performance measures is based on the same percentage and ratings used for the records review. The values are used for records review, observations and interviews. The measures for all three measurement areas are discussed earlier in this chapter.

Employee interviews are used to determine the employee's level of understanding of the required safe work practices, the hazards involved and the controls needed. Employee interviews need to be uniform and consistent with the same questions asked to each employee. Interview questions need to be developed for each safety process being evaluated. Interview questions need to be simple, understandable and easy for employees to answer. Interview questions should require the employee to provide information about the safety process and not just answer yes or no. Example of questions can be: For the lockout process, simply ask the authorized employee to walk you through the lockout process and explain each step and explain why they take the respective action. For hazard communication (Materal Safety Data Sheets (MSDS)), hand them a container and ask them to show you the information available on that material. Interview questions take time to develop to ensure you get a return on your value. Make sure all the people involved with the interviewing, know the questions and the expected answers. Answers do not need to be word for word, but the intent should be met. Interview answers need to be recorded. We use a simple form with the question stated and a column for aware and a column for unaware. Each question should provide you specific information for selected safety areas of each safety process. Not all areas require a question in order to validate the safety process. Areas such as periodic inspection, hazard assessments, and inspections made are a go/no go process. They either exist or they do not.

Interviewers also need to be trained how to interview. Interviewers need to show no expression on their face when answers are given. Whatever an employee states is their perception and must be measured accordingly. We suggest using an introduction to the employee stating, "We are asking employees questions about safety and the more accurately you answer the question, means your plant has done a good job, and if you do not know the answers, it simply means your plant needs to do a better job and is not a reflection of you." This helps take the pressure off the employee. We also ask the questions on the line in their work area rather than in a conference room, as employees often feel intimidated when they are brought into a conference room. We also interview with the knowledge of their supervisor, but without their presence. We inform management of the results as far as a percentage, but do not provide names and answers from those employees.

We get the percentage and rating in a very simple manner. We take the total number of employees interviewed, then divide that number into the number of correct answers. The percent is then correlated to the rating scale. Each question is rated separately. When interviews are completed on separate shift, the total numbers for each shift are added as well as the total number of correct answers. Each shift will have their own percentage and rating, but the plant will have a total that equal the sum of all shifts. This allows the safety professional to identify the shifts that have the least amount of knowledge for each safety process. Using the same questions and training the interviewers will ensure that the measures are consistent and repeatable.

The second method for measuring is observation. We use the same measurement systems as stated earlier. We observe employees involved with a specific safety process and note if the employee completes their assigned task as stated in the written safety process. Example: Using Lockout again, there are seven steps for Lockout and five steps for Release. The names of the authorized employees would be identified in the written safety process. As you observe the authorized employees completing a lockout, watch for the seven steps for lockout and five steps for release. The measures are once again simple. If the authorized employee completes five of the seven steps that equals 71% or a 1 rating. You add the total number of correct actions divided by the total possible actions. This gives you a percentage, which relates to a rating.

Both observations and interviews require a statically acceptable number of observations or interviews for the populace. We generally use 37 to 40% of the affected populace for interviews or observations, unless the group involved is less than a 100 people, then we use 100%. Using the safety standards as the basis, all observations would be consistent, repeatable and based on facts.

Now, how do we decide which safety section requires a records review, observation, or interview? Every company has to decide which measurement media to use for each safety section. Using the lockout example where R = Records review, I = Interviews, and O = Observations:

- Equipment evaluation, RO
- Lockout procedures for lockout set, RIO
- Lockout procedures for release of lockout set, RIO
- Lockout procedures for group lockout set, ROI
- Procedures for lock removal set, RIO
- Procedures set for lockout equipment, RIO
- Procedures for periodic inspections set, RO
- Procedures for temporary test set, RIO
- Procedures for training on lockout set, RO
- Energy source labeled, RO
- Contractor control procedures set, R

- New equipment EID procedures set, R
- Shift change procedures set, RIO
- Certifications completed quarterly, RI

This is an example of only one safety process. Each safety process will have different safety areas and different needs for record review, observations, and interviews. Also, when completing the measures, safety professionals do not have to review each safety process. They can select the targeted safety processes they want to measure.

MEASURES NEED TO BE UNDERSTANDABLE AND RELATE TO A VALUE

The measures stated previously show percentage and a rating. Business leaders understand percentages as long as they have a basis to compare against. The bases for these measurements are 100%, which means all records are in place and effective, all interviewed employees knew their responsibilities to the safety process and that all observations made followed the safe work practice. Celebrate, the safety processes are successful. Business Leaders can also identify the areas that succeeded and which areas need improvement. Even though a plant may gets a 1 rating with a 51% effectiveness, and another safety area records a 1 rating at 97%, management can easily see that the 51% safety issue needs a lot more work. Follow-up measures can also show management if improvements have been made.

Other measures that can be used are:

- Cost of worker's compensation cost per employee, or $100 of payroll
- Severity ratio, or the number of recordable cases before they become a lost time cas
- Total liability cost per safety process.

We will discuss these in detail in the next section.

SAFETY MEASURES DIRECTLY CORRELATE WITH THE BOTTOM LINE

How do we show a cost for each safety measure? We first have to identify the possible liabilities to a company, identify the actual or projected cost for each liability and correlate the cost to the safety performance measures. Liabilities that have a direct relationship to safety and health performance can be broken into three areas:

- Mature cost of worker's compensation
- Third party litigation resulting from wrongful death and other liabilities to the public
- Citations issued from federal or state agencies for violations of safety and health laws

Mature worker's compensation cost must be assigned to a specific safety and health process, based on the type of injury and causation. The principle behind this is that for every work-related injury, a failure in a safety process had to exist. For example, an eye injury occurred because the correct personal protective equipment, which would have prevented the injury, was not worn. Therefore, the safety process that failed was PPE. Another principle to keep in mind is that the greater the efficiency level of a safety process, the lower the liabilities and related cost to the company.

We must develop our liability cost per safety processes so we can correlate safety performance measures to the bottom line. The cost of liabilities per safety process is simply developed. To do so, list all the injuries recorded on the OSHA log for at least a two year period. Identify the cost for each of those injuries, then identify the safety process that failed for that injury to occur. Use a spreadsheet that lists the injuries, injury cost, date of injury, name of employee, and safety process involved. We suggest that all injury cost be mature dollars. Your insurance carrier can provide you an actuarial value to create mature dollars. Sort the spreadsheet by safety process, add the totals, and you have the basic cost for each safety process. You can also add employee surveys to identify safety processes that are not seen by the employees as effective. Cost can be added by using the average cost of an injury for the specific safety process.

We have to develop a relationship of total cost to percentage point of efficient safety performance measures. The percentage points divided into the actual dollar cost will provide you a cost per percentage point of effectiveness. Now let us put real dollar cost to the measures we have discussed. Using Lockout as a safety process, and a cost of implementing that process at $75,000. The $75,000 is the start-up cost and the annual maintenance cost is set at $3,500. The $3,500 pays for new locks and needed training. The $3,500 is fixed cost. The lockout process was evaluated and measured with a percent of efficiency of 59% the first year. The total liabilities as described above are projected at $300,000 for injuries and $75,000 for possible OSHA citations. This equals $375,000 dollars at a 59% efficiency. This amount reflects the variable cost of the safety process. If the efficiency rating was 0, the variable cost would exceed $635,593. If the efficiency rate were 90% the total variable cost would be $63,435 or $6,355 per percentage point of efficiency. The $3,500 is fixed cost and would remain the same no matter what the liabilities or variable cost would be. The measures now reflect dollars that directly relate to the bottom line. The numbers used are only an example and every company must use their actual cost figures. This also gives you a basis for developing a safety business plan, which will be discussed later. Your financial management can help you develop an acceptable formula for your company.

SAFETY MEASURES MUST BE COMPATIBLE
WITH OTHER BUSINESS MEASURES

Business measures focus on the effects they have on the P & L (profit and loss statement). Business measures also focus on quality and efficiency levels The safety performance measures we have discussed, show a direct correlation to the P & L. The safety measures show specific percentage of effectiveness relating to the company standards. The measures also relate directly to the quality of a safety process. Every measure we have discussed is compatible to all other business measures. The liabilities created due to ineffective implementation of specific safety process are real dollars that affect the bottom-line. Every business measure has a direct effect on the company's bottom line.

OTHER VIABLE MEASUREMENTS

The cost of worker's compensation per employee or cost per $100 of payroll can be used as a safety performance measure. However, to direct this cost to a specific safety process, the cost per employee involved in the safety process must be used as well as the number of employees involved. To use the cost for the entire company will yield only an overall picture, rather than a direct focus on a specific safety process.

Severity Ratio allows you to project the cost of future liabilities for each safety process if no actions are taken to improve the process. You are using the average cost of a recordabe and a lost time cases. The cost is then multiplied times the projected number of cases if no actions are taken at the current percentage of efficiency. You can also reverse this process by showing the reduction of cost per percentage point of efficiency. The objective is to create an efficient safety process that reduces the number of injuries and especially the lost time cases.

The cost per safety process as described above will tell management the cost of the total liabilities to the company from a failure in any or all safety processes before improvement actions are taken. Continually measure these liabilities over a rolling twelve-month period to identify the cost reductions after improvement plans are implemented.

Safety performance measures take time to set up and implement. After the process is set up, continual measures take very little time or effort and will yield significantly greater value and information than OSHA frequency rates have ever provided.

CONCLUSION

Traditional safety performance measures focus on outcome results without identifying the factors that produced those results. The measurement process we have described measures the inputs and the outputs, while identifying specifically the areas that require improvements. The process also allows the company to measure the effectiveness of corrective actions taken to improve the safety process.

Safety performance measures need a solid base (see Figure 16.1). We start by developing quality measurable safety and health standards. From the standards we create an effective measurable safety and health process that will serve as the baseline for measures. We then identify the safety areas per safety process that will be measured and select the type of measurement process to use. We can use records reviews, interviews, and observations. All measurement sources help to validate the effectiveness of the safety process. Then by safety process we identify the maximum possible measures that will represent a total quality safety process. Interview questions are set up as well as observation points. The measures are taken and reported to the management. Action plans are established to improve all safety areas that recorded measures of less than 100% or 0 and 2 ratings. As the plans are implemented, measures are taken again to ensure the success of the plan.

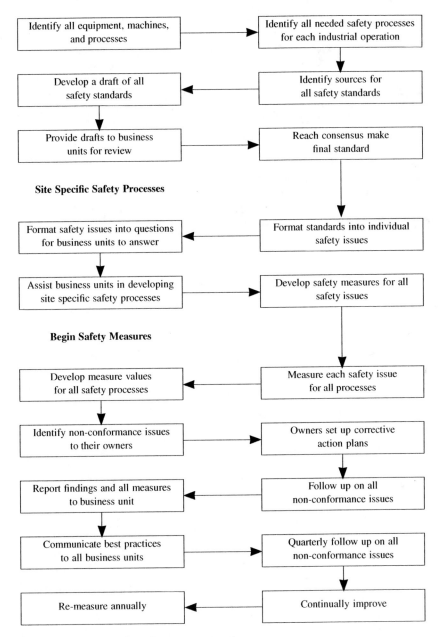

Safety Standards

| Identify all equipment, machines, and processes | Identify all needed safety processes for each industrial operation |

| Develop a draft of all safety standards | Identify sources for all safety standards |

| Provide drafts to business units for review | Reach consensus make final standard |

Site Specific Safety Processes

| Format safety issues into questions for business units to answer | Format standards into individual safety issues |

| Assist business units in developing site specific safety processes | Develop safety measures for all safety issues |

Begin Safety Measures

| Develop measure values for all safety processes | Measure each safety issue for all processes |

| Identify non-conformance issues to their owners | Owners set up corrective action plans |

| Report findings and all measures to business unit | Follow up on all non-conformance issues |

| Communicate best practices to all business units | Quarterly follow up on all non-conformance issues |

| Re-measure annually | Continually improve |

Figure 16.1 Flow chart for safety measurement process.

This process not only measures the effectiveness of safety and health processes but also helps management to focus their resources on the areas that create the greatest liabilities to a company.

REFERENCES

Petersen, D. *Analyzing Safety Performance*. New York: Garland Press, 1980.

Covey, S. *The Seven Habits of Highly Successful People*. New York: Simon and Shuster, 1989.

Huyinl, D. and Westover, C. *ISO 9000*. New York: Irvin Professional Publishing, 1994.

Petersen, D. Integrating safety into total quality management. *Professional Safety*, 28–30, June 1994.

Petersen, D. The four Cs of safety culture. *Professional Safety*, 32–34, April 1998.

Blair, E. Achieving a total safety paradigm through authentic caring and quality. *Professional Safety*, 24–27, May 1996.

Kozak, B. and Clements, R. Building safety into your ISO 9000 program. *Professional Safety*, 31–32, September 1998.

Hansen, L. Re-braining corporate safety and health. *Professional Safety*, 24–29, October 1995.

Rosier, G. OSHA statistics and safety performance measurements, a case for change. *Professional Safety*, 32–34, September 1997.

Manuele, F. A. *On the Practice of Safety*. New York: Van Nostrand Reinhold, 1993.

Crosby, P. B. *Quality Is Free*. New York: McGraw-Hill, 1979.

## CHAPTER 17

Ethics and Professional Conduct within the Safety and Health Profession

Philip E. Ulmer

CONTENTS

INTRODUCTION

In the last quarter of the twentieth century, the safety profession within the United States has experienced an increase in punitive actions being taken against those who are found guilty of professional misconduct, either via the courts, through criminal or civil action or through actions taken by professional societies, licensing boards, and certification bodies. This chapter presents a summary overview of the categories of ethics-related complaints that are filed against individuals in the safety profession. Additionally, a vision for ethics issues that the profession can expect in the early 21st century is offered.

ETHICS AND PROFESSIONAL CONDUCT

Ethics practitioners have been known to say that the mark of a true profession is when it must deal with ethical and professional conduct issues involving those who practice. Indeed, in all fifty states of our country various licensing boards are charged with policing ethical and professional conduct issues involving lawyers, physicians, engineers, and architects to name a few. Ironically, it is a somewhat sad indictment on a profession when a licensing board must exist to protect the public interest. After all, such boards are typically enacted by legislation only after a profession has developed a history of malpractice that has proven harmful to the public. The occupational safety and health profession is being confronted with having to deal with such ethical issues on a more frequent basis than ever before. In the 1990s, there was ongoing public debate in at least eight states regarding the licensing of safety and health professionals. It is outside the scope of this chapter to state unequivocally that the profession should be licensed, but the fact of the matter is that the safety and health profession is being placed under a public and private sector microscope regarding conduct issues.

What, if any, is the difference between ethics and professional conduct? For all intents and purposes, they tend to be one and the same in the eyes of the general public. This is most important when dealing with global issues. The members of the profession itself and even the private sector (corporations) tend to want to split hairs and wish to see a demarcation between the two. For example, a person can be ethically correct but have his or her conduct questioned, such as the manner in which he or she confronted a person. Those who wish to see the two as being different tend to take a very legalistic view of the world versus a practical view where relationships with people should be of profound importance.

Many have met legalistic people such as this — any social structure will have a significant percentage of people who have this characteristic. A professional of this type basically views ethics as what the law allows and disallows and will have tendency to skirt every opportunity to take professional advantage without breaking a law.

Here is a very elementary example. Safety professionals will give public presentations, many times at public registration professional conferences where the audience usually consists of the speaker's peers (i.e., members of the same profession). The agreement between the safety professional and the sponsors of the conference addresses electronic recording of the presentation. For this example, the speaker and the sponsors agree that no electronic recording will be allowed. The sponsor will post notices requesting the audience not record the oral presentation.

As a professional consultant in the field of occupational safety and health, the author agrees that it would be prudent for the speaker to insist on this as a condition of being retained as a speaker. Do sponsors check with each registrant to veryify that they are not concealing recording equipment? No. Would the speaker expect them to necessarily do so? No. Regardless of the agreement with the sponsors, is there a law to prevent a registrant from recording the presentation? Not specifically, because it would be difficult to enforce. There would need to be measures in place to have the registrant arrested by the local authorities if he or she was caught electronically recording the presentation. Ethically, the registrant (again, a peer in the profession) who is recording could argue that there were no laws broken and, therefore, he or she is clear of any wrongdoing in the eyes of the legal and judicial system of our land. Under most codes of ethics for state professional licensing divisions, the professional could not be brought up on charges; however, the registrant's professional conduct would be questioned.

In 1991, an article in a major business journal brought this subject to the forefront when talking about current events such as euthanasia, abortion, and other similar controversial issues. The article addressed the fact that certain types of people will let government define what is legally correct and will let that be the standard from which they wish to be judged as ethically and sometimes morally correct. Of course, the public has a harder time making such a distinguishing judgment.

The first recorded code of ethics, which dealt with the healing profession, was developed almost three thousand years ago by Hippocrates. When one critically observes the early developmental stages of a profession, the first stage is for a group to organize and proclaim and profess to have a special ability. This is where we get the word professional — a group that has made a profession of special abilities. The second stage is when that group dedicates the practice of those special abilities in service to humankind.

In any social environment, the group that is being serviced (i.e., the public) ends up deciding whether this group of practitioners should be recognized as professionals. In most if not all cases, the public makes this pronouncement only after being irrevocably harmed by malpracticing individuals. Before this takes place, there is a presumption by those who are serviced that they should put their faith and trust in the practitioners providing the services. In other words, there coexists a harmonious relationship — until harm is brought to bear on the one that is being serviced.

The interesting thing about this type of social interaction is that, historically, the breakdown and destroyer of that harmonious relationship can be tied to greed, power, lack of compassion and respect for others, laziness, and "trying to get something for nothing."

To be more specific, people who are identified as engaging in professional misconduct fail to recognize a value system that is accepted by those whom they service. Or worse, they recognize the value system and choose to ignore it.

In the safety and health profession, practitioners profess to having special abilities and they dedicate those abilities in the services they provide; however, this is where it gets gray and lesser defined. This leads to the question; Does the safety profession provide services to the public (composed of human beings and their biological and social laws) or to the corporate entity (whether profit or nonprofit in nature)? This writer has asked this question to scores of safety and health professionals throughout the United States in the last ten years. Interestingly, most will initially answer that they serve the public's interest. But when given the opportunity to think more deeply on the question, the vast majority will finally conclude their services are for the furtherance of the corporate good versus the social structure we call the general public. The value system that defines the corporate good may differ at times with the value system of the citizenry of the locality. This should not come as a surprise to the reader.

A profession at some point matures to a level where it defines its value system. Not only that, the safety profession must begin the task of providing a quality control over its own. If it does not, then government will tend to get involved in that aspect of safety professional's lives sooner or later. One of the services of professional societies is to assist a profession in determining its value system and to hold their members to that value system as a condition of membership. This is good, but it creates a gigantic loophole that can promote misconduct.

We can examine the American Society of Safety Engineers (ASSE). It has over 35,000 members and is considered the most recognizable professional society for the safety profession. All who choose to be associated with the ASSE agree to abide by the value system that is defined within that society — the Code of Professional Conduct and Fundamental Canons. This is good for 35,000 practicing professionals. What about the estimated 50,000 to 60,000 safety professionals who choose to not affiliate with ASSE? To what value system do they ascribe and who will hold them accountable? The fact is that no one in the professional community holds those individuals accountable.

Pressures can cause professional misconduct. They emanate from the personality traits of people and the personalities of the corporate entities that employ or retain safety and health professionals.

This writer is in a unique position to see the "bad fruit" that exists within the profession possibly much more than the average reader. Having been the chair of the committee on professional conduct within ASSE, served on the board of directors of ASSE (including chairman of the board and president), and serving on the board of directors (including chairman and president) of the National Institute of Engineering Ethics, this writer shares with the reader some of the issues involving cases that have been studied.

First of all, it will not be a surprise to the reader that there are bad apples within the safety profession. For anyone who has been in the safety profession for just a few brief years, there has probably been some interaction with a peer of this type. Those within the profession who must rely on safety professionals to assist with daily responsibilities may have had some bad experiences.

Behavioral psychologists generally affirm that, in a normal society with a population large enough to be considered statistically significant, one can expect about 60% of the population to meet the normal values set aside by that social structure. An observer can expect that about 20% will not meet the norms simply because of their immaturity in not knowing the difference between right and wrong — that is, babes within the social structure. We can expect 10 to 15% to show qualities exceeding, in a superlative way, the social norms. This leaves anywhere from 5 to 10% of the social structure that will choose to vary from the value system in a way that is distracting, causes harm, provides chaos, or these individuals simply become social outcasts because of their refusal to abide by the social norms.

The safety profession is considered a social group by any definition in the textbooks. Within ASSE and its population, one can expect at least 1,500 members who are or will be problematic from a professional conduct standpoint. Outside ASSE, an analyst would expect up to 5,000 to 10,000 more people who are out there ready to give a black eye to the professional cause. Many corporations and their employees are victims of these people, but these same companies and their employees may be the ones who provide the very foundation for misconduct.

Within the membership of ASSE during the mid-1990s, this writer had seen the tragedy of a member plead guilty in a federal district court to charges brought by federal prosecutors on behalf of OSHA. He was sentenced to probation with a location transmitter locked around his ankle. What did he do to deserve this? While a paid consultant to a corporate client, he falsified his client's OSHA-200 logs as they were filed with the Department of Labor. He got caught during a records audit resulting from an extensive OSHA investigation after a serious injury occurred.

Of cases that have been reviewed, misrepresentation of professional credentials and biographical backgrounds seems to occur at an alarmingly high frequency.

A safety professional in the western United States abused his company credit card privileges and even threatened to physically harm a co-worker if he or she told anyone about it. This person was brought up on criminal charges by the local district attorney and ultimately plea bargained by admitting guilt.

A safety professional brought charges against another ASSE member for flagrantly violating copyright laws. Both individuals were consultants in a small market where they competed against each other for what little business was available. The safety professional who brought the charges developed a very lucrative seminar series and established proof of copyright. A firm retained him to present his seminars internally to its employees. This firm had its own safety department, one member of which had his own secondary consulting business on the side. In his contractual agreement with the firm, the consulting safety professional allowed a clause for limited internal use of the printed materials, including reproduction — this is not unusual in a consultant/client relationship. The firm's safety department was the overseer and enforcer of the terms of the contract. One of the staff professionals allowed a copy of the materials to be given to a friend in the profession who had a successful consulting business. The friend removed the author's name from the document and replaced it with his own name. He then sold the product to his friend in a marketing deal. Fortunately, other safety professionals saw what was happening and alerted the author and owner of the product to the situation.

A manager of a corporate safety group, who managed a staff of more than ten safety professionals, assigned one of his group to analyze a risk situation and recommend proper resolutions. This was to be submitted in a written report by the assigned employee and delivered back to the staff manager. The employee did a fabulous job which was so impressive, that the manager decided to take the results of the document, in verbal and written overview form, up the line to the presidents and vice presidents; however, he decided to take credit for the work himself. The employee brought professional conduct charges against his own manager.

An industrial hygiene (IH) professional brought allegations against another IH for knowingly utilizing the wrong sampling protocol when trying to determine the risk exposure levels to a group of employees. The IH who committed the misdeed acknowledged that he did it on purpose, but only after getting his supervisor's endorsement and approval.

A safety professional bordered on slandering another safety professional in a public forum where both were testifying on a bill introduced in their state legislature. The bill, if passed, would have very narrowly defined who could be licensed to use the title safety professional and who could professionally practice in that state. Although at least eight professional credentials, which require competency by examination, are used in the broad practice of the profession, the bill as written only acknowledged one, which for the purpose of this true story will be identified by a pseudonym of professional credential XYZ. The safety professional who already earned the XYZ designation was a mover and a shaker who got the bill introduced and sent to committee for public comment. Coincidentally, he also was on the payroll of a consulting business that employed only XYZ's. He and his firmed lobbied hard for the bill. In the public testimony, another safety professional's testimony preceded the XYZ professional, and he basically stated that more latitude should be given to other acknowledged credentials for the purpose of determining who should be licensed to practice and use the title. The XYZ professional followed the testimony by attacking the credentials of his peer, using disparaging language because his peer chose to earn a credential other than an XYZ. Ultimately, the bill never got out of committee, and one legislator who witnessed the testimony stated that the safety profession was not a profession in his mind if two members of the profession were so diverse on answering the question of "What is a professional credential?" to the point that one attacked the other for earning a credential that took just as much study and had just as much relevancy to the practice within the profession.

What caused all of these scenarios to take place? In these real-life examples, they are consultants in private practice and safety professionals on corporate payrolls. In the social environment, one could successfully argue that it was greed, laziness, trying to get something for nothing, and lack of compassion and respect for others (i.e., not adhering to the Golden Rule). But in the corporate environment, I believe I can also say that corporate philosophies often place safety professionals in compromising positions.

On the latter, let us take this example. The ABC Corporation has a historical philosophy of 100% regulatory compliance. This is stated in corporate policies and the employer truly feels this is a paramount philosophy of the company. Although this may be a policy that looks good on paper and is impressive to the compliance agencies, it can cause ethical conduct problems for safety and health professionals. Why? The simple answer is that regulations and standards are derived from conflict and compromise discussions at the standards-making table. Typically, this means that in a large enough employee population in the ABC Corporation, there may be a small percentage of employees whom the regulation or standard will not sufficiently protect.

To promote 100% compliance in a case like this may actually bring more risk to an individual employee than should be allowed in the minds of a safety professional. He may be put in a position to recommend noncompliance in order to ensure the proper level of safety for an individual. In most cases this would mean exceeding the regulation with more safety, but some corporations will pressure for minimal compliance, which means follow the letter of the law, rather than meeting the intent.

Another example is the case regarding corporate incentive award programs. The PDQ Corporation has a policy stating that "all injuries, no matter how slight, will be reported to the supervisor." This policy, coupled with an incentive awards program that may have been in place for many years, could place many safety professionals in a very untenable position. Why? In some companies, the death knell to a person's career may occur when they believe it is professionally correct to buck tradition. For example, if a safety professional has seen clear evidence of this fact, he will be reluctant to professionally do the right thing. He hears directly from supervisors that the awards program creates too much pressure for people to not report injuries. He hears it from more than one supervisor and then does his own research of the available literature about the effectiveness of safety incentive awards. He finds credible evidence that the supervisors have a valid point in their views. He also knows that his company's injury and illness database must be accurate in order to be a tool of value for safety management purposes. If people are not reporting injuries, then the database must be false.

He also is now realizing why his company has such a sterling injury record that is reported each year on the National Safety Council (NSC) member company reports and the OSHA-200S forms. He advises his department manager of his findings and professional opinion and his manager, who is also a safety professional says he refuses to believe his findings, and the issue should be dropped entirely. After all, his (the department manager's) annual salary adjustment is predicated on the safety record of the company and he does not wish to jeopardize what has been a history of good pay increases. The PDQ Corporation, surprisingly, also evaluates and rewards the safety professionals based on the company safety record.

Many safety professionals who read the previous example would probably shudder. On the other hand, some non-safety professionals probably think that ought to be the way it is. Regardless, what is the safety professional's ethical responsibility? After all, he doesn't want to be insubordinate of his boss by pressing the issue, but in his heart he knows he is right.

CONCLUSION

Issues like this are tough and they will become tougher as the profession becomes more accepted as a true profession. In corporate life, one never hears about corporate attorneys or corporate physicians being asked to reconsider professional recommendations to their management. Why? Because the management understands that there are well-established codes of professional practice to which individuals who practice medicine and law must adhere. However, this writer can profoundly state that corporations will at times challenge the loyalty of a safety professional by asking him or her to reconsider or even drop issues that must be examined. Sometimes the challenges are done at the supervisory level, where they envision the safety professional as someone who can be manipulated or even ignored.

Although a central repository of records is not kept on safety professionals who are convicted in courts of malpractice for improper conduct or similar deeds, it does not take a rocket scientist to look at leading indicators that point to increasing occurrences. A practicing safety consultant can examine his or her professional liability insurance premium each year and realize there must be some reason that the rates keep increasing — it is because of malpractice claims experience.

In studies on this subject, looking ahead to the first few years of the 21st century, the following are the most obvious areas of ethics or professional conduct allegations that will be seen on the increase — in some cases rather dramatically:

- Gross and willful misrepresentation of professional credentials to clients and employers
- Full-time consultants no longer having an arm's length relationship with clients as they provide services; clients are more likely to report errors of commission, not be confused with errors of omission, of retained consultants
- Consultants bringing allegations against other consultants with respect to unethical business practices in the competitive marketplace, such as grossly stealing or copying materials that have been protected under the copyright laws of the nation
- Safety professionals who will attempt to perform professional services outside their areas of competence; we are seeing in the United States today an attrition in the number of what has in the past been termed safety generalists — specialization is becoming the norm. It is estimated by one analyst that our safety profession may have over 40 specific branches of specialization before the year 2005 (as of the year 2000, there's a mere 32, by one source).
- Emphasis on statistical scoreboard watching placed on safety professionals who do compliance and inspection work, or property/casualty inspections for insurance carriers and brokers, will leave them vulnerable to compromising their professionalism for the sake of meeting the necessary quotas presented to them. Allegations of playing the numbers will increase with those in positions where the line management expects certain numbers of account/client calls, certain number of on-site inspections, and certain numbers of violations found.
- Safety professionals in supervisory positions who will play politics for their career enhancement by gaining credit for work performed by subordinates
- The writer must close this list with the following prediciton that may seem surprising to the reader: One will see an increase in ethics complaints about members who are unprofessional in carrying out their duties as elected or appointed officers in volunteer society work.

The only value systems to which the safety profession can refer are those in professional societies and certification organizations. Whether one is in the safety profession or the recipient of a safety professional's services, it is recommended that the reader consult for further reading a series of essays, *Safety Profession — Year 2000*, published by ASSE. One should focus particularly on chapter three, written by John Russell, on professional practice and ethics.

PART III

Environmental Management Aspects

## CHAPTER 18

Environmental and Hazardous Materials Management Aspects

Marc Bowman and Robin Spencer

CONTENTS

INTRODUCTION

In most companies today, any discussion of safety and health management must consider the several aspects of handling hazardous materials or hazardous wastes in day-to-day operations. Successful companies promote an environmental management program which has been developed to establish and follow corporate environmental philosophy, policies, and procedures to promote compliance and minimize future liabilities. This chapter is intended to provide an overview of environmental management aspects when operating with hazardous materials. The environmental management program can often parallel corporate safety and health programs.

THE IMPORTANCE OF REGULATORY COMPLIANCE

It should be clearly understood by all employees at a facility that compliance with the laws, regulations, and procedures are not only company policy, but that it is expected to be incorporated into every aspect of doing business on a daily basis. The environmental laws and regulations have been established to ensure employee health and safety as well as to protect public health and the environment. In addition, it makes good business sense. The costs of remediation can be enormous from the acute effect of spills and releases or the chronic effect of long-term mismanagement of hazardous wastes and hazardous materials.

If these reasons are not enough, every employee should understand the enforcement provisions for not complying with laws and regulations. While the actual assessment of fines and penalties is a complicated process, as discussed briefly below, they may be very significant. For example, the federal hazardous waste law includes civil penalties of up to $25,000 per day for each violation if the EPA determines that a person has violated or is violating any provision of the law. In addition, individuals may be subject to fines of up to $50,000 per day for each violation and imprisonment for up to two years.

The Department of Justice follows a set of federal guidelines in assessing penalties for cases of environmental noncompliance. These guidelines outline seven basic components for courts to weigh when evaluating a sentence for an environmental crime:

- Line management has to be routinely involved in the environmental compliance program.
- Environmental policies, standards, and procedures must be integrated throughout the organization (up and down the hierarchy).
- An independent auditing, monitoring, reporting, and tracking system that uses terms familiar to all employees should be in place.
- The organization must have regulatory expertise.
- The organization has a system of incentives or rewards for environmental compliance consistent with programs for sales or production.
- The organization has consistently enforced corporate environmental policies using proper disciplinary procedures.
- A way to measure continuing improvement in the organization's effort to achieve environmental excellence. Both internal and external audits as part of an overall Environmental Management Program can measure compliance.

It is clear that a coordinated proactive environmental management program is highly desirable, if not essential.

ENVIRONMENTAL MANAGEMENT SYSTEMS

The environmental management program referred to in the previous section is commonly known as an environmental management system (EMS). An EMS is considered to be an integral part of corporate management structure in most corporations throughout the world today. Numerous environmental management standards or schemes exist throughout the world today, which can be used as a model for developing such a program. These include:

- British Standard 7750
- Eco-Management and Audit Scheme (EMAS)
- Responsible Care
- American Society for Testing and Materials (ASTM) Environmental Management System

However, the most commonly used and most widely recognized environmental management system model is the International Standards Organization (ISO) 14001 standard found in the ISO 14000 series. This series followed the widely used and successful ISO 9000 series, which dealt with quality assurance.

As with any EMS, ISO 14001 standards may be applied to any type or size of organization. In a relatively few pages, the ISO 14000 series outlines specific requirements and principles for environmental management. These standards are voluntary and are not intended to increase or replace existing legal requirements, but rather to help organizations manage environmental obligations (such as compliance with legal requirements).

The ISO 14000 series provides standards for the following areas:

- Environmental management systems
- Environmental auditing
- Environmental site assessments
- Environmental performance evaluation
- Product evaluation

However, ISO 14001 is the only specification standard and the only standard to which an organization can be certified. The primary elements of an ISO 14001-based EMS include:

- Policy
- Planning
- Implementation and operation
- Checking and corrective action
- Management review

These elements involve the classic "Do-Check-Act" cycle of a good management system.

To become ISO 14001 certified, an organization must develop an environmental policy, identify environmental aspects, establish environmental objectives and targets

for significant environmental aspects, develop and implement environmental procedures, commit to pollution prevention, train all personnel whose job responsibilities
may affect the environment, create a hierarchy of auditing and review to ensure the
EMS is functioning properly, and take corrective action to facilitate continual
improvement.

An organization may seek to obtain an ISO 14001 certification which typically
involves a independent, third-party certification audit. An organization may decide
to stop short of a full certification through self-declared conformance to ISO 14001.
In either approach, the organization must comply with all of the requirements of
ISO 14001 and should achieve the same level of environmental management benefits.
The decision to attain full certification is a function of company philosophy and/or
the requirements of customer/supplier obligations.

AUDITING UNDER AN EMS

As described above, an ISO 14001-based EMS follows the classic Do-Check-
Act cycle of a good management system. The checking elements of the EMS involve
auditing. There are typically four types of audits or checks within an ISO 14001
EMS. These include:

- Compliance verification audits (the traditional compliance audit)
- EMS audits
- Management reviews
- EMS certification audits

Compliance Verification Audits

Environmental compliance audits are tools used by most companies to assess
the status of compliance with environmental laws and company policies and procedures. An audit is a snapshot of conditions at the facility at the time of the audit
and provides a starting point for activities to bring the facility into full environmental
compliance if exceptions to compliance are identified. There are three key components of a compliance audit: (1) audit preparation, (2) the site visit, and (3) management of the audit findings. These key components are shown in Figure 18.1 and
are discussed below.

Audit preparation is critical for an effective audit. Prior to visiting the site, the
audit team should establish lines of communication with the facility, corporate staff,
and legal council (if involved). The audit team should then submit to the facility a
pre-audit questionnaire which will help to compile all the information needed to
conduct the audit. Some information is sent to the audit team in advance of the site
visit while other information is made ready for inspection during the site visit. Based
on a review of the completed pre-audit questionnaire, the audit team should then
customize the audit approach in preparation for the site visit.

While at the facility, the audit team will hold an opening conference with site
managers, conduct site inspections, review environmental files and records, interview

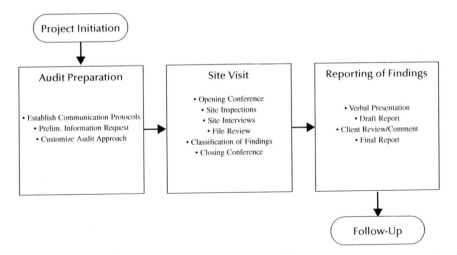

Figure 18.1 Compliance audit steps.

key facility personnel, identify exceptions to compliance, and verbally discuss preliminary audit findings during an exit interview with site managers. The site visit portion of the compliance audit is usually conducted using audit checklists. These checklists may be simple or very lengthy and detailed, depending on the complexity of the issues expected to be encountered at the site.

Management of audit findings is a critical component of the auditing process which involves both the audit team as well as the facility and corporate personnel responsible for environmental compliance. Audit findings can be classified into four issues:

- Compliance Issues — issues that, in the opinion of the auditor, are out of compliance with existing, applicable laws, regulations, or codes
- Management Issues — issues (which are not compliance issues) that do not reflect the current best management practices in similar industries for management of environmental issues
- Significant Liability Issues — issues (which are not compliance issues) that, in the opinion of the auditor, pose significant potential risk to the facility, either as significant environmental liabilities, significant risks to public health, or significant threats to process equipment or production capability
- Punchlist Issues — minor issues that can be readily corrected with minimal resources (usually that day or the next)

Any written report of findings should be objective and without recommendations and should have limited distribution to prevent access by unauthorized persons.

How the audit findings are reported varies widely depending on the needs of the company as well as advice from legal counsel. Of primary importance, however, is the development of a corrective action plan to address all of the audit findings. For each audit issue, the corrective action plan should identify the proposed corrective actions to mitigate the audit finding, the person or team responsible for implementing the corrective action, and the proposed completion date for the corrective action. The corrective action plan should be reviewed at least quarterly thereafter to track the progress of the corrective actions. The status and completion of the corrective actions should be documented and filed with the audit report.

EMS Audits

EMS audits are conducted similarly to compliance verification audits except that they focus on the management systems in place as opposed to regulatory compliance. For example, one area of a compliance verification audit would focus on hazardous waste drum compliance by actually examining each drum for proper handling and labeling. An EMS audit on the other hand would focus on the policies, procedures, and training around drum management to ensure the management systems were in place and were effective. The EMS audit is more of a root cause analysis approach. In some cases, it may make sense to combine compliance verification audits with the EMS audits. Under a good EMS program, less emphasis would be given to compliance verification audits and more emphasis would be given to EMS audits over time.

Management Review

Under an ISO 14001-based EMS, management is required to review the status of the EMS at least annually. This usually involves review of compliance verification and EMS audit results, examination of objectives, and achievement of targets. The management review is intended to keep management actively involved in the EMS.

Certification Audits

The certification audit is usually performed after an EMS has been developed, implemented and in place for one to two years. The certification audit is usually performed by an independent, third-party with experience in certification auditing. The purpose of the certification audit is to assess if the EMS meets all of the requirements of an ISO 14001 EMS.

DEFINITION OF HAZARDOUS MATERIALS AND HAZARDOUS WASTES

Hazardous materials, hazardous substances, hazardous chemicals, and hazardous wastes are legally defined terms in laws such as the Occupational Safety and Health Act (OSHA), Resource Conservation and Recovery Act (RCRA), Toxic Substances Control Act (TSCA); the Transportation Safety Act, Comprehensive Environmental Response, Compensation and Liability Act (CERCLA), and the Emergency Planning and Community Right-to-Know Act (EPCRA). Generally, a hazardous material is any substance or mixture of substances having properties capable of producing adverse acute or chronic effects on human health and safety or the environment. The rule-of-thumb guidance is that if a material has a Material Safety Data Sheet (MSDS), it is probably a hazardous material.

Hazardous wastes are typically hazardous materials that no longer have a beneficial use. According to the RCRA, the law that regulates solid and hazardous wastes, for a waste to be a hazardous waste, it must first be determined to be a solid waste. Solid wastes include solids, semi-solids, liquids, and containerized gases. Generally, solid wastes are materials that are discarded, abandoned, disposed of, disposed of by burning or incineration, accumulated in certain ways, recycled or reclaimed in certain ways or are inherently waste-like.

Once it is established that a material is a solid waste, a hazardous waste determination must be made, and in fact, is required by law. Solid wastes are determined to be hazardous wastes by checking to see if the material is on a list of hazardous wastes and/or checking to determine if the waste exhibits certain characteristics. These characteristics include toxicity, corrosivity, ignitability, and reactivity.

COMMUNITY RIGHT-TO-KNOW

Community involvement in and community knowledge of a facility's operations and the hazardous materials managed there is not only good business practice, but it is required by law. Under EPCRA, the Community Right-to-Know provisions of the Superfund Amendments and Reauthorization Act (SARA) Title III, four types of reporting of hazardous materials may be required:

- Emergency Release Reporting (Section 304)
- Material Safety Data Sheet (MSDS) Reporting (Section 311)
- Inventory Reporting (Section 312)
- Toxic Chemical Release Reporting (Section 313)

SARA Section 304 reporting is triggered when a spill occurs of an extremely hazardous substance in quantities greater than the designated reportable quantity and the spill enters the environment. Extremely hazardous substances are defined in the SARA law and are included in the Extremely Hazardous Substances List of the Comprehensive Environmental Response, Compensation, and Liability Act (CERCLA).

SARA Section 311 reporting is required for any hazardous substance for which an MSDS is required and which is stored on site in quantities greater than 10,000 pounds (or the threshold quantity for extremely hazardous substances). Under this reporting requirement, copies of MSDSs (or a listing of these MSDSs) for each of these hazardous materials stored in quantities in excess of 10,000 pounds must be submitted to the State Emergency Response Commission, the Local Emergency Planning Committee, and the local fire department. The basic purpose is to provide emergency response personnel information about what is stored on site. This is a one-time reporting which must be updated within 90 days upon a change of any hazardous material inventory which triggers reporting.

SARA Section 312 is an annual report which is submitted on Tier I or Tier II forms and provides an inventory reporting for each hazardous material. The reporting requirements are triggered by the same determination for the same hazardous materials as for SARA Section 311 reporting. The purpose of this reporting is to provide emergency response personnel with information about the storage locations for and quantities of hazardous materials.

SARA Section 313 reporting is an annual report of emissions and discharges of certain toxic chemicals. These emissions are reported on the Form R. The purpose of this report is to inform the community of releases of hazardous materials into the environment.

It is also prudent (and required by law for certain hazardous waste activities) to include community leaders and emergency response organizations in emergency planning at a facility. This could include sharing copies of the Emergency Management Plans, familiarization of the site by emergency response organizations and hospitals, participating in community emergency drills, and establishing links of communication with local media.

OTHER IMPORTANT EMS ELEMENTS

Training

Federal and most state laws require employees who handle hazardous materials and hazardous wastes to be trained in their safe handling and their hazards. Typically, the training is provided upon job assignment, then annually thereafter. At times, hazardous materials training overlaps with safety and health training.

Regulations Review

The environmental compliance manager should be aware of new laws and regulations which affect the regulated community. The manager can subscribe to any number of environmental or trade journals which cover hazardous materials. There are update services for regulations and agency policies. Weekly or monthly newsletters keep the manager up-to-date on what is happening in the EPA and elsewhere. If the budget allows, the business can even subscribe to the daily Federal Register. Regulatory agencies offer classes or public hearings for new or proposed regulations. The manager could attend the periodic meetings of local chapters of professional organizations both to gain new technical knowledge and to develop contacts with other professionals with similar interests. One can also attend training classes, seminars, conferences, or university-level classes to receive training on regulations.

Environmental Review for New Projects

Changes are a normal part of daily operations at most facilities. While most of these changes do not have environmental or regulatory consequences, it should be company policy to ensure that all new projects are in compliance with the law, minimize the release of pollutants to the environment, minimize the generation of wastes, minimize the use of toxic substances, and minimize health and safety risks associated with operations. In order to meet this policy, proposed significant changes in any facility should be evaluated to assure that environmental regulatory compliance will be maintained despite the result of the change. Due to the complexity of existing environmental regulations, no one person can be expected to understand all applicable rules and regulations. Therefore, a Management of Change Procedure is typically used. The purpose of this procedure is to ensure that proposed significant changes go through an internal environmental review process.

As with all management decisions, discretion must be used in applying this procedure, keeping in mind that this procedure is intended to protect the public, neighbors, employees, and operations. Typically, significant changes which should be reviewed per this procedure include any and all proposed changes which may have an impact (positive or negative) on:

- Employee health and well-being
- Public or resident health and well-being
- The environment, including air, surface water, groundwater, and soil
- Permits or memoranda of understanding with governmental agencies
- Any air, water, or waste discharges or emissions from any plant/project

At the initiation of a proposed change, an environmental review of the proposed change is performed. This review should be documented using a form similar to Figure 18.2.

PROFESSIONAL ACTIVITIES

The environmental manager and others interested in the field have the opportunity to increase and share their knowledge through participation in workshops, seminars, professional meetings, and as instructors for classes. There are many professional environmental organizations to which to belong or in which to participate: Registered Environmental Professionals, Academy of Certified Hazardous Materials Managers, Groundwater Resources Association, Water Well Association, and others. Many states such as California and Nevada have environmental registration programs. Most of the professional organizations have local chapters that offer periodic meetings, seminars, and national conferences. The knowledge gained by a professional's experience is invaluable for students trying to figure out career paths; if the environmental manager has the opportunity, he should try to be a guest speaker at elementary or high schools, or colleges. Whether corporate or academic, sharing knowledge is the responsibility of the environmental manager or professional. This way, the environmental manager and business can benefit by enhanced public understanding of hazardous materials management.

CONCLUSION

Environmental management of hazardous materials or hazardous wastes can be a full-time job. It is essential that top management supports and commits to making effective environmental policies. An environmental management program will reduce the risks and liabilities associated with facilities operating with hazardous

Environmental Review Form

Location:		Project #:			
Brief Project Description:					
Air Environmental Review			Yes	No	N/A
Will the project create any new point source or fugitive emission?					
Will the project change (increase or decrease) existing point source or fugitive emissions?					
Will there be any dust or VOC emissions from construction or demolition activities?					
Are there any air permitting issues associated with this project?					
Water Environmental Review			Yes	No	N/A
Will the project create any new or different water discharges to the sewer (POTW)?					
Will the project create any new or different water discharges to surface waters or the land?					
Will the project create any new or different waste water discharges to site septic systems?					
Will the project create any new or different waste water discharges to storm drains?					
Will the project create the potential for soil or groundwater contamination?					
Are there any NPDES, storm water, or POTW permitting issues associated with this project?					
Hazardous Material/Solid Waste Environmental Review			Yes	No	N/A
Have outside contractors' wastes/hazardous materials been addressed?					
Will there be any new hazardous materials used or stored on-site resulting from the project?					
Will any haz. materials or wastes remain on-site after completion of the project?					
Will there be a change in the inventories of any existing hazardous materials?					
Will any existing hazardous material storage plant/project be closed as a result of this project?					
Will the change trigger changes to SARA 311, 312, or 313 Right-To-Know reporting?					
Have plans for responding to spills of hazardous materials been adequately addressed?					
Will there be any new or different wastes resulting from the project?					
Will there be any wastes generated during construction or demolition phases of the project?					
Will there be any wastes associated with the project be hazardous wastes?					
Have provisions for storage and disposal of hazardous wastes been addressed?					
Has a reduction in the generation of new or existing hazardous materials been considered?					
Will the project involve any treatment, storage or disposal of hazardous waste on-site?					
Asbestos/PCP Environmental Review					
Will the project involve the removal or demolition of Asbestos containing materials?					
Will all asbestos removal be completed by state-certified, asbestos-trained personnel?					
Have provisions been made for proper packaging, transporting, and disposing the asbestos?					
Will the project involve the removal of any PCB or PCB-contaminated equipment?					
Have provisions been made for proper packaging, transporting and disposing of the PCBs?					

Review Signatures:

_____ _____   _____ _____
Plant/project Environmental Coordinator   Date   Plant/project manager   Date

Figure 18.2 Sample environmental review form.

materials or hazardous wastes. Employees increasingly understand that environmental protection is in the best interest of themselves and their community. Combined with training and general awareness of environmental compliance goals, employees are generally cooperative with management to achieve environmental compliance since they know in order to have a job, the company must stay in business. Also, a proactive approach to environmental compliance is appreciated by regulatory agencies.

FURTHER READING

Books

Blakeslee, H. W. and Grabowski, T. M. *A Practical Guide to Plant Environmental Audits.* New York: Van Nostrand Reinhold, 1985.

Cahill, L. B. Ed. with Kane, R. W. *Environmental Audits.* 7th ed. Rockville, MD: Government Institutes, Inc., 1996.

Cox, D. B. Ed. *Hazardous Materials Management Desk Reference.* New York: McGraw-Hill, 2000.

Hess, K. *E H & S Auditing Made Easy — A Checklist Approach for Industry.* Rockville, MD: Government Institutes, 1997.

Krieger, G. R. Ed. *Accident Prevention Manual for Business and Industry — Environmental Management,* Itasca, IL: National Safety Council, 1995.

Saunders, T. *The Bottom Line of Green Is Black.* San Francisco, CA: Harper Collins, 1993.

## CHAPTER 19

EPA's Risk Management Program

Mark D. Hansen

CONTENTS

INTRODUCTION

Many companies required to comply with the Occupational Safety and Health Administration's (OSHA's) Process Safety Management (PSM) rule (29 CFR Part 1910.119) are also required to comply with the Environmental Protection Agency's (EPA's) Risk Management Program (RMP) (40 CFR Part 68. Also, EPA's RMP requires many companies not required under OSHA's PSM rule to comply with a whole host of new requirements. Even the companies required to comply with OSHA's PSM rule will find the task of complying with EPA's RMP difficult at best.

Table 19.1 Entities Potentially Regulated by EPA's RMP

Category	Examples of Regulated Entities
Chemical manufacturers	Industrial organics and inorganics, paints, pharmaceuticals, adhesives, sealants, fibers
Petrochemical	Refineries, industrial gases, plastics & resins, synthetic rubber
Other manufacturing	Electronics, semiconductors, paper, fabricated metals, industrial machinery, furniture, textiles
Agriculture	Fertilizers, pesticides
Public sources	Drinking and waste water treatment works
Utilities	Electric and gas utilities
Others	Food and cold storage, propane retail, warehousing, and wholesalers
Federal sources	Military and energy installations

Entities potentially regulated by EPA are those facilities that have more than a threshold quantity of a regulated substance in a process. Regulated categories and entities are shown in Table 19.1.

As you can see, many facilities shown above are those most would expect to be compliant with EPA's RMP. However, many facilities that do not traditionally fall into OSHA's PSM rule, do fall into EPA's RMP. As a result, this may cause some facilities considerable effort.

WHAT IS RMP?

The RMP was promulgated to prevent accidental releases of substances that can cause serious harm to the public and the environment from short-term exposures and to mitigate the severity of releases that do occur. The 1990 amendments to the Clean Air Act require EPA to ensure a rule specifying the type of actions to be taken by facilities to prevent accidental releases of such hazardous chemicals into the atmosphere and reduce their potential impact on the public and the environment. In general, the rule requires:

- Covered facilities must develop and implement an RMP and maintain documentation of the program at the site. The RMP will include an analysis of the potential off-site consequences of an accidental release, a five-year accident history, a release prevention program, and an emergency response program.
- Covered facilities also must develop and submit an RMP, which includes registration information, to EPA no later than June 21, 1999, or the date on which the facility first has more than a threshold quantity in a process, whichever is later.
- Covered facilities also must continue to implement the RMP and update their RMPs periodically or when processes change, as required by the rule.

DOES IT APPLY TO ME?

Any facility with more than a threshold quantity of a listed regulated substance in a single process must comply with the regulation. Process, as defined in the regulation, means manufacturing, storing, distributing, handling, or using a regulated substance in any other way. Transportation, including pipelines and vehicles under active shipping orders, is excluded.

On January 31, 1994, EPA promulgated a final list of 139 regulated substances: 77 acutely toxic substances, and 63 flammable gases and volatile liquids. This list included the following 16 substances mandated by the clean air act (CAA):

- Chlorine
- Hydrogen sulfide
- Toluene diisocyanate
- Bromine
- Sulfur trioxide
- Hydrogen chloride
- Phosgene
- Hydrogen fluoride
- Methyl chloride
- Sulfur dioxide
- Ammonia solution
- Anhydrous ammonia
- Ethylene oxide
- Vinyl chloride
- Methyl isocyanate
- Hydrogen cyanide

IF IT DOES, WHAT PROGRAM AM I?

Processes that are required to comply with the RMP Program rule are divided into three tiers or levels of compliance; Program 1, Program 2, and Program 3. Program 1 is the most simplified or streamlined approach and places less requirements on the subject facility. Program 3 is the most intense, and places more requirements on a subject facility than Programs 1 and 2. These program levels were designed by EPA to allow processes that pose different levels of risk to have varying degrees of compliance that are commensurate with the potential risk. This section will provide the tools necessary to determine which program a given process falls into and to understand the different requirements of the three program levels.

It is important to remember that a facility with more than one covered process may have different levels of compliance. For example, if a process meets Program 1 criteria, the facility must only satisfy Program 1 requirements for that process, even if other processes at the facility meet Program 2 or 3 criteria. Therefore, a facility may have processes on one or more of the three RMP Program levels.

To qualify for Program 1, a process must not have had an accidental release with offsite consequences in the five years prior to the submission date of the RMP. Additionally, the process must have no public receptors within the distance to a specified toxic endpoint or flammable endpoint that is associated with the process's worst-case release scenario.

Unless Program 1 eligibility requirements are met, Program 3 applies to processes in the following North American Industry Classification System (NAICS) codes:

- 32211 — pulp mills
- 325181 — chlor-alkali
- 325188 — industrial inorganics
- 325211 — plastics and resins
- 325182 — cyclic crudes
- 325199 — industrial organics
- 325311 — nitrogen fertilizers
- 32532 — agricultural chemicals
- 32411 — petroleum refineries
- 32511 — petrochemical facilities

All processes covered by OSHA's Process Safety Management (PSM) Standard, 29 CFR 1910.119 must also meet Program 3 requirements, unless they meet Program 1 eligibility requirements.

If a covered process does not meet the eligibility requirements for either Program 1 or Program 3, then that process is in Program 2. The following flowchart (see Figure 19.1) walks you through the steps to determine the program level with which your process must comply. The eligibility criteria and a comparison of program requirements are highlighted in Tables 19.2 and 19.3.

EPA was generous enough to determine three levels of compliance depending on several characteristics of your facility. Under EPA's RMP rule, processes subject to these requirements are divided into three tiers, labeled Programs 1, 2, and 3. EPA has adopted the term program to replace the term tier found in the Superfund Amendments and Reauthorization Act of 1986 (SARA Title III) to avoid confusion with Tier I and Tier II forms submitted to the government. Eligibility for any given program is based on process criteria so that classification of one process in a program does not influence the classification of other processes at the source. For example, if a process meets Program 1 criteria, the source need only satisfy Program 1 requirements for that process, even if other processes at the source are subject to Program 2 or Program 3. A source, therefore, could have processes in one or more of the three Programs.

DETAILS OF RMP REQUIREMENTS

If your facility is required to be compliant with EPA's RMP, the following requirements apply:

- Registration with EPA
- Requirements for the development and implementation of a comprehensive RMP
- Requirements for the submittal of a RMP Plan; access will be provided to the implementing agency, State Emergency Response Commission (SERC), Local Emergency Planning Committee (LEPC), and the public

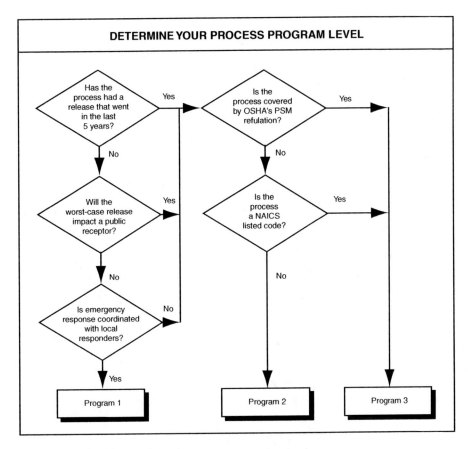

Figure 19.1 Flowchart to determine a process program level.

Table 19.2 Program Eligibility Criteria

Program 1	Program 2	Program 3
No off-site accident history No public receptors in worst-case release scenario circle Emergency response coordinated with local responders	The process is not eligible for Program 1 or 3	Process is covered by OSHA's Process Safety Management Regulation, 29 CFR 1910.119 Process is in NAICS code 32211, 325181, 325188, 325211, 325182, 325199, 325311, 32532, 32411, and 32511.

The EPA rule defines the following three major RMP components for facilities handling extremely hazardous substances:

- Hazard Assessment
- Prevention Program
- Emergency Response Program

Table 19.3 Comparison of Program Requirements

Program 1	Program 2	Program 3
Hazard Assessment		
Worst-case analysis	Worst-case analysis	Worst-case analysis
	Alternative releases	Alternative releases
5-year accident history	5-year accident history	5-year accident history
Management Program		
	Document management system	Document management system
Prevention Program		
Certify no additional	Safety information	Process safety information
steps needed	Hazard review	Process hazard analysis
	Operating procedures	Operating procedures
	Training	Training
	Maintenance	Mechanical integrity
	Incident investigation	Incident investigation
	Compliance audit	Compliance audit
		Management of change
		Pre-start-up review
		Contractors
		Employee participation
		Hot work permits
Emergency Response Program		
Coordinate with local	Develop plan and program	Develop plan and program
responders		
Risk Management Plan Contents		
Executive summary	Executive summary	Executive summary
Registration	Registration	Registration
Worst-case data	Worst-case data	Worst-case data
5-year accident history	Alternative release data	Alternative release data
Certification	5-year accident history	5-year accident history
	Prevention program data	Prevention program data
	Emergency response data	Emergency response data
	Certification	Certification

HAZARD ASSESSMENT

All facilities must assess the worst-case release scenarios. Program 2 and 3 facilities must also assess at least one alternative release scenario for each listed toxic extremely hazardous substance (EHS) and one representing all listed flammable EHSs. The assessment includes dispersion modeling, identification of the area where impacts are above specified toxic or fire/explosion end point criteria, and estimating the potentially affected populations and environmental receptors. The assessment would also include a 5-year accident history involving chemical releases.

The definition of the worst-case release has been perhaps the most controversial aspect of the rules and one that could have a significant effect on public perception and emergency planning. There are specified assumptions for the worst-case release that differ depending upon whether the chemical is a gas or liquid at ambient conditions, whether it is normally refrigerated, and whether it is listed due to toxicity or flammability.

For toxic materials, the worst-case release would usually be the release as a gas of all of the contents of the largest vessel on-site over a 10-minute period. EPA recognizes that the worst-case scenario may represent a highly improbable scenario that may not be the most appropriate for emergency planning. The alternative scenarios are those that may be more credible and more appropriate for emergency planning. Because the results of the worst-case scenario impacts can raise concerns, active communication efforts may be necessary to explain why it may not be appropriate to base emergency management plans and response capabilities on the worst-case. While the worst-case definition poses some problems, it has improved significantly over the initially proposed definition that would have required the assumption of simultaneous, instantaneous release of the total site inventory (including all vessels and locations).

The alternative release scenarios selected should be more likely to occur than the worst-case and reach an end point off-site (if such an alternative scenario exists). The scenario may be based on the accident history or process hazard analysis. While not explicitly stated, the intent is for you to select a credible worst-case that provides the basis for planning emergency response and community response. It should include incidents known to the industry, even if it has not happened at this facility.

The 5-year accident history would include incidents from covered materials that resulted in death, injury, evacuation, sheltering in-place, property damage, or environmental damage.

PREVENTION PROGRAM

EPA's Program 3 prevention program includes most of the elements included in OSHA's PSM program. Both EPA's RMP and OSHA's PSM emphasize the application of performance-oriented management programs rather than the application of specific requirements. The following synopsis outlines key compliance points under these rules:

Management — Only EPA's RMP explicitly requires a documented management system with a clear delineation of overall management responsibility for the program and lines of authority for all personnel assigned responsibility for individual requirements; however, most facilities find that implementation of PSM without a strong management structure is difficult, if not impossible.

Employee Participation — Employers must develop a written plan describing how they will involve employees in hazard analysis and other PSM elements and provide access to all PSM information. Employees generally enjoy participation in process hazard analyses and provide significant input.

Process Safety Information — Maintain toxicity information, permissible exposure levels (PELs), physical data, reactivity data, corrosivity data, thermal and chemical stability data, synergistic effects from mixing process flow, process chemistry, safe upper and lower operating limits, maximum intended inventory, consequences of process deviation, and detailed process design (piping, instrumentation, mass balance, electrical, safety, ventilation, etc.) information. Many facilities do not have accurate and up-to-date drawings and engineering documentation when the program is being initiated.

Process Hazard Analysis — Must be performed in accordance with generally accepted methodologies (i.e., What if Analysis, FMEA, HAZOPS, etc.). This hazard analysis must be revalidated and updated once every five years, and the two most recent analyses must be retained on active file.

Operating Procedures — Detailed process operating procedures must be developed, implemented, and maintained for each hazardous process, including steps for each operating phase (including startup, shutdown, emergency, temporary, etc.), operating limits, safety procedures, and safety and health considerations. These must be more site-specific than the generic equipment O&M manuals that are usually available.

Training (Initial and Refresher) — All employees involved in working with these processes must be thoroughly trained in process and safety (including emergency) procedures. The employer must document that each employee has received the required training. Training and operating procedures are the most frequently cited elements in OSHA PSM inspections.

Contractors — Employers must ensure that contractors working near these processes are informed of all hazards related to the work and the emergency action plan. Employers must evaluate safety performance before contractor selection, log contractor injury and illness, and periodically evaluate performance. Well-designed programs can also reduce incidents and loss-related costs.

Pre-start-up Safety Reviews — Must be performed on new or modified process facilities to ensure that equipment and construction meet design specifications, that procedures and training have been implemented, and that hazard analysis recommendations have been resolved. These reviews also result in more trouble-free startup.

Mechanical Integrity — The mechanical integrity of process equipment will be documented and maintained through inspection and testing programs and through detailed maintenance documentation. Written procedures, training, and quality assurance (QA) programs are also necessary. This program can improve equipment reliability and reduce unplanned shutdowns.

Hot Work Permit — An authorization and permit program for welding and brazing activities near or around these processes shall be implemented and maintained by the employer or his representative. Other safe work practices, such as a line-breaking procedure, are also required.

Management of Change — Implement a program to prevent incidents caused by changes in process chemicals, technology, equipment, or procedures. In manufacturing plants, this is the most difficult element to implement because a strong management structure is required to ensure that production-enhancing changes are reviewed prior to implementation.

Incident Investigation — Incident investigation, reporting, and documentation procedures must be prepared and implemented for any incident that did or could have resulted in an emergency situation.

Compliance Audits — In-house compliance audits must be performed at least once every 3 years to verify that the prevention program elements are in place and that they have been implemented effectively.

EMERGENCY RESPONSE PROGRAM

Both EPA's RMP and OSHA's PSM rely mainly on previous requirements for emergency action plans and training (e.g., 29 CFR 1910.38[a] and 1910.120[a], [p], and [q]). In contrast to the proposed rule, the final EPA rule allows a facility to evacuate and not actively respond to accidental releases, provided that the facility is included in the community emergency response plan developed under U.S.C. 11003 and that appropriate mechanisms are in place to notify emergency responders. While the EPA rule allows the use of plans and programs developed under prior emergency regulations, it stipulates that the plan includes items required under the statute, notably procedures for informing the public and local responders about a release. It also encourages the use of the new EPA Integrated Contingency Planning (or One-Plan) approach.

REGISTRATION AND RMP SUBMITTAL

After completing the Hazard Assessment, Prevention Program implementation, and Emergency Response implementation, a plant must submit the RMP registration and RMP together by June 21, 1999. EPA has yet to develop the format and location of submittal, but is considering the use of a format compatible with electronic submittal. The submittal must include:

1. A copy of the registration form
2. An executive summary of the hazard assessment for each regulated substance
   a. Description of the worst-case release scenario and the alternate-case scenario
   b. Description of the potential offsite consequence analysis methods and results, including distance to end point, public and environmental receptors, and mitigation considered
   c. Compilation of the five-year accident history
3. A description of the major hazards and an overview of the prevention program
   a. Description of the major hazards identified in the PHAs
   b. Description of the consequences of a failure to control for each major hazard
   c. Summary of the actions taken or planned to address these hazards
   d. Description of how accidental releases are prevented or mitigated or the consequences reduced by these actions (action plans with start and target completion dates)
4. A summary of the emergency response plan
   a. Procedures to inform emergency response authorities and the public
   b. Name or position of the point-of-plant contact between the plant and public authorities
   c. Description of coordination with LEPCs
5. A description of the management system developed to implement the prevention program
   a. Definition of the person/position responsible for the overall implementation and coordination of the plan
6. A certification to the truth, accuracy, and completeness by the owner/operator

RMP*SUBMIT

For the computer literate, RMP*Submit is a system developed by EPA to assist in documenting and RMPs required under the Clean Air Act Section 112(r). RMP*Submit allows you to:

- Capture your RMP information and put it on a diskette in the proper format.
- Receive on-line help to assist you in completing your RMP.
- Check your data to identify errors in your RMP prior to submission.

To prepare your RMP, which is the basis for your plan, you will want to consult the following documents:

- List of Regulated Substances and Thresholds for Accidental Release Prevention; Requirements for Petitions under Section 112(r) of the Clean Air Act as Amended (the List Rule) January 31, 1994 (EPA #550-Z94-001) and the List of Regulated Substances and Thresholds for Accidental Release Prevention; Amendments, January 6, 1998 (EPA #550-Z98-001). Both rules help you determine if you have the chemicals and quantities that require you to comply with the risk management program.
- Risk Management Program Rule (RMP Rule) under 40 CFR 68, published June 20, 1996 (EPA #550-Z96-002) is the regulation that outlines the requirements for the risk management program.
- General Guidance on the Risk Management Program (EPA #550B98003) is a comprehensive guidance designed to help a facility implement a Risk Management Program.
- RMP Model Plans simplify your effort to comply with the RMP Rule because it includes compliance information and analyses that are specific to certain sectors.
- The Off-Site Consequence Analysis (EPA #550-B96-014) and RMP*Comp give you modeling information and tables to help you calculate your worst-case and alternative release scenarios.

All of these materials are available, or will be posted as soon as they become available, electronically through EPA's Accident Prevention and Risk Management Program Web page (http://www.epa.gov/ceppo/acc-re.html). They may also be obtained by calling the National Center for Environmental Publications and Information (NCEPI), 1-800-490-9198. For answers to specific compliance questions concerning the RMP, call the Hotline at 1-800-424-9346 (TDD: 1-800-535-0202).

RMP*Submit allows you to electronically enter data as follows:

Section 1: Registration
Section 2: Toxic Worst Case Scenario(s)
Section 3: Toxic Alternative Release Scenario(s)
Section 4: Flammable Worst Case Scenario(s)
Section 5: Flammable Alternative Release Scenario(s)
Section 6: Accident History
Section 7: Prevention Program(s) for Program Level 3 Processes
Section 8: Prevention Program(s) for Program Level 2 Processes
Section 9: Emergency Response Program
Executive Summary

THE ONLY THING LEFT IS THE AUDITING

EPA is proposing that facilities be selected for audits based on the following criteria:

- Accident history
- Accident history of companies in the same industry
- Quantity of regulated substances
- Location of the plant and the proximity to the public and sensitive environments
- Hazards identified in the RMP
- A plan for providing neutral, random oversight

The audit conducted by EPA will focus on the prevention program portion of the RMP.

EPA AND OSHA MEMORANDUM OF UNDERSTANDING

Measures taken by companies to comply with OSHA's PSM for any process are sufficient to comply with the prevention program requirements of all three Programs. EPA will retain its authority to enforce the prevention program requirements and the general duty requirements of CAA Section 112(r)(1). For those who do not already know, this general duty clause is much like OSHA's and provides a similar citation vehicle that OSHA's general duty clause provides.

Further, EPA and OSHA are working closely to coordinate interpretation and enforcement of PSM and accident prevention programs (see the Memorandum of Understanding below). EPA will also work with state and local agencies to coordinate oversight of worker and public safety and environmental protection programs.

CONCLUSION

The Risk Management Plan requires much time and preparation if you are subject to the rule. Hopefully, if you are subject to the rule, this material can help you prepare for the task ahead.

REFERENCES

OSHA 29 CFR Part 1910.119. Process Safety Management of Highly Hazardous Chemicals. *Federal Register*, 57(36): Feb. 24, 1992, Washington, D.C.

Environmental Protection Agency, 40 CFR Part 68. Risk Management Program for Chemical Accidental Release Prevention Requirements, *Federal Register*, 61(120): June 20, 1996, Washington, D.C.

American Petroleum Institute (API), *Management of Process Hazards — API Recommended Practice 750,* 1st ed., Washington, D.C.: API, 1990.

OSHA Instruction CPL 2-2.45A, CH-1. Process Safety Management of Highly Hazardous Chemicals, Compliance Guidelines and Enforcement Procedures, September 13, 1994.

PART IV

Health Program Management Aspects

## CHAPTER 20

Overview of the Key Elements of an Effective Occupational Health Management Program

Barbara K. Cooper, Ellen E. Dehr, and Audrey Lawrence

CONTENTS

1-56670-370-0/02/$0.00+$1.50
© 2002 by CRC Press LLC

INTRODUCTION

This chapter provides practical information needed to plan, organize and implement an effective health management program. As manager of occupational health
for a small company which has never had any health programs, where do you begin?
What programs should receive priority? What constitutes the best use of your
resources? The information presented here can also be used to evaluate and improve
already existing health programs.

Always take a comprehensive approach to the promotion of health, while resisting extreme pressure towards fragmentation. This emphasis on prevention of disease
must permeate the goals and mission of the entire organization at all levels. You
will, therefore, be the consummate politician, as well as the great communicator.
This chapter discusses the steps involved in developing and implementing occupational health programs:

- Conduct a needs analysis.
- Prioritize the health needs of your organization.
- Determine which regulations apply.
- Prepare a plan of action.
- Gain support for your programs.

DEFINITION OF HEALTH PROGRAMS
AND HEALTH PROGRAM MANAGEMENT

An occupational health program has many of the same goals as a safety program,
but has a different emphasis. The profession of industrial hygiene involves the
anticipation, evaluation and control of environmental factors arising in or from the
workplace that may result in injury, illness, impairment, or affect the well-being of
workers and members of the community. The industrial hygienist analyses operations
and materials and recommends procedures to protect human health. The approach

emphasizes prevention of disease arising from the following: noise and vibration, hazardous materials such as lead and asbestos, ionizing and non-ionizing radiation, bloodborne pathogens, psychological stress, heat and cold stress, poorly designed workstations or tools, and other occupational factors. The key to the prevention of health problems lies in the development and implementation of a comprehensive employee health program.

Such a program includes a written plan of action which outlines a schedule of activities and procedures to be followed. These health programs, because of the preventive approach, often involve extensive health education. Other program elements include hazard identification and evaluation, hazard communication, and employee compliance.

Why do health professionals recommend the program approach, and why should these programs be comprehensive in nature? Health professionals often receive direction from those who may not have a clear understanding of health and safety management. Sometimes environmental regulatory compliance constitutes a higher priority for other managers than occupational health issues and employee wellness. Many health professionals find themselves coping with the crisis of the moment.

Having an organized plan for accomplishing the important tasks relating to employee health will help protect you from the changing priorities of those both above and below you within your organization. In addition, the presence of such a program helps demonstrate progress towards regulatory compliance. Finally, a program approach facilitates program documentation and follow-up.

However, health programs, plans and policies will fail if developed without the direct input of others. You cannot know the health priorities of an organization unless you do your homework. Strive to gain support from key individuals while providing education about industrial and occupational health issues. Your programs will succeed if everyone has a stake in the prevention of employee health problems.

Comprehensive programs avoid duplication of effort and competition among different departments. Roles, responsibilities and duties are outlined in detail, eliminating confusion and reducing misunderstandings. Health programs often have similar elements: health hazard evaluation, exposure monitoring, employee training, record keeping, and medical surveillance. Grouping such programs streamlines such difficult tasks as scheduling training sessions or medical appointments. Routine monitoring of the work environment can include measurement of a variety of environmental stressors at the same time.

In addition to these practical considerations, the program approach will relieve you of the pressure of having to cope with a variety of conflicting expectations. Determine your initial goals using a consensus approach. Work with your managers to determine their priorities and concerns. Incorporate these ideas into an overall strategy for health programs. Make each goal realistic. Consider how your organization works as a unit to accomplish tasks.

While the plan is always subject to review and change, you will feel more in control. Those with other priorities involving the use of your time and resources will often respond more favorably once they understand that your programs are part of an overall plan to improve health among your workers and comply with occupational health regulations.

Health, safety, and environmental management programs should report to the highest management level. Executive management must approve and actively work towards the goals of the health management program in order for it to succeed. Otherwise, not all mid-level managers will embrace the changes necessary to promote a healthy work force. Health professionals should have the authority and the executive mandate to carry out their function. If the health programs report to the operating units themselves, a conflict of interest arises.

Grouping industrial hygiene with other technical function such as engineering may prove more successful than placement of this function in human resources or other administrative units, since employees often view those in scientific disciplines as resources rather than adversaries. In addition, scientific units often have better access to laboratory or research facilities.

Obviously, you may have little control over where your unit reports within the organization. After educating managers about occupational and environmental compliance, you may successfully negotiate a dotted-line reporting relationship to the highest levels of management, to ensure communication and cooperation within the organization.

AVAILABLE RESOURCES

Conduct a needs analysis. Beginning a health program in any organization can prove a daunting task. Evaluate past efforts. Technical functions such as engineering may have performed analyses of ventilation equipment or plant design. Often human resources divisions investigate accidents or illnesses and oversee workers' compensation programs. Legal counsel can recommend specific wording for policies, procedures and directives involving health matters. Operating units responsible for production or maintenance often have already established safe work practices.

Interview those within your organization who have the greatest knowledge about employee health. Review the records kept on occupational injuries and illnesses. Compare the injury and illness rates to those within your industry. Make note of the most common types of illnesses and injuries. Review in depth a representative sample of the incidents listed, including the worker's compensation files and medical records.

Visit production and maintenance areas and talk to employees and first line supervisors. What are their concerns? Discuss the injury and illness statistics with them. Conduct brief walk-throughs of production, research and maintenance operations and make flow charts summarizing the processes and materials involved. How many employees work in each area? What types of materials and equipment do they use?

Try to visit all plants and locations for which you are responsible. Every plant has its own culture and idiosyncrasies. You may want to conduct more formal interviews or audits during this information-gathering stage if your area of jurisdiction is wide.

Contact labor union representatives to discuss health and safety concerns. What existing agreements include health and safety requirements? What procedures are recommended for specific operations? Often unions have general operating procedures for the trades they represent. Your human resources department should be able to provide you with the unions representing the employees of your organization.

Interview key managers for their perspectives. Do not neglect those from legal, human resources, and finance. These managers may budget for health and safety equipment, supplies and training, and may manage projects which may involve hazardous materials, such as asbestos in buildings or chemical spills.

Build support for your programs by appearing organized and confident; keep in mind that soon you will be presenting a plan of action for management approval. A few managers will see health programs as your total responsibility. In addition, some may appear defensive about their illness or injury rates. Always present yourself as a resource to these individuals. The function of the health professional involves making recommendations to improve working conditions. The actual implementation of such recommendations and other health policy matters remains a line function.

Prioritize the health needs of your organization. Review the information you collected during your needs analysis. Which operations present the greatest health risk to your employees? Noise-induced hearing loss may threaten some workers due to their work with power tools. Others may use organic solvents without proper personal protective equipment and without adequate ventilation. New chemicals or procedures may present unknown risks to your work force due to the lack of a hazard communication program. Increased workers' compensation costs are often a result of an increase in the severity or number of injuries among employees.

Determine which regulations apply to your operations. Visit the Occupational Safety and Health Administration (OSHA) offices in your area and research your organization's health and safety track record. Introduce yourself to the industrial hygiene and safety compliance officers and consultants. Explain your efforts and discuss any past regulatory violations or incidents. These individuals can provide a perspective on your industry and can provide additional resources.

Obtain and review the relevant regulations. Read carefully the sections pertaining to your priority health risks. Evaluate, to the extent possible, the regulatory status of your operations and make a realistic analysis of the actions required to achieve compliance. State and federal OSHA offices often have resources to assist businesses.

Locate model programs within your industry and summarize their approaches. Check trade and professional journals for articles describing health risks to those in your industry. Contact the authors of these studies for more information. Obtain sample policies and programs (see Figure 20.1). Find out what works.

Become active in business and professional associations. Consider serving on a committee or as an officer. You will make friends, improve your support system, and strengthen your technical knowledge. See Figure 20.2 for a list of organizations.

It is the policy of [organization] to provide a safe and healthful workplace. Line management will be responsible for the recognition and assessment of potential health hazards in the work environment and for the implementation of necessary control procedures and practices to eliminate these hazards or reduce them to the lowest reasonably achievable level. Operating management will also be responsible for ensuring compliance with all government regulations and standards relating to employee health from factors arising from the workplace. The industrial hygiene staff will provide professional assistance and guidance to operating management in achieving these objectives.

Figure 20.1   Sample Industrial Hygiene Program Policy. (From Ross, D. M. *Appl. Ind. Hyg.* 3(5):F30–F34, 1988. With permission.)

American Association of Occupational Health Nurses 50 Lenox Pointe Atlanta, GA 30324 (404) 262-1162 http://www.aaohn.org	American Board of Industrial Hygiene 6015 West St. Joseph, Suite 102 Lansing, MI 48917-3980 (517) 321-2638 http://www.abih.org
American Conference of Governmental Industrial Hygienists 1330 Kemper Meadow Drive Cincinnati, OH 45240-1634 (513) 742-2020 http://www.acgih.org	American Industrial Hygiene Association 2700 Prosperity Ave., Suite 250 Fairfax, VA 22031 (703) 849-8888 http://www.aiha.org
American Public Health Association 1015 Fifteenth St. NW, Suite 300 Washington, D.C. 20005 (202) 789-5600 http://www.apha.org	Institute of Hazardous Materials Management 11900 Parklawn Dr., Suite 450 Rockville, MD 20852 (301) 984-8969 http://www.ihmm.org
National Hearing Conservation Association 9101 E. Kenyon Ave., Suite 3000 Denver, CO 80237 (303) 224-9022 http://www.hearingconservation.org	National Safety Council P.O. Box 558 Itasca, IL 60143-0558 (708) 285-1121 http://www.nsc.org

Figure 20.2 Professional Occupational Health Organizations. (For more associations and other sources of help, see Appendix A in *Fundamentals of Industrial Hygiene*, Plog, B. A., Ed., National Safety Council, 1988.

PLAN OF ACTION

Conduct long-term planning. Prepare a business plan or work plan listing your overall program priorities for the next one to three years. Make the document conform as much as possible to the style used within your organization for reports, audits and plans. List your program goals in order of overall priority, at the beginning of the plan, in a one-page summary. Next outline the objectives or intermediate steps needed to achieve the goals. Finally, list for each objective the activities needed to carry out the objectives. Include a time line or give dates of completion for the objectives. See Figure 20.3.

Carefully consider your staffing requirements and overall budget. Given your resources, what can you accomplish in the time period indicated by your plan? Make conservative estimates of the funding and time needed to complete these tasks; overestimate the amount of time by a factor of three. Your initial program goals must succeed within the time frame and budget indicated.

Gain organizational consensus for your overall plan. Include key staff members you contacted during the needs analysis in the planning process. Route a draft copy of the overall health plan to these individuals and obtain their responses. Present a working version of the plan at senior staff meetings. Begin your presentation by

I. **Summary**
    (List all the primary goals.)

II. **Goal I: Survey of Paint Shop** — to perform a complete industrial hygiene survey of the paint shop and to address the need for a respiratory protection program for these employees.

   Objective 1: Complete an industrial hygiene survey of the paint shop.

   Activities:
   1. Review all operations and material used with shop supervisor. Observe work practices.
   2. Perform air monitoring of all operations involving paint that contains lead.
   3. Perform additional air monitoring of painting, cleaning, and abrasive blasting operations involving toxic materials.
   4. Review sampling results and prepare technical report. Include recommendations for improved working conditions.

Figure 20.3   Sample Work Plan Format.

discussing the results of the needs analysis. Why is this program needed? Show how these programs will benefit the company as a whole. Include a discussion of the costs of these programs and be prepared for extensive questions regarding this subject. Conclude by presenting concrete benefits related to health programs, such as reduced numbers of injuries, illnesses and worker's compensation cases, and increased employee morale and productivity. See Figure 20.4.

Prepare general health policy statements for each of the proposed health programs listed in your work plan. Such policy statements define the scope of the program, and list the responsibilities of those involved. Have these policies officially endorsed and announced by executive management before beginning your work.

What common problems cause health programs to fail? Many health professionals respond that they have too much work, not enough resources, and not enough time to complete projects. However, occupational health programs sometimes suffer from a lack of planning and organization. Time spent planning and positioning yourself within the organization often results in the most efficient implementation of employee health programs. Your goals become the organization's goals. When important health and safety matters arise, you will communicate your ideas effectively.

Unless your programs enjoy the support of executive management, they cannot succeed. These managers control the resources available to the organization, and set the tone for attitudes about worker health. First line supervisors form the critical link between health policy as it appears in training sessions and on paper and actual working conditions. Concentrate on providing quality training and access to technical information to your managers and supervisors. As your employees and supervisors mature in their careers and move into management themselves, they take with them their new attitudes and knowledge about employee health and wellness, making future programs easier to develop and implement.

You and your staff need to devote time to gaining and maintaining the professional skills needed to carry out health management programs. Support your staff in achieving their career goals and in developing themselves as professionals. Frequent attendance at educational seminars, forums and conferences will help these individuals stay current in a rapidly changing field. Subscribe to information services,

I. Introduction

Describe the purpose of the survey: baseline, routine monitoring, health hazard evaluation, or emergency response activity. Identify work area or shop and manager, supervisor, or foreman. Provide names and titles of inspection team.

II. Description of Operations

In tabular form, list employee job titles, work locations, and personal protective equipment. Include duration of all activities and a description of equipment and tools. Describe routine and extraordinary tasks. Describe supervision and training. Itemize all potential health hazards (such as noise, solvent vapors, or dust).

III. Survey Methods

For each potential health hazard, describe method of evaluation in detail. If industrial hygiene sampling was performed, list the sampling strategy (how you determined which employees to monitor). Include the make and model of all monitoring equipment, as well as methods of calibration. Describe how others have surveyed similar operations, and cite journal articles or other references relevant to your methodology.

IV. Survey Results

Tables of monitoring results should include the following information: Name and job title of employee sampled; date; duration of sample in minutes or hours; contaminant or material of interest; volume of air sampled (if applicable); time-weighted average of exposure; 8-hour, time-weighted average; permission exposure limit (PEL); and threshold limit value (TLV) for material of interest. List all health and safety hazards noted during the survey.

V. Discussion

Compare the results of the monitoring to the recommended exposure limits. Are employees overexposed to airborne contaminants or to excessive levels of noise, radiation, heat, or cold? Describe any technical difficulties which could affect the validity of your results. Briefly compare your results to those obtained by others.

VI. Recommendations

Describe changes in operations, materials used, personal protective equipment, or equipment that may reduce employee exposures and improve working conditions. Training and/or engineering controls may provide the best protection for employees.

VII. Appendices

Include sampling sheets, laboratory results, and copies of appropriate regulations.

Figure 20.4   Sample Industrial Hygiene Technical Report.

legislative reviews and professional journals. Require your staff to spend 4 to 5 hours each week reading such literature, and encourage them to take a speed-reading course.

Finally, know your limitations. If budget permits, make use of contract employees or consultants to perform technical evaluations or training. As a manager, you must constantly find ways to accomplish your goals through the work of others. Focus on your own area of technical expertise and always request the assistance of others when necessary. You will develop professionally and serve your organization at the same time.

HIRING CONSULTANTS AND CONTRACTORS

Due to increasing demands on health and safety personnel, many companies and businesses are outsourcing, or hiring consultants and contractors to perform many health, safety, and environmental services. Typically, these services may include:

- Asbestos and lead assessment and abatement
- Program audits
- Specialized training
- Environmental Assessments

Before you consider hiring a consultant for your health program needs, be sure to consider ways you could accomplish the tasks in-house. Some of the benefits of keeping health and safety functions in-house are:

- Availability — A staff person is almost always readily available and able to answer questions and concerns of employees and managers.
- Follow-through — When recommendations are made to improve health and safety, someone must make sure the recommendations are implemented. The end product of hiring a consultant is usually a report with recommendations, not actual changes in the organization.
- Familiarity with the organization — Usually consultants are not around long enough to become very familiar with the organization, including the best ways to find and solve problems.

There are times, however, when it becomes necessary for almost all organizations to hire a consultant. Some of the reasons for hiring a consultant are:

- Expertise — In-house staff may not have the necessary expertise to address all aspects of a health or safety problem.
- Efficiency — Someone who solves particular types of problems all of the time is probably more efficient than someone who rarely deals with the problem.
- Staffing Constraints — In-house staff does not have enough personnel to take on more tasks.
- Legal liability — In some cases, such as asbestos abatement, it is customary to use a consultant not only for expertise and efficiency, but also to have a neutral third party certify that everything is being done according to regulations and best practices.

Prior to contacting consultants and contractors, a resourceful safety and health manager should research other successfully completed projects with as many similarities to the planned project as possible. This research should answer the following questions:

- What are the steps involved in pre-planning?
- What is the cost per task identified?
- What types of problems were encountered?

The best way to find a consultant is to get recommendations from your network of professional contacts. Professional organizations, such as the American Industrial Hygiene Association (AIHA), will usually give you a list of several consultants in your area.

Once you have found some possible consultants, you will need to determine if they will meet your needs. Some things to consider are:

- Expertise in your area of need — Check that they have performed jobs similar to the one you have planned.
- Certification/Licenses — A certified industrial hygienist is desirable for industrial hygiene consulting. Some states require certification or licensing for some types of work, such as asbestos and lead.
- Define your needs/style and make sure consultant matches — Ask about how the consultant would approach your problem to determine if it sounds reasonable to you.
- Solicit proposals.
- Be sure the contract covers your needs — Expectations on costs and end product should be clear to both parties.

For construction-related contractors, such as asbestos and lead abatement contractors, the following also applies:

- Carefully review all specifications.
- Whenever possible, hire a knowledgeable consultant to oversee and monitor the contractor to ensure compliance with all the regulations.
- The contracting process should have several phases, including a request for qualifications, and a pre-bid walk-through, to physically review the scope of work and a follow-up session to ensure all contractual specifications are clearly understood by all parties.
- All hazards and working conditions should be clearly communicated and documented. The recent OSHA regulations governing multiemployer work sites may become an issue if hazards are not clearly stated and controlled.

SPECIFIC PROGRAMS

Incorporate specific programs into an overall health plan that conforms to legal and organizational requirements. Federal regulations may soon mirror the California Code of Regulations, which requires employers to have an injury and illness prevention program (summarized here):

- A written program outlining the scope of the program and designating the person(s) responsible for implementation
- Hazard identification through periodic inspection of the workplace
- Hazard correction procedures for correcting health hazards identified during the inspections
- Hazard communication through use of material safety data sheets (MSDS), labeling, and training of employees regarding occupational hazards
- Investigation of occupational injuries and illnesses
- Employee compliance with health and safety regulations
- Health and safety training that includes both general and specific information regarding occupational health hazards

A general program that includes these elements can form the foundation for all occupational health and safety programs within your organization. The program should also include procedures for medical surveillance, exposure monitoring, and record keeping. Today's occupational health regulations emphasize this comprehensive approach as opposed to the somewhat fragmented standards of the past. This approach also helps document each step as your organization achieves compliance.

Make your program easy to reference. Put your written programs together in a binder for your supervisors and managers. Report on program implementation at staff meetings. Track illness and injury statistics. Consider printing an employee wellness newsletter or contribute regularly to already existing publications.

The following sections summarize each of the specific health and safety programs. Consider this information as a guide, not a comprehensive listing of the required elements of each program. You will find references for additional reading on each particular program. Carefully review the relevant regulations, which may be specific for your geographic area or industry, but may not be discussed in this summary chapter.

Asbestos Safety

A confusing plethora of federal, state, and local regulations govern the handling of this hazardous substance. Extensive media coverage of the health problems caused by exposures to high levels of airborne asbestos fibers has created extreme concern among employees and the general public. High-profile litigation involves many manufacturers of asbestos-containing products and employers of those engaged in asbestos operations. How can you protect your employees from a known human carcinogen but diffuse unnecessary fear of asbestos-related disease? How can you keep abreast of all asbestos regulations but still have time to focus on other health concerns and programs?

Despite the controversy, asbestos exposure remains an important environmental and occupational issue. The health risks relating to occupational exposures to airborne fibers of asbestos have been well documented. Modern environmental and occupational regulations involving asbestos-containing materials include extensive controls to protect worker health and the environment. OSHA promulgated the first asbestos standard in 1971, and has revised the standard several times as new information became available. The permissible exposure limit (PEL) set for this substance by OSHA in 1971 was 12 fibers per cubic centimeter of air (f/cc). On October 11, 1994, a revised PEL of 0.1 f/cc went into effect for all asbestos workers.

In addition to the revised PEL, this standard creates four categories of asbestos work for those involved in the construction or shipyard industries. Employers must use the work practices outlined for each category of asbestos work, regardless of the exposure levels involved. This approach differs from previous standards, which relied upon an action level (AL) of exposure to asbestos. Employers were required to determine if an operation exposed workers to more than the specified AL before utilizing appropriate work practices and controls.

The current OSHA standard incorporates many of the provisions of the Asbestos Hazard Emergency Response Act (AHERA), and the Asbestos School Hazard Abatement Reauthorization Act (ASHARA), enacted by Congress in 1986 and 1990, respectively. This legislation established a program to address the problem of asbestos in schools. AHERA mandated inspections of school buildings for asbestos, and required schools to prepare plans to manage the material if found. AHERA also created extensive training requirements for those involved in performing asbestos work in schools. ASHARA extended the training requirements to persons performing such work in public and commercial buildings. In April 1994, the EPA's Asbestos Model Accreditation Plan (MAP), which clarifies and expands some of these training requirements, went into effect.

Many state governments have established their own asbestos regulations, including accreditation programs for those involved in performing asbestos-related work. Many of these regulations are much more stringent than the federal standard. Other regulations involving asbestos include EPA's National Emission Standards for Hazardous Air Pollutants (NESHAPS) which requires the removal of asbestos prior to demolition or renovation. The EPA and the Department of Transportation regulate the removal and disposal of asbestos as a hazardous waste.

Management Aspects

Determine if you have asbestos-containing products in your building or processes. As with other health programs, begin by identifying the hazardous material involved (in this case asbestos). Building surveys also permit the evaluation of asbestos-containing materials. These surveys must be performed prior to renovation or demolition (under NESHAPS), but OSHA now expects them for all buildings. This is because OSHA expects employers to identify hazards to which employees may be exposed. Protocols for such surveys are outlined in the AHERA legislation, and should be followed. Individuals performing building surveys for asbestos should also have industrial hygiene backgrounds, and should have attended an EPA-approved course in building inspection.

While the recent regulatory emphasis involves asbestos in buildings, you should not overlook asbestos materials in your manufacturing or maintenance operations. Review product MSDS to identify any products containing asbestos in use by your workers. Often roofing materials, flooring, adhesives, gaskets and other products contain asbestos. Find effective substitutes for these products. Phase out the use of asbestos-containing products such as brake shoes during periodic maintenance operations by replacing them with asbestos-free materials.

Develop an abatement plan. Your building survey should identify any damaged materials. You should, with the help of your consultant, determine what materials you want to remove or repair and the priority for abatement. Your priority should be based upon the exposure hazard. Focus on educating your organization about the correct manner of dealing with asbestos, particularly asbestos in buildings. An independent industrial hygienist who holds EPA certification in project design and supervision and is also a certified industrial hygienist (CIH) should plan and oversee all abatement work. This practice is directly analogous to an architect preparing plans prior to the commencement of building construction.

Develop an operations and maintenance plan. The purpose of an operations and maintenance (O&M) plan is to protect employees in buildings where asbestos is left in place. An O&M plan should include procedures for maintaining the condition of asbestos, responding to releases, and reinspecting the condition of asbestos. All asbestos-containing materials should be clearly labeled, and employees should not disturb these materials without training and protective equipment. Establish a method for controlling building contractors and for reviewing all planned building renovations. Compare these plans with the building surveys to determine if asbestos-containing materials will be involved. In many cases, more extensive surveys of building materials will be required. If your organization leases property, you need to ensure that these tenants adhere to your asbestos policy and program when planning and conducting building renovation or demolition.

Train employees. Maintenance and custodial employees need to know how to identify potential asbestos-containing materials. Occupants of buildings containing asbestos materials should receive awareness training. Therefore, your asbestos training program may involve nearly all employees.

During training sessions, provide information in a matter-of-fact way. Your employees may think of asbestos as a dangerous, human-engineered chemical, although asbestos minerals are mined from the ground similar to gold or silver. Although many who worked with asbestos have contracted lung cancer or asbestosis, most of these workers experienced exposure levels many times higher than those permitted by today's asbestos regulations. The actual risk of contracting asbestos-related disease decreases with the level of exposure.

Discuss the latency period associated with asbestos disease, and the synergistic effects of smoking and asbestos exposures. Those working with asbestos should not smoke.

Complying with the asbestos regulations will require an enormous commitment in time and effort from your staff. You may decide to contract out some or all of these responsibilities, while retaining oversight capacity.

Selected Citations and References

For additional information about asbestos safety, consult the EPA publication titled "Managing Asbestos in Place: A Building Owner's Guide to Operations and Maintenance Programs for Asbestos-Containing Materials" (USEPA 2OT-2003).

Confined Space Entry

Many workplaces contain spaces that are considered confined due to their configurations, which may hinder the activities of employees who must enter, work in, and exit such spaces. Additionally, employees working in confined spaces may face increased risk of exposure to hazardous atmospheres, moving parts, entrapment, or other health and safety hazards. Confined space entry (CSE) tragedy ranks among the most tragic and most preventable accidents in industry. Confined spaces are typically manholes, sewage stations, vessels, storage bins, tanks, pits, vaults, silos, vats, and any other spaces that:

- Are not meant for continuous human occupation.
- Have limited or restricted means of entry or exit.
- Are large enough and configured so that an individual can enter to perform work.

The OSHA regulation pertaining to Permit-Required Confined Space Entry can be found in 29CFR 1910.146. The regulation was revised in 1990 following a

National Institute of Occupational Safety and Health (NIOSH) alert focusing on the problem employers had recognizing confined spaces, testing, evaluating, and monitoring confined space atmospheres, and developing and implementing rescue procedures. The report also noted that more than 60% of confined space fatalities occurred among would-be rescuers.

The regulations refer to a required permitting system that may exist if your confined space has one or more of the following characteristics:

- Contains or has the potential for a hazardous atmosphere
- Contains a material that may engulf the entering worker
- Configured such that an occupant could be trapped or asphyxiated by converging walls or a sloped floor
- Contains any other recognized serious safety hazard

Ideally, entering permit-required spaces should be avoided. A little preplanning will usually eliminate the need to enter a space that presents a hazardous atmosphere. Forced ventilation and locking out energized equipment ahead of time may allow you to temporarily re-classify a space as a non-permit-required space, enabling workers to do their jobs more efficiently and unencumbered by personal protective equipment and retrieval systems.

Management Aspects

The cornerstone to any CSE program is recognizing the space, training, communicating the hazards, and ensuring personnel use of all the necessary equipment to prevent exposure to a hazardous environment. The following checklist provides the basic information needed for a CSE program.

- Identify and post danger signs on all confined spaces that may require entry
- Evaluate the hazards of each type of space
- Provide engineering controls such as exhaust fans wherever possible
- Develop a permitting system
- Incorporate a checklist for atmospheric testing, hazard evaluation, and controls
- Provide training for supervisors and workers that includes:
  a. Knowledge of the hazards
  b. Use of gas monitoring equipment
  c. Use of mechanical ventilation
  d. Lockout/Blockout/De-energize procedures
  e. Emergency retrieval equipment and rescue procedures
  f. Personal protective equipment
  g. Hot work procedures (welding, burning)
  h. Special procedure for removal or application of coatings, capping, and purging
  i. First aid and CPR
  j. Use of self-contained breathing apparatus
- Calibration and maintenance of test equipment
- Audit procedures
- Record keeping and filing permits

Confined space programs should also protect employees who work near permit-required confined spaces. The program should emphasize prevention of unauthorized entry into such spaces, and protect authorized entrants from hazards through a permit program. The program should require the employer to control hazards, train and equip affected personnel, document compliance, and authorize any entry operations through written permits. Attendants should be first-aid- and CPR-trained prior to monitoring entry operations. Appropriate precautions for rescuing entrants from permit spaces must be detailed in the program as well. If contractors are hired for entry operations, they must be informed of the hazards identified and any procedures to control them. The contractors should also be required to comply with your confined space program.

Emergency Planning

Many regulations require the employer to prepare emergency action plans, including procedures to follow in case of fire, flood, earthquake, or other disaster. State and local emergency agencies may also require facilities to provide additional information regarding drills and tactics with local resources and community residents. Hazardous materials require special consideration. All employees who routinely work with hazardous materials should know how to cope with small spills, as discussed in the previous section. OSHA regulates Emergency Action Plans under Title 29 CFR 1918.100

Management Aspects

Many local authorities require a business plan that may include information such as:

- Inventory of Hazardous Materials with chemicals, common name, and chemical abstract number (CAS)
- Location of hazardous material
- Physical state of material (solid, liquid, gaseous)
- Standard Industrial Classification (SIC) code
- Site and facility maps with material locations
- Type of storage containers
- Average daily amount of each material stored on-site
- 24-hour emergency response contacts
- On-site emergency response and training plans

Additionally, your plan should include:

- Emergency escape procedures
- Emergency route assignments
- Procedures to be followed by employees who must operate critical plant operations
- Evacuation and a means to account for employees
- Rescue and medical duties
- Preferred means of reporting emergencies

- State emergency management structure
- Designations, duties, and responsibilities, including a primary contact
- List and location of emergency equipment (i.e., eyewash/showers, extinguishers, spill kits, disaster supplies)
- Reference to a storm water pollution prevention plan

A release of an unknown or very toxic material, or of any hazardous material in a poorly ventilated area or any other uncontrolled situation may require immediate evacuation. More acutely hazardous material incidents may also involve community members. Employees should contact the emergency response team. Specially trained units of the fire service often perform hazardous materials response. Clearly post the phone numbers of the appropriate response team in all work areas. In many cases, employees should be directed to call 911.

Organizations that opt to train their own emergency response team fall under Hazardous Waste Operations and Emergency Response (HAZWOPER) Regulations 29 CFR1910.120. Under these regulations, team members require 40 hours of HAZWOPER training, on-the-job training, and an annual 8-hour refresher training course in addition to an investment in personal protective equipment such as gas detection monitors, level A containment suits, self-contained breathing apparatus, decontamination gear, and more. Other specialty training may include water, confined space, tunnel, and mine rescue. Most of the time, it makes sense to contract with a hazardous materials/waste or other specialty companies for these services.

Process Safety Management

If your facility handles highly hazardous materials, you may wish to follow the OSHA compliance guidelines and recommendations for process safety management (Title 29 CFR 1910.119 App C). The objective of such a plan is to prevent the unwanted release of hazardous chemicals especially into locations that could expose employees or others to serious hazard. Such a program requires a systematic approach evaluating the entire process. Basic elements include:

- Process design and technology
- Operational and maintenance activities and procedures
- Nonroutine activities and procedures
- Emergency preparedness plans
- Training programs

Process safety management is the proactive identification, evaluation, and prevention of chemical releases that may occur as a result of failures in processes, procedures or equipment.

Communication

Communication forms a major element of both emergency action and response planning. Factors to consider include:

- How can employees communicate if phone lines do not work?
- Are existing radio communication systems compatible?
- Are there overlapping jurisdictions? (i.e., state, federal, and local agencies)
- Who or what authority is the incident commander? Police? Fire? Operations?
- How will employees be accounted for?

Training and Emergency Drills

After developing a workable program incorporating these points, conduct training, drills and host mock emergencies. In addition to involving emergency responders, you may wish to invite county emergency responders, neighboring businesses and community advocates to participate in your drills. Conduct pre- and post-drill discussions with all involved to optimize your resources. Re-evaluate your program as new processes and hazards are introduced to your facility.

Selected Citations and References

- OSHA Web sites: www.osha.gov — provides standards in full text or a digest; contains a list of OSHA documents, guidelines, preambles (history of the regulation) and reference sources; also contains interpretations of the regulations by way of letters to the Agency.
- OSHA documents: Principal Emergency Response and Preparedness Requirements in OSHA standards and guidelines for safety and health programs (OSHA 3122) — Discusses procedures and requirements for safety and health programs.
- How to Prepare for Workplace Emergencies (OSHA 3088) — Details basic steps for handling emergencies such as accidental release of toxic gases, chemical spills, fire, explosion, and personal injuries.
- Hospital and Community Emergency Response (OSHA 3152) — Outlines principal considerations in emergency response planning to reduce health care workers' exposures to hazardous substances.
- Hazardous Waste and Emergency Response (OSHA 3114) — Highlights components of the Hazardous Waste Standard.
- Process Safety Management (OSHA 3132, 3133) — Outlines and describes OSHA requirements for the management of hazards and strategies to prevent or mitigate the release of highly toxic hazardous chemicals.

Hazard Communication

Management Aspects

Determine how hazardous materials are purchased. An effective Hazard Communication Program involves a careful review of the hazardous materials purchased and used by your employees. Spend time with the Purchasing Department and first-line supervisors to find out how hazardous materials enter the workplace. You may discover that salespersons leave products, with or without MSDSs, for your employees to use on a trial basis. In addition, supervisors or employees themselves may purchase hazardous materials in small quantities for urgent or unplanned activities. This may result in your employees using these materials improperly.

Try to establish the most practical point in the purchasing process for review of the materials used. Approve the use of a material prior to purchase. Manufacturers of hazardous materials will always provide a MSDS for their products, and can send this information by facsimile if necessary. Review the ingredients section of the MSDS and determine whether the product contains any materials considered very toxic or carcinogenic. If so, you will want to work with the supervisor and possibly the supplier to determine a substitute material for the job.

Review the MSDS for completeness. Make sure it meets the regulatory requirements. You could use a checklist (see Figure 20.5). Contact the manufacturer regarding an inadequate MSDS and provide a copy of your checklist indicating the missing information. Sections often found deficient in required information include (1) chemical ingredients, (2) toxicity, (3) personal protective equipment, and (4) date the MSDS was prepared. Occasionally, an MSDS may not have the most basic information, such as the manufacturer's name, address, and telephone number. Do not approve a material for use without the information needed to recommend safe work practices. By requesting this information, manufacturers will improve their MSDS or risk losing business.

Give employees ready access to a file containing the MSDSs after you have approved each one. Each material used by the employee should have an available MSDS. Generic MSDS generally do not meet the requirements of the Hazard Communication Standard. Different brands of the same product may contain slightly dissimilar chemical ingredients. There are internet sites and computerized MSDS systems available, but make sure the system is available to every worker at every work site. Every product should have a label, clearly identifying the material and allowing the employee to reference the relevant MSDS. Each label should contain a brief description of the major health effects associated with the chemical ingredients, as well as information concerning appropriate handling and use. Contact manufacturers regarding inadequate labeling. Most safety equipment suppliers sell labels or label-making kits for any products you have without manufacturer's labels.

What about materials already in use? Follow the procedures described above for new materials after obtaining a detailed inventory from each work area. You will have to provide guidelines for employees responsible for providing such inventories. Provide a brief training session defining hazardous materials and explaining the purpose for the inventory. Provide a format for the inventories. Make use of already existing inventories of stock if the materials listed constitute the actual ones in use. Alternatively, your staff can manually list hazardous materials during a walk-through inspection of each work area. All inventories should note the manufacturer's name, the product name (and number if available), common name, size and type of container, and quantity of material stored. The inventories should also list how these materials are stored and used. Periodic updates of hazardous materials inventories will keep your information current.

Establish a record keeping system. Even small organizations will find their inventories extensive. Assembling corresponding files of MSDS will require extensive clerical support. Cross-reference your file of MSDS by manufacturer, product name, and common name. Using a computer database is helpful. Make it as easy as possible for your employees to find the material, especially during an emergency.

Train employees and supervisors. Develop a flow chart illustrating the system for hazard communication within your organization (see Figure 20.6). Use this chart during general training sessions. Present the basic concepts of hazard communication,

Material Safety Data Sheet Adequacy Checklist

Material Name: _____   Product #:_____

Manufacturer:_____

_____

MSDS Data of Preparation or Revision _____

Required Information	29 CFR 1910.1200(g)(2)	Check if Adequate
I. Product Identification, including 1, 2 or 3:	(i)(A)	
1. If single substance:		
a. chemical name(s),		_____
b. common name(s),		_____
2. If mixture:	(i)(C)	
a. chemical name(s) of ingredients,		_____
b. common name(s) of ingredients,		_____
3. Common name(s) of mixture, if tested as a whole.	(i)(B)	_____
II. Physical and chemical Characteristics, such as vapor pressure and flash point	(ii)	_____
III. Physical Hazards, such as: potential for fire, explosion, reactivity	(iii)	_____
IV. Health Hazard Data, including:	(iv)	
1. Signs and symptoms of exposure,		_____
2. Medical conditions aggravated,		_____
3. Primary routes of entry,	(v)	_____
4. OSHA PEL, ACGIH TLV, and any other recommended exposure limit,		_____
5. Whether listed as a NTP, IARC, or OSHA carcinogen.	(vii)	_____
V. Safe Handling and Use, including: cleanup of spills and leaks	(viii)	_____
VI. Control Measures, such as: Engineering controls, PPE, work practices.	(ix)	_____
VII. Emergency and First-Aid Procedures	(x)	_____
VIII. Date of MSDS Preparation or Revision	(xi)	_____
IX. Name, Address, and Telephone No. of Responsible Party	(xii)	_____

Reviewed by: _____   Date: _____

Figure 20.5 Sample material safety data sheet checklist.

including definitions of hazardous materials, MSDS's, the use of protective equipment, and basic toxicology. All employees should receive this general training, perhaps during regularly scheduled safety or staff meetings.

Training in the use of the specific materials may present the most challenging portion of the hazard communication program. Make use of a number of resources for this training; do not rely on the MSDSs alone. Develop standard operating procedures (SOP) by working with the supervisors and employees. Include detailed instructions for safely using and handling hazardous materials. Glossaries of the

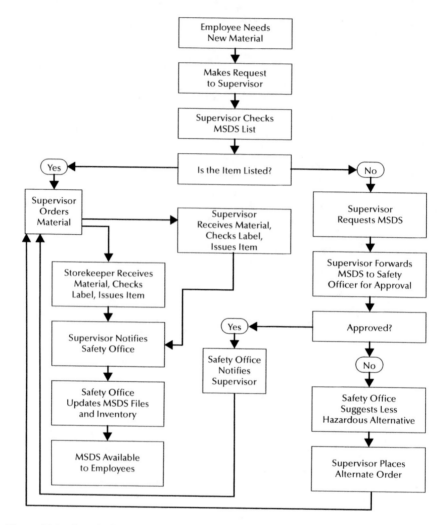

Figure 20.6 Sample flowchart showing hazard communication system.

terms used by toxicologists and industrial hygienists may make it easier to present technical information. Group materials together with similar chemical properties, such as organic solvents. Schedule a number of brief sessions; do not present too much information at one time.

Some organizations effectively use the train-the-trainer approach. The occupational health staff trains first-line supervisors and provides them with the curricula, audio-visual materials, and other supplies needed. This method often works well because employees tend to attach greater significance to information learned from their supervisors. Supervisors may need extensive support from your group, however, in order to feel comfortable presenting technical information.

Occupational Safety and Health Administration (OSHA)   www.osha.gov
- Hazard Communication: Title 29, Code of Federal Regulations, Section 1910.1200 (29 CFR 1910.1200)
- Emergency Action Plan: 29 CFR 1910.38(a)
- Fire Prevention Plan: 29 CFR 1910.38(b)
- Hazardous Waste and Emergency Response: 29 CFR 1910.120
- Chemical Process Safety: 29 CFR 1910.119
- Laboratories: 29 CFR 1910.1450
- Respiratory Protection: 29 CFR 1910.134
- Bloodborne Pathogens: 29 CFR 1910.1030

Environmental Protection Agency (EPA)   www.epa.gov
- Emergency Planning and Community Right-to-Know (SARA Title III): 40 CFR 355, 370, 372
- Underground Tanks: 40 CFR 280, 281
- Pesticides: 40 CFR 152-186
- Hazardous Waste: 40 CFR 240 et seq.

Department of Transportation (DOT)   www.dot.gov
- Transportation of Hazardous Materials and Waste: 49 CFR 107, 171-197

See also:
- Uniform Building Code
- Uniform FIre Code

Note: This is not a complete list. Contact the agencies listed and their state and local counterparts for more information.

Figure 20.7   Selected federal hazardous materials regulations.

Hazardous Materials

Many environmental regulations concerning hazardous materials go beyond the stipulations of the Hazard Communication Standard; Hazard Communication forms a subset of Hazardous Materials Programs. A partial list of hazardous materials regulations appears in Figure 20.7. Many of these overlap.

Management Aspects

Make your inventory versatile. Include in the hazardous materials inventory the information required for public disclosure for each product, including the chemicals ingredients (and percentage contained), toxicity, storage provisions, and emergency procedures. Include detailed maps showing the location of hazardous materials storage and use. During a fire, earthquake, chemical spill, or other emergency situation, this information, as well as the MSDS for all products, should be readily available to emergency responders.

Minimize your hazardous materials. Public interest and environmental action groups will have access to the above information. In addition, your organization must provide the aggregate amount of hazardous materials released to the environment. Substitution of less toxic materials as well as reduction of the overall use of these materials will reduce environmental and occupational health liability.

Store materials safely. Storage requirements are often not clearly spelled out by regulation. General storage requirements are also difficult to apply to specific situations. However, pay particular attention to the storage and handling of the following types of hazardous materials:

  a. Flammable and combustible liquids, such as gasoline, diesel fuel, solvents, paints, and paint removers

  b. Corrosives, including acids and bases

  c. Oxidizers and reactive materials, such as peroxides, chlorine gas, and acetylene

  d. Poisons and toxic materials, inorganic lead and other metals, and pesticides.

  e. Compressed gases

Storage of large quantities of flammable materials require special storage containers or buildings, equipped with fire suppression systems, containment, ventilation, and explosion-proof lighting. Smaller quantities of materials will require flammable liquid storage cabinets or special storage containers. Follow the Uniform Fire Code, National Fire Protection Association (NFPA) standards, ANSI standards, and local fire code for your specific situation.

Combustible and reactive materials should be stored separately from flammable materials as these may provide fuel for a fire. Oxidizers need separate storage. Corrosives require careful handling as these can cause severe burns, and should also be stored apart from other materials.

Conduct periodic walk-throughs of operations involving hazardous materials. Particularly in small operations, you may find incompatible materials stored together, or a complete lack of flammable storage lockers. Containment of these materials in case of a spill is critical, and secondary containment should always be used. Berms and containment dikes separate large tanks of these materials, and provide a collection point in case of spills. Even simple pallets that contain liquids can serve this purpose in smaller operations.

Develop written storage procedures. Many employees will not have an intuitive understanding of the properties of chemicals or the consequences of improper storage of incompatible materials, even after extensive training. Provide detailed procedures for these employees. Begin by grouping materials used in each work area by chemical properties, as listed previously. For each group, chose materials used most often or which are the most representative of these materials as a whole. Next, go to the MSDSs for these materials, and check under the sections on storage, handling, and emergency procedures. If necessary, contact the manufacturer for more detailed information.

Provide control equipment and procedures. Go through work areas or check your inspection records for examples of how these materials are presently used and stored. Break each operation down into steps. Specify the use of personal protective equipment (especially chemical splash goggles and face shields), safety cans for flammable liquids, and a laboratory hood or other ventilation during mixing or transfer. Include provisions for spills by providing spill kits for quick cleanups of small spills of less toxic materials (be sure your training describes which spills employees may clean up themselves). All work sites (including mobile or temporary) where hazardous materials are used should have plumbed eye washes and showers, or other provisions for flushing the eyes or skin with water following an accidental spill.

Employees and supervisors should check the procedures for completeness. To save time, have them prepare the breakdown of the operation into steps, and assist with the incorporation of safe handling procedures. When complete, post the procedures in the workplace, and conduct training.

Hearing Conservation

Definition and Scope

Occupational noise exposures are a major cause of hearing loss. Although noise-induced hearing loss is one of the most common occupational illnesses, it is often overlooked and underrated because there are no visible effects and in most cases, there is no pain. Besides the damage to hearing, noise can result in interference with a person's ability to speak clearly with others, and affect safety and work performance. According to a 1993 EPA report, more than nine million Americans are exposed to daily average occupational noise levels above 85 decibels, a weighted (dBA).

Protecting the hearing of your employees involves physical evaluation of occupational noise exposures as well as monitoring the hearing acuity of those exposed to noisy environments. An effective hearing conservation program seeks to prevent significant shifts in the hearing acuity of employees.

Management Aspects

OSHA 29 CFR1910.95 mandates that "The employer shall administer a continuing, effective hearing conservation program... whenever employee noise exposures equal or exceed an 8-hour time-weighted average sound level of 85 decibels..."

The first step is to enlist the cooperation of management and the support of noise-exposed employees. To assess whether a facility needs a hearing conservation program, Industrial Hygiene professionals should perform an evaluation, which includes personal noise sampling and environmental noise measurements. Ideally, engineering controls are the best long-term solution to decrease noise levels and eliminate the need for a program. However, if engineering controls are infeasible and the OSHA criteria are met, a written program with the following elements should be developed and implemented.

Components of a Hearing Conservation Program

Monitoring. Monitoring is the backbone of an effective program. Both area monitoring and personal monitoring may be used as long as it is representative of the employees exposure. Monitoring should be repeated when there is a change in production, process, or equipment.

Audiometric testing. Audiometric testing is an effective measure of how well your program and controls are working. Changes in hearing levels may indicate noise control, hearing protection or training aspects need further evaluation. Your audiometric service provider should meet OSHA requirements and preferably have certification by the Council of Accreditation in Occupational Hearing Conservation. A baseline audiogram should be established for all employees exposed to 85dBA or above. Note that OSHA is reviewing the feasibility of lowering this threshold to 80 dBA in the near future.

Hearing protectors. Hearing protection should be provided to all employees who are exposed to noise in excess of 85dBA. Employees should be able to choose from several different varieties of hearing protection devices. Custom hearing protectors must be fitted by a qualified individual due to anatomical variations in the size of the ear canal. Other types such as earplugs or inserts, ear caps and earmuffs provide varying levels of noise attenuation or reduction. The noise reduction rating (NRR) should be clearly labeled on each type of protection device. Disposable ear protection should be well maintained or replaced often. Hearing Protection devices should not be given to employees without instructions on proper usage.

Training program. Annual training should be provided to all employees in the hearing conservation program and their supervisors. Training should include:

- Effects of noise on hearing and the importance of hearing loss prevention
- Purpose of hearing protectors, including the advantages and disadvantages and attenuation of various types
- Proper use, maintenance, and storage of hearing protectors
- Purpose of audiometric testing
- Explanation and availability of noise sampling reports and data

Additionally, effective training includes how to avoid noise-induced hearing loss in the workplace and elsewhere.

Enforcement. Because noise-induced hearing loss can contribute to the loss of well-being, comfort, and affect one's lifestyle, disciplinary measures must be established to ensure proper use of hearing protection equipment. These measures should be clear to both supervisors and employees.

Record keeping. OSHA mandates that employers keep employee monitoring measurements for two years, and audiometric tests for the duration of the employee's employment.

Program evaluation. Assess all new equipment and processes. Use medical monitoring results and employee feedback to evaluate your program's effectiveness. Periodic assessment should ensure the program's continued success.

Indoor Air Quality

In recent years, health and safety specialists have recognized that the burgeoning number of complaints associated with indoor air quality (IAQ) require a formalized, systematic approach. Newer buildings designed to efficiently recirculate and condition air have provided an environment that provides little control for its occupants. These buildings share many of the same traits including:

- Limited temperature range
- Inability to open and close windows
- Multiuse facilities
- Improper use of pesticides, sealants, cleansers
- Overcrowding of occupants
- Increasingly complex mechanical systems
- No existing "as-built" blueprints of mechanical systems
- Areas where water or moisture can accumulate

All of these building traits contribute to the increasing frustration of both occupants and building owners in resolving air quality complaints. Additionally, other factors such as ergonomics, psychological stressors, poor lighting, and noise all confound IAQ investigations. Office workers, who tend to have a lower tolerance for nuisance odors or have increased sensitivities to chemicals or allergens, typically generate IAQ complaints. Building owners find it increasingly necessary to implement a set of procedures to investigate air quality complaints. At this time, there are no OSHA regulations governing indoor air quality; however, there are a number of professional organizations that offer guidelines and recommendations to ensure good air quality (ASHRAE and NADUCA).

Management Aspects

Ideally, prior to the building construction, a mechanical engineer should review the design and any modifications to ensure that there is adequate ventilation for all building uses. Certain building related illnesses such as Legionnaire's disease, Pontiac fever, Q fever, and multiple chemical sensitivity have specific symptoms that need to be addressed and investigated immediately. Confirmation of these illnesses requires notification and possible assistance from your local health department. The majority of IAQ complaints, however, are nonspecific and difficult to characterize. The following checklist serves as a guideline to investigate IAQ problems.

1. Occupant interview
   - Location
   - Complaint or symptom timing and spatial patterns
   - Observations
   - Occupant diary
2. Management checklist
   - Assemble team of health and safety, mechanical engineer, building inspector, and maintenance to investigate complaints
   - Heating, Ventilation, and Air Conditioning (HVAC) system design review
   - Inventory complaint locations
   - Visual inspection and walk-through
3. Collection of data
   - Repair information
   - Check negative and positive pressure
   - Check pollutant sources and pathways
   - Check stored hazardous substances
   - Check for moisture, mold or water damage
4. Management plan
   - Confirm that operations and maintenance schedules are sufficient
   - Review pest control procedures
   - Establish IAQ task force or safety committee to address occupant concerns
   - Develop boiler plate specifications to minimize IAQ concerns during renovation and remodeling
   - Schedule work to minimize IAQ problems
   - Isolate work area and provide increased ventilation if appropriate
   - Allow new furnishings, coatings, and materials adequate drying and curing times prior to occupation

Most importantly, inform building occupants of any upcoming work and associated odors, nuisance dust, temperature fluctuations, noise, vibration, and other inconveniences due to the work. Provide a schedule of the work and anticipate their needs by providing earplugs, increasing work breaks and setting up temporary workstations in another location.

Selected Citations and References

- Web sites: www.OSHA.gov; www.EPA.gov — Technical link provides indoor air quality references for recognition, evaluation, control, compliance, and training.
- Government documents:
    Indoor Air Quality Investigation, OSHA Technical Manual
    EPA Fact Sheet — Ventilation and Air Quality in Offices
        www.epa.gov/iaq/pubs/ventilat.html
    The Inside Story — A Guide to Indoor Air Quality
    EPA document #402-K-93-007
    NIOSH Fact Sheet Indoor Environmental Quality
    NIOSH/EPA Building Air Quality (BAQ) Action Plan
        www.cdc.gov/niosh/iaqfs.html
    DHHS (NIOSH) publication #98-123
    Indoor Environmental Quality: A teaching module for clinicians
    Fact Sheet, Mold, HESIS Department of Health Services, Oakland, CA
        www.dhs.cabinet.gov/ohb/hesis/ also www.ohb.org

Lead Safety

Currently, the regulations and standards regarding lead are changing. There has been an increase in public concern and policy making about environmental lead, and thus, is affecting occupational lead programs as well. Congress passed The Housing and Community Development Act of 1992 (commonly referred to as "Title X") to address lead poisoning, and the law is being implemented in stages by various agencies. OSHA has a new standard for the construction industry. The Housing and Urban Development Administration (HUD) and the EPA are developing new guidelines and laws for environmental lead exposure. Many states are also enacting occupational and environmental lead regulations. For example, the lead paint abatement industry is becoming increasingly regulated.

Management Aspects

Determine where and how are you involved with lead. Lead is heavily regulated, and how you manage lead depends on what you do. Do you manage housing or facilities that serve children? Are you involved in manufacturing or construction? You need to gather the regulations that pertain to your industry.

Determine the regulations that affect your organization. When you are first beginning a lead program, you will need to gather information on current regulations and what the regulatory trends are in your region since the regulations are changing quickly. You should check with OSHA, EPA, HUD (if you deal with housing), and local agencies responsible for environmental and occupational safety and health. You will have to ascertain which regulations apply to your operations. Environmental regulations will probably apply to anyone dealing with lead; they will be most applicable to lead abatement.

In terms of occupational health regulations, you will need to decide if you are covered by the general industry standard (29 CFR 1910.1025) or the new construction industry standard (29 CFR 1926.62). You may need to check with OSHA, but generally if a product is made (for example, a battery plant where exposure is fairly consistent) you are covered by the general industry standard. If lead work involves construction, demolition, repair (such as a painting contractor), you are covered by the construction standard. Maintenance activities can be a gray area, and you will have to find out how the standards are being enforced, or follow the most stringent.

Determine who may be exposed to lead. Look at trigger tasks in the OSHA regulation. Is lead work incidental or integral to the function of your organization? The answer to this question will be important in determining your lead management program. Can you contract out some or all lead work? Will you have to train everyone or train a specialized crew?

Purchase basic equipment and personal protective equipment (PPE). The standard includes basic protective measures, such as respirators, PPE, change areas, hand-washing facilities, biological monitoring, and training, be provided during exposure assessment.

Conduct medical surveillance — You will need to have employees tested for initial or baseline levels of lead and zinc protoporphyrin in their blood. Employees will need medical exams to wear respirators and may require medical exams for lead in some cases. The standard requires that employees who have blood lead levels above 50 µg/dL be removed from work involving exposure to lead until their blood lead levels are at or below 40 µg/dL. The standard includes protection of the employee's earnings, rights, and benefits while on medical removal.

Conduct an exposure assessment. Who is exposed over the AL or the PEL? Monitor tasks, but protect employees while you are determining exposure. The standard includes "trigger tasks" with exposure levels that you must assume employees are exposed to until you have completed your assessment.

Develop a written compliance program. The following section outlines the methods for developing a compliance program.

Conduct training. The standard includes a list of topics that must be included in employee lead training for employees exposed above the AL. At a minimum, all employees exposed to lead must be given information about lead hazards in accordance with the Hazard Communication Standard. You should also check your state and local regulations, as well as EPA regulations that may require you to have specialized training and licensure or certification to perform certain lead work (i.e., residential lead abatement).

Compliance Program to Control Exposures

You will need to have a written compliance program that describes your organization's lead activities and the methods you will use to control exposures. The methods include:

Engineering and work practice — This will include the use of HEPA-exhausted tools, HEPA vacuums for cleanup, and removing lead using methods that create less dust (i.e., manual scraping vs. abrasive blasting) where feasible.

Administrative controls — You can limit the exposure by limiting the duration of the lead task. A good example of administrative controls for lead is given in the AIHA *Inorganic Lead Guidance Document*. For example, if you perform a task for 8 hours that has an exposure level of 100 µg/m³, you could assume you would be below the PEL if you performed the task for only 2 hours.

Personal protective equipment — Respirators may be used to supplement other controls. You must ensure that employees are wearing appropriate protective respirators for the tasks they are performing. Other equipment includes coveralls, gloves, hats, faceshields, and goggles, depending on the tasks.

Hygiene and housekeeping — Since ingestion is an important route of entry for lead, hygiene is an important part of your control program. All employees must have a place to wash their hands and face prior to eating, drinking, smoking, applying cosmetics, or going home. Where feasible, employees should shower before going home. Employees need a place to change between work clothing and street clothing. Because lead can be very dangerous for small children, you must make sure that employees do not leave the workplace wearing contaminated clothing or equipment. Employees also need a clean lunchroom and should not enter the lunchroom with dirty clothing or equipment. The work site should be kept as free as possible of lead. This is best accomplished with a HEPA-filtered vacuum.

Respiratory Protection

One of the greatest problems with respirators is that they are used haphazardly and in ways that may be detrimental to the health of employees. You will need to take control of respirator use by determining when they should be worn, what types should be issued, and how they are to be used and maintained. OSHA requires that you formalize your program in written form. Your program should address the selection of proper respirators, training on use and limitations, cleaning, disinfection, storage, inspection, evaluation of work conditions and program, medical surveillance, and approved respirators.

The Respiratory Protection Standard (29 CFR 1910.134) was revised in 1998, so even if you have an existing program, you will probably need to make some changes. Most of the changes to the standard are outlined in the following list. A few other things to note: provisions for voluntary use of respirators (with reduced burdens on the employer) are now in place; new requirements for firefighters have gone into effect; and there will be a separate standard for the use of respirators as protection against tuberculosis.

There have also been changes to the NIOSH certification of respirators. So far the changes affect filters for negative pressure air-purifying respirators. They no longer use the dust/mist, dust/fume/mist, and HEPA designations. These filters have been replaced by nine classes of filters, with three levels of efficiency and three categories of resistance to degradation. The three efficiencies are 95, 99, and 100%. The categories of resistance to degradation are based upon resistance to oil as follows: **N** (not resistant to oil), **R** (resistant to oil), and **P** (oil-proof). In general, you will now use N95, R95, or P95 for dust/mist, and P100 for HEPA. N-series filters should never be used in an environment that contains oil aerosols.

Determine the operations requiring respirator use. Conduct air monitoring of operations, starting with those that you feel may be problematic. If air monitoring is not required by a substance specific standard, you may make an estimate of exposure by the use of objective data (from trade associations, manufacturers, etc.) or by the use of mathematical models. Of course, it is best to conduct personal air monitoring. If the exposures are high, determine the best engineering or administrative control measures, such as product substitution or ventilation. OSHA mandates that you institute other controls before PPE because PPE puts much of the burden of protection on the employees. A respiratory protection program is also more costly than it might seem, and will take a lot of effort to ensure that it is being followed. While control measures are being implemented, you should have employees wear respirators. After implementing control measures, you should conduct more air monitoring. You may find that the control measures are not completely effective and that respirators should continue to be worn. In some cases, such as working with asbestos, the general practice is to always wear some level of respiratory protection, regardless of air monitoring. This is because you want to keep exposures as low as possible and because air concentrations are not uniform or entirely predictable.

Obtain copies of the respiratory protection regulations and standards. The American National Standards Institute has published many standards on respiratory protection, including ANSI Z88.2, which is referenced in the OSHA regulations.

Put your program in writing and get it approved by management. First you must decide who will administer the program. This person will be responsible for purchasing, training, fit testing, and the other elements of your program — it will probably be you. Some of the duties could be given to others in the organization, but make sure that everyone with responsibilities has proper training or professional background, including yourself. Work with management and your human resources department to decide what you will do if employees are unable to wear respirators. What efforts will be made to accommodate them? You should also work with the applicable employee unions.

Determine the types of respirators required for your operations. In selecting respirators consider the contaminant, its form, its concentration, the exposure limit, the possibility of an environment immediately dangerous to life or health (IDLH), flammability, and whether eye or face protection is needed. Familiarize yourself with the uses and limitations of various types of respirators. Presently, the standard does not list assigned protection factors for various types of respirators. You should rely on substance-specific standards or OSHA letters of interpretation where available and applicable. Otherwise, use the protection factors recommended by NIOSH.

Develop a change out schedule for chemical cartridges. The revised standard no longer allows users to rely on the warning properties of contaminants to determine when to change their chemical cartridges for air purifying respirators. Determine a reasonable change out schedule based upon the contaminant concentration and the breakthrough time of the cartridge. Respirator manufacturers may be of help to you in determining the schedule. You should continue to train employees on recognizing breakthrough based on the warning properties of the contaminant.

Send all the employees who work (or could work) in the areas requiring respirators for medical surveillance. Tell the physician what types of respirators you would consider having the employee wear because the examination may vary depending on the type of respirator. Some medical problems may prohibit an employee from wearing a respirator. Respirators can cause physiological stresses due to factors such as restricted airflow. Because being restricted from wearing a respirator can affect the livelihood of an employee, efforts should be made to accommodate physical problems by selecting a different type of respirator. You should work with the physician on this problem. If this fails, rely on the policy you developed earlier.

Fit test employees. Employees who are qualified to wear respirators must be assigned respirators that fit them. Fit testing is now required for all tight-fitting face pieces, regardless of whether they are negative or positive pressure. Fit testing should be carried out at least annually and performed in accordance with standards and regulations. Fit testing is not required for voluntary users or escape-only respirators (which does not include self-contained breathing apparatuses).

There are two types of fit testing, qualitative and quantitative. Quantitative fit testing is the preferred method because it actually measures the amount of contaminant entering the respirator. However, this expensive system may not be practical if you only have a small number of employees wearing respirators. Traditionally, quantitative fit testing has been performed in a booth with a particle generator and counter. However, portable models are now available, which are based on counting the particles in ambient air, that may make quantitative fit testing more reasonable for your situation because they do not require a dedicated location, are easy to use, and less expensive than a booth. Consulting and training firms can also perform fit testing for you.

If quantitative fit testing is not feasible, you can perform qualitative fit testing. This involves finding out if a respirator wearer can taste or smell a test agent. You must follow the protocol carefully for the test to be valid. When qualitative fit testing is used for negative pressure air-purifying respirators, the respirators may only be used in concentrations up to 10 times the PEL, even if it is a full face respirator. For greater concentrations, quantitative fit testing must be used.

Whichever form of fit testing you decide on, you will need to order respirators from various manufacturers and in different sizes so that employees have a selection. One face piece size is not going to be adequate for everyone in your company. Various sizes from just one manufacturer may not be adequate either. Read through the respirator fit testing protocol and obtain the equipment you will need.

Develop maintenance procedures. Respirators will only protect employees if they are properly maintained. If possible, employees should be issued a personal respirator rather than sharing one with others. All respirators must be cleaned and disinfected regularly (right after wearing if shared). Disposable towelettes can be used for cleaning and disinfection, but it is better to wash the respirator in soap and water. There are disinfectant soaps that can be dissolved in water. If the work area does not have a clean sink for this purpose, provide clean buckets. The respirator should be dried and then stored in a clean place. One good way is to put the respirator in a clean plastic container on the top shelf of a clean locker (where it will not get crushed). Respirators should not be left on a hook in the work area, left on the dashboard of a truck, or anywhere that they may become dirty or damaged.

Develop inspection procedures. Before use, respirators should be inspected to see that all the parts are in place and functioning properly. If a part is missing or broken, you will need to order a replacement part from the manufacturer for the particular model of respirator. Respirators are approved as a whole unit and if you use a different part (for example use a safety pin to hold a strap in place) you are no longer using an approved respirator.

Train supervisors and employees. Training for employees and supervisors is a critical part of your program. The success of your program depends upon employees wearing respirators correctly and maintaining them correctly. Both employees and supervisors should be trained and ensuring compliance because you cannot check on all work areas all of the time. You should cover the basics of the types of respirators, their uses and limitations, and all of the aspects of your program. Try to make the training as much hands-on as possible. Have employees bring in respirators and actually clean them, inspect them, and try them on. After your official training, you should try to visit the work areas and check on respirator usage and storage. Talking individually with employees in their work areas and addressing their specific questions or deficiencies can be a good form of additional training. Training needs to be refreshed at least yearly.

Continue to do air monitoring. Periodic air monitoring is a good idea and necessary if conditions have changed. Some regulations for specific chemical hazards specify when monitoring should be done.

Audit the program. It is not enough to supply respirators to employees. Employers must ensure that they are being worn and that they are being used properly. Talk to respirator wearers and inspect the condition of respirators. Evaluate your program and change written procedures if necessary.

Selected Citations and References

- **OSHA Publications:**
  CPL2.120 — Inspection Procedures for the Respiratory Protection Standard
- **Other Publications:**
  United States Department of Health and Human Services. NIOSH Guide to the Selection and Use of Particulate Respirators Certified under 42CFR84. Cincinnati, OH: USDHHS 1996. (NIOSH Publication No. 96-101)
  American National Standards Institute. American National Standard Practices for Respiratory Protection. New York: ANSI, 1991. (ANSI Z88.2-1991)

ACKNOWLEDGMENTS

The authors wish to thank the following individuals, who are current or former employees of the City and County of San Francisco: Richard Lack, Richard Lee, Neva Nishio Petersen, Pamela Reitman, Vickie Wells, Karen Taylor, and Karen Yu. They provided invaluable assistance during the development of health programs for the Port of San Francisco. Their support and friendship is greatly appreciated.

REFERENCES

Books

American Conference of Governmental Industrial Hygienists (ACGIH). *Industrial Hygiene Program Management.* Cincinnati, OH: ACGIH, 1988–1989.

Berger, E. H., Ward, W. D., Morrill, J. C., and Royster, L. H., Ed. *Noise and Hearing Conservation Manual,* 4th ed. Fairfax, VA: American Industrial Hygiene Association, 1986.

Clayton, G. and Clayton, F., Eds. *Patty's Industrial Hygiene and Toxicology.* New York: John Wiley & Sons, 1991.

Garrett, J., Cralley, L. J., and Cralley, L. V., Eds. *Industrial Hygiene Management.* New York: John Wiley & Sons, 1988.

Hearn, L. C. *Inorganic Lead Guidance Document.* Fairfax, VA: American Industrial Hygiene Association, 1995.

Plog, B. A., Ed. *Fundamentals of Industrial Hygiene,* 3rd ed. Chicago: National Safety Council, 1988.

Other Publications

American Industrial Hygiene Association Management Committee. Health and Safety implications of European Community 1992 (EC92). *Am. Ind. Hyg. Assoc. J.* 53(11):736–741, 1992.

American Journal of Public Health (Editorial). Workers' health and safety (WHS) in cross-national perspective. *Am. J. Public Health* 78(7):769–771, 1988.

Andersen, G. H., Smith, A. C., and Daigle, L. L., An approach to occupational health risk management for a diversified international corporation. *Am. Ind. Hyg. Assoc. J.* 50(4):224–228, 1989.

Bardsley, C. A., Lichtenstein, M. E., and Nusbaum, V. A. Industrial Hygiene career planning. *Am. Ind. Hyg. Assoc. J.* 44(3):229–233, 1983.

Belk, H. D. Implementing continuous quality improvement in occupational health programs. II. The role of the corporate medical department — present and future. *J. Occup. Med.* 32(12): 1184–1188, 1990.

Bosch, W. W. and Novak, J. J. Strategies for developing comprehensive occupational health programs. *Water Eng. Manage.* April: 26–30, 1993.

Brandt-Rauf, P. W., Brandt-Rauf, S. I., and Fallon, L. F. Management of ethical issues in the practice of occupational medicine. *Occup. Med. State of the Art Rev.* 4(1):171–176, 1989.

Bridge, D. P. Developing and implementing an industrial hygiene and safety program in industry. *Am. Ind. Hyg. Assoc. J.* 40(4):255–263, 1979.

Brief, R. S. and Lynch, J. Applying industrial hygiene to chemical and petroleum projects. *Am. Ind. Hyg. Assoc. J.* 41(11):832–835, 1980.

Burdorf, A. and Heederik, D. Guest editorial. Occupational and environmental hygiene; more than just a name game. *Am. Ind. Hyg. Assoc. J.* 53(10):A484–A490, 1992.

Colton, C. E., Birkner, L. R., and Brosseau, L. M., Eds., Respiratory Protection: A Manual and Guideline, 2nd ed., Fairfax, VA. American Industrial Hygiene Association, 1991.

Committee report. Scope of occupational and environmental health programs and practice. *J. Occup. Med.* 34(4):436–440, 1992.

Cooper, W. C. and Zavon, M. R. Health surveillance programs in industry. In: *Patty's Industrial Hygiene and Toxicology*, 3rd ed., Vol. 3, Part A. Harris, R. L., Cralley, L. J., and Cralley, L. V., Eds. New York: John Wiley & Sons, 1994.

Corn, M. and Lees, P. S. J. The industrial hygiene audit: purposes and implementation. *Am. Ind. Hyg. Assoc. J.* 44(2):135–141, 1983.

Dalton, B. A. and Harris, J. S. A comprehensive approach to corporate health management. *J. Occup. Med.* 33(3):338–347, 1991.

Dwyer, T. Industrial safety engineering — challenges of the future. *Accid. Anal. Prev.* 24(3):265–273, 1992.

Fallon, L. F. Organizational behavior. *Occup. Med. State of the Art Rev.* 4(1):93–104, 1989.

Fallon, L. F. An overview of management. *Occup. Med. State of the Art Rev.* 4(1):1–10, 1989.

Feitshans, I. L. Potential liability of industrial hygienists under USA law. II. Prescription for reducing potential liability. *Appl. Occup. Environ. Hyg.* 6(3):205–214, 1991.

Geiser, K. Protecting reproductive health and the environment: toxics use reduction. *Environ. Health Perspect. Suppl.* 101 (Suppl. 2):221–225, 1993.

Gough, M. PCIH lecture: Zero Risk or Acceptable Risk. *Am. Ind. Hyg. Assoc. J.* 52(10):A556–A560, 1991.

Hazzard, L., Mautz, J., and Wrightsman, D. Job rotation cuts cumulative trauma cases. *Personnel J.* February: 29–32, 1992.

Henry, B. J. and Schaper, K. L. PPG's safety and health index system: a 10-year update of an in-plant hazardous materials identification system and its relationship to finished product labeling, industrial hygiene, and medical programs. *Am. Ind. Hyg. Assoc. J.* 51(9):475–484, 1990.

Hughes, J. T. An assessment of training needs for worker safety and health programs: hazardous waste operations and emergency response. *Appl. Occup. Environ. Hyg.* 6(2):114–118, 1991.

Jacobs, H. C. Managing safety. *CHEMTECH* July:400–402, 1989.

Johnson, S. L. Manufacturing ergonomics: a historical perspective. *Appl. Ind. Hyg.* 4(9):F24–F29, 1989.

Lichtenstein, M. E. Guest editorial: megatrends and the occupational safety and health professional. *Am. Ind. Hyg. Assoc. J.* 45(11):A17–A18, 1994.

Lichtenstein, M. E., Buchanan, L. G., and Nohrden, J. C. Developing and managing an industrial hygiene program. *Am. Ind. Hyg. Assoc. J.* 44(4):256–262, 1983.

Minerva Education Institute. Management's safety and health imperative: eight essential steps to improving the work environment. *Am. Ind. Hyg. Assoc. J.* 52(4):A218–A221, 1991.

Molyneux, M. K. and Wilson, H. G. E. An organized approach to the control of hazards to health at work. *Ann. Occup. Hyg.* 34(2):177–188, 1990.

Moser, R., Meservy, D., Lee, J. S., Johns, R. E., and Bloswick, D. S. Education in management aspects of occupational and environmental health and safety programs. *J. Occup. Med.* 31(3):251–256, 1989.

Moser, R. Quality Management in occupational and environment health programs — benefit or disaster? *J. Occup. Med.* 35(11):1103–1105, 1993.

Novak, J. J. and Bosch, W. W. Strategies for developing comprehensive occupational health programs for the paper industry. *Tappi J.* 76(9):87–95, 1993.

Occupational Medical Practice Committee of the American College of Occupational and Environmental Medicine (ACOEM). Scope of occupational and environmental health programs and practice. *J. Occup. Med.* 34(4):436–440, 1992.

Reynolds, J. L., Royster, L. H., Royster, L. J., and Pearson, R. G. Hearing conservation programs (HCPs): the effectiveness of one company's HCP in a 12-hr work shift environment. *Am. Ind. Hyg. Assoc. J.* 51(8):437–446, 1990.

Robins, T. G., Hugentobler, M. K., Kaminski, M., and Klitzman, S. Implementation of the federal Hazard Communication Standard: does training work? *J. Occup. Med.* 32(11):1133–1140, 1990.

Roughton, J. Integrating quality into safety and health management. *Ind. Eng.* July: 35–40, 1993.

Roughton, J. Safety and health management through bar code technology. *Appl. Ind. Hyg.* 4(6):F29–F31, 1989.

Sass, R. The implications of work organization for occupational health policy: the case of Canada. *Int. J. Health Serv.* 19(1):157–173, 1989.

Snyder, T. B., Himmelstein, J., Pransky, G., and Beavers, J. D. Business analysis in occupational health and safety consultations. *J. Occup. Med.* 33(10):1040–1045, 1991.

Stanevich, R. S. and Stanevich, R. L. Guidelines for an occupational safety and health program. *Am. Assoc. Occup. Health Nurs. J.* 37(6):205–214, 1989.

Stern, M. B. and Mansdorf, S. Z., Eds. *Applications and Computational Elements of Industrial Hygiene.* Boca Raton, FL: Lewis Publishers, 1999.

Tuskes, P. M. and Key, M. M. Potential hazards in small business — a gap in OSHA protection. *Appl. Ind. Hyg.* 3(2):55–57, 1988.

White, K. Managerial style and health promotion programs. *Soc. Sci. Med.* 36(3):227–235, 1993.

Yarborough, C. M. System for quality management. *J. Occup. Med.* 35(11):1096–1102, 1993.

CHAPTER **21**

Shiftwork: Scheduling Impacts on Safety and Health

James P. Seward

CONTENTS

INTRODUCTION

It is increasingly common to find people working during shifts that differ from the traditional 9-to-5 hours. The term shiftwork loosely subsumes the range of different work schedules, including night work, rotating shifts which change periodically, split shifts (two work periods per day) and other arrangements which vary significantly from fixed daytime hours. Because such schedules, particularly rotating and night work, may interfere with the normal wake-sleep cycles, there has been great interest in their effects on physiologic functioning, overall health, safety, productivity, and social well-being. The design of a work schedule is an important consideration for any organization which requires continuous production or coverage of functions during the evening and night hours. Unfortunately, schedule design is not always a straightforward exercise, and there are trade-offs which must be made between physiologic factors and social preferences.

1-56670-370-0/02/$0.00+$1.50

THE PHYSIOLOGY OF SHIFTWORK

Many human physiologic functions have a circadian periodicity. In addition to the apparent wake-sleep cycle, there is a daily rhythm to a host of other human biologic parameters such as body temperature, release of endogenous hormones, and even lung function. The human temperature curve normally peaks in the late afternoon and reaches a trough in the early morning. A release of glucocorticoid hormones in the early morning hours assists with the resumption of diurnal activity. These endogenous cycles have great implications for human functioning, since humans neither feel nor function optimally when they are disrupted. A sense of well-being and alertness are clearly affected as well as responses to exogenous physiologic stimuli. The metabolism of medications may also be affected by the time of day, and circadian rhythm disruption may require adjustments in medication doses during shiftwork.

These circadian functions can adjust or entrain over time. If an individual maintains a complete change of schedule from day to night functioning over a number of weeks, there would be a reversal of the curves for temperature, glucocorticoid release, etc. However, this reversal is very susceptible to a variety of influences such as light exposure and temporary breaks in routine. Most night workers resume daytime activity on days off and vacations. In addition, most will have significant sunlight exposure while doing errands and commuting from work. These disruptions are usually sufficient to keep night workers from completely entraining their circadian rhythms. As a result, the graphs of these functions would show only a flattening or partial curve reversal.

The biological clock has an additional unusual feature. Without the usual prompts of daily activity (i.e., if a human were secluded for days in a dark cave), the human internal clock would tend to stretch out to a 25-hour day and our biological rhythms would adjust accordingly. Thus, in normal functioning, humans are continually resetting their internal clocks back by an hour each day to stay on a 24-hour schedule. Most shiftwork experts believe that this phenomenon explains why it is generally easier to adjust to a flight westward across time zones than eastward. When crossing time zones westward, the individual essentially has time added to the day and the internal clock is decompressed; the body is allowed to approach its natural 25-hour cycle. By contrast, flying eastward compresses the clock because the individual is required to awake earlier than accustomed, thereby temporarily adapting to a cycle that is 23 hours or less. This phenomenon may also explain why individuals who sleep-in on weekends have trouble getting going on Monday mornings.

When applied to shiftwork, the biological clock findings indicate that successful rotating shifts will progress in the forward direction (from day to evening to night) rather than the reverse. By this principle, workers can take advantage of the natural tendency of their internal clocks to delay the time of waking, rather than forcing the cycle backward. In the ideal situation, shiftwork scheduling would also facilitate the complete entrainment of the body's endogenous biological cycles. A longer period of adaptation on any given schedule (i.e., weeks rather than days) would theoretically facilitate this change; however, both social preferences and other disrupters discussed above limit the achievement of this state.

SHIFTWORK AND HEALTH

The health impacts of shiftwork may be loosely divided into short term and chronic effects. Short term effects clearly include sleep disturbances. The poorer quality sleep of shiftworkers is characterized by reduced sleep time, altered sleep phases with less rapid eye movement (REM) sleep, and more frequent disruptions from the outside (diurnal) world. Gastrointestinal complaints are common with decreased appetite, constipation, and possibly an increase in peptic ulcer disease. The quality of nutrition can be an issue for some shiftworkers, given heavy reliance on caffeine and less availability of a range of foods in many nocturnal work environments.

Mounting evidence indicates that cardiovascular disease is related to shift work; higher rates of heart disease have been found in several studies of shift workers. Other long-term health concerns are the possible effects of schedule changes on existing chronic diseases such as diabetes, asthma, and epilepsy. Individuals with these diseases may require special attention to see that their medical condition remains well-controlled with appropriate adjustments to medication schedule, and in some cases, shiftwork may not be advisable for these people.

SAFETY AND PRODUCTIVITY

Substantial evidence supports a diurnal variation in alertness and the ability to respond accurately to stimuli. The preponderance of evidence indicates that there tends to be an early morning and a mid-afternoon dip in alertness. Studies from meter readers, for example, indicate poorer performance at these times. The literature on the occurrence of accidents is complex and does not give an entirely clear picture as to the contribution of shiftwork. There is some indication of a tendency for accidents to occur at night or in the early morning hours. Despite the relative absence of sound scientific evidence on the issue, there must be a common sense concern for situations in which work scheduling may result in a fatigued individual performing a safety sensitive job.

A link between shift scheduling and productivity has been established. Well known studies of munitions workers during World War I demonstrated that a reduction in the workweek from 53 to 48 hours actually increased production. A study comparing the productivity of workers who change from a reverse shift rotation (day, night, evening) to forward rotation (day, evening, night) showed an increase in productivity on the latter. In general, productivity is less during night work, although the ability to adapt to performing work effectively varies by the type of task.

THE DESIGN OF ROTATING SHIFT SCHEDULES

As indicated in previous sections, when rotating shifts are necessary, it is desirable to rotate them in a forward direction (morning, evening, night) for reasons of the internal biological clock; however, the issue of the duration (number of days) between shift rotations is a matter of controversy. From a physiological perspective, full entrainment of biological rhythms seldom occurs in rotating shift workers even after many weeks due to disrupting factors; nevertheless, there is still an improvement in adaptation over several weeks as compared to a one-week rotation length.

Rotating shifts may be categorized as prolonged rotations (e.g., a change every month), weekly rotating, or rapidly rotating (e.g., change every 2 to 3 days). Family and social factors generally predominate in determining the workers' preferences of shiftworkers. Workers will often prefer a rapid shift rotation. The reason for this is that it provides the worker with a few days of social normalcy every week as opposed to one week out of three (or one month out of three in the more slowly rotating systems). In exchange for this social preference, the worker endures more physiologic disruption and may, indeed, be less alert and less productive while on the night shift. The acceptability of this trade-off may vary by the type of work.

HEALTH AND SAFETY CONSIDERATIONS
IN THE IMPLEMENTATION OF SHIFTWORK

Although shiftwork of all types is common, the evidence suggests that there are significant potential impacts on the individual as well as organizational impacts such as productivity and safety concerns. Careful consideration of the type of shift system and the mitigation of adverse impacts is warranted. Giving workers some voice in how the schedule is arranged makes sense. Where fixed or rotating night shifts are used, employers should consider extra attention to safety in the workplace, the nutritional needs of employees, and, potentially, the provision of quiet sleep areas to assist workers with noisy households. Some evidence suggests that bright lighting in work areas facilitates adaptation to night shift. A list of other recommendations is given in Table 21.1.

There is substantial interindividual variability for tolerance to shift work. Many researchers have noted that existing studies are biased by a survivor effect in which individuals susceptible to ill effects from shiftwork have self-selected out of it. Safety professionals should strongly consider implementing a medical surveillance program for shift workers in which candidates receive medical evaluation and counseling; some individuals might be screened out for overriding health or safety concerns. A part of the program should be training the workers how to best adapt to shiftwork,

Table 21.1 Shiftwork Recommendations

- Rotate shifts in the forward (clockwise) direction.
- Keep the schedule as simple and predictable as possible.
- Avoid early morning starting times.
- Avoid night work if possible.
- Provide at least one full day of rest between shift changes.
- Provide at least 2 consecutive days of rest for every 5 to 7 work days.
- Pay attention to environmental factors such as good lighting.
- Ensure safety measures equally on all shifts.
- Assess and reassess shiftworkers for adequate adaptation to the schedule.

including sleep habits, appropriate avoidance stimulants and alcohol, and other adaptive issues. The program should offer opportunity for periodic contact with health professionals to detect problems at an early stage. Such a program should look for evidence of acute or chronic disease, including sleep disorders and deterioration of preexisting medical conditions.

REFERENCES

Knauth, P. Hours of Work. In *Encyclopedia of Occupational Safety and Health*, 4th ed. Stellman, J., Ed., Geneva: International Labor Office, Chap. 43, 1998.

Scott, A. J. *Shiftwork: Occupational Medicine State-of-the-Art Reviews*. Philadelphia: Hanley Belfus, 1990.

Seward, J. Occupational Stress. In *Occupational and Environmental Medicine*, 2nd ed. La Dou, J., Ed. Appleton and Lange, Stamford, CT, Vol II, Chap. 35, 1997.

Wilkinson, R. T. How fast should the night shift rotate? *Ergonomics* 35(12): 1425–1446, 1992.

FURTHER READING

Coleman, R. M. *The 24-Hour Business*. New York: AMACOM, 1995.

Moore-Ede, M. *The 24-Hour Society: Understanding Human Limits in a World That Never Stops*. Reading, MA: Addison Welsey, 1993.

## CHAPTER 22

Psychological Risk Factors and Their Effects on Safe Work Performance

Barbara Newman

CONTENTS

INTRODUCTION

In 1931, researcher H. W. Heinrich noted that two basic factors can cause accidents: unsafe mechanical or physical conditions and unsafe acts of persons, or the human factor. Many safety programs focus on the physical protection of workers, ignoring the human factors that may lead people to disregard or neglect the procedures that could prevent accidents from happening, yet studies show that employee mistakes account for 50% of all accidents at work.

Psychological risk factors in accidents are the potentially troublesome thoughts, feelings, and behaviors that may occur as a result of stressful events.

PERSONAL
- Martial problems/divorce/end of relationship
- Issues affecting significant other (illness, loss of job, etc.)
- Illness
- Death in family
- Accident
- Moving
- Alcohol/drug use/abuse
- Financial problems
- Marriage
- Birth of child
- School/graduation

OCCUPATIONAL
- New job/job transfer/promotion
- New responsibilities/expectations
- Cutbacks/downsizing/co-workers leaving
- Problem with co-worker or supervisor (harassment, cultural differences)

SOCIETAL
- Natural disaster: Fire, flood, earthquake, etc.
- Man-made disaster: Burglary, robbery/mugging, riot, kidnapping
- Hard times: Recession, depression

CONTAGION: Reaction to an event unrelated to self
- Reading about disaster/television coverage
- Disaster befalling a friend or relative
- Responding to/reporting a disaster

Figure 22.1   The four areas of stress-related events.

These stressors occur in four areas of our lives: personal, occupational, societal, and contagion (Figure 22.1). The events listed under these four categories are examples of many of the stressors we experience. When stress levels rise and the effects of stressors persist, coping skills may become depleted, contributing to a loss of concentration and affecting the ability to perform tasks safely.

LEVEL OF STRESS

In 1967, the Life Change Index was formulated by T. H. Holmes and R. H. Rahe. Given a list of 43 life events, individuals are asked to check the number of occurrences of each event over the past year. Every occurrence is ranked according to its numerical stress-value impact, from a low of 11 to a high of 100. Adding up the score provides the total amount of stress experienced. The Index contains many of the stressful events that can become strong contributors to the human factor in work-related accidents. However, the primary determinant in evaluating the impact of stressors on behavior will be the reaction of each individual to the stressful situations that occur.

DURATION OF STRESS

In estimating how long a stressor may affect an individual, consideration must be given to the event, its after-effects, and the individual's response to the circumstances. An earthquake or flood may not last very long as an actual stressful incident, but the emotional consequences may last much longer and be more disruptive. On the other hand, an individual caring for someone with a chronic illness may experience more consistent stress, but the caregiver may adjust well by learning healthy coping mechanisms and calling on resources for assistance. The post-traumatic stress syndrome resulting from the emotional reaction to the earthquake or flood, consequently, would be more significant in terms of becoming a psychological risk factor in accidents.

As individuals with distinctive personality styles, each of us will respond to a given stressor in an idiosyncratic way. In general, however, short-term problems will not have as great an impact on performance levels as longer-term concerns. Being involved in a minor car accident may create a temporary distraction of attention. Ongoing, more severe issues, such as drug or alcohol addiction, may create a more critical loss of concentration and present a more profound risk to safe work performance.

ABILITY TO COPE

Each person differs in how well he or she handles stressful situations and the accompanying internal stress messages (see Figure 22.2). The most notable factor that interferes with successfully managing a stressful event is the sense of loss of control. The more an individual feels a loss of control, the greater the level of stress. This out-of-control feeling may lead to confusion and preoccupation with the stressor, resulting in increased negative self-talk and anxiety. A rising sense of frustration and an accompanying loss of concentration may lead either to taking unnecessary risks or to overcompensation for the lack of attentiveness by becoming overly cautious and hesitant. Indecision, uncertainty, and anxiety levels may then escalate further, precipitating an unhealthy stress cycle and contributing to increased accident risk.

Because many people feel uncomfortable discussing the problems that may be plaguing them, a method of assessment is needed to appraise the factors involved in an employee becoming a risk for an accident. The most effective assessment comes from observation of behavior or performance level.

When someone is distracted, performance and productivity go down. Stressful events trigger behavioral changes. Some of these changes may be overt and obvious, others less so. It takes a willingness to get to know employees and to learn their habits in order to notice changes in functioning, many of which are listed in Figure 22.3. The longer these changes in conduct continue, the more at risk the individual becomes.

Once you have identified an employee as being an accident risk, the issue must then be addressed directly. There are three levels of confrontation: personal, performance, and policy (see Figure 22.4).

MESSAGES
- Generalized fear and anxiety
- Confusion
- Preoccupation with problem, loss of concentration (loss of productivity)
- Hypervigilance (overcompensation for distractedness)
- Grief process
- Negative self-talk
- Alienation
- Isolation
- Fear of reprisal
- Fear of showing weakness

REACTIONS
- Post-traumatic stress disorder, including
  - fatigue, low energy
  - irritability, agitation
  - anxiety
  - apathy, emotional numbness
  - headache, nausea, dizziness
  - minor illness
  - difficulty concentrating, distractedness
  - confusion, forgetfulness
  - insomnia
  - nightmares
  - increased drug and alcohol use
  - jumpiness
  - recklessness
- Fearfulness
- Anger
- Feeling overwhelmed
- Nervousness
- Sleeplessness

Figure 22.2   Internal stress messages and emotional/physical reactions.

LEVELS OF CONFRONTATION

Personal

If you are willing to leave the realm of your professional capacity, try approaching the individual as a friend who is concerned and who cares. Ask if there is a problem and what you can do to help. Attempt to become an active support system for the individual if he or she desires it. The effort must be sincere, of course. The personal level of contact may be very beneficial for people who can appreciate your sincerity, know that confidentiality will be respected, and trust you to listen, empathize, and help. The risk you run is that the problem may be denied, and you may be rebuffed as being intrusive. Extreme tact is necessary if you decide to intervene on a personal level. You may prefer to remain in the professional arena.

- Change in affect (demeanor)
  - Mood changes, mood swings
  - Teariness
  - Withdrawal
  - Forced cheerfulness
  - Bursts of anger; overreaction to situations
  - Agitation
  - Sluggishness
- Distractedness, daydreaming
- Lateness; no excuse
- Excessive use of sick days or personal days
- Long lunches; returning from lunch with smell of liquor on breath
- Personal telephone calls; conduction personal business
- Workaholism
  - Working late or overtime, with no previous pattern
  - Adding to already taxed work load
  - Unwillingness to leave work
- Frequent breaks and bathroom time
- Forgetfulness, passiveness
- Performance drop — more mistakes than usual; minor mistakes
- Erratic performance
- Secretiveness
- Denial of problem when confronted

Figure 22.3 Behaviors that contribute to accident susceptibility.

PERSONAL
- Awareness of gender differences and employees' behavior patterns
- Approachability ("open door" policy)
- Willingness to take a risk and intervene
- Concern for safety (must be sincere)
- Privacy must be ensured
- Increased cooperation, decreased defensiveness
- Confidentiality
- Openness, willingness to listen
- Availability

PERFORMANCE
- Performance reviews
- Professional, impartial evaluation of conduct
- Behavioral modification alternatives
- Negotiation, mediation skills
- Cross-training of employees

POLICY
- Written policy
- Training and wellness programs, classes, information
- Guidelines for safe behavior

Figure 22.4 Levels of intervention.

Performance

From a professional stance, your tactic will be to discuss performance level and behavior changes. Note the problem areas and address them specifically. Try to find the reason behind the decline in productivity or change in conduct. If you encounter denial that a problem exists, keep the conversation focused on specific behavior, informing the employee how he or she can improve performance and how that improvement will be documented. Express your concerns regarding safety issues. Cross-training of employees will ensure that job duties are covered if it should become necessary to remove someone from a potentially hazardous situation.

You may notice gender differences occurring at the personal and performance levels. Women in general, are more likely to be willing to discuss personal problems and accept solutions and offers of help. They are inclined to respond positively to compassion and an honest desire to listen. Men, who are often raised in an atmosphere that emphasizes problem solving and privacy, may believe that acknowledgment of difficulty or stress is an admission of weakness. They will usually respond more positively to an intervention at the performance level, with clear expectations and guidelines for improvement.

Policy

Policy and performance go hand in hand. Written expectations of safe behaviors, with posted policy guidelines, are a legal requirement. A specific policy, combined with a comprehensive training program, prepares a company to take action when conduct crosses the line and becomes unsafe. Along with safety regulations, policy can address other behavioral issues such as lateness, personal and sick days, personal telephone calls, and lunch and break times. A good written policy fosters safe behavior.

Naturally, the best course of action is prevention. The best case scenario is for accidents never to occur. If an accident should take place, immediate debriefing is necessary to prevent the contagion that can create an atmosphere conducive to further accidents (see Figure 22.5). Open honest discussions of the event are essential; secrecy breeds anxiety. Although the need for counseling is not always immediately apparent, sharing of feelings helps to contain the sense of helplessness. Additional safety training with respect to the accident will help other employees feel more aware and in control. There is no such thing as too much information.

- Debrief to prevent contagion; provide counseling if necessary
- Promote open, honest discussion and education
- Share knowledge and information — secrecy breeds anxiety
- Ensure confidentiality

Note: A written program should be in place — the need for debriefing is not always apparent.

Figure 22.5 When an accident occurs.

Successful intervention in controlling accidents depends on the nature of the corporate culture. Dissemination of information, thorough training, and wellness programs provide opportunities for the corporate culture to encourage the values that promote safe performance. Emphasis on communication, accountability, involvement, flexibility, priorities that put employees first, and a positive perception of the workplace all contribute to preventing emotional issues from adversely affecting performance. When coupled with keen observation of behavior and timely intervention, healthy corporate values create a safe and responsible environment.

REFERENCES

Holmes, T. H. and Rahe, R. H. Life change index. *J. Psychosomatic Res.* 11:213–218, 1967.

Jones, J. W. Breaking the vicious stress cycle. *Best's Rev.* 88(3): 74, 1988.

Scherer, R., Brodzinski, J. D., and Crable, E. A. The human factor (human failings as a main cause of workplace accidents). *HR Mag.* 38(4): 92, 1993.

FURTHER READING

Books

Blanchard, M. and Tager, M. *Working Well: Managing for Health and High Performance,* New York: Simon & Schuster, 1985.

O'Donnell, M. P. and Ainsworth, T., Eds. *Health Promotion in the Workplace.* New York: John Wiley & Sons, 1986.

Other Publications

Anonymous. Sleeping on the job. *Small Bus Rep.* 16(1):27, 1991.

Borofsky, G. Pre-employment psychological screening. *Risk Manage.* 40(5):47, 1993.

Halgrow, A. A trauma response program meets the needs of troubled employees. *Personnel J.* 66(2):18, 1987.

Krause, T. R. Hidley, J. H., and Hodson, S. J. Measuring safety performance: the process approach. *Occup. Hazards* 53(4):49, 1991.

Reher, R., Wallin, J. A., and Dubon, D. I. Preventing occupational injuries through performance management. *Public Pers. Manage.* 22(11):301, 1993.

Van Wagoner, S. I. Stress's many disguises. *Across the Bd.* 24(3):51, 1987.

Verespej, M. A. Wired for disaster: drug and alcohol abuse can wreak havoc on a safety program. *Occup. Hazards* 52(4):107, 1990.

CHAPTER **23**

Ergonomics for Managers — A Practitioner's Perspective

Donald L. Morelli

CONTENTS

1-56670-370-0/02/$0.00+$1.50
© 2002 by CRC Press LLC

INTRODUCTION

For most safety professionals, ergonomics has been considered primarily a tool to improve workplace safety and health. Although these are important factors, the core of the proper application of ergonomic principles is to enhance operational and organizational efficiency and effectiveness. An understanding of the basic philosophy of ergonomics, along with readily usable approaches to apply it, will allow the engineer or manager to make ergonomics an integral part of any plant layout, job design, or business decision.

Regardless of a manager's title — safety, environmental, production, or human resources — many forces shape the decisions that must be made every day. These forces include operational performance, compliance with laws and regulations, judicious use of limited — and often sparse — resources and, not the least of which, acting in accordance with professional and personal ethics.

Managing has long been defined as those activities necessary for planning, staffing, organizing, directing, and controlling the actions of groups of people to attain some desired outcome. It is in defining the desired outcome that safety interests and production interest often clash: safety is too often seen as inhibiting production or work processes. Understanding and applying the principles of ergonomics is one strong approach to bridging the gap between these opposing views.

However, ergonomics is not a magic pill, and it is not capable of miracles. These ideas must be dispelled before any meaningful ergonomic intervention or improvement is attempted. Inflated expectations can doom future efforts even if the initial improvements are successful. Application of ergonomics will make jobs easier and less stressful on workers. Work output would be expected to improve, potentially both in quality and quantity where warranted. Equipment utilization should be better, tool breakage diminished. These outcomes should be expected as long as the big picture of ergonomic application is embraced. The conflict between production and safety outcomes arises when *only* injury reduction or *only* increases in production quantities are defined as the expected results of ergonomic improvements.

ERGONOMIC PERSPECTIVE

To understand ergonomics and its application, one must appreciate its basic approach. This is vital because ergonomics is more than an area of interest or field of study. It is not just a science in terms of a collection of rules and data. Ergonomics is more a philosophy to understanding the world and how it functions. Although once considered a derivative of the industrial engineering and industrial psychology disciplines, it continues to evolve and now represents a further development of, and an enhancement to, these disciplines.

The ergonomic perspective considers a work setting to be comprised of four components:

Human — the person or group of people engaged in a purposive activity
Task — the series of actions necessary of the human(s) to accomplish the activity

Ergonomic System Model

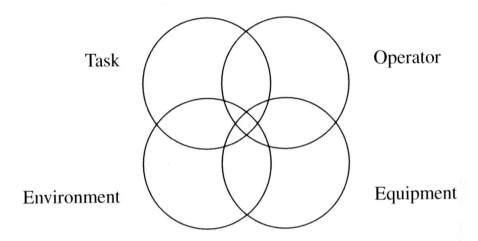

Objectives: • Maximize system performance
• Minimize mismatches between
system elements

Figure 23.1 The ergonomic system model, which illustrates the physical, organizational, and psychological factors that can affect human performance.

Machinery/Equipment — the hardware and devices provided to the human(s) to assist in the performance of the activity

Environment — the overall arena in which the purposive activity takes place, to include not only the physical factors of temperature, lighting, and noise, but the organizational and psychological factors that can affect human performance.

Any ergonomic evaluation considers the interaction of these four components as a human/task/machinery/environment system. The primary objective of ergonomics is to attain and maintain balance in the system and thereby maximize the performance of the system. Analysis of this system seeks to determine if imbalances exist between the elements of the system and to identify the sources of the imbalance. This is done by examining if the demands of any of the components of the system require performance beyond the capabilities of any other component.

In most cases of physical work, and where the risk of worker injuries is a focus, the system assessment tends to focus on whether the demands of the task require capabilities the human worker either cannot perform (resulting in a traumatic incident) or cannot sustain (potentially resulting in a cumulative incident), without undesirable results. This imbalance will lead to a system failure, which can manifest itself as injuries, reduced productivity, quality problems, and/or, more likely, a combination of all three.

In some instances, the equipment provided to aid in performing a task may be the source of demands that can also lead to system failures. Tool handles may be too large for smaller workers to use effectively. Complex systems may require extensive training and experience to operate properly, limiting the people who should be allowed to work with these systems but also limiting the number of people available for these jobs.

The work environment may present demands beyond worker capabilities, such as downsized organizations expecting fewer workers to do the work of the entire previous staff, with no improvements in work design or methods. In a more common way, the temperature or humidity of a work setting may increase the physical demands of performing strenuous work, or a noisy setting may inhibit one-on-one conversation that may be important to timely completion of necessary tasks.

In these ways, the problems that are routinely faced in organizations are looked upon not as single-issue events (i.e., human error), but the results of the interaction of the components of the workplace. The ergonomic perspective also looks beyond behavioral and simple training factors as the so-called root causes of undesirable incidents.

Inattention to these often complex interactions within the work system can result in design induced error, a concept very different from the classic, and outdated, ideas concerning unsafe acts and unsafe conditions. Design induced error is often related to large scale disasters (i.e., the incident at Three Mile Island) where the operators responded to inaccurate information regarding the status of the system. But many more day-to-day incidents are also the result of design induced problems.

If a worker must grasp a work piece with a hand in an awkward and statically held position, the muscles performing that grasp will fatigue rapidly. When the work piece is dropped and damaged, this is not the unsafe act of a clumsy worker. It is the foreseeable outcome of muscle fatigue. Computer software that presents inconsistent placement of command buttons, dialog boxes, or other screen elements will prove frustrating to the user. Software that uses inconsistent terminology or complex syntax can lead to erroneous inputs, which can have results that are more serious than just frustrating.

ERGONOMIC PRINCIPLES

Human Characteristics

Once the basic systems approach of ergonomics is accepted, it is necessary to understand human capabilities and limitations — human characteristics — in order to apply the system analysis technique. In other words, fitting the task to the worker can only be done if the characteristics of the worker are known so that they can be assessed in relation to the task demands, the demands of the equipment, and of the work environment.

Many sources for information on human characteristics exist. Several excellent texts are listed in the references. Discussing all the human characteristics pertinent to safety considerations would be beyond the scope of this entire text, let alone this individual chapter. But the essential human characteristics in a given situation can be identified in terms of the type of task demands required by the specific work elements.

Size, Variability, Reach, and Clearance

Human characteristics vary, pure and simple, from one person to the next. People are different heights, weights, have different sized feet or hands, and have differing capacities for physical or mental workloads. Although this variability is very wide, it can be addressed in quantifiable terms because the variability of human characteristics is normally distributed.

All the statistical rules and tests that apply for normally distributed data can be applied to human data. This allows design criteria and dimensions to be based on collections of population data, and not on data collected from a small, and possibly skewed, set of workers.

Design criteria to address size differences among people generally seek to set task requirements so that they are acceptable to the largest practical population segment. This usually translates to attempting to design for the 5th percentile worker at the small end of the range in reach demands or the 95th percentile worker at the large end of the range for clearance dimensions. This approach attempts to disadvantage the fewest people in the foreseeable worker population.

Typical dimensions that are important to workstation design and layout are given in Table 23.1 and Figures 23.2 and 23.3.

The proper use of this data is to design reaches to accommodate the smallest workers and to design clearance dimensions to accommodate the largest workers. More importantly, this data can also be used to design equipment, tools, and work environments that can be adjusted over the proper dimensional ranges to allow the work setting to accommodate many different workers.

Strength and Material Handling

The assessment of force requirements is especially important in reducing the probability of musculoskeletal injuries. Research going back over twenty years and practical experience from today suggests that strength requirements set such that at least 75% of the worker population can perform them, can reduce the risk of injury from 30 to 50%. Some strength requirement suggestions are offered in the Guidelines section of this chapter.

Where material handling is the routine task demand, two main measures relate to the risk of injury: compressive loads in the back and energy expenditure. Compressive loads in the tissues of the back are generated by the contraction of the back muscles while performing the material handling or lifting. This muscle contraction can be very forceful, even when the lifting technique used adheres to the generally accepted — and truly misnamed — safe lifting. The major variable that results in high compressive loads in the back is the horizontal distance from the center of gravity of the load handled to the low back of the worker. This is true and independent for the most part, of the lifting technique used.

Table 23.1 Selected Body Dimensions for Females and Males

	5th Percentile (Japanese/Asian Female)		95th Percentile (U.S. Male)	
	Inch	Cm	Inch	Cm
Standing Dimensions[a]				
Stature	57.0	145.0	74.6	189.5
Eye height	53.1	135.0	69.4	176.3
Shoulder height	42.3	107.5	61.9	157.2
Elbow height	35.2	89.5	47.8	121.4
Knuckle height	24.4	62.0	32.7	83.1
Seated Dimensions[b]				
Seated height	31.5	80.0	38.3	97.3
Eye height	27.2	69.0	33.2	84.3
Shoulder height	20.1	51.0	25.4	64.5
Elbow height	8.5	21.5	11.3	28.7
Buttock-knee length	19.1	48.5	25.6	65.0
Popliteal height	12.8	32.5	19.5	49.5
Knee height	17.5	42.0	23.8	60.5
Reach				
Standing vertical grip reach	66.1	168.0	84.0	213.4
Sitting vertical grip reach	39.2	99.5	53.4	135.6
Forward grip reach	24.0	61.0	31.0	78.7

Note: All dimensions are measured from the floor and include a one-inch allowance for shoe height.

[a] Standing dimensions are measured from the floor (see Figure 23.2).
[b] Most seated dimensions are from the surface of the seat (see Figure 23.3).

From Pheasant, S. *Body Space.* London: Taylor and Francis, 1989. With permission.

If the structures of the back are chronically exposed to high compressive loads, the end plates of the vertebrae can become damaged (i.e., literally suffer compression fractures). This damage to the end plate affects the nutrition of the discs, which act as shock absorbers between the vertebrae. With diminished nutrition, the disc can shrink, which can result in shifting of the vertebrae and potentially pinching nerves which run from the spinal column.

High compressive loads have also been implicated in damage to the structural tissues around the joints between the vertebrae. This damage can weaken the containment of the discs, potentially leading to a slipped disc problem. The slipped disc, which, most often, actually is the protrusion of disc tissue outside its protective capsule, can also pinch nerves producing significant and potentially disabling pain.

For most work settings where high strength tasks or repetitive material handling are present, the sources of the risk of physical injury generally involve:

- Postures required
- Force demands
- Frequency of activity
- Duration of activity

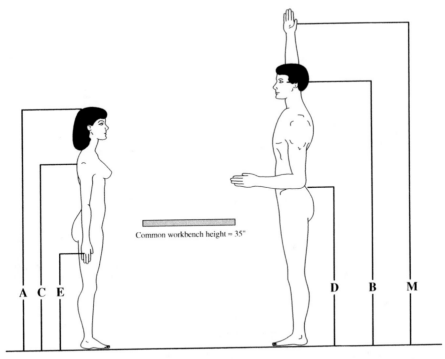

Figure 23.2 Selected 5th percentile female and 95th percentile male standing dimensions.

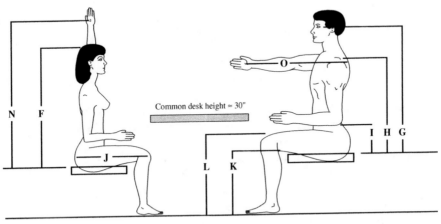

Figure 23.3 Selected 5th percentile female and 95th percentile male seated dimensions.

These terms relate to the following:

Postures — the body positions assumed to perform the work element
Force — the requirement to exert strength to perform the task
Frequency — the repetitious nature of the activity
Duration — the length of time a work demand, either dynamic or static, is maintained

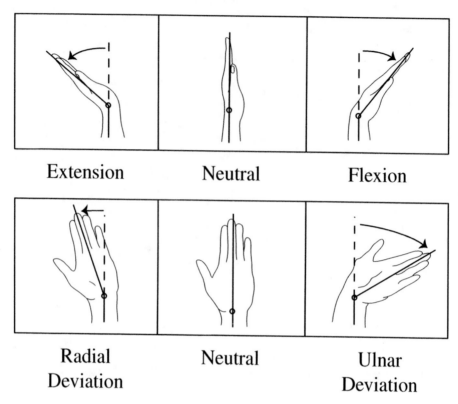

| Extension | Neutral | Flexion |

| Radial Deviation | Neutral | Ulnar Deviation |

Figure 23.4 Postures of the hand and wrist. (Used with the permission of the Travelers Insurance Company.)

Postures

Postural considerations associated with work generally focus on neutral vs. non-neutral postures. Neutral postures can be defined in many ways. A generally accepted definition of a neutral posture is one where, with no externally applied load, there is minimal stress in the tissues, the muscles, and connective tissues associated with the body parts and position. For example, neutral posture of the forearm, by this definition, would be with the elbow bent roughly 90°. For the wrist, the neutral posture would have the hand bent neither toward the palm (flexion), toward the back of the hand (extension), toward the thumb (radial deviation), nor the little finger (ulnar deviation). For most body parts and joints of the body, the neutral posture generally can be considered the position at the mid-point of the range of movement of the involved joint.

Posture during manual work, especially material handling, is the subject of continuing debate. However, a universally agreed upon risk factor is the horizontal distance from the workers lower back to the center of gravity of the load being handled. The guiding principle is to keep this distance as small as possible.

This is where the classic squat lift can be advantageous, if the load handled is small enough to fit close to the body or between the knees. If the load is large or cannot be grasped tightly to the body, then virtually all of the research and analytical tools suggest that the overall body posture assumed is not so important — the task will probably present a high risk of injury. Large distances from the center of gravity of the load to the worker's back create large moment arms resulting in high compressive loads in the lower back, which have been correlated to incidence of back injury.

The position of the hand and wrist is also a postural concern during gripping activities. Here again, flexed or extended wrist positions, as well as awkward or pinching type grips, reduce the maximum available force a worker can generate. Therefore, to perform what may appear to be a simple task, but in an awkward posture, may actually require a very high percentage of the worker's available hand strength.

This will lead to high forces and loads in the connective tissues of the hand, wrist and arm (the tendons and ligaments) and more rapid fatigue of the muscles involved. The high loads in the tendons and ligaments increase the risk of sprain or repetitive motion disorder (if the frequency is high) and the muscle fatigue will diminish performance.

Force

The strength required to perform a task is dependent on several key factors:

- Size or bulk of an object handled or held
- Posture or body position assumed by the worker
- Size or configuration of hand tools used
- Frequency of the exertions

It should be apparent that force, frequency, and posture are interrelated factors, as suggested previously. To illustrate this concept, consider a task demand to pick up a cassette of silicon wafers out of a dip tank with one hand, hold it over the tank to drain, and then move and insert it into a water rinse tank. The weight of the cassette obviously has to be such that the worker can lift it. However, the arm position — determined by the horizontal reach to the tank — is also vital. The functional forward reach of the 5th percentile (female) worker is about 27 inches, so the tank must be located no further than that from the operator.

But this is not the full solution. If the tank is placed requiring a 27-inch reach, that will mean the small worker will have to lift the cassette with a fully extended arm. The available strength of the 5th percentile (female) worker in this posture is roughly only half of the maximum strength for such a task demand. Therefore, the cassette will have to be very light in weight or the tank should be moved closer to the worker, which permits a more desirable arm posture, and a maximum upward force can be generated.

How frequently the cassette is dipped and how long it is held to drain are also important. If the holding time and the repetitiveness of the activity do not allow for adequate recovery time for the muscles involved, overall performance will suffer.

Frequency

The frequency of activity determines the factors that must be analyzed. For tasks performed infrequently, strength limitations, as outlined previously, generally take precedence. In tasks that are performed continuously or for relatively long time periods (i.e., unstacking a full pallet of items all at one time), energy demands and appropriateness of recovery times become most important. Energy demands and recovery time calculations are discussed below.

Duration

The length of time a work demand, either dynamic or static, is maintained is also important to assessing the risks involved. Long-duration tasks generally require greater energy and, therefore, should be provided with greater recovery time. The duration of seemingly light tasks, such as keyboard work, is just as important. Several of the upper extremity disorders associated with this type of work have been linked to overactivity of the flexor and extensor muscles of the forearm. Additional evidence suggests that chronic fatigue in the tissues of the forearm can lead to formation of free radical molecules which are capable of attacking the nerve, connective, and muscle tissues.

Energy Demands and Recovery Times

Repetitious work or work of long duration is physiologically demanding. The body's response to such work will involve an increase in heart rate, oxygen consumption, respiration rate, blood pressure, and body temperature. The elevation of these physical measures clearly suggests the systemic changes that are the normal responses to heavy physical work. Inattention to the risks to workers associated with physiologically over-stressing the body can, at the low end, reduce productivity and, at the high end, could result in work induced heart attack.

To minimize the risks, proper job design in terms of work and recovery times is necessary. Here, recovery time is not synonymous with rest or break time. Rest or break time suggests inactivity (i.e., leaving the work setting and performing no work). Recovery time, on the other hand, can be productive time. Recovery time should represent low energy, that is, alternative work activities that allow adequate recovery from the high demand tasks while still being productive. An example would be paperwork necessary to track pallet loads of goods that have been manually stacked. The material handlers can have the paperwork included as a part of their jobs to provide low-energy, recovery time work, which would be in addition to any regular or contractually required rest breaks.

Energy demands and expenditure can be expressed as kcal/min and guidelines have been developed for energy demand and recovery times based on a kcal/min basis. Caution should be used with such guidelines because the maximum levels are generally based on 35 to 40 year old healthy males. Adjustments are suggested to apply these limits to female and older workers.

Table 23.2 Suggested Percentage of Recovery Time Needed
for Differing Energy Demands and Task Durations

Energy Demand (kcal/min)	Work Duration (in minutes)	Recovery Time (% of Work Duration)
4.0	30	20%
4.0	60	30%
5.2	30	40%
5.2	60	>50%

From Eastman Kodak Company, *Ergonomic Design for People at Work*, Vol. 2, 11, 206–227, 1986. With permission.

Research suggests that for a healthy adult male, an energy demand of roughly 5.2 kcal/min is a reasonable upper limit for continuous work. When adjusted for female and older workers, the continuous work energy demand is reduced to roughly 4.0 kcal/min. (This difference should not be interpreted to mean that women are not capable of significant physical work. The main variable in determining the amount of energy expended during physical work is the body mass of the worker. Therefore, the ≈4.0 kcal/min energy expenditure for women represents the same level of demand as the 5.2 kcal/min expenditure for the adult male.)

The low-energy, recovery times suggested with these energy expenditure levels is dependent on the duration of the work. Some general ranges are given in Table 23.2.

COMPUTER WORK

The approach to assess and analyze computer work is no different from any other ergonomic effort. What needs to be known is what tasks are to be completed by the worker, what equipment is best suited to support the performance of those tasks, and what environmental factors can enhance or inhibit the task performance.

Stressing perfect posture or a particular keyboard as a panacea for problems associated with computer work is analogous to teaching safe lifting without ever investigating the load or the frequency of the lifting. Such efforts are short term band-aids that will not produce long-term results.

Computer work is most often associated with repetitive motion disorders, primarily carpal tunnel syndrome, as well as other repetitive motion ailments, such as epicondylitis and tendinitis. General pain and stiffness in the hands, wrists, shoulders, and neck are also often attributed to computer work.

Computer work presents two major factors that can be linked to these physical disorders:

- Long duration and high frequency of similar activity (i.e., keystroking)
- Long duration of static or fixed postures

The desired outcomes of any efforts to improve the ergonomics of computer work settings should be very simple and straightforward. Because the major sources of risk, as presented previously, both involve long-duration activities, the best outcomes are often associated with increased flexibility in tasks and postures. That suggests that jobs must be structured to include more than just keyboard work. Take advantage of peoples' experience, creativity, and abilities to learn to contribute in a greater sense to your organization.

Increased flexibility also suggests that the work setting be readily adjustable and even movable, where possible, to allow the worker to exercise greater control and input to how the work setting is put together to best support the tasks that must be performed. This sense of control can improve not only the comfort experienced by the worker, but also contribute to lessening the impact of psychological and physiological stressors in the work environment.

For the situations where physical shortcomings of computer work settings must be dealt with, Table 23.3 presents a set of typical potential problems and possible interventions. This list is not exhaustive, but it represents the most common potential problems found in long duration computer work. Before applying any of the interventions presented in the table, one should gain a high comfort level with the redesign requirements presented in the next section, to make sure that these interventions are understood as the product of that approach, and not quick fixes to what are often complex situations. See Table 23.3.

REDESIGN REQUIREMENTS

Once system imbalances have been identified, it is often too easy to jump to a "fix" to get a problem solved. This temptation is understandable in our rush-rush world, but it should be avoided. Instead, the information regarding what may be out of balance should be turned into redesign requirements that state clearly, but in broad terms, what factors need to be addressed and how. At this point, input should be solicited from every viable source to construct an intervention, improvement, design change, or process upgrade. Workers, engineers, operating supervisors and managers, and your equipment vendors all should have input to the decision process. Some general redesign requirements, upon which to model your future thought processes, are as follows:

- Reduce forces required.
- Reduce frequency of repetitive activities.
- Improve postures.
- Limit the duration of high demand tasks.

Force — redesign requirements call for the reduction of the:

- Weights handled.
- Force exerted in awkward postures.
- Time the force is exerted.
- Gripping force when the wrist is bent.
- Time an item or body part is held unsupported.

Table 23.3 Fundamental Problems and Potential Interventions for Computer Work Settings

Observe Potential Problems	Possible Interventions
Elbows raised from sides (abducted)	Lower work surface/keyboard Raise chair — footrest if needed Lower armrests
Shoulders raised or tensed	Lower work surface/keyboard Raise chair — footrest if needed Lower armrests Telephone use — headset
Hand or arm resting against hard surfaces for long periods	Pad surfaces Forearm supports Height of work surface/keyboard
Wrists extended or flexed for long periods	Palm/wrist rest (for posture avoidance) Forearm supports Change height or angle of keyboard
Hands suspended over keyboard when not keystroking	Raise keyboard Forearm supports Habit retraining
Feet not supported	Chair too high Work surface too high Chair seat pan too deep Footrest as a last resort
Torso not supported	Chair too high Chair seat pan too deep Work surface too low Computer monitor too far for user's vision
Forward (peering) posture of torso, head/neck	Monitor image not sharp Monitor too far for user's vision Task lighting inadequate Back support of chair inappropriate Chair seat pan too long
Head/neck flexed forward (i.e., chin down to chest)	Monitor or documents too low Back support of chair inappropriate Chair seat pan too long
Head/neck extended backward	Monitor or documents too high Chair too low Bifocal wearer — may need single vision glasses
Eyestrain complaints	Screen glare Too much/too little illumination Color selections inappropriate
Computer monitor too close to user's face	Monitor image not sharp User's vision uncorrected Task lighting inadequate Work surface depth inappropriate
Long-duration, near-focus work	Work redesign Workplace alteration providing long focus or outside views More frequent breaks Habit retraining
Multiple visual targets at different viewing distances	Holder for documents at monitor distance Illumination of targets may be uneven

Table 23.3 (continued)   Fundamental Problems and Potential Interventions for Computer Work Settings

Extended reaches to use mouse	Locate mouse on same surface as keyboard Habit retraining
Congestion of work around computer monitor	Work surfaces too high (only foot support at monitor) Work surfaces inappropriate for volume of paper work File/storage space inadequate Training in organization of work
Rapid or sustained keystroking	Work redesign Work enlargement, variety More frequent "mini-breaks"
Forceful keying	Alternative keyboards Forearm supports Habit retraining (limited impact)

Frequency — redesign requirements call for the reduction of the:

- Number of times an item is handled or lifted.
- Occurrences of extreme body positions.
- Repetition of identical work elements.

Posture — redesign requirements call for the reduction of:

- Deviations from 'neutral' body positions.
- Torso inclinations, especially while lifting.
- Shoulder abduction and adduction.
- Forward arm reach to grasp, hold, or dispose of material.
- Wrist deviations while gripping.

Duration — redesign requirements including the following:

- Provide adequate recovery time for repetitive work.
- Reduce energy demand by assisting manual work.
- Reduce the amount of time high-energy work is performed.

By fulfilling the challenge stated by the redesign requirements, the risks associated with the tasks at hand can be reduced and the work performance of the worker enhanced. For example, if postural risks are identified as a prime source of concern in a light assembly job, then in attempting to fulfill the first redesign requirement, the reduction of the deviations from "neutral" body positions, several options could exist:

- Explore different hand tools, fixtures, or jigs to improve the working hand and wrist postures.
- Adjustable height work surfaces would allow more erect torso postures while performing standing work.
- Improving color and illumination contrast between the work pieces and the work surface could improve head and neck postures, as well as increase productivity and quality.

By first generically defining what needs to be changed via the concept of redesign requirements, more options can be considered, rather than a quick fix. This approach takes some practice and discipline but, when worked through, it can yield interventions that fundamentally improve the work and the way it is performed.

JOB STRESS

Job stress is an increasing source of concern and cost for many organizations. A consistent theme in the research literature suggests that job satisfaction and job stress are *inversely* related.

Job satisfaction can be linked to:

- Greater degrees of worker control.
- Autonomy.
- Input and involvement in decision making.

Indicators of job stress are associated with:

- Lack of control or input.
- Repetitive and monotonous jobs.
- Environments from which the worker gains little or no respect.
- Jobs that have high levels of physical or mental workload.

Performance studies of typical quality control/inspection tasks suggest that very low and very high levels of stimuli (e.g., the number of nonconforming products to be detected) both correlate to decreased performance.

Research in job satisfaction vs. **job complexity** suggest a similar pattern of affect: *Low complexity and high complexity are both correlated with low satisfaction and reduced efficiency.* Both low- and high-complexity tasks also show the biochemical waste products signatures associated with increased levels of stress.

Part of the stress response within the body is the release of various hormones that, in the short term, are very helpful. In the long term (i.e., when there is chronic stress arousal), these hormones can have undesirable effects. Research done over 20 years ago in Sweden indicated that men in high-stress occupations could be at a seven times greater risk for heart disease than men in low-stress occupations. Some of the stress hormones also inhibit immune response and the body's ability to deal with soft tissue inflammation.

For these reasons, and many more, stress is an important factor to be addressed in assessing organizational demands in a system analysis. It continues to play a larger role in undesirable safety-related outcomes.

REGULATORY ACTIVITIES

State and federal regulations have been in the works since the 1980s regarding ergonomics and workplace design. In 1997, the State of California passed a regulation addressing repetitive motion injuries. This regulation has, as of this date, withstood several court challenges. The essence of this regulation requires employers to institute programs including work site evaluations, control of exposures, and employee training.

The U.S. Federal Government, through the Department of Labor, has explored the creation of a regulation to address ergonomics, but these efforts have not yet yielded legislation that has advanced to consideration for enactment. In early 1999, a new draft regulation from Federal OSHA surfaced, which primarily addressed musculoskeletal injuries and their prevention. The main elements of this draft regulation were:

- Management leadership and employee participation
- Hazard identification and awareness
- Job hazard analysis and hazard control
- Training
- Medical management
- Program evaluation

GUIDELINES

The following guidelines provide more depth and some quantifiable information to the ideas and approaches developed in this chapter. While not absolutes, they are the product of much research and practical experience and should prove helpful when applied in the context of the ergonomic perspective laid out in this chapter.

Task and Work Setting Design

Whole body force — lifting: ≤250 kg compressive load in the lower back (posture and range of motion dependent); refer to NIOSH's *Work Practices Guide for Manual Lifting*, 1981, and the revised guide, published in 1991.
Arm strength — lifting: ≤25 lbs
Arm strength — pulling: ≤20 lbs
Gripping force: ≤25 lbs (actual force generated in the hand, not the load lifted by the hand)
Static holding force: ≤40% of the maximum force available in the working posture
Static holding time: ≤20% of the maximum possible holding time for the working load and posture
Forward reach tasks: repetitive, 18 inches; occasional, 24 inches
Dimensions for keyboard interface:
    Height (standing): range, 39 to 44 inches; fixed, 41 inches
    Height (seated): range, 25 to 28 inches; fixed, 26 inches (if thigh clearance permits)
Height of adjustable chairs: seat height range should be 15 to 20 inches
Height of computer screen: The top of screen should be just below eye height.

Information Display Design

A major reference for display design and specification is Military Standard (MIL STD) 1472D (Human Engineering Design Criteria for Military Systems, Equipment, and Facilities). The following is a synopsis of the display design criteria in MIL STD 1472D. These criteria also apply to the design of a software interface, whether for desktop applications or computer controlled equipment.

- The content of the information displayed should be limited to what is necessary to perform specific actions or make decisions.
- Information should be displayed only to the precision necessary.
- The format should be in a directly usable form so the operator does not need to transpose, compute, interpolate, or translate it into other units or forms.
- Failure of a display or display circuit should be immediately apparent.
- Unrelated information should not be displayed along with error or failure messages.

The format of information displayed, as suggested previously, can be the source of error in interpretation or action by the operator. Alphanumeric information presents certain advantages and disadvantages. The greatest advantage is that a virtually unlimited number of combinations of letters and numbers can be used. Letters can also be used to create clear language messages (e.g., REMOVE WORKPIECE) and numbers used to display quantitative values (e.g., 150 seconds, millimeters, etc.) that generally do not have to be learned. However, when letters and numbers are used in coding systems, a necessary learning component exists.

Color Coding of Controls

Color coding conventions for lights (either control or alarm displays) is also provided in MIL STD 1472D. The most important and frequently applied conventions are provided below. Consistent coding of controls and alarms can improve operator response, lessen the probability of error in responding, and thereby reduce risks to the operator and to the equipment or process.

Color	Meaning
Red	Malfunction; action stopped; failure; stop action; error
Flashing Red	Emergency condition (immediate action necessary)
Yellow	Caution; delay; check/recheck; condition marginal
Green	Go ahead; within tolerance; acceptable; ready; power on
White	Status indication (no right or wrong); action in progress
Blue	Advisory or alternative status

These color conventions can and should be applied to color monitors as well as to more familiar control panel lights or lighted controls.

TOMORROW'S CHALLENGES

The basic concepts and applications of ergonomics can help deal with problems faced by virtually every business or organization today. However, the forces shaping tomorrow's work world are even farther reaching than those that gave birth to the Industrial Revolution and the scientific method.

The seemingly constant efforts to reduce the size of organizations will force many able and eager people into new professions. How are we to adapt to such constant change and aid each other in taking command of our careers? How are the people left in downsized organizations going to perform their own work as well as the work of all who have been let go?

Mega-mergers signal an end to many jobs that are seen as redundant in the new structure. How can other organizations better themselves from this increase in experienced talent cast out into the market place? Do we have the creative job design ideas to build opportunities for these people that will be win-win for them and for a future employer?

The growing interest in the total quality of work life and reducing stress while increasing job satisfaction can be viewed as a variation of the concept of maximizing our system performance, with the performance measure being a person's overall well-being. If we do not learn to address our system failures in the workplace, how will we begin to deal with them on a social, or societal basis?

To design workplaces and manage workers in the future, we will face greater differences among our workers and in the work environment. How will we specify our tasks and equipment (the two elements of the system we can readily affect) to minimize injuries and attain maximum performance?

We are facing a work force that is constantly increasing in average age. What capabilities will our workers have as they work well into their sixties, seventies and beyond? What type of work can use the best that these people have to offer? How can we retain the knowledge and skills of experienced workers?

The United States continues as a land of opportunity to many from other countries. As the demographics of the work force evolves, do we understand the demands placed on a small Asian woman to use tools designed for a large North American male? Are we willing to find out and make the necessary changes to assure her a safe work situation and one that may enhance her performance?

The apparently increasing gap between skilled and unskilled workers may present a requirement to design jobs and equipment very differently for each group. Are we willing to do this so the less skilled can have jobs, support themselves and contribute to society?

The ergonomic system approach provides a framework to enhance performance today and avoid problems tomorrow. Ergonomics simply must be an integral part of the overall skills and decision approaches of every person involved in creating safe places to work.

FURTHER READING

American National Standards Institute/HFS 100-1988. *American National Standard for Human Factors Engineering of Visual Display Terminal Workstations*. Santa Monica, CA: Human Factors and Ergonomics Society, 1988.

Baily, R. *Human Performance Engineering*, 2nd ed. Englewood Cliffs, NJ: Prentice-Hall, 1989.

Eastman Kodak Company. *Ergonomic Design for People at Work*, Vol. 1, New York: Van Nostrand Reinhold, 1983.

Eastment Kodak Company. Ergonomic Design for People at Work, Vol. 2, New York: Van Nostrand Reinhold, 1986.

Hutchingson, R. D. *New Horizons for Human Factors in Design*, New York: McGraw Hill, 1981.

Kroemer, K. H. E. and Grandjean, E. *Fitting the Task to the Human,* 5th ed. London: Taylor and Francis, 1997.

Military Standard-1472D, *Human Engineering Design Criteria for Military Systems, Equipment and Facilities*, Washington, D.C.: U.S. Department of Defense, 1989.

National Institute for Occuational Safety and Health (NIOSH), *A Work Practices Guide for Manual Lifting*, Tech Report No. 81-122, Cincinnati, OH: NIOSH, 1981.

Pheasant, S. *Bodyspace*, London: Taylor and Francis, 1988.

Putz-Anderson, V., Ed. *Cumulative Trauma Disorders, A Manual for Musculoskeletal Diseases of the Upper Limbs*. Washington, D.C.: Taylor and Francis, 1988.

Rodgers, S. H. Recovery time needs for repetitive work, *Seminars in Occupational Medicine* 2(1): March 1987.

Rodahl, K. *The Physiology of Work*. London: Taylor and Francis, 1989.

Waters, T. R. et al. Revised NIOSH equation for the design and evaluation of manual lifting tasks, *Ergonomics*, 36(7): 749–776, 1993.

CHAPTER **24**

Principles of Risk Communication

Lucy O. Reinke and Michael L. Fischman

CONTENTS

INTRODUCTION

Given the increased level of concern expressed by employees and the public in recent years about occupational and environmental hazards, risk communication has become an increasingly important topic for industry and policy makers. After many failures reflecting poor or no communication about environmental and occupational hazards, industry and government agencies are looking to other approaches. Recently, recognition has been given to the need to address not only the scientific data but also the concerns and perceptions of employees and the public, in part by adopting a more democratic process in which members of these groups serve as active participants in the collection and communication of information and the decision-making process. Effective risk communication requires an understanding of multiple disciplines including medicine, science, sociology, and psychology. Although risk communication typically occurs in the aftermath of an untoward event, one would intuitively expect that such communication preventively, as part of a regular dialogue with employees or the community, would foster trust and be more effective. Risk communication is fundamentally similar, whether addressed to community residents or employees, with a few exceptions. Occupational exposures tend to be higher and more prolonged than environmental exposures, increasing the likelihood of demonstrable hazards to health. On the other hand, employees are generally more familiar with the processes, hazards, controls, and safety philosophy at their workplace and have a greater economic investment in the organization, which may tend to reduce their perception of risks.

Although risk communication efforts attempt to convey scientific information to groups of concerned individuals, it is clear that risk communication itself is not yet a science. Literature searches in this area do not identify articles that have scientifically evaluated the impact or effectiveness of different approaches to communicating risk information. Certain authors, most notably Sandman (1986 and 1988) and Chess (1988), have developed some very useful yet theoretical constructs, by which we can try to understand the layperson's reactions to hazards and thereby adjust our approach to communicating information about these hazards. They support their publications with numerous instructive examples, but controlled studies in this field are not available for citation. Apart from an understanding of this body of work and thorough technical knowledge about the hazard, one cannot be successful in risk communication without possessing the skills of any good communicator — good listening skills, common sense, sincerity, and sensitivity to the audience.

THE ROLE OF RISK ASSESSMENT AND RISK MANAGEMENT

Before one can communicate risk, one must understand it. The process of risk assessment must precede risk communication, although a good risk assessment does not ensure good risk communication. Risk assessment may be a qualitative or quantitative process. Although a full discussion of risk assessment methodology is beyond the scope of this chapter, risk assessment basically involves four steps — hazard identification, dose-response assessment, exposure assessment, and risk characterization. Hazard identification involves two steps — identification of exposures and risk identification (identifying the adverse impacts of an exposure, as observed in epidemiologic studies or animal toxicology tests). The dose-response assessment attempts to determine the doses required to elicit particular adverse effects and the no observed adverse effect level (again, from epidemiologic studies or animal toxicology tests). Exposure assessment involves a determination of the exposures to the group of interest (e.g., by industrial hygiene monitoring). Risk characterization integrates the above information, in order to make an assessment as to whether significant risk exists.

It is important to recognize that the risk assessment process (and the estimated magnitude of risk derived from it) is not a completely objective process. Risk assessment typically relies on models and theories, assumptions, estimates of exposure, and extrapolations, which have some but incomplete scientific support. The choice of different models or assumptions by the risk assessor involves subjective judgments and beliefs (Slovic, 1999). The communication of risk estimates, as derived from the risk assessment method, often ignores or downplays these subjective elements, while inappropriately discounting the input of concerned community members or employees because it is not based upon objective analysis.

Management of identified risks is also a process separate from risk communication, though risk communication will likely not be well received without a plan for managing identified exposures and risks. If the risk assessment process identifies a likely hazard, efforts should include reduction of exposure (by industrial hygiene or environmental control measures) and consultation with the risk assessor or other appropriate experts as to the likely impact of prior exposures. In the workplace setting, employees should be notified of their exposures and the significance of these exposures. In some cases, consideration may need to be given to recommending removal of the employee from exposure, allowing voluntary transfers, or modifying duties. In these situations, it is best to involve the employee(s) in the decision-making process.

DIFFERING PERCEPTIONS OF RISK

There is often a polarization in attitudes toward health risks associated with environmental and occupational exposures. Individuals often react angrily toward these risks when they become aware of them. Many individuals subscribe to a notion that private industry puts profits or production ahead of safety. From the point of view of many organizations, the public is thought to consist of laypersons who are unable to understand science, do not want to listen, and are afraid of the wrong risks. In this view, the public has a tendency to ignore some serious risks, while it frequently gets alarmed at significantly less serious ones. The layperson's response to risk not surprisingly reflects his or her perception of that risk, not necessarily the actual magnitude of the risk. As part of a risk communication effort, it is essential to accept these different viewpoints. While it may be appropriate to attempt to persuade the audience that there is no risk based upon scientific analysis, one must recognize that it may not always be possible to change their views.

Sandman (1986) explains differences between experts and the public. To the experts, Sandman points out, risk is the death rate (or a disease or injury rate). The public, however has a broader view. Sandman has defined the death or illness rate as the hazard and the other factors as outrage. He then defines risk as the sum of hazard and outrage. Although the public does not pay much attention to the hazard, industry or agencies do not pay much attention to the outrage. Thus, we have two entities in dialogue about two fundamentally different issues. Ignoring the issues that concern the public just because the data do not support their concerns will only multiply their outrage and at the same time increase their distrust.

THE PERSPECTIVE OF THE AUDIENCE

Whether from enlightenment, regulation, or fear of liability, there is a growing tendency for companies to involve affected citizens, employees, and/or agencies in environmental and occupational health decisions. Although there are different ways of doing this, it is important to understand who your audience is and how they view or perceive risk. In *Improving Dialogue with Communities: A Short Guide for Government Risk Communication,* Chess outlines, as quoted below, some reasons why public risk perceptions may not correlate with the scientific evidence. While focusing on community concerns and governmental organizations, these factors apply equally well to employee concerns and employers:

> Voluntary risks are accepted more readily than those that are imposed. When people don't have choices, they become angry. Similarly, when communities feel coerced into accepting risks, they tend to feel furious about the coercion. As a result, they focus on government's process and pay far less attention to substantive risk issues; ultimately, they come to see the risk as more risky.
>
> Risks under individual control are accepted more readily than those under government control. Most people feel safer with risks under their own control. For example, most of us feel safer driving than riding as a passenger. Our feeling has nothing to do with the data — our driving record vs. the driving record of others. Similarly, people tend to feel more comfortable with environmental risks they can do something about themselves, rather than having to rely on government to protect them.
>
> Risks that seem fair are more acceptable than those that seem unfair. A coerced risk will always seem unfair. In addition, a community that feels stuck with the risk and little of the benefit will find the risk unfair — and thus more serious. This factor explains, in part, why communities that depend on a particular industry for jobs sometimes see pollution from that industry as less risky.
>
> Risk information that comes from trustworthy sources is more readily believed than information from untrustworthy sources. If a mechanic with whom you have quarreled in the past suggests he can't find a problem with a car that seems faulty to you, you will respond quite differently than if a friend delivers the same news. You are more apt to demand justification, rather than ask neutral questions of the mechanic. Unfortunately, on-going battles with communities erode trust and make the agency message far less believable.
>
> Risks that seem ethically objectionable will seem more risky than those that do not. To many people, pollution is morally wrong. As former Environmental Protection Agency (EPA) Assistant Administrator Milton Russell put it, speaking to some people about an acceptable level of pollution is like talking about an acceptable number of child molesters.
>
> Natural risks seem more acceptable than artificial risks. Natural risks provide no focus for anger; a risk caused by God is more acceptable than one caused by people. For example, consider the difference between reactions to naturally occurring radon in homes and the reactions to high radon levels caused by uranium mine tailings or industrial sources.
>
> Exotic risks seem more risky than familiar risks. A cabinet full of household cleansers, for example, seems much less risky than a high-tech chemical facility that makes the cleansers.
>
> Risks that are associated with other memorable events are considered more risky. Risks that bring to mind Bhopal or Love Canal, for example, are more likely to be feared than those that lack such associations.

Slovic (1999) adds other factors, such as uncertainty, values and dread to the list of additional elements that affect the public's conception of risk. As originally described by Thompson and Dean and discussed by Slovic, the contextualist conception of risk places more objective attributes, such as probabilities, alongside more subjective attributes, such as voluntariness, fairness, and perceived controllability. While these subjective factors do not speak to whether an individual is at risk, they cannot be ignored if risk communication is to be successful.

Social science research, particularly through surveys, has been helpful in understanding factors that affect perception of risk by the public and others. In a survey study of over 1,500 Americans, Slovic found that white males, compared to white females and non-white males and females, perceived consistently lower magnitude of risk for a variety of diverse hazards, such as pesticides in food, nuclear waste, and cigarette smoking. Further analysis demonstrated that the lower perceived risk in white males could be attributed to about 30% of the sample of white males, who had higher education levels, higher income, and were politically more conservative. The author suggests that sociopolitical factors, such as power and trust in organizations, account for these observed differences. Such differences also extend to surveys of scientists. For example, Slovic cited a study of members of the British Toxicological Society, in which female members were much more likely than male members to rate various societal risks as moderate to high.

Utilizing a survey of over 800 individuals, MacGregor et al. (1999), similarly reported that the public's perception of what constitutes exposure may be very different from that of the risk assessor or communicator. Dependent upon the manner in which the question was asked, the agent and its familiarity, or the consequences (e.g., cancer and the associated negative emotions), exposure might mean one-time contact, extensive and regular contact, or contact sufficient to cause a health effect. It is important that the risk communicator understand the audience's potential interpretations of exposure and attempt to clarify the meaning and significance of exposure in any given situation.

Slovic also reported studies, which demonstrated the role of emotion in the development of risk perceptions. He described the immediate sense of like or dislike, which he termed affect, conjured up by a phrase describing a hazard or activity. This response leads to perceptions of risk and benefit, which are independent of analytic thinking. For example, swimming at the beach might trigger a positive immediate emotional response (despite obvious hazards), while going to the hospital or living near a hazardous waste site might trigger a negative response. Survey studies have demonstrated an interaction between affect and perceived benefit and risk, such that a negative affective response tended to result in low perceived benefit and high perceived risk. In turn, information that increases the perception of a benefit from an activity (e.g., nuclear power) can reduce the negative affective response and lead to reduction in perceived risk.

Finally, Slovic discussed the role of trust or lack of trust in determining the success of risk communication. He noted the difficulty in and time required for building trust and the ease and speed with which trust can be destroyed by a mistake or a negative event. Further, negative events or stories, which tend to destroy trust, have greater visibility and tend to be more readily accepted as true or accurate than positive events. While there are no easy solutions to dealing with differences in risk perception that result from differences in socio-political views, affective responses, and willingness to trust, it is important to recognize these differences when planning and delivering risk communication. The addition of more scientific data and concepts alone will not address these other elements of perceived risk. Slovic argues for greater participation of concerned individuals in the risk assessment and risk management process, resulting in greater trust in the organization and acceptance in the decisions that result.

Employee and community groups deserve credit for being capable of understanding the scientific aspects of risk assessment. Many groups include some professionals, who work as physicians, engineers, chemists, lawyers, etc. Acknowledgment of their capabilities and, in some cases, using them as intermediaries for the communication of information lays down a foundation for trust and effective interaction.

Understanding the audience, including its demographics, employment, education, and prior experiences with health risks, is a crucial part of this process. A presentation on the risk of a hazardous gas leak will differ significantly when addressed to the employees of a facility vs. members of the surrounding community. Similarly, the presentation will differ significantly when the audience is primarily scientists from the EPA vs. a group of concerned citizens. Once one establishes the nature of the audience, it is important to gather information as to what the audience members already know, in addition to assessing their needs and concerns. In order to do this, one may need to do some research by making phone calls or interviewing representative members of the audience. Make sure your presentation reflects their needs as well as your needs.

PRESENTING INFORMATION

It is then important to establish those crucial pieces of information that are most important to communicate. Often, the issues in these situations are quite complex; from this, one must distill out the main messages that must be conveyed, in order to avoid overwhelming the audience. One must identify main ideas and gather relevant data and supporting details. Throughout the presentation, be attentive to signals from the audience that may indicate confusion or frustration. The audience may merely want a definition of a term or explanation of a more complicated issue. Expect that there may be questions that you will not be able to address or answer. If other speakers will be addressing this issue, let the audience know it. If you cannot answer a question, let the questioner know that you do not know the answer at present but that you will research it and provide a response in a timely manner. It is helpful to allow questions throughout the presentation and, at the same time, to permit time for a question and answer period at the end.

The use of a team of professionals from appropriate disciplines to present information and serve as a panel permits each team member to present information that he or she has gathered or analyzed (and with which he or she is most familiar). The team may consist of an industrial hygienist, a safety or environmental engineer, a facilities manager, an occupational health nurse, a toxicologist or risk assessor or an occupational medicine physician, as indicated. Relevant industrial hygiene or environmental monitoring data should be presented, to support subsequent conclusions regarding exposure. It is important for one team member, often the occupational physician, to integrate the medical, toxicologic and exposure assessment information into one coherent picture. Since no one professional is likely to be able to address all questions, the presence of the team strengthens the presentation and inspires greater confidence in the audience. It is critical that someone on the team be very familiar with the relevant literature (e.g., epidemiologic data, animal toxicology studies) in order to promptly respond to concerns and to appear to be competent to provide information and reassurance. It is not difficult for an audience to recognize that a speaker has not prepared for his or her presentation, which may further any feelings of mistrust. Many individuals attending these meetings have researched the subject, in some cases using similar resources and reference materials as were available to the speakers. It is prudent to never underestimate your audience.

Ethical issues and obligations regularly arise in the planning and delivery of risk communication efforts. Samuels noted several ethical imperatives, including the right of affected individuals to understand their exposure and risk and to participate in decision-making. Johnson (1999) listed several additional ethical obligations, which should underlie decisions about risk communication, including to do no harm, to increase understanding, to tell the truth, to keep promises, and to treat groups fairly. He then described a number of ethical issues, which must be confronted, including:

- Sharing fully (or not fully) with the audience the goals and motivations for a presentation
- Determining whether the goal is to inform or persuade the audience
- Determining the amount of effort to expand on understanding the health risk (including acceptance of information from other parties or experts)
- Deciding to include all concerned parties in a meeting (including those who might be disruptive or highly critical)
- Determining how much information (and uncertainty) and which topics to share (balancing the benefits of full disclosure against the downside of overloading or confusing the audience and recognizing the concern that selective disclosure, if later identified by the audience, may lead to erosion of trust)
- Choosing to present contrary viewpoints regarding the hazard

Although definitive or universal solutions to these issues are not available, they need to be considered by risk communicators with the ethical obligations in mind. The language of the presentation should be such that it can be understood by lay people but should not be condescending. Practicing the presentation by rehearsing it prior to the meeting will aid in the delivery. If rehearsed in front of a small group, the ultimate presentation can reflect their constructive feedback and prepare one for possible questions. During the presentation, it is important to acknowledge any concerns or symptoms experienced by audience members, giving credence to them even if not accepting that they are connected to the exposure. One should respond directly to the concerns expressed by attendees. When possible, be definitive, don't waffle. If the data supports a conclusion with a high degree of probability, one should not convey an equivocal answer, because there remains some very small amount of uncertainty. If you are not definitive, you will not be successful in reassuring your audience, even though the data indicate that they should feel reassured.

It is very important to put the concerns into perspective with regard to severity, the potential for long-term consequences, and the limits of uncertainty. For example, with an indoor air quality problem, symptoms may be quite uncomfortable and may even be temporarily disabling. However, a statement, supported by scientific literature, that such symptoms have been commonly observed in problem buildings, universally resolve after leaving the building, and do not result in permanent disability should help to ease some of these concerns. With regards to the limits of uncertainty, individuals will often state to an occupational health professional, "Science doesn't know everything there is to know about that chemical; therefore, how do you know that this exposure won't cause cancer?" Although the degree of toxicologic and epidemiologic data for different chemicals varies considerably, it is most often possible to address these concerns by discussing the low dose, the transient nature of the exposure, the requirements for chemical carcinogenesis, and the existing knowledge base.

When significant uncertainty persists because of lack of scientific data about the health effects of a chemical, one needs to be open about the existence of the uncertainty, while utilizing all available information to address the likelihood of a hazard (e.g., physical properties, exposure data, and nature of the exposure environment). Because of the lack of developmental toxicity information for many chemicals, this situation may arise when discussing the potential risks to an unborn child with a pregnant employee. Polifka et al. (1997) report that a layperson facing such uncertainty will often attempt to deny it, either by minimizing the uncertainty or grossly exaggerating the perceived risk, requesting a leave of absence from work during the pregnancy. The ability to ensure optimal control of the hazard or to arrange job modifications or other accommodations can help to resolve these situations, which result from scientific uncertainty.

Another common misconception may need to be dispelled: "They didn't think that asbestos was harmful in the past. How do we know that what you're assuring us now won't be proven wrong in a few years?" Indicating that the health risks from asbestos were known but concealed in the past and focusing again on the limits of uncertainty should help to dispel this notion.

One should avoid characterizing symptoms as being psychological in origin, even if it is clear that psychological or psychosocial factors are contributing to the development of symptoms. Some individuals will view this suggestion as an implication that they are crazy or that they are to blame for the problem. The audience will likely lose trust in the speaker and experience more outrage. However, there are non-toxicologic phenomena, which may account for symptom development in some situations. Neutra et al. (1991) and Shusterman et al. (1991) from the California Department of Health Services have demonstrated, in a series of epidemiologic studies conducted near hazardous waste sites, that odor and environmental worry are powerful risk factors in the induction of common symptoms such as headache and nausea:

Retrospective symptom prevalence data, collected from over 2,000 adult respondents living near three different hazardous waste sites, were analyzed with respect to both self-reported environmental worry and frequency of perceiving environmental (particularly petrochemical) odors. Significant positive relationships were observed between the prevalence of several symptoms (headache, nausea, eye and throat irritation) and both frequency of odor perception and degree of worry. Headaches, for example, showed a prevalence odds ratio of 5.0 comparing respondents who reported noticing environmental odors frequently vs. those noticing no such odors and 10.8 comparing those who described themselves as very worried vs. not worried about environmental conditions in their neighborhood. Elimination of respondents who ascribed their environmental worry to illness in themselves or in family members did not materially affect the strength of the observed association. In addition to their independent effects, odor perception and environmental worry exhibited positive interaction as determinants of symptom prevalence, as evidenced by a prevalence odds ratio of 38.1 comparing headache among the high worry/frequent-odor group and the no-worry/no-odor group. In comparison, neighborhoods with no nearby waste sites, environmental worry has been found to be associated with symptom occurrence as well. Potential explanations for these observations are presented, including the possibility that odors serve as a sensory cue for the manifestation of stress-related illness (or heightened awareness of underlying symptoms) among individuals concerned about the quality of their neighborhood environment (Shusterman, et al., p. 30).

There was neither an excess of serious health effects observed in residents near these waste sites, nor did air measurements document any toxicologically significant exposures. Experimental studies have documented headaches, irritation, and other symptoms in volunteers exposed to the odors of solvents, present in very low, toxicologically insignificant concentrations (Otto, et al., 1990). In addition to the physiologic effects induced by odors (e.g., nausea), environmental odors may serve as a sensory cue for the manifestation of autonomic or stress-related symptoms (e.g., headache and nausea). According to Shusterman et al., several studies document a 4 to 10 percentage point increase in the reporting of symptoms associated with stress. These studies document a mechanism for symptom induction that is not mediated by toxicologic effects or psychological factors. Under some circumstances, it may be appropriate to convey the findings of these studies to employees or community residents to help in allaying their concerns about symptoms. In addition to odor, other cues sensed in the environment (e.g., alarm sirens, emissions from smoke-stacks, or visible soot particles) may lead to perceptions of increased health risk and to anxiety among community residents. Elliott et al. (1999) observed such findings in a large survey of an urban community.

Not uncommonly, in relation to exposures, employees or community residents report apparent clusters of illness, which may or may not be true clusters of like illnesses aggregated in space and time. When the apparent cluster consists of several individuals with a serious illness, particularly cancer, the level of concern may be quite high. Involvement of the concerned individuals in the investigation and attention to risk communication principles throughout the investigation are critical to a successful resolution of the concern (Drijver and Woudenberg, 1999).

In some situations, typically involving employee populations, there may be the potential for significant health effects related to poorly controlled handling of chemicals or hazards (e.g., lead exposure in a radiator shop). In these settings, risk communication involves sharing information about the hazard and its consequences as well as approaches to prevent exposure and illness. Tan-Wilhelm et al. (2000) reported the effect of a worker hazard communication program in a beryllium machining plant on attitudes and behaviors. The program utilized high threat messages designed to convey the serious health risks associated with beryllium exposure and take-home beryllium. It also used high efficacy messages, indicating that the recommended practices, such as hand washing and car vacuuming, can be readily accomplished (self-efficacy) and eliminate the threat (response efficacy). The authors found statistically significant favorable changes in both attitudes and protective behaviors in workers in the intervention plant compared with the control plant.

CONCLUSION

Despite the fact that occupational and environmental exposures are, in general, much better controlled now, the level of concern about such real or perceived hazards has increased dramatically in recent years, coincident with increased media attention and the aftermath of a number of well-publicized environmental disasters. The principal tasks of environmental, health, and safety professionals remain in hazard reduction. However, effective communication of actual risk is essential in order to avert fear and anger associated with concerns, to maintain good employee and community relations, and to avoid unnecessary disruptions in production or other key activities. Risk communication is not and probably will never be amenable to rigorous scientific methodology, even if the informational content of the message is scientifically demonstrable. We need to approach our audience, whether employees or public citizens, with concern, sensitivity, honesty, respect, and a willingness to understand their point of view. Risk communication must be a two-way street to be accepted and effective. The combination of a well-conceived risk assessment and appropriate and sensitive risk communication can, at least in many cases, permit resolution of vexing employee and community concerns to the satisfaction of all parties.

REFERENCES

Chess, C., Hance, B. J., and Sandman, P. M. *Improving Dialogue with Communities: A Short Guide for Government Risk Communication.* Environmental Communication Research Program. Division of Science and Research, New Jersey Department of Environmental Protection:Trenton, NJ, 1988.

Drijver, M. and Woudenberg, F. Cluster management and the role of concerned communities and the media, *European Journal of Epidemiology* 15:863–869, 1999.

Elliott, S. et al., The power of perception: health risk attributed to air pollution in an urban industrial neighbourhood, *Risk Analysis* 19(4):621–634, 1999.

Johnson, B. Ethical issues in risk communication: continuing the discussion. *Risk Analysis* 19(3):335–348, 1999.

MacGregor, D., Slovic, P., and Malmfors, T. How exposed is exposed enough? Lay inferences about chemical exposure. *Risk Analysis* 19(4):649–659, 1999.

Neutra, R., Lipscomb, J., Satin, K., and Shusterman, D. Hypotheses to explain the higher symptom rates observed around hazardous waste sites. *Environmental Health Perspectives* 94:31–38, 1991.

Otto, D. et al., Neurobehavioral and sensory irritant effects of controlled exposure to a complex mixture of volatile organic compounds. *Neurotoxicology and Teratology* 12(6):649–652, 1990.

Polifka, J., Faustman, E., and Neil, N. Weighing the risks and the benefits: a call for the empirical assessment of perceived teratogenic risk. *Reproductive Toxicology* 11(4):633–640, 1997.

Sandman, P., *Explaining Environmental Risk*. Washington D.C.: Office of Toxic Substances, United States Environmental Protection Agency, 1986.

Shusterman, D., Lipscomb, J., Neutra, R., and Satin, K. Symptom prevalence and odor-worry interaction near hazardous waste sites. *Environmental Health Perspectives* 94:25–30, 1991.

Slovic, P. Trust, emotion, sex, politics, and science: surveying the risk-assessment battlefield. *Risk Analysis* 19(4):689–701, 1999.

Tan-Wilhelm, D. et al. Impact of a worker notification program: assessment of attitudinal and behavioral outcomes, *American Journal of Industrial Medicine* 37:205–213, 2000.

FURTHER READING

Fisher, A. et al. One agency's use of risk assessment and risk communication. *Risk Analysis* 14(2):207–212, 1994.

Hance, B. J., Chess, C., and Sandman, P. M. *Industry Risk Communication Manual: Improving Dialogue with Communities*. Boca Raton, FL: Lewis Publishers, 1990.

Samuels, S. Communicating Risk to Workers: History and Ethics, in Tinker et al. eds., *Communicating Risk in a Changing World*. Solomons Island, Maryland/Beverly Farms, MA: The Ramazzini Institute/OEM Press, 1998, pp. 119–134.

Sparks P. J. and Cooper M. Risk characterization, risk communication and risk management. The role of the occupational and environmental medicine physician. *Journal of Occupational Medicine* 35(1):13–17, 1993.

## CHAPTER 25

Practical Application of Occupational Health Programs

Neva Nishio Petersen

CONTENTS

INTRODUCTION

In this section, I plan to address some of the obstacles that you may encounter as you plan and implement your occupational health programs and to discuss some ideas to help keep your program on track.

First, let us look at some of the problems that you may encounter — this is by no means an exhaustive list, but hopefully it will provide some insight into real-world issues that you may encounter.

- There is not enough time or staff to adequately address projects and/or complete a needs assessment. This is a problem with which most of us are already familiar, and one that will probably continue indefinitely as corporations and governments try to control costs.
- Your planned programs meet resistance due to the cost. This includes the cost of the equipment and the cost of personnel to attend training or implement the safety procedures. Often, compliance with safety and health programs will mean that employees need additional time to complete the job. Job preparation and setup may take longer (as in the case of asbestos work); employees may find it slower to work with personal protective equipment; additional paperwork (such as obtaining a confined space permit) may take time; and cleanup after a job may also require additional time (time to clean and store safety equipment such as respirators).

- Apathy may exist on the part of managers and supervisors. Often, safety is perceived as the safety department's problem, and managers and supervisors may look to you to fix the problem rather than take ownership themselves.
- Manager and supervisors try to reset your priorities based on their perceptions of the issues or on information from the other sources.
- Employees are enthusiastic about the new programs, but interest wanes and the program dies.
- A cluster of occupational injuries or illness resets your priorities.

So, how can you overcome these and other obstacles that you may encounter?

One of your most valuable tools is your occupational injury and illness record, that is, your OSHA 200 log. By tracking your injury and illness frequency and severity rates, you can track your progress and success. By breaking down your log into categories — how many of each type of injury have you had — you can identify the areas and programs that should be a priority for attention.

For example, if you look over that last three years and you discover that you have had several chemical exposures, but only one cumulative trauma injury, you should prioritize your respiratory protections and hazard communication programs, and ergonomics may be lower on your list. This is a quick and dirty way to set priorities if you do not have time to complete a full needs assessment. (Please keep in mind, however, that a needs assessment would be preferable if possible.)

Tracking injury and illnesses may also help justify the priorities you have set, and can help to convince other managers and supervisors that your priorities are the right occupational health priorities for your organization. However, as you set these priorities, you must keep legal compliance in mind — if you have a high rate of cumulative trauma injuries, you will still need to ensure that your organization meets respiratory protection and hazard communication regulations. Often, your programs will be driven by both factors.

The cost of industrial injuries and illnesses and the cost of regulatory noncompliance (fines and liability) can help demonstrate the cost effectiveness of occupational health and safety programs. Workers' compensation carriers often can provide the direct costs of your injuries and illnesses — medical cost and disability — and you can estimate the cost of lost time or lost productivity due to injuries. The total injury cost can then be compared to the estimated cost of the occupational health program — necessary equipment, training time, and any additional job time. Usually, the cost of compliance is lower than the cost of the injury.

However, once you start tracking injuries and illnesses, do not forget to look beyond the statistics as the root cause of the incident. Your policy and program should be designed to address the problem so that the injuries and illnesses decrease. Also, feedback to each department is a critical step in giving the department ownership of any issue. Departments need to know of any problems so they can correct them. They must also be involved in identifying the fix for the problem, and they must be a part of implementing the fix. This way, they will readily accept ownership of the program, and if they have ownership they will follow through and continue with the program.

For example, you begin implementing a respiratory protection program in response to all your chemical exposures. If the department is aware of the problem and is involved in developing the program (Where do they want to store respirators? When is training most practical for them? Who in the department is best suited to ensure compliance in the department?), they are more likely to implement the program and follow through to ensure continued compliance. And without continued compliance, you will not see any reduction in chemical exposures.

To help ensure that your programs continue, you should also provide continued support and feedback — let managers and supervisors know about their successes in implementing a program or reducing injuries. Alternatively, if necessary, provide continued information on problem areas and continue to solicit ideas for improvement as well as suggest your own ideas to correct problems.

Finally, be flexible. Your priorities and your programs will need to change and evolve. As new regulations are passed, you will need to update your programs and develop new ones. A cluster of occupational injuries or illnesses may reset your direction. However, if you keep your focus on your overall goals, you will be able to successfully develop, implement, and maintain an effective occupational health program.

FURTHER READING

Plog, B. A. et al. *Fundamentals of Industrial Hygiene*, Itasca, IL: National Safety Council, 1988.

## PART V

Safety and Health Management: Regulatory Compliance Aspects

## CHAPTER 26

Safety and Health Management — Regulatory Compliance Aspects

Gabriel J. Gillotti

CONTENTS

INTRODUCTION

The cost of preventable workplace injury, illness, and death in the United States is almost incomprehensible to the average person. In economic terms, it is in the billion-dollar range every year — a magnitude to which the average worker whose well-being is at stake cannot relate to in any terms.

Data gathered from state workers' compensation agencies and the National Safety Council (NSC) are very revealing with respect to the costs associated with the prevention of deaths, injuries, and illnesses. Workers' compensation figures indicate that in 1983 approximately $17.6 billion was paid out to employees, whereas in 1993 the figure was $50.2 billion. These are payouts to employees — actual costs to employers were about 30% higher, or $83 billion.

The NSC claims that the true cost to the nation for work-related deaths and injuries is greater than that of compensation alone. The NSC estimate as to the cost for 1999 was $123 billion. In other words, each worker in the country must produce an additional $96 in goods and services to offset the $123 billion.

Every human, compassionate being is for improved safety and health — whether a worker or employer, union or nonunion, or Republican or Democrat. The mission of the Occupational Safety and Health Administration (OSHA) is the same in 2001 as it was at the time of the passage of enabling legislation in 1970: to save lives and to prevent deaths, injuries, and illnesses in order to protect the American working man and woman.

OSHA's task is difficult in large part because of changes in the work environment, including the associated technology, worker–employer relationships, impact of priorities set by society, marketplace conditions, and the impact of government priorities; all have varied enormously over the 30-year life of OSHA. It has been extremely difficult for OSHA to maintain its focus on this moving target — the changing workplace.

Unavoidably, OSHA's effectiveness at meeting its mandate is frequently challenged. One or more measures are used to argue the point, for example, the number of inspections conducted annually, the level of monetary penalties, the negative impact on the small-business community, the negative impact on specific industries such as agriculture or chemical processing, and others.

The available data bear out the fact, however, that OSHA's enforcement of its standards has resulted in a fulfillment of its mission, at least for 18 of its nearly 25-year existence. According to the Office of Technology Assessment of the federal government, the rules on lead and cotton dust, as examples, have dramatically reduced workers' exposures and substantially reduced illness rates. Also, the federal Bureau of Labor Statistics data indicate that in 1975, 9.4 workers per 10,000 were killed on the job, while in 2000 the number was 4.3 per 10,000 workers. In 1973 there were 11 injuries or illnesses per 100 full-time workers, while in 1999 the number had fallen to 6.3.

For a variety of reasons, both political and administrative, OSHA's approach has varied dramatically as well. The results of one approach (i.e., strong commitment to and emphasis on enforcement) have been measured at least once. After studying illness/injury data from 6842 manufacturing plants, Wayne Gray and John Scholze concluded in 1991 that when OSHA inspects and imposes penalties for violations, there is a measurable injury reduction in those workplaces following the inspection. It is nearly impossible to measure the other way (i.e., training and consultation) that prevention is achieved, either directly or indirectly, as a result of OSHA action.

Because the law was interpreted to have the Congressional intent that prohibits its federal OSHA from providing direct on-site consultation and assistance, a few years after the law's passage contracts with state and other agencies to deliver such services were initiated. The results of the efforts of those OSHA agents have not been measured in any credible manner in the past 23 or so years. To what extent the current reductions are attributable to that approach and, in what proportion consultation and enforcement go hand in hand to prevent injuries and illnesses, are indeterminate. How does one measure the number of accidents prevented, lives saved, etc., with any confidence? The Bureau of Labor Statistics' OSHA data are but a snapshot of the universe of workplaces and are not a measure of success.

I believe that the efforts of responsible employers, particularly those who never have the occasion to be face to face with an OSHA representative, are difficult to measure also. The enlightened owners and managers who view OSHA's standards as a minimum, and who have the compassion and concern for the well-being of their workforces are responsible for a significant portion of the reduction. Their efforts are essential and must continue if the numbers are to continue to improve.

Having commented on roles and on some of the results, let us address the extent to which relationships between employers, workers, and government bring about success, or the lack thereof.

NEW ERA FOR LABOR–MANAGEMENT RELATIONSHIPS

In December 1994, a report was issued on workers' opinions about their jobs. The year-long study, coordinated by a Harvard economist and a University of Wisconsin law professor, polled 2048 workers about how they felt about job satisfaction, loyalty, and particularly about participation. Although a single study does not make a case, it is timely to this discussion because labor–management committees are viewed as the new era mechanism, essential to the establishment of effective safety and health programs in all workplaces.

What did the study conclude? Over half of the workers polled want participation and influence in decision making and nearly the same number feel loyalty toward their employers. That combination suggests that labor–management joint committees in all probability could be successful in addressing safety and health issues. The axiom "Who knows better what hazards exist in a given work situation than the exposed worker?" is self-evident and certainly worthy enough for management to solicit opinions from that worker.

The 1994 attempt at a reform of the OSHA law by the Democrat-controlled Congress prompted extensive debate and controversy on a provision that mandated labor–management safety and health committees. Currently, 10 of the 24 states that administer the OSHA program require workplace accident prevention programs, and 7 of the 10 require joint committees. The effectiveness of this provision in those seven states has not been measured such that a convincing case for adoption by other states and federal OSHA has been made.

However, at least two NLRB cases, Electromation and Dupont, clearly "dampen" the efforts of employees, employers, and government to encourage the formation of joint committees. Some people believe that it is virtually impossible to establish committees for any purpose, including safety and health, without violating the National Labor Relations Act (NLRA). Although in the two cited cases — Electromation, a company that has no unions, and Dupont, which is a union workplace — the fact situations are different and are specific to that workplace, and the decisions are being badly interpreted to disallow committees in any workplace.

An example of such an interpretation is that, if a committee is established and becomes effective, the employees will consider themselves a labor organization and not attempt to exercise their rights to join a union. The NLRB views such a situation as an effort by the two parties, employer and employee, to deal with one another via bilateral discussions that include an exchange of ideas. That is what a labor union engages in under the NLRA, particularly if the safety and health committee gets into issues that affect manning and production, which is often the case.

Until the NLRB gives a green light to all parties, employers risk violating the law (NRLA), however constructive their intent. OSHA and unions are equally frustrated because they see committees as a potentially new and progressive mechanism for empowering workers and preventing injuries and illnesses.

Currently, there are efforts in Congress to amend the NLRA to lessen the effect of the recent decisions and to encourage the creation of committees.

Despite what has been discussed to this point regarding committees and programs and the hurdles that impede the effort, OSHA has been working with companies for about 19 years in establishing about 583 voluntary protection programs (VPPs). The VPPs include committees and accident prevention programs and are proving to be very successful based on site-specific data. I will elaborate on the VPP effort further on in this chapter.

UNION AND NONUNION WORKPLACES

Clearly, the driving force behind the initial enactment of the OSHA law was the organized labor establishment. Without their persistent effort to represent their members by bringing their voice to the agency, other interests may have prevented the agency from staying the course and protecting workers.

One truly appreciates labor's efforts when one realizes that about 85% of the country's workers have no representation and that what labor does for one of its members benefits about six other workers who have no voice but their own. We all know how difficult it is for a lone voice to be heard when addressing the government rule makers on issues that affect all of us.

Are there any real differences between union and nonunion workforces when it comes to safety and health programs, accident prevention, problem resolution, or whatever else necessitates positive interaction between employers and workers? If I am an employee whose supervisor practices a style of management that relies on intimidation and absolute control, I will have at some time a compelling need for a union to represent and protect me using their knowledge, the contract, labor laws, and any legitimate means available. The nonunion employee has no such protection, although he or she may have the good fortune of having an enlightened, compassionate manager who does not put productivity before safety.

The labor unions have been very aggressive in their support of the OSHA law reform provisions for labor–management committees as well as OSHA's efforts to promulgate a proposed accident prevention program enforceable standard, presently scheduled for release in late 1995. Other interactions with OSHA include participation on advisory committees, at public hearings on standards, as commenters on proposed rules, etc. Again, the direct voice of the majority, the nonunion workplace, is virtually silent but for the effort of the unions. In the end, the forces of both will prevail if the concept of joint labor-management problem resolution is accepted as a valued method.

As for conflicts in unionized workplaces over what is subject to collective bargaining and what is a safety and health issue, they will arise and must be addressed quickly to avoid compromising either the negotiated contract or the joint committees' efforts. All too often, conflicts have been labeled as safety or health matters when they have not been, and care and effort should be made to separate the agendas unless and until labor and management choose to include the safety and health program and procedures as an integral part of the contract. Many unions have chosen this approach, and in many cases it works very well.

BEHAVIOR VERSUS MECHANICAL FIX

Over the years, veterans in the safety profession have repeated an old axiom, "80% of hazards can be changed only by changing human behavior, 20% by instituting mechanical fixes." It is not evident when one reviews federal or state OSHA standards that drafters of standards have sufficiently believed that axiom. They still specify the mechanical solution to the exclusion of any consistent mention of the need to direct employees toward assuming a greater share of responsibility for themselves, and their fellow worker, while carrying out their daily tasks.

Leaders of government programs in many foreign nations have for years put greater emphasis on efforts to modify employee behavior to become more safety conscious. Some nations make it the focus of their efforts through extensive training programs and by strongly supporting labor-management cooperative efforts. Workers who have a stake in determining and instituting a preventive measure will surely change behavior to benefit co-workers and themselves.

Psychologist E. Scott Geller, Ph.D., has spent over 20 years lecturing about the need to have employees not only involved in safety; he stresses that they must become a force behind making safety a positive, proactive practice. People care about the well-being of family members and may develop interpersonal relationships at work that can be nearly as strong. His emphasis is on teaching workers a learned optimism so that they expect things to be safe and improving rather than believing that, no matter what they do, accidents will happen.

Dr. Geller also points out that the same classic management principles that apply to safety apply to other workplace issues, i.e., seeking employee input and leadership on key issues, keeping people informed, creating feedback systems, auditing progress, setting short- and long-term goals, and rewarding employees for safe behaviors, something they can control. As for control, the reward and acknowledgment system should be designed to recognize employees for that which they can control, not for those changes about which they have little or no individual impact or control.

The federal government's effort to affect employee behavior is contained largely in numerous required training standards. There are approximately 158 specific federal standards that call for training on both narrow and broad topics. Of the total, 83 apply to general industry, 41 to construction, 32 to the maritime industry, and 2 to agriculture. The topics range from personal protective equipment (PPE) to electrical hazards. Some training programs require a minimum number of hours, while others are performance oriented and the employer is expected to design and deliver the information. For the most part, the training is designed to sensitize the employee to the hazard and to alert him or her to the proper work procedure, work practice, or equipment safeguard.

As is true in most cases, there is no follow-up to determine whether the lessons were learned, the safe practices were applied or injuries/illnesses decreased because of training.

The standard that specifically calls for employee training to be conducted in the construction industry by the employer is 29 CFR 1926.21(b)(2), which reads as follows:

> The employer shall instruct each employee in the recognition and avoidance of unsafe conditions and the regulations applicable to his work environment to control or eliminate any hazards or other exposure to illness or injury.

Over a 1-year period (1999 to 2000) 1053 inspections addressed this standard, and 89% of the time, the employer was cited with an average penalty of $800. Obviously, the 64 training requirements applicable to the construction industry are not being adhered to if this level is indicative of employer effort.

ACCIDENT PREVENTION PROGRAMS

OSHA's direct experience with the subjects of programs and committees has been primarily in three areas, the VPPs initiative, the issuance of the Safety and Health Program Management Voluntary Guidelines in January 1989, and enforcement of 29 CFR 1926.20(b), *Accident Prevention Responsibilities*, in the construction industry.

The VPP concept can be described as follows:

- The concept of the voluntary protection programs is based on the realization that workplace compliance with OSHA standards alone will never completely accomplish the goals of the Occupational Safety and Health Act. Good workplace safety and health programs which go well beyond OSHA standards can provide a system of rules which can be set and enforced quickly, rewards for positive action, and safety and health improvements in ways simply not available in OSHA.
- Voluntary protection programs are intended to supplement OSHA's enforcement effort by identifying employers committed to protecting workers through internal systems so that limited enforcement resources can be directed to workplaces where the most serious hazards exist; they are also intended to encourage voluntary protection systems.
- All approved participants in voluntary protection programs are required to have implemented safety programs and meet all relevant OSHA standards. Participation

will not diminish either employer or employee rights or responsibilities under the Act.
- OSHA will verify qualifications; remove approved participants from routine scheduled inspection lists; provide any necessary technical support; investigate formal, valid complaints and serious accidents; and evaluate the programs.

Employee participation is an essential and required element in these programs, and many of the participating companies have very successful and effective committees. Not only have employees clearly bought into being part of the solution and not part of the problem, but the enthusiasm and commitment is very evident. OSHA, the government partner, creates an incentive for companies to apply for approval by exempting the workplace from scheduled inspections. Both the employers and employees have made safety and health concerns as much a part of day-to-day operations as production, maintenance, etc.

OSHA continues to encourage companies to apply for VPP recognition and approval, not only because it results in the institutionalization of the program in their company over the long term, but because it also enables OSHA to direct its resources to other high-hazard establishments.

Although specific, quantifiable measures of VPP labor-management committee contributions do not exist, other than reduced rates that may be a consequence of a number of factors, one has only to meet and discuss the important intangible consequences of employee involvement to know how employee behavioral changes have occurred. The workers themselves clearly change their behaviors because they have a stake in the success and well-being of the entire group.

The Safety and Health Program Management Voluntary Guidelines of 1989 have proven to be one of the most useful documents published by OSHA since its inception. These guidelines are particularly useful to small employers who want to have an effective program. As stated earlier, responsible and enlightened employers do want to protect their workers and to reduce their compensation and insurance costs, and they need the guidance that this document provides. The Guidelines will be discussed in more detail later in this chapter.

The enforcement of the 29 CFR 1926.20 standard in construction has produced interesting results. The standard reads as follows:

Accident Prevention Responsibilities. It shall be the responsibility of the employer to initiate and maintain such programs as may be necessary to comply with this part.

Such programs shall provide for frequent and regular inspections of the job sites, materials, and equipment to be made by competent persons designated by the employers.

The use of any machinery, tool, material, or equipment which is not in compliance with any applicable requirement of this part is prohibited. Such machine, tool, material, or equipment shall either be identified as unsafe by tagging or locking the controls to render them inoperable or shall be physically removed from its place of operation.

The employer shall permit only those employees qualified by training or experience to operate equipment and machinery.

During 1999 and 2000, there were 916 inspections that involved the identification of an accident prevention program as a significant issue at a workplace. The lack of the program resulted in serious violation citations, with an average penalty of $602. That clearly indicates that this long-standing requirement has not become an integral part of the construction industry's effort to comply with the standard and recognize the importance of the intent and purpose of the standard.

It should be mentioned here that employees who participate in these programs, as a committee member or in any other respect, are protected from recrimination by employers for that participation. The drafters of OSHA's law foresaw the need to protect those workers who do exercise their rights to be committee persons, to file complaints, and to report hazards by including the 11(c) provision. OSHA expends considerable resources and effort investigating employees' allegations of recrimination, in many cases where unions are nonexistent and grievance procedures are not available to the employee. Many employees have been made whole as a result of OSHA's 11(c) action.

OSHA has attempted to learn more about the value of accident prevention programs as part of this effort to promulgate both the 1989 Guidelines and the enforceable standard presently being drafted.

In an attempt to get beyond typical anecdotal evidence which tends to focus on larger employers, federal OSHA had a report prepared by outside experts on the subject of the benefit to construction companies who institute effective worker protection programs. The evidence in this report released on June 9, 1994 demonstrates major gains in injury reduction when individual firms undertake safety and health programs. The report also estimated that a net savings to the entire construction industry, if programs were instituted, could be as high as $16 billion annually due to the prevention of deaths and injuries.

The report cited the successful, nonanecdotal U.S. Army Corps of Engineers program to reduce construction injuries and fatalities. Results have been compiled to indicate that, between 1984 and 1988, contractors working for the Corps registered an average lost workday case rate of 1.34 to 1.54 per 100 full-time workers compared with the nationwide construction industry average of 6.8 to 6.9 per 100 full-time workers. For that period and since, the Corps has required a written safety and health program and has required the contractor to conduct work-site hazard analyses, to implement hazard prevention and control measures, and to provide safety and health training.

In 1994, the U.S. General Accounting Office (GAO) released a report titled "Occupational Safety and Health: Differences between Programs in the U.S. and Canada" (GAO/HRD-94–15S). As a result of this study comparing the United States as a whole with the three most populous subdivisions of Canada — Ontario, Quebec, and British Columbia (because the authority rests with provinces in Canada) — the GAO found differences in three major areas: operations and funding of safety and health programs; the extent of worker involvement in policy making and accountability; and enforcement.

With respect to programs and worker involvement, in Canada there must be a joint worker/employer health and safety committee at all sites with 20 or more workers, with at least one half of its members being chosen by workers. Workers can refuse to do hazardous work until the condition is corrected, and joint committees can shut down a hazardous work process until the hazard is abated (Ontario). The emphasis is placed upon the internal responsibility of the joint committee rather than on the number of provincial government inspectors whose presence is all that determines the level of protection. Unlike the U.S. statute, workers can be held accountable and fined by the government as well.

Unfortunately, few data demonstrate the success of the provincial programs and to compare with the U.S. data. On the other hand, there are few critics of the philosophy and approach taken by the provincial government according to the GAO report.

FEDERAL OSHA GUIDELINES

As recently as July 1994, OSHA hosted a "stakeholder" meeting of industry, labor, state agencies, and others to discuss the OSHA agenda for the future. Predictably, a major issue with broad support was the matter of OSHA stimulating effective safety and health programs with top management commitment and meaningful employee involvement. The workplace-level, bottom-up solution to problems was the notion promoted by attendees provided that employee empowerment should be accompanied by stronger OSHA 11(c) whistleblower protection for workers who agree to committee involvement. Top-down effort is essential also, and OSHA's burden was to develop strategies for enhancing awareness on the part of our nation's business leaders and then holding them accountable. The goals of safety and of high performance or productivity are compatible, and top management must be convinced of this.

The stakeholders also shared their ideas as to what elements must be integrated into a safety and health program. The bottom line to OSHA was that OSHA model programs and self-help materials and other stimuli were necessary to get the employers engaged.

Even prior to the stakeholder session, as far back as the institution of the VPP policies in the early 1980s and at various OSHA top level management meetings prior to 1989, the need for published guidelines was discussed. On January 26, 1989, OSHA published a document titled *Safety and Health Program Management Guidelines; Issuance of Voluntary Guidelines.* As a result of extensive staff work and the input of 67 commenters from a broad spectrum of individuals, labor, trade associations, professional associations, etc., the guidelines were published. This document is limited in its approach to all industries except construction because of the uniqueness of construction work sites.

Convinced that there were a number of basic principles that underlie an effective program, OSHA began its Guideline language by stating that (1) the program must provide for systematic policies, procedures, and practices that are adequate to recognize and protect employees from hazards; (2) systematic identification, evaluation, and prevention or control of general, specific, and potential hazards must be included; (3) the employer's effort must go beyond the minimum the law requires, must prevent injuries and illnesses regardless of whether OSHA has an enforceable standard to force compliance; and (4) effectiveness is the measure, not the fact that it is in writing.

Given those principles, OSHA identifies four elements fundamental to an effective program: (1) management commitment and employee involvement, (2) work site analysis, (3) hazard prevention and control, and (4) safety and health training. Let us expand and clarify these elements.

Management Commitment and Employee Involvement

It must be clearly articulated in writing so that all personnel with any responsibility on site, or in a support function elsewhere, understand the priority of safety and health in relation to other organizational values. By making their own management practices visible to the entire organization, employees will get the priority message.

As is true with the setting of production goals, a goal for the safety and health program which sets forth the results desired and the measures of effectiveness is a crucial first step.

Employees will commit to the program only if they buy into the responsibility that comes from involvement in decision making. Realizing, as stated earlier, that employees have the most intimate knowledge of the job they perform and the hazards presented, the commitment is there if management values and respects the employee's perspective.

The employee's role can take various shapes, including, but not limited to, being involved in actual inspections for hazards and recommending abatement methods, being engaged in analyzing job hazards, providing input to the drafting of the work rules, training co-workers when his or her expertise is relevant, being active at safety meetings, and assisting in accident investigations. Performance of these tasks as an employee can be in the context of joint committees, labor committees, rotational assignments, or volunteer opportunities presented by the employer.

In place in successful organizations is the clear understanding of assignments and the constant practice of communicating responsibility to managers and employees. With that must be the delegation of adequate authority and resources so that the assigned responsibility can be met.

Accountability, after the above actions are in place, must be instituted so everyone knows he or she must carry his or her share of the load. By conducting comprehensive reviews and audits to confirm success, note surface deficiencies, and institute program improvements, the loop will be closed on what is intended by this first element. This is not constructive, however, if done with an eye toward punishment, but it may be necessary to institute major changes in roles and responsibilities, which can be healthy for the program.

Work Site Analysis

The hazards, actual and potential, cannot be acted upon unless comprehensive workplace surveys are conducted at least annually or as necessary. When major changes occur, such as the installation of new equipment or processes or the purchase of new material such as chemicals, a hazard analysis, however small in scale, must be conducted.

The policy should provide for a continuous flow of information on perceived hazards from employees, with the understanding that a timely and thorough response will be provided by management.

Two other forms of analyses that must occur as part of the program are (1) investigations of accidents, and of near misses, focusing on accident cause; and (2) the periodic review of injury/illness data for identifying patterns and causes.

Hazard Prevention and Control

Engineering control is the place to start when determining how to prevent an incident. Designing safe equipment and equipment features is the most effective control. Additionally, there must be safe work procedures and provisions for purchasing and providing appropriate PPE. Administrative controls, such as reducing the hours an employee may be exposed, are another acceptable means if engineering the prevention is not achievable. A combination of two or all three of the engineering, administrative, and PPE controls may be appropriate in some cases.

A comprehensive facility and equipment maintenance program will prevent many accidents. All too often it is not preventative, but is reactive, which can be costly in terms of injuries, illnesses, and equipment.

Every workplace must exercise control with respect to emergency situations by training employees how to act quickly and avoid harm's way.

If an incident occurs, medical response — in order to minimize and control harm — should be adequate by virtue of a well-instituted medical program.

Safety and Health Training

All employees must be trained to understand how they may be exposed while doing their particular job and while working with others. That way they will understand and implement the program's policies, procedures, and the established protection. Training that enables an employee to recognize hazards, particularly if he or she is a member of a safety and health committee that conducts walk-through inspections, is crucial.

Managers must be trained how to analyze the work under their supervision in order to identify hazards, how to carry out their roles within their assigned managerial function, including, but not limited to, instituting and maintaining controls.

Often it is more effective to incorporate the safety and health training within job practices training so that the employee has an even better understanding of how safety and health are an integral part of this work.

The need that is most often overlooked is the need to ensure that those trained comprehend the material. Various approaches such as observing their work, oral questioning, and even formal testing should be undertaken. Traditionally, workplace training is conducted such that background and comprehension skills are overlooked and all employees are treated as equals in these regards. The verification of the understanding of the message will ensure, to some degree, a measure of success.

Our discussion is now complete — the government's challenge is to create the momentum necessary to shift the exercise of responsibility from government to employers, but it has not yet been met. In these times of lesser government intervention into the business community due to Congressional mandates, the roles of the partners in injury/illness prevention — namely, the government, employers, and employees — is shifting in terms of their relative magnitude and importance. We therefore, in a constructive fashion, must each fulfill our roles effectively, however differently. Accident prevention programs may well be the vehicle that enables that to happen. In fact, we may learn that it more closely meets the intent of the OSHA law than our past practice. It may also result in a greater appreciation by the industrial community of what can be achieved in reducing the loss of life and the injuring of their workers, without government involvement.

FURTHER READING

Freeman, R. and Rogers, J. Worker Representation and Participation Survey. [City?]: Gannett News Service, [year?].

OSHA 29 CFR 1910. OSHA Regulations applicable to General Industry. July 1, 1994.

LeBar, G. Activating Your Greatest Safety Resource. *Occupational Hazards Magazine*, January 1994.

OSHA publication 2254 (Revised 1992). Training Requirements in OSHA Standards and Training Guidelines.

GAO Report (GAO/HRD-94–155), 1994. Occupational Safety and Health: Differences between Programs in the U.S. and Canada.

OSHA Guidelines (January 26, 1989). Safety and Health Program Management Guidelines; Issuance of Voluntary Guidelines; Notice.

## CHAPTER 27

The Regulatory Viewpoint — California Standards: A Regulator's Perspective on Safety and Health Management

Thomas A. Hanley

CONTENTS

OVERVIEW

California Senate Bill 198 required California employers to comply with the most stringent and far-reaching legally mandated occupational safety and health program in the United States. California employers were required to develop, implement, and maintain an effective injury and illness prevention program (IIPP) with seven mandated, integrated components. Employers responded to the requirement with varying motivational perspectives: from recognizing the value of the concept for loss control to apprehension of substantial penalties following a Cal/OSHA inspection. Those employers, irrespective of motive, who in good faith developed, implemented, and maintained an effective IIPP, are a real asset to California and its workers.

Real, direct benefit has come to these responsible employers as well. Not only have these employers experienced the bottom-line benefits of loss control, but they have also found that effective injury and illness prevention tends to place an employer in an advantageous posture with respect to Cal/OSHA by minimizing both the likelihood of a Cal/OSHA inspection and the impact of any Cal/OSHA inspection that does occur. Furthermore, the process of maintaining an effective IIPP tends to make an employer familiar with the organization and structure of Cal/OSHA, as well as the services available through Cal/OSHA and the employer's rights under the Cal/OSHA program. I will develop this position from the perspective of one Cal/OSHA regional manager.

Those employers who have developed, implemented, and maintained effective IIPPs, based on internal motivations or in response to concern about potential sanctions following a Cal/OSHA inspection, tend to be the subject of fewer Cal/OSHA inspections than other employers. Despite the recent establishment in California of a High Hazard Compliance Inspection unit, the predominant focus of Cal/OSHA compliance assets is on employee complaints of unsafe conditions and reports of serious occupational injury and illness. Cal/OSHA, unlike federal OSHA, is largely complaint driven. Employee complaints of unsafe conditions and mandatory employer reports of serious injury accidents and occupational illness still motivate most Cal/OSHA inspections.

Employers with effective injury and illness prevention programs have given serious attention to establishing and committing resources to the required communication component of the IIPP. Responsible employers have established one or more means to communicate with employees and to facilitate employee communication with safety and health management. Whether these employers have selected postings of written material, distribution of written material, meetings of safety and health committees, or some combination thereof to communicate expectations and requirements of their safety and health programs is not as important as the fact that some form of effective communication does take place.

The communication component, in addition to providing and reinforcing a general awareness of the structure and organization of the company's program and safety rules, invites and encourages employees to communicate their concerns about safety and health matters to the employer. For these responsible employers, a valuable employee is one who reports his or her concerns about potentially unsafe and unhealthful working conditions. In implementing the two-way communication component, the responsible employer clearly delineates one or more routes by which employees can relay their concerns. Often this includes directing employees to report their concerns to the line supervisors, an identified safety and health committee person, or the safety and health coordinator. Frequently, the employer will provide an anonymous route by which concerned employees may express their concerns without any fear of discrimination.

The required sanctioning component in the responsible employer's IIPP ensures that employees are motivated to report their concerns through one of several reporting channels. Most responsible employers not only institute a system of graduated discipline in compliance with the requirement that they maintain a system to ensure conformance to the requirements of the company's IIPP, but also have a system of positive inducements. The responsible employer in California is confident that concerned employees know how to report, know what to report, and are motivated to report.

The responsible employer, through supervisory staff, the safety and health office, or other means, investigates employee-reported problems and commits resources to control and abate the hazard when it is substantiated. Even when the employee-reported condition is determined not to represent a hazard, employees are briefed and informed as to the investigation and results. The reporting employees, in observing or learning of their employer's response to their concerns, are motivated to continue the practice and are confident that the employer wants to maintain a safe and healthful workplace and can accomplish such a goal with its own assets. The safety and health effort receives information on potential hazards not only from employees, under the communication component, but also from its own in-house inspection program, under the hazard identification component, and by isolating the causes of in-house accidents under the IIPP requirement to investigate accidents.

When employees have confidence in their employer's own system of complaint reporting, as a result of the employer's response to employee complaints, employees continue to report their concerns in-house and do not complain to Cal/OSHA. Since the Cal/OSHA program is largely complaint driven, the employer who has maintained an effective IIPP is subject to fewer Cal/OSHA inspections. An effective IIPP solicits, accepts, and resolves complaints in-house.

Apart from other economic benefits that result from effective loss-control measures, employers with effective IIPPs who solicit, investigate, and resolve employee safety and health complaints demonstrate that they are good corporate citizens who are sensitive to the limited resources of government, especially the regulatory arm of Cal/OSHA. To the extent that Cal/OSHA field compliance resources are not assigned to inspect a business with an effective IIPP, they can be focused on hazardous settings that deserve attention. The responsible employer, as a good corporate citizen, is preserving both governmental and societal assets.

In addition to requiring inspections in response to employee complaints, the California Labor Code requires that all industrial accidents resulting in serious injury or illness be investigated, contrary to the mandate of federal OSHA, which limits its focus to fatal and catastrophic events. The requirement to investigate all serious injury accident and illnesses is the other primary motivator that brings the Cal/OSHA inspector to an employer's door in California.

An effective IIPP prevents and limits the severity of destructive and injurious health and safety incidents and again limits the likelihood of Cal/OSHA having to respond to the responsible employer's place of business. The net result of the efficient identification of safety and health hazards, and their subsequent control or elimination, is fewer severe hazards to cause accidents or illnesses. To the extent that the implemented IIPP of the responsible employer motivates the investigation of near-miss events, minor accidents, serious accidents, and incidents of serious occupational illness, the likelihood of serious injury or illness in the future is reduced along with the likelihood of a visit from Cal/OSHA.

The responsible employer assigns a competent and qualified employee or supervisor to investigate the accident or near-miss event with an eye to identifying causes, particularly preventable causes. The accident causes are shared through report and discussion with the safety director, coordinator, safety and health committee, and line supervisors. Such discussion tends to serve as a quality control mechanism for the in-house investigation. Through such a process, there is a greater likelihood that all relevant factors and causes are considered and the abatement actions, appropriately selected, are prioritized and implemented.

For the responsible employer, such a process decreases the frequency of occurrence of accidents and occupational illnesses and generally reduces the severity of such incidents that do occur. Reducing the occurrence of serious injury and illness-producing events simultaneously reduces the frequency of Cal/OSHA visits. Accidents are also reduced by the general inspection process required of the IIPP. The mandated periodic inspections conducted by responsible employers are not just pro forma. The inspection results are reported and discussed; the hazards are identified, prioritized, and corrected under the IIPP's abatement requirement.

Because Cal/OSHA enforcement scheduling is so sensitive to employee complaints and the requirement to investigate serious occupational injury and illness events, effective employer activity, which results in preventing accidents and motivates employees to report their complaints and concerns in-house, tends to minimize or prevent Cal/OSHA inspections and investigations. Fewer Cal/OSHA inspections means less time disruption to an employer's workplace and less potential for citations and civil penalties.

Although keeping the Cal/OSHA enforcement arm away from the employer's door is one of the positive results of an effective safety and health management and genuine IIPP compliance, the same responsible employers tend to be the employers who acquaint themselves with the structure and organization of Cal/OSHA and the Division of Occupational Safety and Health, the employer's rights under the Cal/OSHA program, the benefits and services available through the program, and Cal/OSHA inspection procedures. With this knowledge, responsible employers can avail themselves of opportunities to further reduce the chances for an inspection, to influence the outcomes of any inspections that do occur, and to impact the development and amendment of California's occupational safety and health standards.

Whereas California State Assembly Bill 110 (workers' compensation reform legislation) has mandated that Cal/OSHA refer problematic employers with high experience modifications to the loss control services of their workers' compensation carrier or the expanded Cal/OSHA Consultation Service unit, responsible employers have already discovered the benefits and services of the Cal/OSHA Consultation Service and have used this service as another resource in their overall loss control effort. It is paradoxical that the employers who need it the least are the ones inclined to utilize the Cal/OSHA Consultation Service more often. The responsible employer tends to be the employer who has learned about the Cal/OSHA Consultation Service and has utilized the service to receive technical guidance and information, publications, consultative inspections, and safety and health videos from the video library.

The process of developing, implementing, and maintaining an effective IIPP leads an employer, through its own efforts or those of its consultants, to secure the Cal/OSHA orientation publications: *The Guide to Cal/OSHA*. Guides to developing IIPPs, Model IIPPs, Cal/OSHA Policy and Procedures, and contact information for the local Cal/OSHA Compliance Office and Consultation Service office. Useful information for the public is now available online at www.dir.ca.gov/DOSH/doshl.html. The responsible employer knows how to plug in to Cal/OSHA and knows how to contact the local Cal/OSHA supervisors and managers. These same employers are the ones who request Cal/OSHA representatives to deliver speeches and make presentations (to the extent Cal/OSHA resources permit). These employers have the opportunity to bring personnel responsible for IIPP support activities to such presentations and, in so doing, stay abreast of changes, trends, and Cal/OSHA policy interpretations. The responsible employer stays ahead of the loss-control power curve by maintaining an effective IIPP, by minimizing Cal/OSHA inspections, and by embracing available resources and knowledge available through Cal/OSHA on a proactive basis.

Responsible California employers are finding not only that the resources expended to maintain an operative IIPP are cost effective, but that an effective IIPP acts as a force multiplier. The number of personnel actively or partly, but meaningfully, focused on safety and health issues increases because all supervisors and employers are mindful of, and encouraged to do, their part in accord with the IIPP. The responsible employer includes consideration of compliance with IIPP and loss control efforts in performance appraisals of employees, supervisors, and managers. The performance evaluation is a suitable vehicle for motivating employees, supervisors, and managers to do their parts under the IIPP. When efforts in support of a company's IIPP are recognized in performance evaluation, occupational safety and health becomes as locally significant as production. Even the language of IIPP, when used by senior supervisors and managers, motivates subordinate supervisors to understand that continued success and advancement within the company requires effort and support of the IIPP.

The IIPP acts as a force multiplier that literally increases the number of people focused on safety and health issues. The more eyes, ears, and noses focused on safe and healthy work operations, the better. The more eyes, ears, and noses involved, the sooner potential hazards, whether they be physical, chemical, biological, or human, are reported, investigated, and controlled. Effective safety and health management mobilizes more than just a company's safety and health personnel. Nearly everyone is mobilized into the effort. IIPP effectiveness spawns additional resources to build on its own effectiveness.

It is my experience that effective IIPPs tend to increase overall employee morale and, consequently, production. Effective IIPPs create a culture of respect for the knowledge, opinions, and experience of a company's employees. Employees are encouraged to report their ideas and concerns about safety and health. Demonstrated employer response reinforces the employee's belief in the employer's respect for the employee's individual contribution to safety and health and the overall company mission. From explaining how procedures, methods, tools, and processes can or should be modified to achieve safer operations, it is but a short distance to employee suggestions as to how to improve the quantity and quality of production. Mutual respect creates loyalty and high morale. High morale feeds self-respect and pride, which in turn feeds efficient and improved production. Employers with effective IIPPs tend to be productive, competitive companies.

The IIPP can serve as an umbrella or integration mechanism for other specialized safety and health program requirements driven by particular safety and health standards relative to the nature of an employer's operation. The emergency action plan, fire prevention plan, hazard communication program, respiratory protection plan, bloodborne pathogen program, and other plans and programs are incorporated under the general IIPP as specialized components of the plan. By incorporation into the IIPP, the particular substance and requirements of a specialized plan are strengthened by the various components of the IIPP. So, the emergency action plan is the subject of the communication component, compliance component, hazard identification component, and training component of the IIPP, which enhances its effectiveness and efficiency. The same is true of the other programs. The language of the IIPP standard reinforces the language of the specific plans. The most effective and vital IIPPs reference and incorporate subordinate plans and programs.

The effective IIPP will tend to minimize the occurrence of those events that motivate Cal/OSHA inspections — complaints and accidents. But even the most sophisticated employer with a very effective IIPP cannot keep all employees happy all the time. The disgruntled employee can generally find some issues to raise in a call to the local Cal/OSHA office to generate a valid complaint, which will be assigned for field inspection. However, the responsible employer with an effective IIPP will consistently minimize any adverse impact of a Cal/OSHA inspection.

Responsible California employers with an effective IIPP typically know Cal/OSHA inspection procedures and will react proactively to facilitate the inspection in their favor. When the Cal/OSHA field compliance person arrives in the reception area of the responsible employer, the receptionist will know which office or person to call. The safety and health coordinator or director will know in advance which person to assign to greet the inspector and represent management. The plan is already in place. The Cal/OSHA inspector will be greeted positively and graciously in a way that acknowledges that Cal/OSHA cannot make appointments for initial responses and must conduct unannounced inspections. The representative for the responsible employer takes a tack similar to, "How do you do, Mr. [or Ms.] Jones. I am Herb Gibson, assistant safety and health coordinator here at Business, Inc. If you will show me your identification, we can step into my office and discuss the parameters of the opening conference and inspection." Contrast this with, "Mr. Callanan is the person responsible for safety and health here, and he is on vacation. He will be back next week. Why don't you make an appointment then?"

The assistant safety and health coordinator of the responsible employer with an effective IIPP and knowledge of the Cal/OSHA program knows that Cal/OSHA is not mandatorily required to conduct a comprehensive wall-to-wall inspection in response to every complaint. Cal/OSHA policy provides sufficient latitude to conduct partial inspections. District Office scheduling and individual workload often limit inspections to partial inspections.

An assistant safety and health coordinator for the responsible employer knows that the Cal/OSHA inspection is often a response to a confidential complaint. The assistant safety and health coordinator knows that the Cal/OSHA inspector cannot provide, in addition to the confidential complainant's name, the specific and exact nature of the complaint, out of caution that the specific nature of the complaint may tend to identify the complainant relative to the number of workers/employees associated with specific processes, tools, or parts of the operation.

Our model assistant safety and health coordinator typically proceeds to sketch and delineate the units, departments, and sections with their associated missions and the approximate number of employees engaged in each. The assistant safety and health coordinator from our model company knows that this approach will have a reasonable chance of generating the following response from the Cal/OSHA inspector. "As you know, I must always preserve the right to conduct a comprehensive inspection, but at present I intend to inspect the warehouse, including forklift operation and the spray coating operation. When I conclude my inspection of those areas, I will determine which, if any, additional areas to inspect."

During the opening conference, our model assistant safety and health coordinator takes advantage of opportunities to demonstrate the effectiveness of the employer's IIPP and the good faith of the employer with respect to safety and health issues generally. During the opening conference, the Cal/OSHA inspector provides evidence and input contributing to the judgment about the effectiveness of the IIPP and the good faith of the employer including: the IIPP being in writing; the presence of the necessary supplemental plans and programs; the active presence of the required IIPP components in the written plan; the maintenance of training and inspection records; the impressive demonstration of supervisory knowledge of the IIPP; the expression of full cooperation with the pending inspection; the stated inclination to improve the program and correct any hazards that may be found; and the Cal/OSHA Log 200 suggesting no serious unreported accidents or unmitigated trends. Even before the inspection has started, the responsible employer with an effective IIPP, through the assistant safety and health coordinator, has gone a long way toward limiting the impact of the Cal/OSHA inspection. After the opening conference, only two other criteria will be considered in the Cal/OSHA inspector's final determination of IIPP effectiveness — the overall results of the inspection in terms of the nature and number of hazards identified, and the impact of the IIPP on employees.

Assuming that the Cal/OSHA inspector does not find an excessive number of violative conditions or plain-view serious violations, and that employee contacts reveal employee knowledge and involvement with the IIPP, the inspector will find the IIPP fully effective and determine a high degree of good-faith on the part of the employer. A high good faith rating significantly reduces any civil penalties that may be subsequently issued. Furthermore, the Cal/OSHA inspector will likely conclude that expanding the inspection beyond the partial inspection already conducted, in view of the employer's very effective IIPP and sophisticated assistant safety and health coordinator, is not an effective utilization of the Cal/OSHA inspector's time, given his or her backlog of pending assignments. An effective IIPP communicated by a knowledgeable safety and health representative can save time, money, and production resources.

The sophisticated employer also knows an employer's rights under the Cal/OSHA program and readily takes advantage of such rights. In the current scenario, if the Cal/OSHA inspector has issued any citations and proposed any civil penalties, the employer has the right to an informal conference by appointment with the Cal/OSHA district manager and the Cal/OSHA inspector at the local Cal/OSHA office. The district manager has the authority to amend the citations and civil penalties in response to the employer's presentation of new credible evidence, not yet considered, or in response to supported and reasonable interpretations of existing evidence.

In the event that the sophisticated employer does not prevail at the informal conference, but still considers its position persuasive, that employer can initiate the Cal/OSHA appeal process with a phone call and subsequent completion of a simple form within 15 days of citation service. The sophisticated employer also knows that abatement of the cited violative conditions and payment of the associated civil penalties are stayed, pending appeal. Further, an appeal hearing will be held locally, before an administrative law judge. Although the process is sufficiently formal, to the extent that a record is maintained and general rules of evidence apply, the sophisticated employer knows that it is not necessary to retain an attorney.

It has been my experience that occupational safety and health directors, coordinators, and production supervisors and managers, who are capable managers of IIPP activity within their purviews, are very effective advocates at Cal/OSHA Occupational Safety and Health Appeals Board hearings. They understand the evidence burden Cal/OSHA must present to sustain any citations and civil penalties. They know how to present contrary evidence with respect to the significant elements, particularly with respect to employer knowledge of the violative conditions cited and employee exposure to the violative conditions. Even after the inspection, the responsible employer with an effective IIPP can continue to minimize the overall impact of a Cal/OSHA inspection.

The Division of Occupational Safety and Health, as the unit of California state government exclusively responsible for the enforcement of occupational safety and health, recognizes the vital role of employers who take occupational safety and health seriously and develop effective IIPPs. These employers are doing their part to protect the safety and health of California workers. As a Cal/OSHA regional manager, I salute the efforts of responsible California employers and offer my congratulations for their concurrent rewards of reduced losses, high morale, competitive production, and minimized adverse Cal/OSHA impact.

FURTHER READING

Books

National Safety Council. *Accident Prevention Manual for Industrial Operations: Administration and Programs,* 9th ed. Chicago, IL: National Safety Council, 1988.

Other Publications

1994 California Labor Code. Carlsbad, CA: Parker Publications, 1993.
Guide to Cal/OSHA. San Francisco: State of California, Division of Occupational Safety and Health, 1993.

CHAPTER **28**

The Employer's Viewpoint — Standards and Compliance

Peter B. Rice

CONTENTS

INTRODUCTION

Occupational safety and industrial health issues have become increasingly prominent in recent years. Employers' viewpoints are changing. This is an exciting time in which employers are abandoning the belief that programs like safety and health are hindrances to production. Many companies have recognized the importance of safety and health; however, for many reasons, mostly financial, there are thousands of employers who are just now recognizing their importance.

In response to the directives of the Occupational Safety and Health Administration (OSHA) and various state OSHA plans, employers are recognizing the importance of strong, effective occupational safety and health programs and regulatory compliance. Most employers recognize that the primary purpose of complying with OSHA regulations is to prevent occupational injuries and illnesses and to provide a safe and healthy workplace. Employers are becoming increasingly aware of other important reasons to recognize and comply with OSHA standards.

The progressive employer recognizes that OSHA compliance actually allows him/her to be more competitive in the marketplace, with a strong impact on productivity and profitability. With OSHA compliance, how can this make an employer more competitive, more productive, more profitable? The first issue to consider is the impact of an injury/illness on the bottom line. Injuries/illnesses and noncompliance mean more workers' compensation claims. Workers' compensation costs are rising every year, and claims translate into direct dollar losses. Noncompliance with the regulations can also lead to more frequent OSHA inspections and the potential for thousands of dollars in penalties.

The astute employer should be aware of not only the consequences of an OSHA inspection but, in cases of intentional negligence, the potential for criminal charges as well.

Failure to comply with OSHA regulations can also have a big impact on other, more intangible factors such as employee morale. What employee is going to be motivated to work for an employer who has little or no regard for safety and health in the workplace? On the other hand, when employees see that an employer is genuinely concerned with compliance with the regulations, setting the example for safety, employees recognize this and are also positively affected.

Public and community relations are becoming increasingly important to employers. Failure to recognize the impact of public relations can produce dramatic negative effects.

Employers also need to maintain strong client and customer relationships. Many clients will choose not to purchase goods or solicit services from a company with a poor track record in safety and health.

OSHA + EMPLOYER = A PARTNERSHIP

Maintaining good relations with the OSHA Compliance Office and its representatives makes good sense and is strongly encouraged. OSHA may appear bureaucratic in trying to enforce the hundreds, if not thousands, of safety and health rules and regulations currently in place. There is, however, a strong positive advantage of working with OSHA. OSHA is also made up of knowledgeable, often highly educated professionals with varied backgrounds such as engineering, safety management, chemistry, physics, industrial hygiene, construction, and public administration.

The vast majority of OSHA compliance personnel are strongly motivated to establish solid working relationships with employers. They recognize that cooperation, rather than enforcement, is much more effective in creating these relationships. The result has been such programs as the Voluntary Protection Programs, which are designed to recognize and promote effective safety and health management.

Whether or not an employer participates in any of these voluntary programs, developing an effective partnership with OSHA is still possible and very much encouraged. Employers should establish contact with the local OSHA office, meet the area manager, and tour their facilities. OSHA is generally very receptive to this approach and is usually eager to review an employer's existing safety and health program to evaluate the employer's level of compliance. Other means for strengthening employer/OSHA relationships include (1) meetings or conferences featuring OSHA representatives as speakers or attendees, (2) panel discussions with OSHA representatives, and (3) direct telephone contact with representatives with regard to OSHA compliance questions.

Making use of an OSHA consultation, a free service funded by the U.S. Department of Labor to assist employers (priority is still given to the small employers) in their safety and health programs, can help in the development of a partnership between OSHA and the employer. The addresses and phone numbers of the OSHA consultation programs are available at www.osha.gov/oshdir/consult.

EMPLOYERS' OBLIGATIONS

As specified in the Occupational Safety and Health Act (OSHA), an employer is responsible for providing a safe and healthy workplace. This implies that the employer not only must comply with OSHA requirements (which really are minimum standards), but also must know which ones apply to their situation. Employers and their representatives (supervisors and managers) are required to ensure compliance with safety and health standards in facility and process design, equipment selection and setup, and normal operations. It is important to recognize that all safety and health hazards need to be identified, evaluated, and controlled, regardless of whether or not there is a corresponding OSHA regulation. General safety and health hazards are listed by category in Table 28.1.

Management is required to maintain records and conduct employee training with regard to general and specific hazards and safe work practices in the workplace. Employers are required to provide safety and health services and resources, including personal protective equipment, first aid supplies, and fire protection equipment. An employer's program must be documented with clearly defined policies and procedures since OSHA regulations are in part generic and are written for a wide range of industries or businesses. The employer needs to examine these regulations and apply them to the specific needs of their facilities, operations, and processes. The employer also needs to establish a hierarchy for implementing and enforcing OSHA regulations within their company.

Table 28.1 Example Classification of Occupational Safety and Health Hazards

Chemical	Physical	Biological	Ergonomic	General Safety
Fumes[a]	Cold stress	Insects (e.g., ants, bees, scorpions, spiders)	Circadian rhythm (e.g., shift work/rest cycles)	Construction[a]
Gases[a]	Heat stress[b]	Microbes (e.g., bacteria [tuberculosis[b]], parasites, viruses [hepatitis B[a], HIV[a], others])	Fatigue (extended work hours)	Maintenance[a]
Liquids[a]	Ionizing radiation[a] (e.g., alpha-, beta-, gamma-, X-rays)	Toxic plants (e.g., poison oak)	Hand tools[a]	Electrical[a]
Mists[a]	Noise[a]	Reptiles (e.g., snakes)	Manual material handling[b] (e.g., biomechanics, lifting, pushing, pulling, carrying)	Emergencies[a]
Particulates/ dusts[a]	Nonionizing radiation[a] (e.g., lasers, radio frequencies, microwaves, ultraviolet light)	Sanitation[a] (e.g., drinking water, hygiene facilities)	Mental task overload	Environmental conditions[a]
Vapors[a]	Pressure	Small mammals (e.g., dogs, rodents, skunks)	Stress (occupational and nonoccupational)	Fires/explosions[a]
	Vibration	Potentially violent people[b]	Substance abuse	Mechanical and machinery systems[a]
	O_2 deficiency		Work station design[b] (e.g., dials, controls, signals, labeling, office [computer] workstations)	Motorized equipment[a]
			Repetitive physiological stress	Pressurized systems[b]

Note: All hazards must be addressed in employer's overall safety and health program if present at work site.

[a] Employee safety and health regulations adopted.
[b] Employee safety and health regulation in development.

UNDERSTANDING OSHA

Occupational safety and health issues have been a concern to employers for more than 2000 years. Hippocrates recognized and documented the hazards of lead to miners. Laws to enforce worker protection surfaced in the 18th and 19th centuries. The implementation of these laws coincided with the start of the industrial revolution. In the United States, the first workers' compensation law was passed in 1911 in

Wisconsin, and today all states have various forms of workers' compensation and OSHA-type regulations.

Approximately half of the states in the United States have their own federally approved OSHA programs. The remaining states rely on federal OSHA for both regulations and enforcement. All of the OSHA-type programs consist of (1) standards generated from a standards development process, (2) enforcement activities generally conducted by OSHA enforcement officers, and (3) an appeals process. Employers may appeal any or part of an OSHA enforcement action that results in a penalty or citation.

OSHA INSPECTIONS AND THE EMPLOYER'S RESPONSE

Inspections generally are the result of (1) imminent danger of serious violation, (2) catastrophes and fatal accidents, (3) employee complaints, (4) programmed inspections (generally of high-hazard industries), and (5) reinspections.

Although OSHA inspections cannot be totally avoided, they certainly can be minimized by having a strong safety and health program. It is worth noting that the number one cause of an OSHA inspection is an employee complaint. While it is illegal to prevent an employee from complaining to OSHA, minimizing these complaints can be achieved by implementing an internal mechanism whereby employees can share their complaints and grievances without fear of employer reprisals. Once a complaint is received, the employer should investigate the situation and take the necessary control and other follow-up actions. Above all, take action and let the employee(s) know what has been done.

OSHA INSPECTIONS — THE EMPLOYER'S ROLE

OSHA compliance officers generally have free access to investigate and inspect any workplace during regular working hours and at other reasonable times if necessary to enforce safety and health compliance.

What should the employer expect and even require?

- The compliance officer must present appropriate credentials to the employer.
- The investigation or inspection must be carried out within reasonable limits and in a reasonable manner.
- During the course of any investigation or inspection, the OSHA compliance officer may obtain any statistics, information, or physical materials in the possession of the employer that are directly related to the purpose of the investigation or inspection.
- The OSHA compliance officer may conduct any test necessary to the investigation or inspection.
- The OSHA compliance officer may also take photographs. Any photograph taken during the course of an investigation or inspection is to be considered confidential information. Also, it is important to note that any employer information that might contain or reveal a trade secret is protected.
- Because certain requirements prevent OSHA from giving advance warning or notice of an inspection, except under specific conditions, it is important for an employer to have a plan outlining procedures for an OSHA inspection of the facility.

The employer should have a standing policy and procedure plan or manual, such as "How to Prepare for a Compliance Inspection." Employees should be informed of the elements of the policy and procedure, which should be readily available to all those who may be in the position of greeting an inspector (e.g., receptionist, supervisor, safety officer).

If the employer is suspicious of the inspector's credentials, the employer should contact the inspector's supervisor or manager for verification. Under most circumstances, it is prudent to allow the inspector entry. However, there are cases where an employer may refuse entry. Of course, the OSHA inspector may, with probable cause, request and obtain a search warrant. An employer may be able to effectively narrow the scope of an inspection by refusing entry and requiring a search warrant. An inspection warrant generally cannot permit an OSHA compliance officer to search beyond the scope created by the declarations upon which the warrant is based. Once admitted to a workplace by the employer, the OSHA compliance officer may not inspect beyond the scope of the warrant.

Some tips for handling an inspection include:

- Consent to any inspection can be given by any person with ostensible authority. The employer's plan to deal with an inspection should specify who or what position (i.e., safety director) should allow entry. Having a knowledgeable person involved at the time of the OSHA compliance officer's visit is desirable. An employer's representatives should be familiar with the processes, hazards, safe work practices, and overall safety and health programs.
- The employer's representatives need to be knowledgeable and careful in expressing the employer's position, since anything said during such discussions can be used against the employer at an appeals hearing, assuming an appeal is filed.
- The employer's representatives should keep careful and detailed notes of such meetings to document both what they say and what the OSHA compliance officer(s) say, because their comments can also be used at an appeals hearing.
- An employer may be able to avoid a citation altogether if the employer is able to convince the OSHA compliance officer that no violation, and no hazard, exists at the time of the inspection.
- At the closing conference, required by law, an employer may be able to convince the OSHA compliance officer that the proposed citation is erroneous and should be dropped or modified.
- An employer can request an informal conference to discuss the matter with other OSHA management personnel (e.g., the district manager).

Practical Considerations of an OSHA Inspection

- In deciding whether to require an inspection warrant, employers should consider the impact on the company's relationship with OSHA. Apart from easing the inspection and penalty process, maintaining a good relationship with OSHA is beneficial from the standpoint of the wealth of information resources available from the agency.
- If the employer is not prepared to grant the OSHA inspector access to its facility, and it is expected that the OSHA compliance officer may request a warrant, the employer may notify the appropriate court and agency officials issuing the warrant. The presiding judges of the courts in which the employer is located should be

advised of the employer's desire to be present in the event that the OSHA compliance officer applies for an inspection warrant. In most cases, an employer does have the right to participate in the warrant issuing process and may have the opportunity to have a warrant declared invalid.

- The employer should designate an individual to act as the key person for OSHA inspections (e.g., a safety director). Upon arrival, the OSHA compliance officer should be referred to the key person. If that key person is unavailable, the compliance officer should be informed that the company program requires the presence of that person and that they are currently unavailable. Also, the inspector should be asked to wait until the key person is available. Generally speaking, OSHA compliance officers will wait for a reasonable period of time.
- The key person should understand the nature of an inspection and the rights of both OSHA and the employer. If the key person is unsure of the employer's rights, he or she should ask the compliance officer.
- Nearly all OSHA inspections will entail an evaluation of the employer's compliance with occupational safety and health standards that apply to all employers. These standards, to which all employers must comply, include:
  - An emergency action plan to address emergency response capabilities and procedures to address anticipated emergencies; types of emergencies would include those that are reasonably foreseeable (e.g., chemical spills, power failures, adverse weather conditions)
  - A fire prevention plan that addresses fire hazards and proper procedures for responding to fire emergencies
  - An employee exposure and medical records plan that provides for the access and maintenance of such records
  - A hazard communication plan to address the labeling of hazardous substances in the workplace, employee training, and access to material safety data sheets
  - Compliance with general requirements for workplaces and work surfaces
  - An injury and illness prevention plan — required in some states, a written plan to identify, evaluate, and control occupational safety and health hazards, including employee training, accident investigation, identified responsible person(s), and record keeping
- During the inspection, the employer's representative should be present and take notes throughout. If the OSHA inspector takes a photograph, the employer's representative should be able to identify any trade secret information and so inform the OSHA representative. Details of tests (e.g., air monitoring) or samples collected by an OSHA inspector should be noted. Also, the employer may want to collect their own companion samples or tests.
- Inspection of any records other than those specifically described in a warrant (assuming a warrant was used to gain entry), or beyond those records which employers are required to maintain under the OSH Act (such as the log and summary of occupational injuries and illnesses, the supplemental record, and the annual summary), should not be permitted.
- If possible, to minimize disruption, try to schedule the inspection outside of working hours or during a lunch break. Generally speaking, if these concerns are made known to an OSHA compliance officer, he or she will be open to rescheduling the inspection.

Because it is the objective of the OSHA compliance officer to evaluate the employer's safety and health program, it is important for the employer's representative to share with the inspector all that the employer is doing to promote safety and health — in a sense, to sell their program.

Finally, all permits, licenses, and previous reports from compliance agencies should be maintained on-site or be readily available.

CITATIONS AND PENALTIES

Following an inspection, the OSHA compliance officer and his or her management will make a determination as to what citations, if any, will be issued and what penalties, if any, will be imposed. If a citation is issued, the employer will receive it by certified mail. The employer also has the obligation to post a copy of each citation at or near the place where the violation occurred for three days or until the violation is abated, whichever is longer.

Although the types of citations and penalties may differ between state plan programs and the federal OSHA program, they both follow the same pattern and include the categories Other Than Serious Violations, Serious Violations, Willful Violations, and Repeat Violations. All types may carry financial penalties. The willful and repeat violations, because of their nature, are assessed at a much higher rate.

In addition to financial penalties, criminal actions may be considered and taken against employers and employer representatives for violations of OSHA regulations. Generally, these actions are taken against employers that are intentionally negligent in carrying out their responsibilities of providing a safe and healthy workplaces for employees. Criminal prosecution is becoming a more widely used tool by local law enforcement agencies to hold employers accountable.

APPEALING A CITATION

When issued a citation or notice of a proposed penalty, the employer has the right to appeal. The employer may appeal either the citation itself, the violation type, or the penalty.

Why appeal a citation? Several reasons must be considered when appealing a citation. However, before an appeal is submitted, the employer should request an informal meeting with OSHA management responsible for the citation. As more and more employers are appealing citations, and in order to avoid legal disputes and prolonging the abatement and legal process, OSHA management is oftentimes motivated to settlement agreements that revise citations and penalties.

Should you appeal? That depends. The employer may consider the following:

The possibility of repeat citations — Having a repeat citation down the road for the same condition is possible. As noted earlier, a repeat citation carries with it much greater penalties.
Other legal ramifications — There may be legal ramifications other than OSHA to consider. These include the potential impact of civil litigation, workers' compensation, and criminal prosecution.

The potential for setting a precedent — For example, an employer does not want to become an easy target for OSHA compliance activity by acquiescing to every citation that is issued regardless of its merits.

Cost of abatement — The cost of abatement may be more than the employer can afford at the moment. Having to abate the violation may outweigh the cost and hassles of the appeal.

Cost of appeal — Depending on the type of citation and whether an attorney is put on your appeal team, appeal costs could be substantial.

Should the employer be represented by an attorney in the appeal process? Generally yes, because OSHA compliance representatives have usually handled dozens or hundreds of appeal cases, and without the representation of an experienced OSHA attorney the employer would be at a serious disadvantage.

Potential defenses to a citation that an attorney familiar with OSHA law might use are:

- Evidence does not support a violation
- Evidence does not support the characterization of the citation
- Wrong safety order was cited
- Independent employer action occurred

If the employer decides to contest either the citation, the time set for abatement, or the proposed penalty, they have 15 working days from the time the citation and proposed penalty are received in which to notify the OSHA area office in writing. An orally expressed disagreement will not suffice. This written notification is called a Notice to Contest, and forms can be obtained by calling the local OSHA area office.

Once received, the OSHA area director will forward the case to the Occupational Safety and Health Review Commission (OSHRC) for action. The OSHRC will assign the case to an administrative law judge for review. More often that not a hearing is scheduled, in which the employer and even the employees have the right to participate. The OSHRC does not require that the parties be represented by an attorney; however, depending upon the case, because it is a legal hearing, an attorney familiar with the process can be instrumental in defense of the employer and the employer's rights.

States with their own occupational safety and health programs have a state system for review and appeal of citations, penalties, and abatement periods. The procedures are generally similar to federal OSHA's, but cases are heard by a state review board or equivalent.

RECORD KEEPING AND REPORTING

Safety and health records must be maintained since they are required in some cases and can be very meaningful over the long term. These records may serve as compliance and may be admissible in legal proceedings. Also, records are essential for an effective safety and health program. They aid in identifying and evaluating problem areas and accident causes and assist management in measuring overall safety and health performance. Certain records are required by law (i.e., Material Safety Data Sheets, OSHA log of occupational injuries and illnesses, employer's accident records) and are retained for specified periods. It is advisable to keep these required records indefinitely.

Most OSHA-related records are available to the OSHA inspector during an inspection. The types of records that the employer must generate and maintain are many. What records are kept is in some cases dependent upon the size of the company and its type.

What types of records are generally required to be developed and maintained by the employer to satisfy the OSHA compliance?

- Injury and illness records
- Log of occupational injuries and illnesses
- Reporting a work-related death or serious injury
- Medical and exposure records
- Safety and health inspection checklists

Injury and Illness Records

Injury and illness records differ by state, and the OSHA office with jurisdiction should be contacted for identification of the necessary records and their requirements. In general, injury and illness records include:

- Employee's claim for workers' compensation benefits
- Employer's report of occupational injury or illness
- Doctor's first report of occupational injury or illness

Log of Occupational Injuries and Illnesses

With limited exceptions, every employer must complete the Log and Summary of Occupational Injuries and Illnesses, also known as the OSHA log. When an employee in the course of his or her employment suffers an injury or illness requiring medical treatment (generally other than basic first aid), that injury or illness must be recorded on the log. Information to be included on the log includes the employee's name, occupation, department, description of the injury or illness, and the number of days of restricted work activity or lost workdays. The log is kept for a calendar year and posted from February 1 to April 30 in the year following the one for which the log was maintained.

With limited exceptions, the employer maintains this log, even if no recordable incidents have occurred. In the event of an OSHA inspection, odds are the log will be requested for inspection by the OSHA compliance officer.

The logs must be maintained for 5 years in most cases (it is recommended that the logs be indefinitely maintained as they serve as a good source of record). In some cases, the employer may be exempt from keeping the OSHA Log. The employer should check with OSHA to determine whether it meets the small employer exception or other exception by industry type (certain industries are exempt based upon their standard industrial classification).

Reporting a Death or Serious Injury

With limited differences, OSHA regulations require an employer to submit a report to the nearest OSHA compliance office if an employee is seriously injured on the job or in connection with the job, suffers a serious job-related illness, or dies on the job or in connection with it. The report must be made as soon as practically possible, but not longer than 8 hours (federal OSHA) after the employer knows or otherwise becomes aware of the death, serious injury, or illness.

Although differences in reporting requirements exist between certain state plans and federal OSHA, the employer should check with the local OSHA compliance office and identify the applicable requirements. Incorporate those requirements into your program and make sure that employees who have safety and health program responsibilities are informed of the reporting requirements.

Medical and Exposure Records

Records that the employer has acquired for employees pertaining to medical conditions (e.g., physician and employment questionnaires or histories, laboratory results, medical opinions, diagnoses, treatments, etc.) and exposure records (e.g., air monitoring data) must be:

- Maintained by employer
- Preserved by the employer
- Made available to employees and, in some cases, to OSHA

They can also be a wonderful source of helpful safety and health information for the employer.

OSHA AND THE INTERNET

Using the Internet as a tool to allow us to accomplish safety and health objectives, makes our jobs as safety and health professionals and managers more efficient. We can gather a tremendous amount of information to help us make responsible, informed decisions. Also, the Internet can be used to train employees who would otherwise not receive the type of training they deserve and we are required as employers to provide. Using the internet can make safety administration and training easier by allowing us to focus our attentions where we can be most effective and actively involved with the recognition, evaluation, and control of workplace safety and health hazards.

OSHA has provided us with an excellent Web page with many resources to assist employers. By accessing OSHA's Web page at www.osha.gov, as well as state-equivalent Web sites, the safety and health professional may find: OSHA statistics; publications about OSHA; safety and health hazards recognition, evaluation and control techniques; assistance with topics such as ergonomics, process safety management, safety inspections, and needle sticks in health care; OSHA hot topics; conferences and meetings; technical links; record-keeping information; the regulations and standards; posters and outreach information; and much more.

CONCLUSION

Compliance with OSHA regulations has advantages far exceeding simply the avoidance of OSHA citations and penalties. Also, developing a positive relationship with OSHA and its representatives can bring rewards to the employer. In the experience of this author, putting a little energy and effort into a safety program can oftentimes bring the employer significant rewards in the form of fewer injuries and illnesses, fewer workers' compensation claims, improved production and morale, fewer regulatory and legal hassles, and, in some cases, more clients and an overall improved product and image.

FURTHER READING

Books

National Safety Council (NSC). *Accident Prevention Manual for Business and Industry: Administration and Programs.* Itasca, IL: NSC, 1992.
National Safety Council. *Supervisors Safety Manual.* 8th ed., Itasca, IL: NSC, 1993.
Blosser, F., *Primer on Occupational Safety and Health*, Bureau of National Affairs, Washington, D.C., 1992.

Other Publications

Dufour, J. *The Cal/OSHA Handbook.* California Chamber of Commerce, 2000.
Occupational Safety and Health Administrtion, *All About OSHA*, Washington, D.C.: U.S. Department of Labor, OSHA 2056, 2000.

PART VI

Legal Aspects

## CHAPTER 29

An Overview of Regulatory Compliance and Legal Liability Issues

J. Michelle Hickey and Melissa M. Allain

CONTENTS

1-56670-370-0/02/$0.00+$1.50
© 2002 by CRC Press LLC

INTRODUCTION

This chapter provides an overview of some of the key laws, regulations, and policies governing two major areas:

- Occupational health and safety in the workplace
- Environment and public health and safety[1]

Regulation of occupational safety and health as well as environmental protection expanded rapidly over the last thirty years. The first generation of environmental compliance, beginning with the National Environmental Protection Act of 1969 (NEPA), focused on the environmental impacts of proposed development projects on federal lands or involving federal funds,[2] and on implementing stricter controls on discharges to air and water. The Council on Environmental Quality was established to administer NEPA and, in addition, the U.S. Environmental Protection Agency (EPA) was established to administer federal water and air laws and regulations. Existing laws were amended to set stricter limitations on discharges and emissions, to expand the number of regulated substances, and to mandate the use of best pollution-reducing equipment, practices, and technologies.

During the same period of time, Congress passed the Occupational Health and Safety (OSH) Act of 1970,[3] discussed in detail in the following sections in this chapter. The Occupational Safety and Health Administration (OSHA) was created within the U.S. Department of Labor. Similar to the environmental programs, OSHA over the last three decades has developed numerous new standards designed to protect workers. Penalties later were increased, including criminal sanctions, particularly for willful, egregious, or repeated misconduct.

As state programs developed during this period, OSHA and EPA delegated their authority to administer their programs to those states with programs that were at least as strict as their federal counterparts. Over time, new deterrents were developed, including stiff fines of $25,000 or more per day, with each day of a continuing violation potentially a separate offense. Criminal prosecutions increased dramatically at federal, state, and local levels, implicating the local plant manager as well as the president and the business entity. In addition to hefty fines, businesses faced probation, restrictions on operations, mandatory contributions to environmental projects, and continued reporting to agencies. At the same time, officers and managers were subject to arrest and incarceration.

Beginning in 1980, both federal and state legislatures enacted broad programs to clean up abandoned industrial sites and landfills, imposing strict liability on innocent property owners as well as the businesses whose operations contaminated surface or ground waters, or the soil. Superfund site became a dreaded harbinger of long-term — some would argue perpetual — liability. To prevent future Superfund sites, solid waste laws were expanded to regulate hazardous waste from cradle to grave (see the discussion on the Comprehensive Environmental Response, Compensation, and Liability Act (CERCLA) and Hazardous Waste Storage, Handling, and Disposal later in this chapter).

In the late 1980s, government turned its attention to emergency planning for major spills or releases of hazardous substances in order to prevent or reduce widespread injuries and fatalities such as those that occurred from toxic releases at the Union Carbide plant in Bhopal, India.[4] Under these right-to-know laws, businesses were required to develop written emergency plans identifying the hazardous materials at their facilities and the procedures for handling an emergency. In addition, they became obligated to report spills and other unauthorized releases immediately, to keep records of their emissions, and to provide annual reports to federal, state and local agencies (see the discussion on Right-to-Know Laws later in this chapter).

A second generation of environmental compliance has emerged during the last decade, with a growing number of regulatory incentives for businesses to develop and implement voluntary compliance programs that go beyond the legal requirements. In addition, the regulatory agencies have implemented incentive programs encouraging businesses to find new ways to reduce or eliminate hazardous emissions, discharges, and waste (so-called source reduction). This trend toward self-policing encourages businesses to develop effective compliance programs and conduct periodic audits of their operations (see the discussion on legal implications of Compliance Management and Audits later in this chapter).

The OSH and environmental programs overlap to a certain extent on the regulation of hazardous materials, emergency planning, and certain safety practices. In addition, the U.S. Department of Transportation also regulates the transport of hazardous materials. Generally, OSH programs focus on workplace and employee safety, while the environmental laws address adverse effects to public health and the environment, including water and air.

The increased regulation of the workplace and the environment creates complex legal problems and risks for employers, their managers, and their employees, and requires substantial compliance planning and management. Although space will not permit an in-depth survey, the resources in the footnotes at the end of the chapter provide additional guidance on obtaining further information.

LABOR CODE

The following sections provide a summary of three major areas relevant to safety managers:

- Occupational safety and health
- Workers' compensation
- Substance abuse

Occupational Safety and Health Requirements

The federal Occupational Safety and Health (OSH) Act of 1970[5] was designed to reduce illnesses and injuries caused by unsafe working conditions in private industry. The OSH Act excludes the public sector, mining industry, and agriculture, although federal agencies are required under the Act to operate programs consistent with OSHA standards, and a separate program applies to mining.[6]

OSHA imposes two categories of duties on employers:

- Duty to comply with specific safety standards promulgated by the federal and state agencies[7]
- General duty to provide workers with a safe and healthful working environment, free from recognized hazards likely to cause serious harm (General Duty Clause)[8]

Each state is authorized to enact occupational safety and health legislation so long as it is at least as protective as federal law. Over twenty states and territories[9] have adopted OSHA plans to date. California's OSHA plan exemplifies a more stringent program than the federal government's, particularly as to criminal enforcement (see the discussion on Criminal Penalties later in this chapter).[10] Occupational Safety and Health agencies may initiate enforcement actions on their own initiatives in response to notice of an accident, worker complaints,[11] or through a routine inspection.[12]

Specific Safety Standards

OSHA has promulgated numerous regulations establishing specific safety standards for a wide array of potential workplace hazards, from airborne grain dust to ladders and from trenches to noise. The agency may issue standards generally applicable to all industries, the so-called General Industry Standards, or standards for specific industries, such as construction or health care. In addition to standards, OSHA has issued less formal guidelines. State agencies have followed suit, issuing similar specific safety standards and guidelines. We will focus on one set of safety standards — those dealing with toxic substances in the workplace, as a general example of specific safety standards. OSHA's guidelines on preventing workplace violence will be discussed briefly as well.

OSHA has established two types of specific safety standards for toxic and hazardous substances. Some standards set limits on a worker's exposure to a toxic substance in a specified period of time. Other standards, in addition to setting such limits, require the employer to implement an extensive protective program including medical examinations, training, and record keeping. OSHA generally prefers that employers use reasonable engineering controls to reduce exposure below the prescribed standard.[13] If engineering controls do not sufficiently reduce exposure, personal protection equipment and other protective measures must be implemented.

OSHA has promulgated more than 25 specific comprehensive standards for recognized hazardous substances, which include arsenic, asbestos, benzene, formaldehyde, lead, and vinyl chloride.[14] OSHA has also set exposure limits for more than 375 additional hazardous substances, including acetone, acetylsalicylic acid (aspirin), aluminum, carbon dioxide, chlorine, chromium, copper, cotton dust, ethanol, nicotine, silica, starch, wood dust, and zinc.[15]

Example of General Industry Standard — Asbestos

The safe handling of asbestos-containing materials (ACM) remains a challenge for safety managers. While historical industrial uses such as shipbuilding and the concomitant exposures have abated, aging buildings with ACM pose an increasing risk of exposure to the occupants and workers who may perform repairs and maintenance. In addition, in certain industries such as automotive repair, construction and asbestos mining and manufacturing, workers continue to face exposure to asbestos.

A naturally occurring mineral fiber, asbestos was used for most of the 20th century as a fire retardant in over 3600 building materials, including roofing, floor and ceiling tiles, pipe insulation, wallboard, grout, mastic, and cement. Asbestos was phased out of most building materials beginning in 1980;[16] however, certain ACMs remain in use.[17] Asbestos fibers can be released into the air when ACM is damaged, for example, by sanding or breaking up floor tiles, or when it deteriorates to the point that the asbestos becomes friable, that is, the material can be pulverized, crumbled, or turned into powder when dry with hand pressure. Over time, the fibers may accumulate in the lungs in significant concentrations and cause lung cancer, particularly mesothelioma, a specific type of cancer caused exclusively by exposure to asbsestos, and asbestosis, scarring of the lung tissue.[18]

Employers risk violating the OSHA standard for asbestos by not checking records for ACM before commencing renovations or routine installations such as cable or phone lines that may disturb ACM.[19] Employers and owners of buildings constructed no later than 1980, are required to treat thermal insulation, asphalt and vinyl flooring, and troweled-on surfacing materials as presumed ACM.[20] Building and facility owners must determine the presence, location, and quantity of ACM in the building, and notify their employees as well as employers (contractors) who will perform housekeeping activities in the building.[21] A written management plan is required to manage friable or presumed ACM.[22] Employers must also comply with OSHA's hazard communication standard for asbestos, and post warnings if ACM is present.[23]

In addition to building materials, asbestos is still used in brake and clutch parts for automobiles.[24] Employers in auto repair shops where brakes and clutches are repaired or serviced must institute engineering controls and work practices to reduce employee exposure to asbestos by using a specific vacuum system or low pressure/wet cleaning method.[25] Alternative methods may be used if fewer than five pairs of brakes or clutches are inspected, disassembled, repaired or assembled on a weekly basis.[26] Similar to other standards, OSHA has established a permissible exposure level for asbestos; however, safe work practices are required for asbestos regardless of the exposure levels.[27]

Independent of OSHA, EPA regulates ACM under several statutory programs, which involve airborne and other pathways of exposure, for example, as a water pollutant.[28] Moreover, EPA may pursue criminal sanctions for the improper handling of ACM.[29]

Businesses face liability not only from the regulatory agencies, but also from lawsuits brought by employees and other private parties.[30] In sum, it is critical for a safety manager unfamiliar with this complex intertwining of several programs regulating asbestos to consult with a qualified asbestos management expert and determine whether a maintenance program is needed, and whether any precautions are needed before engaging in any demolition or renovation.

Record Keeping

Many of OSHA's safety standards impose record keeping requirements on employers. An employer who is required to monitor employee exposure to asbestos, for example, must keep the related records for thirty years, and notify OSHA before going out of business to permit the agency to retain the records.[31] In addition, OSHA has promulgated general record keeping rules governing all employers, except those who operate within designated standard industrial classifications.[32] These rules require, among other things, that the employer:

- Maintain a log of all workplace injuries, including type of injury and number of days lost or restricted
- Post the log annually
- Retain and make the log available for employee and OSHA inspection for 5 years
- Promptly report certain types of injuries, including those involving fatalities, to the local OSHA office
- Maintain medical and exposure records and retain them for 30 years from the time the employee leaves the workplace[33]

An employer's failure to maintain records properly is a separate statutory violation.[34]

Hazardous Materials

In 1983, OSHA promulgated the Hazard Communication Standard[35] requiring employers to give specific information and training to employees who handle chemicals considered to be hazardous (e.g., combustible, explosive, flammable, unstable, or water-reactive). The keys to compliance include:

- Maintaining, and making available on employee request, complete and updated Material Safety Data Sheets (MSDS)
- Developing a written hazardous materials communication program and list of hazardous chemicals used in the workplace
- Training all workers on the safe handling of the materials at issue

The manufacturer or distributor of a regulated product must provide the purchaser with a copy of the MSDS.[36]

In conducting inspections throughout the country, the lack of training and lack of a written communications program are among the top ten violations OSHA has identified every year from 1996 to 2001.[37] For businesses with high employee turnover, training can be challenging. Under the applicable OSHA regulations, training is required:

- Before an employee is exposed (essentially, within 30 days of hiring)
- Whenever a new product is introduced into the workplace
- As a refresher, usually on an annual basis

Many businesses rely on professional trainers, videos or computer-based training; however, these programs should be tailored to the particular business.

Employers who handle hazardous materials regulated under OSHA should also be alert to emergency response procedures. Under certain OSHA standards, employers are required to develop emergency response plans, site characterization and analysis, site control, training, medical surveillance, engineering controls and personal protection monitoring, information programs, and decontamination procedures. In addition, facilities using certain listed highly hazardous toxic and reactive materials may be required to meet additional standards, such as process safety management, designed to minimize the risk of explosions and other catastrophic events.[38]

Inspections

OSHA officials may inspect a workplace either as part of a routine administrative program of inspection or in response to a reported accident or to an employee complaint.[39] Employers may find themselves inadvertently triggering an inspection by the notice they are obligated by law to give OSHA whenever there is an accident resulting in a fatality or the hospitalization of five or more workers.[40]

Ordinarily, OSHA will arrive to conduct an inspection without advance notice.[41] If the OSHA inspector wants an employer representative to be present or wants to conduct the inspection after business hours, it may give advance notice, but in no event longer than 24 hours. Once on the premises, OSHA inspectors may test equipment and interview employees to document plainly observable violations. If an employer initially resists an inspection, OSHA is authorized to obtain search warrants and to issue administrative subpoenas to compel interviews with employees. Although employers generally cannot prevent an audit and are unwise to try, usually the employer is permitted to and should accompany the OSHA inspector during the inspection; an employee designee may do so as well.[42] Prudent employers ensure that whoever represents them during an inspection is knowledgeable about OSHA rules and regulations, and the company's policies and operations.[43]

Whether to conduct an internal audit, inherently a complex endeavor, was further complicated by one court ruling in 1992 that OSHA may subpoena an employer's internal self-audit of its own compliance with OSHA standards.[44] In July 2000, OSHA published a self-audit policy, under which the agency will not routinely seek the production of an audit report upon commencing an investigation or to identify potential hazards (see Protecting Self-Evaluations and Audits from Disclosure, this chapter). [45]

General Duty

In addition to the specific safety standards described above, both federal and state occupational safety and health agencies impose upon employers a "general" duty to each employee to provide a workplace free from recognized hazards that are likely to cause death or serious injury.[46] OSHA has not provided a precise definition of an employer's obligations under what has come to be known as the "General Duty Clause," nor has the agency imposed any minimum requirements, except on an after-the-fact case-by-case review.

Typically, the court or other trier of fact will examine whether:

- A recognized hazard existed
- The employer knew of the potential danger or should have known because it was generally known in the particular industry
- Feasible measures were available to abate the hazard
- The employer in fact failed to abate the hazard[47]

Allowing equipment to be operated without safety devices and failing to provide a meaningful safety program, for example, have been found to be violations of the General Duty Clause.[48]

In the next section, we discuss briefly the Agency's longstanding efforts to develop a specific standard on ergonomics.

Basis for Recent Regulation — Ergonomics

In November 2000, OSHA issued the final version of a controversial general industry standard on ergonomics.[49] The standard was designed to improve workplace design and reduce the incidence of musculoskeletal disorders, such as tendinitis, back injuries, and carpal tunnel syndrome. These disorders can be caused by, among other activities, performing similar repetitive motions over time, such as factory assembly work or typing, or by exerting too much pressure on the same part of the body.[50] The ergonomics standard would have required employers to begin applying the standard to their businesses by October 2001.[51] In March 2001, however, in the early days of the Bush Administration, Congress voted against the standard, utilizing a law passed in the 1990s that permitted Congress, after relatively brief hearings, to repeal regulations issued by federal agencies.[52] One of the primary concerns in the regulated community was the difficulty in determining whether a physiological disorder was caused by workplace activities or personal activities.

Although implementation of the ergonomics standard was suspended, OSHA still has authority to regulate ergonomics problems under the General Duty Clause. The agency has indicated that it intends to prosecute egregious situations, and may issue a modified version of the standard.[53] In addition, some states such as California operate their own ergonomics program.[54]

Basis for New Guidelines — Workplace Violence

Employers and regulators alike are grappling with an alarming increase in workplace violence. More than 1000 workers each year since 1993 have been the victims of homicide in the workplace, far surpassing work related vehicular accidents, fires, and construction falls.[55] OSHA reports that nearly 2,000,000 assaults have occurred annually during the same period.[56] Although much of the publicity has centered on the disgruntled employee who commits murder in the workplace, over 90% of the homicides are related to robbery, for example, at a convenience or jewelry store, or in a taxicab or motel.[57]

Businesses face a growing number of lawsuits claiming negligence, for example, in hiring violent-prone individuals, in retaining such individuals after an incident has occurred, in failing to provide adequate security, and in allowing weapons in the workplace.[58] In addition to retail clerks and taxicab drivers, employees in hospitals and other medical institutions are particularly vulnerable to attacks by patients.[59] Large residential complexes have also been subject to violent confrontations between tenant and landlord or visitors.[60]

To help alleviate the problem, OSHA has published guidelines for managing workplace violence for specific industries such as healthcare and social service workers.[61] In addition, the agency will prosecute serious omissions under the General Duty Clause.[62] Similarly, Cal-OSHA published general guidelines in 1995,[63] specific guidelines for the health care industry,[64] and a model prevention program.[65] Several other state or municipal agencies have issued similar guidelines, including New Jersey, Washington, and Chicago.[66]

In contrast to formal standards, the federal and state agency guidelines usually are advisory and do not impose binding obligations on any business.[67] Nonetheless, OSHA relies on the General Duty Clause to encourage employers to develop workplace violence prevention programs. In addition to potential liability under the OSH Act, an employer could be vulnerable to lawsuits by private parties for negligence if the employer took none of the preventive measures set forth in the guidelines, and it was feasible to do so.

Businesses are increasing preemployment screening to evaluate the propensity for aggressive or violent behavior. In addition, management and human resources personnel can be trained to detect and deal at an early stage with potentially violent workers. Providing employee assistance programs, counseling, grievance procedures, and job placement counseling after termination have proven to be effective means of managing the problem.

Potential Penalties

Administrative Penalties — Civil and Criminal

Federal and state OSH agencies are authorized to issue citations based on findings of alleged violations of either specific or general compliance standards. Citations include a statement of the alleged violation and the proposed length of time to set for abatement. The employer is obligated to post a copy of the citation for 3 days or until the violation is abated, whichever is longer, at or near the place where the violations occurred.[68]

In addition, agencies are authorized to initiate civil as well as criminal penalties depending on the type of violation allegedly found. There are five possible categories of violation in the federal system:[69]

1. A general violation (a violation that has a direct relationship to job safety and health, but not likely to cause death or serious physical harm) is punishable by a fine of up to $7000 per violation per day. The fine is discretionary with the agency, and may be reduced depending on the employer's good faith, history of compliance, and size of business.

2. A serious violation (a violation where there is a substantial probability that death or serious physical harm could result and the employer knew or should have known of the hazard) is punishable by a mandatory penalty of up to $7000 per day per violation. This penalty may be reduced depending on the employer's good faith, history of compliance, gravity of alleged violation, and size of business.

3. A willful violation (an act done intentionally and knowingly) is punishable by a mandatory penalty of up to $70,000 per day per violation, with a minimum penalty of $5000. This penalty may be reduced by business size and compliance history but not good faith. A willful violation may also trigger criminal prosecution that can be punished by a criminal penalty of up to $500,000 per violation for a corporation and $250,000 per violation for an individual and or 6 months of imprisonment.

4. A repeat violation (a violation for substantially similar reasons received upon reinspection following final citation) is punishable by a penalty of up to $70,000 per day per violation.

5. Failure to correct a prior violation is punishable by a civil penalty of up to $7000 per day from the time the violation was received through the abatement date.

The administrative agency is authorized to impose civil penalties. Criminal penalties, including imprisonment, are imposed through the courts.

When the employer has failed to take preventive action and an injury occurs, the violation can be elevated to the willful category.[70] Even civil penalties can be overwhelming in particularly egregious cases or where the agency concludes that the employer has not moved swiftly or extensively enough to abate the violation. One of the largest administrative penalties, over $800,000, was assessed in 1999 against an oil refinery in the San Francisco Bay area, after four employees were killed and several others injured in a fire. The issuance of penalties for violations at the same facility two years previously was a factor in finding that the company had acted willfully in violating OSHA regulations.[71] A multiagency team, including US/OSHA and Cal/OSHA, as well as other federal, state, and local agencies participated in the investigation (see the discussion on Preparing for Inspections later in this chapter).

Appeal Process

In response to a citation, employers may request an informal meeting with OSHA's area director, who is authorized to execute settlement agreements revising citations or penalties. Employers receiving citations must correct the cited hazard within the prescribed time period, or contest the citation or the abatement date by filing a written Notice of Contest within 15 working days of receiving the citation.[72] There is no specific format for the notice, but it must clearly identify the citation, the proposed penalty, the abatement period, and the employer's basis for contesting. Employers must give a copy of the notice to the employees' authorized representative or post it in a prominent place.[73]

The area director then forwards the matter to the Occupational Safety and Health Review Commission, where an administrative law judge hears the case.[74] The Commission is a federal agency that operates independently from OSHA or the Department of Labor. The administrative law judge reviews the contest and then sets the matter for public hearing, where both employer and employees may participate. Any party may request that the full Commission further review any ruling of the administrative law judge. In turn, a party may appeal the Commission's rulings to the appropriate U.S. Court of Appeals.[75]

Criminal Penalties

Every corporate employer, as well as every individual manager responsible in any manner for health or safety issues, should be aware of the potential criminal liability for health or safety violations, specifically including individual criminal liability. First, the specter of manslaughter charges arises whenever a workplace death results from negligent or willful conduct. Where the federal agency may impose criminal sanctions at its discretion in those circumstances, Cal/OSHA must prosecute such conduct as a misdemeanor or felony.[76] Penalties under the Cal/OSHA program can be ten times as much as those under federal law.

Second, in California, a business that is subject to an inspection should not assume that a Cal-OSHA inspection is merely civil or administrative. The courts have found that such inspections are inherently criminal in nature.[77] Consequently, the manager should try to clarify the inspector's purpose and intent before the inspection begins to avoid any inadvertent waiver of applicable rights or privileges, such as an individual's privilege against self-incrimination.

Workers' Compensation

Both federal and state worker's compensation laws have been enacted.[78] Although these laws may vary, they generally conform to the following scheme. Under workers' compensation systems fault is irrelevant, except in the most extreme circumstances.[79] The employee generally receives a relatively quick, although limited, award for injuries arising from employment.[80] In exchange, employees and their families surrender their common law right to sue employers and others for damages.[81] On the other hand, employers are liable for a greater number of awards under a workers' compensation system, but the amount of such awards are less than what might be obtained by the employee through litigation.[82]

Workers' compensation can be costly; however, there are ways to manage those costs. For example, workers' compensation insurance is available,[83] and some insurance companies provide loss control services to their insureds to reduce the rate of injuries and the associated costs.[84] In addition, some states have established programs to reduce injuries in certain hazardous industries.[85] Such programs and the safety self-audits discussed later in this chapter can help to reduce overall safety risks and the injuries and costs associated with such risks.

Substance Abuse: Drugs and Alcohol

It is undeniable that employee drug and or alcohol use substantially increases the risks of both accidents and disease, with concomitant increases in employer's costs. Some employers have responded by instituting drug and or alcohol testing programs. Certain features of those programs, however, may violate the employee's constitutional privacy rights or illegally discriminate against the employee. The following is a brief discussion of some of the federal laws regarding drug and alcohol testing and the termination of the employee that abuses drugs or alcohol.

The Americans with Disabilities Act (ADA) protects physically or mentally disabled individuals from employment discrimination (see details on ADA later in this chapter).[86] An individual who is currently engaging in illegal drugs is not disabled for purposes of the ADA.[87] On the other hand, the ADA protects an individual who has successfully completed a drug rehabilitation program and is no longer engaged in the use of illegal drugs or is wrongfully accused of engaging in the use of illegal drugs.[88] Unlike the illicit drug user, the Act does not automatically exclude alcoholics from coverage under the ADA and an alcoholic may be protected under the ADA.[89] However, even if an alcoholic or recovering drug user is covered as disabled under the ADA, an employer can hold such alcoholic or recovering drug user to the same performance standards as other employees and terminate the alcoholic or recovering drug user if he or she does not meet such standards.[90] This is true even if the reason for not meeting the standards was due to alcoholism or the illegal drug use.[91] For example, an employer can fire an alcoholic who misses work or fails to show up for work timely because of drug or alcoholism.[92]

The ADA does not prohibit testing for illegal drugs or alcohol.[93] In fact, such testing may be legally required, for example, under applicable regulations of the U.S. Department of Transportation.[94] However, drug testing by an employer may violate the ADA if it fails to reasonably accommodate known physical or mental limitations of an otherwise qualified individual with a disability or if it discriminates against an individual because of that individual's disability.[95]

The information gathered from drug testing should be kept confidential, particularly information not related to the use of alcohol or illegal drugs. Releasing this information could be a violation of the employee's privacy.[96] For example, a drug test could reveal that the employee takes properly prescribed drugs for an unrelated medical condition.[97]

State laws may have different regulations regarding drug and alcohol testing; therefore, the safety professional should become familiar with any applicable state laws prior to instituting a testing program.

ENVIRONMENTAL LAWS

Federal, state, and local governmental authorities have promulgated a myriad of environmental laws and regulations, making it difficult for safety professionals to ascertain and comply with such laws and regulations. However, the civil and criminal penalties and potential legal liability for damages associated with failing to comply make it imperative that every company devote adequate resources and commitment to environmental compliance.

This section will summarize the following federal environmental laws and regulations: (1) the Clean Water Act and the Safe Drinking Water Act's regulation of discharges of waste to surface water and groundwater, (2) the Clean Air Act's regulation of discharges of pollutants to the air, (3) the Resource Conservation and Recovery Act's regulation of the handling, treatment, storage and disposal of hazardous waste, (4) the Environmental Protection and Community Right-to-Know Act's emergency planning and reporting requirements, and (5) the Comprehensive Environmental Response, Cleanup and Liability Act's statutory scheme and common law tort causes of action to recover cleanup costs and other damages caused by environmental contamination. The last section of this chapter will discuss the issues surrounding environmental, as well as worker safety, compliance self-audits, and government inspections.

Unfortunately, the scope of this chapter will not allow us to examine all of the other numerous federal,[98] state, and local environmental laws that may apply to a company. States and local authorities promulgate and enforce the great majority of the environmental laws and regulations that affect day-to-day management of companies. This is in part due to the fact that many federal environmental statutes allow states to enforce their own laws and regulations in lieu of such federal environmental laws and regulations, as long as the state's program is as strict as the federal program and has been approved. Accordingly, the following discussion only provides an introduction to some of the major federal environmental laws which are either enforced in your jurisdiction or which form the basis of the environmental laws enforced in your jurisdiction. Safety and environmental officers and managers must ascertain and comply with the specific environmental laws and governmental authorities that regulate their company.

Regulation of Discharges of Waste to Water

Because of the devastating environmental effects of allowing untreated waste, including hazardous waste, to be discharged into waters, the federal government has enacted laws and regulations designed to reduce and ultimately eliminate such discharges.

National Pollutant Discharge Elimination System Permits (NPDES)

The Clean Water Act (CWA), as well as similar state laws, prohibits anybody from discharging pollutants through a point source into a water of the United States unless they have a NPDES permit.[99] The term pollutants has been broadly defined under the CWA and includes dredged soil, garbage, rock, sand, and industrial, municipal, and agricultural waste.[100] The term point source means "any discernible, confined and discrete conveyance, including but not limited to any pipe, ditch, channel, tunnel, conduit, well, discrete fissure, container, rolling stock, concentrated animal feeding operation, or vessel or other floating craft, from which pollutants are or may be discharged" except agricultural stormwater discharges and return flows from irrigated agriculture.[101] The term waters of the United States has also been broadly defined under the CWA and includes navigable waters, interstate and intrastate waters, tributaries to such waters, wetlands, and the territorial oceans of the United States.[102]

NPDES permits (general and individual) are issued by the EPA or by states that have been delegated that authority by EPA.[103,104] General permits are issued for multiple facilities within a specific category while an individual permit is tailored to a particular facility.[105] It may be more economical and expedient for a company to obtain a general permit.

A NPDES permit generally provides effluent limits for discharges of pollutants, monitoring and reporting requirements and standard and special conditions.[106] These conditions control all aspects of discharges and, thus, directly affect the operation of the source business. Permits are subject to public review and comment prior to being issued and must be periodically renewed.[107]

The EPA is in the process of implementing a program that would require NPDES permits for storm water discharges as mandated in the 1987 amendments to the CWA.[108] The EPA has established a two-phased approach to addressing storm water discharges.[109] Phase I has been implemented and requires permits for storm water discharge linked with industrial activity, including construction activity disturbing at least five acres of land, and municipal separate storm water systems serving large and medium sized communities with over 100,000 inhabitants.[110] Under Phase II, the EPA will require permits for construction activity disturbing between one and five acres of land and municipal separate storm water systems for small communities.[111] Permit applications for individuals and entities of Phase II regulated construction activity and municipal separate storm water systems for small communities will not be required until early 2003 unless an earlier date is set by a state permitting authority.[112]

Pretreatment Regulations and Discharges to Publicly Operated Treatment Works (POTWs)

Many plants discharge to publicly operated treatment works (POTWs) through public sanitary sewers. A plant or other entity or person discharging into a sewer of a POTW generally does not need a NPDES permit but must comply with pretreatment standards and other regulations. Pursuant to the CWA, the EPA, states and local authorities set limits for the discharge of pollutants from nondomestic, such as industrial, sources to POTWs.[113] The goal of the pretreatment regulations is to protect the operations of the POTW and the environment from pollutants that pass through the POTW.[114]

The two sets of pretreatment standards are categorical pretreatment and prohibited discharge.[115] The categorical pretreatment standards are technology-based limitations on pollutant discharges to POTWs that apply to specified process wastewaters of particular industrial categories.[116] Prohibited discharge standards prohibit the discharge of wastes that pass through or interfere with the operations of the POTW.[117] Such prohibited wastewaters include wastes that are ignitable, explosive, corrosive, obstructive, or interfere with the operations of the POTW.[118]

In order to encourage compliance, the CWA imposes severe civil and criminal penalties for failing to comply with the provisions of the act, including the NPDES and pretreatment regulations. For example, a person who negligently violates any permit condition or pretreatment standard shall be punished by a fine of up to $25,000 per day of violation or by imprisonment for up to one year.[119] Moreover, a person, who does these same acts even though he or she knows that another person may be placed in imminent danger of death or serious bodily injury, is subject to a fine of up to $250,000 and or imprisonment of up to 15 years, while an entity could face a fine of up to $1,000,000 for such knowing endangerment.[120] A person who knowingly makes false material statements on a permit application or other document is subject to a fine up to $10,000 and or by imprisonment for up to two years for the first offense and subject to a fine up to $20,000 and or by imprisonment for up to four years for any subsequent offenses.[121] Citizens can also enforce the CWA through citizen suits.[122]

Regulation of Subsurface Discharges to Groundwater

The federal government primarily regulates subsurface discharges that affect groundwater pursuant to the Safe Drinking Water Act (SDWA).[123] Under the authority granted it by the SDWA, the EPA has implemented the Underground Injection Control (UIC) program to regulate subsurface discharges to prevent underground injection that endangers drinking water sources.[124] While some activities such as petroleum exploration are exempt from UIC regulation,[125] disposal of effluent via deep wells is generally subject to UIC permits.[126]

As with many other federal environmental programs, a state may operate its own UIC programs in lieu of the federal program, as long as the state program meets the minimum requirements set forth under the SWDA and the EPA has approved such state program.[127] Moreover, state or local authorities may have stricter regulations including prohibitions on subsurface discharges. Accordingly, safety and environmental officers and managers should check with state and local agencies to determine local regulatory schemes.

Regulation of Emissions to Air

Under the Clean Air Act (CAA), the EPA sets limits on the amount of certain pollutants that can be present in the air in the United States.[128] The CAA sets limits for the common or criteria air pollutants, commonly known as National Ambient Air Quality Standards,[129] and limits for hazardous air pollutants (HAPs).[130] The common or criteria air pollutants include ozone, Nitrogen Dioxide (NOX), carbon monoxide (CO), particulate matter (PM-10), sulfur dioxide (SO2), and lead.[131] HAPs are chemicals that cause serious health and environmental effects, such as cancer, birth defects, and death.[132]

The CAA requires states to develop state implementation plans (SIPs) that explain how each state will attain and maintain the National Ambient Air Quality Standards[133] and allows a state to submit a plan for approval to implement and enforce the regulation of HAPs.[134] If the EPA approves a state's SIP, the state can take over enforcement of the CAA for the EPA in that state.[135]

The EPA has set primary and secondary standards for criteria air pollutants.[136] The primary standards protect health while the secondary standards prevent environmental and property damage.[137] An area that meets the primary standards is an attainment area and an area that does not is called a non-attainment area.[138] Since the CAA mandates that the states attain the primary standards,[139] the regulation of the discharge of a pollutant in a non-attainment area may be more stringent than in areas of attainment.

The CAA regulates both mobile sources of pollution, such as cars, and stationary sources of pollution by setting air quality standards and issuing permits. Numerous industries have had to change the way their products are made to meet the CAA standards and regulations. For example, car manufacturers have been required to build cleaner cars and gasoline manufactures have had to reformulate gasoline to be cleaner burning.[140]

The CAA regulates sources of pollution in part through a national permit system.[141] The permits are issued to certain sources of air pollutants by the approved states or the EPA for states that have not been approved to enforce the CAA.[142] A CAA permit will, among other things, set forth the quantity of a pollutant that may be released, the steps the source owner is taking to reduce pollution and the monitoring requirements.[143]

The CAA encourages companies to reduce and eliminate air pollution. In fact, safety and environmental officers and managers can create a valuable commodity of marketable emission allowances by reducing air emissions.[144] The air emission allowances can be traded or sold to other companies or used to offset another source of emissions.[145]

Similar to other environmental statutes, the CAA is enforced through civil and criminal penalties. For example, the CAA imposes civil penalties of up to a $25,000 a day per violation and criminal penalties which in certain circumstances can include up to 15 years imprisonment for first time offenders and 30 years for repeat offenders.[146] Moreover, an organization who knowingly releases a hazardous air pollutant and who knows at the time that he thereby places another person in imminent danger of death or serious bodily injury shall, upon conviction, be punished by a fine of up to $1,000,000 for each violation.[147] Citizens may also bring citizen suits to enforce the CAA.[148]

Hazardous Waste Storage, Handling, and Disposal

The Resource, Conservation and Recovery Act (RCRA)[149] was designed to regulate solid and hazardous waste from the point of generation through storage, transportation and treatment to disposal or, in other words, from "cradle to grave." The following is a discussion of the general requirements of the federal RCRA program. However, RCRA allows a state to administer its own hazardous waste regulation program that is as strict as or stricter than RCRA in lieu of RCRA.[150] Accordingly, it is important for managers and safety officers to be familiar with any applicable state hazardous waste program's requirements.

Companies that generate hazardous waste in certain quantities must comply with RCRA's requirements for generators discussed below.[151] First, RCRA requires generators of hazardous waste over a certain quantitative limit to identify hazardous wastes.[152] A hazardous waste under RCRA is a solid waste, or combination of solid wastes, which because of its quantity, concentration or physical, chemical, or infectious characteristics may cause or significantly contribute to an increase in mortality, an increase in serious irreversible, or incapacitating, reversible illness, or pose a substantial present or potential hazard to human health or the environment when improperly treated, stored, transported, disposed of or otherwise managed.[153]

The EPA has established several lists of hazardous wastes and assigned numbers to each hazardous waste listed.[154] In addition to the listed wastes, RCRA regulates wastes that have any of the following characteristics:

- Ignitability
- Corrosivity
- Reactivity
- Toxicity[155]

Accordingly, wastes that are not listed may still be regulated by RCRA. Managers and safety officers should be aware that RCRA also regulates the following in certain circumstances:[156]

- Solid waste that has become mixed with hazardous waste
- Waste that is generated by the treatment, storage or disposal of hazardous waste
- Material that has been contaminated by a hazardous waste
- Some recycled and reclaimed hazardous waste

A generator must keep any records, such as test results, used to determine that a waste is hazardous for three years.[157] However, as discussed in the subsection on liability under the Comprehensive Environmental Response, Compensation and Liability Act (CERCLA) and common law, a company should keep important environmental records indefinitely.

Under RCRA, the EPA also encourages generators to reduce the volume and toxicity of their hazardous waste streams. In fact, the Uniform Hazardous Waste Manifest requires that the generator certify that it has a program to reduce the volume and toxicity of waste it generates to the degree economically practicable and has selected the method of treatment, storage or disposal that minimizes the threat to human health and the environment.[158]

A generator that has a hazardous waste regulated under RCRA must obtain an EPA identification number in order to transport, treat, store, or dispose of such hazardous waste.[159] In certain circumstances, generators of hazardous waste may accumulate up to 55 gallons of their own hazardous wastes near the point of generation in containers that are properly marked and maintained prior to moving such waste into storage without a RCRA permit.[160] In addition, generators may store hazardous wastes for 90 days prior to transportation without a RCRA permit provided that certain standards are met.[161] Finally, generators of small amounts of hazardous wastes may be allowed under certain circumstances to store hazardous wastes for a longer period without a RCRA permit.[162]

Generators also are responsible for manifesting, properly labeling and packaging hazardous waste for off-site transportation.[163] The generator fills out the Uniform Hazardous Waste Manifest, which is used to track the transportation, treatment, storage and disposal of the waste.[164] The generator must describe the waste, list the names and EPA identification numbers of each transporter and treatment, storage, and disposal facility that the waste will be handled by, on the manifest and certify that the waste has been properly packaged and labeled.[165] The generator must file an exemption report with the EPA if the generator does not receive a properly executed manifest in a timely manner.[166] Generally, a generator must keep a copy of the final signed manifest for three years but should keep it longer.[167]

In addition, companies that generate hazardous waste will be considered transporters if they move wastes from the site even for a short distance.[168] A transporter must obtain an EPA identification number prior to transporting any hazardous waste.[169] Among many requirements, a transporter must also keep the manifest with the hazardous waste, obtain the signature of the next transporter or operator of a permitted treatment, storage, or disposal facility prior to delivering the hazardous waste to such transporter or operator and maintain such executed manifest for three years.[170]

RCRA provides for civil penalties of up to $25,000 per day for each day of noncompliance and criminal felony penalties of up to $50,000 and 5 years in prison for knowing RCRA violations and up to $250,000 for individuals and $1,000,000 for organizations and 15 years in prison for knowing endangerment violations.[171] Knowing violations include omitting information or providing false information in reports or filings.[172] Knowing endangerment involves violations that place another person in imminent danger of death or serious bodily injury.[173]

In 1984, RCRA was amended by the Hazardous and Solid Waste Amendment Act (HSWA). Pursuant to this act, the EPA obtained for the first time the authority to compel generators of hazardous waste to remediate past releases of solid waste and hazardous waste through a RCRA Corrective Action.[174] The RCRA Corrective Action process is faster than the Superfund site remediation process.

In summary, to avoid such steep civil and criminal penalties and the other liabilities associated with improper handling and disposal of hazardous wastes, safety managers and professionals must (a) accurately determine if the wastes are hazardous, (b) safely handle hazardous wastes, and (c) properly dispose of hazardous wastes.

Community Right-To-Know Laws

The federal and many state and local governments have enacted community right-to-know laws. The federal law, Title III of the Superfund Amendments and

Reauthorization Act of 1986 (SARA), also known as the Environmental Protection and Community Right-to-Know Act (EPCRA),[175] will be discussed below. The myriad of state and local community right-to-know laws and their additional requirements are beyond the scope of this book. However, safety and environmental officers and managers must be aware of and comply with any applicable state and local community right-to-know laws.

EPCRA has four major parts with each part having different reporting requirements: emergency response planning,[176] emergency release notification,[177] community right-to-know reporting,[178] and toxic chemical release inventory reporting.[179]

Under the emergency planning section, EPCRA requires states to set up a state emergency response planning commission and emergency planning districts and local emergency planning committees in each of the districts.[180] Businesses that produce, use or store certain extremely hazardous substances (known as Section 313 chemicals) in threshold quantities[181] must notify the state emergency response planning commission and designate a facility emergency response coordinator to participate in the local emergency planning process[182] and provide information necessary for developing and implementing the emergency response plan.[183]

EPCRA requires owners or operators to immediately report a release of a reportable quantity (RQ) of CERCLA defined extremely hazardous or hazardous substances.[184] Generally, the report would be made to the local and state emergency response planning commissions, as well as the National Response Center.[185] These reporting requirements may overlap with the reporting requirements mandated under other environmental statutes, such as the CAA[186] or CWA,[187] or permits issued under such environmental statutes. Safety and environmental officers and managers should ascertain the governmental authorities to whom it is responsible for reporting releases to prior to such releases in order to insure compliance with all applicable regulations and permits. In addition, safety and environmental officers and managers should develop a plan for insuring that all reportable releases are timely reported and properly remediated. Such a plan should include training all individuals who work with hazardous substances to notify the individual(s) responsible for reporting releases to the proper governmental authorities and providing both work and home phone numbers and other emergency contact information for such responsible individual(s). In light of the civil and criminal penalties associated with the release of hazardous substances, a company should carefully chose the individual(s) that will report the release and interact with the response agencies and regulators regarding the releases and remediation.

In order to provide the community with information regarding chemicals located nearby, EPCRA requires that owners and operators submit information regarding the chemicals that they use, produce or store at their facilities in certain circumstances.[188] The required information may include material safety data sheets (MSDSs) and annual inventory, usage, and manufacture information about such chemicals.[189] Companies can obtain limited trade secret protection from these EPCRA reporting requirements in certain circumstances.[190] EPCRA also requires owners or operators to report the amounts of listed toxic chemicals their facilities release into the environment, including toxic chemicals released to the air, water, land and POTWs.[191] This information is then made available to the general public through EPA's Toxic Release Inventory.[192]

EPCRA provides for the issuance of civil and criminal penalties for failure to comply.[193] For example, EPCRA imposes a civil penalty of up to $75,000 per day of violation and criminal penalties of up to $50,000 in fines and or five years imprisonment for a violation of the emergency release notification requirements.[194] Citizens can also enforce EPCRA by bringing citizen suits.[195]

Liability under the Comprehensive Environmental Response, Compensation and Liability Act (CERCLA) and Common Law

CERCLA or superfund law was enacted in 1980 to enable the cleanup of the numerous contaminated sites located in the United States, including many sites where the parties actually responsible for the disposal or release of hazardous substances were no longer in existence or did not have the adequate financial resources to pay for the necessary remediation. To accomplish this task, CERCLA has been interpreted to impose strict[196] and joint and several[197] liability on all potentially responsible parties (PRPs). PRPs include past and present owners and operators of the contaminated site, generators who arranged for the disposal or treatment of a hazardous substance at the site, and transporters who transported a hazardous substance to the contaminated site.[198] In addition, CERCLA has been interpreted to impose liability on successor corporations,[199] parent corporations,[200] and shareholders[201] of the PRPs in certain circumstances.

Accordingly, CERCLA can hold responsible parties that did not have any direct involvement with either the disposal or release of hazardous substances at the contaminated site. For example, under CERCLA, a present owner that did not own the site at the time of the disposal of the hazardous substances and did not cause the hazardous substances to be disposed at the site may be held jointly and severally liable for all the contamination.[202] CERCLA also can hold a generator of hazardous waste lawfully transported and disposed at a disposal facility liable for all the cost to remediate such facility. For example, the EPA frequently seeks recovery of costs for the clean up of landfills, such as the Casmalia Disposal landfill, from any party that transported hazardous materials to such landfill.[203]

Many states have enacted superfund-type statutes.[204] These state superfund statutes may differ from CERCLA in their requirements and procedures but generally attempt to impose similar stringent liability on PRPs for the clean up of contaminated sites.[205]

In addition to lawsuits brought pursuant to CERCLA and the other federal and state environmental statutes, a company could be sued for the recovery of damages caused by environmental contamination or exposure to hazardous substances under common law causes of action. These common law causes of action may include nuisance, trespass, ultra-hazardous activity, negligence, negligence per se, equitable indemnity, and equitable contribution.[206] In addition, if the plaintiff purchased, leased or leases the contaminated property, the plaintiff may be able to bring causes of action for breach of contract, fraudulent misrepresentation, negligent misrepresentation, and or a waste.[207] Moreover, a plaintiff, who is exposed to a hazardous substance, can bring a cause of action for personal injury. For example, numerous plaintiffs have sought to recover personal injury and property damages allegedly caused by asbestos.[208] The common law causes of action can be used to recover damages for substances not generally covered by the environmental statutes. For example, plaintiffs have recently recovered personal injury and property damages allegedly caused by exposure to mold.[209]

To avoid being caught in the web of liability created by CERCLA and other federal and state laws and regulations and the common law, safety professionals and managers should take the following steps, among other precautions and compliance measures. First, wherever possible avoid the potential future liability associated with sending waste to a landfill or other disposal site by reducing the amount of hazardous substances and wastes used by choosing nonhazardous or less hazardous products. Second, recycle or incinerate any hazardous wastes that are generated instead of sending such wastes to a landfill or other disposal facility.

Third, if you have to send hazardous waste to a landfill or other disposal facility, do your due diligence to determine that the landfill or disposal facility is authorized to handle the specific hazardous waste you are disposing of and that such facility is being operated properly. Do not rely on assurances from state or federal regulatory agencies for your due diligence. Visit the landfill or disposal facility and verify such items as financial assurance for closure, employee training and waste management practices. The EPA named the State of California as a PRP in two superfund sites (Casmalia Resources and Stringfellow Acid Pits) because the state had issued operating permits to both facilities. The State of California paid $111 million for its share of the response costs for these Superfund Sites in settlement with the EPA. In addition, be wary of the disposal site that offers below market disposal fees.

Fourth, keep all environmental records, including manifests, tank tests, permits and waste analysis. Often numerous operators have used the same site, or neighboring companies have used similar hazardous substances. Accordingly, companies that have records that show environmental compliance, such as passing tank tests or remediation of releases, are in a much better position to defend themselves against a lawsuit based on CERCLA, other environmental statutes or common law. Fifth, consider performing a compliance self audit. The pros and cons of performing such a self-audit are discussed later in this chapter.

Finally, prior to acquiring any property or any company, the acquiring company or individual must conduct due diligence to determine if any environmental conditions exist. CERCLA provides an innocent landowner's defense for a party that purchased a previously contaminated site if such party can establish that it did not know and had no reason to know of the contamination when it purchased the site.[210] In order to be protected by the innocent landowner's defense, a company must conduct adequate due diligence.[211] As to acquisition of a company, you should not only look at the present hazardous substance use and hazardous waste handling and disposal practices, but also the past uses and disposal practices. If the property or company that your company wishes to acquire has an environmental condition, such liability should be reduced or allocated through deal structure, indemnity agreements, and insurance.

CIVIL RIGHTS LAWS

Beginning in the 1990s, safety managers have had to deal with additional statutory programs under the civil rights laws. We will discuss two areas that are presently emerging for companies and their managers — disability rights and environmental justice.

Disability Rights and Safety Requirements

Managers in the course of hiring and managing employees must take into account the federal and state prohibitions of employment discrimination based on disabilities. Seemingly well-intended efforts to achieve a safe workplace can backfire. For example, to reduce the risk of illness and injury and the related costs, it may seem medically justifiable to refuse to hire or assign persons with prior back injuries to positions where they may suffer further injury. Applying these selectivity factors, however, can run afoul of the federal ADA of 1990,[212] which contains broad prohibitions of discrimination against individuals with disabilities.

All aspects of employment such as recruitment, hiring, and fringe benefits (including insurance), and employment-related activities, fall within the scope of the Act. It applies to private and public employers of 15 or more employees, except for the federal government, Indian tribes, and certain tax-exempt private membership clubs.[213] Federal agencies face similar obligations, however, to provide reasonable accommodations to qualified applicants or employers under the Rehabilitation Act of 1973.[214]

Employers may violate Title I of the ADA if they discriminate against qualified individuals (a) with protected disabilities, (b) with a record of impairment, or (c) who are regarded as having such an impairment and who can perform the essential elements of the particular job (with or without reasonable accommodation).[215] Under Title III of the Act, access to public transportation, accommodations, and services must be provided.[216]

The rights of the physically and mentally disabled to employment may be protected under certain conditions, which are determined on a case-by-case basis. A disability is defined as a physical or mental impairment that substantially limits one or more major life activities.[217] A reasonable accommodation is a modification or adjustment to a job or work environment that would allow the qualified disabled person to perform the essential job functions. It can include restructuring a job, modifying work schedules, acquiring or modifying equipment, providing readers or interpreters, and the like. The duty imposed by the ADA to provide reasonable accommodation is limited, however, when it causes an undue hardship on the operation of the employer's business. Whether or not the undue hardship exception applies is determined on a case-by-case basis, taking into account the nature and cost of the accommodation, the financial resources of the employer, and the type of operation.

Under the ADA Act, employers may restrict workers from posing a direct threat or significant risk to the health or safety of themselves or others. If the threat or risk can be eliminated or lowered by reasonable accommodation to an acceptable level, the employer must do so.[218]

Although the Equal Employment Opportunity Commission (EEOC) has provided guidelines,[219] many of these key terms have been the subject of extensive litigation: what is a reasonable accommodation, what is a disability, and what constitutes undue hardship to an employer remains in dispute.[220] During the six-year period beginning in 1992, when the EEOC began enforcing Title I of the Act, the agency received over 110,000 charges based on the ADA.[221] The most frequent complaints involved the alleged failure to provide reasonable accommodations, harassment, and unlawful discharge.

In sum, although the ADA has spawned a whole new area of litigation and EEOC charges, the congressional objectives of reducing discrimination and increasing the employment of the disabled have yet to be achieved. In the meantime, employers are still struggling to understand the extent of their obligations to provide reasonable accommodations to the disabled individuals.

Environmental Justice

Public concern began growing in the 1980s over the apparent concentration in predominantly low-income or minority communities of large-scale industrial facilities such as landfills, waste incinerators, sewage treatment plants, and bulk chemical factories.[222] For example, from 1986 to 1996, permits were issued to construct seven landfills in Delaware County, Pennsylvania, five of which were clustered in a single low-income, predominantly African-American community along the Delaware River.[223] Residents and others questioned whether decisions to construct such facilities in their communities were motivated by racial discrimination, giving rise to the concept of environmental racism.

The movement, also known as environmental justice or equity, gained significant recognition in 1994 with the execution of an Executive Order by former President Clinton. The order triggered the mandatory review by every federal agency of the effects of its policies, programs and activities, including proposed federal projects, on low-income or minority communities.[224]

In response, EPA,[225] agencies in the U.S. Department of Transportation,[226] and the Council on Environmental Quality[227] have issued guidelines to address environmental justice concerns. EPA defines environmental justice as the fair treatment for people of all races, cultures, and incomes, regarding the development of environmental laws, regulations, and policies.[228] EPA created the Office of Environmental Justice and a national council to advise the agency on environmental justice. The agency's Civil Rights Office also assists in evaluating the potential effects of a federally funded project, and its formal policy extends to state permits.[229] From New Jersey and Pennsylvania to California and Tennessee, task forces charged with policy development at the state level have been established; whose members include residents of the affected communities.[230]

Skeptics have questioned whether disproportionate concentrations or increased health problems have occurred, and whether development in low-income or minority communities was based improperly on racial discrimination.[231] Regardless of the debate, businesses would be well served to engage the local community early in the proposed development of an environmentally sensitive project, and to make reasonable accommodations.[232] Failure to do so could cause costly delays or worse, completely derail a project. Consider, for example, the ruling in 2000 of a federal appeals court that effectively suspended a 25-year, $19 billion transportation plan for Atlanta, which EPA had approved.[233] Many other development projects have been subjected to legal challenges under the civil rights laws.[234] While these challenges may not always succeed on the merits, settlement frequently has required the developer to modify the project and alleviate the impacts to the local community.

One unintended consequence of the environmental justice movement may be to stifle or hamper redevelopment of blighted inner cities and other urban areas. Long-standing federal, state, and local government efforts to redevelop these areas, commonly known as brownfields,[235] may need modifications to take into account the potential for disproportionate concentrations of environmentally sensitive industrial facilities, as well as adverse health effects on surrounding minority and low-income communities. Another consequence has been increased governmental review of the compliance of operations of companies located in minority and low-income communities. For example, EPA Region IX recently cited 36 of the 43 companies located in minority and low-income communities that were inspected pursuant to the Los Angeles Environmental Justice Initiative.[236]

OPERATIONAL COMPLIANCE MANAGEMENT

Safety and Environmental Compliance Auditing

The enactment of the Federal Organizational Sentencing Guidelines in 1991[237] prompted most if not all of the larger businesses in the United States to develop formal corporate compliance programs. Under the Guidelines, a business that has implemented an effective compliance program may receive more lenient treatment in the event that a violation of law occurs. The effectiveness of a compliance program is frequently measured by, among other factors, the extent to which the company conducts periodic compliance audits.[238]

Although the Sentencing Guidelines do not apply to violations of environmental laws,[239] the EPA developed a formal policy to encourage businesses to self-police its operations and compliance with laws.[240] Similar to the U.S. Sentencing Commission, EPA expects businesses to perform periodic audits as an essential component of an effective compliance program. The EPA has defined an environmental audit as a systematic, documented, periodic and objective review by regulated entities of facility operations and practices related to meeting environmental requirements.[241]

Whether the EPA or other enforcement agencies will grant leniency for violations depends on several other factors, including the company's voluntary and timely disclosure of the violation(s) to the appropriate enforcement agency, and prompt redress of the problem.[242] Consequently, advance preparations to address potential findings from an audit are critical, including the allocation of adequate resources.

Since 1982, OSHA has developed several programs to encourage voluntary compliance efforts, which include periodic audits. Upon accepting a company's application to one of OSHA's Voluntary Protection Programs, for example, the

Agency will provide a team to conduct an inspection of a facility and cooperate with the business to redress any deficiencies.[243] Although OSHA normally would not seek a business's self-audit reports, the Agency may need to examine them to determine eligibility. Assuming the deficiencies have been corrected, the business then can enjoy a two-year reprieve from routine inspections; however, OSHA will reassess the business every one to three years, depending on the particular program. To avoid unexpectedly large problems and concomitant costs, a business considering entry to the star program should be prepared to surmount a high hurdle of eligibility: OSHA accepts only facilities that are striving to excel by using flexible and creative strategies that go beyond the requirements to provide the best feasible protection for their workers.[244]

Compliance auditing can include checking whether a business is fulfilling its duties under worker safety and environmental laws,[245] its permits for discharges and emissions, its labor and employment agreements, and agreements with government agencies. An audit or self-evaluation can also identify ways a company's management system can promote compliance and reduce risk. One added benefit to the company over time is the reduction in accidents and injuries, and the preservation of assets, which can result in significant savings.

Protecting Self-Evaluations and Audits from Disclosure: The Law

Many federal and state laws require immediate or prompt reporting upon discovery of a reportable quantity of a release the regulated substance, even from historical activities by former unrelated owners or operators.[246] Substantial penalties (up to $25,000 per day) may be imposed for failing to give timely notification.

A business may need to consider whether it should make voluntary disclosure of a violation to the EPA, OSHA, or other government enforcement agency. Such voluntary disclosure may be a mitigating factor that could reduce penalties and other sanctions. EPA normally requires companies to disclose such violations within 21 days after discovery.[247]

Compliance auditing often reveals violations of law that, if promptly reported and remedied, can be resolved with governmental agencies. The findings can be based on highly sensitive and confidential information. The company should treat written product of such audits (particularly the thought processes and analyses of company managers and their counselors) as carefully as trade secrets and employee medical records.

One way to protect such information is by understanding the law. If compliance auditing is conducted so that the company can receive legal advice, the company may be able to preserve the confidentiality of the information. Preserving confidentiality can be very important in defending against government investigations and third-party lawsuits (citizen or private attorney general lawsuits, and those by competitors).

If the company's attorney provides legal counseling based on the information developed by a safety professional and or the environmental consultant, the company may be able to invoke three evidentiary privileges to preserve confidentiality of the information. These privileges include the attorney–client confidential communication privilege, the attorney work product doctrine, and a developing self-evaluation privilege. In addition, approximately 25 states have enacted laws that recognize an environmental audit privilege.[248]

The attorney–client privilege is similar to other communication privileges developed by the courts and included in state and federal codes of evidence law.[249] Information can be protected if it is compiled for the purpose of securing an opinion of law.[250] The attorney work product doctrine protects information gained and communications in anticipation of litigation. Routine audits or self-evaluations may not qualify.[251]

Under a self-evaluation or self-critical analysis privilege, it is possible for a business to assess its compliance with regulatory requirements without creating evidence that may be used against it in future litigation.[252] However, the history of the confidentiality of such self-evaluations in the face of government prosecutions is not so positive.[253] The state legislation, mentioned earlier, creates a form of self-evaluation privilege. However, the statutes generally only provide a qualified privilege; for instance, the company must promptly remedy any violations revealed in the audit. The EPA is presently studying the issue, and federal legislation is being considered that would not allow a privilege, but that would allow mitigation of penalties if an audit had been conducted.

The EPA has had a policy since 1986 that encourages companies to develop, implement, and periodically upgrade environmental auditing programs.[254] However, the policy also includes the agency's position that audit reports are subject to disclosure to the EPA. Likewise, the U.S. Justice Department's guidance on investigating for and prosecuting environmental crimes offers no safe harbor.[255] Under OSHA's self-audit policy adopted in 2000, agency inspectors will not routinely request audit reports during inspections, or use the reports to identify potential violations to investigate during an inspection.[256] Similar to the EPA, OSHA may consider an employer's good faith efforts in addressing a hazard to reduce the level of violation and the resultant penalties.

Preparing for Inspections

In addition to the numerous reporting requirements, governmental regulators rely on inspections to enforce compliance with environmental and safety laws and regulations. Moreover, environmental protection and OSHA agencies are turning to multimedia inspections.[257] Multimedia inspections combine the review of record keeping, employee training, discharges and emissions, and disposal of wastes. The inspectors, who may be cross-trained (EPA often assigns trained investigators to state-led teams), also look for physical evidence of violations, such as spills, stains, faulty secondary containment, and the inability of workers to articulate safety and environmental compliance procedures and policies.

It is crucial that a company be prepared for a regulatory inspection at all times since inspectors have a habit of dropping by unannounced. The following steps should be taken in preparation of an inspection. First, make sure the company's operations are in compliance before the inspector arrives. It is generally easier and cheaper to correct a problem before, than after you are served with a citation, a notice of violation or a complaint bringing civil and criminal charges against the company, its officers, and managers. Ultimately, the company will have to comply with the laws and regulations. Complying beforehand allows the company to be in charge of choosing and implementing compliance measures and potentially saves the company the cost of legal representation and fines and penalties. Moreover, a company with a good compliance record is less likely to be inspected or to be scrutinized when it is inspected.

Second, have all records organized and in one area if physically possible and in compliance with permit and regulatory requirements.[258] You do not want the inspector waiting around (and therefore looking around) while you try to locate documents or taking an unsolicited tour of the entire facility because documents are located in numerous places. Moreover, you do not want the inspector to think that your compliance efforts are disorganized and haphazard based on poor record keeping.

Third, keep a neat house. Clean up spills and stains even if not required by law. Don't keep trash, junk, empty or unlabeled barrels around. Even if the barrels do not contain regulated substances, such as hazardous chemicals or waste, the inspector will most likely require you to prove that they do not contain such regulated substances. Again, you don't want the inspector hanging around because he wants to make sure that a junk pile or unlabeled barrel is not a violation, and you don't want the inspector to jump to the conclusion that your compliance program is in shambles because your house looks that way. In sum, first impressions count.

Finally, appoint an individual and a backup to be responsible for interacting with the inspector. An inspection can lead to severe civil and criminal penalties and thus, it is important to assign a knowledgeable, levelheaded person to represent the company. This person will need to decide such important issues as whether to cooperate with government requests for inspection, whether to seek advice from counsel prior to speaking with governmental officials, and whether a privilege or protection should be asserted to the inspection of certain information.

In order to make these difficult decisions, the persons appointed by the company to interact with the government inspector should have a solid understanding of the inspector's rights to investigate and any privileges and protections the company may have the right to assert. Governmental regulators have broad rights to inspect pursuant to laws and permit conditions. In fact, regulators have the authority to impose civil fines on individuals or entities for refusal to comply with an inspector's request for information under certain laws.[259] On the other hand, inspectors do not have unlimited authority to search or inspect. The Fourth Amendment of the Constitution protects individuals and entities from unreasonable search and seizure and the Fifth Amendment protects individuals from self-incrimination. In addition, certain protections are afforded to attorney–client communications, attorney work product, self-evaluations and audits,[260] and trade secret or confidential business information.[261] EPA's Region 4 manual on investigation procedures recognizes that the EPA's authority is subject to the unreasonable search and seizure provisions of the Fourth Amendment to the Constitution and adopts policies to protect trade secret and confidential information, including but not limited to, a policy that branch personnel not accept confidential information when conducting a plant evaluation, inspection, or reconnaissance. Also, written requests for information must provide a statement allowing a facility to designate all or part of the information as confidential.[262]

CONCLUSION

A facility officer and or manager must deal with a complex legal regime for health, safety, and environmental compliance. Inadvertent or well-intentioned mistakes can lead to civil and criminal penalties where employee safety, public health, and the environment are concerned. However, a solid grasp of the regulations, good communications internally and with the regulatory community, a commitment of management to compliance, and common sense can go a long way to avoiding such penalties and allowing the manager to succeed in today's regulatory environment.

NOTES

1. The discussion in this chapter is not intended to substitute for legal advice applicable to specific circumstances; the authors encourage the reader to seek legal counsel to review a particular situation. Laws and regulations are subject to legislative amendment or repeal, and judicial and administrative interpretation. The reader will find acronyms throughout the discussion, which are offered solely as a guide because most of the professional literature in this area includes such references.
2. 42 USC §§ 4321–4370 (e.g., NEPA requires the Council and other federal agencies to review the environmental impacts of projects on federal land or supported by federal funds). *See* discussion on Environmental Justice in this chapter for a more recent application of the Act.
3. *See* 29 U.S.C. §§ 651–678.
4. *See* David J. Abell, Emergency Planning and Community Right to Know: The Toxics Release Inventory, 47 SMU L. REV. 581 n.3 (1994) (explaining that Congress enacted the Emergency Planning and Community Right-to-Know Act in response to the Union Carbide disaster in Bhopal, India). *See also* Thomas N. Gladwin, A Case Study of the Bhopal Tragedy, *Business Matters* 223, 227–234 (Charles S. Pearson ed., 1987); D. Weir, The Bhopal Syndrome 36 (1987).
5. *See* 29 U.S.C. §§ 651–678.
6. Under 29 U.S.C. § 652(5), the definition of an "employer" excludes the United States or any state or political subdivision of a state. *See also* 29 U.S.C. § 668(a) (requiring each federal agency to establish and maintain an OSH program). Similarly, the occupational, safety and health standards of other federal agencies take precedence over OSHA regulations. *See* 29 U.S.C. § 653(b) (1). Consequently, farm workers are protected from pesticide exposure under the Federal Insecticide, Fungicide and Rodenticide Act, (FIFRA). *See* FIFRA § 25; 7 U.S.C. § 136w; 40 C.F.R. § 170.8. In addition, workers in the mining industry are protected under the Mine Safety and Health Act. *See* 30 U.S.C. §§ 801–962, particularly 30 U.S.C. § 811; 30 C.F.R. §§ 1–104. An alternative federal program has been implemented for employees of the Nuclear Regulatory Commission under the Atomic Energy Act, 42 U.S.C. §§ 2011–2021.
7. *See* 29 U.S.C. § 654(a) and (b).
8. *See* 29 U.S.C. § 654(a) (1).
9. The States and territories are: Alaska, Arizona, California, Hawaii, Indiana, Iowa, Kentucky, Maryland, Michigan, Minnesota, Nevada, New Mexico, North Carolina, Oregon, Puerto Rico, South Carolina, Tennessee, Utah, Vermont, Virgin Islands, Virginia, Washington, and Wyoming. Connecticut, New Jersey, and New York have

also enacted OSH programs for public employees. *See* Conn. Gen. Stat. §§ 31-367–31-385; N.J. Stat. § 34:6A-29; N.Y. Lab. Law §§ 27 and 27-a.

10. *See* Cal. Labor Code §§ 6300–6719.

11. *See* 29 C.F.R. § 1903.11(a); 29 C.F.R. § 657 (f) (1). Workers may file complaints with many agencies through the Internet. A complaint that an OSHA violation has occurred may be filed at http://www.osha.gov/as/opa/worker/eComplaintForm.html.

12. *See* 29 C.F.R. § 1903.14(a).

13. *See, e.g.,* 29 C.F.R. § 1910.1001(f) (best engineering controls/practices should be implemented where feasible to keep exposure to asbestos under the permissible exposure limit and excursion limit).

14. *See* 29 C.F.R. §§ 1910.1000–1910.1450.

15. *See* 29 C.F.R. § 1910.100, Tables Z-1, Z-2 & Z-3.

16. EPA banned the use of asbestos-containing building materials in July 1989 pursuant to section 6(a) of the Toxic Substances Control Act, 15 U.S.C. § 2605(a). That ban against commercial applications later was overturned. *Corrosion Proof Fittings v. Envtl. Protection Agency*, 97 F.2d 1201 (5th Cir. 1991).

17. Under section 112 of the Clean Air Act, 42 U.S.C. § 7412(c) and (d), certain applications of ACM previously were prohibited, such as spray-applied insulation materials. These materials now may be used, however, if encapsulated to prevent the release of asbestos fibers. New uses of asbestos remain prohibited, as well as the use of asbestos in corrugated paper, rollboard, commercial paper, and flooring felt. *See, e.g., Technical Amendment in Response to Court Decision on Asbestos,* 59 Fed. Reg. 33,208 (June 28, 1994).

18. *See, e.g., OSHA Guidance Documents at* http://www.osha-slc.gov/SLTC/asbestos/index.html.

19. *See, e.g., Secretary of Labor v. Southern Nuclear Co. Inc.,* No. 97-1450, available at 199 WL 569136 (O.S.H.R.C. 1999) (serious violation found for facility owner's failure to determine presence of ACM or presumed ACM, and to notify contractor of presence).

20. 29 C.F.R. § 1910.1001(b).

21. 29 C.F.R. § 1910.1001(j) (2) (i).

22. 29 C.F.R. § 1910.1001(f) (2).

23. 29 C.F.R. § 1910.1001(j).

24. Construction, 29 C.F.R. § 1926.1101, and shipbuilding industries are regulated separately, *id.* § 1915.1101.

25. 29 C.F.R. § 1910.1101(f) (3).

26. *Id.*

27. *See, e.g., Secretary of Labor v. Southern Nuclear Co. Inc.,* No. 97-1450, available at 199 WL 569136 (O.S.H.R.C. 1999) (citing 59 Fed. Reg. at 40974 (Aug. 10, 1994)).

28. EPA regulates asbestos under several statutory programs, including (1) the Clean Air Act (CAA) as a hazardous air pollutant (42 U.S.C. § 7412, 40 C.F.R. § 61.145 (standard for demolition and renovation); (2) the Clean Water Act (CWA), 33 U.S.C. §§ 1311, 1119 (maximum contaminant level (MCL) and the maximum contaminant level goal (MCLG) for asbestos set at 7 million fibers per liter); (3) the Comprehensive Environmental Response, Compensation, and Liability Act (CERCLA), 42 U.S.C. § 9601, *et seq.* (reportable quantity (RQ) of 1 lb. has been established for asbestos); (4) the Resource Conservation and Recovery Act (RCRA), 42 U.S.C. § 6901, *et seq.*; and (5) the Toxic Substances Control Act (TSCA), 15 U.S.C. §§ 2641–2656 (prohibiting the manufacture and use of asbestos in certain products). In addition, EPA has promulgated standards covering asbestos abatement project personnel not covered

under OSHA standards. The Food and Drug Administration (FDA) also regulates, under the Food, Drug, and Cosmetic Act, the use of asbestos in indirect food additives, adhesives, components of coatings, and polymers. FDA has restricted the utilization of asbestos filters in the manufacture of drugs and drug ingredients. For guidance on demolition and renovation activities, *see, e.g.,* EPA Region 4, "Common Questions on the Asbestos NESHAP," available at http://www.epa.gov/region 4/air/asbestos/asbqa.htm; EPA Region 4, "Asbestos/NESHAP Regulated Asbestos Containing Materials Guidance," available at http://www.epa.gov/region 4/air/asbestos/asgmatl.htm. For an overview of cross-agency guidance, *see* OSHA's Web site on asbestos compliance, http://www.osha-slc.gov/SLTC/asbestos/compliance.html. Unlike OSHA, EPA assumes no safe level for asbestos.

29. *See, e.g., U.S. v. TGR Corp.,* 171 F.2d 762, (2nd Cir. 1999) (affirming criminal conviction under Clean Water Act for discharging ACM into federal navigable water without permit). In *U.S. v. Ho,* No. H00-133 (5th Cir. 2001), the court of appeals upheld the conviction of an employer who was sentenced to six months in a halfway house and subject to more than $1 million in civil fines under the Clean Air Act for exposing employees to excessive levels of asbestos in a demolition project, and for falsifying records.

30. *See, e.g., Cantrell v. GAF Corp.,* 999 F.2d 1007 (6th Cir. 1993) (ACM manufacturer held liable to employees for intentional continued exposure of employees to asbestos, and liable for emotional distress caused by reasonable fear of contracting cancer).

31. *See, e.g.,* 29 C.F.R. § 1910.1001(m)(1) (iii) and (m)(6)(ii).

32. *See* 29 U.S.C. § 657(e); 29 C.F.R. § 1904.0. *See* 29 C.F.R. § 1904.2 (setting forth partial exemption for establishments in certain industries). OSHA published updated record keeping regulations in 2000, which are under review by the Bush Administration as this chapter goes to press. *See* discussion in OSHA National News Release USDL:01–25, Jan. 18, 2001.

33. *See* 29 C.F.R. § 1904.2-1904.8.

34. *See* 29 C.F.R. § 1904.9(b)

35. *See* 29 C.F.R. § 1910.1200.

36. *Id.*

37. OSHA publishes the most frequently violated standards on its Web site, http://www.osha.gov/oshstats/, which includes an analysis of violations issued most frequently to specific industries, according to their standard industrial classifications.

38. *See* 29 C.F.R. § 1910.119.

39. *See* 29 U.S.C. § 657(a) & (f).

40. *See* 29 C.F.R. § 1904.8.

41. 29 C.F.R. § 1903.6(a).

42. *See* 29 U.S.C. § 657(e); 29 C.F.R. § 1903.8.

43. For a more complete description of the inspection process, see Chapter 29.

44. *See Martin v. Hammermill Paper Div. of Int'l Paper Co.,* 796 F. Supp. 1474 (S.D. Ala. 1992).

45. *See* Final Policy Concerning the Occupational Safety and Health Administration's Treatment of Voluntary Employer Safety and Health Self-Audits, 65 Fed. Reg. 46,398 (July 28, 2000) (hereinafter, *OSHA Self-Audit Policy*).

46. 29 U.S.C. § 654(a) (1).

47. *See, e.g., Caterpillar, Inc. v. Occupational Health and Safety Review Commission,* 122 F.2d 437 (7th Cir. 1997) (upholding Commission's ruling that employer willfully violated General Duty Clause due to heightened awareness of risk of injury from repairing a press).

48. *McKie Ford, Inc. v. Secretary of Labor,* 191 F.3d 853 (8th Cir. 1999) (lack of safety features in freight elevator car made risk plainly obvious to employer).

49. *See* 65 Fed. Reg. 68,261 (Nov. 14, 2000).

50. *See id.*

51. *See id.* at 68,262.

52. OSHA officially withdrew the Ergonomic Program Standard effective April 17, 2001, 66 Fed. Reg. 20403 Feb. 23, 2001. *See* Steven Greenhouse, *House Joins Senate in Repealing Rules Issued by Clinton on Work Injuries,* N.Y. Times, March 8, 2001, at A1; Helen Dewar, Cindy Skrzycki, *House Scraps Ergonomics Regulation,* Wash. Post, March 8, 2001, at AO1; Steven Greenhouse, *Senate Votes to Repeal Clinton Workplace Injury Rules,* N.Y. Times, March 7, 2001, at A1; Helen Dewar, Cindy Skrzycki, *Workplace Health Initiative Rejected,* Wash. Post, March 7, 2001, at AO1.

53. OSHA's ability to impose penalties for egregious violations of the General Duty Clause has been challenged successfully. *See, e.g., Reich v. Arcadian Corp.,* 110 F.3d 1992 (5th Cir. 1997) (holding that OSHA lacked authority to apply egregious case policy to General Duty Clause violations).

54. *See* Cal. Labor Code § 6401.7; 8 Cal. Code of Reg. §§ 3202-3203 (known as the Illness and Injury Prevention Program).

55. *See* U.S. OSHA, *Workplace Violence* at 1 (revised Aug. 18, 1999) (citing U.S. Dept. of Justice, G. Warhol, *Workplace Violence, 1992-96. National Crime Victimization Survey (report No. NCJ-168634).*

56. *Id.*

57. *Id.*

58. *See, e.g.,* S. Kaufer, et al., *Workplace Violence: An Employer's Guide.* Workplace Violence Research Institute (Palm Springs, CA 1998), available at http://www.nowork-violence.com.

59. *Id.*

60. *See Sec. of Labor v. Megawest Fin., Inc.,* 17 O.S.H. Cas. 1337, found at 1995 WL 383233 (Occupational Safety & Health Review Comm'n 1995).

61. California State Department of Industrial Relations. *Cal/OSHA Guidelines for Workplace Security,* revised Mar. 30, 1995, available at http://165.235.90.100/DOSH/dosh_publications/worksecurity.html (hereinafter, "Cal/OSHA Guidelines for Workplace Security").

62. *McKie Ford, Inc. v. Secretary of Labor,* 191 F.3d at 856 ("[S]imply failing to address a known hazard will not support a willful violation" but finding willful violation due to a plain indifference to employees' safety where employer allowed elevator to be operated for years without repairing or replacing safety devices).

63. *Cal/OSHA Guidelines for Workplace Security, supra,* note 61.

64. Cal/OSHA Guidelines for Security and Safety of Health Care and Community Service Workers (1995), available through Cal/OSHA's Consultation Service Area Office at (415) 703-4341.

65. Cal/OSHA Model Injury and Illness Prevention Program for Workplace Security, available through Cal/OSHA's Consultation Service Area Office (see Web site address at n. 85 *infra* for the telephone number of local offices).

66. New Jersey Public Employees Occupational Safety and Health, *Guidelines on Measures and Safeguards in Dealing with Violent or Aggressive Behavior in Public Sector Health Care Facilities.* Washington State Department of Labor and Industries (1990). *Late Night Retail Workers Crime Protection Act. WAC 296-24-102, 296-24-10203.* Metropolitan Chicago Healthcare Council, *Guidelines for Dealing with Violence in Health Care.*

67. OSHA *Guidelines for Preventing Workplace Violence for Health Care and Social Service Workers* at 2 (OSHA 3148 1998) (noting that the failure to implement the guidelines is not a violation of the General Duty Clause).

68. *See* 29 U.S.C. § 658(b); 29 C.F.R. § 1903.16.

69. *See* 29 U.S.C. § 666; OSHA Field Inspection Reference Manual, ch. IV.E, available from agency's Web site at http://www.osha.gov.

70. *See, e.g., McKie Ford, Inc. v. Secretary of Labor,* 191 F.3d at 856.

71. Cal/OSHA Citation and Notification of Penalty issued Aug. 4, 1999, CSHO ID No. F9862, to Tosco Refining Co. and its succcessors.

72. *See* 29 U.S.C. § 659(a); 29 C.F.R. § 1903.15.

73. *See* 29 C.F.R. § 1903.16(b)-(c).

74. *See* 29 U.S.C. § 659(c); 29 C.F.R. § 1903.17.

75. *See* 29 U.S.C. § 659 (c) (providing for Commission review over Secretary of Labor); 29 C.F.R. § 1978.110; and 29 U.S.C. § 660(a) (providing for judicial review).

76. Cal. Labor Code § 6425(a). Under California AB 1127 passed in 2000, officers of a company may be individually liable for up to $2 million.

77. *See Salwasser v. Occupational Safety & Health Appeals Bd.,* 214 Cal.App.3d 625, 262 Cal. Rptr. 836 (1989).

78. *See* MARK A. ROTHSTEIN, ET AL., EMPLOYMENT LAW §§ 6.1–6.4 (2nd ed. 1999) (hereinafter "ROTHSTEIN, EMPLOYMENT LAW") (identifying various federal and state workers compensation laws); Workers' Compensation Guide § 1.02[1], sub-chapter 7.02 (West 1998).

79. *See* ROTHSTEIN, EMPLOYMENT LAW § 6.3 (noting that "[i]njured workers no longer ha[ve] to establish negligence attributable to their employer in order to obtain legal redress" and only "[must] demonstrate that their conditions arose out of and during the course of their employment"); Workers' Compensation Guide § 1.01[3] (West 1998). *See also* http://www.dir.ca.gov/dwc/basics.html.

80. *See* ROTHSTEIN, EMPLOYMENT LAW § 6.3 (explaining that workers compensation laws provide employers several benefits, including imposing limited liability for workplace injuries and precluding compensatory or punitive damage awards to employees for such injuries); Workers' Compensation Guide § 1.01[3] (West 1998). *See also* http://www.dir.ca.gov/dwc/basics.html.

81. *See* ROTHSTEIN, EMPLOYMENT LAW 6.3 (explaining that "by creating the exclusive remedy for employee injuries, [workers compensation] laws [give] employers immunity against most tort actions arising from the employment relationship"); Workers' Compensation Guide § 1.01[3] (West 1998). *See also* http://www.dir.ca.gov/dwc/basics.html.

82. *See* 29 U.S.C. § 658(b); 29 C.F.R. § 1903.16; Workers' Compensation Guide § 1.01[2] (West 1998). *But see* ROTHSTEIN, EMPLOYMENT LAW § 6.3 (suggesting that limited liability features of workers compensation statutes and "statutorily-mandated social insurance coverage financed by private companies" limit total employer exposure and allow employers to distribute costs of the inherent risk of workplace accidents to shareholders, consumers, and employees).

83. *See* 29 U.S.C. § 659(a); 29 C.F.R. § 1903.15; Workers' Compensation Guide § 3.02[1] (West 1998); Cal. Lab. Code § 3700. *See also* ROTHSTEIN, EMPLOYMENT LAW § 6.3 ("The limited liability concept and the extension of tort liability to industrial accidents [make] it possible for employers to estimate their annual exposure and to obtain insurance policies to cover individuals who sustained work-related injuries.").

84. *See* Workers' Compensation Guide § 3.07[8] (West 1998); 9 Mass. Prac. Workmen's Compensation § 13.9 (2000 ed.) (TREATISE SUPPLEMENT) (noting that 1991

Massachusetts workers compensation reform encourages "employers [to] work closely with the loss control departments of their workers' compensation insurers to provide a safer workplace"); Cal. Labor Code § 6354.5 (providing that an insurer desiring to write workers' compensation insurance must maintain or provide occupational safety and health loss control consultation services to places of employment posing significant preventable hazards to workers).

85. *See, e.g.,* Cal/OSHA's Workplace Injury and Illness Prevention Program, which includes a hazard assessment and control component at http://www.dir.ca.gov/dosh/dosh_publications/iiphihzemp.html. *See also* http://www.dir.ca.gov/DIRNews/2001/IR2001-01.html (discussing decrease of injuries and illnesses in high-hazard industries such as agriculture and construction due in large part to the Cal/OSHA inspection programs targeting such industries).

86. *See* 42 U.S.C. § 12112(a). *See generally* Americans with Disabilities Act of 1990, 42 U.S.C. §§ 12101–12213.

87. *See* 42 U.S.C. § 12114(a) (providing that a "'qualified individual with a disability shall not include any employee or applicant who is currently engaging in the illegal use of drugs, when the covered entity acts on the basis of such use"). *See also* EEOC Regulations to Implement the Equal Employment Provisions of the ADA, 29 C.F.R. § 1630.3(a).

88. *See* 42 U.S.C. § 12114(a) & (b); *Herman v. City of Allentown*, 985 F. Supp. 569 (E.D. Pa. 1997).

89. *See* EEOC and U.S. Dept. of Justice Civil Rights Division booklet titled "Americans with Disabilities Act Questions and Answers" which can be found on the internet at http://www.usdoj.gov/crt/ada/qandaeng.htm (stating that "While a current illegal user of drugs is not protected by the ADA if an employer acts on the basis of such use, a person who currently uses alcohol is not automatically denied protection. An alcoholic is a person with a disability and is protected by the ADA if he or she is qualified to perform the essential functions of the job. An employer may be required to provide an accommodation to an alcoholic. However, an employer can discipline, discharge or deny employment to an alcoholic whose use of alcohol adversely affects job performance or conduct. An employer also may prohibit the use of alcohol in the workplace and can require that employees not be under the influence of alcohol."); *Mararri v. WCI Steel, Inc.*, 130 F.3d 1180, 1183 (6th Cir. 1997) (noting that the plain language of Section 12114 does not automatically exclude alcoholics from ADA coverage since alcohol is not an illicit drug).

90. *See* 42 U.S.C. § 12114(c) (4); *Mararri v. WCI Steel, Inc.*, 130 F.3d at (noting that employer may terminate an employee for violating reasonable rules of conduct even if related to employee's handicap or alcoholism); *Nielsen v. Moroni Feed Co.*, 162 F.3d 604 (10th Cir. 1998) (explaining that employee's unsatisfactory conduct caused by alcoholism or illegal drug use is not protected by the ADA); *Carroll v. Illinois Dep't of Mental Health & Developmental Disabilities*, 979 F. Supp. 767 (C.D. Ill. 1997) (finding that alcoholism-related discharge is not a violation of the ADA if the conduct caused by employee's alcoholism merits discipline); *Salley v. Circuit City Stores, Inc.*, 160 F.3d 977 (3rd Cir. 1998) (holding that employer offered legitimate, nondiscriminatory reason for terminating store manager due to, among other things, the manager's drug-related absences); *Brown v. Lucky Stores, Inc.*, 246 F.3d 1182 (9th Cir. 2001) (holding that employer is not required to make reasonable accommodation for employee to miss work to attend rehabilitation program following alcohol- and drug-related arrest).

91. *See id.*

92. *See id.*

93. *See* 42 U.S.C. § 12114(d).

94. *See* 42 U.S.C. § 12114(e) (providing that the ADA does not restrict the lawful authority of entities, pursuant to statutes and regulations implemented by the Department of Transportation, to test employees in "positions involving safety-sensitive duties for the illegal use of drugs and for on-duty impairment by alcohol"). *See also* 49 C.F.R. § 392.5(a)(1) & (a)(2).

95. *See* 42 U.S.C. § 12114(d) (providing that "a test to determine the illegal use of drugs shall not be considered a medical examination" and that the ADA does not "encourage, prohibit, or authorize the conducting of drug testing for the illegal use of drugs by job applicants or employees or making employment decisions based on such test results"); *Buckley v. Consolidated Edison Co.*, 127 F.3d 270, 24 A.D.D. 758, 7 A.D. Cas. (BNA) 794 (2d Cir. N.Y. 1997) (holding that drug testing by employer does not violate the ADA unless it fails to reasonably accommodate known physical or mental limitations of otherwise qualified individual with disability or if it discriminates against qualified individual because of that individuals disability).

96. *See* EEOC Regulations to Implement the Equal Employment Provisions of the Americans with Disabilities Act, 29 C.F.R. § 1630.16(c) (3); EEOC and U.S. Dept. of Justice Civil Rights Division booklet entitled "Americans with Disabilities Act Questions and Answers" which can be found on the internet at http://www.usdoj.gov/crt/ada/ qandaeng.htm (stating that "If the results of a drug test reveal the presence of a lawfully prescribed drug or other medical information, such information must be treated as a confidential medical record.").

97. *Id.*

98. For example, this chapter does not cover the following federal environmental statutes in detail: (1) the National Environmental Policy Act of 1969, 42 U.S.C. §§ 4321-4370f, (2) the Toxic Substances Control Act, 15 U.S.C. §§ 2601-2692, (3) the Federal Insecticide, Fungicide, and Rodenticide Act, 7 U.S.C. §§ 136-136y, and (4) the Endangered Species Act, 16 U.S.C. §§ 1531-1544.

99. Section 301 of the CWA, 33 U.S.C. § 1311, prohibits the discharge of such pollutants as mine drainage, logging debris, agricultural runoff, cooling water from power plants, alteration of stream flows by sand or gravel mining, silt. Section 402 of the CWA, 33 U.S.C. § 1342, allows for discharge pursuant to a NPDES permit notwithstanding the Section 301, 33 U.S.C. § 1311, prohibition against discharge. In addition, Section 404 of the CWA, 33 U.S.C. § 1344, allows for the discharge any fill materials into waters of the United States pursuant to a permit issued under the joint administration of the U.S. Army Corps of Engineers and the EPA. An example of a comparable state law is Section 5650 of the California Fish and Game Code, which makes it unlawful to deposit in, permit to pass into or place where it can pass into the waters of this state certain pollutants.

100. 33 U.S.C. § 1362(6).

101. 33 U.S.C. § 1362(14).

102. 40 C.F.R. § 122.2; 33 U.S.C. § 1362(7).

103. 33 U.S.C. § 1342(b) & (c).

104. *See* EPA's NPDES Permit Writer's Manuel, Chapter 3, Section 3.1 *Types of Permits.* The NPDES Permit Writer's Manuel can be reviewed and downloaded for free at EPA's Office of Wastewater Management's Web site http://www.epa.gov/ow-owm.html/ permits/pwcourse/manual.htm.

105. 40 C.F.R. § 122.28.

106. EPA's NPDES Permit Writer's Manuel, Chapter 3, Section 3.2 *Major Components of a Permit*. The NPDES Permit Writer's Manuel can be reviewed and downloaded for free at EPA's Office of Wastewater Management's Web site at http://www.epa.gov/owm/permits/pwcourse/manual.htm.

107. For information on public review and comment requirements *see* 33 U.S.C. § 1251(e); 40 C.F.R. §§ 124.10-124.12, 123.25(a)(28)-(30) and for information on expiration and renewal of permit application deadline. *See* 40 C.F.R. § 122.21(d).

108. Section 402(p) of the CWA, 33 U.S.C. § 1342(p), exempts "discharges composed entirely of stormwater except for discharges that (a) were permitted before February 4, 1987, (b) are associated with industrial activity, (c) are from a municipal separate storm sewer system serving a population of 250,000, (d) are from a municipal separate storm sewer system serving a population of 100,000 or more but less than 250,000 and (e) contributes to a violation of a water quality standard or is significant contributor of pollutants as determined by the EPA or the state. The EPA is required to establish regulations setting forth the permit application requirements for (b)-(d) pursuant to 33 U.S.C. § 1342(p)(4). For a discussion of the implementation of the Phase I and Phase II of the NPDES Storm Water Program go to the EPA Office of Wastewater Management's discussion in "NPDES Permit Program — General Information" at its Web site http://www.epa.gov/owm/gen2.htm.

109. *Id.*

110. *Id.*

111. *See* EPA Office of Wastewater Management's discussion in "Phase II of the NPDES Storm Water Program" at its Web site http://www.epa.gov/owm/sw/phase2/index.htm.

112. *Id.*

113. 33 U.S.C. § 1317(b). For additional information on pretreatment standards for POTWs see EPA's Introduction to the National Pretreatment Program, which can be downloaded for free from the Office of Wastewater Management's Web site at http://www.epa.gov/owm/permits/pretreat/final199.pdf.

114. 40 C.F.R. § 403.1(a).

115. 40 C.F.R. §§ 403.5, 403.6. For additional information on pretreatment standards see EPA's Introduction to the National Pretreatment Program, Chapter 3, which can be downloaded for free from the Office of Wastewater Management's Web site at http://www.epa.gov/owm/permits/pretreat/final199.pdf.

116. 40 C.F.R. § 403.6.

117. 40 C.F.R. § 403.5.

118. 40 C.F.R. § 403.5(b).

119. 33 U.S.C. § 1319(c)(1) (A).

120. 33 U.S.C. § 1319(c)(3).

121. 33 U.S.C. § 1319(c)(4).

122. 33 U.S.C. § 1365.

123. *See* discussion in EPA's Office of Water's "Summary of the History of the UIC Program" which can be downloaded for free from the Office of Water's Web site at http://www.epa.gov/safewater/uic/history.html. The EPA also cites numerous other statutes as giving it authority to protect groundwater, including the Resource Conservation and Recovery Act, the Comprehensive Environmental Response, Compensation and Liability Act, the Federal Insecticide, Fungicide, and Rodenticide Act, the Toxic Substance Control Act, and the Clean Water Act. *see* EPA's Office of Water's guide entitled "Citizen's Guide to Groundwater Protection," which can be found at the Office of Water's Web site at http://www.epa.gov/safewater/protect/citguide.html.

124. *Id.;* 42 U.S.C. §§ 1421(b)(1), 1421(d)(2).

125. 42 *see also* U.S.C. § 1421(b)(2)(A) & (B).

126. 42 U.S.C. § 1421(b)(1)(A); 40 C.F.R. § 144.12.

127. 42 U.S.C. §§ 1421 & 1422. Currently, the majority of states have EPA approved UIC programs. However, more than fifteen states do not have an EPA fully approved UIC program. In these states, the EPA has primary or joint responsibility for the UIC program. *See* EPA's Office of Water's article "State UIC Programs" at the Office of Water's Web site at http://www.epa.gov/safewater/uic/primacy.html.

128. *See* EPA's Office of Air Quality Planning and Standards publication "The Plain English Guide to the Clean Air Act" for an overview of the CAA. This guide can be found at the EPA's Office of Air Quality Planning and Standards' Web site at http://www.epa.gov/oar/oaqps/peg_caa/pegcaa02.html. Other useful information can also be found starting at the home page for the EPA's Office of Air Quality Planning and Standards at http://www.epa.gov/oar.

129. 42 U.S.C. § 7409.

130. 42 U.S.C. § 7412.

131. 40 C.F.R. §§ 50.1–50.12.

132. 42 U.S.C. § 7412; 40 C.F.R. Part 61.

133. 42 U.S.C. §§ 7407(a), 7410.

134. 42 U.S.C. § 7412(l).

135. 42 U.S.C. §§ 7407(a), 7410, 7412(l).

136. 42 U.S.C. § 7409(a) (1) & (2).

137. 42 U.S.C. § 7409(b) (1) (A) (i).

138. 42 U.S.C. § 7407(d) (1) (A) (i).

139. 42 U.S.C. §§ 7407(a), 7410.

140. 42 U.S.C. §§ 7521, 7545.

141. 42 U.S.C. §§ 7661a–7661f.

142. 42 U.S.C. § 7661a; 40 C.F.R. Parts 70 & 71.

143. 42 U.S.C. § 7661c(a)-(c).

144. *See* EPA's Clean Air Market Programs Web site at http://www.epa.gov/airmarkets/index.html; Emissions Trading Handbook at the Emissions Trading Education Initiative's Web site at http://www.etei.org/handbook_changed.htm.

145. There are numerous emission credit websites that provide information and online services relating to marketing emission credits, including the Emission Marketing Association's Web site at http://www.emissions.org/ and The Cantor Fitzgerald Environmental Brokerage Services emissions trading Web site at http://www.cantor.com/ebs/.

146. 42 U.S.C. § 7413(b) & (c).

147. 42 U.S.C. § 7413(c) (5) (A).

148. 42 U.S.C. § 7604.

149. 42 U.S.C. §§ 6901–6992k.

150. 42 U.S.C. § 6926.

151. 42 U.S.C. § 6922.

152. 40 C.F.R. § 262.11.

153. 42 U.S.C. § 6903(5).

154. 40 C.F.R. §§ 261.31, 261.32, 261.33. The non-specific source hazardous wastes are given numbers starting with the letter "F" and the site specific source hazardous wastes are given numbers starting with the letter "K." Commercial chemicals that must be treated as a hazardous waste if discarded or spilled are given numbers starting with the letter "U" and if deemed acutely hazardous waste when discarded or "P."

155. EPA Office of Solid Waste's RCRA Orientation Manual, Section III, Chapter 1, pages III-24 to III-25, a copy of which can be found at the EPA's Web site at http://www.epa. gov/epaoswer/general/orientat/. *See* 40 C.F.R. §§ 261.3, 261.20–261.24.

156. EPA Office of Solid Waste's RCRA Orientation Manual, Section III, Chapter 1, pages III-7, III-27 to III-29. *See* 40 C.F.R. §§ 261.2(c), 261.3(a) (2), 261.3(b), 261.3(f).

157. 40 C.F.R. Part 262.40.

158. 42 U.S.C. § 6922(b); 40 C.F.R. Appendix to Part 262 — Uniform Hazardous Waste Manifest and Instructions (EPA Forms 8700-22 and 8700-22A and their instructions).

159. 40 C.F.R. § 262.12(c).

160. 40 C.F.R. § 262.34(c).

161. 40 C.F.R. § 262.34(a).

162. 40 C.F.R. §§ 262.34(d)-(f), 261.5.

163. 42 U.S.C. § 6922(a) (2).

164. 40 C.F.R. §§ 262.20, 262.23. *See also* 40 C.F.R. Appendix to Part 262 — Uniform Hazardous Waste Manifest and Instructions (EPA Forms 8700-22 and 8700-22A and their instructions).

165. 40 C.F.R. Appendix to Part 262 — Uniform Hazardous Waste Manifest and Instructions (EPA Forms 8700-22 and 8700-22A and their instructions).

166. 40 C.F.R. § 262.42.

167. 40 C.F.R. § 262.40(a).

168. 40 C.F.R. § 260.10 definitions of "on-site" and "transporter."

169. 40 C.F.R. § 263.11.

170. 40 C.F.R. § 263.20.

171. 42 U.S.C. §§ 6928(c), 6928(d), 6928(e), 6928(g).

172. 42 U.S.C. §§ 6928(d) (3), 6928(e).

173. 42 U.S.C. § 6928(e).

174. 42 U.S.C. § 6973.

175. 42 U.S.C. §§ 11001–11050.

176. 42 U.S.C. § 11003.

177. 42 U.S.C. § 11004.

178. 42 U.S.C. § 11022.

179. 42 U.S.C. § 11023.

180. 42 U.S.C. § 11001(a)-(d).

181. 42 U.S.C. § 11002(a) (3); 40 C.F.R. Part 355, Appendix A and B; 40 C.F.R. § 355.30(e).

182. 42 U.S.C. §§ 11002(b) (1), 11002(c), 11003(d)(1); 40 C.F.R. § 355.30(a)-(c).

183. 42 U.S.C. § 11003(d) (2); 40 C.F.R. § 355.30(d)(1)-(2).

184. 42 U.S.C. § 11004(a); 40 C.F.R. § 355.40(a)(1).

185. 42 U.S.C. § 11004(b) (1); 40 C.F.R. § 355.40(b)(1).

186. 42 U.S.C. § 7401- 7671q.

187. 33 U.S.C. § 1251–1387.

188. 42 U.S.C. §§ 11021, 11022.

189. 42 U.S.C. §§ 11021, 11022.

190. 42 U.S.C. § 11042(a)-(i); 40 C.F.R. §§ 350.5-350.25.

191. 42 U.S.C. § 11023.

192. *Id. see also* EPA's Toxic Release Inventory: Community Right-to-Know Web site at http://www.epa.gov/tri/.

193. 42 U.S.C. § 11045.

194. 42 U.S.C. § 11045(b); 40 C.F.R. § 355.50.

195. 42 U.S.C. § 11046(a).

196. Courts have consistently held that CERCLA imposes strict liability on potentially responsible parties. *See, e.g., United States v. Colorado & E.R.R.*, 50 F.3d 1530, 1535 (10th Cir. 1998); *Bedford Affiliates v. Sills*, 156 F.3d 416, 423 (2nd Cir. 1998); *B.F. Goodrich v. Betkoski*, 99 F.3d 505, 514 (2nd Cir. 1996);*United States v. Mexico Feed & Seed Co.*, 980 F.2d 478, 484 (8th Cir. 1992); *Dedham Water Co. v. Cumberland Farms Dairy, Inc.* 889 F.2d 1146, 1152-53 (1st Cir. 1989); *New York v. Shore Realty*, 759 F.2d 1032, 1044 (2nd Cir. 1985).

197. *See, e.g., OHM Remediation Servs. v. Evans Cooperage Co.*, 116 F.3d 1574, 1578 (5th Cir. 1997); *Uniroyal Chem. Co. v. Deltech Corp.*, 160 F.3d 238, 242-243 (5th Cir. 1998); *United States v. Stringfellow*, 661 F. Supp. 1053 (C.D. Cal. 1987); *O'Neil v. Picillo*, 682 F. Supp. 706 (D.R.I. 1988), *aff'd* 883 F.2d 176 (1st Cir. 1989).

198. 42 U.S.C. § 9607(a) (1)-(4). CERCLA uses the term facility which for simplification and ease of understanding we have referred to as the contaminated site.

199. *See, e.g., B.F. Goodrich v. Betkoski*, 99 F.3d 505 (2nd Cir. 1996); *North Shore Gas Co. v. Salomon Inc.*, 152 F.3d 642, 649 (7th Cir. 1998); *Smith Land & Improvement Corp. v. Celotex Corp.*, 851 F.2d 86 (3d Cir. 1988).

200. *See United States v. Bestfoods*, 524 U.S. 51 (1998).

201. *See, e.g., Donahey v. Bogle*, 129 F.3d 838, 845 (6th Cir. 1997); *State v. Markowitz*, 710 N.Y.S. 2d 407, 411-12 (Sup. Ct. 2000); *United States v. Northeastern Pharmaceutical & Chemical Co.*, 579 F. Supp. 823 (W.D. Mo. 1984), *aff'd in part, rev'd in part*, 810 F.2d 726 (8th Cir. 1986); *Kelley Michigan v. ARCO Industries Corp.*, 723 F. Supp. 1214 (W.D. Mich. 1989).

202. *See, e.g., Uniroyal Chem. Co. v. Deltech Corp.*, 160 F.3d 238, 242-243 (5th Cir. 1998); *New York v. Shore Realty Corp.*, 759 F.2d 1032 (2nd Cir. 1985).

203. A look at the various former disposal facility sites that are listed as superfund sites on the EPA Superfund Web site http://www.epa.gov/superfund/ will confirm this.

204. Susan M. Cooke, *Law of Hazardous Waste — Management, Cleanup, Liability and Litigation* § 13.02[1] (2001) Matthew Bender.

205. Susan M. Cooke, *Law of Hazardous Waste — Management, Cleanup, Liability and Litigation* § 1.02[1] (2001) Matthew Bender.

206. *See, e.g., Lewis v. General Elec. Co.*, 37 F. Supp. 2d 55, 61 (D. Mass. 1999) (stating that plaintiff's inability to sell her property because of defendant's interference with public health and environment represents special injury enough for a public nuisance claim and that interference with use and enjoyment of land sets forth a theory of private nuisance upon which plaintiff may recover); *Peters v. Amoco Oil Co.*, 57 F. Supp. 2d 1268, 1278-79 (M.D. Ala. 1999) (stating that allegations regarding defendants' contamination of plaintiff's properties were sufficient for a trespass claim); *Crawford v. Nat'l Lead Co.*, 784 F. Supp. 439, 442-43 (S.D. Ohio 1989) (noting that strict liability applies in situations with abnormally dangerous conditions and activities); *Kolnick v. Fountainview Ass'n*, 737 So. 2d 1192 (Fla. Dist. Ct. App. 1999) (alleging negligence against a housing association for lack of maintenance that resulted in harmful exposure to mold); *Moore v. Texaco, Inc.*, 244 F.3d 1229, 1233 (10th Cir. 2001) (explaining that plaintiff needs to show proof of violations of certain pollution statutes for negligence per se); *Vandenberg v. Superior Court*, 88 Cal. Rptr. 2d 366, 21 Cal. 4th 815 (1999) (alleging equitable indemnity where an individual installed and operated waste oil storage tanks that were the source of oil contamination); *Pacific Indemnity Co. v. Bellefonte Ins. Co.*, 95 Cal. Rptr. 2d 911, 80 Cal. App. 4th 1226 (2000) (holding that equitable contribution requires proration of costs when the defendant's insured breaches its duty to protect a bay).

207. *See, e.g., Lousiana-Pacific Corp. v. ASARCO, Inc.*, 24 F.3d 1565, 1571 (9th Cir. 1994); *Old Republic Ins. Co. v. Superior Court*, 66 Cal. App. 4th 128, 77 Cal. Rptr. 2d 642 (1998).

208. *See, e.g., Carter-Wallace, Inc. v. Admiral Ins. Co.*, 712 A.2d 1116, 1121 (N.J. 1998).

209. *See, e.g., Blaine Constr. Corp. v. Insurance Co. of N. Am.*, 171 F.3d 343 (6th Cir. Tenn. 1999); *Kolnick v. Fountainview Ass'n*, 737 So. 2d 1192 (Fla. Dist. Ct. App. 3d Dist. 1999); *Centex-Rooney Constr. Co. v. Martin County*, 706 So. 2d 20 (Fla. Dist. Ct. App. 4th Dist. 1997). *See also* Van Voris, *Sexy it's Not, but Mold is Real Hot*, The National Law Journal, June 4, 2001 which can be found on the internet at http://www.law.com/cgibin/gx.cgi/AppLogic+FTContentServer?pagename=law/ View&c=Article&cid=ZZZRJBBCGNC&live=true&cst=1&pc=0&pa=0&s=News& ExpIgnore=true&showsummary=0.

210. 42 U.S.C. § 9601(35) (A) (i).

211. 42 U.S.C. § 9601(35) (B).

212. 42 U.S.C. §§ 12101–12213. Employers, for example, may not ask, on an application form or during an interview, whether and to what extent an individual is disabled. The employer only may describe a job and ask whether the applicant can perform the essential elements of the job. If the applicant's disability is obvious, or is known to the employer, the employer may ask the applicant to describe or demonstrate how (with or without reasonable accommodation) the applicant can do the job. The ADA also prohibits pre-employment physical screening until an offer has been extended. *Id.* § 12112(d) (2), 29 C.F.R. 1530.13(a); 1630.14(a), (b). If a disability is discovered, the employer cannot withdraw its offer until it meets the "reasonable accommodation" standard (i.e., a modification to the job that allows the disabled applicant to perform the essential job functions). U.S. Equal Employment Opportunity Commission, Enforcement Guidance: Pre-employment Disability-Related Questions and Medical Examinations (Oct. 19, 1995); *see* U.S. Equal Employment Opportunity Commission, Enforcement Guidance on Reasonable Accommodation and Undue Hardship Under the ADA (1999).

213. 42 U.S.C. § 12131.

214. 29 U.S.C. § 791.

215. 42 U.S.C. § 12111.

216. 42 U.S.C. § 12182.

217. 42 U.S.C. § 12102.

218. 42 U.S.C. § 12113(b); *see* ROTHSTEIN, EMPLOYMENT LAW § 2.44.

219. 29 C.F.R. §§ 1600–1699.

220. *Sutton v. United Air Lines, Inc.*, 527 U.S. 471, 119 S.Ct. 2139, 144 L.Ed.2d 450 (1999); *Murphy v. United Parcel Service*, 527 U.S. 516, 119 S.Ct. 2133, 144 L.Ed.2d 484 (1999); *Albertsons, Inc. v. Kirkingburg*, 527 U.S. 555, 119 S.Ct. 2162, 144 L.Ed.2d 518 (1999).

221. U.S. Equal Employment Opportunity Commmission, Americans with Disabilities Act of 1990 Charges (last modified Jan. 18, 2001), available at http://www.eeoc.gov./ stats/ada.html.

222. *See, e.g.*, U.S. General Accounting Office, Siting of Hazardous Waste Landfills and Their Correlation with Racial and Economic Status of Surrounding Communities, GAO/RCED-83-168 (1983); Dominique R. Shelton, *The Prevalent Exposure of Low Income and Minority Communities to Hazardous Materials*, 32 BEVERLY HILLS B. ASSOC. J. 1 (1997).

223. *See, e.g., Chester Residents Concerned for Quality Living v. Seif*, 132 F.3d 925 (3rd Cir. 1997), *vacated*, No. 97-1620 *U.S. Lexis 4604 (1999)*.

224. Executive Order No. 12898 directs federal agencies to collect, analyze, and assess data on environmental health risks in disadvantaged communities. 59 Fed. Reg. 7629 (Feb. 11, 1994). The order has had a substantial impact on the EPA's regulatory programs for clean water, clean air, community right-to-know, and hazardous wastes control — as well as the exercise of its prosecutorial discretion. The order also has resulted in lawsuits challenging expansion of landfills and incinerator projects. *See, e.g., Chester Residents Concerned for Quality Living v. Seif*, 132 F.3d 925 (3rd Cir. 1997), *cert. granted*, 118 S. Ct. 2296 (1998), *vacated, U.S. Lexis 4604 (1998).*

225. *See, e.g.,* EPA Office of Solid Waste and Emergency Response, Environmental Justice Action Agenda, May 1995; Elliott P. Laws, "Integration of Environmental Justice into OSWER Policy," memorandum dated Sept. 1994. Both documents are available at http://www.epa.gov/swerosps/ej.

226. U.S. Dept. of Transportation Environmental Justice Strategy, 60 Fed. Reg. 33896, June 21, 1995; U.S. Dept. of Transportation Order on Environmental Justice to Address Environmental Justice in Minority Populations and Low-Income Populations, 62 Fed. Reg. 18377 (April 15, 1997). Both documents are available at http://www.epa.gov/swerosps/ej. *See also* U.S. Dept. of Transportation, Federal Highway Administration, FHWA Actions to Address Environmental Justice in Minority Populations and Low-Income Populations (DOT Order 6640.23) (Dec. 2, 1998).

227. *See* Council on Environmental Quality, Environmental Justice Guidance Under the National Environmental Policy Act (Dec. 1997).

228. Office of Fed. Activities, U.S. Envtl. Protection Agency, Final Guidance for Incorporating Environmental Justice Concerns in EPA's NEPA Compliance Analyses 1.1 (Apr. 1998).

229. EPA Interim Guidance for Investigating Title VI Administrative Complaints Challenging Permits, Feb. 1998.

230. *See, e.g.,* Hillary Gross, et al., ENVIRONMENTAL JUSTICE: A REVIEW OF STATE RESPONSES 2 (2000) (unpublished, U. Cal. Hastings College of Law) (on file with author) *Tennessee Receives an Environmental Justice Grant*, 10 TENNESSEE ENVTL. L. LETTER (Nov. 1998), available at http://www.lexis.com.

231. *See, e.g.,* Christopher J. Foreman, Jr., THE PROMISE AND PERIL OF ENVIRONMENTAL JUSTICE, Brookings Institute (1998); Jimmy White, *Environmental Justice: Is Disparate Impact Enough?*, 50 MERCER L. REV. 1155, 1159 n.16 (1999) (explaining that studies to determine whether the disproportionate distribution of waste treatment facilities in minority communities is a result of race-based siting decisions or "a result of minorities 'coming to' the hazard due to lower property values" have been less than conclusive) (citing Richard J. Lazarus, *Pursuing "Environmental Justice:" The Distributional Effects of Environmental Protection*, 87 Nw. U. L. REV. 787, 802 n.56) (critiquing 1987 United Church of Christ Commission Study on the distribution of waste facilities because it relies upon "present demographic data rather than the demographic data pertaining to the time that the initial siting decision may have been made").

232. *See, e.g., Road Design and Design and Construction: Where Philosophy Meets the Road; Environmental Justice*, 2 PUBLIC WORKS (Feb. 1, 2001) (describing a successful highway project in which public comment was solicited and key suggestions followed), available at http://www.lexis.com.

233. The U.S. Court of Appeals for the 11th Circuit issued a stay on July 18, 2000 prohibiting the U.S. Department of Transportation (DOT) from determining that the Atlanta metropolitan area was in conformity with Clean Air Act standards and approving Atlanta's transportation improvement program and long-range transportation plan

using a projected vehicle emissions budget. The DOT responded by making a conformity determination based on 1996 vehicle emissions figures.

234. *See, e.g., Rozar v. Mullis,* No. 95-8227 (11th Cir. June 17, 1996) (environmental justice claim dismissed because no discriminatory intent shown); *NAACP, et al. v. Engler,* (E.D. Mich. 1995).

235. U.S. Environmental Protection Agency, Brownfields Tax Incentive Economic Redevelopment Initiative, included in Taxpayer Relief Act, P.L. 1005-34, title IX, subtitle E (Aug. 5, 1997). *See also* http://www.epa.gov/swerosps/bf/. The EPA defines "brownfields" as "abandoned, idled, or underused industrial and commercial facilities where expansion or redevelopment is complicated by real or perceived environmental contamination." Office of Solid Waste and Emergency Response, U.S. Environmental Protection Agency, Brownfields Glossary of Terms, available at http://www.epa.gov/swerosps/bf/glossary.htm (last modified Sept. 30, 1997). *See also* U.S. ENVIRONMENTAL PROTECTION AGENCY, LIABILITY AND OTHER GUIDANCE FACT SHEET (April 1997) (defining a "brownfield" as a site, or portion thereof, that has actual or perceived contamination and an active potential for redevelopment or reuse").

236. *See* EPA Environmental Justice Program Results in Large Fines for 10 Los Angeles Facilities, *California Environmental Insider,* April 16, 2001, 14(21):8.

237. U.S. Sentencing Commission, GUIDELINES MANUAL, ch. 8 – Sentencing of Organizations.

238. Other key components of an effective compliance program include proper communication to and education of employees of the company's ethics code and compliance policies, providing adequate personnel to oversee and operate the program, and providing employees with access to report perceived violations of company policy or law.

239. *See* U.S. Sentencing Commission, GUIDELINES MANUAL § 8C2.1.

240. EPA Final Policy Statement on Incentives for Self-Policing of Violations, signed April 3, 2000, available at EPA's Web site, http://www.epa.gov/oeca (hereafter, "EPA Final Audit Policy").

241. EPA Final Audit Policy, *supra.*

242. *Id.*

243. OSHA, *Voluntary Protection Programs,* 47 Fed. Reg. 29025, July 2, 1982.

244. OSHA, *Revisions to the Voluntary Protection Programs to Provide Safe and Healthful Working Conditions.* 65 Fed. Reg. 45649, July 24, 2000.

245. And including the civil rights laws, e.g., the Americans with Disabilities Act, supra.

246. *See, e.g.,* CERCLA § 103, 42 U.S.C. § 9603 subdiv. (a) (and state counterparts, e.g., Cal. Health & Safety Code § 25359.4).

247. *See* EPA Final Audit Policy, *supra.*

248. The states include **Alaska** (ALASKA STAT. §09.25.450–490 (1997)), **Arkansas** (ARK. CODE ANN. §8-1-301–312 (Michie 1997)), **Colorado** (COLO. REV. STAT. ANN. §13-25-126.5 (1997) (Privilege); COLO. REV. STAT. ANN. §25-7-122 (Immunity) (1997)), **Illinois** (ILL. REV. STAT. ch. 5, para. 52.2 (1997)), **Indiana** (IND. CODE. ANN. §13-11-2-69 (1997); IND. CODE. ANN. §13-28-4-1~10 (1997)), **Iowa** (IOWA CODE §455K.1–.13 (1998)), **Kansas** (KAN. STAT. ANN. §60-3332–3339 (1996)), **Kentucky** (KY. REV. STAT. ANN. §224.01- 040 (1997)), **Michigan** (MICH. STAT. ANN. §13A.14801–14810 (1997)), **Minnesota** (MINN. STAT. ANN. §114C.20–.31 (1997)), **Mississippi** (MISS. CODE ANN. §49-2-71 (1997) (Privilege); MISS. CODE ANN. §17-17-29 (1997) (penalty mitigation)), **Montana** (MONT. CODE ANN. §75-1-1201–1206 (1997)), **Nebraska** (NEB. REV. STAT. § 25-21, 254-264 (2001)), **Nevada** (NEV. REV. STAT.

§445C.010–.120 (1997)), **New Hampshire** (N.H. Rev. Stat. Ann. §147-E (1996)), **New Jersey** (N.J. Rev. Stat. §13:1D-125–132 (1996)), **Ohio** (Ohio Rev. Code Ann. §3745.70–.74 (1997)), **Oregon** (Or. Rev. Stat. §468.963 (1997)), **Rhode Island** (R.I. Gen. Laws §42-17.8-1–8 (1997)), **South Carolina** (S.C. Code Ann. §48-57-10–110 (1997)), **South Dakota** (S.D. Codified Laws Ann. §1-40-33–37 (1997)), **Texas** (Tex. Rev. Civ. Stat. Ann. art. 4447cc (1997)), **Utah** (Utah Code Ann. §19-7-101–108 (1997) (Privilege); Utah Code Ann. §19-7-109 (1997) (Immunity); Utah R. Evid. 508), **Virginia** (Va. Code Ann. §10.1-1198–1199 (1997)), and **Wyoming** (Wyo. Stat. §35-11-1105–1106 (1997)). **Idaho**'s statutory provision, formerly codified at Idaho Code § 9-801–811, expired in December 1997.

249. These include such protected communications as those between spouses, doctors and patients, clergymen and penitents, domestic violence victims and counselors, etc. *See* Calif. Evid. Code, Ch. 4.

250. *See*, for example, *Olen Properties Corp. v. Sheldahi Inc.*, No. 9l-6446-WDK-JWM, 38 ERC 1887, 1994 U.S. Dist. Lexis 71725 (C.D. Calif., April 2, 1994). A supervisor testified in an affidavit that he prepared environmental audit memoranda to gather information for company in-house attorneys to assist the company in evaluating compliance with environmental laws.

251. The attorney work product doctrine can be overcome if the opposing party demonstrates substantial need and undue hardship and expense in obtaining the same information from other sources.

252. This privilege was originally recognized in *Bredice v. Doctor's Hospital, Inc.*, 50 F.R.D. 249 (D.D.C. 1970), *aff'd without opin.*, 479 F.2d 920 (D.C. Cir. 1973), where doctors' peer review records were protected from disclosure to survivors of a deceased patient in a medical malpractice case. The court held that such peer review was necessary to improve health care and would not be conducted if discussions were discoverable. In *Reichhold Chemicals, Inc. v. Textron, Inc.*, No. 92-30393-RV, 1994 U.S. Dist. Lexis 13806 (N.D. Fl. Sept. 20, 1994), documents prepared by a company during its investigation of ground-water contamination are immune from discovery because of the public interest in companies candidly assessing their compliance with past regulations.

253. *See U.S. v. MacDonald & Watson Waste Oil Co.*, No. CR-88-0032 (D.R.I. 1990), *affd in part, vacated in part*, 933 F.2d 35 (1st Cir. 1991) (internal environmental audits can be used to prove that violations were "knowing"; *U.S. v. Hammermill Paper, Inc.*, 796 F. Supp. 1474 (S.D. Ala. 1992) (OSHA subpoena of voluntary safety audits upheld, against a claim of self-evaluative privilege).

254. *Environmental Auditing Policy Statement*, 51 FR. 25004 (July 9, 1986). EPA's view of an effective environmental compliance auditing program contains seven elements: (1) top management support and commitment to follow-up; (2) independence; (3) adequate staffing and training; (4) explicit objectives, work, commitment of resources, and regular audits; (5) sufficiency of the process; (6) candid and clear written reports of findings, corrective actions, and schedules for implementation; and (7) quality assurance procedures, training, and accountability.

255. *Factors in Decisions on Criminal Prosecutions for Environmental Violations in the Context of Significant Voluntary Compliance or Disclosure Efforts by the Violator*, July I, 1991. In fact, some of U.S. Department of Justice "factors," such as immediately turning over to the government information and surrendering the names of all involved individuals to the prosecuting agency will be regarded by companies as an unwarranted intrusion into corporate management.

256. OSHA Self-Audit Policy, *supra*, n. 44.

257. *See, e.g.,* EPA's Office of Regulatory Enforcement Multimedia Division's internet site at http://es.epa.gov/oeca/ore/med/; U.S. Environmental Protection Agency News Release on multimedia inspection of National Steel in Detroit Michigan which can be found at http://www.epa.gov/Region5/news97/97opa267.htm (stating that Region 5 April 22 -May 3, 1996 multimedia inspection of National Steel resulted in the filing of 31 separate counts of environmental violations, including: 13 PCB-related violations of the Toxic Substances Control Act, 9 hazardous-waste related violations and 3 underground storage tank violations of the Resource Conservation and Recovery Act, 4 benzene-related violations of the Clean Air Act, and 2 violations of the Emergency Planning and Community Right-to-Know Act. *See also* Office of Inspector General Audit Report *Enforcement EPA's Multimedia Enforcement Program* Report No. 2000-P-000018 (June 30, 2000) (results of audit performed by Office of Inspector General because of concern that EPA was not taking a multimedia approach to enforcement in areas where it would be most beneficial and making recommendations to assist EPA in implementing more multimedia inspections), a copy of which can be found on the internet at http://www.epa.gov/oigearth/audit/list900/MMfinal1.pdf.

258. Certain environmental and OSHA documents must be kept in certain areas pursuant to law or permit conditions and, thus, the original or a copy of such document may need to be located in another area as well as the central location. For example, an employer must ensure that a copy of any applicable Material Safety Data Sheets ("MSDSs") "are readily accessible during each work shift to employees when they are in their work area(s)" pursuant to 20 C.F.R. § 1910.1200(g) (8). *See* OSHA Standards Interpretation and Compliance Letters 12/07/1999 — Employee access to MSDSs required by 1910.1200 vs. 1910.1020 a copy of which is posted on the internet at http://www.osha-slc.gov/OshDoc/Interp_data/I19991207A.html.

259. *See, e.g., United States v. Charles George Trucking Co.,* 823 F.2d 685, 688-89 (1st Cir. 1987) (upholding lower court's ruling that failure to timely furnish information requested by inspector was a violation of Section 3007(a) of RCRA, 42 U.S.C. § 6927(a), and was subject to the civil penalties set forth under Section 3008(g) of RCRA, 42 U.S.C. § 6928(g) and imposing a fine of $20,000.00 to each individual for late submission of information); *U.S. v. Barkman,* 784 F. Supp. 1181 (E.D. Pa. 1992) (holding that individual unreasonably failed to timely comply with information request by EPA in violation of Section 104(e) of CERCLA, 42 U.S.C. § 9604(e), and imposing a fine of $55 for each of the 700 days the submission of information was late for a total of $38,500.00 on individual pursuant to Section 104((5) (B) of CERCLA, 42 U.S.C. § 9604(e) (5) (B)).

260. *See* discussion and footnotes in the preceding section entitled "Protecting Self-Evaluations and Audits from Disclosure: The Law."

261. *See, e.g.,* 42 U.S.C. § 11042(a)-(i) and 40 C.F.R. §§ 350.5-350.25 (providing trade secret protections under the federal Community Right to Know law or EPCRA).

262. *See* EPA Region 4's *Environmental Investigations Standard Operating Procedures and Quality Assurance Manual,* §§ 2.3.3 (entitled "Unreasonable Search and Seizure") and 2.3.4 (entitled "Requesting Information") (May 1996), a copy of which can be found on the internet at http://www.epa.gov/Region4/sesd/eisopqam/eisopqam.pdf.

CHAPTER **30**

An Overview of Product Liability

Kenneth M. Colonna and Steven M. McConnell

CONTENTS

FOREWORD

The writings in this chapter do not focus on the legal aspects of product liability. Discussions of the legal mechanics of product liability are reserved for texts that focus on business law and are beyond the expertise of the authors.

The chapter does present the basis for product risk assessment within the framework of a product safety management program. It is intended to provide the readers guidance for assessing and reducing risk in the design, manufacture, and distribution of their company's products.

1-56670-370-0/02/$0.00+$1.50
© 2002 by CRC Press LLC

CASE STUDY AND INTRODUCTION

The following is a typical case study of product liability. Judy worked for your customer, a large food-packaging company, for over 16 years. In all her time on the job, she always worked on the can–packaging line and was an experienced operator who knew each piece of equipment on the line. Her normal job at the time of the accident was label operator. In her capacity as label operator, she ran the packaging equipment that applied labels to finished sealed containers of her company's product. Traditional practices dictated that Judy and her co-workers rotate to each piece of equipment during a shift for relief of the normal operator during breaks and as a means of training. At the time of her accident, Judy was operating the can-filling machine. The can filler, a large rotating piece of equipment capable of filling 10-ounce cans of food products at a rate of 1100 per minute, is not difficult to operate but requires certain skills acquired only from experience. It was common for operators to make minute adjustments to the filler during the day. This practice is known as fine tuning the run. At the time of her accident Judy was fine tuning the filler, which had become necessary because cans were jamming in the filler as they moved off the rotating machine and onto the conveyor for sealing. As was common at this plant, operators would often reach into the area of the machine where filled cans discharged in order to remove jams or product that had spilled from the filler and accumulated in this area. This process was typically done while the machine was running. To accomplish this task, the operator must reach over a nylon gear approximately one foot in diameter that acts as a guide for the cans as they discharge from the machine. Judy had reached across the gear to remove a can that had jammed in the machine. She was wearing a long-sleeved uniform shirt with the cuffs unbuttoned. As she reached into the machine, her shirt caught on the nylon wheel, pulling her arm and eventually her upper torso into the machine. As a result of her accident, Judy's left arm was amputated, and she suffered permanent brain damage as a result of oxygen deprivation that occurred when her chest was compressed by the machine, rendering her unable to breathe for several minutes.

Judy's employer paid her the workers' compensation benefits allowable under the state statutes where she lived. The settlement for permanent total disability, however, was barely enough to cover all the expenses that a single mother of two children incurs, especially since Judy requires assistance in managing many everyday tasks such as bank accounts, bills, and normal expenses that are more difficult for her to understand today.

Judy's attorney has decided to pursue a product liability suit against your employer. The basis of the suit revolves around your company's design of the equipment and the machine guards. Her attorney alleges that your employer built a machine with known hazards and failed to follow generally accepted industry practices in designing machine guards for the machine. Additionally, the attorney alleges that, since your company employed professional engineers for design purposes, your company's engineers violated their professional duty by failing to exercise due care relative to the design of the machine.

The suit asks for damages in the amount of $5 million. It names both your employer and individual officers as defendants. You have been requested to aid in the investigation and in preparation of a defense. You will be working with your company's legal counsel and the company's insurance carrier in the investigation.

While this scenario is a compilation of different events known to the authors, it reflects the very real exposure all companies face in developing and selling products or services. As a safety professional you have an obligation to your employer and your profession to participate in the assessment of your company's products and services. In today's evolving workplace, where downsizing is commonplace, the safety professional must assume a broader role in all phases of risk assessment. Where should the safety professional begin?

Product safety management is viewed differently from company to company. The level of importance that management gives to product safety is dependent on the type of products or services provided and their related hazards. Therefore, while some companies pay relatively little attention to product safety, others implement a comprehensive product safety management (PSM) program.

Customer misuse of products is difficult to eliminate completely. The manufacturer can, however, take certain steps to minimize the likelihood of liability suits. The manufacturer's best opportunity to prevent claims is to produce and sell products (including instructions when necessary) that are reasonably safe. The PSM concept can help minimize a company's exposure to product liability because safety is incorporated into the design, manufacturing, and marketing phases of the product.

A typical PSM program consists of people from a variety of departments, including design engineering, manufacturing, marketing, and distribution. Other team members represent quality assurance, purchasing, legal, insurance, human resources, safety, and the service organization. Each department is important to the overall success of the program and, therefore, to the overall success of the company.

It is important for management to demonstrate support for the program through commitments made in strategic or tactical plans, policy statements, and other objectives of the company that support the company's mission statement. Management must set the goals and objectives related to product safety and liability and designate overall responsibility for the program to an individual. The safety professional is often delegated the responsibility in small- to medium-size companies. In larger companies, and when multiple product lines are involved, the assignment may be given to someone in engineering. In some cases, the company hires a product safety director. Whatever the organizational structure, the safety professional will play an important role in the successful program. The safety professional's broad base of knowledge and experience will allow him or her to offer valuable insight across departmental lines. The safety professional's understanding of applicable regulatory requirement will help shape the control measures implemented by the company to minimize product liability. The safety professional can and should be involved in each phase of product's life.

PRODUCT DEVELOPMENT

From the moment an idea is brought forth within your company, the safety professional plays a crucial role in the development of the new product or service.

Usually, the process begins at the conceptual stage where discussions of products or services under consideration take place. The safety professional in any organization should take an active role in these discussions, offering informed input into the product or service being developed. Regulations that could offer barriers to the development of the product or service should be fully explored. Too often, companies overlook the significance of safety and environmental regulations that may affect their goods or services. These regulations should be considered both from the impact they will likely have on your company and from the impact they could have on your customers. A product is not worth developing if its use exposes your customers to increased risk of regulatory intervention. This is especially true in the professional services industries where certification or licensing requirements may exist. Often these requirements are not uniform between states and could create problems for your company or your customer.

For example, in many states, businesses confronted with regulations affecting underground storage of petroleum products may use licensed geologists to conduct assessments of possible leaks. Although this is appropriate in most states, some states, such as North Carolina, require that certain assessment documents be signed and stamped by a professional engineer, which is a discipline that may not be employed by many consulting firms. A company that offers consultation services to a customer in one state may find its efforts and its customer's resources wasted in another. Although this may not result in a product liability lawsuit, it does threaten your company's standing within the industry, where reputation is as important as the service provided.

MATERIALS AND MANUFACTURING

Once an idea has passed from the drawing board to the prototype stage, it becomes important for the safety professional to be involved in the process. Are the materials selected for use in manufacturing the product suitable? Has the engineering group considered possible uses of the product other than those intended by your company? Have assessments of possible failure modes and their effects been performed?

Materials selected for use in your product are critical to your company. Although the engineering department is more suited to materials selection, the safety professional can offer valuable input. Regulations such as hazard communication, process safety, and confined space entry should all be considered in the process of selecting the materials for construction. One large manufacturing company recently embarked on a multimillion–dollar project to install pollution control technology, only to discover that a material used in the pollution control process, ammonium hydroxide, would require the company to implement PSM techniques and install further pollution control technologies. This critical error went undiscovered by the professional engineers who were hired to design the process. The safety professionals employed by the company quickly recognized this problem early enough in the design process to allow for redesign, thereby avoiding a potentially costly error.

Materials selection alone is not the only important aspect of reducing risk for your company's product. Suitability for use is another. Here the safety professional can assist his or her company by employing proven techniques for evaluating the process or the product. Fault tree analysis, failure mode and effect analysis, and what if analysis are examples of hazard identification/analysis techniques available to the safety professional. It is beyond the scope of this chapter to explore these techniques in detail; however, the safety professional can find ample information on these methods and their application in texts and professional journals.

PRODUCTION

The manufacturing department is responsible for turning a reasonably safe and reliable design into a finished product. The safety professional can provide assistance in manufacturing by performing preliminary and specific hazard analysis, including the analysis of the process and materials used to complete fabrication of the product. The need for additional operating permits (e.g., air emissions permits) must be evaluated well in advance of the production process. Similarly, employee training and educational needs should be assessed and implemented prior to the initial trial run of the product.

Working closely with the quality assurance staff, the safety professional can assist in the development of quality techniques and measurement tools that will help track the manufacturing process. Although product rejects may be indications of failures or restraints in the manufacturing process, they may also help identify product design deficiencies and other product-related deficiencies.

PRODUCT PACKAGING

Although often overlooked in the process, product packaging is an important part of risk assessment for the safety professional. Errors in product packaging can lead to accidents and injuries on the user's end and can trigger product liability claims.

The safety professional can help determine if product packaging is acceptable. Will the packaging prevent damage to the product during shipment? Will the packaging prevent product contamination, corrosion, and deterioration during shipment and while on a retail outlet's shelf? The adequacy of the packaging should be assessed for suitability through the entire distribution channel. This assessment should consider conditions during manufacturing, transit, warehousing, and display.

The manner in which a product is presented through labeling and graphics can be interpreted by the end user as representation for use or suitability for a particular purpose not intended by your employer. Although labeling and graphics decisions are made by the marketing and product development groups and most often undergo legal scrutiny, the safety professional should also review the material and provide insights about the product packaging. It helps avoid embarrassment later if the safety professional discovers that the graphics on the product contain items that are clear violations of safety and health regulations.

A producer of industrial training films recently offered the authors an opportunity to preview a new safety video training program that was in the final stages of development. The production company was quite proud of the new product and its marketing materials as well as its ability to produce the product ahead of the competitors, which was a clear marketing advantage. They were somewhat embarrassed to learn that, after a great deal of time, effort, and editing, a critical segment of the video showed actors engaged in activities that were correct and in compliance with the regulation the topic addressed, but clearly in violation of other longstanding safety regulations. This example relates not only to a product-related problem, but it also illustrates the need for the safety professional to evaluate product packaging. The use of clip art as a graphic in packaging promotion, for example, could communicate a message not intended by the producer of the product.

OPERATING MANUALS AND OTHER MATERIALS

If your company produces a product that includes or requires operating instructions, maintenance manuals, or related training materials, you as the safety professional should be engaged in the review and development of these documents. Much like the packaging and labeling issues, manuals and other materials developed by your company should be reviewed to identify instructions or directions that fail to warn of possible hazards or that may mislead the user of your product to engage in activities that put him or her at risk.

Instruction or operating manuals related to the product should repeat hazard warnings and address how those hazards can be reduced or avoided by the user. Instructions for product inspection, assembly, and maintenance should be addressed. Any potential hazards associated with troubleshooting should be indicated, and preventive maintenance guidelines should include a discussion of pertinent safety issues. The safety professional is usually not responsible for drafting these documents, but a concise review could prevent unnecessary and avoidable problems after the product is sold.

Many courts have held that a manufacturer has a duty to warn customers about any reasonably foreseeable use of a product beyond the purposes for which it was designed. An adequate warning must advise the user of the hazards involved in the product's use (and in some cases, possible misuse), avoidance of these hazards, and potential consequences of failure to abide by those warnings.

MARKETING AND SALES

Understanding the market for your company's products or services is necessary for success in risk assessment, even though this concept is addressed during the early stages of a product or service's life cycle. Knowing who the target market is, what channels of distribution will be used to get your product to the consumer, and whether the product will be pushed to the consumer, or the marketing department will pull the consumer to the product, will help the safety professional develop a more comprehensive assessment.

It is also important to understand your company's sales strategy. How will the sales force sell the product? What type of training does the sales force receive? Is there a clear understanding of how the product is intended to be used? Do sales personnel understand the product's limitations? What type of sales support material is provided? Will your company offer free trials? If your product will be sold to the retail consumer through a distribution channel of retail outlets, how have the retailers trained their sales force? Do the retailers understand the proper uses and limitations of your company's products? If your company is selling services, does the consumer or customer understand the scope of services that your firm is capable of providing and what limitations exist on those services?

PSM includes a review of sales brochures and advertising materials. These materials should be evaluated by the PSM team to ensure that the product's features and capabilities are accurately depicted and illustrate only safe operating and maintenance procedures. The company's legal counsel should be included in this review cycle.

PRODUCT RECALL

Despite a company's best efforts, circumstances may arise that require a recall of all or a portion of the products it has sold. Other situations may require field modifications to products already in the hands of the end consumer. Here again, the safety professional may play a critical role. Does your company have a contingency plan to recall products? Is the chain of command during these crisis situations clear? Who will address the press during this period? What will be your company's approach in dealing with the public? A successful PSM program will address these questions.

Companies initiate product recalls or field-related modification programs when a substantial performance or safety defect exists in one or more of its shipped products. Such programs may also be considered when a company either is made aware of or has developed a substantial product improvement or change.

Product recalls and modification programs can prove costly to a company — both in dollars and in public perception. Either effort requires a substantial commitment of resources. It should be recognized that the basis for implementing either product recall or modification can often be traced to inadequate PSM efforts.

It is easy to understand why your company should develop a contingency plan for rapid implementation of product recalls or field modifications. Quick action can often minimize the costs and improve a company's image during times of adversity. McNiel-PPC, Inc., the maker of Tylenol®, is an excellent example of a manufacturer with well-planned and well-implemented strategy for product recall. However, the reader should recognize that a successful PSM program strives to identify and eliminate the product hazards and misuses and, therefore, the need for most product recalls and field modification programs.

CONCLUSION

Regardless of size, any company producing products or supplying services should assess its exposure to potential product liability issues and take proactive steps to ensure that a PSM program of appropriate size and effort is in place to reduce or eliminate associated risks.

The interaction of representatives from many departments is essential to the success of the PSM program; the safety professional plays a significant role through involvement in many program areas. The safety professional must utilize his or her expertise in the field of safety and health and participate assertively in the program.

As with a conventional safety and health program, senior management must actively demonstrate support for the product safety management program. Senior management must undertake the effort to ensure that the program is set up properly and is sufficiently funded. Finally, senior management must either establish or approve the goals and objectives associated with the program, and actively communicate these goals and objectives to all employees.

REFERENCES

Allison, W. *Profitable Risk Control.* Chicago, IL: American Society of Safety Engineers (ASSE), 1986.

Anderson, R., Fox., I., and Twomey, D. *Business Law, Principles, Cases, Legal Environment.* Cincinnati, OH: South-Western Publishing, 603–604, 1987.

Coughlin, G. *Your Handbook of Everyday Law.* New York: HarperCollins Publishers, 341–350, 406–407, 1993.

Jablonski, J. *Prosper through Environmental Leadership: Succeeding in Tough Times.* Albuquerque, NM: Technical Management Consortium, 55–62, 1994.

National Safety Council (NSC). *Accident Prevention Manual for Business and Industry: Administration and Programs,* 10th ed. Itasca, IL: NSC, 1992.

Porter, M. *Competitive Strategy: Techniques for Analyzing Industries and Competitors.* New York: The Free Press, 3–52, 1980.

Schonberger, R. *World Class Manufacturing: The Lessons of Simplicity Applied.* New York: The Free Press, 3, 4, 7, 124–133, 201–203, 1986.

FURTHER READING

Books

Hammer, W. *Product Safety Management and Engineering.* Chicago, IL: American Society of Safety Engineers (ASSE), 1994.

National Safety Council (NSC). *Product Safety Management Guidelines.* Itasca, IL: NSC, 1989.

Petersen, D. *Techniques of Safety Management.* Goshen, NY: Aloray, 1989.

Slote, L. *Handbook of Occupational Safety and Health.* New York: Wiley-Interscience, 1987.

Other Publications

Colling, D. Materials and product safety. *Prof. Saf.* 36(4): 17–19, 1991.

Kitzes, W. Safety management and the Consumer Product Safety Commission. *Prof. Saf.* 36(8): 25–30, 1991.

Nassif, G. Products liability and the economic community. *Prof. Saf.* 36(1): 21–23, 1991.

PART VII

Risk Management Aspects

## CHAPTER 31

Risk Management: Its Application to Safety and Health

Gerard C. Coletta

CONTENTS

INTRODUCTION

Risk management is a term used to describe various operational and financial activities aimed at minimizing the impact of accidental or unplanned loss. As a process, risk management focuses on:

- Identifying and monitoring an organization's exposures to possible loss; such exposures range from natural perils and manufacturing hazards to kidnapping/ransom and contractual omissions.
- Assessing the possibility (the risk) that each identified exposure will result in losses and then quantifying the likely financial and operational impacts.
- Controlling or reducing these risks through administrative, procedural, and engineering methods (a direct tie-in to safety and health).

1-56670-370-0/02/$0.00+$1.50
© 2002 by CRC Press LLC

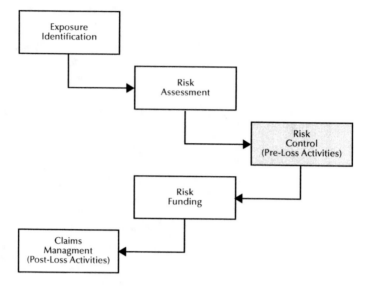

Figure 31.1   Schematic of the risk management process.

- Funding the losses that actually do occur by use of internal resources such as self-insurance or self-assumption, by risk transfer through a commercial insurance program, or by other formal financing options.
- Managing claims aggressively to contain and minimize the overall cost of incurred losses.

The five steps are illustrated as a flowchart in Figure 31.1.

The risk management process is obviously much broader than the purchase of insurance and, subsequently, the processing of claims. Many businesses are focusing more and more on the third segment — risk control (i.e., employee and third-party safety, property protection, emergency preparedness, and contract stipulations) — as the principal method for reducing both the frequency and severity of losses over the long term.

Such a broad approach to risk management is not a new concept. It has, however, gained increasing attention and sophistication over the past few years. Restrictions on the availability and changeable prices of commercial insurance, a move toward higher deductibles or assumed limits, and heightened responsibility for risk control imposed upon senior managers by employees, by governmental regulators, and by society at large have all provided impetus for this change.

Many organizations, regardless of whether or not they fund losses through commercial insurance or private insurance programs such as self-insurance or captive insurance companies, are finding the costs of many risks to be strategically unacceptable because of the potential for unexpected, substantive disruptions in operations.

The five technical segments of the risk management process should be developed equally as part of a tightly integrated system. Each of the segments is discussed more fully in the following sections and, where appropriate, illustrative action items are presented. This discussion concludes with brief comments on overall program management. A focus on planning and communications is the best approach to ensuring that the program is integrated into the usual day-to-day operations of a business.

EXPOSURE IDENTIFICATION

Most organizations should be identifying exposures to possible loss on a regular, in-depth basis. This includes looking at single-incident exposures as well as multiple-incident exposures that could lead to accumulation of significant aggregate losses.

Key exposure categories are:

- Property, equipment, and facilities
- Restricted use of assets and resources
- Adherence to contractual/legal obligations
- Health, safety, and availability of employees
- Use of hazardous/toxic materials
- Dependence on single or sole-source suppliers
- Synergy or integration of individual operations
- Geographical and business competitiveness
- Errors and omissions

Each of these categories should be cross-matched against various perils to identify whether an exposure actually exists for a particular business. This cross-matching, illustrated by Figure 31.2, is subjective and is not intended to quantify the risk. Instead, it is intended to "flag" categories that should be explored more fully and in detail.

The process is best accomplished in multidiscipline discussions. For example, in reviewing exposures to employees, representatives from product groups, maintenance, security, and risk management should all be involved. Initial discussions should be as broad as possible.

Although the exposure identification process can be conducted formally on an annual basis, it also can be structured to easily accommodate changes throughout the year. The risk manager would be an active participant in the addition of new business segments, major changes in operating procedures, planning a safety and health program, and monitoring security practices.

RISK ASSESSMENT

This step should actually be viewed as a continuation of exposure identification and represents the quantification of the identified exposures. It is intended to estimate the value of losses under various conditions and events deemed probable or substantive by management. Both financial and legal consequences should be considered. This dual approach is especially important when assessing compliance with (1) government regulations such as those promulgated by the Occupational Health and Safety Administration (OSHA) and the Environmental Protection Agency (EPA), (2) societal expectations such as planning for business continuity, and (3) good business practices such as reducing overall loss costs and process disruptions.

Exposure Category*	Key Perils								
	A	B	C	D	E	F	G	H	I
Property, Equipment, and Facilities	X	X		X				X	
Use of Assets and Resources	X	X		X				X	X
Adherence to Contractual/Legal Obligations			X						
Health, Safety, and Availabillity of Employees	X	X			X				
Work with Hazardous/Toxic Materials					X		X		
Synergy or Integration Operations	X	X		X			X		
Geographical and Business Competitiveness	X					X	X	X	
Errors and Omissions							X		X

* Categories can be subdivided into individual exposures for a more thorough analysis.

Legend

A = Natural Disaster (flood, earthquake, windstorm)

B = Fire and Explosion

C = Injury and Illness

D = Theft, Dishonesty, Vandalism

E = Negligence

F = Economic/Business Conditions

G = Project Design Deficiencies

H = Contractual Oversight

I = Litigation

Figure 31.2 Exposure identificatio matrix.

Figure 31.3 presents a second matrix that illustrates a method for assessing the order of magnitude of losses and, thereby, prioritizing risks whose impact should be considered in much more detail. The risk assessment process should be taken to enough depth to produce an accurate view of potential losses. This, in turn, provides the roadmap for both risk control activities and, later, risk funding options.

RISK CONTROL

Risk control is in many ways the most important part of the risk management process. If done well, it permits a business to effectively anticipate and prevent many, if not most, material losses. Many organizations do not include risk control as an element in strategic planning. Such an exclusion represents a failure to recognize continuing growth of loss costs, tightening insurance market, and stringent regulatory liabilities.

Risk control should focus on a number of issues:

- Employee safety and health
- Hazardous materials management
  - Environmental affairs
  - Storage and use practices
  - Industrial hygiene
  - Waste generation
- Property protection
  - Buildings and facilities
  - Equipment (including computers)
  - Product and supplies
  - Cargo
- Security systems
  - Buildings and facilities
  - Employees
  - Information and data
- Business continuity planning
  - Emergency preparedness
  - Incident management
  - Operations recovery
- Contractual liability
  - Subcontractor oversight
  - Appropriateness of commitments

Employee safety and health provides an excellent model for describing benefits that can be derived from aggressive risk control. A safety and health program should be viewed as a function that addresses the interests of employees, officers, directors, and shareholders in preserving this segment of a company's assets (employees should be considered a primary asset). Such a program can be an important element in productivity and quality improvement efforts. But most of all, effective safety and health should be looked upon as a way to prevent bottom-line dollars from slipping

Exposure Category**	Financial Loss*						Non-Financial Loss*			
	to $1M	to $10M	to $25M	to $50M	to $100M	above $100M	Loss of Key People	Damage Reputation	Legal Action	Loss of Market Share
Property, Equipment, and Facilities					X					X
Use of Assets and Resources						X				X
Adherence to Contractual/Legal Obligations			X				X	X	X	
Health, Safety, and Availability of Employees			X				X	X	X	
Work with Hazardous/Toxic Materials			X					X	X	X
Synergy or Integration of Operations				X			X			X
Geographical and Business Competitiveness						X				X
Errors and Omissions					X		X	X	X	X

* Both single incidents and multiple incidents should be considered.

** Categories can be subdivided into individual exposures for a more thorough analysis.

Figure 31.3   Estimate of possible losses.

away. For example, stress and repetitive motion claims are still a major drain on many U.S. businesses and now represent a significant portion of many workers' compensation bills.

A strong safety and health program should have four primary goals:

- To prevent accidents that result in injury or illness and in related damage to property and equipment.
- To manage regulatory exposures and potential liabilities created by an organization's duty to exercise due diligence in worker safety and health.
- To integrate safety and health into day-to-day activities and management responsibilities.
- To cast safe performance as a strategy for efficient and productive operations.

Many aspects of safety and health fit well as part of quality improvement efforts like those included within the ISO 9000 protocols and under Baldridge Award guidelines.

RISK FUNDING

Risk funding naturally follows exposure identification, risk assessment, and risk control, all of which are preincident, proactive sets of activities. Risk funding represents a reactive activity; it sets up financing for losses after they have already occurred.

The most effective risk funding activities are based on a number of factors, not the least of which is an organization's posture in setting deductibles and self-assuming various levels of risk. The process includes use of commercial insurance, self-insurance, captive insurance companies, and other (in some cases, creative) mechanisms to finance those losses that do occur.

CLAIMS MANAGEMENT

Claims management becomes important in those loss areas with high frequency and or high severity rates. Again, employee safety and health via workers' compensation provides an illustrative model for this process. It is important to note that the model can be adjusted to fit most property and casualty claims — the model is not limited only to workers' compensation or, more broadly, only to insured claims.

Workers' compensation claim costs can be brought under control by aggressively pursuing five key activities:

1. Diligent case management — For those accidents or situations that do occur and become claims, effective case management requires through investigation, medical utilization review, realistic case reserving, and aggressive negotiation for proper settlements.
2. Monitoring employee lost time — Successful control of indemnity costs requires a team approach to managing an injured or ill employee's lost time. The team approach, backed by specific target dates for return to work, is used to ensure that

individual activities are coordinated. This means close cooperation and open communications between the injured employee, the treating physician, other medical care providers, and the claim technician.

3. Reviewing medical treatment — Workers' compensation should pay only for related and necessary medical treatment at reasonable fees. Every medical charge should be reviewed to confirm its relationship to the compensable injury, its reasonableness, and its necessity.

4. Overseeing legal services — Costs in workers' compensation are not limited to indemnity and medical issues. As workers' compensation claims have become more complex and their values have risen, fees for outside defense attorneys have also grown substantially. As with the other aspects of case management, case litigation oversight requires methods and procedures aimed at controlling associated costs.

5. Regular use of light duty — The final element in a workers' compensation cost containment package is light duty. Regular use of a well-planned light duty program can generate meaningful savings by returning an employee to a productive position. This is the alternative to paying for stay-at-home, do-nothing time. Care must be taken to ensure that light duty is clearly defined and meaningful.

Consistency in managing claims can be highly effective. Reductions of 25% or more have been demonstrated by following four actions:

- Assigning clear responsibility for various aspects of the claims process.
- Fostering controls to avoid "double dipping," such as can occur with workers' compensation benefits and health benefits.
- Supporting a comprehensive accident investigation process tied to the safety and health function.
- Providing risk management monitoring for all activities relating to claims.

PROGRAM MANAGEMENT

Integration of all segments of the risk management process is important in building an effective, productive program with long-term stability. Centralized coordination and oversight are appropriate in many organizations.

For risk management to be effective, it must be structured to quickly address changes in both internal and external business environments. Most organizations benefit from risk management activities that in one way or another penetrate to every part of the business and motivate staff to actively manage the business' risks of doing business.

To stimulate a suitable level of involvement, several steps may be necessary:

- Convene a senior-level steering committee to oversee all aspects of the risk management process. This does not have to be a large group, just one in which key business interests are represented.
- Establish a focused mission statement that addresses all strategic aspects of the process.

- Foster ongoing communications among all departments that benefit from the risk management process.
- Establish a centralized system for information collection, analysis, and reporting (especially to senior management).
- Develop a company-wide risk management manual with specific, but concise, chapters on the five segments:
  - Exposure identification
  - Risk assessment
  - Risk control
  - Risk funding
  - Claims management

These steps provide a solid foundation for an effective risk management program that can successfully minimize an organization's cost or risk. An overall template for the process is provided in Figure 31.4, along with a reiteration of the action items suggested throughout this chapter.

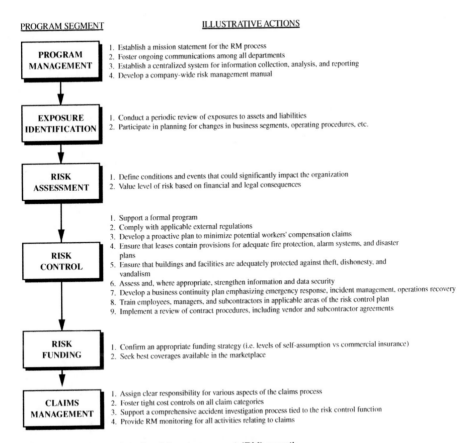

PROGRAM SEGMENT   ILLUSTRATIVE ACTIONS

PROGRAM MANAGEMENT
1. Establish a mission statement for the RM process
2. Foster ongoing communications among all departments
3. Establish a centralized system for information collection, analysis, and reporting
4. Develop a company-wide risk management manual

EXPOSURE IDENTIFICATION
1. Conduct a periodic review of exposures to assets and liabilities
2. Participate in planning for changes in business segments, operating procedures, etc.

RISK ASSESSMENT
1. Define conditions and events that could significantly impact the organization
2. Value level of risk based on financial and legal consequences

RISK CONTROL
1. Support a formal program
2. Comply with applicable external regulations
3. Develop a proactive plan to minimize potential workers' compensation claims
4. Ensure that leases contain provisions for adequate fire protection, alarm systems, and disaster plans
5. Ensure that buildings and facilities are adequately protected against theft, dishonesty, and vandalism
6. Assess and, where appropriate, strengthen information and data security
7. Develop a business continuity plan emphasizing emergency response, incident management, operations recovery
8. Train employees, managers, and subcontractors in applicable areas of the risk control plan
9. Implement a review of contract procedures, including vendor and subcontractor agreements

RISK FUNDING
1. Confirm an appropriate funding strategy (i.e. levels of self-assumption vs commercial insurance)
2. Seek best coverages available in the marketplace

CLAIMS MANAGEMENT
1. Assign clear responsibility for various aspects of the claims process
2. Foster tight cost controls on all claim categories
3. Support a comprehensive accident investigation process tied to the risk control function
4. Provide RM monitoring for all activities relating to claims

Figure 31.4   A template for risk management (RM) growth.

FURTHER READINGS

Books

Head, G., Ed. *Essentials of the Risk Management Process.* Malvern, PA: Insurance Institute of America, 1989.

Head, G., Ed. *Essentials of Risk Control,* Vols. 1 and 2. Malvern, PA: Insurance Institute of America, 1989.

Other Publications

The Risk Report, a monthly publication of the International Risk Management Institute, Dallas, TX.

Risk Management, a monthly publication of the Risk and Insurance Management Society, New York.

PART VIII

Asset Protection Management Aspects

CHAPTER **32**

Security

James S. Cawood

CONTENTS

INTRODUCTION

Security is the function in a company that is responsible for preventing intentional actions directed against the company's interests or assets. Safety is the function in a company that is responsible for preventing unintentional or accidental actions that would harm the company's interests or assets. Both disciplines are composed of a vast number of discrete, technical, information segments comprising the day-to-day thinking and activities of each professional. As with all conceptual frame works which are this broad, it is impossible to identify every bit of knowledge required to practice, as a professional, under the framework. However, we can cover, with broad and sometimes controversial strokes, the major categories of security knowledge that a safety practitioner might find useful for their daily work. I will also attempt to provide a philosophical framework that will be useful to the practitioner in understanding what the goal is to be accomplished in each of these areas. Because business entities differ in management, culture, industry, and geography, applying any of this information to a specific company or set of circumstances would require a more detailed text. However, my hope, as a security professional, is that you will understand the variety of questions that need to be asked and have enough information so you can proceed in gathering additional information to address your specific needs.

MINDSET

It has been my experience that security professionals and safety professionals have a distinctly different mindset, or professional viewpoint. Understanding this fundamental difference allows us to work together to protect more area of our employer's interests, however, it can also be a stumbling block in situations where turf issues or program funding appear to put interests in conflict.

To illustrate the difference let me tell you a very short story. When I was a young law enforcement officer, I overheard in the police academy the phrase "if you want to be liked, be a fireman." This was an acknowledgement that fire fighters were almost universally seen as good guys who did nothing controversial, while police officers were in the business of inserting themselves into the most difficult of society situations (e.g., domestic violence, drug use, sex for hire, robbery, murder). I went to the academy in the late 1970s when public safety agencies were being created, combining the law enforcement and firefighting functions at the municipal level. My academy had several firefighters being cross-trained to be police officers, and upon return to their cities they would be rotated between the role of firefighter and police officer. It was evident to me, even being new to the field, that the firefighters I observed were approaching the work of law enforcement from a distinctly different, and less aggressive, mindset then the more traditional recruits. They appeared to be more cooperative with their fellow cadets and on confrontational exercises were more even-tempered and slower to show aggressive action. It also meant that, in those exercises, they were the ones who ended up being killed more often. It appeared that they wanted to be liked more than the other recruits and that their mindset, in confrontational situations, made them more vulnerable to attack. The concept of combined fire and police operations never did blossom into more than an experiment and, years later, studies by the United States National Institute of Justice involving officer safety have shown that officers that are perceived to be nicer by their fellow officers are the ones most likely to be killed in the line of duty.[1]

My point is that occupational mindset is the most important difference to understand between the security and safety professions. Security professionals are primarily engaged in thinking about how to preserve company assets and interests from intentional acts that have to be presumed to involve actively circumventing observable controls. Safety professionals are primarily engaged in thinking about how to preserve company assets and interests from unintentional or accidental acts where it is assumed that the individuals are not actively attempting to circumvent controls. This is a vastly different perspective, and not understanding the difference can lead to tragedy.

A dramatic example of this type of tragic difference in perception occurred on June 21, 2000 in the small town of San Leandro, California. Federal Department of Agriculture and state agricultural inspectors were visiting a small sausage factory that had been operating illegally. They had visited the factory, after providing notice, and were in the process of shutting it down because the processing being done there was allegedly risking the health of the public. The owner, on videotape, shot to death three of the inspectors and chased the fourth one down the street shooting. In the aftermath, it was learned that these were the first Department of Agriculture inspectors to be murdered in the history of the agency. They were not armed and were only looking out for health and safety of the community. Yet, they were acting like law enforcement personnel, enforcing the federal agricultural regulations, with none of the training or equipment of other federal law enforcement officers. One of the inspectors had called the police, minutes before the shooting, asking that an officer come and stand by because of a possible confrontation, but they still proceeded with their inspection. Thinking like safety professionals in a law enforcement or security situation can lead to avoidable deaths.

Intentional acts are always more difficult to guard against then unintentional acts, though the results of the acts themselves, in many cases may be the same. It will be important to keep that distinction between intentional and unintentional acts or accidents clearly in mind when you are asked to provide security consultation or manage security operations.

BASIC SECURITY MANAGEMENT

Security programs and attendant hardware, at any level of application, are meant to first, divert someone from committing an unsafe or harmful act, and then, if diversion is unsuccessful, to delay the progress of the individual in committing the undesirable act until such time as trained individuals are able to be notified and respond to nullify the problem. All effective security programs work with the assumption that, at any level of problem, an effective response by properly trained personnel will occur. An example of this is the case of threats of workplace violence. In a well-planned response, at a pre-determined threshold point, correctly trained, armed, personnel will need to handle whatever is developing or is in progress. In some workplace violence situations these responding personnel may be the local police, however, based on their average response time to your location and the amount of prior notice which your company may have been given as the situation developed, properly qualified private personnel may be the only defensible option.

Once it is recognized that diversion and delay of the instigator and notification of the appropriate responding personnel are keystones to any successful security program, the questions become:

- What cost-effective security system components should be available to facilitate this process of diversion, delay, and notification?
- Will these system components also facilitate the ongoing assessment of the situation and an effective response by the appropriate responding personnel?

RISK MANAGEMENT

Professional safety and security require an understanding of the balance between risks to the company and costs to preserve or protect a particular asset. Central to determining this balance is a risk analysis or risk assessment that identifies all the assets to be protected, identifies the historic and possible ways the asset could be lost or damaged, the probability of each loss type, the cost to the company of each lost type, and the cost of securing against each loss type. Then, through analysis and discussion, each business makes a determination about how much they are willing to pay to secure against each type of loss taking into account the value of the asset and the probability of loss. A loss that would be catastrophic to the business and is very likely to occur would take priority over something that would happen infrequently and or have no discernable impact on the business. Conducting competent risk assessments is a topic for a book-length discussion and has been covered in the professional literature quite well.[2] As you might imagine, this process of risk assessment can be very difficult for a company to work through, as, in the best case, it forces the company to grapple with different valuations, business philosophies, and interests within the company. However, this risk assessment process establishes the framework that will be used to make decisions about what priority to assign each of the distinct areas that will be explored next.

PERSONNEL SECURITY

The most critical area in any company is the selection, retention, and protection of its living assets — its personnel. Security is involved directly or indirectly in all of these processes. Concerning the selection of personnel, the security function could become involved with the designing the application forms to allow for the gathering of the type of information that will speed background investigations. It may get involved in working with human resource professionals in designing interview procedures and training interviewers to more effectively screen applicants for positions. It may design and run the background investigations process to validate the credentials and experience of applicants, vendors, and contractors. For more information concerning a basic framework for each of these issues refer to a prior chapter on Personnel Screening.[3]

Concerning the retention of personnel, Security may become involved in investigating interpersonal conflicts and incidents in the workplace. These can include harassment, intimidation, threats, violence, conflicts of interest, and theft. The correct handling of these types of investigations can have a tremendous impact on the company, not just from a liability perspective, but from the positive or negative effect it can have on workgroup productivity and morale.

An excellent basic document on the prediction and interdiction of potential violent behavior, titled *Protective Intelligence and Threat Assessment: A Guide to Law Enforcement*,[4] is available on the United States Secret Service Web site (www.treas.gov/usss/NTAC) under its National Threat Assessment Center or on the United States National Institute of Justice Web site. For learning about how to conduct other types of sensitive personnel investigations, consult the Society of Human Resource Management (SHRM) at www.shrm.org for articles in *HR Magazine* or obtain referrals on investigative texts from your organization's employment or employment law counsel. See more information on this topic in the Investigations section of this chapter.

An executive protection, or active threat personnel protection, program could also be a part of some security programs. This type of program could make recommendations concerning security at the homes of targeted individuals, train personnel and their families in threat awareness and crisis response, including protection while driving, and educate them concerning personal information security. It is always important when establishing a program like this that consideration is given to the legal ramifications of such training. If steps are not taken to address the legal issues involved, the company may extend its liability for the safety of its employees into their personal lives. This additional exposure should only be considered in the most extreme cases and after careful consideration.

PROPERTY PROTECTION

The most obvious application of security is for the protection of a company's physical assets. Security, like safety, is achieved through the interweaving of many different control, monitoring, and response systems that, collectively, provide protection to the assets marked for preservation. Once a risk analysis has been conducted to identify the value of any given asset to the company's ongoing viability, then a protection plan can be formulated to address the level of protection that is justified.

Regardless of the level of protection which is justified, there is one constant in the protection of property — inventory control. Companies must be able to identify what assets they own and where they are located. Consequently, an accurate inventory control system is a fundamental necessity for any security program. An accurate inventory control system depends on identifying, marking, documenting each asset as it is acquired (computers, vehicles, buildings, etc.) or created (ideas, processes, computer code, etc.), and tracking the locations of these assets, as close to real time as possible. This is a very difficult program to implement, especially in very rapidly changing environments. For physical assets, every point of intake for the company must be monitored for new equipment and the process needs to begin immediately.

Ideally, all departments are tied into the system so that they know how much equipment they are responsible for, where it is located, and how much it is worth. Value, or worth, needs to be tracked both from a tax perspective (depreciated value), as well as a protective perspective (real value to the business, value to be insured, value to be protected by security). An example of an often overlooked department that needs access to the inventory security system is human resources. When any individual is hired, transferred, promoted, or released the human resources department should be aware of what business property that person is assigned or controls and they should be responsible to work with the company security function to maintain the assets of the corporation as the personnel changes are made.

In the following sections, I have listed the basic components of a protection plan that will generally protect both property and personnel: Access Control, Lighting, Close Circuit Television Systems (CCTV), Alarm Systems, Guard Force Management, and Investigations. These are the threads that are interwoven to provide the fabric of a protective security shield.

ACCESS CONTROL

Physical access control is the process of channeling individuals to points of access to a site, building, or area and then determining which individuals should be allowed into those areas at particular times or under particular circumstances. Much of the security hardware being used today is related to this control function.

The most benign of this hardware is signage. Signs are used to communicate boundaries, procedures, access levels, and warnings. As safety professionals know, a wide range of shapes, colors, typesets, and symbols are used in signs to attract attention and communicate information. The correct use of signage, from a security perspective, can communicate a perception that the site, building, or area is being protected, not neglected. An understanding and well-planned utilization of perceptual security is a critical success component in the area of access control. If an individual or group intent on mischief perceives that an area is well managed, whether they are located inside or outside the company, they will take longer to plan any inappropriate action or will seek a target that is easier to attack.

Another obvious element in access control is the control of a site's perimeter. It may involve fencing and gates, or it may involve the hardening of the external walls and windows of facilities having very little offset from the public sidewalks and roads. The easiest place to gain information about all the issues involved in choosing between all the options in fencing, gates, wall construction, and window glazing is to read through a detailed security manual, such as POA Publishing, LLC's *Protection of Assets Manual*. This type of manual can speed your understanding of all types of hardware and provide enough information to begin making decisions about what type of hardware might be best for your facility.

Once entry to the site is allowed, how does the company identify who the people are and control their movement inside the perimeter of the site. This raises the issue of entry logs, visitor and contractor badges, employee identification badges, and escorted or unescorted access to various parts of the facility. The important thing to remember when making decisions about these types of procedures and hardware is that what your company requires people to do should be something that will be done consistently, not an ideal. Any process which is done on a consistent basis is better than a more comprehensive system that is ignored because it is too cumbersome. It might help to consider security policies and procedures from a viral perspective. Your success rate will increase if your company thinks about incorporating security into all discussions of development or improvement. Over time, this will infect employees with an understanding of the importance of security, that over time will hopefully increase their acceptance and compliance with security procedures. This is important when you realize that security, like safety, is strongest when every employee takes responsibility for its implementation in the company.

Access control also involves access to a company's information assets. This involves access to its internal networks (e.g., local area networks (LAN), wide area networks (WAN), and virtual private networks (VPN)) and its external interactive sites, including e-commerce sites and Web sites, but also its nonelectronic information assets such as filing cabinets, private offices, desks, information archives, and other information sites. After personnel, information is the second most important asset to any company or organization. Regardless of the industry in which you work, large physical assets are becoming less important, while the information that is critical to the competitive advantage of a business is becoming more vital.

Access control to information assets can be addressed on individual computers by something as low-tech as activating the BIOS (Basic Input Output System) level password protection feature which is available on any modern computer — both desktop and laptop or by adding layers of passwords and data encryption. Effective solutions are found in using a blend of hardware and software which match the type of hardware and software used in the various parts of your company's operations. Certainly, the use of office door, file cabinet, and desk locks should be encouraged. BIOS level passwords and locking computers into their environments are also low–tech and low–cost solutions. The consideration of more sophisticated solutions like automated log-offs for unused computers, pinging computers to assure they are on the physical network, sounding an alarm for computers removed from the network, and encryption of all data are more sophisticated and costly solutions. Other technologies, such as firewalls, are essential for off-site attempts to intrude into your company's networks or data management operations.

LIGHTING

The concept of perceptual security is also seen in the use of lighting for security purposes. The better the lighting scheme, the more difficult it is for individuals to avoid detection, both inside and outside the company. This tends to influence individuals who are trying to make a target selection to choose a target that is less likely to allow for identification and possible capture.

The decisions about lighting fixtures, luminences, and coverage should always be made in recognition that standards are beginning to emerge, both in litigation and in practice. The Illuminating Engineering Society of North America has published detailed standards for the average level of light they suggest for all sorts of locations including walkways, parking lots, the exterior of buildings, the interior of various parts of businesses, and private roadways. Not considering these standards in your decisions about what level of light you place in and around any business site will not only increase the probability of loss, but also the exposure to litigation that may follow.

Once the issue of lighting is raised, the normal course of action is to conduct a lighting survey in the area of concern. This consists of having a knowledgeable person take light readings at regular intervals (3 to 5 feet), at ground level, over the entire area of concern. The type of light meter used should allow for readings in foot-candles. An example of this type of meter would be the Gossen Panlux Electronic2 light meter. The average luminescence can then be calculated and compared to the lighting standards mentioned above and the lighting could be adjusted as needed.

CLOSED CIRCUIT TELEVISION (CCTV)

CCTV technology is often used without proper thought or planning. Having the presence of cameras can be a perceptual security enhancement in certain areas. It is most effective when used to record images of incidents in progress for later analysis and evidentiary reasons. It is less effective when deployed to provide a real-time oversight of security areas. In most facilities the number of cameras already being used outstrips the ability for human operators to monitor them. Even with the use of motion-activated or alarm monitoring only, a human operator can only effectively monitor about 10 cameras for 15 minutes without some form of break. Since most monitoring shifts are for eight hours with two 15-minute breaks and a half-hour break for lunch, the effectiveness of the monitoring is significantly reduced.

An additional complicating factor is cameras panning a large area on a constant basis. This practice, which was meant to maximize the coverage area of certain cameras, actually reduces efficiency by guaranteeing that, the majority of the time, a camera will be out of the position when something is happening.

The technology for recording video input is constantly improving. A quantum improvement has been made in the recording of time-lapse images. Instead of using multiple video tape recorders on a multiplexed basis, up to 15 cameras can be recorded by a single digital recorder in time slices as small as a frame every half second. This technology not only allows for quicker searches of activity on a particular camera, but it also allows for more images to be captured per second than can be recorded by a video tape based system.

Prior to the installation of any system of this type, a thorough, professional, assessment and design should be conducted. Selection of equipment that does not address the issues of climate, lighting, distances to monitor, response time of the guard or police force to the site, and the expectations of the personnel at the site will not only produce substandard results, but could significantly increase the legal exposure of the company should an incident occur.

An example of increased legal exposure is the classic mistake of installing fake cameras in a location. These cameras are meant to deter crime by providing the perception of security. The idea is that someone will choose not to commit a crime because it might be documented and they would be caught. However, several rapes and robberies have occurred on sites with fake cameras and the victims have successfully sued for damages. The suits hinged on the idea that the company knew they might have a security problem (constructive notice) or they would not have installed the fake cameras. The company then compounded the problem by having the victim believe that they were being protected by the cameras and no one responded to help them. Consequently, the fake cameras not only did not deter the crimes from occurring, they multiplied the exposure of the company to liability.

ALARM SYSTEMS

Alarm systems are meant to alert designated personnel to a condition being monitored. In the case of security alarm systems they are used in a variety of ways, depending on the level of security needed. At a minimum, a security alarm system should have the following components and setup:

- The alarm should be monitored by personnel who will either respond or send response elements (security or law enforcement) to the scene of the alarm for verification.
- Each protected element (door, window, zone of area protector, access control card reader, panic button, etc.) should be on its own zone.
- Each protection device (door contact, pressure pad, passive infrared detector, etc.) should be home run wired to the alarm subpanel or main panel. This will initially cost more, but will allow easier repair and maintenance, which will more than pay for the initial increase in cost.
- Each zone, when alarmed, should designate itself with a plain English identifier so that during an emergency response no one has to remember what zone is tied to what physical location for response.
- Alarm panels should be placed at strategic locations throughout the facility so that during an alarm designated people will be able to quickly check, and avoid, alarm areas.
- If the alarm is to be monitored at a remote location, consider digital dialers for the primary alarm with cellular phone as a secondary system. Some systems may even allow real-time monitoring over the company's WAN or the Internet.
- Always provide adequate battery backup for your alarm system. A minimum of 24 hours of battery power should be built into the system in case of intentional or unintentional power loss.

Consider piggybacking other functions onto the security alarm system, such as building monitoring systems or vice-versa, particularly if you are planning to monitor the systems with company personnel.

GUARD FORCE MANAGEMENT

This component of the security program is the one that will provide the most feedback from the organization. If the security officers in the program are selected, trained, and managed well, they will prove to be the most effective and critical aspect of the security program. Conversely, any failings that they have will be quickly noticed, commented on, and affect the overall perception of the security program.

The art of selecting, training, and managing a guard force is a subject worthy of a book-length essay.[5] What is initially important to an organization is realizing that security officers must be able to identify emerging security problems and be trained to quickly and effectively respond to solve them. Organizations must decide how much of the responsibility for the selection, training, and staffing they want to control given the identified risks at any site or facility. This question of control usually breaks down to a question of who is the employer of the security officers. Some organizations, those with a need for maximum control (regardless of cost), have a proprietary security force, one in which all of the security personnel are their employees. Some organizations, with a need for specialized control, might use a blended security force. A blended security force has certain management and specialized personnel (e.g., security director, investigators, executive protection staff, and information security personnel, etc.) as direct employees of the organization, while security officers and their supervisors are employed by a contract guard company and assigned to their site. The last option would be to hire a completely contractual security force where all security management and personnel are employed by a contract security firm.

Many questions will need to be raised and answered by any organization concerning the necessary qualifications for their security management or officer personnel. One question that is often asked is whether former federal, state, or local law enforcement personnel automatically have the high-end qualifications that should be considered for selection of key personnel. The answer is no. Law enforcement and security are two different professions, just like safety and security. Though law enforcement is also in the business of handling intentional acts, the vast majority of any law enforcement officer's experience comes from responding to crimes, not proactively attempting to prevent crimes. Also, the vast majority of law enforcement officers have limited experience in interacting with and working in the complex organizational environments that are present in the private sector. This means that any organization hiring for a security position should closely consider the actual experience, specialized training, and accomplishments of the individual candidates for a fit to the specific organization and job involved, and not allow assumptions based on perception to inappropriately influence the hiring decisions. Most doctors would not be effective emergency medical technicians, most lawyers would not be effective judges, and the majority of former law enforcement personnel would not be effective private security personnel.

INVESTIGATIONS

Investigation is the careful search, detailed examination, and systematic inquiry[6] of any matter. The investigation of anything requires a complex set of skills that take time and experience to acquire. Examples of these skills are the ability to suspend personal assumptions and emotions; interact with a wide range of people on both an intellectual and emotional level; provide accurate extemporaneous documentation of the investigative process; conduct sophisticated analysis of a wide range of collected information; and have the ability to deliver effective oral and written communications; as required. In the state of California, just to qualify to take the examination to become a licensed private investigator, a candidate must have 6000 hours of compensated (paid) investigative time, which is the equivalent of three years of full-time employment. This time can be acquired as a paid apprentice to another licensed investigator or by being designated and compensated as an investigator by a company or law enforcement agency. This is the minimum amount of experience that California believes is required to acquire the skills necessary to provide a competent, third-party, investigation in the private sector. The fact that California only requires psychologists who wish to be licensed to have 3000 hours of supervised practice might raise the issue about which profession California considers the most complex to master.

Once an individual has been identified to conduct an investigation, the next step in providing valid and defensible results is a well-constructed strategy which is designed to reach the established investigative goals within the time and cost constraints that are present. This requires that the investigative goals have been identified, the most basic of which is to understand, from the outset, whether the investigation might lead to the need to present the results in a criminal action. This goal needs to be defined because of the difference in the level of evidence needed in different venues (arenas) of adjudication. In the world of civil litigation the normal level of certainty required to prove an issue is a preponderance (generally 51% or more) of the evidence. For a criminal matter the level of certainty has to be beyond a reasonable doubt, which is a much higher standard. Consequently, if at the outset it is possible that an investigation could lead to a criminal trial, a higher level of documentation and evidence handling (chain of custody) should be established as a part of the investigative strategy, because the quality of these investigative elements is almost impossible to raise in a viable way in mid-investigation. Obviously, higher quality levels of documentation and evidence handling require more time, which means they require more money.

In the area of post employment investigations, some civil investigations an organization might have to conduct include sexual harassment, discrimination, misuse of company assets (time or property), and conflict of interest. Some potential criminal investigations include threats, assault, sexual assault, theft, commercial bribery, rape, and robbery. Remember, just because the police are involved in investigating some of these criminal matters the organization may not be absolved from also conducting some form of investigation as well.

The purpose of the material in this section is to inoculate any reader of this book from assuming that, without proven experience, he or she understands the complexity of the investigative process and, therefore, underestimates the need to obtain professional investigative support when the need arises. No safety professional is qualified in every specialty of safety, and no security professional is qualified in all aspects of security. Investigation is a complex area of special security skills that is often underappreciated with the predictable, poor results.

POLICY AND PROCEDURE DEVELOPMENT

This area is one of the most discussed, yet under supported, areas in the security world. Too often, policies are written as reactions to an event that has scared personnel, usually at the management level. Careful consideration should be used, instead of quickly copying a policy or procedure from another organization (e.g., benchmarking, best practices, etc.) which often leads to creating more problems than are solved. An example of this is the current fad to have zero tolerance policies in many areas of safety or security. In the area of workplace violence, as in some other critical areas, zero tolerance policies can work against the interests of the organization, not for them. To the workforce, zero tolerance means termination. Consequently, in potentially serious cases, reporting can be delayed, which can put lives at risk, because the employees know that if they report the activity of Johnny or Janey then Johnny or Janey, will be fired and the employees are not certain they want the responsibility for that result. This may be for several reasons, but the statements I hear the most are:

- "If Johnny or Janey find out I turned them in they will be mad at me and may hurt me."
- "I know Johnny or Janey's family and they will not be able to stand the loss of the job."
- "I know what they are doing is wrong, but I will not help management take their job."

Also, in most employment-related situations, it is necessary at some point in the investigative process to let Johnny or Janey tell us their side of the story. When we approach them on an issue that they know is a zero tolerance issue, they may instantly do the math, and in many cases it will alter their truthfulness with us. In the case of a seriously destabilized person, however, it may cause them to act immediately because they have nothing to lose and time is running out.

Aside from that general observation and warning about security policies, policies can be developed in many areas, but it should always be done with the understanding that a certain amount of flexibility should be built in for the benefit of the organization. Policies could be considered for the minimum following areas:

- Alarm system operation, including response
- Access control
- Key control
- Background investigations for personnel screening
- Internal investigations
- Information security
- Personnel security (workplace threats and violence); include three parts as a minimum to this policy:
    - Mandatory reporting
    - Multiple reporting paths (supervisor, manager, head of HR, head of security)
    - Mandatory disclosure of restraining orders sought or received with have the worksite as a protected area

- Weapons policy
  In this area, consider using the penal code statute in the state of operations as a base level or defining weapons. Remember to build in the statement such as "Weapons can be authorized by _____ to be brought onto the company premise in connection with the protection of the safety of employees." This allows for armed individuals on the premise for protection in certain instances without breaking the policy.
- Natural disasters such as fire, flood, earthquake, tornado, or hurricane
- Major disruptions of operations such as power loss, water loss, or loss of chemical containment; the safety plan needs to have a security element that should be the security policy/procedure in this area
- Civil unrest

Procedures are a series of steps to take in implementing the intent of the policies. Often, the two are confused because they are used together in the phrase "policies and procedures." Actually, I coined a term about eight years ago to describe a document I wrote for a client on the process of managing workplace violence. I called it a "poli-cedure" because it had a policy front end and a procedural back end. Policies should describe the intent of the company in the given area; procedures fulfill the objectives of the policy by delineating actions to make the intent of the policies concrete. The best procedures give a general outline of the process of implementation and then some specific steps that are constantly reviewed and updated as the practices are used over time. A yearly review of procedures to validate their current applicability would be a fine preventative model to facilitate this concept.

CONCLUSION

In this brief chapter, I have attempted to encapsulate an overview of the security for the safety professional. I hope that discussion of mindset and the purpose of security have stimulated some interest in further exploring the field of security and its relationship to safety. The references provided and the attached reading list should provide a good start on the process. Another source would be to contact the American Society of Industrial Security (ASIS) in Arlington, VA on the Web (www.asisonline.org) or contact them at (703) 519-6200 to request the reading list used in the preparation for the Certified Protection Professional Exam. That reading list is more exhaustive than mine and should serve to further increase the scope of your exploration. I wish you the best of luck in your journey of interdisciplinary exploration.

REFERENCES

1. Pinizzotto, A. and Davis, E. Cop killers and their victims, killed in the line of duty. *FBI Law Enforcement Bulletin,* December 1993.
2. Broder, J. F. *Risk Analysis and the Security Survey,* 2nd ed. Woburn, MA: Butterworth-Heinemann, 1999.

3. National Safety Council (NSC). Personnel Screening. In *Accident Prevention Manual for Business and Industry.* Itasca, IL: NSC, 1997.
4. Fein, R. and Vossekuil, B. *Protective Intelligence and Threat Assessment Investigations: A Guide to Law Enforcement,* July 1998, NCJ1706/2.
5. Canton, L. G. *Guard Force Management.* Woburn, MA: Butterworth-Heinemann, 1995.
6. *Webster's New Twentieth-Century Dictionary,* Unabridged, 2nd ed. New York: Simon & Schuster, 1972.

FURTHER READING

Canton, L. G. *Guard Force Management.* Woburn, MA: Butterworth-Heinemann, 1995.

Dalton, D. R. *The Art of Successful Security Management.* Woburn, MA: Butterworth-Heinemann, 1997.

Lack, R. W., Ed. Accident Prevention Manual for Business and Industry: Security Management. Itasca, IL: National Safety Council, 1997.

Lyons, S. *Lighting for Industry and Security: A Handbook for Poviders and Users of Lighting.* Woburn, MA: Butterworth and Heinemann, 1993.

Post, R. S. and Schachtsiek, D. A. *Security Managers Desk Reference.* Woburn, MA: Butterworth-Heinemann, 1986.

Purpura, P. P. *Security and Loss Prevention: An Introduction,* 3rd ed. Woburn, MA: Butterworth-Heinemann, 1998.

Sennewald, C. A. *Effective Security Management,* 3rd ed. Woburn, MA: Butterworth-Heinemann, 1998.

Stees, J. D. *Outsourcing Security: A Guide for Contracting Services.* Woburn, MA: Butterworth-Heinemann, 1998.

Tyska, L. A. and Fennelly, L. J. *150 Things You Should Know about Security.* Woburn, MA: Butterworth-Heinemann, 1997.

Walsh, T. J. and Healy, R. J., Eds. *Protection of Assets Manual.:* P.O.A. Publishing, LLC, Los Angeles.

CHAPTER **33**

Fire Protection: Pressure Vessels, Boilers, and Machinery

Peter K. Schontag and Robert A. Panero

CONTENTS

INTRODUCTION

Property loss control, once referred to as fire protection or fire prevention, has been expanded and covers a much wider scope. This function includes not only fire protection and prevention, but also boiler, machinery, and natural hazard risks. Boiler and machinery risks broadly relate to the malfunctioning, deterioration or overloading of production machinery. Natural hazard risks are those presented by flood, wind, snow, ice, earthquake, and other natural-hazard-related events.

Although the loss potential from fire and boiler and machinery risks is mostly property related, there are also life and safety exposures. In 1997 alone, 120 civilian deaths and 2600 injuries were caused by nonresidential structure fires, as estimated by the National Fire Protection Association. During the period from 1992 through 1996, an average of 16 civilian deaths and 414 injuries were caused each year by fires solely in industrial and manufacturing occupancies. The Association's latest analysis shows a very diverse spread of fire causes, with the leading ones heavily influenced by the equipment specific to the type of operation at the facility. Metal or metal products, wood, furniture, paper, and printing products manufacturing led with the largest percentages of structure fires and property damages. The National Board of Boilers and Pressure Vessel Inspectors (National Board) reported a total of over 2400 incidents resulting in 75 injuries and 18 deaths in 1997. It is clear that, in addition to property loss, a heavy toll is taken in terms of injury and death for hazards that have not been properly managed.

The loss exposures from boiler and machinery related incidents are usually underestimated. The failure of a single critical mechanical or electrical component in a process can have profound operational effects on a business, be the catalyst for a much larger property loss, and result in negative life safety experience. Given the high probability of failures with mechanical equipment and the potential resulting consequences, proactive boiler and machinery loss control is a very important component of any overall risk management program.

PROPERTY LOSS CONTROL MANAGEMENT

In Chapter 31, Overview of Risk Management and Its Application to Safety and Health, the importance of risk control is pointed out. Yet, far too often, the risk control or loss prevention program is the weakest link in the risk management effort.

Many property loss control programs are reactive, rather than proactive in nature. The insurance company makes loss prevention recommendations and then the safety manager is asked to take care of them. This is not an approach that follows the principles of good management because it is reacting to past events rather than anticipating, preventing, or minimizing future incidents.

This chapter covers the fire protection and boiler and machinery aspects of property loss control. The protection of facilities and equipment is a technical field that requires a high degree of education and experience if mitigation strategies are to be sufficient and cost effective. Only qualified personnel should be employed to evaluate risk and identify loss mitigation strategies. Qualified in this context means not only qualified in the field of fire protection engineering or boiler and machinery, but also proficient in the specific area they are retained to evaluate.

After the loss mitigation steps have been taken and the appropriate protection is in place, there are two other areas which need constant attention. They both involve the human element.

The first is inspection, testing, and maintenance of the protection systems and safeguards. Many losses occur in protected property or to safeguarded equipment due to a breakdown in the program established to make sure the protection is in service and functional. It takes a very high degree of diligence to keep this from happening.

The second human element area is prevention. It requires that employee training be adequate to assure safe operations. In regard to the prevention of fires, it has been documented that employee awareness is a major factor in reducing incidents; when employees are given fire extinguisher training, the number of fires decreases. In cities where people knew telephone service was out, preventing the reporting of fires, the number of fires decreased dramatically. After the telephone service was restored, the fire rate returned to the norm. In both cases, the situation did not decrease the hazard level, but the awareness of the people involved had increased. For the same reason, frequent employee reminders of the need for fire safety are required for an effective program.

Experience has shown that, as in many other areas, a fire and boiler machinery loss control program will not be effective without the commitment, support, and involvement of senior management. As with safety, property loss control will be given short shrift by managers who are measured solely on short term results.

To provide a strong level of management control over the property loss control effort, the following steps should be taken:

- Develop loss control objectives that correlate with the company's risk management philosophy.
- Ensure that loss control surveys provide a high level of loss exposure identification.
- Have exposures quantified as to magnitude and probability of occurrence.
- Establish a formal program to develop and implement changes to eliminate or mitigate the exposures.

FIRE PROTECTION

Existing Facilities

The individual(s) responsible for fire protection for facilities should first establish a general policy that includes the programs, standards and guidelines needed to establish the desired level of protection. Procedures should be developed to outline the steps to be taken to maintain that level during operations. Also included in the procedures should be steps that meet Occupational Safety and Health Administration (OSHA) requirements, especially Subparts E, H, and L. The specific responsibilities of all involved individuals and departments should also be detailed. Some of the positions that are typically responsible for fire protection include: facilities manager, engineer, fire/safety supervisor, and maintenance supervisor.

A system should be established to report and document both losses that occur and near misses. An analysis of the resulting data will often show areas or operations that need greater protection or increased prevention efforts. If it is to be effective, it is important that the system be reward based, without retribution.

An inspection program is critical to assuring that safe conditions are maintained. Fire suppression equipment and systems should be inspected and tested (see next section for a sample fire inspection form used). The fire inspection should also cover prevention items such as housekeeping, storage, hazardous operations including flammable liquids/gases, open flames and hot surfaces, as well as welding and hydraulic oil systems. Access to fire exits should be checked regularly. Items that do not always get attention are the safeguards on fuel burning equipment such as furnaces, boilers, ovens, kilns, etc. Combustion controls and temperature limit switches need regular testing by qualified personnel.

When a fire alarm, detection, or extinguishing system is out of service or impaired for any reason, the situation should be reviewed to assure all measures have been taken to offset the reduced level of protection. Also, it is very important that a unique program (not an operational clearance system) is in place to monitor the impaired system until it is returned to service. It has been frequently demonstrated that unless unique impairment programs are created, out of service protection features may not be returned to service. Impaired systems should be returned to service in the most expeditious manner possible. Most insurance companies have an impairment control program that can be adopted to your facility.

Fire brigades are now referred to as emergency response teams or plant emergency organizations. Requirements established by OSHA must be addressed in the organizational plan for your team (including Subpart E, Emergency Action Plan), in addition to the following factors:

- A team should be established for each operating shift.
- Training should be done on a regular basis and good records of the training kept for review.
- Team members must be provided with the appropriate protective clothing and safety gear, as well as medical exams.
- Contact should be made with the local fire department and joint drills held at least once a year so both the facility team and the fire department are clear on who is responsible for each action to be taken in an emergency. This joint training drill is often a good time to review the hazards at your facility with the fire department.

Many facilities have security officers for security and access control. They perform the important function of property patrols during nonoperating periods. All security officers should be well trained in how to report fires and other types of emergencies. The necessity that this be done promptly before any other action is taken must be made clear to all security officers on a regular basis. Figure 33.1 presents an inspection checklist for a facility fire prevention inspection.

New Facilities

The design objective of a fire protection plan for new facilities should assure they can withstand both fire and natural hazards with minimal damage or interruption of operations. Included should be provisions for safe evacuation of personnel from all areas that meet OSHA Subpart E requirements.

All state and local building code requirements should be addressed, with each item either being met or offset by an equivalent or better solution. Insurance companies should also be given a chance to review the plans as they often can help with items that might affect insurance costs, as well as earthquake and windstorm risks.

For facilities with very high values and or unusual design features, it is wise to have a fire protection engineering consultant included in the design team. Engineering firms used by most architects are often weak in fire protection engineering, so it is essential that their qualifications be verified carefully.

The facility and the planned operations should have a fire hazard analysis developed to identify operationally critical systems, hazards, and protection needs. Included in the analysis should be the arrangement of operations, processes, and systems to minimize fire hazard risk and the possible business interruption exposure which can result from a fire or explosion. Some items to cover include:

- Separation or isolation of critical operations
- Separation of backup fire protection systems
- Use of noncombustible or fire-resistant materials
- Designs to control and protect storage of flammable gases and liquids
- Arrangement and protection of electrical and ventilation systems
- Layout of building exits
- Combustion controls and safeguards for fuel-burning equipment
- Special protection for large hydraulic systems

BOILER AND MACHINERY

Scope of Loss Potential

Traditionally, the concern for safety and property preservation evolved from insurance carriers who bore the responsibility of underwriting boiler failures. Because of inconsistent design practices, materials, construction and operation, the incidence of failure (which sometimes resulted in injury and death), was extremely high before and during the early twentieth century. The establishment of construction and operational standards, initiated by the insurance industry, has evolved to individual states and ultimately to the American Society of Mechanical Engineers (ASME) and the National Board.

The field of boiler and machinery has taken on the complexity and diversity of the industrialized world and encompasses hazards from boilers and pressure vessels to all types of complex mechanical and electrical equipment. The role this equipment plays is directly linked to the financial success of a business. Causes of losses range from the obvious, such as a tube failure in a boiler or a blade failure on a turbine, to the less obvious, such as a chiller system designed to remove heat from an electronic data processing center. Loss scenarios from events like these are only limited by the nature of the operation and the imagination of the person conducting the loss control evaluations. Equipment fails, as do the people who operate it, which makes the proper management of boiler and machinery loss control programs vital to a successful business.

Inspection Checklist

Facility Fire Prevention Inspection

Facility _____ Location _____ Date _____

Sprinklers, Piping, and Water Supply

Sprinkler system main valve open?	Yes ___ No ___ *
Sealed or chained and locked open?	Yes ___ No ___ *
Sprinklers and sprinkler piping OK?	Yes ___ No ___ *

Deficiencies to look for:

• Storage 18" or less from sprinkler head deflector • Sprinkler piping leaking • Sprinklers turned sideways • Sprinkler piping hangers missing • Sprinkler head deflector or fusible link painted • Sprinkler heads or piping corroded •

Verify pressure on gauge on sprinkler riser(s)   Yes ___ No ___ *

• Use lower gauge if two provided on riser • Low or no pressure indicates impaired water supply •

Fire department connection OK?   Yes ___ No ___ *

• Caps missing - Replace caps if missing after checking for debris in pipe •

Note: If "No" is indicated on any item, make comments

Comments _____

Fire Hoses and Nozzles

Hose attached?	Yes ___ No ___ *	Not obstructed?	Yes ___ No ___ *
Neatly racked?	Yes ___ No ___ *	Nozzle attached?	Yes ___ No ___ *

Has the hose been hydrostatically tested in past year?   Yes ___ No ___ *

Note: If "No" is indicated on any item, make comments

Comments _____

Fire Extinguishers

General condition Good_____ Fair_____ Poor_____

Current date on tag (within last year)?   Yes_____ No_____ *

Seal in place?   Yes_____ No_____ *

Pressure gauge in proper range?   Yes_____ No_____ *

Obstructed?   Yes_____ No_____ *

Comments _____

Note: If "No" is indicated on any item, make comments

Housekeeping

List all areas where housekeeping was not satisfactory _____

Deficiencies to look for:

• Blocked aisles/exits • Trash on the floor • Improper storage of packing materials • Enforcement of "No Smoking" area, check for butts on the floor • Oily rags, located mainly in the garage area (

Flammable Liquids

Flammable liquids stored in proper safety cans?   Yes_____ No_____ *

• Any flammable liquids over 1pint should be in a Safety Can •

Flammable liquids/aerosols stored in proper cabinets?   Yes_____ No_____ *

• Any amount of Flammable Liquids over 10 gallons required to be in Flammable Liquids Cabinet • Cabinets should be approved for storage of flammable liquids • Any amount of aerosols should be stored in at least a cabinet constructed of at least 18-gauge steel or plywood 1-inch thick •

Flammable liquids cabinets in good condition?   Yes_____ No_____ *

• Doors shut and locked, flame arrestors in place, shelves in good condition, bungs in place

If mixing and dispensing is done, is area designed properly?   Yes_____ No_____ *

• Proper lighting, grounding, curbing, ventilation

Note: If "No" is indicated on any item, make comments

Comments _____

Alarms

Is the Power light on showing power to the alarm panel?   Yes _____ No _____ *

• The power indication light indicates that there is power to the Alarm Panel so alarms can be received and transmitted •

"Trouble" or "Alarm" lights OK?   Yes _____ No _____ *

• If any trouble or alarm lights are showing, bring it to the facilities manager attention and make recommendations to repair •

Note: If "No" is indicated on any item, make comments

Comments _____

General Comments and Recommendations

List all necessary repairs, replacements, unusual conditions, and suggestions for additional fire protection for limiting fire hazards

_____

Inspected By _____   **Reviewed By** _____   **Date** _____

Figure 33.1   Facility Fire Prevention Inspection Checklist.

Loss Control Program Elements

Hazard Evaluation

A successful loss control program identifies loss potentials, evaluates their importance, and provides mitigation strategies that either prevent or minimize their effect. The identification of system or process components and potential failure modes is the fundamental step in this process. Next, the material condition of the component needs to be evaluated. Is it operating correctly or vibrating? Is it operating outside of its design limits or is it deteriorated by corrosion or cracking? Would a catastrophic failure of a piece of machinery or pressure vessel compromise an adjacent redundant component and force an outage of the entire operation?

The key to a successful hazard evaluation is the utilization of a qualified loss prevention engineer teamed with knowledgeable plant personnel, so that credible what if scenarios can be developed and analyzed. Once failure modes are identified they can be ranked in order of importance based on mission criticality and protection methods developed to minimize the potential or consequences of failure.

Predictive and Preventive Maintenance Programs

One of the most important strategies in minimizing loss potential in the boiler and machinery field is the implementation of effective predictive and preventive maintenance programs. Predictive maintenance programs utilize state of the art technologies such as vibration analysis, thermography, and ferrography. These tools, if consistently and properly used allow equipment owners to develop operational signatures or trends on a piece of machinery or electrical equipment so that changes in state (which could be indicators of impending failure) can be identified, analyzed and corrected. Preventive maintenance is generally based on the original equipment manufacturer's recommendations, supplemented by plant experience, and is intended to prevent failure due to the deterioration of system components. Corrective or breakdown maintenance is the consequence of ineffective predictive and preventive maintenance programs.

Performance trending of equipment provides a tool that helps establish whether a piece of equipment is running in a malfunctioning or overloaded condition. Records of maintenance trends and performance should be reviewed frequently by supervisors and aberrant conditions should be monitored closely. The compilation of data without this review serves no purpose, but is a common practice. There is normally plenty of time after the failure to review this data and observe that it would have predicted the occurrence.

A management commitment to these programs is essential to their success. In times of tight dollars, they are typically the first programs to be cut since there is not always a direct relationship between their implementation and production goals. While not immediate, this is a false economy that will ultimately result in reduced production and reduced revenues. If it is necessary to consider a reduction in programs, it should only be done with input from qualified operators and maintenance personnel.

Business Interruption Studies

As previously noted, in addition to property losses and safety issues, boiler and machinery related events can have profound effects on the continuity of operations. Business interruption can represent a larger catastrophic loss potential than the actual property damage, since relatively small losses to inadequately backed up or hard to obtain components can result in long down times.

As part of the hazard evaluation process, process flow diagrams should be developed to identify mission critical components so that strategies can be implemented if the component fails. Are there spares available for hard to get, long lead pieces of equipment, or are there manufacturers or equipment suppliers who can provide timely replacements? These questions should be asked and answered before the incident so that proper contingency planning can be accomplished. Thorough planning, together with hazard evaluations, are the keys to minimizing the loss potential.

Management and the Human Element

The implementation of an aggressive boiler and machinery loss prevention program is only as effective as the management commitment to that program. Human element influences on a safe and reliable operation far outweigh any design features intended for loss mitigation. Since inadequate maintenance is the leading contributor to losses, predictive and preventive maintenance programs must be supported by management, with maintenance personnel responsible for the programs they implement.

An integral part of any maintenance program is the testing of safety systems. Because safety systems and process control interlocks can go unused for years, their operational status can never be taken for granted. For this reason, they require methodical and frequent testing to verify their operational status. New systems should have complete system check outs, or acceptance tests, to verify that they are operational and perform their intended function.

Operators, properly trained and empowered with the authority to make prompt corrective action decisions, including emergency shutdown, are considered the first and last line of defense against catastrophic failures. For this reason frequent training and clearly established performance expectations should be provided for operators. Written, practical operating procedures should supplement training and be updated periodically to reflect changing design and operational conditions.

A review of loss history demonstrates the importance of the human element aspects of any loss control program. Simple reliance on engineered systems, without proper testing and maintenance, has proven that losses will continue to occur and are generally much more severe. Senior management has a responsibility to ensure that the human element programs are implemented in a manner that promotes safe and reliable operation and fosters a sense of commitment and responsibility in employees for the overall success of the business. If a loss does occur, cause needs to be determined and loss control, maintenance, and operations programs should be modified to prevent reoccurrence.

Jurisdictional Inspections

One aspect of a boiler and machinery program, which relates to loss control but is generally a separate function, is the performance of regulatory-required (jurisdictional) pressure vessel inspections. These inspections are required by each individual state to ensure the safe operation of boilers and other pressure vessels and result in state issued permits to operate. Requirements and scope of vessels to be inspected vary from state to state and inspectors are required to be commissioned (licensed) by each state in which they work. These inspections provide a level of independence from the owner and are intended to assure that equipment is not operated in a degraded condition. In addition to meeting regulatory requirements, an owner of a pressure vessel is required to verify by independent inspection that any repair or alteration is conducted in accordance with various codes (e.g., ASME, National Board, and state codes.)

REFERENCES

Global Property Loss Prevention Datasheet 9/2000. Norwood, MA: Factory Mutual System.

American Society of Mechanical Engineers (ASME). New York, ASME Boiler and Pressure Vessel Code.

National Board of Boiler and Pressure Vessel Inspectors. *National Board Inspection Code.* Columbus, Ohio, 2001,

National Fire Protection Association (NFPA). *National Fire Codes.* Quincy, MA: NFPA,

National Fire Protection Association (NFPA). *Sax's Dangerous Properties of Industrial Materials.* Quincy, MA: NFPA, 1996.

Occupational Safety and Health Administration (OSHA). Hazardous Materials, Subpart H. Washington, D.C.: OSHA,

OSHA. Air Receivers, Subpart M. Washington, D.C.: OSHA,

OSHA. Fire Protection, Subpart L. Washington, D.C.: OSHA,

OSHA. Means of Egress, Subpart E. Washington, D.C.: OSHA,

OSHA. Personnel Protective Equipment, Subpart I. Washington, D.C.: OSHA,

OSHA. Storage and Handling of Anhydrous Ammonia, Subpart H. Washington, D.C.: OSHA,

State Fired and Unfired Pressure Vessel Safety Orders. Requirements promulgated by individual states to govern the safe operation of pressure vessels.

FURTHER READING

Insurance Institute of America. *Readings in Property Protection,* 2nd ed., 1990.

National Fire Protection Association (NFPA). *Fire Protection Handbook,* 18th ed. Quincy, MA: NFPA, 1997.

National Fire Protection Association (NFPA). *Industrial Fire Hazards Handbook,* 3rd ed. Quincy, MA: NFPA, 1990.

Society of Fire Protection Engineers (SFPE). *SFPE Handbook of Fire Protection Engineering,* 2nd ed. Bethesda, MD: SFPE, 1995.

PART IX

Professional Management Aspects

## CHAPTER 34

How to Be a Successful Safety Advisor

Robert A. Lapidus

CONTENTS

1-56670-370-0/02/$0.00+$1.50

INTRODUCTION

Technical expertise does not guarantee success. You can know all there is to know about a given subject, but if you do not know how to build long-lasting, credible relationships, no one will listen to you. Your expertise will be of no use to anyone. This lesson is one that many safety people fail to learn.

We fail to learn this lesson because we believe that our technical safety knowledge is the reason we have been hired. In many cases, what we believe is the truth. We have been hired for our technical expertise.

The typical impetus for instituting new safety programs and putting someone in charge of the safety function is management's desire to reduce losses, prevent serious injuries, and comply with governmental mandates. To implement such programs requires people who know the standards, regulations, techniques, and methods so that programs can be designed for specific needs.

What the new employer fails to tell us, and probably does not consciously recognize, is that they also expect us to work with others in the organization to achieve safety success. Consequently, when we begin to espouse our knowledge and lay down our expectations, the workforce, those people who have to apply our remedies to their daily work lives, start to feel put upon. We may find doors closed to us if we push too hard to achieve our goals.

Accordingly, to do our safety advisory jobs in a professional manner, we must have current technical knowledge and skills, but we cannot assume that such expertise will in itself get safety incorporated into the work environment. That takes other knowledge and skills — that which is associated with how we relate as human beings and how we seek to meet expectations. This chapter discusses 18 steps that can help you become a more successful safety advisor.

STEPS TO SUCCESS

Advisory success comes from giving yourself to others.

Step 1: Develop a Trusting Relationship from the Beginning

Establish a base of management to learn to trust you. Discuss your role with management. Listen to what they want to accomplish, and massage your goals to fit their goals.

In safety, specific actions need to be taken when we start to create a safety program. These actions are usually nonthreatening to anyone in the organization because they are simply information retrieval, and they include the identification of the

- Causes of why losses are occurring
- Exposures to loss that could result in additional losses, including compliance requirements
- Current activities being done to target the organization's losses and exposures
- Organization's style of management and culture so that future programs will fit

---

ADVISORY SUCCESS STEPS

1. Develop a trusting relationship from the beginning
2. Ask questions
3. Get input before taking action
4. Keep communication open
5. Tailor your advice
6. Be prepared to explain why
7. Put management's priorities first
8. Suggest alternatives
9. Avoid the enforcement role
10. Stay behind the scenes
11. Be timely with all activities
12. Be responsive to all requests
13. Be positive in your daily communications with others
14. Apologize
15. Learn to foresee problems before they happen
16. Avoid "I told you so"
17. Recognize others' successes
18. Show you are proud to be on the management team

Figure 34.1   Advisory success steps.

Obtaining this information can be an excellent opportunity to start to build relationships within the organization. Begin by having dialogue with others by sitting down with top, middle, and first-line managers, employees, and union representatives and talking with them about the current safety program, the future as they see it, and what they believe would work in the organization.

Developing a trusting relationship means that the advisor needs to listen. In the beginning, refrain from giving your opinion on every subject that is discussed. Build your knowledge base by gathering the knowledge, opinions, and feelings of others.

Step 1 should be initiated at the commencement of a new opportunity in safety, but can also be initiated after an advisor has been in the job for some time. The latter situation requires the advisor to set aside ego and make a concerted effort to reinitiate the advisor/management relationship. To accomplish this mission, we need Step 2.

Step 2: Ask Questions

Both Steps 1 and 2 are part of the building of personal credibility. Safety advisors need to be seen by others as people who are easy to talk to, are open to new ideas, and can be trusted to hold confidences.

No single person has all the answers. Therefore, probably one of the most effective means to obtain buy-in on the part of all managers, supervisors, and employees is to gain their involvement in the creation and implementation of the safety program. Instead of doing an "information dump" (one-way communication) on everyone else, ask questions to discover what is going on in the organization and how others see the safety program being designed. Upon starting a new job, ask others: "What are your expectations of me and my function?" If you are in a current job and have never asked this question, ask it now.

After three or four months of working on those expectations, return and ask, "Do you feel that I am meeting your expectations?" If they do not feel that your are achieving their expectations, find out what you can do to change that. Keep pursuing this line of questioning.

Those to whom you give advice are your customers. Treat them with respect. Let them know that their opinions, comments, and ideas are important to you. They need to know that you will recommend the implementation of safety programs and activities that are based upon the input you have received from those who will be affected.

Step 3: Get Input before Taking Action

Others may be privy to information crucial to your job. Getting input about what you plan to do before you actually do it will strengthen your relationship through a mutual sharing of information.

Get input on all programs that are going to affect other people. Such input should come from top and middle management, first-line supervision and employees, and special interest groups that are represented in the organization.

Step 4: Keep Communication Open

Continually apprise management of what your objectives are and how you are going to attain them. Use the communication tools that management prefers. Communication techniques may include face-to-face contacts, short, simple communiqués, formal periodic reports, or simply telephone contacts anytime something needs to be communicated. In the latter case, management may then decide what kind of follow-up documentation they need. The key is to use the mode of communication that your recipients want, not what you feel more comfortable using.

Keep a dialogue going with as many parties as possible. No one wants to be blindsided with surprises. Get input and provide input.

Step 5: Tailor Your Advice

Avoid canned answers or traditional approaches that worked for other organizations, but that do not fit the reality of your own. Be creative. Whatever advice you provide, be sure to tailor it to the climate of the organization, to the nature of the way things are. You should also seek creative new ideas from others. Those who actually do the main line of work have a pretty good idea of what will and will not work.

Step 6: Be Prepared to Explain Why

"Because" and "the law requires it" are not the answers that most of us appreciate receiving to the question of why something has to be done. Whenever you provide a suggestion, have as many facts as possible at your fingertips. Describe how your suggestion will affect the nature of the activity that is being done and how the suggestion will improve the workings of the organization. Provide actual examples of situations where the failure to have your recommended action in place resulted in a loss, such as an occupational injury or a property damage incident. Obtain backup data and real-life anecdotes from internal and external sources that will support your propositions. Know how your recommendations might provide a positive financial reward.

Step 7: Put Management's Priorities First

Conceptually, certain organizations may put safety first as the most important goal, but in reality safety is not first. Safety should be an integral part of how the work gets done, but a perfectly safe environment rarely exists. Everyone has to establish for themselves their own level of safety, what they consider to be and acceptable level of risk.

In this regard, sometimes management does not recognize certain elements of the safety advisor's work as being necessary. Management may not see your activities fitting within their priorities.

If your objectives appear to be important for the welfare of the organization, educate management on the importance of the objectives or activities that need to be done. Plant seeds (suggestions) in a manner that, in time, management may see the importance of the concept or need.

If you're not getting the response from management that you are seeking, you may wish to back off. Tell management, "I guess I should rethink my ideas." Then return to your office and rethink what you are trying to do and how you are trying to do it.

You may be on the right track, but not explaining it well. Then again, this time may not be right to pursue this particular activity. It may be necessary for you to let go of some portion of your job, objectives or some recommendation for change if management will not allow you to do it or will not support you in doing it.

One of the most difficult concepts for safety advisors to accept is that sometimes it is better not to force an issue for the time being, but to wait until the time is right to reinstate the activity or recommendation.

Step 8: Suggest Alternatives

Refrain from coming across as if you are the only one with *the* answer. Seek to be a true advisor (helper) by avoiding playing a game of gotcha.

Working in a technical profession, safety advisors have to balance the need to comply with regulatory requirements to the need of maintaining a working environment that permits employees to have input into their daily activities. The inspector mentality tends to cite the regulatory standards and require compliance regardless of the impact on the work. The helping or servant mentality tends to cite the regulatory standards, but then discusses with others how best to comply while maintaining the ability to get the job done. Be straightforward in your comments from a helping standpoint. Ask people for ideas about how problems might be solved.

Step 9: Avoid the Enforcement Role

If at all possible, avoid the role of enforcer. Instead of you playing the role of police officer, train line people to play that part. Work with them to develop the technical information they need to comply with external and internal mandates; then let line managers, supervisors, and employees enforce their own rules and regulations. Let them establish their own parameters and enforce proper compliance. They would then be doing their own self-monitoring.

Try to avoid the enforcement role by doing positive things instead:

- Seek to observe people doing their work in a safe manner, both in terms of safe practices and maintaining a safe environment. Let people know that you appreciate their positive efforts.
- Through line management, publicly acknowledge successful activities and results.

Step 10: Stay behind the Scenes

Advisors should avoid putting themselves in the limelight, either as the receiver of praise or the giver of recognition. Allow management the privilege of taking the public credit and giving appropriate commendation to their people. You will get advisor gratification by seeing your recommended programs implemented by operational management. When your advice is taken and others are successful, you will also reap the fruits of success.

Step 11: Be Timely with All Activities

Agree to mutually acceptable timetables for getting things done. If you do not do this, the parties who are expecting you to respond to them will normally have a shorter deadline in their minds than you will have in your mind. You will then be in trouble. If you think you will not be able to meet the deadline or target date that has been mutually agreed upon, inform the expecting party of the problem prior to the established date. Again, no one likes surprises.

Step 12: Be Responsive to All Requests

Keep those who have asked you for something informed as to the status of their request and your response. Let them know you have received their request. Work out a mutually acceptable timetable, and be responsive. Do not let a request drop into a black hole. Your credibility will be greatly hurt.

Step 13: Be Positive in Your Daily Communications With Others

The advisor needs to be seen as a helper, someone whom others feel comfortable coming to and obtaining support from. To achieve this positive status in the organization, avoid sarcastic or abrasive comments. Be careful about joking around; someone could easily take offense. Guard your temper. Do not lose your self-control, no matter how right you think you are.

Step 14: Apologize

There may be situations in your work life where you will do or say something that hurts another's feelings or that hurts the final outcome of a project. Be prepared at all times to repair relationships by apologizing for what you said or did. You may think you were correct in what you did, but if the relationship is suffering, be the first to apologize and seek to make things right.

Step 15: Learn to Foresee Problems Before They Happen

Advisors, especially safety advisors, are supposed to protect their organizations from potential problems. Keeping up to date on the latest new regulations is critical for knowing what type of compliance will be required. Keeping posted on accidents and incidents that have occurred in similar types of organizations will help you prevent those things from happening in your organization.

Step 16: Avoid Saying "I Told You So"

When management has a problem, and you had previously recommended a way to have avoided it, be there to help out and rectify the problem. Avoid saying "I told you so" in any way whatsoever. This includes side comments to others, facial expressions, body language, and direct statements that show that those in charge did not listen to you.

Step 17: Recognize Others' Successes

When other people succeed, recognize their success. Be on the lookout for anyone who has done something special. Keep congratulatory and thank you cards in your desk, and send them out in a timely manner. Those who receive your notes will appreciate you and your thoughtfulness even more.

Step 18: Show That You Are Proud to Be on the Management Team

Let management know you are proud to be part of their team, working together to achieve mutual goals. Tell them. Write thank you letters. Be there to participate in celebrations and special events. If you are not proud to be on the team, if you awake each workday with a headache or stomach ache, and dread going to work, move on, get out of that situation. You are not doing yourself or management any favors by staying around.

SUMMARY

Seek to be a credible resource. We need to have the knowledge necessary to help other solve safety-related problems. We need to be current on all subjects that relate to our work, but such knowledge does not guarantee success. In support of this knowledge we need to have a reputation for being dependable, honorable, reliable, and trustworthy; that is, we need to be credible.

The best way to achieve this credibility is to follow the Golden Rule: *Do unto others as you would have them do unto you.*

How do you want to be treated? Most of us have similar answers to this question. We want to be treated respectfully, honestly, and as if our ideas are worthy. We want to be treated fairly and be recognized for doing a good job. We want to have input into how our jobs and work activities will be done.

This latter desire is probably the most important element of safety advisory success, as has been described throughout this chapter. If we are to be successful advisors, we need to facilitate the involvement of others in the development and maintenance of the safety program itself. Everyone will then have an opportunity to become part of the safety effort, buy into the safety program, be treated respectfully, and look to the safety advisor as a credible resource.

FURTHER READING

Books

Barker, J. A. *Discovering the Future — The Business of Paradigms.* St. Paul, MN: ILI Press, 1985.

Blanchard, K. and Peale, N. V. *The Power of Ethical Management.* New York: William Morrow & Company, 1988.

Kouzes, J. M. and Posner, B. Z. *Credibility.* San Francisco, CA: Jossey-Bass, 1993.

Petersen, D. *Safety Management — A Human Approach.* Goshen, NY: Aloray, 1988.

Other Publications

Greenleaf, R. K. *The Servant as Leader.* Indianapolis, IN: The Robert K. Greenleaf Center, 1991.

CHAPTER **35**

Developing Effective Leadership Skills — How to Become a Successful Change Agent in Your Organization

Paula R. Taylor

CONTENTS

INTRODUCTION

For millennia, scholars and ordinary folks have been trying to define leadership. The result is a multitude of definitions and no consensus. One reason may be that leadership is demonstrated in many ways and in many places. There is personal leadership, political leadership, artistic leadership, scientific leadership, world leadership, and organizational leadership. A cry often heard today is that there is a lack of leadership in all areas. Perhaps the tremendous amount of change that is occurring requires more or stronger leadership than ever before. Our specific concern is the leadership role of the health and safety manager and how that manager can be a positive force in times of change and transition.

As organizations and the nature of work change, leaders are needed more than ever. Leaders demonstrate specific characteristics. Fortunately, each of these characteristics has associated skills and these skills can be learned. As each characteristic is discussed you will have the opportunity to assess your current strength in it, identify where you need to develop your skill, and learn specific steps you can take to increase your ability to lead and to be perceived as a leader. No matter where you are in your organization, or your life, you can make a difference.

Use the assessment checklist in the next section to measure your current strength in each characteristic. If you want additional feedback other than your self-perception, ask members of your team, peers, or employees to complete an assessment checklist on you. Use the checklist again in six months and at various intervals to check your progress.

ASSESSMENT CHECKLIST — LEADERSHIP CHARACTERISTICS

As you read through each leadership characteristic, assess your current level of strength or that of the person for whom you are completing the assessment. Use a 1 to 10 rating scale. A "1" indicates a deficiency, no strength at all, and a "10" means fantastic, could not be better, and no need for improvement.

Current Strength Leadership Characteristics   **Rating**

    1. Envision the Future
       Have a clear vision of the future that is different from the present   _____
    2. Communicate the Vision
       Am able to clearly communicate that future vision to others   _____
    3. Motivate and Inspire Others to Participate
       Am able to bring others on board to work toward the vision   _____

4. **Recognize and Take Advantage of Opportunities**   ———
   Can see opportunities for making changes and am willing to take
   advantage of them
5. **Empower Others to Act**   ———
   Am able to let go and let others take responsibility and authority for actions
6. **Model Behavior**   ———
   Am able to consistently model the behavior desired of others
7. **Provide Recognition and Rewards**   ———
   Consistently provide recognition and rewards to those who deserve them
8. **Possess and Demonstrate Character**   ———
   Consistently demonstrate qualities of respect, honesty, and trust to others

The first four of the eight characteristics are creation characteristics. These generate action and ideas. Characteristics 5, 6, and 7 are implementation characteristics. They make things actually happen. The eighth characteristic, the ability to demonstrate character, is in a category of its own. It is typical to be naturally stronger in either the creation characteristics or the implementation characteristics. However, both categories of skill are required.

Before leadership characteristics and skills are discussed in more depth, it is important to understand the environment in which today's manager operates and what is expected of him or her in that environment. The environment is one of change, and the manager is expected be a leader and a change agent. Managers are expected to move their organization through change, which is likely to be continuous and constant, and to create a creative, productive, effective group that will look different in structure, output, and style than anything they have known or imagined in the past.

Organizational change is, and will continue to be, the norm, not the exception. Some will find it exciting; many will find it unsettling and frightening. What should a leader know about change?

UNDERSTANDING CHANGE

Change, in all aspects of our lives, whether perceived as positive or negative, causes stress and anxiety. Change means leaving the past behind and going into an unknown future. Regardless of how much the past was disliked, it was familiar, and familiar is comfortable. Change feels uncomfortable and unfamiliar. Since organizational change is usually imposed upon us, it often feels as though our past is being taken away. This feels like a loss. Most people go through five stages as they make the transition in a changing environment.

Five Stages to Move to the Future

Denial — Regardless of what is said or done, people believe nothing will change or, if it does, it will have nothing to do with them. "They don't mean it." "It's just a rumor; we've heard this before." "Even if it happens it won't affect me (or this group)."

Resistance — At this stage people are very emotional. They are angry, depressed, resentful, self-pitying, and blaming. "Why me?" "After all I did for the company, I can't believe I'm being treated this way." "No way. I'm not changin' nothin'!"

Bargaining — People start making deals with themselves or God. "Okay, just let me stay until my retirement in 2 years." "I'll even take a pay cut; just don't move me."

Depression — People start to feel sad and anxious. Productivity declines.

Acceptance — The final stage is acceptance of what is happening and the beginning of a future orientation. "Hey, this is really happening. What do I need to do now?"

Actions for the Leader

Recognize and accept your own movement through the five stages — Leaders are not immune from passing through the five stages. If you anticipate and expect that you will experience the same feelings as everyone else, you will move through the stages quicker, will understand the process, and will be more empathetic to others.

Open and expand the communication channels — Keep communicating. Provide every possible means of getting out information — from group meetings to newsletter to hotlines — even if the information is constantly changing. The void of no information will be filled with rumors and speculation.

Be the first to give out the news — Do not allow your employees to hear about what is happening to them from the *Wall Street Journal*, the local newspaper, or another department.

Communicate with honesty and clarity — Make every effort to be clear and honest. This is not the time for ambiguity and "corporate-speak". If you do not know, say you do not know.

Encourage others to speak — Allow and encourage others to speak their minds, express feelings, and ask questions. Accept feelings, answer questions, and do not get defensive.

LEADER FOR CHANGE

In the midst of a major organizational change, strong leadership is more important than ever. Consider the eight characteristics of leaders and what you can do to strengthen your ability as a leader.

Characteristic 1: Envision the Future

All leaders without exception possess this characteristic. They are able to envision the future. They can see possibilities; they can imagine what could be. Leaders often see the future as they would create it or improve it. Much has been written about our ability as human beings to think in future terms. It is this vision, this hope of a future — a better future — that keeps us alive and keeps us going. Once we have passed through the acceptance stage, we ask, "What can I do? What will it be like?"

How strong is your ability to envision the future? If you want to develop skills in this characteristic, here are some specific steps you can take.

Steps to Increase Your Ability to Envision the Future

Spend quiet time alone — Finding time is often the hardest thing to do these days. It is important to set aside 20 minutes a day, an hour a week, and half a day or more once or twice a year.

INSTANT EXERCISE

STOP! Stop right now and make that commitment. Take out your calendar. Give yourself 20 minutes every day for the next month. The best time is usually in the morning, right after you wake up, or right before you go to sleep. If that does not work for you, take it out of your lunch hour. (Yes, take time for lunch.) What are you going to do in that time with yourself? You are going to sit alone in a quiet place, get relaxed and calm, and go inside your head and find out what is going on in there. Then you move to the next step.

Ask yourself what you would like to be different in your life, your work, and your organization — Spend time just thinking about what you would like to be different. This is personal brainstorming. The rule is no censorship. After you have considered for a while, write your thoughts in a notebook. This will be a valuable reference book for you. You are now ready for the final step.

Ask yourself what this vision would look like if it were in place — Close your eyes, stay relaxed, and imagine the future. The more vividly you can imagine the outcome the more clear your vision will be. Again, do not censor your vision. Make it big. Imagine you in your role; imagine how you would feed, how others would feel. Imagine the accomplishments of the organization, the impact on people's lives, the additional dollars made or saved by the organization. Include as much detail as possible. Engage all your senses. Finally, write the details in your notebook. Imagine first; write second.

Every dream and vision, every change starts in the mind of one person. This vision comes from a leader. That brings us to the second characteristic.

Characteristic 2: Communicate the Vision

Leaders can communicate their vision. Having a good idea and not being able to communicate it is as useless as having no idea or vision at all. Being able to articulate your vision is critical. Nothing produces more frustration in a group, particularly during times of change, than a leader without a vision or one who cannot communicate it. The result is the same. Think back on your own experience. Don't you find this to be true?

The vision is communicated verbally, in writing, or by example. Great leaders use all three. The skill required here is one of communication.

Steps to Increase Your Ability to Communicate the Vision

Write out your vision or mission statement — The very act of taking pen and paper and writing out what you see and really want for yourself and your organization will have the effect of breathing life into your vision. It will become more real and more possible to you, and ultimately to others.

Say it out loud. Speak into a tape recorder — Tell your mirror. Articulate your vision before a video camera. Communicate your vision out loud in a meeting. People want to know how you envision the future. Remember how frustrating it is not to know.

Walk your talk — If you imagine an organization where people speak out, try out new ideas and support each other, behave as you speak.

Assess your communication — As you listen to the tape recorder, watch yourself in the mirror or on the videotape, or read what you have written, ask: "Is my vision concise, clear, enthusiastic, and congruent?" If it is not, people will not understand you or will not believe you. If possible, get feedback from a good friend or a colleague. Does he or she understand what you are trying to say?

INSTANT EXERCISE

Take a few minutes right now. Take out a piece of paper or an index card. Write out your vision for your organization or group. Make your first draft as long as you need it to be. Then edit it down to no more than three sentences. Use this first write-out as you follow the steps listed above. The more you articulate and assess your vision, the clearer it will become to you and to others.

Having a vision and being able to communicate it are still not enough. Leaders possess the next characteristic.

Characteristic 3: Motivate and Inspire Others to Participate

Leaders have followers. How well do you move and motivate other people? Are you able to get people on board, to buy into your ideas? The special communication skill of influencing plays an important role with leaders who are strong in this characteristic.

Each person is motivated and inspired by something different, but great leaders are able to appeal to a wide range of people. The ability to influence others goes back to how well, how clearly and concisely you communicated your vision initially. The more dramatically you present your vision, the more passion and belief you demonstrate in your ideas, the more people you will bring on board with you. People respond strongly to leaders who show confidence, passion, and commitment.

Many potential leaders are unable to motivate and inspire others because they are unwilling to share their ideas completely. If you give only partial information, if you hold back your enthusiasm, if you think you have an idea that is too good to share, you may be left with an idea and nothing else. Your desire to have other people join you must be genuine.

Steps to Increase Your Ability to Motivate and Inspire Others

Own your vision, but be willing to share it — Once people are inspired, they share the vision. They feel it is theirs. This is when things happen, when people are working on a common goal. A sense of shared vision is critical.

Communicate the benefit of being involved — Since each person is motivated by something different, the benefit to pursuing the shared vision will be different for everyone. Understand and be able to communicate the benefits of being involved with you and your organization.

INSTANT EXERCISE

Reread your vision statement. Take out a sheet of paper and draw a vertical line down the center. Label the top "Benefits." Above the left column write "For Me." Above the right column write "For Them." List the benefits to you of accomplishing the vision. List all the benefits you can think of for others involved in working on the vision. Benefits can be both tangible and intangible. Keep these benefits in mind and communicate them as you are discussing your vision.

Moving the vision and talk into action brings us to the last of the *creation characteristics* and the fourth leadership characteristic.

Characteristic 4: Recognize and Take Advantage of Opportunities

Leaders notice what is going on. They are willing to act and willing to take a risk. Leaders are not careless and they are not reckless. They are action oriented. Strong leaders are up-to-date on their company, their industry, and the external environment. They also frequently have interests in seemingly unrelated events and organizations. Often opportunities and ideas come from an atypical and unexpected source.

Leadership strength in this characteristic is built by communication, participation, education, self-assessment, and risk. If this is an area in which you would like to be stronger, here are some things you can do.

Steps to Increase Your Ability to Recognize and Take Advantage of Opportunities

Communicate and participate — In order to identify opportunities you have to be out there listening and talking to people. The more people you know and who know what your vision is, the more possibilities will come your way.

Practice listening attentively to what other people are saying — As you hear what other people need and want, you will recognize where your plans and ideas fit. You will hear benefits for you and them.

Be prepared to act — Know your risk threshold. Think about just what you would be willing to do if the opportunity presented itself. Would you be willing to change departments, jobs, or companies or move to another city to see your vision in place? If you have a sense of what you are willing to do when the opportunity presents itself, you can take action or you can create your own opportunity. We all have stocks we should have bought, things we should have said, actions we should have taken; leaders minimize the if onlys and act. Dr. Maxwell Maltz expressed it well: "Often the difference between a successful man and a failure is not one's better abilities or ideas but the courage that one has to bet on his ideas, to take a calculated risk — and to act."

Take a moment to review the first four characteristics of leaders, the *creation characteristics*. These leadership skills sow the seeds necessary to create the future. They begin the process of change, the process of creating the future.

INSTANT EXERCISE

Consider your own strength in the creation characteristics. Identify the characteristic in which you are the weakest. What has not happened as a result of your not being as strong as you might be in that characteristic? Make a list of the specific things you will do differently that will make you more effective in that characteristic.

It is important to remember that although these characteristics are necessary in all leaders, each individual has a different style and a different approach. Some leaders will be noisy and flamboyant; others will be quiet and steady. One style is not inherently better than another. Each person must operate within his or her own personality while consciously doing what is necessary.

Let us look now at the implementation characteristics of leaders. If you can envision the future but can do nothing else, you may be known as a visionary. If you can communicate your vision, inspire others to join you, and seize every appropriate opportunity, you may be known as a group of visionaries. Visionaries are needed during times of change, but in order to make the changes work, a manager must be able to keep employees on board and moving in the right direction. If you cannot do that, the vision and the desired changes for the future will never happen.

The manager and the leader as a loner is not the model for today. The age of the leader as a loner is gone. A manager who cannot create a team, be part of a team, and keep that team together will not succeed in today's changing world. That brings us to leadership characteristic number five, the first of the implementation characteristics.

Characteristic 5: Empower Others to Act

A real leader today empowers other people. The leader shares the vision and shares the planning. As things are changing around them, people are comforted by having a hand in the solution. The leader not only asks the question, "How are *we* going to make this happen?" but listens to the answers.

Empowering others means sharing responsibility and authority. People must be given the authority to decide what is to be done and then must face the resulting consequences of their decision. The manager provides the team with the tools to do the job, with training in skills such as problem solving, listening, and presenting, with the latitude to accomplish something, and with support and assistance. Although time often seems like a commodity in short supply in transitional periods, it is the wise leader who finds the time and money to develop the team. It pays off in the end.

A leader also knows his or her team and knows the strengths and weaknesses of every team member. A leader who is capable of empowering others to act utilizes and encourages the strengths of each person and ignores or works around their weaknesses. Look for what people do well: use it, praise it, benefit from it. Allow and encourage people to learn new skills and improve weak areas, but go for the strengths.

INSTANT EXERCISE

Create a strength profile of your group. Use a separate sheet of paper for each member of your team. Write their name at the top. On the side list their strengths, on the other their weaknesses. Draw an "X" through the weaknesses. List how the strengths can benefit the group at this time. Praise the strengths, utilize them, and make assignments as appropriate. Variation: A variation on this exercise, and one that would be very beneficial to the team, is to share with the group what you want to do. In the whole group (if the level of trust is high) or with each individual, ask them to list their strengths and weaknesses. You and/or the other team members can add to each person's list. Again, put an "X" through the weaknesses — this is about strengths — and identify how each person can best benefit the group at this time. It may be that people decide to work together on tasks they once did separately, or they may shift duties altogether.

Steps to Increase Your Ability to Empower Others to Act

Communication is the key here; that means talking, listening, providing feedback, developing trust.

Create goals of "how we get there" as a group — Although it is the responsibility of the leader to present the vision, it is the responsibility of the group to figure out how to accomplish it. Being part of the solution helps people feel more involved and more committed to the changes that will take place.

Create clear expectations of each person's responsibility — Clarity is critical. Change produces enough ambiguity. The job of the leader is to make things as clear as possible.

Trust others to represent the best interests of the team — If the vision, goals, and roles are clear, people can represent each other. Everyone does not have to attend every meeting.

Utilize and encourage the strengths of the group members — When in doubt, refer to your list.

Allow freedom to make mistakes — Mistakes will be made because they always are. Changing times require new approaches, and this often means trial and error. Do not look for whom to punish, but to what can be learned. If people are allowed to make mistakes then they will try new things. That is what is needed.

Help people stretch to the next level — The natural tendency during change is to become more cautious and to withdraw when what is needed is to stretch and expand. Encourage others (and yourself) to go to the next level in ideas and action.

Goethe said, "Treat people as if they were what they ought to be and you help them to become what they are capable of becoming."

Characteristic 6: Model Behavior

A manager today must be not only the leader we are describing, but also a member of the team. In many ways that means modeling the behavior that you desire in others. The leader is not the exception to the rule but the rule itself. How a manager communicates, solves problems, spends money, manages time, and treats other people is observed by the group. People see what you do and take their lead from that regardless of what you say.

Steps to Increase Your Ability to Model Behavior

Do what you want others to do — It is rather simple: do not do anything you would not want a member of your group to do. You all represent each other.

Keep the end in mind — It is the responsibility of the leader to keep the vision alive. Remember the desired outcome and keep reminding the group so you all stay on track.

Stay positive — Things may get tough, but if you — the manager, the leader — do not stay positive, who will?

INSTANT EXERCISE

By yourself, or better yet with your group, create a list of norms of behavior. Ask the following series of questions, answer them, and post the answers as the norms of the group:

- How do we communicate with each other?
- How do we approach problem solving?
- How do we manage our time?
- How do we spend our dollars?
- How do we treat other people?
- How do we run our meetings?
- What do we do when things go wrong?

Characteristic 7: Provide Recognition and Rewards

Everyone needs recognition for what they accomplish. Leaders never forget this. Creating ceremonies and celebrations, particularly during times of change, helps remind people that they are important and valued. These rewards can be both public and private, monetary and nonmonetary. They can be for individuals and for the entire team.

Steps to Increase Your Ability to Provide Recognition and Rewards

Reward the behavior you want — Reward the behavior and accomplishments you want or movement toward it. This encourages people to keep going.

Share the wealth — If your group is recognized for an accomplishment, share the glory. Praise the team and all those involved in the success. If you do not you may not get many other opportunities.

Shoulder the blame — If there is a mistake or a problem, the leader takes responsibility. Do not point the finger or shift the blame. In private, with the team, review the problem and learn from the mistake.

Give the recognition and give it yourself — Good intentions will buy you nothing. Do it. Do not just plan to do it. Give the recognition. Also, do it personally. As the leader your feedback and praise mean a great deal to people. Make it personal.

INSTANT EXERCISE

Make a list of the people who deserve recognition for a job well done, a favor, a great idea, etc. Next to their names write what you are going to do (call, write a note, send a gift, etc.) and the date by which you will do it. Then follow through with the action.

Characteristic 8: Possess and Demonstrate Character

Leaders have character. Although each person may have their own definition of character, it is the essence of an individual. It has to do with the respect they give to people and the respect they receive. Character is demonstrated by ethics, honesty, and sincerity. Character is the bedrock on which all the other characteristics of a leader rest.

Two traits that may best demonstrate leaders of character are that they trust people and that they possess empathy. Strong leaders have a positive view of humankind. They trust that people, given the opportunity, will do a good job. Their view is not "everyone is stupid except me," but rather "I know they can do it." Strong leaders also have empathy. They remember what it was like when they were in a similar position, or they have the capacity to imagine how someone in a given situation may be feeling. It is also their ability to communicate trust and empathy to others that results in the respect and loyalty required for a leader to lead a team through change, toward its vision.

INSTANT EXERCISE

Think back among the many people in your life to someone you really admire. This may be a friend, a relative, a business acquaintance, or someone you met for just a moment. Think about how that person makes you feel when you are with him/her, what he/she does or says that impresses you, what he/she has done that you admire. List the specific characteristics he/she possesses that makes him/her special to you. Put a check next to those that you would like to possess. Decide what you need to do differently and take action.

SUMMARY

In order to be a change agent in your organization, your role as a manager requires you to be a strong leader capable of making the difference in your organization. Strong leaders (1) envision the future, (2) communicate the vision, (3) motivate and inspire others to participate, (4) recognize and take advantage of opportunities, (5) empower others to act, (6) model behavior, (7) provide rewards and recognition, and (8) possess and demonstrate character. Each of these characteristics has associated skills that you as a manager can learn and utilize.

FURTHER READING

Bass, M. *Leadership and Performance beyond Expectations.* New York: The Free Press, 1985.

Batten, D. *Tough-Minded Leadership.* New York: AMACOM, 1989.

Belasco, A. and Stayer, C. *Flight of the Buffalo.* New York: Warner Books, 1993.

Bennis, G. and Nanus, B. *Leaders.* New York: Harper & Row, 1985.

Bennis, G. and Nanus, B. *Leaders: The Strategies for Taking Charge.* New York: Harper Perennial, 1985.

Byham, C. and Cox, J. *Zapp! The Lightning of Empowerment.* New York: Faucet Columbine, 1988.

Cleary, T. *The Book of Leadership and Strategy: Lessons of the Chinese Masters.* Boston, MA: Shammbala, 1992.

Covey, S. *The Seven Habits of Highly Effective People.* New York: Fireside, 1990.

Covey, S. *Principle-Centered Leadership.* New York: Summit Books, 1990.

Cox, A. *Straight Talk for Monday Morning,* New York: John Wiley & Sons, 1990.

Cox, D. and Hoover, J. *Leader When the Heat's On.* New York: McGraw-Hill, 1992.

Cribbin, J. *Leadership.* New York: AMACOM, 1981.

DePree, M. *Leadership is an Art.* New York: Dell Publishing, 1989.

DePree, M. *Leadership Jazz.* New York: Doubleday, 1992.

Dilenschneider, L. *A Briefing for Leaders.* New York: Harper Collins, 1992.

Gardner, N. *On Leadership.* New York: The Free Press, 1990.

Kotler, P. *A Force for Change.* New York: The Free Press, 1990.

Kouzes, M. and Posner, B. Z. *Credibility.* San Francisco, CA: Jossey-Bass, 1993.

Kouzes, M. and Posner, B. Z. *The Leadership Challenge.* San Francisco, CA: Jossey-Bass, 1987.

Leavitt, J. *Corporate Pathfinders.* Homewood, IL: Dow Jones-Irwin, 1986.

Loden, M. *Feminine Leadership, or How to Succeed in Business without Being One of the Boys.* New York: Times Books.

McLean, J. W. and Weitzel, W. *Leadership: Magic, Myth, or Method.* New York: AMACOM, 1991.

Nanus, B. *Visionary Leadership.* San Francisco, CA: Jossey-Bass, 1992.

Peters, T. and Austin, N. *A Passion for Excellence.* New York: Random House. 1985.

Phillips, T. *Lincoln on Leadership.* New York: Warner Books, 1992.

Richardson, J. and Thayer, S. *The Charisma Factor.* Englewood Cliffs, NJ: Prentice-Hall, 1993.

Tichy, N. M. and De Vanna, M. *The Transformational Leadership.* New York: John Wiley & Sons, 1986.

Yeomans, W. N. *1000 Things You Never Learned in Business School.* New York: Signet, 1985.

## CHAPTER 36

Interpersonal Communications

Laura Benjamin

CONTENTS

INTRODUCTION

Perhaps you've heard it said, "It's not so much what you say as how you say it!" I strongly disagree. I believe they both count! Interpersonal communication skills are now more important than ever for the safety and health practitioner's professional development because of the following factors:

- Technological advances, which require more-sophisticated, subtle communication skills.
- Cross-cultural influences, which require us to work with people who are not always just like us.
- Mergers and acquisitions, which introduce new ways of doing business overnight.
- Multigeneration workers, which have broadened the age range of those we work with.

More than ever before, your ability to play well with others is critical to your success as a leader, a mentor to others, or the member of a highly functional team. It will influence not only the professional connections you make within your industry and organization, but also your personal development and career success.

FACTORS THAT INFLUENCE INTERPERSONAL COMMUNICATION

Have you noticed that men communicate differently from women, younger generations seem to have a language all to themselves, and one co-worker is easier to connect with than another? A number of factors influence how we express ourselves, both verbally and nonverbally, including:

- Culture
- Family background
- Career history
- Personality or behavioral style
- Job demands
- Education
- Physiology
- Socioeconomic levels
- Religion
- Voice tone

The list could go on and on! No one communicates in exactly the same manner. Therefore, our challenge becomes not only how to express ourselves in a meaningful way, but also how to increase our understanding of others.

Tip — See how many other influencing factors you can add to the list above. It will increase your awareness of just how challenging effective communication can be!

FIRST IMPRESSIONS COUNT

This is the key to our survival as a species. Ancient people needed to make rapid choices of whether they should run or fight to stay alive. They counted on their senses to draw conclusions of whether the person approaching was friend or foe, and if they made the wrong choice or ignored the data their bodies received, it could mean the difference between life and death.

In more modern society, we make the same choices and must recognize that others also draw rapid conclusions about us. Within 3 to 5 seconds of meeting

Table 36.1 Body Language

Behavior	Interpretation
Brisk, erect walk	Confidence
Standing with hands on hips	Readiness, aggression
Arms crossed on chest	Defensiveness
Hand to cheek	Evaluation, thinking
Rubbing the eye	Doubt, disbelief
Hands clasped behind back	Anger, frustration, apprehension
Rubbing hands	Anticipation
Open palms	Sincerity, openness, innocence
Tapping or drumming fingers	Impatience
Patting/fondling hair	Lack of self-confidence
Steepling fingers	Authoritativeness
Stroking chin	Trying to make a decision
Biting nails	Insecurity
Pulling or tugging at ear	Indecision

someone new, we can usually tell whether they are friend or foe. When approaching a manager or co-worker, we can often sense if they are going through a difficult day or are relaxed and open to conversation. What sensory data helps us draw those conclusions? We usually make accurate decisions based on nonverbal clues or body language as outlined in Table 36.1.

Paying attention to the nonverbal cues that you receive from others will help you evaluate your own behavior and the impact it's having on those around you.

Tip — Ask people you trust to give you feedback regarding the nonverbal cues you send. Avoid arguing or justifying, and ask them why they chose the cues they did to discuss with you.

COMMUNICATION STYLES

There are almost as many ways people communicate as there are different personality types! To make it easier to learn interpersonal strategies that work, consider the following basic categories.

Direct: They Seek Action and Results

Direct communicators want to get results or achieve a goal. They communicate in a very direct way. They will "cut to the chase" and get right down to business without spending any time with the niceties like, "How is your day going?" It's easy to misinterpret their direct approach as unfriendly, aloof, or demanding.

Tip — To be more effective with direct communicators, speak with confidence, focus on the deliverables, and reference specifics vs. vague generalities. Let them know what you will achieve vs. how you will do it. Appreciate their abilities as change-agents.

Rapport Builders: They Seek Acceptance

Rapport builders want to increase understanding and build friendships. They usually start a conversation with the niceties and can spend the majority of their time on social talk before getting down to business. It's important for them to know you like them as individuals prior to conducting business. It's easy to misinterpret their social approach as frivolous, or unfocused.

Tip — To be more effective with rapport builders, temporarily set business objectives aside long enough to address their social needs. They respond well to positive nonverbal cues like smiles and appropriate physical touch. Appreciate their ability to influence others.

Quality Controllers: They Seek Accuracy

Quality controllers want to do it right. Their communication style centers on clarifying, correcting, and ensuring accuracy. They will focus on getting the facts, down to every detail and nuance. They may grow impatient if you avoid their questions or respond with vague, general statements. It's easy to misinterpret their need for accuracy as picky or controlling.

Tip — To be more effective with quality controllers, do your homework and have your facts in order before approaching them. Don't try to wing it since you will lose their respect if they find you out. Appreciate the value they bring in identifying gaps and holes in process or procedure.

Team Builders: They Seek Cooperation

Team builders want to create consensus. They strive to find areas of agreement and commonality to gain cooperation from others. They tend to be the peacemakers and avoid contentious communication. It's easy to misinterpret their need for agreement as "wishy-washy."

Tip — To be more effective with team builders, use a "win/win" approach and build options and choices into your conversations. Since they will automatically look for areas of agreement, avoid making either/or statements.

COMMUNICATION CHALLENGES

Occasionally, situations arise where you wish you had a tool to handle a challenging set of circumstances. The following are strategies to use in times when you are not quite sure how to approach the situation.

Conflict Communication: The Four F Method

The most difficult type of communication occurs when you must confront people regarding their behavior. The following tool may be used to coach someone to better performance, identify inappropriate behavior, or draw personal boundaries to manage an unhealthy relationship:

State the *facts*: This is the when, where, how, and what of the circumstances that occurred. It's advisable to have the facts written down to ensure accuracy and avoid misstatements. For example: "Bob, I've noticed your safety checklist has been incomplete 4 of the past 5 weeks."

State your *feelings*: Indicate how you feel about Bob's lack of follow-through. For example: "Bob, I feel you're not as committed to the quality of your job as you were when we hired you 6 months ago." (This may or may not be necessary, depending on the situation.)

State the *fallout*: This is the result or consequences of what will follow should the behavior continue. It lets the other person know exactly how you will respond to a recurrence. For example: "Bob, if you submit another incomplete safety checklist in the next 3 months, it will disqualify you for your next pay raise."

Encourage *Feedback*: This component has a dual purpose. First it allows Bob the opportunity to respond right away, *or* to reconnect with you at an appointed date to respond to your concerns. This is especially helpful if emotions make it difficult for Bob to respond appropriately at that time. It gives you access to important information you may not have known, which could influence your understanding of the facts. Second, if Bob makes an improvement in his behavior, it reminds you to follow up with him to recognize and commend him on his efforts.

Tips

- Use "I" statements when making observations or stating facts. This prevents the other person from becoming defensive and shutting down.
- Avoid going below the "water line" by assuming you know how the other person feels or what the person's motivation might have been. Think of the iceberg that sunk the *Titanic*. Only a small portion of the iceberg was visible above the waterline. You can only deal with what you see because what lies below the waterline may be totally different from what it appears!

Talking Sideways

The best way to define talking sideways is to picture two sports fans sitting side-by-side watching a televised event. Their conversation is periodic, abbreviated, and interspersed between plays. This strategy is helpful when direct conversation may be interpreted as too threatening, intimidating, or formal. It's best done while you are engaged in an activity alongside the person you would like to communicate with, for example, while setting up a computer, clearing a meeting room, or inventorying equipment.

Nonstop Talkers

To end a conversation with someone who is monopolizing your time, drop your voice tone down at the end of a sentence to indicate finality. You may observe those who end their statements with a questioning tone. This implies lack of confidence and invites participation. To discourage further communication, end your sentence firmly with your voice at least one to two tones deeper than your normal range.

Obtaining More Information

To obtain more information, ask more questions! As the old saying goes, we have two ears and one mouth, so we should listen twice as often as we talk! To get people talking so you obtain more information it is important to know how to ask appropriate questions. There are two types of questions:

Closed ended — These are questions that can be answered with either a yes or a no response. They will not help you get much information and usually start with, do you, can you, will you, have you, etc.

Open-ended — These questions begin with how, what, when, where, or why. They encourage the respondent to provide details, tell a story, or discuss surrounding issues leading to a particular result. The best questions, however, begin with how or what since the others (when, where, or why) often put people on the defensive.

CONCLUSION

In summary, interpersonal communication is an art form that takes practice and requires that both participants be actively engaged in the process. It may come easier for some people than for others, but it is the responsibility of both parties to bridge the gap toward productive, respectful workplace relations. As our labor force diversifies, technology changes shape, businesses merge, and new workers join our ranks, the demand for highly skilled communicators will increase proportionately. It is worth your time, your career, and the relationships that will grow and develop through this investment.

FURTHER READING

Bayan, R. *Words That Sell*, Caddylak Systems, Inc., 1984.

Bramson, R. *Coping with Difficult People*, Bantam Doubleday Dell Publishing. New York, 1981.

Cialdini, R. *Influence: The Psychology of Persuasion*, New York: William Morrow and Company, Inc., 1993.

Keirsey, D. *Please Understand Me II*, Prometheur Nemesis Book Company, Delmar, CA, 1998.

Levinson, J. C. and Smith, M. S. *Guerrilla Negotiating*, John Wiley & Sons, New York, 1999.

## CHAPTER 37

Techniques to Make You a Successful Negotiator

Dale A. Gray

CONTENTS

1-56670-370-0/02/$0.00+$1.50
© 2002 by CRC Press LLC

INTRODUCTION

Negotiating is one of the least-understood skills in human affairs. How often do you negotiate? Never, often or seldom? Like it or not, you are a negotiator. Negotiation is a fact of life. You negotiate all the time. Sometimes we don't even realize it. You discuss a raise with your boss. You discuss a design or safety issue with management or the union. You negotiate with OSHA over the type of citation, penalty assessed, or abatement date. You try to agree with a stranger on a price for his house. Two lawyers try to settle a lawsuit arising from a car accident. You negotiate with your spouse about where to go for dinner and with your child about when the lights go out. Negotiation is a basic means of getting what you want from others. It is back-and-forth communication designed to reach an agreement when you and the other side have some interests that are shared and others that are opposed.

All people want to participate in decisions that affect them; fewer and fewer people will accept decisions dictated by someone else. People differ, and they use negotiation to handle their differences. Standard strategies for negotiation often leave people dissatisfied, worn out, or alienated — and frequently all three. As a negotiator you have an obligation to help the other party achieve a win/win outcome. You should look for mutual gains whenever possible, and where your interests conflict, you should insist that the result be based on some fair standards independent of the will of either side. Good negotiation is firm on the merits of the issues, but easy on the people. It employs no tricks and no posturing. Professional negotiation allows you to obtain what you are entitled to and still be decent. It enables you to be fair while protecting you against those who would take advantage of your fairness.

Application of the following techniques will add a new dimension to your personal and business life.

NEGOTIATION DEFINED

Webster defines *negotiation* as mutual discussion and arrangement of the terms of a transaction or agreement; others say it is a field of knowledge that focuses on gaining the favor of people from whom we want things. Still others say it is the timely use of power and information to affect behavior within a web of tension. *Mediation*, on the other hand, involves a third party to effect an agreement or reconciliation between the two or more negotiating parties. *Arbitration*, is the hearing and determination of a dispute or the binding settlement of differences between the parties by a person or persons chosen or agreed to by them.

HOW GOOD A NEGOTIATOR ARE YOU?

Like any management skill, negotiation can be learned, practiced, and mastered. Here is a questionnaire on the personal characteristics necessary to be a great negotiator. It will help you determine where you have strengths as a negotiator and where you may need improvement. Circle the number that best reflects where you fall on the scale. The higher the number, the more the characteristic describes you. When you have finished, add up your numbers and put the total in the space provided.

1. I enjoy dealing with other people, and I am committed to creating a win/win outcome.

   1 2 3 4 5 6 7 8 9 10

2. I enjoy solving problems and coming up with creative solutions.

   1 2 3 4 5 6 7 8 9 10

3. I am willing to ask as many questions as it takes to get the needed information to make the best decision.

   1 2 3 4 5 6 7 8 9 10

4. I do not take my counterpart's strategies, tactics, and comments personally.

   1 2 3 4 5 6 7 8 9 10

5. I like to uncover the needs, wants, and motivations of my counterparts — so I can help them achieve their goals.

   1 2 3 4 5 6 7 8 9 10

6. I am able to think clearly under pressure.

   1 2 3 4 5 6 7 8 9 10

7. I have good self-esteem and tend to have a high level of aspiration and expectation.

   1 2 3 4 5 6 7 8 9 10

8. I recognize the power of strategies and tactics and use them frequently.

   1 2 3 4 5 6 7 8 9 10

9. I am willing to compromise to solve problems when necessary.

   1 2 3 4 5 6 7 8 9 10

10. I am a good listener.

   1 2 3 4 5 6 7 8 9 10

Grand Total: _____

Scoring

90+:   You have the characteristics of a great negotiator. You recognize what negotiation requires, and you are willing to apply yourself accordingly.

80–90: You should do well as a negotiator, but you have some areas that need improvement.

70–79: You have an understanding of negotiation but can improve in some important
areas.

0–70:   Go over the questions again. You may have been hard on yourself, or you may
have identified some key areas that you will need to concentrate on when you
negotiate.

THE FIVE POSSIBLE OUTCOMES OF NEGOTIATION

A negotiation will end in one of five possible outcomes: lose/lose, lose/win,
win/lose, win/win, no deal (no negative or positive consequences). In most situations,
you should strive for a win/win outcome.

Lose/lose — Lose/lose is the outcome when neither party achieves his or her needs or
wants and is reluctant to negotiate with the same counterpart again.

Lose/win; win/lose — The second and third possible outcomes of negotiation are the
win/lose and the lose/win. The difference between the two is which side of the
fence you end up on.

Win/win — In almost all negotiations, you should strive for a win/win outcome, in
which the needs and goals of both parties have been met. Both parties will walk
away with a positive feeling and will be willing to negotiate with each other again.
There are four keys to creating a win/win outcome:
1. Do not narrow your negotiation down to one issue.
2. Realize that your counterpart does not have the same needs and wants you do.
3. Do not assume you know your counterpart's needs.
4. You need to believe point number two in your heart.

No deal — The fifth possible outcome is no outcome: neither party wins or loses.

THE IMPERATIVE ELEMENTS OF NEGOTIATION:
TIME–INFORMATION–POWER–TRUST

The four crucial elements in negotiation are:

Time — The period over which the negotiation takes place.
Information — The more you have, the better.

Power — Great or marked ability to act and accomplish.... To exercise strength and
control over people, events, and situations in many ways.

Trust — Reliance on the integrity, strength, ability, and surety of a person.

People need time to accept anything new. Resistance to change is universal. It
takes time to become used to ideas that are foreign or unpleasant.

Most people think of negotiation as an event that has a definite beginning and
ending. Nothing could be further from the truth. Most negotiations, like life, are a
continuous process.

Time plays a critical role in negotiations. Most often, negotiations will conclude in
the final 20% of the time allowed. This aspect of negotiation follows an interesting rule
that seems to apply to life in general. It's called the 80/20 rule, or Pareto's law. It states:

Twenty percent of what you do produces 80 percent of the results; conversely, 80%
of what you do produces only 20% of the results.

Here are a few suggestions that will help you bring time to your side of the
negotiations.

- Have patience.
- If there are benefits to resolving the negotiation quickly sell your counterpart on
  the value added to him or her.
- Realize deadlines can be moved, changed, or eliminated.
- In most negotiations you are better off if you know your counterpart's deadline
  and your counterpart does not know yours.
- Remember that generally you will not achieve the best outcome quickly.

Information

Remember, information is power. Information is the heart of the matter. Before
you can use information, you must know where it is, you must have access to it,
and you must be organized to receive it.

Most often, the side with the most information will receive the better outcome
in a negotiation. Why do people fail to obtain adequate information prior to a
negotiation? Because they tend to perceive the negotiation as the actual interaction
between two parties. People seldom think about the information they need until this
fact hits them in the actual negotiation.

A negotiation is not an event, it is a *process*. It starts long before the face-to-
face encounter. Your chances of obtaining information during the actual negotiation
are relatively remote.

The earlier you start, the easier it is to obtain information. Advanced preparation
is critical. The more information you have, the better off you will be. Also, it will
be in your best interest to have a clear set of goals going into the negotiation. You
should set your goals based on the information you collect.

Power

Power is based on perception — if you think you've got it, you've got it. If you
think you don't have it — even if you've got it — then you don't have it.

Power is a concept with ugly connotations. It implies a master-slave relationship. Knowledgeable people complain about power because they don't like the way it is being used, or they don't approve of the goal of the power.

There are ten types of power that can influence the outcome of a negotiation. The word can is emphasized because if one has power but doesn't use it, the power adds no value to the negotiation.

Position — Some measure of power is conferred based on one's formal position in an organization.

Legitimacy — Certain positions of authority confer legitimate power.

Knowledge or expertise — People who have knowledge or expertise can wield tremendous power. Knowledge in itself is not powerful. It is the use and application of knowledge and expertise that confers power.

Character — The more trustworthy individuals are, the more power they will have in negotiations.

Rewards — Those who are able to bestow rewards or perceived rewards hold power.

Punishment. Those who have the ability to create a negative outcome for their counterpart have the power of punishment.

Sex — Dealing with someone of the opposite sex can confer power.

Behavior style — Everyone most likely displays one or a combination of the following behavioral styles:

   a. Analytical, process oriented, methodical
   b. Driven, task oriented, goal directed, bottom-line focused
   c. Supportive, relationship oriented, focused on feelings

No power — In some instances giving up all power can be very powerful.

Crazy — This may sound funny, but bizarre or irrational behavior can confer a tremendous amount of power.

Most people have more power than they think. I believe there is a link between self-esteem and the amount of power you believe you have.

Trust and Building Trust

The more your counterpart trusts you, the easier you will find it to negotiate a win/win outcome. If for whatever reason your counterpart considers you untrustworthy, you will find it difficult to obtain even minor concessions. Think about it.

Here are some suggestions to help build trust with your counterpart:

- Do what you say you are going to do.
- Go beyond the conventional relationship.
- Overcommunicate.
- Be honest ... when it costs you something to be honest.
- Be patient.
- Accept honest mistakes.
- Safeguard for fairness.

In a continuing relationship, the more trust you place in others, the more they will justify your faith. Convey your belief in their honesty and you will encourage them to live up to these expectations.

TRAITS OF GOOD NEGOTIATORS

Negotiation is one of the most difficult jobs a person can do. It requires a combination of traits not ordinarily found in the business world or professions. The process of negotiating demands not only good business judgment but also a keen understanding of human nature. Nowhere in business does the transmission of power, persuasion, economics, motivation, and organizational pressure come together in so concentrated a fashion and so narrow a time frame. Nowhere is the potential return on investment so high. To be professional in negotiation, certain characteristics are imperative. Among them are the following.

Integrity

Integrity is the uncompromising adherence to moral and ethical principles; it is a soundness of character.

The absence of integrity cannot be offset by intelligence, competence, or tight legal phraseology. Without this quality, no deal, however carefully written, is worth much. Built into the transaction must be a high sense of values, the generosity to resolve subsequent difficulties equitably, and a commitment to meet the intent of the agreement.

Trust

No one is empowered with trust. Trust is earned. Trust takes time to build. One is trustworthy or not. It is the reliance on the integrity, strength, ability, and surety of a person. Without trust, agreement can seldom be reached.

Patience

Patience is the most powerful tactic in negotiation, more powerful even than deadlock or threat. Patience, persistence, and determination can make up for inadequate resources. Patience fulfills the basic mission of negotiation. Patience has another payoff. It gives both parties a chance to find how best to benefit each other.

Patience works: this is what it can accomplish for you:

- It divides the opponent's organization.
- It can lower the other party's expectations.
- It leads to concession after concession.
- It forces a new look at priorities.
- It separates wishes from reality.
- It gets other people involved.
- It can cause a change in leadership.
- It provides new information.

Other Traits of a Good Negotiator

- An ability to negotiate effectively with members of their own organization and win their confidence.
- A willingness to negotiate anything.

- A willingness and commitment to plan carefully in advance and to know the rules and the alternatives.
- Good business judgment. An ability to discern the real bottom-line issues.
- An ability to tolerate conflict and ambiguity.
- The courage to commit to higher targets and take the risks that go with it.
- A willingness to get involved with the opponent and the people in the opposing organization, that is, to deal in personal and business depth with them.
- An ability to listen with an open mind.
- The insight to view the negotiation from a personal standpoint, that is, to see the hidden personal issues that affect outcome.
- Self-confidence based on knowledge, planning, and good intraorganizational negotiation.
- A willingness to use team experts.
- Stability, the quality of one who has learned to negotiate with himself or herself and laugh a little of one who doesn't have too strong a need to be liked, of one who doesn't have to be smarter than everyone in the room.

Can anyone fit a bill as tough as this? Probably not. However, when the stakes are high it pays to find the right person. Common sense and research tell us that skilled negotiators come home with better settlements.

LISTENING SKILLS

A successful listener must keep an open mind and strive to be free from bias and preconceived notions.

Few negotiators know how to be good listeners. Statistics indicate that the normal, untrained listener is likely to understand and retain only about 50% of a conversation. This relatively poor percentage drops to an even less impressive 25% retention rate 48 hours later. To be a good listener, you must attempt to be objective. The best negotiators almost always turn out to be the best listeners as well.

Three Pitfalls of Listening

Negotiators tend to encounter three pitfalls that hinder effective listening. First, many think that negotiating is primarily a job of persuasion and, to them, persuasion means talking. Second, people tend to overprepare for what they are going to say and to use their listening time waiting for their next turn to speak. Third, we all have emotional filters or blinders that prevent us from hearing what we do not want to hear.

Types of Listening Skills

Great listening does not come easily. There are two major types of listening skills, attentive and inter-active. The following attentive skills will help you better receive true meanings your counterparts are trying to convey.

Attentive Listening Skills

- Be motivated to listen.
- If you must speak, ask questions.
- Be alert to nonverbal cues.
- Let your counterpart tell his or her story first.
- Do not interrupt when your counterpart is speaking.
- Fight off distractions.
- Do not trust your memory. Write everything down.
- Listen with a goal in mind.
- Give your counterpart your undivided attention.
- React to the message, not to the person.
- Don't get angry.
- Remember, it is impossible to listen and speak at the same time.

Interactive Listening Skills

The second type of listening skills are those used to interact with the speaker. These skills help ensure that you understand what the sender is communicating, and they acknowledge the sender's feelings. Interactive skills include clarifying, verifying, and reflecting. Clarifying is using facilitative questions to clarify information, get additional information, and explore all sides of an issue. Verifying is paraphrasing the speaker's words to ensure understanding and to check meaning and interpretation. Reflecting is making empathetic remarks that acknowledge the speaker's feelings.

To be empathetic, negotiators need to perceive the content of the message accurately. Second, they need to pay attention to the emotional components and the unexpressed core meanings of the message. Finally, they need to attend to the feelings of the other, but remain detached.

When you want to improve your listening skills, a good rule to remember is that you have two ears and one mouth and you should use them in respective proportions.

Listening is the easiest way to recognize needs and discover facts. Why don't we listen? Of the 11 reasons below, only the first is the responsibility of the speaker. The rest are self-inflicted by the listener.

1. Most people speak before they think. Their speech is disorganized and difficult to listen to.
2. We have a lot on our minds that cannot be switched off at a moment's notice.
3. We tend to talk and interrupt too much.
4. We are anxious to rebut the other person's arguments.
5. We dismiss much of what we hear as irrelevant or uninteresting.
6. We tend to avoid listening to difficult material on the basis that it is too technical or detailed.
7. We allow ourselves to be distracted and don't concentrate.
8. We jump to conclusions before all the evidence is in.
9. We try too hard to remember everything so that the main points are lost.
10. We dismiss some statements because they come from people whom we don't consider important.
11. We tend to discard information we don't like.

PREPARATION, AGENDA, AND OPENING REMARKS

Preparation

Negotiating does not start and stop on a certain date or when a contract expires. It starts long before and lasts long after. Preparation is imperative. Following are tips for success.

- Do your homework.
- Determine the number and kinds of people that will compose your team:
  - Single negotiator
  - Team negotiators
- Establish the meeting site.
- Establish seating arrangements and amenities.
- Make caucus rooms available.
- Establish objectives.
- Determine major and minor issues.
- Reveal minimum and maximum position.
- Determine counterparts maximum position.
- Establish behavior rules.
- Prepare overall game plan.
- Determine where and when to begin.

Agenda

Prepare the agenda, if possible. Those who control the agenda formulate the questions and time the decisions. A good agenda can clarify or hide motives. The person who controls the agenda controls what will be said, and, perhaps more importantly, what will not be said. Always try to negotiate an agenda before talks begin. It will help you keep the initiative. The following guidelines are pertinent:

- Do not accept the other person's agenda without thinking through the consequences.
- Consider where and how issues can best be introduced.
- Schedule the discussion of issues to give yourself time to think.
- Study the opponent's proposed agenda for what it deliberately leaves out.
- Be careful not to imply that your "must" demands are negotiable. You can show your resolve early by not permitting such items into the discussion.

An agenda is a plan for discussion. It is not a contract. If either party does not like the format after talks begin, the party must have the courage to change it. Neither party can afford to treat the matter lightly.

Opening the Negotiation

There are no strict rules on opening or conducting a meeting, but experience suggests the following:

- Come in like velvet, not like coarse sandpaper.
- Arrange for proper introductions.

- Begin with an irrelevant topic.
- Use humorous stories to lighten tension.

BREAKING AN IMPASSE

Too many negotiations break down for the wrong reasons. There is nothing wrong with deadlock in itself. A seller has every right to prefer no deal to one at too low a price. A buyer may prefer impasse as a tactic for reaching its objective. What is of concern is how to break the deadlock we do not want.

These moves are useful in averting or breaking an impasse:

- Change a team member or the team leader.
- Change the time shape of uncertainty. This can be done by postponing some difficult parts of the agreement for renegotiation at a later time when more information is known.
- Change the time shape of risk sharing. A willingness to share unknown losses or gains may restore a lagging discussion.
- Change the shape of future satisfaction by recommending grievance procedures or guarantees.
- Change the bargaining emphasis from a competitive mode to a cooperative problem-solving mode. Involve engineers with engineers, operators with operations people, bosses with bosses.
- Change the base for a percentage. A smaller percentage of a larger base or a larger percentage of a smaller but more predictable base may move things back on track.
- Call a mediator.
- Add options of a real or apparent nature.
- Set up a joint study committee.
- Tell a funny story.

The reason that impasse breakers work is that they reinvolve the opponent in discussions with the opponent's team and other members of the organization. These ice breakers help create a climate in which new alternatives can be developed. Surprisingly, the introduction of new alternatives sometimes has the effect of making old propositions look better than ever.

Breakdowns are not always caused by world-shattering issues or great matters of economics. Many are the result of simple things like personality differences, fear of loss of face, troubles in the organization, a poor working relationship with the boss, or the sheer inability to make a decision. Any consideration of how to break an impasse must take into account the human factor. It is not what you do but how you do it that may be the critical factor.

CLOSING THE DEAL

How do you bring negotiations to a close? We know closure happens when each party believes that the other has conceded all that it will and that further efforts are not likely to be too productive. At closure the parties insulate their minds to further information and make a final decision.

The reason for the final decision may be based on fact or intuition. That is immaterial. What is important is that the expectations of both parties come together and that closure follows. The 12 techniques below are designed to nudge an opponent into agreement:

1. Start with a positive attitude about closing.
2. Don't talk too much in requesting closure.
3. Keep asking the other party what the problem is if no agreement is reached.
4. Repeatedly assure the other party that it is wise to settle.
5. Don't be afraid to assume that things are already settled.
6. Walk the opponent into settlement by talking about details like the wording of a clause or when the opponent wants to start. Act as though agreement has been reached on the main issues and price.
7. Take a physical action toward closure.
8. Emphasize the possible loss of benefits now available if agreement is not soon reached.
9. Provide a special inducement for closing that cannot be offered later.
10. Tell a story legitimatizing the deal.
11. Make closure a real and desirable experience.
12. Take the initiative to write the final document.

These techniques have been used by negotiators throughout the world. They seek to nudge the other party into an agreement by projecting an attitude that is pleasant, persistent, and positive without being overbearing. At that point whatever you've been talking about is a deal.

The crisis and closure mistakes made at the end of a negotiation happen so fast that they are not recognized until after the session is over. The mistakes below happen to anybody under pressure:

- Never make the error of assuming that an impasse on an issue will result in deadlock in the overall negotiation.
- Do not be intimidated by the opponent's last and final offer.
- Deadlock is unpleasant for both parties, not only for yourself.
- Do not be afraid to admit that you have made an error in coming to an agreement.
- Do not make a last and final offer until you have evaluated precisely how the statement will be made and how discussions will be continued if it is not honored.
- Do not fail to prepare members of the home organization for deadlock or threat tactics by the opponent.
- Never be panicked into a final agreement by a time deadline.
- A final agreement is not necessarily fair or reasonable.
- People will not succeed in winning their objectives if they try to be liked in this final phase of negotiation.

SEVENTEEN RULES OF NEGOTIATION

1. Assume that everything is negotiable.
2. Aim your aspirations high.
3. Never accept the first offer.

4. Deal from strength if you can, but create the appearance of strength, regardless.
5. Put what you have agreed on in writing.
6. Recognize that the other party is probably holding back valuable information.
7. Flinch to create doubt in the counterpart's mind and to add value to a concession.
8. Find out what your counterpart wants. Do not assume that his or her wants and needs are the same as yours. Concede slowly and call a concession a concession.
9. Keep your counterpart in the dark about your strategy and find your stake in the deal.
10. Try to make your counterpart lower his or her level of aspiration.
11. Ask questions if you do not understand what is going on. Do not let your counterpart deliberately confuse you.
12. Answer questions with a question to avoid giving away information needlessly.
13. Invoke the higher authority to buy more time.
14. Information is power — obtain as much as possible.
15. Verify anything you are told that you do not know to be a fact.
16. Be cooperative and friendly. Avoid abrasiveness, which often breaks down negotiations.
17. Use the power of competition. Remember that power can be real or imaginary.

SELLING YOUR VIEWPOINT

Safety professionals are negotiators. They have points of view and want to convince the other party that it is correct. The buyer has one set of beliefs and the seller another. An exchange of viewpoints is a negotiation, a very tough one because ideas are like possessions. People don't like to part with them.

There are eight points to remember if your viewpoint is to prevail. They are so easy to follow that you won't want to forget them.

Point 1 — Talk less and listen more. Other people want to express themselves. If you keep quiet they will talk. They will reciprocate by being more attentive when you talk.

Point 2 — Don't interrupt. Interruptions make people angry and block communications.

Point 3 — Don't be belligerent. There is good reason that we respect the soft-spoken individual. It is harder to be firm in a moderate, self-controlled way than to be harsh, loud, or derisive. Soft-spoken individuals, by their manner, encourage the same treatment from others. An argumentative attitude changes no opinions.

Point 4 — Don't be in a hurry to bring up your points. As a general principle, it is best to listen to the other party's full viewpoint before expressing yours.

Point 5 — Restate the other person's position and objectives as soon as you understand them. People like to know that they have been understood. It is an inexpensive concession to make. There is another hidden benefit in forcing yourself to restate an opponent's position. It will cause you to listen better and help you phrase your points in the opponent's terms.

Point 6 — Identify the key point and stick to it. Avoid the tendency to overwhelm with arguments. Cover one point at a time.

Point 7 — Don't digress from the key point and keep the other person from digressing. There are three ways to minimize digressions: agree on some nonessential point temporarily for argument's sake, agree to discuss it later, or try to treat the intrusion as being somewhat off the point.

Point 8 — Be for, not against, a point of view.

People prefer cooperation to conflict. The ideas above lead in the right direction. When you think about it, the problems of selling your viewpoint and of winning an argument have much in common.

CONCLUSION

This conclusion is unlike any other written. Much of what you have read in this chapter you already knew. When it comes to successful negotiations what follows is worth remembering.

Peace is not made at the council table or by treaty, but in the hearts of men.

— Herbert Hoover

The question should never be who is right, but what is right.

— Glenn Gardiner

A person who is angry on the right grounds, against the right persons, in the right manner, at the right moment, and for the right length of time deserves great praise.

— Aristotle

The art of being wise is the art of knowing what to overlook.

— William James

Our conduct is influenced not by our experience, but by our expectations.

— George Bernard Shaw

The best and most beautiful things in the world cannot be seen or even touched. They must be felt with the heart.

— Helen Keller

Everything that is done in the world is done by hope.

— Martin Luther

Courage is doing what you're afraid to do.
There can be no courage unless you're scared.

— Eddie Rickenbacker

You can observe a lot by watching.

— **Yogi Berra**

If you think you can or you can't, you're always right.

— **Henry Ford I**

Never get angry. Never make a threat. Reason with people.

— **Don Corleone,** *The Godfather*

The meek shall inherit the earth — but not its mineral rights.

— **J. Paul Getty**

The secret of walking on water is knowing where the stones are.

— **Anonymous**

It ain't over till it's over.

— **Yogi (again)**

Worry never lessens the sorrow of tomorrow, it only saps the strength of today.

— **Dale Gray**

FURTHER READING

Cohen, H. *You Can Negotiate Anything*. Secaucus, NJ: Carol Publishing Group, 1996.
Karrass, C. *Give and Take*. New York. Thomas Crowell, 1974.
Neirenberg, G. I. *Fundamentals of Negotiating*. New York. Hawthorne/Dutton, 1973.
Schaffizin, N. *Negotiate Smart*. New York. Random House, 1997
Stark, P. B. *It's Negotiable*. San Francisco. Jossey-Bass/Pfeiffer, 1994.
Ury, W. and Fisher, R. *Getting to Yes*. New York: Penguin Books, 1991

## CHAPTER 38

The Role of a Safety Manager and Safety Leader

Michal F. Settles

CONTENTS

> "Managers do things right ... leaders do the right thing."
>
> — **Warren Bennis**

OVERVIEW

The safety industry has faced continuous challenges to meet existing and newly passed legislation in recent years. What is needed by the industry to address these demands? Some say more management is needed. I ask you to consider leadership roles as the safety profession undertakes existing and newly developing challenges. Safety professionals face catastrophes, technology changes, injury control, environmental issues, and ergonomic influences, to name a few. The role of management with any of these issues is, of course, of grave concern.

ROLE OF MANAGERS

Management, as defined in Mondy's *Supervision*, is the process of getting things done through the efforts of other people. Managers are vital to the success of any organization. Without managers, an organization would be faced with a vacuum in service, execution, and delivery. Successful managers are results oriented and focus on goal attainment. A safety manager ensures that the details are not missed or overlooked. Many organizations have moved from the title of manager to a team leader or coach.

ROLE OF LEADERS

Leadership can be defined as getting people to follow your example because they want to. So how would you describe a safety leader? When I think of leadership, some characteristics that come to mind are:

- Is a visionary — Leaders see what the potential possibilities are and what can be.
- Takes risks — Leaders are willing to act on their visions (turn them into realities).
- Sees the big picture — Leaders view a world without limitations and see the relationship to other things in the environment.
- Taps available resources — Leaders maximize all existing potential resources (human and capital).
- Is creative — Leaders are constantly challenging existing paradigms.

People usually respond to leaders because they elect to do so. Successful leaders usually have a clear purpose and are associated with improvements.

MANAGER OR LEADER

What is the difference between leadership and management? At a recent seminar, participants brainstormed this question. The quotations that surfaced were classic:

Managers are selected for their positions while leaders rise to the occasion.

People respond to leaders because they want to. They respond to managers because they have to.

Managers have a function while leaders have a purpose.

The books of Dr. Stephen Covey (*The Seven Habits of Highly Effective People*) and Dr. Warren Bennis (*Leaders: The Strategies for Taking Charge*) are major works for the study of organizational leadership. Dr. Covey's book highlights effective leadership factors. The seven identified habits are:

1. Be proactive.
2. Begin with the end in mind.

3. Put first things first.
4. Think win/win.
5. Seek first to understand, then to be understood.
6. Summarize.
7. Sharpen the saw.

Dr. Bennis' book is considered a classic. Bennis' work focuses on four leadership strategies:

1. Attention through vision.
2. Meaning through communication.
3. Trust through positioning.
4. Deployment of self.

One of the most important questions answered by Bennis is, "Where do leaders get their ideas, visions, and dreams?" Both books are insightful and describe in detail the power of leadership. Anyone interested in the topic of leadership should avail themselves of these resources.

The American Society for Training and Development (ASTD) document *Fundamentals of Leadership* distinguishes a manager from a leader:

Manager	Leader
Knows how to follow directions from above	Inspires people to cooperate
Takes modest risks	Teams with others, provides tools for success, then gets out of the way
Sees people as employees, not partners	
Creates and realizes a vision	Is followed by people out of choice
Uses power cautiously	Takes risks
Is committed to the organization	
Is proficient at planning, organizing, and controlling	

Nine leadership competencies were also identified:

1. Technical and tactical proficiency (know your job and the jobs of others).
2. Communication (speaking, writing, and listening).
3. Professional ethics (do the right thing).
4. Planning (forecasting).
5. Use of available systems (technology and people systems).
6. Decision making (make tough choices and involve others).
7. Teaching/counseling (train your successors).
8. Supervision (to plan, direct, evaluate, coordinate, and even control, if necessary).
9. Team development (share the knowledge, involvement, and responsibility freely).

SUMMARY

It is my contention that a successful organization needs both managers and leaders. Can an individual be both a leader and a manager? Absolutely! Can most people accomplish both with ease and success? For most people, achieving both can be very difficult and complex at best. The continuous visionary abilities of leaders allow an organization to move from a short-term day-to-day focus to a long-term focus. Managers, on the other hand, are the key to making the vision a reality. Organizations need both (leaders and managers) to succeed. An organization without one or the other will not be a healthy organization for long.

Safety professionals have an opportunity to shape their organizations in both capacities. Which role will you play?

FURTHER READING

Bennis, W. *Leaders, the Strategies of Taking Charge,* New York: Perennial Library, 1985.
Covey, S. R. *The Seven Habits of Highly Effective People,* New York: Simon & Schuster/Fireside Books, 1990.
Mondy, W. *Supervision,* New York: Random House, 1983

## CHAPTER 39

Essentials of Effective Influence

John D. Adams

CONTENTS

INTRODUCTION

We live today in a world of frequent, unpredictable changes. New technologies spring up daily that pose new challenges and new opportunities for safety and health professionals. As organizations restructure and as the marketplace changes, we must always be ready to update the safety and health policies in our workplaces.

Safety and health professionals seldom have the formal authority to implement changes in the programs and policies they oversee. Rather, we must use whatever personal and political skills we have to express our ideas to others. This chapter provides a checklist and a road map for improving our effectiveness in establishing win/win influential communications with others.

Most of us already know much of what we should do to be effective in situations calling for influence, but we frequently fail to put our knowledge into practice. For example, have we not already learned that we should listen actively, set clear and mutually agreed upon goals, establish clear agendas, and so on?

Much has been written about effective influence and negotiation techniques, but most people do not employ these techniques with any regularity. Instead, we assume that if we would just try harder with the techniques that are not working, we would eventually break through the resistance and be successful.

People's learned styles of operating and ways of processing information have a lot of inertia. The more often a thought or behavior is repeated, the more difficult it becomes to replace it with a new thought or behavior that would be more effective. It takes a great deal of commitment and discipline to reprogram our autopilots.

SIX ESSENTIALS

I have been collecting information on the ingredients of successful, win/win influence engagements for over 15 years. In the course of this work, I have found that the essentials for effective influence can be nicely sorted into six categories: mind-set, preparation, awareness, skills, models, and stage management. You can use the following paragraphs as a self-assessment and a checklist for preparing yourself for significant influence engagements.

Mind-set

Versatility of thinking and behavior — Versatility of thinking means being able to hold multiple perspectives simultaneously (e.g., immediate and long-term focuses). Versatility of behavior means having a repertoire of styles and skills from which to draw.

Relationship — It is becoming more and more apparent that things happen in organizations through effective relationships much more rapidly than via the hierarchy. Trust takes a lot of effort to establish and is easily lost forever.

Optimistic outlook — The self-fulfilling prophecy never takes a break! Obtain a clear sense of how you would be at the end of a perfect influence and adopt that "posture" from the outset.

Continuous learning — Plan to learn something new about your influencing style every chance you get.

Be yourself — Never put on an act or play a role. The other person can always spot it.

Stay centered and relaxed — Be sure to relax yourself and keep breathing normally throughout influence engagements.

Preparation

Attend to all details.
Clearly envision successful, positive outcomes.
Update your knowledge of the subject matter.

Style — Reflect on the style of the person(s) you plan to influence. Always be ready to align with their focus and their energy. Do not force the other person to adopt your focus at the outset.

Skill practice — Always identify a skill that you will practice as part of your preparation. This is an extension of the continuous learning point above.

Awareness

Attend to emotions and procedure as well as the task — Emotional and procedural issues will always block work on the task.

Emotions are sometimes more important than the task.

Reflect on your own influence assets and liabilities — What are you good at, and where do you often shoot yourself in the foot when you are engaged in influencing someone?

Fear — Remember that the person you want to influence probably has fears that are the same or similar to your own.

Skills

Active listening — This especially includes paraphrasing, summarizing, and asking open-ended questions.

Flexibility — This means being open to feedback, willing to be influenced, and able to let go of no-win situations.

Language — Less effective influencers use more qualifiers (i.e., "sort of ..."), self-discounts ("I'm only a...."), intensifiers ("Wow, super!!!"), and gotcha questions ("Don't you think...?").

Assertiveness — This should be coupled with patience and tolerance. It is not the same as aggressiveness.

Style differences — It is normal for each of us to want others to set their priorities and see the world in the same way that we do. You will have much more success in your influence activities if you first gain an awareness of the priorities, focus, and world views of the person you want to influence and then match or support those positions regardless of how different they are from your own. For example, if your priority is urgency and getting on with things, you are going to have trouble establishing a positive influence relationship with someone whose priority is having all the data and being deliberate, unless you are willing to shift from the viewpoint of hurry up to one of systematic analysis.

Models

Systematic — Having a systematic model of the influence process guides the development of the interaction and ensures that you will not overlook anything essential to your success.

In the section that follows, a systematic model for stage managing your influence interactions is developed. With practice, this model becomes an automatic guide that you do not have to think about to make use of. The best way to make it automatic is to practice the various elements in the model, one step at a time. After some period of practicing the bits and pieces, you will find that you are automatically using the whole thing.

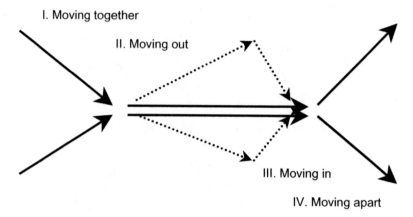

I. Moving together

II. Moving out

III. Moving in

IV. Moving apart

Figure 39.1   Stages for exerting infuence. (Adapted by John D. Adams from a model developed by Dennis Kinlaw.)

Stage Management

The diagram shown in Figure 39.1 provides a graphic portrayal of a systematic way to stage manage your influence interactions. There are four aspects to the model, which should not be taken as linear or sequential. In other words, your goal is eventually to learn to monitor all four of the aspects all of the time.

Stage I. Moving Together

This aspect is about contacting and contracting, and it has to do with building the necessary relationship with the person you want to influence. Neglecting to establish and maintain rapport and not to establish a clear contract for working together are the most frequently overlooked aspects of the influence process. When these things are overlooked, a successful influence interaction is unlikely.

Whenever there is tension in an interaction, it is easy to fall out of rapport. It is essential that you pick the right time and place, ensure that there is enough time to talk, eliminate interruptions, and do whatever is necessary to ensure that both you and the other person are paying attention to each other.

An easy way to establish a working contract with someone is to remember the acronym TEA, which stands for time, expectations, and agenda. A single sentence can establish a solid working agreement. "I understand that you have 10 minutes available right now [time] and we both understand how important the elimination of the toxic fumes is [expectations], so let's focus right now on identifying the source of the fumes and clarifying our roles relative to getting them stopped [agenda]."

Stage II. Moving Out

This aspect is about diverging and diagnosing, and it has to do with exploring the content of the influence. Most of us move to taking action too quickly. We also expect the other person to understand what we are after and to buy in far more quickly than is usually possible. It is essential that you maintain an exploratory, divergent stance for longer than you think necessary. Always ask, "Is there anything else we need to consider?" at least once more than you think is necessary.

There are five elements that are useful to know about in this aspect if you are expecting there to be any resistance to your influence. When these five elements are developed fully, most nonpolitical resistance will dissipate. You may find them to be useful guidelines for doing homework prior to the influence meeting.

Belief — what your are asking is both desirable and possible
Sufficient dissatisfaction — with the status quo or present situation
Goals — for change that are understood and agreed upon
First steps — guaranteed to be successful
Other people — who needs to be informed, included, excluded, negotiated with, etc.?

Stage III. Moving In

This is the stage of deciding among the alternatives and taking action. Little needs to be said about this aspect if the other three stages are attended to sufficiently. In our culture we are all too often in a hurry to get to this aspect, at the expense of the other three — and at the expense of success!

Stage IV. Moving Apart

This aspect is about finishing and following up, and it has to do with sustaining the relationship you have built with the person you are influencing. Neglecting to summarize agreements, to repair damages to the relationship, and to establish specific follow-up accountabilities are the second most frequently overlooked aspect of the influence process. When these are overlooked, a successful influence interaction is unlikely.

As the influence process progresses, many agreements are likely to be reached. These are often lost if they are not summarized from time to time, especially at the close of the interaction.

One reason that agreements are not carried out is that they often require the person to act in nonhabitual ways. When this is the case, it is exceedingly easy to forget or to gradually regress. Following-up is essential to keep this from happening.

CONCLUSION

In preparation for a crucial influence interaction, look over the above sections and select one or two points to concentrate on. Use these ideas to build your own personal plan for the event. While working at improving your general ability to influence others, do not try to adopt every point on the list simultaneously. Pick one or two key ideas to work on in a conscious way in every influence situation into which you enter. When those items become a natural part of your approach, select a few more. After a few months, you will have dramatically improved your ability to influence others in a positive, win/win fashion.

FURTHER READING

Bellman, G. M. *Getting Things Done When You Are Not in Charge.* San Francisco, CA: Berrett-Koehler Publishers, 1992.

Deep, S. and Sussman, L. *What to Say to Get What You Want.* Reading, MA: Addison-Wesley, 1992.

Laborde, G. Z. *Influencing with Integrity.* Palo Alto, CA: Syntony Publishing, 1987.

McCann, R. *How to Influence Others at Work.* Oxford, U.K.: Heinemann, 1988.

Northstine, W. L. *Influencing Others.* Los Altos, CA.: Crisp Publications, 1989.

## CHAPTER 40

Creative Problem-Solving Techniques

Anne Durrum Robinson and Holland A. Young

CONTENTS

MENTAL ENERGY: MIRED IN THE PROBLEM OR MOVING TOWARD THE SOLUTIONS?

In Chapter 11, Richard Lack addresses problem-solving techniques for safety and health professionals. This particular chapter narrows the focus to an encouraging look at the globally practiced approach offered by the Creative Education Foundation through its often-repeated Creative Problem-Solving Institute (CPSI) in Buffalo, New York. The institutes bring participants from every continent and many countries for a creative field day (actually almost a field week) of basic instruction in the CPSI process and hundreds of enriching adjunct sessions on many phases of creative thinking. As one adherent so aptly expressed the general feeling, "It's like being a kid in a creativity candy store!"

1-56670-370-0/02/$0.00+$1.50
© 2002 by CRC Press LLC

Because of the confines of the chapter, we will address the process rather directly, covering roles, rules, advantages, and process steps. We will illustrate the approach to the process with concrete but limited examples. Many of its enriching capabilities must, of necessity, be left to readers to seek for themselves from the various books listed in Further Reading section at the end of the chapter.

To make the chapter more pertinent for safety and health professionals we are taking the dual view that, often, specific safety and health measures for individuals stand or fall on the basic health of an entire organization. So we are combining Anne Robinson's knowledge of the creative problem-solving process with Holland Young's expertise as planning and environmental manager for the Austin (Texas) airport.

Executives at the Austin facility face rough-rider challenges that can be compared with those of circus performers who attempt to ride two steeds at once, with one foot on the back of one horse and the other foot on the back of a different animal. After a long, involved (and often bitter) struggle among civic factions, a decision was made to move the present airport from its in-town position to the rather recently closed Bergstrom Air Force Base. Bergstrom is located on the edge of Austin proper. Complete relocation from the present site will be a very complex process that will take many years. The Austin Department of Aviation's New Airport Project Team has relocated to Bergstrom to begin the process of building new airport facilities. Questions continually arise about what to keep as is, what to transform or renovate, what to demolish, etc. Obviously, there are multiple distractions that consume energy and affect the safety as well as the mental and physical health of employees at both locations. A department that heretofore has needed only to operate one airport now must meet the challenge of building a new airport while continuing efficient and safe operations at the existing airport.

The first caveat to any problem solvers is to caution them to erase the word *problem* and, instead, substitute

- Challenge
- Opportunity
- Learning event

or any other word that connotes a positive outlook, a belief that "there are solutions and all we have to do is find them!" Such an approach conveys a much more encouraging message to the subconscious mind, which can then work more effectively with the conscious mind to uncover productive directions of effort.

Research has shown that any tried-and-true problem-analysis or problem-solving process offers a much greater possibility of successful outcome than repeated gnawing on the problems themselves. As our chapter heading indicates, with such a process we move the energy from the mire of problems toward the higher (and safer) ground of beneficial solutions.

There are any number of proven process, including Synectics (Cambridge-based but world renowned) and Kepner-Tregoe (also globally taught and recognized). Challenge meeters will find that some processes work best in some instances, other processes or techniques in others. In this chapter, we dwell only on the CPSI process and, as indicated, primarily on its structure. The CPSI process involves three principal roles:

1. The *client* (or decision maker), who has the challenge, knows the facts, and can have *some* impact on the outcome.
2. The *leader* (or facilitator), who knows the process and is adept at leading client and participants through it, but who has the least possible stake in the process outcome. (This neutral attitude is essential. There must be no feeling that the leader is manipulating the process or the people involved.)
3. The *participants*, who have gathered to help the client (or clients) meet the challenge.

The process can be effectively used by an individual who conscientiously follows the steps, by partners (one of whom takes the role of leader and participant while the other takes the role of client and participant), or by a reasonable-sized group (of perhaps eight people). Anne Robinson has given CPSI-process demonstrations to much larger audiences (even one where some of the participants were blind and some were visually impaired). She also mentors a process-learning group from a large state organization that often approaches 15 or more attendants. However, she finds a group of six to eight most effective. A group of six allows for a client, a leader, and four helpers; a group of eight for a client, a leader, and six helpers. Such a group is recommended during the initial learning of the process. Once one is comfortable using and leading the process, one can profitably move to creative individual or partnership problem-solving efforts.

Necessary supplies include a flip-chart easel and paper (perhaps the 3M® chart paper that adheres to a wall or paper similar to Wall-Right), black and colored markers, possibly Post-it® notes or colored sticky dots, paper, and pencils or pens for participants.

Participant materials should include a description and expectations of the roles, a listing of the rules, a listing and explanations of the process steps, and a sample of process-step pages (so learners can duplicate those for their own future use and guidance).

Where a team or group is involved on a regular basis, individuals should rotate roles as much as possible. Thus each participant will learn to lead, to participate, and to act as an effective client.

Experienced leaders can, when necessary, handle more than one client. However, the process works more cleanly and moves more quickly when only one decision maker is involved at a time.

The usual brainstorming rules are:

- All contributions are honored and recorded exactly the way the donor wants them.
- Suggestions may be embellished or built on.
- No criticism or editing (direct or implied) is allowed.
- Attention is focused on the leader and the flip chart. No side conversations are permitted. Experience has shown that these take the focus away from the mainstreamed funnel energy off into wasteful tributaries.
- Participants may contribute or pass if they wish.
- Steps in the process are taken in order — no leaping to ideas or conclusions before the proper time.
- Complete records of the process are kept on the flip chart up-front.

- These records are given to the client at the close of the session.
- Participants are encouraged to have fun with the exercise as this fosters a more relaxed and creative atmosphere.

Steps in the process include:

- The *fuzzy mess* or *possible problem* — This is a one-sentence statement of the worry or wonder or concern or challenge or opportunity, as it now seems.
- *Fact finding* — The client states the concern or challenge as he or she sees it, gives all the pertinent facts, says why it is a group or personal problem, and states all the things he or she has tried or thought of to resolve it. In this step participants may ask questions for clarification. After this step participants are discouraged from asking further questions. Too often a question is, in itself, a safety measure. The participant wants to be safe in making a suggestion. Creative problem-solving encourages the making of suggestions without fear of being ridiculous or wrong.
- *Problem finding* — In this step, participants (and the client) look at various aspects of the problem and state them in several ways:
  - How to (H/2) ...
  - In what ways might (IWWM)...
  - I wish (I/w)...

  Participants may have been thinking of these problem analysis statements as facts are being given. The leader may encourage random suggestion of such statements or follow some order of the leader's choosing. A leader may also involve a scribe to help with fast posting so momentum is maintained. Posting can be done with Post-it® notes (large ones). A leader continues to press for more problem statements because these give more range for later ideas. Often a group's creativity is enhanced toward the end of a process step. When the group has exhausted its problem finding (or the time it can allot to that), the client is asked to select the problem analysis statements to which he or she wants ideas addressed. The client may combine statements and may select as many as appeal. This is the client's decision alone; the client need not explain a choice as criticism. Sometimes, a client knows why such a suggestion or problem statement is not feasible.
- *Idea finding* — Now participants suggest specific, behavioral ideas that may help solve or improve the problem statements selected by the client. Anyone (other participants or the client) may build on or add to the suggested idea. Participants are not there to get credit or shoot down other people's ideas. They are there to help the client move toward solutions of the problem aspects. Again, the leader presses for the greatest quantity. Quality can be determined later. Any type of judgment (verbal or implied) slows the idea flow. A group (or a person) will tend first to offer more practical, obvious, standard ideas. As the flow increases and the group warms to its collective creative task, ideas tend to become more imaginative, more free. Here is where the leader may inject some of the ideation techniques listed below. Such techniques encourage the thinking to move across the corpus callosum into the right brain's more playful, risk-taking, visionary, and intuitive realm. When group trust is high, these ideas come more freely. When the idea finding has apparently run its course (or when the client signals that he or she has a sufficient number of ideas), the group stops. The client then again selects (without explanation or excuse) the ideas that the client wishes to incorporate in the action plan.

- *Solution finding* — This step involves making a decision table with a list of a number of favored options (identified by key words), which the client will rate against a chosen list of criteria. Options are listed vertically on the left. Criteria (musts and wants) are listed across the top. Each option is judged against the same criterion in sequence. Against the *must* criteria, the options are checked only to see if they pass that necessary hurdle. Against the other criteria, each option is rated on a scale of 1 to 10, with 10 as the best. Again, each option is measured against the same criterion in sequence. After all the options are measured against all the criteria, the leader totals the ratings. If the ratings tend to cluster too closely together, the client can weight the criteria by assigning a relative numerical importance, multiply ratings by weights, and get new weighted totals. The decision table is not a table of stone. It simply indicates how the client subjectively feels about the various options and which he or she might want to pursue first.
- *Action finding* — Once the client has defined preferred options, he or she should record a definite action plan, stating which option should be done by whom at what time or by when.
- *Acceptance finding* — This step allows the client to look at the action plan and determine who might help with it, who might hinder, and how to follow up with either help or hindrance. The client might take the action plan through a who, what, where, when, how, and why catechism and alter or improve the plan accordingly.

We have taken this process outline and — in the confines of this chapter — attempted to show how it might be employed to remedy some of the dilemmas faced by the Austin Aviation Department as previously discussed. Please bear in mind that actual use of the process would go much further in the problem analysis and idea-finding steps. It would also be greatly enriched by the addition of such ideation stimulators as those listed in the Methods of Ideation section.

SIMULATION

Fuzzy Mess or Task

The original problem statement offered by Holland Young was the following: We have a big problem with communication.

Fact Finding

- Aviation department employees have problems with intradepartment communication.
- They also have problem with airlines and other businesses that operate in the airport (interdepartment communication).
- We are not all informed as to what others are doing.
- We often are not aware of where other people are focusing their efforts — what they are trying to accomplish.
- The lack of communication cuts into the ability to run an efficient and cost-effective operation.
- Personal friction evolves when different parties do not understand what others are doing and why.

- We have done some things with the tenants:
  - Have held regular meetings
  - Instituted a series of mail boxes for tenant communication
- Because we are building a new airport we have part of our staff in two different physical locations.
- People tend to establish a territory and do not know that they might be stepping over the boundaries of others' territory. This is related to organization as well as to communication.
- We do have regular staff meetings, but they are not well organized.
- We have been through the total quality management (TQM) process and are very deeply involved in TQM implementation.
- A management consultant is now reviewing the organization.
- We are in the process of building a new airport, but our organization was set up to run our existing airport.
- Because of the change involved in taking on a project like the new airport, the lack of communication leads to some fear among staff members that their individual positions might not be needed. This really is a problem for us. It is siphoning off energy and saps people's initiative and their desire to do a good job.
- Some people see the group that is building the new airport as the "elite" group, one which gets favored treatment.
- The very fast change and fast-paced nature of the business demands that we all know what we are all doing for us to work well.
- Generally speaking, we have a very high level of individual excellence in the department.
- There has never been any sort of formal or informal communication process other than departmental newsletters.
- Holland is doing a new airport newsletter, *NEW AIRPORT NEWS,* which is distributed to the public. People need to be informed and we need to work toward consensus.
- We have established a computer network between the two locations, with everyone linked by e-mail. It is a slick system — easy to use, can print, can attach documents. It has already eased the situation somewhat.

Problem Finding

- I wish (I/w) that everyone had more clearly defined areas of responsibility.
- I/w there were only two people in the department so it would be easier to talk. (*Note*: This is an example of a problem statement where the person making it knows it is an impossible wish but hopes something might come out of this perception.)
- I/w a solution would fly in on a plane. (*Note*: This is another absurdity that might open some thinking avenues.)
- How to (H/2) give training in clear communication to all involved staff.
- I/w people would not first jump to conclusions without getting the facts.
- H/2 make sure all the people get the right facts at the right time.
- I/w people would take a little more time to communicate with one another.
- I/w people would be a little more sensitive to how their information affects others.
- I/w each airport could define its own vision.
- I/w the two airports could define a shared vision.

- H/2 keep people aware of the existing vision.
- H/2 translate the existing vision into concrete action.
- I/w we could define the best ways for staff to communicate — media? verbal? written? electronic?
- I/w the entire staff could be taught a good problem-solving system.
- H/2 get teams to work more rapidly and/or efficiently.
- H/2 teach teams idea-finding methods and H/2 see that they use them.
- I/w all levels of employees would be involved in the problem-solving experience, especially for those problems that affect them.
- H/2 conserve personal energy for the direct demands of each person's current job.
- I/w people would be less negative and focus more on the good things we have and how to improve them.
- I/w the airport would buy some helpful books and audio/video tapes for its staff to use on their own time or in groups.
- I/w that people would realize the high quality of others among us.
- I/w the most productive relationship between the two airports could be envisioned and built on.
- I/w people had a clearer understanding of their career paths in the department.
- I/w everybody involved had a clear picture of the perfect outcome.

Holland Young was then asked to check over the problem-finding (problem analysis) statements to determine which ones he wanted addressed. He was reminded that ideas should be specific and behavioral. We both acknowledged that, obviously, problem finding could have gone on much longer. However, in the interest of brevity, we decided that we should begin the judgment/focusing part of the process. Holland Young then decided that, inasmuch as many of the ideas might be grouped, and inasmuch as there were two of us instead of a larger group of participants, he would like ideas addressed to the entire list of problem statements.

Idea Finding*

- Redefine the organization.
  - Have a new chart end product.
  - Look at how the department functions, where logical work units are, and how they are related.
  - Try to involve staff in developing a new organization so people have some control of their own destiny and see what is happening and why.
  - Have a contest among staff for new organizational suggestions.
  - Give awards for helpful suggestions.
  - Have a contest for best overall organizational chart.
  - Appoint a committee to judge the contest.
- Encourage wider use of e-mail.
- Add training on use of e-mail.
- Increase e-mail implementation. (You can copy other people on e-mail.)
- Encourage people to communicate more quickly and more often when they have something that is important.

* Both Holland Young and Anne Robinson submitted ideas.

- Encourage creative ideas for improvements in any area.
- Have a regular schedule for publication of good suggestions.
- Be more accurate in the information that circulates.
- Determine the level of communication skills the staff possesses.
- Develop a program to enhance speaking/writing/listening skills.
- Train staff in meeting skills.
- Train staff in facilitating skills.
- Check on communication consultants.
- Make sure people receive facts at the proper time by:
  - Formal communication procedures such as newsletters.
  - Update papers.
  - Hiring someone for communication control.
- Have a routine process for looking at each employee and celebrating his or her best skills. (Solicit employee suggestions about how this can be done.)
- Place feature stories in the local paper. (Contact Lifestyle editor.)
- Define human assets available at each airport.
  - Tap someone from outside.
  - Tap a staff member.
  - Investigate a way to place communication specialists on staff.
  - Assign this to the communication division.
- Implement regular and structured management/staff communications.
- Redesign employer-supervisor performance.
- Develop a clear outline of the department career paths.
- Develop imagineering abilities in all employees in a series of workshops.
- Reiterate the visions of both airports.
- Simplify and integrate those into culture, perhaps with:
  - Ball caps.
  - Letterhead.
  - Bottom of the newsletter.
  - T-shirts.
  - Everywhere.
- At fun gatherings, wear "rumperstickers" with pertinent slogans.
- Integrate staff into the development of a shared vision for the two airports.
- Stress that visions of both airports are of equal importance.
- Translate vision into specific goals and objectives.
- Encourage each employee to devise a course of individual action leading toward the goals.
- Ask all employees to define what they think the perfect position for them would be.
- Study and analyze how individuals want to communicate. Find what they prefer and cater to that.
- Have monthly lunch meetings for project managers:
  - At Bergstrom.
  - At the existing airport.
- Analyze what teams need to do to improve.
- Keep teams focused.
- Improve facilitators' skills.
- Train more facilitators.
- Make teams aware of excursions for getting ideas.
- Allow latitude to be on the team of one's choice.

- See that team suggestions are given sincere consideration.
- Define what empowerment means at the airports.
- Teach employees how to work smart.
- Advertise the positive attributes and the successes. Find fun and clever ways to do that.
- Establish a commendation process for positive leaders and practice it.
- Encourage informal approbation of positive performance and outlook.
- Determine the best print/audio/video materials from available sources and encourage employees to use them.

As with problem finding, idea finding could go on much longer, with Anne Robinson and Holland Young experimenting with the rich possibilities of the idea-finding "excursions" on the following page. Groups are often led through many of these escaping-from-the-thinking-rut techniques with extremely fruitful results. Such seemingly absurd approaches can give much wider glimpses of possibilities and encourage participants to seek answers from all parts of their limitless minds.

Methods of Ideation — CPSI Method

Pocket articles — Each participant takes an article from a purse or pocket and suggests an idea based on that article. Suggestions can be practical, wild, funny, etc.

Imagery — Participants close their eyes and relax to music. Each uses the first image that comes to mind to serve as the basis for an idea.

Personal analogy — Each participant writes a short paragraph about what he or she would think or say if he or she were some element in the problem: a person, an inanimate object, a concept, a situation, whatever. The group then attempts to derive ideas based on the analogies.

Sensory check — Each participant searches his or her senses for responses that might trigger ideas.

Attribute listing — The group works together to list attributes of the problem or elements in the problem to see what comes to mind.

Role play — Volunteers from the participant group role-play aspects of the problem to see if ideas arise from this dramatic approach.

Other worlds — The leader or the participants suggest some other world (such as the world of insects or the world of space) and try to get ideas from comparisons of the problem world to the selected other world.

Random related words — One participant suggests a word related to the problem. The next participant immediately says a word the first word suggests. In the proper order participants suggest random related words. Then each picks a word or words to stimulate ideas.

Back to the sun — The group attempts to take some part of the problem back to its origins to find ideas.

Morphological charting — The group lists various types of attributes of a problem. It then assigns these lists to various sides of an imaginary stack of cubes. The leader then draws one of the imaginary cubes and participants try to force-fit the sides (the words on the various sides) to see what ideas are produced.

Forced relationships — Participants call out random words for a vertical, list, then for a horizontal list. They then force-fit vertical 1 to horizontal 1, etc. to try for ideas.

Alphabet: what else can I do with this? — The participants go through the alphabet, asking, for example, "What else can I do with this which begins with A?" "With B?" "With C?"

Scamper — Participants take for first letters the word S-C-A-M-P-E-R. (Substitute, Combine, Alter or Adapt, Modify or Magnify or Minimize. Put to other uses; Elevate or Eliminate, Reverse or Rearrange). They attempt to see if these words give them any ideas.

Purge sheet — The leader or the participants keep a purge sheet on which is put thoughts or ideas that occur at improper times.

In-out sheet — Participants keep a personal in–out sheet, showing what thoughts they may have when their attention wanders. This helps access the right hemisphere of the brain.

Solution Finding

In determining the options for judging, Holland Young chose the options that he thought represented the best possibilities for improving departmental communications. It was apparent that several were related and would perhaps be implemented together; thus, they were judged as one. Options chosen were:

- Redefinition of the organization.
- E-mail enhancements (includes encouraging wider use, training, and implementation).
- Training (includes communication/meeting skills/facilitator training).
- Vision enhancements (includes vision development and integration).
- Employee/supervisor performance/career tracks.
- Training library.

Criteria given were:

- Must be able to be implemented.
- Want it to be relatively inexpensive.
- Want potential for fast results.
- Want solution(s) to be easy for people to do.

Options and criteria were organized as in Table 40.1. In reviewing the totals in Table 40.1, three options were obviously the highest ranked: vision enhancements, e-mail enhancements, and performance evaluations. Holland Young felt that this made sense given that all three were strongly related to communication. He believed that all three areas should receive immediate attention.

As can be seen from the example, the CPSI process has tremendous potential in the development of solutions for very complex problems.

SUMMARY

In employing the CPSI creative problem-analysis, problem-solving method, remember that:

Table 40.1 Solution-Finding Matrix

Options	Must Be Implementable	Want It to Be Inexpensive	Want Fast Results	Want It to Be Easy	Total
Redefine the organization	✓	5	5	5	15
Enhance e-mail	✓	8	8	10	26
Employee training	✓	3	3	3	9
Vision enhancements	✓	10	8	10	28
Performance evaluations	✓	8	8	8	24
Training library	✓	7	5	5	17

- It can be used by individuals, partners, or groups (with the most effective groups numbering six to eight).
    - *Leader/facilitator*, who has least stake in the outcome but is experienced in the process.
    - *Client (or decision maker)*, who is familiar with the problem and can in some way influence the outcome.
    - *Participants*, who are there to assist the client in solving the problem.
    - *Scribe*, who helps the leader with recording all of the process on the flip chart.
- It comprises seven steps which should be followed in order:
    1. *Fuzzy mess/task* — One-sentence statement of apparent challenge.
    2. *Fact finding* — Statement of pertinent facts by the client, including what he or she has tried or thought of. Only in this stage may participants ask questions.
    3. *Problem finding* — Multiple statements by all concerned of alternate statements of the challenge or parts of the challenge. Everyone (including the leader) may take part. Problem or challenge statements begin with I/W, IWWM, or H/2.
    4. *Idea finding* — Specific, behavioral ideas offered after the client has selected the problem statements to which he or she wants ideas directed. Ideas can be added to or built on. No criticism (stated or implied, verbal or visual) is permitted. No judgment is allowed at this stage.
    5. *Solution finding* — Key ideas selected by client are listed on a decision table, then rated and weighted by the client to discover the most promising ones. In neither this nor the selection of problem statements does the client explain choices.
    6. *Action finding* — Client offers an action list, indicating who will do what, by when in carrying out highest-rated ideas.
    7. *Acceptance finding* — Client checks possibilities of acceptance or hindrance in projected actions.

The CPSI process offers opportunities for focused and random awareness, for quantity of suggestions, and for timely judgment about quality. It centers attention and offers equal opportunity for participation to everyone involved. It also stimulates all parts of the fertile brain/mind.

FURTHER READING

Ackoff, R. L. *The Art of Problem Solving.* New York: John Wiley & Sons. 1978.

Adams, J. L. *The Care and Feeding of Ideas: A Guide to Encouraging Creativity.* Reading, MA: Addison-Wesley, 1986.

Albrecht, K. *Brain Power: Learn to Improve Your Thinking Skills.* Englewood Cliffs, NJ: Prentice-Hall, 1989.

Barrett, F. D. *10 Techniques for Creative Thinking: Bionics, Synectics, Morphology, Brain-Storming, Transcendental Meditation, Forced Association, Work Improvement, Attribute Listing, Value Engineering, Scenarios.* Canada: Management Concepts Limited, 1972.

De Bono, E. *Serious Creativity: Using the Power of Lateral Thinking to Create New Ideas.* New York: HarperCollins, 1992.

Firestien, R. L. *Unleashing the Power of Creativity: The Key to Teamwork, Empowerment and Continuous Improvement.* 41-minute video, copyright by Firestien, with workbook, 1994.

Grossman, S. R., Rodgers, B. E., and Moore, B. R. *Innovation, Inc.: Unlocking Creativity in the Workplace.* Plano, TX: Wordware Publishing, 1988.

Kepner, C. H. and Tregoe, B. B. *The Rational Manager: A Systematic Approach to Problem-Solving and Decision Making.* Princeton, NJ: Kepner-Tregoe, 1965.

Koberg, D. and Bagnall, J. *The Universal Traveler: A Soft-Systems Guide to Creativity, Problem-Solving and the Process of Reaching Goals.* Los Altos, CA: William Kaufmann, 1976.

Osborn, A. F. *Applied Imagination: Principles and Procedures of Creative Problem-Solving,* 3rd revised ed. New York: Charles Scribner's Sons, 1963.

Parnes, S. J. *A Facilitating Style of Leadership.* Bearly Limited in association with the Creative Education Foundation, Buffalo, New York, 1985.

Parnes, S. J. *Visionizing: State-of-the-Art Processes for Encouraging Innovative Excellence.* East Aurora, NY: D.O.K. Publishers, 1988.

Parnes, S. J. *Source Book for Creative Problem-Solving.* Buffalo. NY: Creative Education Foundation Press, 1992.

Prince, G. M. *The Practice of Creativity: A Manual for Dynamic Group Problem-Solving.* New York: Harper & Row, 1970.

VanGundy, A. B. *Techniques of Structured Problem-Solving.* New York: Van Nostrand Reinhold, 1981.

VanGundy, A. B. *Idea Power: Techniques and Resources to Unleash the Creativity in Your Organization,* New York: AMACOM, 1992.

Wenger, W. and Wenger, S. *Your Limitless Inventing Machine.* Gaithersburg, MD: Psychegenics, 1979.

CHAPTER **41**

Effective Writing: Guidelines for Clear Communication

Yvonne F. Alexander

CONTENTS

1-56670-370-0/02/$0.00+$1.50
© 2002 by CRC Press LLC

INTRODUCTION

This chapter is about writing effectively. It is a chapter that might surprise you — it is not filled with grammatical rules and parts of speech, fancy labels, or laborious technicalities. It is about saying what you mean and meaning what you say, because if you don't, the health and safety of your readers could be at stake.

For example, imagine that you are the pilot of a large jet plane. You have just completed what you think is a successful take-off. Suddenly you hear the air traffic controller saying, "Flight 1486, you've lost an engine. Make a left turn back to runway 32." As you begin to turn, you hear, "1486, a left turn, a left turn! You have lost an engine! Any runway, 1486, any runway!"

How will the air traffic controller's instructions determine what you do next? What do you think has happened to the engine? Is the plane gushing fuel because an engine has fallen off, or is the plane flying too slowly because it has lost power in an engine?

CLARITY, BREVITY, SIMPLICITY

Clarity, brevity, and simplicity are the three goals of effective communications. Think of them as the larger context in which you will find the other guidelines and techniques of good writing. Let us begin by defining these three concepts; then I will explain how you can use them to make your writing more effective. At the end of the chapter you will find a list of writing guidelines.

Over 2000 years ago, Quintilian, a Roman rhetorician, defined clear writing as "writing that is incapable of being misunderstood." Nowhere is this guideline more important than in health and safety management. Clarity means there can be no ambiguity in your writing. When writing is clear, there is no room for lost engines.

The famous American writer Mark Twain wrote to a friend: "Sorry for the long letter. I didn't have time to write a short one." Brevity does not mean writing the perfect first draft; it means editing to cut repetitious and unnecessary information and ten-dollar words that might alienate your reader. Brevity means using as few words as possible to convey your meaning. If you can delete a word without changing the meaning, do so. For example, does very add meaning to very hungry? Of course, because you can be mildly hungry or extremely hungry. On the other hand, you cannot be very famished.

Simplicity means using easy words and short sentences to convey your meaning. So when you can, try to use one and two-syllable words. Many times, when you use words with three syllables or more, you force your reader to work too hard. It is far better to write, "Grandmother, what big eyes you have" than to tell granny that her ocular implements are of an extraordinary order of magnitude. And do not worry — you will not sound like a simpleton. The fact is that simple writing is a sign of clear thinking and hard work. Writing that is wordy and rambling is a sign of a writer who does not care or does not know any better.

Using simple words does not mean you should limit your vocabulary. On the contrary, with a large vocabulary you can express yourself clearly and precisely — in as few words as possible. A good vocabulary lets you use one word to convey your meaning rather than several words to define what you are saying. With a strong vocabulary, you can say tolerate instead of put up with and imminent rather than likely to occur at any moment.

FOGGY WRITING

This chapter might have been called "Effective Written Communication." However, isn't writing (one word, two syllables) the same thing as written communication (two words, seven syllables)? Pretentious, overblown writing is also called the official style, weasel words, and fog. Writing becomes foggy when we use too many words or use big words inappropriately: medication for medicine, utilize for use, purchase for buy.

Foggy writing is imprecise and ambiguous, and it can be costly or dangerous if your reader misunderstands your meaning. Compare these two examples. Which is easier to understand? What might happen if your reader does not follow the directions because they are unclear?

- These special procedures for manual door operation must be continued and performed prior to each flight until electrical restoration and operation are resumed and reinspection of the lock sectors has been accomplished.
- Until the electrical locking system is repaired, the doors must be manually locked before each flight.

Why do we write in this complicated, official style? Often it is because we think we are supposed to. The truth is that readers have a difficult time understanding writing that is unnecessarily complex. If your reader has to work too hard, your writing is ineffective, and the results will be poor communication and wasted time on everyone's part. It is your job as the writer to make reading easy. Do not try to impress your reader with fancy words; write to express, not impress. *Remember that the more technical or complicated something is to explain, the simpler the writing needs to be.*

READABILITY FORMULA

Readability is the term used to describe how difficult something is to read. It might surprise you that an appropriate level for most business writing is the eighth to tenth grade. For scientific or highly technical writing, you might write at the 12th-grade level. Anything beyond grade 13 will be too difficult for most readers. Most major metropolitan newspapers, for example, are written at the sixth-grade level.

How can you improve your readability level? Simplify, simplify, simplify. Avoid too many big words (three syllables or more), and keep your sentences short (an average of 15 words per sentence; a maximum of 25).

To measure your readability level, use the grammar checker on your computer or do the math yourself. Using a passage of at least 100 words, add the average number of words per sentence to the percentage of words with three or more syllables. (Treat the percent as a whole number.) Multiply the total by 0.4. For example:

> 12   average number of words per sentence
>
> +10   percentage of big words
>
> 22   total × 0.4 = 9th grade

How can you practice reducing fog? Just imagine that you are talking on the phone rather than writing a memo. On the phone, we usually speak in a clear, simple style. "Bring your umbrella because it might rain." Often when we write, we obscure the message with fog: "It is highly recommended that you take into consideration the utilization of appropriate foul weather gear due to the fact that precipitation is anticipated."

ASPECTS OF GOOD WRITING

Now that we have covered how to simplify and clarify your message, let us look at the other aspects of good writing that you will need to understand.

Appearance

First impressions are crucial because your writing is judged by its appearance before the message is even read. Carefully select letterhead, logo type, ink, and paper. Use an appropriate size and style of typeface. Always use the computer spell checker if you have one; proofread slowly and carefully.

Seize Your Reader's Attention

STOP! DANGER! KEEP AWAY! One of the most critical aspects of health and safety writing can be getting your reader's attention quickly and conveying a sense of urgency. You can do that by using words that appeal to your reader's self-interest and are filled with emotion. So you might consider using no, don't always, never, caution, stop, danger, or beware.

Focus

Unless you have bad news, get to the point immediately to get your reader's attention. If you ramble, you risk losing your reader's interest.

IDENTIFYING YOUR PURPOSE

Effective writing means that you accomplish your goals. To do so, before you begin writing, it is important that you identify your purpose, your reason for writing. Unfortunately, identifying your purpose is not always as easy as it sounds. Are you trying to explain, persuade, promote goodwill, analyze, warn, justify? Knowing your purpose will help you organize your ideas before you begin to write.

Your writing will seldom have only one purpose. Since you will usually have a combination of purposes, you will need to clarify and separate them first. For example, perhaps you want to promote goodwill by thanking your customer for the new order for the fire extinguishers and expressing your appreciation for the business. You might also want to remind your customer about the importance of following the safety rules for using the extinguishers. In addition, maybe you want to persuade him or her to take advantage of your volume discount. After you have separated your purposes, you will need to prioritize them; doing so will help you organize your thoughts before you begin writing.

In a letter or memo, it is a good idea to start by telling your reader what your purpose is: "This letter is to let you know that your order for the fire extinguishers will be delivered on Friday, May 2." Do not make the mistake of thinking you are merely conveying information. Always ask yourself what purpose the information serves. Also, do not avoid identifying the purpose altogether; chances are if you do not know where you are going, you will end up somewhere else.

TARGETING YOUR AUDIENCE

Audience means the reader. As with the purpose, we often have more than one audience, which makes writing even more difficult. Keep in mind your principal audience and target your message to it. To identify your audience, you will need to answer these questions:

- What does she know? Is she familiar with the policies, procedures, rationale, methods, industry jargon, technical concepts? What is her background? Is she a generalist or a specialist? Will she need just the big picture or the details?
- How does he feel? Is he open-minded, resistant, skeptical, enthusiastic, hesitant, reasonable? How much of your writing will have to be persuasive? Will you need to be especially diplomatic or tactful, or can you be blunt?

Point of View

Write from your reader's point of view, not your own. You know why you are writing this, but why should the reader read it? What is in it for the reader? How will the reader benefit? How will the reader's health and safety (or that of his or her employees or customers) be improved or assured?

Tone

Tone conveys an underlying feeling to your reader. It is often the writing between the lines. Like facial expressions or gestures in speech, tone often sends a message that the actual words might not. Match the appropriate tone with the circumstances. Do not sound friendly when you want to reprimand. Do not sound official when you want to be cordial. Be sure to sound professional; remember that anger and disrespect are inappropriate in business writing. If you question the appropriateness of your tone, ask a colleague for his or her opinion.

You will notice that my tone is warmer and more personal than you might expect from the average textbook or desk reference. To get this tone, I use a lot of personal pronouns: I, you, we. I chose to write this way to get your attention and encourage you to read about a topic that otherwise might not hold your interest.

Tone is particularly important in health and safety management. You will want to identify your desired tone before you begin writing. Perhaps you want to use a tone that conveys a sense of urgency, immediacy, caution, or alarm. Notice how the meaning changes when tone differs:

- Use caution on left turns
- Left turns not advised
- Turn left at your own risk
- No left turn
- Never turn left

First decide what meaning you want to convey to your reader; then choose the words that match the tone with your meaning. Remember that you cannot please all the people all the time with any particular tone. What one reader considers straightforward and matter-of-fact might be seen by another reader as blunt or even abrasive.

Jargon

Jargon is a specialized vocabulary known to those in a particular profession. For example, in medical journals, cardiologists sometimes refer to heart attacks as events. Jargon is good shorthand and can save time as long as your reader understands it. For example, "Training is handled through the 550 process at the OPM."

The problem is that most of us use far too much jargon (and too many abbreviations and acronyms). So use jargon only if you are sure it will clarify rather than obscure your message.

Cultural Diversity

Be sensitive to communications styles in other cultures. Americans tend to be direct, to-the-point, and informal — qualities that can be offensive in some cultures. You might need to modify your style to make it appropriate for your reader. For example, "I think you made a mistake" might be too harsh; instead you can write "It appears there has been an error." Personal pronouns (I, you, we, they) can create closeness, so avoid them if you want distance, and use them cautiously when you have bad news. In some situations you will want to avoid using personal pronouns; they can seem excessively personal, informal, harsh, or inappropriate in many cultures.

WRITING CLEAR DIRECTIONS

Writing directions can be one of the most important aspects of health and safety management; yet directions are some of the most difficult things to write well. They should be written in simple language, in a clear style, and in chronological order. Be sure that the logic of what you are writing is clear to your reader. Does your reader want to know only what to do? Or does he or she need to know how to do it or why it should be done this way? Be sure the directions are complete; do not assume your reader will know your meaning if you leave something out. When you say push the button, do you mean push and let go or push and hold? Sometimes things that are simple or obvious to you may be confusing to your reader.

STYLE

Style means putting words together in a way that is unique to you. When you like a certain writer's work, you usually mean that his or her style appeals to you. Style is not quite the same thing as tone, which you can change, depending on your message. It is hard to change your style, which is much like your signature. Think of your style as your voice.

We all have our own style, and we are entitled to sound like ourselves, as long as the style is appropriate for the topic and audience. There is no ideal style; some people are terse, others verbose. Although some uniformity in a company or department is desirable, people should not be expected to sound alike.

EDITING

We are all sensitive about our writing, which is an extension of ourselves. So if you edit someone else's work, do not change his or her style unless it is inappropriate. Also, be sure not to change the meaning. Sometimes changing a word or inserting or deleting punctuation can change the writer's intent. It is important to put your ego aside when you edit another's work and not change something just because it is different; the change must be better. Try to make as few changes as possible when you edit someone else's work.

A LIFELONG JOURNEY

Keep in mind that writing is a process, not a goal. By using these guidelines you can make your writing clearer and stronger every time you write. Think of writing not as a destination but as a journey. And have a pleasant trip.

WRITING GUIDELINES

- Always be clear.
- Be brief.
- Use simple words.
- Avoid fog.

- Keep the readability level below grade 13.
- Limit the number of big words (three or more syllables).
- Write short sentences (average 15 words per sentence, maximum 25).
- Make appearances count.
- Seize your reader's attention.
- Get to the point immediately.
- Identify your purpose.
- Target your audience.
- Write from your reader's point of view.
- Use the proper tone.
- Avoid jargon.
- Be sensitive to other cultures.
- Write directions in a clear, simple style.
- Respect other people's writing styles.
- Edit another's writing with care.
- Think of writing as a process, not a goal.

FURTHER READING

Books

Bernstein, T. *The Careful Writer.* New York: Atheneum, 1965. A 500-page guide to usage that clarifies words such as *assume* and *presume.*

The Chicago Manual of Style. Chicago, IL: University of Chicago Press, 1993. Primarily for publishers and editors. Answers esoteric questions about writing. Nice to have, but not a necessity.

Gordon, K. E. *The New Well-Tempered Sentence.* New York: Tichnor & Fields, 1993. A grammar and punctuation handbook with a refreshing, whimsical approach.

Rico, G. *Writing the Natural Way.* Los Angeles, CA: J. P. Tarcher, 1983. Using right-brain techniques to release your expressive powers.

Strunk, W., Jr. and White, E. B. *The Elements of Style.* New York: Macmillan, 1979. An absolute must! Fewer than 100 pages. Read it once a year.

Venolia, J. *Write Right!* Woodland Hills, CA: Periwinkle Press, 1988. A small book, easy to use, covers punctuation, grammar, and style.

Zinsser, W. *On Writing Well.* New York: HarperPerennial, 1994. An informal guide to writing notification.

Writing to Learn. New York: Harper & Row, 1988. How to write and think clearly about any subject.

Other Publications

Dictionaries

Dictionaries vary considerably. The word *Webster* is a generic term like aspirin. It tells you nothing about the quantity because anyone can use the word. Be sure to get a dictionary that has the derivations of words so you can build your vocabulary by learning roots. Here are my favorites:

Webster's New World Dictionary, 2nd college ed.
American Heritage Dictionary
Random House Dictionary

Thesaurus

I like Roget's hardback desk version.

Word Guide/Speller

An electronic spell checker is ten times faster than using a dictionary for checking the spelling and syllabication of words.

CHAPTER **42**

Managing Diversity in the Workplace*

Sondra Thiederman

CONTENTS

INTRODUCTION

"It drives me crazy," one frustrated manager said, "No matter how often I tell my staff to ask for help if something is too heavy to lift, many will still go ahead on their own. I've lost count of the number of injuries I've seen because of this sort of thing over the last few years." A safety officer at a large company in Southern California had another complaint. His main concern was the fact that a high percentage of his employees refused to wear the new safety masks that the plant purchased some months before. He would get comments like, "We did fine before without them. Why do we have to start wearing them now?"

---

* The material in this chapter is based on that found in Sondra Thiederman's books: *Bridging Cultural Barriers for Corporate Success: How to Manage the Multicultural Work Force*, San Francisco: Lexington Books, 1990, and *Profiting in America's Multicultural Marketplace: How to Do Business across Cultural Line*, San Francisco: Lexington Books, 1991.

Because this chapter is about how to manage employees who have come to the United States from other countries, you might be wondering why I have begun with these particular examples. Why, for example, didn't I start out with something to do with language differences or perhaps differences in values or etiquette? The reason I chose these more subtle examples, both of which involve immigrant employees, was to illustrate the fact that cultural factors can come into play in the most unexpected ways. In these instances, for example, it was later discovered that the first complaint grew out of a value shared by the manager's Mexican employees. That value was the culturally rooted desire to appear strong and masculine in front of others. In the second case, employees — most of whom were from Vietnam — explained their reluctance to change with comments such as, "In my country we believe that when something works well, there is no need to change it. We did fine without the masks before — my friends and I just don't see the point."

When working with a culturally diverse workforce, good cross-cultural management skills become synonymous with good safety management skills. As most of you have discovered, when morale is up, safety claims are down. When people feel respected and valued, cooperation, satisfaction, and concentration increase and, in turn, safety improves. The purpose of this chapter is to examine two key values that managers should understand to create this type of atmosphere. The values, varying attitudes toward authority, and the importance of saving face are central to the confusion that has arisen in workplaces all over America.

ISSUE I: ATTITUDES TOWARD AUTHORITY

The Challenge

American managers have been complaining for years that employees do not have enough respect for them. "It's not like the old days." "It would sure be nice to be looked up to once in a while like my father was." The paradox of this complaint is that these same managers are now confused by the amount of respect that they are seeing in many of their immigrant employees. What many fail to realize is that, with almost 1 million immigrants entering the country each year, workplace values are changing. One of these values is respect for authority. Most cultures throughout the world are hierarchical in structure. Showing respect for one's boss is considered a virtue. In the United States, on the other hand, such behavior can be perceived as kissing up or, at least, as evidence of a neurotic lack of self-esteem.

On the surface, this kind of respect may sound like good news to many of you. The problem is that the current climate of participative management can make this degree of respect impractical and frustrating. Employees, for example, from authoritarian cultures tend not to complain to superiors about faulty equipment, inefficient procedures, or confusion over safety regulations. This reluctance stems not, as we might first assume, from passivity, stupidity, or lack of interest in the job, but more likely from the belief that the boss has, as a Thai proverb clearly states, "been bathed in hot water." In short, the manager knows what is right and to disagree with him

or her is to communicate disrespect by implying the employee is more knowledgeable than the superior.

The Solution

One of the fascinating things about cultural differences is the way that seemingly
complex problems often have simple solutions. In the case of what seems for native-
born managers to be excessive respect for authority, the answer lies in the very
hierarchical structure that creates the problem in the first place. Because of the
respect for hierarchy that so many immigrants have, there is apt to be, somewhere
among the workers, an individual who has been informally designated the leader of
the group. This individual functions as an informal leader, a role model, and, in the
best of workplaces, as a liaison between manager and staff.

Most of you have informal leaders in your workplaces. They are the persons to
whom others turn for advice, to whom they go if they do not understand an instruction, and who provide general guidance to those who are less experienced in the
American workplace. The following tips will help you identify these leaders:

- They tend to be bilingual. This is important because language means that the
  individual can communicate directly — without the aid of an interpreter — to
  both you and foreign-born staff.
- In a group of mixed gender, the leader will generally, although not always, be male.
- Informal leaders are usually the eldest of the group. Seniority and respect for age
  are dominant values among many immigrant groups and are manifested most
  clearly in their choice of leadership.
- Frequently, informal leaders held a prestigious position in the homeland. A leader,
  for example, might have been a military officer in Vietnam, a government official
  in South America, or a physician in India.

It is a simple matter to approach the informal leader with the goal of gaining
cooperation in achieving compliance with safety regulations. These steps will help
you overcome any initial hesitation you may feel.

- Acknowledge your respect for their status. This means to state clearly that you
  understand the importance of the individual's leadership position.
- Explain clearly what you want. If you need employees to wear their goggles, to
  follow safety procedures more carefully, or to let you know if a piece of equipment
  is faulty, spell out details so that there is no risk of misunderstanding.
- Explain why your request is important. One of the ways is which cross-cultural
  management can fail is by asking for behaviors that seem nonsensical to employees. This is not to say that the request is nonsensical, but simply that it involves
  new concepts or approaches that require explanation. This applies, I believe, more
  to safety than almost any other area. When people have worked, either in their
  homeland or in the United States, for years without having to worry about safety
  procedures or special equipment, it takes some convincing to persuade them what
  you are asking is more than a bureaucratic whim.

- Ask the leader to pass your explanation on to the group. Do not be surprised if his or her message is far more persuasive than yours. Do not take this personnally. Sharing the culture and the language of the employees as well as being in a position of respect is a magical combination that can be very effective in getting the message across.
- Ask the leader to model the behavior that you desire. Words are one thing, but action is far more powerful. When employees see their leader complying with the rules that used to seem so silly, you will be amazed how rapidly they begin to cooperate.

ISSUE II: THE IMPORTANCE OF SAVING FACE

The Challenge

Saving face is another value that can interfere with cross-cultural safety management. Although traditionally associated with the Far East, this value is also found in Hispanic, Middle Eastern, and Southeast Asian cultures. Saving face may seem like a complex philosophical concept, but it is really very simple. The doctrine of saving face places priority on seeing to it that no one in a relationship is being embarrassed or humiliated. Here is a brief list of the ways in which saving face can be manifested in the workplace.

Reluctance to disagree with the boss — Although sometimes employees fail to disagree simply because they are afraid or do not care, more often, especially in the multi-cultural workplace, this hesitancy has to do with saving face for the boss. The concern is that if the employee disagrees, the boss will be embarrassed and suffer loss of face, especially if that disagreement takes place in front of others.

Hesitance to admit lack of understanding — We have all felt this one. How often have you pretended to understand something said, for example, by your tax accountant or auto mechanic, just because you did not want to look stupid? It is not that you actually were stupid, or that you did not care about learning — more likely you just wanted to save face by pretending you knew something that you did not.

Reluctance to ask questions — In America we are taught that there is no such thing as a stupid question. You might be surprised to learn that this view is not shared by much of the rest of the world. In schools and workplaces all over the globe, asking questions is more apt to be considered a sign of ignorance than of intellectual curiosity. Saving face also affects the immigrant's willingness to ask questions. For many, to ask a question of a superior not only makes the asker appear stupid, but it also risks loss of face for the boss if he or she is unable to answer the question.

Reluctance to admit mistakes out of fear of appearing inadequate — This may apply to instances like losing a pair of protective glasses, performing a procedure wrong, or carelessly lifting something that is too heavy.

Sensitivity to being reprimanded — Discomfort when being reprimanded or coached is certainly worse when conducted in front of others and can result in extreme loss of face and emotional withdrawal.

We have all felt these emotions at one time or other. For those, however, from cultures in which saving face is a dominant value, the discomfort is apt to be far worse. The Chinese proverb, "If there is no face, there is no life," points out the depth of importance that this value holds for many immigrant groups. It is not to be taken lightly and, as I am sure many of you would agree, is probably the dominant value that distinguishes many immigrants from native-born Americans.

The Solution

As in the case of respect for authority, the challenges resulting from the value of saving face can best be solved by using the value itself. Saving face is probably the most deeply rooted value you are apt to encounter — it affects the ability to sustain good communication in the face of language barriers; it influences your ability to coach and counsel effectively; and, most tragically, it can lead to the loss of good employees because of hurt feelings and humiliation.

It is, however, because of how deeply rooted this value is that you can use it to your advantage. When you feel that an employee is not giving you what you need — information, admissions of errors, etc. — because of fear of losing face, the following steps will help.

- Express an understanding of their point of view. One of the big mistakes that managers of multicultural employees make is to believe that it is wrong to acknowledge that a cultural value exists. As long as it is done with respect, mentioning cultural differences can be an effective management strategy. In the case of losing face, it is perfectly acceptable to say that you understand how embarrassing it can be to disagree or admit lack of understanding and that you have often felt the same way and know how difficult it can be.
- Tell employees that you are concerned about losing face for yourself. Point out that you cannot do your job effectively if they do not give you the information you need, admit lack of understanding, or let you know if a piece of equipment is faulty. In other words, you will be embarrassed in front of your superiors and colleagues — you, too, will lose face — if the employee does not help you out. What you are doing here is stating a simple truth, but using terms — the terms of the employee's central value — to which he or she can readily relate.
- Reinforce the behavior once it happens. This does not mean that you further embarrass employees by overdoing it, but simply thank them when they bring you a problem, admit a mistake, or make a suggestion. I mention this rather simplistic step because it is all too easy to forget to reinforce behaviors that we regard as easy. For most Americans, for example, asking questions is fairly straightforward — we would never expect anyone to reward us for what seems to us to be a basic tenet of proper employee behavior. The challenge comes in recognizing that what is easy for one person can be difficult for another, and we need to stay alert to what behaviors require, and deserve, our reinforcement.

CONCLUSION: TAKING IT TO THE WORKPLACE

Managing cultural diversity can be a daunting task. Language barriers, differences in values, varying ideas about proper employee behavior — all of these can render even the most experienced manager frustrated and confused. It will help if you remember that beneath the cloak of culture all human beings are after the same things: human dignity, physical comfort, and social support. By building on this commonality we can work to find the compromises that will make even our most multicultural companies productive and safe places to work.

FURTHER READING

Note: The emphasis in this list is on those items that address the issue of cultural and ethnic diversity and how this diversity affects the way we manage and do business. Although several of the books deal with international business, much of the material contained within them is applicable to those ethnic and immigrant cultures found within the borders of the United States.

Presses Specializing in Diversity Publications

Amherst Educational Publishing, P.O. Box 6000, Amherst, MA 01004-6000, (800) 865-5549.
Brigham Young University, David M. Kennedy Center for International Studies, 280 HRCB, Provo, UT 84602, (801) 378-6528.
Gulf Publishing Company, P.O. Box 2608, Houston, TX 77001, (713) 529-4301.
Intercultural Press, P.O. Box 700, Yarmouth, NE 04096, (207) 846-5168.
Sage Publications, Inc. P.O. Box 5084, Newbury Park, CA 91359, (805) 499-0721.

General Works on Managing Workplace Diversity

Baytos, L. *Designing and Implementing Successful Diversity Programs*. Englewood Cliffs, NJ: Prentice-Hall, 1995.
Fernandez, J. P. *Managing a Diverse Work Force: Regaining the Competitive Edge*. New York: Lexington Books, 1991.
Jamieson, D. and O'Mara, J. *Managing Workforce 2000: Gaining the Diversity Advantage*. San Francisco: Jossey-Bass, 1991.
Loden, M. *Implementing Diversity*. New York: McGraw Hill, 1996.
Loden, M. and Rosener, J. B. *Workforce America!: Managing Employee Diversity as a Vital Resource*. Homewood, IL: Business One Irwin, 1991.
Thiederman, S. *Bridging Cultural Barriers for Corporate Success: How to Manage the Multicultural Work Force*. New York: Lexington Books, 1991.
Thomas, R. R. *Beyond Race and Gender: Unleashing the Power of Your Total Work Force by Managing Diversity*. New York: Amacom, 1991.
Wilson, T. *Diversity at Work*. Etobicoke, Ontario: John Wiley & Sons Canada, Ltd., 1996.

General Works on Cross-Cultural Business

Axtell, R. E. *Do's & Taboos around the World*. New York: John Wiley & Sons, 1985.
Chesanow, N. *The World-Class Executive: How to Do Business Like a Pro around the World*. New York: Bantam Books, 1986.
Fisher, G. *International Negotiation: A Cross-Cultural Perspective*. Yarmouth ME: Intercultural Press, 1980.

Foster, D. A. *Bargaining across Borders: How to Negotiate Business Successfully Anywhere in the World.* New York: McGraw-Hill, 1992.

Kennedy, G. *Doing Business Abroad.* New York: Simon and Schuster, 1985.

Morrison, T. and Conaway, W. A., and Borden, G. A. *Kiss, Bow, or Shake Hands.* Holbrook, MA: Adams Media Corporation, 1994.

Thiederman, S. *Profiting in America's Multicultural Marketplace: How to Do Business across Cultural Lines.* New York: Lexington Books/Macmillan, 1991.

Books Focusing on Specific Cultures or Countries

Note: The volumes listed here are a sampling of the dozens that are available on specific countries and cultures. Many of the books mentioned elsewhere in this list also include extensive material on specific groups.

American Culture

Stewart, E. *American Cultural Patterns: A Cross-Cultural Perspective.* Yarmouth, ME: Intercultural Press, 1972.

Arab Culture

Nydell, M. K. *Understanding Arabs: A Guide for Westerners.* Yarmouth, ME: Intercultural Press, 1987.

Asian Cultures

Engholm, C. *When Business East Meets Business West.* New York: John Wiley & Sons, 1991.

Black Culture

Kochman, T. *Black and White Styles in Conflict.* Chicago: University of Chicago Press, 1981.

Chinese Culture

Seligman, S. D. *Dealing with the Chinese: A Practical Guide to Business Etiquette in the People's Republic Today.* New York: Warner Books, 1989.

European Cultures

Hall, E. T. and Hall, M. R. *Understanding Cultural Differences: Germans, French, and Americans.* Yarmouth, ME: Intercultural Press, 1990.

Platt, P. *French or Foe?* London: Culture Crossings Ltd. 1994.

Filipino Culture

Gochenour, T. *Considering Filipinos.* Yarmouth, ME: Intercultural Press, 1990.

Hispanic Culture

Condon, J. C. *Good Neighbors: Communicating with Mexicans.* Yarmouth, ME: Intercultural Press, 1985.

Knouse, S., et al. *Hispanics in the Workplace.* Newbury Park, CA: Sage Publications, 1992.

Japanese Culture

De Mente, B. *How to Do Business with the Japanese.* Lincolnwood, IL: NTC Business Books, 1991.

Moran, R. T. *Getting Your Yen's Worth.* Houston: Gulf Publishing Company, 1985.

Rowland, D. *Japanese Business Etiquette: A Practical Guide to Success with the Japanese.* New York: Warner Books, 1993.

Korean Culture

Leppert, P. *Doing Business with the Koreans: A Handbook for Executives.* Chula Vista, CA: Patton Pacific Press, 1987.

## CHAPTER 43

Total Quality Management
and Safety and Health

Richard W. Lack

CONTENTS

INTRODUCTION

Being a serious student of the art and science of management, I am naturally interested in the total quality management (TQM) process. Since its inception, I have attended numerous conferences, seminars, lectures, and meetings at which this subject has been discussed. I have also read many articles, papers, and books describing the process and results achieved at various organizations and facilities. In order to gain some perspective, I invite you to join me in a short journey of reflection on the whole issue of TQM.

Throughout history, the achievement of high-quality goods and services and a need to satisfy the customer were of paramount concern. Artists and artisans alike shared this concern. One has only to visit museums or art galleries or tour the wonders of ancient civilizations to immediately realize that TQM has been with us for a very long time.

I share Dr. Wayne Dyer's view that every human being since the beginning of time is a part of an intelligent system and that each of us has an heroic mission, as he terms it, to accomplish while we are on this planet. The difficulty, of course, is

1-56670-370-0/02/$0.00+$1.50
663

that few are born with this knowledge and skill. Most of us have many lessons to learn on our road to peak achievement and inner peace. Translate this situation into any human organization, be it a social club, restaurant, hotel, municipality, factory, or giant corporate business, and you are dealing with a group of people each on their various paths toward getting to their lives' purposes.

No doubt the pharaohs faced this problem when building the pyramids, and every undertaking, now and in the future, will probably do so, until a system is found whereby people will more easily be able to find their personal life mission. Undoubtedly, progress is being made, when you consider the developments of the industrial revolution. The safety and health profession itself was born from the concern for preventing injuries and illnesses in what frequently were inherently dangerous workplaces.

In more recent times, the constant need to improve production efficiency and reduce costs, especially as automation was introduced, produced people with skills to serve these needs. Sometimes called efficiency experts or time and motion studiers, these were your industrial engineers of the 1950s and 1960s. This profession has become far more sophisticated with the advent of the computer.

Quality inspection and control also grew up with the mass-production era, especially in the U.S. Frequently, the quality control and industrial engineering functions were, and still are, merged into one group serving their respective organizations.

The difficulty with this relentless search for production efficiency is that it tends to dehumanize the organization. People feel they have lost control, that their efforts are not recognized, and the inevitable result is a buildup of stress which finds outlets in ways that are frequently unproductive.

As has been discussed in other chapters in this book, a number of approaches and systems have been developed in recent times to help address this problem. These range from reengineering the organization to the team approach and behavior-based motivation programs. On the horizon are techniques developed by such luminaries as Scott Peck, Tom Melohn, Margaret Wheatley, Peter Senge, and Peter Block.

Another approach is the technique of intropreneuring developed by Gifford and Elizabeth Pinchot. This approach is designed to help organizations transform their bureaucratic management systems into systems which make more use of their employees' intelligence and help them achieve their personal missions.

I was recently intrigued to receive a brochure on a special 3-day conference organized by the Association for Quality and Participation entitled *Practical Applications of Leading Edge Theory*. The three presenters at this course are Margaret Wheatley, Peter Senge, and Peter Block. Need I say more?

THE TOTAL QUALITY MANAGEMENT PROCESS

Readers interested in studying the specific elements of TQM in depth are encouraged to review the books and other publications listed in Further Reading at the end of this chapter.

The techniques developed by such pioneers as W. Edwards Deming, Joseph Juran, and others were built around creating a more empowered workforce and providing them with tools so they could improve their production efficiency and maintain customer satisfaction. This movement also helped to demystify the statistical process, and a number of simple techniques for pinpointing potential production problems were developed. Examples are:

Flowcharts — these are a useful way of outlining the steps in a procedure or process. There are many applications in safety management. For example, the chart could define the steps for confined space entry or lockout-tagout. This technique can supplement or in some cases may replace the traditional job safety analysis approach.

Fishbone diagrams — cause and effect charts. This technique can be a useful aid in accident investigation.

Pareto charts — to help determine priorities (the 80–20 rule). This graphic technique can help with accident analysis, e.g., type of injury and the immediate causes of this injury.

Histograms — to chart how frequently something occurs. This could be used to track the shift, day, or time that accidents occur.

Run or trend charts — to identify trends over a period of time (example, 12-month total case incidence rate for injuries and illnesses).

Control charts — establish upper and lower control limits and track the actual results. This technique could be used to track injury experience results.

Problem solving in the TQM process has a variety of approaches, especially now that computer have come to our aid. They all involve the following broad steps:

- Define the problem.
- Analyze the facts as to its cause and effect.
- Generate alternatives.
- Select feasible solutions.
- Take action.
- Evaluate the results.

Brainstorming to help generate ideas for improvement or to solve a situation is now commonly used in successful organizations.

The Six Sigma concept, as adopted by several corporations (the best known being Motorola), has created much interest. Six Sigma is in effect a measure of goodness in that it measures defects or mistakes. Each group in the organization has to identify their customers (internal and/or external) for their service and find ways to mistake-proof their service and ensure continuous improvement.

Achieving Six Sigma, for example, equals 3.4 defects per million opportunities for error. Translating this to safety would mean no more than four incidents per million hours worked. Note that by incident I mean four mistakes or unsafe work practices that could have caused an injury.

APPLYING TOTAL QUALITY MANAGEMENT
TO SAFETY AND HEALTH

In my opinion, the TQM approach is a natural for safety and health practitioners because it is a proactive preventive method. It will help you to identify your customers and find better ways to serve them.

TQM is also concerned with measuring performance to help ensure that results will continuously improve. This concept has already been discussed in several other chapters of this book.

The analytical techniques involved with TQM, discussed earlier in this chapter, can be used to help identify trends or root causes of accidents, incidents (near miss), or hazards (either unsafe work practice or unsafe condition).

In addition to the tools and techniques already described, readers are referred to the section on accident investigation in Chapter 6. This outlines a source and cause analysis approach. The key is to reach actionable root causes by asking "Why?" (ask "Why?" five times). Then select the root cause with the greatest probable impact.

Finally, the safety and health practitioner will need to help his or her organization set up measurement systems that the employees and work teams can use to track their own performance. The old approach of the safety department preparing and distributing the reports on supervisory and team performance can then be phased out as TQM really takes hold in your organization.

CONCLUSIONS

For the reasons stated in the introduction, there is no doubt in my mind that every organization will eventually shift its management style to one that helps its people achieve a pattern of high performance. This, in turn, will open up a unique opportunity for safety and health practitioners to help adapt their unit's safety and health goals and systems to the unit's overall production goals and systems.

If you are not already involved in this process, I recommend that you take immediate steps to prepare yourself by reading books, attending courses, and joining organizations in the TQM and professional training fields.

I see our role as helping the people in our organization find ways to achieve error-free peak performance. Prevention is essential in order to ensure continuous improvement, and this will only be possible when the people are given the necessary tools and training and are provided with a supportive community in which they can do their jobs safely.

REFERENCES

Here is a list of sources that are referred to in the chapter which are related to but not directly concerned with the TQM process.

Books

Block, P. *Stewardship — Choosing Service Over Self Interest*. San Francisco, CA: Berrett-Koehler Publishers, 1993.

Dyer, W. W. *Real Magic — Creating Miracles in Everyday Life*. New York: Harper Collins Publishers, 1992.

Fletcher, J. L. *Patterns of High Performance — Discovering the Ways People Work Best*. San Francisco, CA: Berrett-Koehler Publishers, 1993.

Melohn, T. *The New Partnership — Profit by Bringing Out the Best in Your People, Customers and Yourself.* Essex Junction, VT: Oliver Wright Publications, 1994.

Peck, M. S. *A World Waiting to Be Born — Civility Rediscovered.* New York: Bantam Books, 1994.

Pinchot, G. and Pinchot, E. *The End of Bureaucracy and the Rise of the Intelligent Organization.* San Francisco, CA: Berrett-Koehler Publisher. 1994.

Senge, P. *The Fifth Discipline Fieldbook — Strategies and Tools for Building a Learning Organization.* New York: Doubleday, 1994.

Wheatley, M. J. *Leadership and the New Science — Learning about Organization from an Orderly Universe.* San Francisco, CA: Berrett-Koehler Publishers, 1992.

FURTHER READING

Books

Adams, E. E. *Total Quality Management — An Introduction.* Des Plaines, IL: American Society of Safety Engineers, 1995.

Camp, R. C. *Benchmarking.* Milwaukee, WI: American Society for Quality Control, ASQC Press, 1989.

Gitlow, H. S. and Gitlow, S. J. *The Deming Guide to Quality and Competitive Position.* Englewood Cliffs, NJ: Prentice-Hall, 1987.

Sashkin, M. and Kiser, K. J. *Putting Total Quality Management to Work.* San Francisco, CA: Berrett-Koehler Publishers, 1993.

Walton, M. *The Deming Management Method.* New York: Perigee Books–Putnam Publishing, 1986.

Other Publications

Deacon, A. The role of safety in total quality management. *Saf. Health Pract. (U.K.)* 12(1), 18–21, January 1994.

Lack, R. W. More on the safety-quality management issue. *Insights Manage.* 3(1), February 1991.

Manuele, F. A. Make quality the watchword of your safety program. *Saf. Health* 148(4), 106–108, October 1993.

Motzko, S. Variation, system improvement, and safety management. *Prof. Saf.* 34(8), 17–20, August 1989.

Petersen, D. Integrating safety into total quality management. *Prof. Saf.* 39(6), 28–30, June 1994.

Vincoli, J. W. Total quality management and the safety and health professional. *Prof. Saf.* 36(6), 27–32, June 1991.

## CHAPTER 44

Integrating Quality, Health, Safety, and Environmental Issues into One Management System

Mark D. Hansen

CONTENTS

INTRODUCTION

It seems that every time safety professionals become proficient at one set of standards, the standard either undergoes a major revision or new regulation is published. This requires safety professionals to literally start all over again. An example involves the OSHA 1910.119 Process Safety Management (PSM) and the Chemical Manufacturers' Association (CMA) Responsible Care Program, which contains a Process Safety Code. These standards are somewhat redundant, but are different enough to require much more documentation and coordination by the safety professional. The same can also be said of the Environmental Protection Agency (EPA) Risk Management Program (RMP), the OSHA 1910.119, and the CMA Employee Health and Safety Code, Community Awareness and Emergency Response Code, and Process Safety Code. As soon as we become proficient with these codes and with integrating them with the quality requirements in International Organization for Standards (ISO) 9000:1994, then ISO 14000 and ISO 9000:2000 were issued. All these standards are similar but different enough to require much more work by the safety professional. This is true of numerous standards, including ISO 9000:1994, ISO 9000:2000, and ISO 14001, OSHA 1910.119, EPA RMP, EPA Clean Air Act (CAA), and the Food and Drug Administration (FDA) Good Manufacturing Practices (GMP) standards. All these standards are different in their own ways, but they are also similar in many aspects, such as management of change, documentation, security and integrity of the review, and design and implementation process.

It seems that those generating standards clearly don't talk to each other to provide a consolidated and consistent direction in the area of standards. However, at the end of the day, if a company doesn't make money, it won't be in business. I'm not saying that being in compliance isn't important, but that staying in business is just as important. Quality, health, safety, and environmental (QHSE) professionals should not function merely as document generators to produce the appropriate paperwork for compliance.

STANDARDS EVERYWHERE

Everywhere we turn, standards emerge, requiring more redundant work for safety and health professionals. Since the standards identified have diverse functions, this chapter focuses on the effect on the safety profession. Specifically, this chapter focuses on how key elements of the OSHA PSM program map to the other standards.

OSHA PSM

The key elements of the OSHA PSM program include general information, process safety information, process hazard analysis, management of change, operating procedures, safe work practices, training, mechanical integrity, pre-start-up safety reviews, emergency response and planning, incident investigation, auditing the process, contractors, capital projects, human factors, standards, codes and regulations, waste management, and references. These elements were chosen for their application across many industries and their frequent appearance in the other standards presented here.

Because it is not the purpose of this chapter to describe in detail each of the standards discussed, only those areas applicable to the OSHA PSM will be detailed. Table 44.1 presents the standards and the referenced sections of each standard where similar activities apply and some form of redundancy occurs.

EPA Clean Air Act Amendments of 1990

The 1990 Amendments to the CAA make comprehensive changes with respect to air quality permitting and enforcement, operations in non-attainment (dirty air) areas, control of utilities and acid rain, regulation of toxic air pollutants, motor vehicle and fuel requirements, stratospheric ozone protection, and other critical areas. See Table 44.1 for references to the EPA CAA. Of particular note regarding the convergence of standards is Title III, General, Sections, 301 Administration, and 304, Citizen suits.

EPA Risk Management Program

Under the EPA RMP rule, processes subject to these requirements are divided into three tiers, labeled Programs 1, 2, and 3. The EPA has adopted the term *Program* to replace the term *Tier* found in the Superfund Amendments and Reauthorization Act of 1986 (SARA Title III) to avoid confusion with Tier I and Tier II forms submitted to the government. Eligibility for any given Program is based on process criteria so that classification of one process in a Program does not influence the classification of other processes at the source. For example, if a process meets Program 1 criteria, the source need only satisfy Program 1 requirements for that process, even if other processes at the source are subject to Program 2 or Program 3. A source, therefore, could have processes in one or more of the three Programs.

Program 1 is available to any process that has not had an accidental release with offsite consequences in the 5 years prior to the submission date of the RMP and that has no public receptors within a specified distance to a specified toxic or flammable end point associated with a worst-case release scenario. Program 3 applies to processes in Standard Industrial Classification (SIC) codes 2611 (pulp mills), 2812 (chloralkali), 2819 (industrial inorganics), 2821 (plastics and resins), 2865 (cyclic crudes), 2869 (industrial organics), 2873 (nitrogen fertilizers), 2879 (agricultural chemicals), and 2911 (petroleum refineries). Program 3 also applies to all processes subject to the OSHA PSM standard, unless the process is eligible for Program 1. Companies are required to determine individual SICs for each covered process to determine whether or not Program 3 applies. All other covered processes must satisfy Program 2 requirements.

American Petroleum Institute Recommended Practice 750 Facility Siting

The American Petroleum Institute (API) Facility Siting document addresses the location of buildings in processing facilities that are generally addressed in the PSM program in the hazard analysis and emergency response portions.

Table 44.1   Standards Convergence Evident in International Standards Development

Title	API RP 750 Sect	OSHA PSM Para	AIChE CCPS Chap	CMA's Responsible Care			MBNQA Para	CMA TQC CCI Para	ISO 9000 Para	EPA RMP Para	FDA GMP Part 211 Para	1990 CAAA Sect
				PSC Tabs	EHS Tabs	CAER Tabs						
General	1	a,b,c	123	123	1	1	1.1	1	9001-4.1	68.2	B	301, 304
Process Safety Information	2	d	4	5.5	6	7	2.1	4C	9001-4.5.1, 4.16, 9002-4.5, 4/6	68.3	D, E,F, G, J	301, 304
Process Hazard Analysis	3	e	46	5.7	7, 8	7	1.3		9001-4.4, 4.14, 9002-4.139004-15	68.2	C, F	301, 304
Management of Change	4	l	7	5.10.8	56	2		4A, 4C	9001-4.9, 4.5.2, 9003-3.2,6.16.2	68.4	C, F	304
Operating Procedures	5	f	4	7	4	6	5.1	4A, 4C	9001-4.2, 4.4, 9002-4.8.1, 4.13	68.3	F, I	301, 304
Safe Work Practices	6	k	4	7.4	41213	6		4F	9001-4.2,4.9,		I	301, 304
Training	7	g	10	7.7	17, 18	8, 9	4.3	3	9001-4.18, 9002-2.4, 9003-3.4, 11, 9004-4.2,5.3	68.3	C, G, H, K	301, 304
Assuring Quality & Mechanical Integrity of Critical Equipment	8	j	8	6.1	12, 13	10	5.2	4	9001-4.14, 9004-3.11.3	68.3	C, D	301, 304
Pre-Start-up Safety Review	9	i	5	13	5	2		4B	9001-4.9, 9004-2.6, 2.7	68.3		304
Emergency Response & Control	10	n	11	6.16	16	34		4C	9001-4.9			301, 304
Investigation of Incidents	11	m	11	4.1	14		5.2	4	9001-4.14, 9002-2.4, 9003-3.4, 11, 9004-4.2, 5.3	68.4		301, 304
Audit of Process Hazards Management Systems	12	o	13	6.7	5		1.3, 5.5	4	9001-4.17, 9002-4.16, 9004-5.4	68.4	J	301, 304
Contractors		h		7.13	3		5.4		9001-4.6, 9004-9.1, 9.3		B	304
Capital Projects			5									
Human Factors		f, d	9	7.9	4		5.1	4A, 4C	9001-4.2, 4.4			
Standards, Codes & Regulations			12	6.4, 12			1.2, 4.5	23	9001-4.12.1			304
Waste Management				2.5-2.7	1						H, K	
References	13	Apps A,B,D	14	9	78						A	

CMA = Chemical Manufacturers Association; PSC = Process Safety Code; EHS = Employee Health and Safety; CAER = Community Assistance and Emergency Response; MBNQA = Malcolm Baldridge National Quality Award; ISO = International Organization for Standards; OSHA = Occupational Safety and Health Administration; PSM = Process Safety Management; TQC = Total Quality Council; CCI = Criteria for Continuous Improvement; FDA = Food and Drug Administration; GMP = Good Manufacturing Practices; CAAA = Clean Air Act Amendment; AIChE = American Institute of Chemical Engineers; CCPS = Center for Chemical Process Safety; API = American Petroleum Institute; RP = Recommended Practice

American Institute of Chemical Engineers Center for Chemical Process Safety

The Center for Chemical Process Safety (CCPS) was established in 1985 as a directorate of the American Institute of Chemical Engineers (AIChE) and focuses on engineering and management practices that help prevent or mitigate catastrophic process safety accident. This Center published a text in 1989 entitled, *Guidelines for Technical Management of Chemical Process Safety.* Many of the criteria are illustrated and mapped to other standards in Table 44.1.

CMA Responsible Care Initiative

The CMA Responsible Care® Initiative was designed to improve public perception and to encourage chemical manufacturers to implement management systems to ensure the safe use of their chemicals from cradle-to-grave. There are six codes as follows: Employee Health and Safety Code, Process Safety Code, Pollution Prevention Code, Community Awareness and Emergency Response (CAER) Code, Distribution Code, and Product Stewardship Code. Each code requires a written management system to ensure that the code is properly implemented.

International Organization for Standards 9000

The publication of the ISO 9000:1994 series and the terminology standard (ISO 8402) in 1987 attempted to bring some harmony on an international scale. It has supported the growing impact of quality as a factor in international trade. The ISO 9000 series has been adopted by many nations and regional bodies, and is rapidly supplanting prior national- and industry-based standards.

The ISO 9000-9004 and the American National Standards Institute (ANSI)/American Society for Quality Control (ASQC) Q90-94 series documents contain information relevant to systematic management for product and service development, design, production, and installation activities, including safety, health, and environmental aspects.

These standards represent an international movement to establish worldwide quality system standards for hardware, software, and processed products and services. Although strongly driven by the European Community originally, the Total Quality Management (TQM) vision represented by these standards will most probably drive U.S. business to become compliant and require certified practitioners (e.g., certified safety professionals (CSPs)) in the near future if a business wishes to conduct business internationally.

CMA Total Quality Council Criteria for Continuous Improvement

The Total Quality Council (TQC) is a CMA CHEMSTAR business council dedicated to promoting the total quality culture in the chemical industry. The council saw an opportunity in 1991 to advance the understanding of total quality by demonstrating that its practices and techniques could serve the practical needs of managers charged with implementing the Responsible Care Codes of Management Practices. At the same time, the CMA Engineering and Operations Committee, sponsor of the Process Safety Code (PSC), reached the same conclusion as code implementer.

These conclusions were supported by the observation that CMA member companies with mature total quality systems were implementing the codes more smoothly and rapidly than companies with traditional management systems. A subsequent survey of the membership of the association revealed a strong need for guidance in linking total quality initiatives such as ISO 9000 to process safety programs such as the PSC and the OSHA PSM Rule.

As a result, a joint task group of TQC and Engineering and Operations Committee members was commissioned to examine the practical relationships between PSM and total quality management. The task group then developed the Criteria for Continuous Improvement (CCI) shown in Table 44.1.

The FDA GMP 21 CFR Part 210 and 211, Current Good Manufacturing Practice in Manufacturing, Processing, Packing, or Holding of Drugs is a standard that identifies the minimum requirements for drug processing to ensure the safety, identity, strength, and purity of commercial manufacture of drugs. These practices are identified as GMPs that parallel PSM, ISO, and other standards. The applicable criteria have been identified and mapped in Table 44.1.

COMPLYING WITH MULTIPLE STANDARDS BY WORKING SMARTER

The problem is that health and safety professionals and managers already spend an inordinate amount of time digging through multiple regulations and filing reports rather than focusing on doing the job of health and safety. There is literally no time to worry about special processes and procedures unless you have excess time on your hands.

Health and safety should be key elements of process improvement and quality and environmental enhancement. With all of the inherent similarities, it is difficult to understand why companies fail to integrate their quality and environmental management systems and fail to include health and safety in the process. Many U.S. companies have chosen not to seek ISO certification but still implement a management system that is consistent with ISO requirements without the burden of documentation or administrative costs required by some registrars. The goal here is to illustrate the benefit of integrating your health and safety management systems into your quality and environmental management systems. The result is that you now have one management system for QHSE rather than one for quality, one health and safety, and one for environmental management. Because all of these systems have a common management and process engineering foundation, it is not leap but it is a major step. The key for future process improvement and financial success is to implement a QHSE management system that integrates all three disciplines into one overarching management system.

If you want to streamline your whole process, you should consider using ISO systems to establish strong process quality management based on ISO 9000:2000 and 14000. One reason is that it is already developed so you do not have to reinvent the wheel. Second, it has already proven successful. And third, it is already accepted in the worldwide market. Your primary focus should be on improving performance, and not on ISO certification and regulatory compliance. If you are still stuck on regulatory compliance, a management system merely adds to the list of things to do. Management systems, when implemented correctly actually reduce the burden of developing documentation and records while regulatory conformance is being accomplished. Your system should first of all make sense from an operations and process improvement perspective and not be driven by concerns over your historical audits. By working with operations you will determine the implementability of the management system and tailor it to your needs and business objectives. Your objective is to reduce risk, not to establish a huge paper trail based on audit results.

In addition, organizations can work with their information system (IS) departments to develop the record-keeping and data management systems, policy, and procedure templates that will need be shared between various business units. Using your company's intranet is a tremendous technological advantage in providing resources in the field.

A well-thought-out management system that integrates QHSE will significantly improve process performance and reduce risk and the liabilities associated with nonconformance over time, even if certification is not a priority. The decision to implement a certified management system should be a second priority and not the focus of the implementation effort. A key to improved performance is system design and tapping the resources that exist with the first-line supervisors and employees who perform the daily critical tasks of the organization. If employees are your most valuable resource, then they must be heavily involved in the QHSE process improvement system.

This chapter should help you see the power of working together to improve process performance at all levels. The tools and concepts that are common to ISO, can become part of an overall decision-making strategy to help manage and mitigate risks.

Deming drew his experiences from World War II. The systems used by manufacturers supporting the war effort focused on statistical control of mass production to supply the military. Lives were at stake, the fate of the world was at stake, and quality on-time delivery of goods and services was critical. Deming then used these methods to help him with the postwar reconstruction of Japan. For some reason, American manufacturers believed that the postwar production of products did not require the same rigor as the military–industrial output. This thought was probably attributed to arrogance over having the only significant industrial infrastructure remaining in the world, and to a return of men to production jobs who were not familiar with these systems.

The boom years of postwar America focused on quantity, and taking advantage of increasing market demand. The Japanese, however, were starting over and had focused on developing products and the systems to deliver them, which balanced both quality and quantity.

Process management is becoming very important for companies not only in the United States, but also in Europe and in developing countries with burgeoning economies. Integrating quality and environmental management with your health and safety management systems can only make things run more smoothly in your organizations. This objective is now possible by using ISO 9000:2000 and ISO 14000 as definitive models to build this integrated management process.

Historically in the United States, the development of an overarching occupational health and safety management system has met with resistance by industry. However, recently the ANSI Z10 committee has drafted the Occupational Health and Safety Management System standard. Internationally, this kind of effort is already well under way, as ISO is developing a draft occupational health and safety management system standard OHSAS 18001. This draft standard will most likely become part of a future ISO standard. It is interesting to note that in the United Kingdom, management systems have been required by standards such as BS 8800 for several years. OSHA is developing a best practices standard that may have an impact on management systems development. However, the process for OSHA could be years away.

Planning is the foundation for designing and successfully implementating an overarching management system. Clear roles and responsibilities, as well as a process for systematically reviewing process and resource inputs to the system such as the procurement procedures, training, maintenance, and procedural controls, must be established.

The concepts of multiple root-cause analysis, risk management, risk mapping, process incident reports/nonconformances and accident costs, should be introduced as part of the QHSE management system. These are tools to help prevent incidents and nonconformances and to improve performance from lessons learned. Processes have primary or critical process inputs as shown in Figure 44.1, that, if substandard in any way, will ultimately be the root cause of non-conformances and accidents/incidents.

It is our responsibility as QHSE professionals to evaluate all primary process inputs to ensure that they are adequate and that they do not become the root cause of an incident. We are also responsible for learning from our mistakes and evaluating process nonconformances and incidents to identify the multiple root causes of the process failure, including the establishment of incompatible goals. These responsibilities must include taking a systems approach to ensure that failure does not happen again, including sharing of lessons learned with managers and employees so that they to can avoid recommitting the incident.

The necessary elements and implementation of an integrated total process management system include the following:

- Employee involvement
- Training
- Communications
- System documentation and document control
- Records
- Design control
- Purchasing
- Hazardous materials/waste management
- Management of contractors and vendors

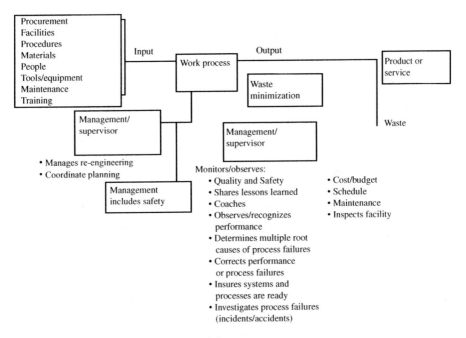

Figure 44.1 Industrial work process model.

- Emergency preparedness and contingency planning processes
- Use and maintenance of equipment, tools, and the facility

Positive operational control demands emphasis on the use of procedures to control the work at the process resource input and design stages. Critical to understanding process improvement and incident/accident prevention is understanding the management system and the elements of the system that are causing process nonconformance and incidents.

The concept of multiple root-cause analysis and the tools for process nonconformance or incident investigations, including the need to document the total costs (direct and indirect costs), will need to be part of any credible management system. Until management understands the financial impact of substandard process design and execution and the benefits of process change communicated in financial terms, such as return on investment, payback period, very little will be accomplished. Otherwise, remedial efforts such as retraining and disciplinary action will be applied to the process enhancement or corrective action effort.

Identifying the causes of process incidents and nonconformance and sharing lessons learned to correct and prevent recurrence is one of the most important tools supervisors and managers will have on the path to continual improvement. If you know what went wrong and why, you can improve the process and prevent a recurrence. Audits, and the follow-up to completing corrective actions, are critical to successfully implementing any management system.

We need to get away from the way we've always done things. We need to ask ourselves the following questions:

- What can we do to improve the process, reduce rework, breakage, and worker's compensation claims while increasing productivity, efficiency, profitability, and customer, management, and employee satisfaction?
- How can we involve employees and management in creating a process that saves money and increases profitability while reducing waste, rework, and losses due to accidents, as well as process nonconformance to requirements?

Over the past 15 to 20 years managers have been bombarded with a variety of management programs and processes, as shown in Table 44.1. If you look to the European Union, and the United Kingdom in particular, you see regulatory bodies adopting the ISO framework for QHSE. This trend can be seen in other countries around the world such as New Zealand and Australia, and it clearly reflects a different approach to our prescriptive regulatory approach taken by the U.S. EPA and OSHA. An example of a management system approach can also be seen in the draft OHSAS 18001 standard.

A management system is intended to help facilities establish processes, procedures, and metrics that allow them to perform tasks right the first time, within budget, and on time, and with no rework, redundant procedures, injuries, or environmental liability.

An integrated QHSE management system will be critical to the success and competitive posture of companies. In the future, management must align its resources (people, tools/equipment, facilities and maintenance, procedures, materials, vendors, contractors, training, and supervision) to optimize QHSE performance.

Oftentimes, programs or processes have their roots in the works of Deming, or in systems safety, which originated with the Department of Energy (DOE) and National Aeronautics and Space Administration (NASA). For those unfamiliar with systems safety, this process is not focused simply on worker safety. Systems safety includes an assortment of risk assessment tools that help management identify and, to the extent possible, quantify potential process risks and to ensure operational readiness. This process could include evaluating business risks, as well as the risk of environmental contamination or injury or illness to employees, customers, or contractors.

Another problem in some organizations is their tendency to focus on regulatory compliance vs. process management. Some organizations have implemented a management system only to become frustrated with it or question its value as they focus on documentation vs. process management.

This assessment does not mean that you should ignore regulations or documentation. You must conform to both; however, regulatory and documentation conformance should be a by-product of good process management and not the primary focus of the management system.

Unfortunately, we have too often created confusion in the workplace by not consistently sticking to a chosen plan for process or quality improvement. We have jumped from Deming to Crosby's cost of quality, TQM, and or reengineering. Some versions of these quality programs have been adopted internationally in the ISO quality and environmental standards.

The reason ISO has become so popular globally is that managers of multisite or global operations have recognized that having multiple systems is dysfunctional.

Also, corporations may create additional liability by not consistently applying corporate policy. The biggest problem for managers is that common management measurements and language are needed. Too often we tend to adopt metrics and language to describe various programs that are not consistent, or we tend to fragment process management into artificial compartments such as quality, environmental, or occupational health and safety. Another reason management systems have become popular is because they are seen as a way to control the escalating costs of process nonconformance. Six Sigma, the newest and most rigorous of systems, has a goal of defect elimination, or no more than 3.4 defects per 1,000,000.

All accidents nationwide cost ~3% of the gross national product (GNP). When this loss is expressed in medical expenses, lost wages, insurance claims, production delays, and equipment downtime, it significantly reduces business productivity and profitability. A more costly problem that would often be related to the same management control issues is rework, which would easily be more than double the injury losses.

Potentially more costly problems, which can often be related to the same management control issues, are quality or rework, and environmental liability that can again easily be more than double the losses due to workplace injuries. Indirect costs associated with incidents range from four to ten times the direct costs of wage loss and medical cost, greatly increasing the significance of this type of process failure.

The problem is often attributed to a lack of understanding of process management, coupled with our tendency to compartmentalize process quality, health and safety, and environmental management. Who benefits by this fragmentation? The line supervisor, production management, or the specialists who manage these departments? Too often these program managers consider these processes to be the center of their corporate universe, and their peers' goals to be less important, failing to take advantage of the opportunity to work together.

We need to step back and look at work processes from our internal customer's perspective, as well as the perspectives of the line supervisor and work center employee. As a senior manager or manager of a quality system or QHSE program, ask yourself the following questions. What is the line manager or supervisor's primary mandate or concern? What is senior management holding the supervisor accountable for? At the end of the day is management accountable not only to its external customers and shareholders but also to its internal customers, supervisors, and community stakeholders?

True understanding and appreciation of their internal and external customers and the summation of incremental improvements over time are what sustain companies and their profit margins. "Sustained focus on these values and objectives" must be the management mantra for the new millennium. Too many companies lack focus on process management and have conflicting objectives, as is evident in redundant and multiple management systems for QHSE. The other problem contributing to lack of focus is when leaders delegate responsibility and are not actively involved in the process enhancement system. Too often it is believed that the QHSE manager or the quality manager can cover the same territory as hundreds of supervisors.

Why is it that QHSE programs often appear to be the responsibility of the quality or QHSE department or a distant corporate entity vs. the production manager who controls the work process? Why is it that workplace process control instruction or designs often fail to include QHSE? Why aren't quality managers and QHSE working more closely together?

Does the concept of continuous improvement apply only to production rework? Why wouldn't you consider environmental damage or employee injuries a process failure? Quality failures often have safety and environmental implications. After all, aren't our employees our most important resource, or do we see them as disposable and easily replaced? Through painful experience and financial consequences many companies have learned that environmental quality control is necessary to eliminate potential liabilities that can cause the financial failure of a company, not to mention potential criminal liability. In fairness to operations managers, too many safety or QHSE managers have not understood how to address process issues effectively, especially in financial terms.

These questions may seem overstated, but they are intended to highlight the dangerous attitudes exhibited by some managers and supervisors who do not understand process control. If any company in today's global market expects to be competitive, it must be willing to consider integrating its QHSE management systems. QHSE managers must also learn more about quality management and communicating with production and financial managers in terms that they understand. Changing metrics and reporting losses or risks in financial terms, or considering returns on investment when recommending process controls or process changes, is critical if we expect to bridge the gap between production, quality, and QHSE management.

What are we trying to do with an integrated QHSE Management System?

Managers of most mature companies understand the need for a quality management process. Industrial managers and developers by now know the risk associated with mismanagement of hazardous materials and waste. The difficulty has been how do you do it, which process TQM — Deming or Crosby.

In many ways environmental issues appear to be the most difficult because of the technical complexity associated with remediation projects or the design of new chemical processes. Then there is occupational health and safety that many managers think is out of their control. They believe many injury claims are fraudulent but difficult to prove, or that the employee was careless. If they only had more conscientious employees the safety problem would go away! In reality, 80 to 85% of the process failures, including injuries, are caused by common processes and multiple root causes, variations built into the process by the existing management system.

The typical company spends one week per employee to implement a TQM program. The biggest mistake is a failure to focus on a critical few issues, and too much emphasis is focused on documentation and regulatory compliance vs. process and task management.

One of the objectives here is to demonstrate that there is tremendous benefit to integrating the management process for controlling these risks. Using ISO 9001 and ISO 14001, you can clearly see that a common process management system QHSE integrated and built on ISO can be very powerful. The ISO standards and their various derivatives provide a framework consisting of a number of interconnected process management tools and services organized to integrate quality and QHSE management.

The ultimate goal is to reduce the time needed to perform tasks and to create added efficiency while reducing costs, waste, and personnel injuries. Qualitative risk assessments and detailed fault tree or systems analysis are needed, but in many instances they should be used only after passing through the sieve of the qualitative risk assessment. A qualitative risk assessment involves employees and other stakeholders and in most instances allows management to quickly identify, assess, and establish a plan to eliminate or control process risks.

William B. Smith, Vice President for Quality Assurance for Motorola, pointed out the value of employee involvement (e.g., quality circles, and process enhancement teams or in processes such as the qualitative risk assessments). It should also be noted that, like the statistical quality control process used widely by the military–industrial complex during World War II, quality circles were also widely used; however, like the quality improvement process it did not survive postwar industrialization in the United States but was reinvented in Japan. In Smith's speech to the National Private Truck Council, he stated:

In the Motorola experience it was found that 91% of problems were hidden from general management. The general manager was aware of only 4% of problems on the production floor. General supervision was only a little better off, being aware of 9% of the problems. The largest gap in the information flow appeared between the production supervisors and general supervision. Line supervision was credited with being aware of 74% of the problems. Not surprising, the workers were found to be aware of 100% of the problems.

Smith's statement is a compelling reason to ensure that your most valuable resource, your employees, is heavily involved in the QHSE process.

California implemented legislation 10 years ago that attempted to require some elements of a process management system for safety. The California Occupational Safety and Health Administration adopted legislation (Senator Bill Greene's Accident Injury Program Legislation SB-198) on October 2, 1989, requiring management to evaluate processes, perform risk assessments, and develop management plans to mitigate the identified risks.

The proposed OSHA safety management standards and voluntary protection program (VPP) reflect a growing understanding that QHSE management system approaches to Safety 101 are much more effective than prescriptive-type regulations. In addition, countries around the world such as Australia and New Zealand have adopted an occupational health and safety (OH&S) management systems based on the U.K. BS 8800 standard. We also have a draft OH&S management system standard that was developed with input from several counties and organizations using the BS 8800 standard as a model. This draft could ultimately become the model for a QHSE standard, into which the ISO standard or elements of it could be incorporated.

European Union requirements mandate management audits and risk assessments of facility programs, and ultimately the design of programs to reduce or eliminate risk. In Europe the regulators expect companies to evaluate their facility and process design and to control their processes through well-thought-out risk assessments and mitigation plans that can form the foundation of an OH&S management system. In Australia, some states require self-insured companies to implement a process called safety map, which is harmonized with ISO 9000. OSHA is promulgating an OH&S programs standard 29 CFR 1910.700 with provisions similar to the OSHA voluntary guidelines. The standard requires employers to identify and control work-site hazards and to involve employees in all phases of the program. Marthe Kent, Director of the OSHA Office of Regulatory Analysis, stated:

> If OSHA had it to do over again, this would probably be the first standard we would promulgate because it truly provides employers and employees with the foundation for workplace safety and health.... Our evidence suggests that companies that implement effective safety and health programs can expect reductions of 20% or greater in their injury and illness rates and a return of $4–$6 for every $1 invested.

Companies with ISO 9000 or ISO 14000 programs are starting to evaluate harmonizing worker safety and health into these programs and, to the extent possible, integrating the management aspects of these quality and environmental management programs.

There is really nothing terribly significant about ISO management system standards, nothing new in them, except the fact that there is a consensus on the elements contained in the standard. The value of the ISO standards is that they give companies a common foundation to discuss, and evaluate the minimum expectations in a competently designed and executed system.

The common elements establish the benchmarks for expected quality and QHSE performance. It is for these reasons that I am an advocate for including occupational health and safety in the systems and, to the extent possible, integrating them. This in no way means that you will not have a need for specific environmental, quality, or safety requirements, or technical resources, because you will. What it does mean is that these resources will be used more effectively. Your focus will be on prioritizing "enterprise risk" and in controlling the systemic causes of process nonconformance, which will in most cases eliminate or reduce quality, environmental liability, and safety risks by correcting the multiple root causes of process failures. All management systems including Deming, Juran, and Crosby have common principles that are identified below, and which are discussed and compared in more detail in this chapter.

Table 44.2 is provided to show how the quality improvement principles defined by Deming are consistent with ISO Standards and good QHSE management principles.

FIVE PRINCIPLES

The five principles detailed below are specifically discussed in the U.K. standard BS 8800, Australian/New Zealand Occupational Health and Safety Management System Draft Standard (1996). They are also covered in ISO 14001 and found in a different context in ISO 9001, TQM, and the multitude of quality and QHSE management system standards emerging around the world. They are the foundation of any good management system.

Table 44.2 Quality Management TQM (Key Deming–Crosby Steps) vs. Safety Program Management

Quality	HSE
Management commitment and organizational constancy of purpose	Same
Adopt new management philosophy including Crosby's four absolutes:	Same
1. Quality is conformance to requirements (defined by the customer, internal and external, and defined in process requirements, training, etc.).	
2. The system of quality is prevention.	
3. The performance standard is zero defects, 100% conformance to defined standard or process requirements.	
4. The measurement of quality is the price of nonconformance (direct and indirect cost) and frequency rates.	
Must have a management system for continuous improvement. Constantly improve the system for product or service delivery. Do not depend on inspections alone to achieve quality.	Same
Institute formal training and on-the-job training (OJT), coaching, and mentoring; self-improvement for process improvement, which includes quality, safety, and environmental requirements.	Same
Leadership: Create open, trusting environment.	Same
Drive out fear.	Same
Break down barriers between departments, and internal customers understand the concept of which your customer is.	Same
Understand and implement process for error cause removal; identify the multiple root causes of process failure.	Same
Employee involvement: Quality action or improvement teams.	Safety committee or quality improvement team
Eliminate slogans.	Same
Put everybody in the company to work to accomplish the transformation.	Same
Other Deming requirements include:	
End the practice of awarding business on the basis of price alone	
Eliminate management by objective	
Remove barriers that rob the hourly workers their right to pride of workmanship	
Remove barriers that rob people in management their right to pride of workmanship	

Principle 1. *Commitment and Policy*

An organization should define its process control policy to ensure commitment to the management system. Execution of management policy should be evident in internal documents and procedures, including management, supervisors, and staff training. The policies should include conformance to regulatory requirements, systematic risk assessments, and process control evaluations.

Principle 2. *Planning*

An organization should plan to fulfill its process control policy, goals, and objectives. This plan should include review of facility, tool, and equipment process readiness or applicability. It should also include the review or development of well-defined process control instruction for hazardous operations. Process management requires evaluation of assigned supervisors and staff to ensure competency to perform work or that they have available and that they use correct procedures, tools, etc. Roles and responsibilities must also be clearly defined to ensure implementation and adherence to policy and standard operating procedures (SOPs). Inherent in this stage is the need to perform risk assessments that enable prioritizing and focusing the planning process.

Principle 3. *Implementation and Operational or Risk Control*

For effective implementation, an organization should develop the capabilities and support mechanisms necessary to achieve its OH&S policy, objectives, and targets. The policy should be posted and clearly communicated to all managers, supervisors, and employees. Action plans and target dates or milestones must be established and monitored to ensure plan execution. Controls start with systems review and risk assessments at the design and planning stages and include all resources applied to the process: people, purchasing, supervision/management, tools, equipment, facilities, raw materials/parts or components, maintenance, communications, including training, and procedures. These key categories of process resource input provide the tools and create the process or work environment for producing goods and services. They are also the source of the multiple root causes of process failure creating nonconformance and forcing employees to adopt substandard behaviors that ultimately lead to on-the-job injuries, process failures, and rework. Increasingly, operational management must evaluate the capability of contractors or subcontractors to deliver needed products or services. Also, managers will need to develop emergency response and contingency plans to eliminate or reduce the potential for losses.

Principle 4. *Measurement and Evaluation*

An organization should measure, monitor, and evaluate its process performance and take preventative and corrective action. This approach must include identifying the multiple root causes of nonconformance accidents or process failure, determining appropriate process control and corrective actions, and sharing the lessons learned with other parts of the facility or organization. This practice could include behavioral job observations, audits, inspections, and safety and industrial hygiene surveys or systems reviews, including job–task analysis or systems safety reviews. Behavioral performance management or safety observations can be a critical element of this process, providing proactive measurements of systems performance vs. reactive measurements such as those from accident reports or employee–customer complaints.

Benchmarking, measurement, and evaluation are essential for program success. Benchmarking compares performance against target industries, companies, and processes internal or external to the company. It is the standard against which managers measure their operations performance. Benchmarking is always important, and there are various ways to approach this aspect. One way is to define your own meaningful

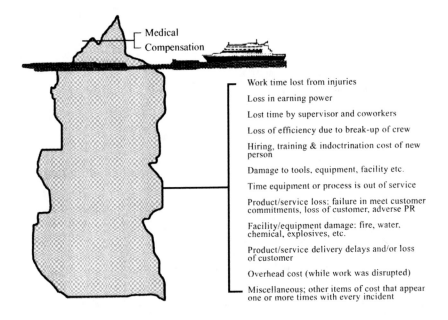

Work time lost from injuries

Loss in earning power

Lost time by supervisor and coworkers

Loss of efficiency due to break-up of crew

Hiring, training & indoctrination cost of new person

Damage to tools, equipment, facility etc.

Time equipment or process is out of service

Product/service loss: failure in meet customer commitments, loss of customer, adverse PR

Facility/equipment damage: fire, water, chemical, explosives, etc.

Product/service delivery delays and/or loss of customer

Overhead cost (while work was disrupted)

Miscellaneous; other items of cost that appear one or more times with every incident

Figure 44.2   The hidden cost of accidents.

Example: At 10% Profit Margin

$15,000 loss (direct cost)

4:1 ratio of indirect to direct costs, causing $60,000 indirect costs

A facility will need $75,000 in additional sales to make up this loss!

Figure 44.3   Impact to profits.

benchmark, for example, benchmark against yourself, other facilities, or your industry. However, in establishing benchmarks, it is important that the costs of nonconformance or accidents be much more accurately documented, including direct and indirect costs associated with losses or rework (Figures 44.2 and 44.3). Indirect cost must be accounted for, as they will almost certainly exceed direct cost and are often not recognized.

Principle 5. *Management Review and Improvement*

An organization should regularly review and improve its process management system, with the objective of continual process improvement, progressive risk reduction, and compliance with relevant standards or guidelines. If the system is not being fully executed or is not generating expected results, then additional management attention and changes in process direction or design will be necessary to improve the system, including the establishment of new or revised goals and objectives.

Elements of ISO 9000/14000 are consistent with Deming's 14 Principles Management Program: Responsibility and Control (management commitment, including resources):

- Quality system, policy, objectives
- Documentation and document control
- Operations control
- Design control and process review
- Purchasing
- Maintenance
- Material handling, storage, and packaging, including waste minimization
- Emergency preparedness
- Monitoring
- Nonconformance evaluation and corrective actions (includes rigorous follow-up)
- Audits (monitoring/measurement)

ISO 9001/14001 HARMONIZED MANAGEMENT SYSTEMS

ISO harmonized systems provide for continuous improvement and are sufficiently flexible to include safety.

- They are applicable to companies of different sizes.
- They are appropriate for the hazards and risks specific to a given organization.
- They are compatible with company or local culture, including integration (harmonization) of existing environmental or quality management system (QMS).

The ISO was established in 1947 to create international standards for the manufacturing of products that would be used by customers in different countries. International managers realized that the cost and duplication of effort by not having standardized methods could be significant. In the late 1980s ISO decided that this same principle applied equally to the management of quality and environmental systems.

Effective QHSE or QMSs have four stakeholders with respective needs or corporate goals for improvement. These typically include:

- Shareholder: profitable growth
- Society: benefit — value, positive economic and environmental impact
- Customer: higher product/service quality at a lower cost
- Employee: contribution, dignity, and quality of life

Because ISO 9000 and ISO 14000 draw from the same quality principles taught by Deming and Juran, it is relatively easy to integrate them. These systems make it much easier for multi-site and, more important, multinational companies to execute effective QHSE and QMSs consistently. Companies cannot afford the redundancy and its associated costs caused by multiple and fragmented systems around the globe. In addition, most corporate councils will advise their boards that inconsistent implementation of policy will create potential liability for them in the event of a serious accident.

It can be reasonably expected that more companies will integrate their QHSE programs as they become more familiar with their ISO management systems and see the opportunity to include workplace safety more efficiently as an integral element of their system. Internationally, regulators and corporations are regarding the American (U.S. OSHA) prescriptive approach to standard enforcement, although necessary in some cases, as not as effective as performance improvement based on a management system that allows managers to better understand and control process risks.

One of the primary difficulties created by the fragmentation of these process management systems is competition for scarce resources and management or involvement in the systems. Process control fragmentation ultimately creates three different teams (safety, quality, environmental) evaluating critical work processes. Each of these separate process control teams requires the time commitments of supervisors and work centers for process improvement and support of systems for QHSE risk reduction. This requirement results in poor utilization and ineffective or duplicated efforts of facility resources. Many times this inefficiency frustrates supervisors who feel the pressure of budget and schedule demands, and it results in passive support or in some cases malicious compliance for processes they don't believe directly support this focus.

Most organizations need to address these questions as part of their quality or QHSE planning efforts. What do the supervisor and manager need? How can we make QHSE a seamless component of process management and quality control? What can employees do to reduce rework, waste, duplication of material or package handling, package damage or loss, and injuries? What tools are needed? What are the life cycle risks associated with a product, including the cost of postuse disposal?

Multinational corporations are demanding consistent global facility management, and at the same time they expect their organization to improve its competitive position in the marketplace. Thoughtfully designed systems or process management, including the integration of QHSE, can help companies realize this goal.

It can be reasonably expected that more companies will integrate their QHSE programs as they become more familiar with their ISO management systems and see the opportunity to include workplace safety more efficiently as an integral element of the system. The big three automobile manufacturers' QS 9000 QMS based on ISO 9000 illustrates how seriously some large companies are taking ISO management systems. This standard is required for all vendors providing parts and services to these manufacturers. In addition Ford and GM both announced in December 1999 that vendors would be required to comply with ISO 14001.

The emphasis of ISO-harmonized process management systems is an attempt to develop controls that do not rely excessively on the behavior of people. More emphasis is placed on process management or systems safety, ensuring that the work environment and all of its critical elements outlined in Figure 44.1 are designed and executed in a way that ensures process conformance. In addition, an underlying expectation is that, in the event of nonconformance or process failure, managers will identify the multiple root causes of the failure or accident before developing corrective and preventive actions.

Managers analyze and evaluate the supervisor's process from their biased expertise. The ultimate result is that the supervisor becomes frustrated and returns to focusing on more immediate and certain pressures such as budget and schedule. Process control fragmentation caused by disjointed QHSE program implementation will tend to cause supervisor overload and weak commitment to QHSE. This tendency is especially true when the supervisors have not received adequate training. Training of all management and employee levels must be supported and should include technical, system, and managerial training. The supervisor is a very important element of process control and is in effect the gatekeeper for process control management. The supervisor is both the internal and external provider of customer service, which includes employees.

Table 44.2 outlines the common elements of most QMSs, including those of Deming, Juran, Crosby, and — more recently — TQM. Table 44.2 also compares some of the major elements of most QMSs, with similar requirements in any well-developed health and safety management system.

As you can see the requirements outlined here are very similar to what is required in ISO management systems and TQM. The lesson to be learned is that by integrating their quality and QHSE systems, management will significantly leverage the effectiveness of its efforts and resource utilization.

CONCLUSION

Understanding process management and systems safety is critical to identifying process or business risk. I will once again emphasize that systems safety is a process for understanding work processes and that it establishes a system for identifying the risks associated with that process. Please note that this emphasis does not mean that systems safety is strictly a safety tool; it is a management tool, and does not and should not be considered a tool only for the corporate safety engineer or safety consultant.

Operations managers must manage the process and rely on the QHSE managers or technical consultants; systems safety can be an invaluable tool for helping them understand their processes or system risks. Experience performing only OSHA-type regulatory audits will not be sufficient to qualify an auditor to perform competent management system audits. In addition, the use of audits is critical to understanding the level of systems, process, or program conformance. To be successful, however, an auditor must understand process management, how to ask the right questions, and how or what to appropriately review as documentation. Your best auditors will have good technical and managerial skills.

From a more simplistic perspective, think of the efficiency and increased cooperation that can be realized among supervisors by discontinuing the fragmentation of process management into quality, environmental, and safety. After 25 years I have never investigated or reviewed an investigation of a serious process failure, injury, or fatality that did not involve multiple root causes in the management system. The investigations also demonstrated such an incident result at times generated environmental liability. The direct and indirect costs of these related impacts are many times not well documented and therefore do not attract the management attention they deserve.

An effective health and safety program is good management and should be viewed as part of the employer's overall quality assurance program. Real productivity is the balance between striving for excellence and protecting human resources. The chairman of the board, president, and general manager of the company or facility must give priority to environmental health and safety control programs. Managers must take this responsibility as seriously as they do meeting schedules, producing superior products, or controlling corporate losses.

Integrated Process Quality and Health, Safety and Environmental (HSE) Program Benefits

- Reduces costs which increases profit margins
- Increases competitiveness
- Facilitates injured employees to return to work
- Reduces incident frequency and severity rates or lost time
- Reduces damage to equipment, inventory or product loss, and generation of hazardous waste
- Increases companies' regulatory compliance
- Integrates process quality and safety
- Improves employee and public relations

One of the key considerations is that the goals of an OH&S, QHSE, or QMS are all the same. All three are trying to establish a system of performance standards, and first time/every time conformance to those standards, thus ensuring customer and stakeholder satisfaction, maximization of profits, and minimal process risk. Organizations with effective integrated QHSE management systems should be better able to control process risk and ensure customer satisfaction. Decision making will be improved, and the level of management participation in identifying and controlling process and production risks significantly enhanced.

Integrated QHSE management systems offer organizations the opportunity to improve, among other things, business effectiveness, as well as QHSE performance. The process of implementing a system, however, will require significant management commitment and organizational effort.

To advance, management systems must require that companies' establish or document processes to review/investigate process failure, rework, and incident/accidents. These review procedures should include process nonconformance or incident multiple root cause analysis, systems readiness, cost of nonconformance, and system for ensuring that appropriate corrective actions are implemented and followed up, including sharing of lessons learned with stakeholders. In addition, effective performance measurement tools and management systems are necessary.

The bottom line is that the smart process or quality management system cannot treat safety as a separate issue. Therefore, effective programs will integrate quality and QHSE. One of the objectives of this chapter, and of my book, is too show managers that all of these processes are under their control, and that there is tremendous benefit to integrating the management process for controlling these risks. Using ISO 14001/9001, I hope to convince you that a common process management system QHSE integrated and built on ISO can be very powerful.

FURTHER READING

Nolde, M. T. *Organizational Mastery with Integrated Management Systems: Controlling the Dragon.* New York: John Wiley & Sons, 2000.

International Standard Organization. ISO 9001: Quality Systems — Model for Assurance in Design/Development, Production, Installation and Servicing.

International Standard Organization. ISO 9002: Quality Systems — Model for Assurance in Production and Installation.

International Standard Organization. ISO 9003: Quality Systems — Model for Assurance Final Inspection and Test.

International Standard Organization. ISO 9004: Quality Management and Quality System Elements — Guidelines.

International Standards Organization. ISO 14001: Environmental Management Systems Specifications with Guidance for Use, 1996. Available from American National Standards Institute, New York.

Baum, D. Client/server development tools for Windows, *Computerworld*, April, 73–75, 1993.

Hansen, M. D. Catching COLD is not just a workplace illness anymore, instead it is the hottest new development in long-term document storage and retrieval. *Plant Services*, 16(5), 47–50, 1995.

Hansen, M. D. ISO 9000 and safety. *Plant Services.* 16(3), 70–76, 1995.

Hansen, M. D. ISO 9000: The effect on the global safety community. *Professional Safety, The Journal of the American Society of Safety Engineers*, June, 44–47, 1994.

Hansen, M. D. *ISO 9000: the effect on the global safety community,* presented at 33rd Annual Professional Development Conference & Exposition of the American Society of Safety Engineers, June 1994.

Hansen, M. D. Environmental safety & health compliance in the 1990s. *Professional Safety, The Journal of the American Society of Safety Engineers,* February, 29–32, 1994.

O'Lone, E. J., Datapro, document imaging systems, management issues, *Client/server Computing,* June, 1–8, 1993.

Books

Hansen, M. D. Environmental safety and health in the 1990s, using computer technology to manage the paper chase. In: *Essentials of Safety & Health Management*, CRC Press, Boca Raton, FL, R. Lack, Ed. 1996.

Kozak, R. J. Krafaisin. Safety Management and ISO 9000/QS-9000: A Guide to Alignment and Integration, New York: Quality Resources, 1997.

Sheldon, C. ISO 14001 and Beyond: Environmental Management Systems in the Real World. London, U.K.: Greenleaf Publishing, 1997.

Waite, D. A., Heartz, W. T., and McCormack, W. D. Integrated Performance Assurance: How to Combine Your Quality, Environmental, and Health and Safety Management Systems, New York: Quality Resources, 1998.

Weinstein, M. B. Total Quality Management and Auditing. Boca Raton, FL: Lewis Publishers, 1997.

Zwetsloot, G. Joint Management of Working Conditions, Environment and Quality. Amsterdam, The Netherlands: Dutch Institute for the Working Environment NIA, 1994.

## CHAPTER 45

Beyond Teamwork: How the Safety and Health Practitioner Can Foster the New Workplace Community

Carolyn R. Shaffer

CONTENTS

OVERVIEW

When the members of a home-visit-based program at a large, urban hospital went out on the job, they did not know what they might find. Their mission was to provide emotional support and developmental training to families with severely disabled children. One hour a staff professional might be in a run-down, flatlands apartment counseling a cocaine-addicted mother with a child unable to walk. The next hour she might be visiting a luxurious, 12-room home in the hills to provide support to two usually high-performing, in-control parents, reduced to sobs by the frustration of caring for their deformed and unresponsive infant.

In most front-line health-care programs like this, burnout and staff turnover run extremely high, with some staffs turning over completely in the course of a year. The staff of this child disability program confounded their colleagues by working with an increasingly high-stress population for 11 years without a single staff member leaving. All in the program, including the secretary, loved their work. Today, staff members agree that a major ingredient in this success was the sense of community that they had developed. "We felt more like a family than a work team;" says one.

Although the staff of this health-care program did not analyze data on the effect that this sense of community had on their safety and health, other workplaces experimenting with community-generating styles of management have gathered impressive quantitative evidence. A bakery division of a major supermarket chain reversed its downward slide into unacceptable rates of absenteeism and accidents and became a model for safety and health practices. It did so by shifting to a management approach that bore several striking resemblances to the above hospital program. According to Robert H. Rosen's report on this success story in The Healthy company, employees in the bakery division of Safeway Stores in Clackamas, Oregon had felt powerless, discouraged, and unimportant under the earlier iron-fisted management style of the bakery manager. When the manager changed his style to give employees more chances to connect with one another and more say in shaping their work environment, the negative indicators began to fall. Among the dramatic results: absenteeism decreased from 8 to 0.2%, accidents plummeted from 1740 to 2 workdays lost in a year, and turnover dropped from 100% in some jobs to less than 10%. The flood of grievances and discrimination cases also turned into a trickle, decreasing from 75 to 80 a year to 1 grievance and no discrimination cases in 5 years. Besides improving the health and well-being of the employees, the Safeway bakery successes contributed significantly to a healthy bottom line for the company. Safeway calculated that, in real dollars, for every dollar invested in the new programs it generated an estimated return of $15, saving between $700,000 and $800,000 a year.

By fostering community — and the sense of commitment inherent in it — in the workplace, safety and health professionals can improve the bottom lines of their companies while saving the lives, limbs, and immune systems of employees. To do this effectively, safety and health professionals need to know what community is, how it benefits health, how it generates commitment in employees, and how it can be created in the workplace. This short chapter can only touch on some of these topics and will refer readers to other sources of information — although much of the literature on the kind of workplace community that goes beyond teamwork has yet to be written. What is well documented are the health benefits of positive social connectedness (see Further Reading, especially Shaffer and Anundsen, *Creating Community Anywhere*, chap. 2). The primary focus here is on drawing practical lessons from success stories that safety and health professionals can apply in their organizational settings.

First, a word about community, a term as amorphous and overused as love. In *Creating Community Anywhere,* my co-author and I chose to define the term strongly. Here is our short version:

> Community is the dynamic whole that emerges when a group of people commit themselves for the long term to their own, one another's, and the group's well-being.

In essence, community is about commitment. We included a long-term time element because commitment tends to manifest over time. Our long definition spells out the practices and sense of collective identity that flow from, and are an essential part of, this commitment. When people are in community, they:

- Do things together
- Depend upon one another
- Make decisions together
- Identify themselves as part of something larger than the sum of their individual relationships

The members of the bakery and the hospital communities described above engaged in these practices and developed a sense of collective identity that withstood the test of time.

The professional staff of the child disability program, which grew from 8 to 11 members over the 11 years, met weekly for lunch at a local restaurant before going into their staff meeting. They shared stories about their lives as well as the joys and frustrations of working with their patient families. In the staff meetings, the director generated an atmosphere of mutual respect and empowerment. Without giving over her power or responsibilities, she encouraged consensus around important issues. She made sure the staff felt aligned around new hires and any changes in policy before implementing them. Occasionally, she and the staff arranged for a day-long retreat. Although the work group was technically hierarchical in structure, it felt egalitarian. Members liked and respected one another even when they disagreed, and they disagreed often in passionate discussions on the way to reaching a consensus. At the same time, they supported one another in practical as well as emotional ways. They adjusted their schedules to meet each other's parenting needs and substituted for one another when necessary. When one member was diagnosed with breast cancer, the others helped her continue to work — as she desired — while she received chemotherapy treatments. The team members also encouraged each others' individual gifts to flourish. "We became wildly creative," says a former member. "We invented new forms, set up offices, and created new programs." Through all this, the program produced results. Every grant the group applied for, year after year, it received. Not only did the members deem themselves productive and a success, but the hospital and the funding agencies agreed. Their collective identity included being like family and engaging in important, effective work. Even today, several years after the program and the personnel began to change and the sense of community to fade, members of the original group stay in touch and several remain close friends.

At the supermarket bakery, when the manager loosened his control, he generated a sense of connection and commitment by encouraging employees to meet in problem-solving groups focused on such common concerns as sanitation, safety, and ergonomics. He also let workers know that he wanted suggestions from them. Once employees were convinced that management was listening and responding positively, they submitted hundreds of suggestions. Soon the employees began expressing their commitment and their newly unleashed creativity in more significant ways. They built a small employee fitness center. This led to a wellness program and eventually to a 5000-ft^2 fitness center not just for employees but for their families as well. The center began sponsoring classes and special fitness events that brought employees into even greater positive connection with one another and with the larger community.

In both workplaces, managers increased commitment and creativity by loosening control and inviting participation in the decision-making process. Both also created vehicles for employees to come together to talk about matters of importance to them — the weekly lunches, occasional retreats, problem-solving groups, and wellness classes. Just as important was the sense of positive collective identity that emerged. For the hospital staff, the work itself created a strong bonding. "We were very dedicated to our patient families and our work," says one member, "and were convinced that we were making a difference in people's lives." The bakery work became more than a job for employees when it began involving their general wellness and their families. They developed a fitness event called "Buns on the Run" at which T-shirts were handed out to regular exercisers. Activities such as this led to the development of a sense of camaraderie and belonging, which in turn led to a positive collective identity.

Creating community requires trust and vulnerability. These are difficult to develop and sustain in any setting, but the workplace presents special challenges. The hierarchical structure and competitive climate in most places of work are hardly conducive to self-revelation, honest feedback, and interdepartmental cooperation. A person could become frozen out of the information loop or fired for engaging in such behavior. It took the employees of the Safeway bakery some time to learn that it was safe, as well as effective, to make suggestions. Nonetheless, it is possible to create a community in the workplace that is both honest and committed, as the staff of the child disability program demonstrated. These women — during that 11 years, the director and staff were all women — not only shared their deep fears and joys with one another, but they also argued with and challenged one another. They had learned that the director would not arbitrarily fire, or hire, anyone, although she had the power to do so. They also came to know that no one else in the program would use the personal information a member shared against her, nor would small groups of members form factions and engage in malicious gossip or secretive plotting. If someone were having a problem with others in the group, she brought it to the group — an agreement that any collection of people intent on becoming a community should consider making.

One of the first steps toward creating the trust that leads to community in the workplace is to be open and honest about where the power lies and who wields it. If the director of the hospital program had engaged with the staff in a process that she labeled consensus, but then made decisions on her own without regard to their input, she would have undermined any germinating sense of community and plunged the group into betrayal-induced cynicism, effectively immunizing them against attempts at community building in the future.

Another key step is to encourage the members to align with a common vision or purpose — one strong enough to help them hold together and support the larger enterprise through the inevitable differences and disagreements ahead. This is where a well-facilitated group retreat can be effective. To maintain the openness and sense of mutual support that such a retreat can generate, the organization needs to create or encourage vehicles for ongoing conversations about matters of importance to the group. It also needs to clarify what role these groups play and be ready to follow up on decisions or recommendations that come from them. However, the vehicles are not enough in themselves. Participants also need to develop group agreements and collaborative attitudes and skills. These include agreeing on how and when members meet, make decisions, and deal with differences and conflict. A healthy group will make sure everyone has a chance to contribute, will appreciate the varying gifts of each member, and will welcome differences as opportunities for learning and generating creative solutions to problems. Its members will be willing to learn to communicate well, including giving and receiving negative feedback, and to work effectively with conflict. However, even the best groups will eventually grow stale or become stuck unless the members regularly take time out to renew their common purpose and agreements, looking honestly at what worked and what could be done better and celebrating their accomplishments.

Fostering community in the workplace is not easy, but it is possible. Generated skillfully, with care and honesty, workplace community can contribute significantly to the health and safety of employees — primarily by reducing stress and stress-related accidents and by boosting the immune system. It can infuse a business with vitality and creativity and lead to increased productivity, lower turnover, and healthier profits.

A final word to safety and health professionals who may think this sounds fine but cannot imagine it happening in their organization. Even if all these benefits cannot help you convince top management to support community-generating programs, you need not give up. You can still educate managers and employees about the health value of community and provide them with information on how they can create this outside the profit centers of the organization — in their neighborhoods through peer support groups, labor unions, and professional associations and even at the workplace through volunteer activities. Perhaps the best way to begin is by giving yourself the gift of community.

FURTHER READING

Books

Autry, J. *Love and Profit: The Art of Caring Leadership.* New York: Morrow, 1991.

Harrington-Mackin, D. *The Team Building Tool Kit: Tips, Tactics, and Rules for Effective Workplace Teams.* New York: AMACON/American Management Association, 1994.

Helgesen, S. *The Female Advantage: Women of Leadership.* New York: Doubleday/Currency, 1992.

Nirenberg, J. *The Living Organization: Transforming Teams into Workplace Communities.* Burr Ridge, IL: Irwin, 1993.

Peck, M. S. *The Different Drum: Community-Making and Peace.* New York: Simon & Schuster, 1987.

Rosen, R. *The Healthy Company: Eight Strategies to Develop People, Productivity, and Profits.* New York: Tarcher/Putnam, 1991.

Senge, O. M. *The Fifth Discipline: The Art and Practice of the Learning Organization.* New York: Doubleday/Currency, 1990.

Shaffer, C. R. and Anundsen, K. *Creating Community Anywhere: Finding Support and Connection in a Fragmented World.* New York: Tarcher/Putnam, 1993.

## CHAPTER 46

New Directions in
Safety Communication

Bonita B. Zahara

CONTENTS

OVERVIEW

"Plane canceled due to mechanical problems!" You can imagine the angry passengers crowded in the airport terminal complaining in unison about the heat and the inconvenience. Yet I had just come from the 1994 American Society of Safety Engineers (ASSE) conference, I could only smile and try to get the waiting grumblers to realize how lucky we were that the ground crew was so efficient that it found a serious problem and prevented an accident. "Consider the alternative!"

"What," you may ask, "can possibly be new in safety communications?" I would like to invite you to look outside your standard box of safety tools and back at some basic communication skills. No, communication skills are not new, but what is new is the attitude that we must bring to the importance of the safety messages we communicate every day. The tools of safety professionals are powerful and, for the most part, very well utilized. I would like to invite you to take a new look at communication skills and how they might influence the safety message you have to deliver.

You may recall the opening scenes of a long-running television series where the police sergeant would close his evenings roll call with the empathetic charge. "Be careful out there!" Somehow the same old message, given week after week, had to come across as new. The new messages are the same old messages when it comes to safety. Familiar phrases like "Safety Doesn't Hurt," "Safety First," and "Safety Saves Lives" are as true and important today as they ever were. The challenge to safety professionals, as well as to the general populus, is: How do we keep the simple principles of safety heard in an information society overloaded with messages and instructions?

Shifting attitudes does not seem like a big deal, but in fact our attitudes shape almost everything in our lives. Attitudes to safety consciousness are critical to the success of a safety program. While attitude awareness might not seem a new direction for safety professionals, I would like to suggest that there are lessons to be learned from a better understanding of communication skills — in particular, understanding whole-brain thinking.

There is nothing new under the sun, yet the futurists and marketing gurus all make their fortunes by forecasting the new trends. I would like to make a more humble new claim and state up-front that much of my thesis is indeed good old-fashioned common sense. They are the kinds of things safety professionals used to do as a routine part of their work. That new/old thing is to communicate the importance of life.

When I was a child, safety was an everyday part of our lives. My father was a personnel director and safety manager for an aluminum casting plant and rolling mill. While other families had backyard barbecues, we showed safety films and learned CPR. Mind you, the CPR practice came in handy during our teenage years! We knew Dad was serious, because we had a swimming pool in the backyard and he instilled in us the responsibility that was inherent in owning a pool. We all had to learn how to save lives.

My professional training is in the area of whole-brain learning skills. I became fascinated with how the brain/mind system worked and how we could increase human productivity. Peter Drucker's simple statement, "Effectiveness must be learned," was part of my battle cry as I set about to champion the underutilized infinite capabilities of the human brain. As a regional vice president for Evelyn Wood Reading Dynamics, I became fascinated with the brain's capacity to learn and remember. As computers and ergonomics overtook people's interest in old fashioned reading and we moved into the age of information accessing, a powerful and simple safety message reappeared:

> If we understand how people think and how they remember, they will better remember our safety message.

A simple understanding of the human brain is helpful for anyone interested in improving his or her communication skills. Scientist Richard Sperry won a Nobel prize for his research in split-brain development, and since that time extensive studies have repeatedly supported the idea that the capabilities of the brain are infinite. There are some simple aspects of human dynamics that, if understood, can help us to remember and help others to remember the messages we want to get through to them.

"When do accidents happen?" They happen when we are not paying attention, or when we lose concentration, or when our minds wander. As I studied the human brain, some simple exercises for stimulating the learning abilities seemed to be applicable to many professions. The confusion about left and right specialization, and which side of the brain does what, can be eliminated if we focus on being whole brained and put the emphasis on getting it all to work together.

In working with whole-brain stimulation to increase our comprehension of written information, I became aware of how simple and powerful it was to truly wake up both hemispheres of the brain. Simple visual exercises, like following one hand at a time as you draw a large infinity sign in front of you, literally stimulate the dendritic connections in the outermost reaches of the brain. The brain wakes up with this simple stimulation exercise, and by so doing memory in enhanced. In many of the workshops I conduct across the country I teach juggling as a form of whole-brain stimulation. Kevin Strupe, the teacher and educator who introduced me to this simple brain exercise, has had enormous success working with students with learning disorders. By stimulating and integrating both hemispheres of the brain, reading levels and test scores have jumped enormously.

If we can learn, as individuals, how to increase our brain's effectiveness, as Peter Drucker has challenged, then our effectiveness as safety professionals will also be improved. And if we can begin taking that message home to our families and sharing not only how to wake up, but also how important some of our lessons are, then we can begin to make a strong and powerful difference in the world. It is a sad truth that seven times more accidents happen at home than in the workplace. It is important that the profession of safety begins to look beyond the workplace for ways to contribute. Learn communication skills, for they will help you not only at work, but also at home. Then take the big step into the third realm, out into your community and the world at large.

I suggest the following new directions to safety professionals for applying whole-brain learning theory.

- First, Understand how *you* think and remember.
- Second, Take it home. Take your safety knowledge and communication skills, along with the basic knowledge of safe practices, home with you and spread the message.
- Third, Spread the word to your communities.

In my work across the country I often ask individuals to share what makes them proud. In about 90% of the cases family is part of the answer that is given. Family truly is at the heart of safety. As you learn more about yourself, you can share that with family and also at work. Work is, in fact, a form of family. Bette Friedan has written an excellent book, *The Fountain of Age*. There are two major points in this book: first, that we must stop treating age as a disease and second, that we are evolving as a culture that will create families of choice. Work is a sort of family, and we need to learn more about communication skills to share not only with those with whom we live, but also with those with whom we spend 60% of our waking time, our fellow workers.

Stephen Covey's *The Seven Habits of Highly Successful People* has become a classic in a very short period of time. Covey's seventh habit is to "sharpen the saw" — to learn; to study. My new direction for safety professionals is to learn as much as you can about the human brain, about how people think and how we can aid memory, and then use that knowledge. Use it on yourself, share it with your family, and then take it out into the world. The message of safety is sorely needed in our world today. Safety professionals for too long have been the quiet, gentle spirits with big hearts and a bad tag of safety cops. I challenge you as safety professionals to become guardians of the future. Come out loud and strong with your message of safety and hope and humanity. While you will forever continue to make the workplace safer, I beg you, for our children's sake, to also use your skills and knowledge and make the world safer. Let us be sure that safety does not become a message of fear, but instead a message of hope. "Make it safe out there!" You are the guardians of the future!

FURTHER READING

Buzan, T. *Use Both Sides of Your Brain.* New York: E.P. Dutton, 1983.
Covey, S. *The Seven Habits of Highly Effective People.* New York: Simon & Schuster, 1989.
Friedan, B. *The Fountain of Age.* New York: Simon & Schuster, 1993.
Joyce, M. *Ergonomics: Humanizing the Automated Office.* Carlsbad, CA: Southwestern Press, 1989.
Robbins, A. *Unlimited Power.* New York: Fawcett Columbine, 1986.

PART X

Training Aspects

CHAPTER 47

Managing Workplace Safety and Health Training

Michael-Laurie Bishow

CONTENTS

1-56670-370-0/02/$0.00+$1.50
© 2002 by CRC Press LLC

INTRODUCTION

Today, risk exposure, liability, new technologies, media exposure, regulation, and mandates create challenges even for organized and dedicated safety training officers.

The future is limited for courses tracked by sign-in; repeating refresher/awareness courses are also endangered. Hazardous waste, chemical spills, bloodborne pathogens, environmental management, and public awareness create a volatile yet exciting training opportunity. This chapter helps you to organize your thinking with practical tips and a plan that anticipates your workforce needs. *Anticipation* of changing demands and organization objectives marks the difference between a stellar training program and a pedestrian one.

Traditional Program Planning Model vs.	New Program Planning Model
Employee request	Organization mission/objectives
Feel good participant exit	Gained a skill before it was needed
Individual registration	Team registration
Off-the-shelf training plans	Customized training plans
Individual gains	Team gains, competitive advantage
Low accountability, short term	High accountability, long term
Reactive training (request-demand)	Proactive training (anticipate need)
Accountable to supervisor, manager	Accountable to the community/and the stockholders
Market: employees, individuals	Market-business/community partners
Risk and incidents drive safety training	Organization development drives training
	Training reduces risk

FASTER ORGANIZATIONS REQUIRE NETWORKING

Top-down organization once meant that training topics were selected by executive order. Often the safety training topics were based on existing mandates and laws that were known to be enforced by government. At other times, a work accident created recognition of an acute need for antidote training — after the fact. Prevention was often ineffective because of these firefighting responses and accompanying funding constraints on the safety training program.

Today, our organizations are more lean and mean; our executives are held to a higher accountability with added regulatory enforcement. Safety and health training programs are now expected to respond to a mind-boggling array of factors related to employees' safety practices. Ultimately we seek to reduce accident rates, workers' compensation costs, and health care costs, as well as comply with regulations, laws, and mandates. Today, regulatory requirements are more specific and may even define training content or length (at least 200 regulations require training at this writing). Psychosocial factors may create more demands on safety programs, such as when communication breakdowns lead to accidents; thus safety communication may become part of your program responsibility. Similarly, low worker literacy rates may lead to safety literacy when warning signage procedures use a specific vocabulary.

The increasing use of teams in organizations may require critical thinking skills for work groups. Downsizing and hiring temporary workers further burden the safety officer. How can we deal with these expanded responsibilities? And where can we draw the line with our limited resources? To develop answers, network with managers and safety officers!

The Safety Committee

The best safety officer cannot stay on top of it alone — so the safety committee is now a survival tool (rather than a pain in the neck). Be careful to select committee members who approve resource allocations and also to select employee representatives. Your safety committee members are an internal source for expertise on training, marketing, and forecasting. Members can become your advocates to integrate safety training early in the process of technical change.

For written references, consult texts such as the National Safety Council Occupational Safety and Health Series, which includes the volumes entitled *Administration and Programs, Engineering and Technologies, Fundamentals of Industrial Hygiene, and Environmental Management*. Anticipate exposure by attending to media coverage of incidents in your industry around the world. Your organization will immediately turn to safety when an incident such as Chernobyl occurs. Workers may suddenly realize they are at risk when an accident occurs half a globe away. I hope you have anticipated this training need so employees who watch the news will say, "Yeah, we just had that safety training last week!" Now that is a stellar program.

SAFETY TRAINING ALIGNED FOR IMPACT

Our accountability is higher than ever, and we are accountable to a wider constituency. Building program credibility depends on our attention to the concerns of all our stakeholders: relevance to our employees, cost-effectiveness for our managers, short/long term learning for our participants, and consistency with the mission and objectives of the organization.

Grassroots Relevance

No matter how many mandates you meet, your common personal appeal must be maintained. If you doubt this, ask a safety officer how he or she would be received if an accident occurred and he or she said, "We don't have that training because there isn't a state mandate." By listening and responding to the complaints and scuttlebutt of the people doing the work, you can establish credibility for your program. To obtain this informal information, you may want to leave your office for a safety walkabout. For the grassroots of our organizations, anticipation means valuing the wisdom of people who may not give you a sophisticated technical description of risks. Your best recommendations are conveyed by word of mouth. Even if the training is mandated, people vote with their feet. High no-show rates or early departures can indicate a need to redesign your training.

Cost-Effectiveness for Managers

Resistant managers often say they cannot afford to let people go to training on work time. Your first strategy is to develop policies that take managers off the hook by executive order; that is, employee hours are set aside for mandated safety training. The safety committee sets annual training priorities and helpful policies, such as negotiating that training occur on paid time in union–management agreements.

If executive edict is not available, try documenting the cost of a related accident and compare that to the cost of prevention. Be accurate about the cost of training; include the employee time, training materials, trainer costs, and program development. Divide by the number of people trained to get a per-person cost. For example (without indirect costs):

Class: 3.5 hours Bloodborne pathogens	Class Size: 20 to 25 (2 sessions per day)	Materials: Booklet/certificate... $8.00 per person	Contract Trainer: $1500 per day (2 classes)
Administrative Costs: Registrar: $75. Snack Setups: $80 (AM/PM)	Completion Rate: (85%) 34 completed	Total Materials: (40 sets) = $ 320	Cost per Person: (34 complete) $58 per participant (add indirect costs)

You can sometimes reduce costs by using in-house trainers and writing your own courses. Be careful to calculate development costs; it may cost more than you anticipate and you may not meet deadlines. A safer approach is to open your training to outside participants from other facilities or companies. This can be a win/win collaboration for small companies or departments. Some internal training functions have a department charge-back system with preapproved annual safety training budgets. This may help to stabilize your budget. The safety committee can advocate on your behalf to the executive and other companies as long as you have internal credibility.

Usually the cost per participant is reasonable; our public sector institute charges about $65.00 per half-day training per participant. This includes some costs (development, videos, equipment) that are recovered by repeating programs and updating older topics.

Learning for Participants

Changes in behavior after the training session are the reason for taking the time. If your program was short and intended to be a wake-up call or to raise awareness, your constituency might call it a waste. Unless the event is an after-dinner speaking engagement or a stockholders' luncheon, you might want to avoid awareness objectives without behavioral outcomes.

Sometimes the pressure is on to schedule 1- or 2-hour sessions. If you cannot avoid this, ask yourself, "what is do-able within the time frame?" Give your managers a choice of several objectives or a few critical behaviors to practice; restrict classes to ten participants. Your calculations showing how many months it will take to train 400 employees may motivate managers to approve a longer session. If not, cut the exposition from your course or cut the history; use one or two theoretical concepts/scientific principles with practices and demonstrations that address the dramatic consequences of ignoring just one principle. For example, oil is lighter than water (extinguish a grease fire chemically). In short, only deliver training that results in increasing the probability of safe behavior. If you cannot achieve an impact, avoid doing ineffective training by citing regulations on course length, researching courses in similar companies, or just saying no.

Participants in safety training should always complete a written evaluation. Even if you thought the class was a dud, ask for suggestions about content, delivery, materials, exercises, and practices. Be sure the trainer gets a copy and reads the evaluations every time. This is particularly important for in-house trainers you want to develop into top-notch presenters. Be sure to provide pretraining practice sessions and coaching for your trainers to reduce stress and ensure training quality.

Promoting Organizational Objectives

Politics! Well, that's real! From a political standpoint, the safety officer should make the organization look good in a crisis. Also, safety program planning should anticipate needs generated by new organizational objectives. Let us say your company plans to change from aluminum alloy castings to extruded plastics. Current training reduces the probability that an employee will pour a 50-pound bag of zirconium into an open-air vat.*

This training will be obsolete in 6 months; consequently, it is time to research airborne plastic gas ignition. If the new extrusion process is still a secret, schedule multiple sessions of a generic training program, e.g., "Preventing Production Pitfalls," to reserve the space and the employee time. Prepare impressive marketing for your new training and schedule the publicity release for the same day the new extrusion process is to be announced. With luck, an executive will be announcing your new training. Above all, do not wait until the secret is out to design the new training. Be on the cutting edge!

In addition to new technologies, new plant locations, hiring surges, reengineering, or management turnover can create an opening for safety and health training topics. If possible, have the safety committee stay on top of trends and provide a heads up service for the executive. Of course, always bring a training solution with the heads up problem.

ORGANIZING THE TRAINING PROCESS

How far in the future are your rooms and training scheduled? Two months? Six months? Let us talk about the process of managing a stellar program: needs assessment,

---

* Zirconium is a metal that explodes and burns, as it does in flashbulbs. One manufacturing facility provided training for employees in the handling of this metal demonstrating that exposure to air and friction tend to ignite the material. A night shift employee who did not follow training precautions dumped a bag of zirconium alloy into the top of an open vat. He was seriously burned and the flash was seen all over the plant. This incident led to a rush of safety reeducation sessions for all shifts.

program funding and flow, designing and conducting courses, program evaluation, and documentation. **Caution:** This is a parallel process, not a linear process. Expect to do multitasking.

Needs Assessment

Who do you talk to? Who has the need? How do you weight the input from various sources? A common pitfall is to rely only on participants' input. If you do that, you may not prepare employees for new technologies, you may repeat popular programs too often, and you may use up allotted training hours (or manager goodwill).

Participants like to recommend entertaining programs to their friends. Of course, your programs should keep people awake, but you take the hit if people use risky practices because you did not prepare them in time. I recommend that participants' input influence 80% of the course structure and delivery but only account for 20% of future program planning.

Another pitfall is to ask only managers what their people need. If you do this, managers will often think practice is too fluffy and will favor cognitive classes with tests to document that the participants got something out of it. The safety officer may become a kind of enforcer, an extension of the strong arm of the law. You can serve the organization by educating the managers; for example, document the reduction in downtime if participants use specific safety behaviors. Managers also relate to reducing hassle factors such as endless report writing and defensive meetings. I would weight manager influence on program planning at 30%. Gather manager input by hosting an annual luncheon with a training needs assessment process (high budget) or send a survey to all managers (low budget).

A final pitfall is to rely too heavily on the safety committee for forecasts about the future of your industry and the organization. Most committee members are not high enough in the organization to forecast accurately. I would weight the safety committee input at about 30%. The safety officer should meet regularly with a visionary in a power position (preferably the CEO/CAO) and follow the technology trends carefully.

Another training needs source is a conscientious objector in the ranks. This person notices internal problems and has high ethical expectations of your company. He or she often seeks options from people in other divisions/companies to suggest specific solutions using design and process changes. I would look for this person in research and development or sales/marketing and listen carefully. Often objectors do not have enough power to correct the problem and will see the safety officer as an ally seeking to improve safety practices. Vendors will often alert you to the safety hazards in the competitor's equipment, but do your own research after that kind of news flash. I would weight your information from objectors and vendors at 20% of future training plans.

Program Objectives: Funding and Cash Flow

After multisource input and your own research, forecast the top priorities for the next 2 years and sort them into 6-month reaches. The near reach lists your current programs; the midcourse reach finalizes which programs in development will go to pilot stage/early trials; and the far reach forecasts needs to the executives to inform them, to gain approval, and to access resources for training. Anticipate new organizational training demands resulting from new regulations, production changes, or accidents in your industry.

The 2-year plan may be too far-reaching for some organizations, but in our training institute we may need 2 years to get funding to certify our trainers in national training programs. We then have the course ready to go for our next funding cycle (which is annual). We project a budget and plan cuts in case we lose 10% or even 20%. This is depressing, but it is good discipline to clarify our core program and our bottom lines. If possible, we devise methods to generate revenue or cut costs by sharing resources with other companies or departments.

In your annual presentation for the budget, connect activities with funding for each reach, such as "350 employees will successfully complete the BBP training in the next 6 months at a per-person cost of $58." Document training needs by quoting your managers' planning luncheon, safety committee recommendations, employee survey results, and new government regulations. With high engagement, people who decide on the budget would have been consulted already and would be well informed. Your chances of approval are much better with key people on-board before your annual presentation.

If you have to reduce your budget, ask vendors if they will cut prices for volume; then form organizational partnerships to raise your participant numbers. Use out-of-state safety conference presenters while they are in town to reduce travel expenses. Locate college internship programs (human resources, educational technologies, adult education, and business) that offer team teaching and course design internships in safety and health training. By the way, students are older and more experienced these days; the average age at some state universities is 28, and mature students often have full-time work experience.

Program Delivery

The size of your program has great bearing on how formally organized you should be. Let us discuss a program for a medium-sized organization consisting of between 1000 and 3000 employees. This is your trainee population; your total annual attendance will be about 2000 participants in 100 class sessions per year. You personally deliver two topics twice each month, in half-day segments, and handle the administration. You share two trainers with Human Resources (HR); they deliver most of the safety and health training. A university intern helps with course research and design. Three managers on the safety committee deliver technical safety training topics relevant to their professions. Fortunately, your staff is competent; all trainers earn evaluations above 80% in their courses. Your training clerk is skilled with desktop publishing, registration software, and spreadsheet programs for budgets.

Your primary instructional unit is the 3.5-hour segment with typical courses lasting one half-day or one 7-hour day. Course designs include a balance of 30% cognitive input (lecture/discussion, reading), 30% individual work, and 40% practice and demonstration. Participants are expected to demonstrate competency during training practice to earn a certificate. Often pre- and post-tests are used both to assess the difficulty level of the class and to document retention of technical information. All participants complete an anonymous course evaluation at the end of the last session, and 85% receive a certificate of completion. When necessary, passing test scores are verified for licenses. All current courses are certified for continuing education units (CEUs) through the state provider number of the HR department.

Course content is considered the property of the safety/HR function if it is developed there. Other courses belong to the vendor/contractor. Time needed to develop courses yourself may be estimated with these rules of thumb:

- New and technical topics/ground-up design: 4 hours per class hour.
- Intern prepared/desktop publishing booklets: 3 hours per class hour.
- Well-known subject/updating existing course: 2 hours per class hour.
- Off-the-shelf/contract trainer: 1 hour per class session (setup, briefing).

The in-house process of course design is usually done by an individual trainer until materials are in final draft form; ideally, all trainers meet to review the content and walk through the training design. These practice sessions are festive and comic as well as a beneficial training for trainers. The final design is approved by the safety committee and any certification review groups. After designing course materials, there will also be duplication costs by the page, plus binding. The service center requires 7 working days from pickup to delivery. Check costs for commercial copy work with faster turnaround in case you have emergency changes. If possible, negotiate a contracted lower rate.

Program Evaluation and Documentation

The same targeted groups contacted in the needs assessment process should also have a voice in the evaluation. Participants who completed course evaluations contribute to improved course delivery, especially if they add to lively class discussions about unmet training needs. Since you surveyed the managers, you can determine if they see an adequate training impact on the job. Also, managers and executives may be asked a hypothetical question to help the safety officer select annual priorities. Here is a sample format for a manager survey:

Dear Manager: You have a $10,000.00 training allocation for next year. We offer some "hot topics" on which to spend your allocation, but add any others you think relate to our current or future safety and health training needs. Be sure to spend all your allocation! Thanks for your Safety Program planning advice. [List current and projected classes; use "other" blanks.]

Managers' feedback seems more focused when they are asked to distribute a sum, thus choosing between favorite training topics, needs, and government requirements. Return a summary to all managers for their information. PC-based survey software is available to help format, send electronically, and tabulate results with a minimum delay.

Reporting to Executives

Organization level accountability might be easiest to document by using measures of success such as:

- Number of participants who completed courses within the time frame.
- Percentage of planned courses that were conducted (cancellation rate).
- Cost of courses (or per participant) within projected budget.
- Percentage of participants who passed competency or were licensed.
- Average of participant evaluations (or percentage at excellent/satisfactory).

Automated registration is necessary at some point to avoid long delays and to track your success measures. Whatever software program you choose should generate rosters, keep ongoing class lists, produce individual transcripts, and print 5-year histories for department training. The registration program should also easily merge class lists with a confirmation letter, a certificate of completion, placard, or other personalized mailings.

If you need to justify the cost of registration software, calculate the labor/materials needed to produce reports, rosters, certificates, and budgets. True, if you purchase software, the initial data entry is tedious. However, you will probably realize that it is a blessing when you first print a list of all department employees who have completed a mandated course in the last 3 years. Now you can convert your file space to archiving the history of course content, handout masters, and evaluations. Individual records are now in the automated registrar, which you back up every night. The training clerk can run the course signup process paper-free, straight from the phone to the screen roster or online.

SELECTING YOUR TRAINING TEAM AND TECHNOLOGY

For some extroverts, training is a perpetual joy. If you are careful to pick people who enjoy groups, they will train for the fun of it; however, many technically qualified people are detail oriented and not very delighted to be in front of a group. Course development might be the best contribution for the high-detail trainer; even then, during new course development, high-detail trainers will grieve for the loss of every fact and figure. The safety officer must determine the level of detail which participants can comprehend in the allotted time frame.

To maintain a stable training staff, your task is to provide rewards that each trainer will appreciate. Trainer gatherings to talk about training problems over free food seem to be popular. Selecting a balanced trainer team will help blend different training styles into an excellent program. Rotating or trading topics can be appreciated if trainers become bored; trading may raise sagging evaluations caused by low trainer energy. Forming a cadre of managers who like to train, but have limited time, provides a motivated and fresh resource for your organization.

Wherever you get the talent, be sure to represent multiple ethnic backgrounds, ages, levels of physical ability, and genders. Diversity enhances participant learning, conveys fairness, and adds to the safety officer's knowledge of human factors in the organization.

Recruiting Internal Trainers

Recognize there is a trainer mind-set. Recruit people with a problem-solving focus, people who resist the temptation to blame people. Great trainers think well on their feet, especially under group pressure or resistance. Look for optimism and patience with human attitudes and fears. Be sure to avoid anyone who blames or credits by reference group (i.e., "you women know all about that."). Thinking in stereotypes is not compatible with excellent training.

A spontaneous and even slapstick sense of humor helps a trainer smooth the rough spots. I have observed that most trainers have low resistance to change and high tolerance for ambiguity. I saw this clearly when I attended a national workshop for trainers. The learning groups quickly changed the exercises, rules, roles, and seating. They even provided some ambiguity of their own by adding a bouncing beach ball into the mix: this added uncertainty, whimsy, and more performance pressure — for fun! Trainers often enjoy the class process so much that they almost become a self-generating energy system. These trainers do not burn out.

Selecting Contract Trainers

Selecting contract trainers adds a cost factor into your decision making. A quality trainer who has an excellent program is willing to design to your participants' needs. Wherever you work, there is a going rate for contract trainers. If you are unsure about contractors or compensation, ask your local American Society for Training and Development (ASTD) chapter for a recommendation. Your trainer candidates should have respectable resumes with professional memberships related to their specialty and to adult learning. Be sure to ask for recommendations and call previous clients on the phone for a more candid report. Also ask for course materials and evaluations (you may have to look at these during an interview due to copyright concerns).

If a candidate seems acceptable, have the candidate do a 10-minute audition for the safety committee or your training cadre, using an evaluation form to give you written feedback from your staff. Write the first contract for a short term, perhaps two to four class sessions with the possibility of an extension, even if you must pay a bit more for short term. That way you can stop a disaster if you have misjudged. In any event, be there at the first class to introduce the trainer, to watch at least part of the training, and to close the session with evaluations. Stay to debrief the trainer: provide more good news than corrections whenever possible.

High-Technology Training

People-free learning, as some trainers say, reduces the humanity of learning. The pace of change has encouraged us to consider rapid and easily changed methods.

We know humans, both adults and children, need relevance and application to understand. Online learning presents an opportunity to increase speed, update easily, and reach large numbers at many locations without a travel expense. The safety officer needs to consider the CD-ROM and the modem as training tools. As with any other training, we match the objectives to the tools: information retention is one of the objectives that fits electronic learning.

Some trainers are concerned about the lack of human contact and group reinforcement that is lost without classroom interaction. Consider the dilemma of airtime in a big class. Participants may drift off, rather than lift off. Some entertain themselves with interpersonal hassling, while you try to hear everyone. The picture is not 100% pure learning even in the best situation. Another problem is the schedule; classes occur as if all participants learn best between 10:00 a.m. and 3:00 p.m. We know people pick very diverse learning times if left to their own resources. There are advantages to distance learning.

Consider the Ford Motor Company mechanics courses. Mechanics study individually with CD-ROM to retain basic facts about the engine and its systems. After passing the online tests, mechanics earn access to hands-on classes. The Ford method has advantages since the mechanics have already established their basic knowledge. Classes are more efficient with motivated participants who value access to face-to-face instruction. I speculate the instructors are also happy with the pace and quality of the group.

Combining technologies is more and more common, and safety officers can save time and money on some knowledge-based training. Methods of individualized learning save resources for topics requiring experience and practice such as first aid, CPR, teamwork, public speaking, and media relations. Safety officers should become the resident expert on available CD-ROM, satellite, online, and distance learning safety programs.

SUMMARY

Developing the safety and health program organizes the effort to protect employees from risk. In the faster network organization, employees may experience increased stress and reduced job security. Your program may be the biggest investment your company makes in the employees. Thus, your job is to provide a stellar training program that communicates concern, caring, and investment in employees. You can plan well, select an excellent staff, develop internal training resource people, stay on top of the breaking needs of your organization, and enjoy the process.

In fact, take a moment now to reward yourself with a short retrospective. Think back. What have your safety and health training accomplishments been over the last 2 years? Write them down and save them to read on a low-energy day. Now, think ahead. Plan your stellar training program for the next 2 years, remembering to set a few goals for your personal satisfaction.

FURTHER READING

Books

Buckley, R. and Caple, J. *The Theory and Practice of Training*. San Diego: University Associates, 1990.

Friedman, P. and Yarbrough, E. *Training Strategies from Start to Finish*. Englewood Cliffs, NJ: Prentice-Hall, 1985.

Janov, J. *The Inventive Organization: Hope and Daring at Work*. San Francisco: Jossey-Bass, 1994.

Johnson, S. and Johnson, C. *The One Minute Teacher — How to Teach Others to Teach Themselves*. New York: William Morrow, 1988.

Other Publications

Gilbert, S. R. Company specific safety information. *Professional Safety*, June, 44–47, 1993.

LeBarr, G. New rules for safety training. *Occupational Hazards*, December, 27–30, 1993.

Pfieffer, J. W. *The 1995–2001 Annuals*. Volume 1: *Training*; Volume 2: *Consulting*. San Diego: Pfieffer & Company, 1995–99.

Note:

Current journal issues are helpful: *Training and Development Journal (astd. org), Training Magazine, and Personnel Journal.*

Web search using keywords: training, employee development, methods/strategies, OSHA/Safety/Management, virtual learning.

CHAPTER **48**

Safety and Health Employee Training Sessions

Michael-Laurie Bishow and Richard W. Lack

CONTENTS

INTRODUCTION

In this overview of the planning, delivery, and evaluation of effective training, you will be encouraged to think from the participant's point of view. Training objectives are always about behavior change. Information is provided to reduce barriers, provide reasons, or inform about the risks of not following protocol. Adults (and children) learn more permanently by doing, and for this reason training must include practice and activities. Safety topics are often mandatory, which presents a particular problem about free will. This chapter discusses handling resistance.

Evaluation may make you nervous, but training evaluations can support your budget and course changes without threatening your personal future. Yes, some people will write brutal comments; still most write thoughtful and constructive suggestions. The alternative, no feedback from participants, is infinitely more dangerous to your professional future and personal credibility.

The Training Environment

Employees are often not exactly enthusiastic at the prospect of attending a presentation by the safety department. The subject, most likely required by law, may be viewed as dull and uninteresting. The presenter may approach the topic in negative terms, i.e., what not to do in the work setting. Finally, past training might have been in lecture format or other modes that restrict participant learning and retention. No doubt readers are familiar with the following figures from studies showing how people learn and retain instructional material. Nevertheless, we believe they are worth repeating because as trainers we need to always keep this in mind.

Retention of Instructional Material: It is believed that participants retain:

Only 10% of what they READ
20% of what they HEAR and 30% of what they SEE
50% of what they SEE and HEAR
70% if they can SAY what they have learned by talking it over with others
90% if through physical and verbal means they can DEMONSTRATE learning
95% if they TEACH someone else what they have learned

To reinforce this point, trainers should keep in mind that, while in days past having documentation as proof that your employees completed the required training may have been sufficient to satisfy regulatory agencies, now the agency officials are likely to randomly observe and talk to employees to verify retention, understanding, and compliance with the training.

PLANNING SAFETY AND HEALTH TRAINING

The safety and health practitioner should develop an annual plan for training. This will identify all mandatory training and the groups of employees who need the skills. Setting out an annual schedule in this way will minimize the stress and pressure of preparation of training material, finding speakers, scheduling attendees, etc. Naturally, other subjects can be included during the year as specific needs are identified. Safety and health practitioners are frequently required to develop presentations that are portable, that is, that can be delivered in many different settings. For example, managers may want health and safety information delivered as part of staff meetings. In the interests of production efficiency, employees may not be able to leave their workplace, and you may have to literally carry your training to them and meet in a lunchroom or work break area. These situations require the safety and health practitioner to be creative and to develop effective mechanisms for delivering the required information.

Making Arrangements

Here again, planning is key. We have provided a useful checklist at the end of this chapter as a start to your personal planning tools. This pretraining session checklist provides preliminary information, steps in presentation development, and administrative activities that must occur prior to the training session.

On the subject of planning, consider your stress level and Murphy's law, "If anything can go wrong, it will." Divide you planning checklist into time-sensitive sections and allow for Murphy time. You can be sure that checking the training room the day before and talking to the crew who will do the setup to be sure the work order is correct will save many headaches. Similarly, anticipate duplication problems; double your turnaround so upside-down copies on pea-green paper can be redone. Your courteous behavior is easier to access when you allow enough time.

On the subject of time, we strongly recommend that you avoid scheduling training any later than the fifth hour of any shift. For example, on the day shift in the late afternoon, attendees are often inattentive because of fatigue, because of their commute, or because of other obligations after work. Also you can increase positive regard for training if you can end 15 to 30 minutes before the usual end of shift. This helps participants concentrate on learning without the worry that they will be late to take their children to soccer practice. Be sure to allow sufficient time at the end to complete the evaluation; participants will not believe you want to hear from them if the evaluation spills over the closing time.

In selecting a site for the training, here are some important conditions to consider:

- Noise level and general distractions (sounds, visuals, and smells).
- Access to restrooms (both men and women).
- Electrical outlets (available where you need the audiovisual setup and the coffee pot).
- Adequate ventilation (for the group size and the class length for each room).
- Room temperature (can be controlled adequately).
- Comfortable furniture (chairs with soft seats influence participant satisfaction).
  - Access to the room for setup at least 1 hour before the first arrivals.
  - Politics of the location: Does the location affect the credibility of your training?

The seating in the room will vary according to the size of the group, configuration of the room, style of the training presentation, and audiovisual equipment. Small groups up to 15 may be seated around a conference table if you expect equality and conversation. A horseshoe, square U, or fan arrangement of tables works better for groups of 25 to 40 if the trainer anticipates speaking more than 50% of the time or maintaining a formal style. Theater seating with chairs can increase numbers in any room, and reduce participant interaction with each other. Theater seating is common for groups exceeding 75, but in the new training methods, small-group rounds with an open mike setup are very pleasing to participants. We recommend tables called restaurant rounds that are six or eight feet across and seat eight to ten comfortably. In some cases, you will use a fixed-seat theater or auditorium with built-in audio-visual equipment. Inattention may be a demand characteristic of this setting. You will need to work harder to keep the group active.

Selecting Training Methods

When developing a training presentation, consider using a variety of methods to present the information. Use visuals, including photographs or slides, and especially involve members of the group to help generate interest and reinforce the training points. Assessment of the reading level of the attendees will be critical in the preparation of written material. Role-playing and group exercises may increase attention, retention and learning. If your audience is at a management level, a concise summary of the key issues with action items is desirable. Any group of participants requires an introduction including a purpose statement, and the rationale or relevance to them.

Methods	Advantages	Limitations
Demonstration	Increases interest, makes the topic real if class helps	May not be visible in a large venue, less technical control
Discussion, question and answer (Q&A)	Helps pace the session, increases fit to participants	Airtime limits; some people may ask boring questions
Lecture or panels	Known to have the highest information density, group compliance	Known to have the lowest information retention rate; difficult to reach all trainees
Practice	Participants more likely to do the recommended action, highest retention	Smaller classes required to get to everyone, may be seen as favoritism if not
Role play, situation practice	Faster than going on site, can tell if participants get it	Some are embarrassed; monitor optional activity
Video models with discussion	Safer than role plays, real on-site examples of both mistakes and positive actions	Equipment: screen and sound adequate for room size; longer video may reduce attention (TV)
Small-group exercises	High interest, vast increases in airtime, team building	Time must be flexible; group control lower during activity
Interactive computer programs (CD, Internet, touch screen, etc.)	Pace individualized, learning of information documented, high retention of technical information	No or low personal contact, not suitable for topics related to cooperation, human factors
Individual reading and self-assessment inventories	Popular activity, highly structured, quiet; may yield valuable awareness	Instrument must be research based to be valid and reliable; variable time

Selecting and Using Audiovisual Aids

As already discussed, visual aids definitely increase comprehension and retention of your training material. You will select from many methods such as overheads, stationary or recording charts, handouts, PowerPoint-type presentations, keypad participation, videotapes, group projects, props, specimens, or science demonstrations. Add sensory channels to keep interest. Courtesy of Dr. Robert Fish, here are some important guidelines to keep in mind whatever visual aid you select:

- Use color, and also graphics — pictures and diagrams — not just words. Use several colors.
- Make them visible. Place only a few words on a line and only a few lines on a page. In PowerPoint presentations: six lines and six words maximum.
- Have only one main idea per screen. Speak only to that idea when using the visual.
- Use upper- and lowercase lettering because it is easier to read than all uppercase.
- Proofread. Double-check your visuals for errors. Admit errors if found.
- Practice and rehearse using the audiovisual equipment to discover technical difficulties in each venue.
- Face and speak to the participants, not to your prop or visual aid.
- Turn off the power or cover the aid when not in use to focus attention.
- Use sparingly. Aids are not a substitute for participant activity.
- Have a backup plan in case the equipment fails (handouts, alternate activity on same skill).

Handouts

A handout is an essential item for any training presentation. Besides providing the attendees with material for further reference, it can also include copies of your visual aids so that the trainees can make notes as they follow your presentation. The quality of your handout will be one of the factors influencing how the audience evaluates your training, so our advice is to always go first class! If your handout is a complete text of the paper you are presenting, you may choose to hold your handouts for distribution at the end of the session to minimize people drifting away in midsession.

Color handouts help participants find your references (be sure they are all the same color). Be sure to provide an outline of your visual aids to facilitate note taking with numbered pages. Tell the participants the page you are on to keep people from shuffling through the contents when they refer to the particular subject being discussed.

Considerations When Planning a Useful Learning Event

Whatever you select should help participants understand and retain your subject matter. In planning, consider the impact of your methods on your credibility, the relevance of the subject, the attention of the participants, and the available time, space, or authority. Your role is usually one of serving the group; rarely do you manage them. Therefore, limits exist on your ability to deliver effective training.

For every session, do you have or do you have the authority to get:

- Working equipment needed in the room before each session time?
- Coffee and tea in the morning sessions (a matter of respect)?
- Support materials needed: pens, tape, foam-core mounting, VCR, window shades, etc.?
- Sufficient copies with a few extras for *all* participants on site on time?
- Transcription if you use a whiteboard or charts; copies for all participants after the session?
- Resources to pay for nifty up-to-date tools in time to practice using them before the session?
- Approval for registering the appropriate participants for each training event?
- Approval for setting and limiting class size, for scheduling enough time to learn?
- Documentation of the credibility of the training methods, this class, and your delivery?

DELIVERING THE TRAINING

In planning your delivery, consider your training subject and the goals of your presentation. To deliver an effective presentation, what should your role be? When should you behave like a trainer? An instructor? A facilitator? Here are some practical definitions for these very different techniques:

- Instructing — An instructor is primarily an educator whose focus is to impart information. The goal of a good instructor is to motivate people to retain the information provided.
- Training — A trainer is a coach whose focus is skill building and application. The goal of a good trainer is to help people develop necessary skills/actions/behaviors, to teach them how to use the knowledge provided to act differently on the job.
- Facilitating — A facilitator is part discussion leader/part coordinator whose focus is group process and problem solving. The goal of a good facilitator is to help the group generate ideas, respect each member's contribution, and guide process with methods such as brainstorming and root-cause analysis. As a facilitator, you guide the process, ease around barriers, point out opportunities to help the group reach its goal. This type of a meeting can be used to tap into the group members' creativity and also help them reach solutions to mutual problems in their organization. In classic facilitation, the role does not involve expressing a content-related opinion. Facilitators who judge members or contribute a solution to the technical problem are out of role and should step down for a moment to regain their content neutrality. A phrase to remember is, "Steer the car, not the people!"

When preparing yourself for a training class, consider the objectives of each time block. Is it information retention, skill building or group process? A training class will be more effective if you build your training outline around specific observable skills. The participants should be able to demonstrate their knowledge by doing the skill. If the group requires knowledge to understand when or how to do the skill, teach a section of the training as the instructor with this information. Then move on to a skills practice exercise with you in the trainer role.

Avoiding Training Trouble

When delivering training, a useful acronym to remember is PEDPER:

Prepare — **E**xplain — **D**emonstrate — **P**erform — **E**valuate — **R**eview

Some other important delivery steps are as follows:

- Whenever possible, set up the night before so you will have time to deal with snags.
- Be there ahead of your attendees so you can greet them informally. You are the host; your friendly greetings will relax the group.
- Provide coffee and tea in morning sessions and consider providing other refreshments such as candies or cookies. This helps to show you care about the people attending and builds rapport.
- In your opening remarks, discuss breaks, note restroom locations, and explain the purpose, materials, length, and general process for the session.
- Remind attendees that no one has all the answers and that you will be relying on them to contribute their knowledge and experience to the training. Encourage attendees to ask questions at any time.
- Dress neatly and appropriately; don't distract the participants!
- Solicit suggestions for improvement of the presentation. Use an anonymous evaluation sheet. Use brief numerical ratings and short answers such as "If I could improve the program, I would ..."
- You may chart comments at the end of each session using a divided chart "+ and Δ" (what is positive about the training, what should be changed).

Some questions are essential for analysis purposes since the trainer should continuously improve the content and quality of presentation. Be sure to thank attendees for their input and participation.

Handling Resistance in the Group

As a trainer and instructor for many years, Richard Lack has found that sessions will rarely get out of hand as long as the trainer demonstrates respect for the people and practices the "Golden Rule" when dealing with them. You must be honest with your group at all times. If someone asks a question and you do not know the answer, do not waffle or guess; say you do not know and ask the participants if anyone in the group knows. If not, say you will find out and get back to the participants who are interested. If someone asks a tough or controversial question, turn it back to the group and let them deal with the question. Alternatively, you can say you will seek the appropriate person to investigate.

Michael-Laurie Bishow adds, group culture can predetermine that the trainer will be challenged, sometimes in the first 2 minutes of the session. Try to remember this is not a personal attack; they hardly know you! Handle the problem with a review of ground rules for the session. A chart on the wall before anything happens allows the trainer to walk to the chart and review it as if this were an ordinary daily introduction. Ground rules should be customized to each group; ask if any other guidelines should be on the chart and add the suggestions. The ground rules also govern you; a common rule "start and end on time" applies to your time management.

Power dynamics are affected by your demographics and the makeup of the group. Characteristics such as gender, age, position, title, class, ethnicity, or tenure with the company may lead to perceptions about your competence on the topic. Often stereotypes lower your credibility until you confront them directly with real information about why you know about your topic. Above all do not reinforce the stereotype or play like you don't really know your material. Some groups will see that as a sports challenge in a game "Chase the Trainer." Again this is not a personal attack unless you allow the stereotype to persist.

Here are some tips to help reduce the likelihood of negative interactions. First, avoid use of the following "red flag" words that tend to accelerate conflict and increase stress:

Red Flag Words	What They Convey
Ought, should	Blame
Can't, couldn't, have to, must	Victim, helpless
Never, always	Absolutes, rigidity
It, them, they	Projection, not responsible
(Person)____ is ____(type)	Labeling/stereotyping

Instead, choose the following "green flag" words to decelerate conflict and stress:

Green Flag Words	What They Convey
Choose to, decided to	Self-determined
Want, wish, hope	Nonrestrictive, intent
Some, many, several, few, seldom	Conditional
Usually, often	Speculative
Seems, notice	Value-free
(action)____ suggests you____	Neutrality, informed choice

Evaluating Training Outcomes

On the subject of evaluation, Richard Lack remembers a phrase Homer Lambie frequently used: "The value of any meeting is measured in terms of what the group attending the meeting will do differently after the meeting." This point is really the bottom line in any evaluation of a training event. Perhaps some aspects of your presentation were not perfect, but if your message got across and the group will put the message into action, that is what really counts. Apart from observable behavior change, other methods for validation of training effectiveness include:

- Questionnaires at the end of the training
- Performance assessments during or after training
- Interviews of participants or their managers
- Observation checklists for skills in the workplace

Evaluating training is rather like problem solving. If your problem is specific, it is solvable. If it is too general, it will be unsolvable. If your training objectives were specific, you will be able to measure the results. If your training objectives were too general, or described in vague terms, evaluating the training will be equally vague or general.

Accountability today requires cost-effective training. Documenting more than good exit impressions is necessary to establish the worth of your programs. Followup in several months with participants' supervisors to check on safety compliance. Drawing statistical relationships between the injury or accident rate and training participation is also useful. Many evaluations ask participants, "How likely are you to begin using this safety protocol when you return to work?" or other specific questions about intent. See Chapter 47 for other ideas about establishing the cost-effectiveness of training.

Summary

Making effective training presentations may not be something that comes naturally to many safety and health practitioners. However, as part of their professional development, safety and health practitioners should attend courses and seminars on delivering effective training, group facilitation, and persuasive communication. Developing these abilities will enhance your professional credibility and open new opportunities for growth and promotion.

FRIEDLAND'S TRAINING TIPS

These tips are drawn from a talk given by Steve Friedland, management consultant and trainer, at the 1994 Annual Conference of the Northern California Human Resource Council.

1. Schedule at least one 15-minute break each 90 minutes. Allow one hour for lunch.
2. Pay attention to how you dress — no distractions!
3. Play music that fits the program to help the ice-breaking process.
4. Form small groups for better participation — a maximum of five or six.
5. Have more rather than less material.
6. Your handout reflects you. Include references and graphics. Use colors.
7. Training feedback is important. Attendees need to evaluate the program and you for:
   • Delivery
   • What was learned
   • What can be improved
8. Study the evaluations and then let go of them. Some people are very judgmental.
9. Ask attendees to tell you what they expect. Record requests and address them.
10. Use anecdotal stories to support your message. Talk about life experiences that relate to the training content.
11. Openings and closings are very important. This is the heart of your presentation.
12. Transitions between content areas are the heart of comprehension.
13. Be very careful with humor. Test out jokes. Do not ever joke about ethnicity, religion, gender, or any other legally protected human categories.
14. Be careful of copyright issues. There are some things you should not reproduce or have as overheads without written permission from the author or publisher.

15. Do not call upon people who don't want to participate. People will volunteer.
16. Use "round robin" small groups. Your objective is to get as many to talk as possible.
17. Do not rely too much on visuals. Remember: you are the message. You need to be out there and keep eye contact. Less is more when it comes to visuals.
18. LISTEN carefully. Paraphrase to help the class understand the participant's answer.
19. Do not confront people while in the group — do it privately.
20. Always build the audience up, never down.
21. SMILE!

SAFETY AND HEALTH TRAINING PROCESS CHECKLIST

Administration Issues

☐ Check site areas for smoking, eating, parking, public transportation
☐ Travel arrangements made for the trainer(s)
☐ Select site: Room and equipment reserved
☐ Lesson Plan and funding established, contracts signed/updated
☐ Check for training policies, compensation, and labor/management agreements
☐ Name and address of key person for assistance on site
☐ Determine if security badges or special arrangements are needed for access to site

Three to Six Weeks before the Training

☐ Survey the site for size, lighting, ventilation, furniture needs
☐ If off-site, check on transportation arrangements for equipment
☐ Send announcement/reminder letter, e-mail, or memo to potential participants
☐ Contact appropriate person about opening and closing site
☐ Order the necessary number of chairs and tables
☐ Breakout rooms, contingency plan for larger group
☐ Have handout materials printed and available
☐ Notification to attendees sent out, maps with directions
☐ Locate and reserve audiovisual equipment and chart easels with paper
☐ Room setup, request copy of the work order if off site

One Week before the Training Program

☐ Confirm number of participants, verify supervisor permission to attend
☐ List of attendees/sign-in sheet available/print name tags or tents
☐ Confirm order for refreshments, delivery time
☐ Obtain direction signs indicating designated room
☐ Locate restroom facilities, disabled access/parking permits
☐ Locate emergency copy-center, phone number, and charges
☐ Obtain felt-tip markers, trainer toolbox fully stocked
☐ Obtain pencils or pens and notepads for participants
☐ Assemble class setups for each participants, boxed and ready
☐ Recheck equipment arrangements; check for spare bulbs, batteries
☐ Have a "dress rehearsal;" test all equipment.

Day of Training Session (night before and one hour before session)

- ☐ Organize materials on the trainer's table and check all visual aids
- ☐ Make sure furniture is set up, correct as needed
- ☐ Check room temperature settings, door locks, restrooms
- ☐ Verify that an available clock is on time and working
- ☐ Tape electrical cords to the floor and verify extension cords are working
- ☐ See that refreshments are delivered, coffee is perking
- ☐ Set the tables: booklets, handouts, notepads, pencils, name tags
- ☐ Test equipment and visual aids again

TRAINING MATERIALS AND EQUIPMENT CHECKLIST

Jog your memory with the following list for trainers.
Modify to include items that apply to your own situation.

To Project Your Images

TV-VCR	Video camera	Fader switch setup
Overhead projector	Slide projector(s)	Screens
Sound system (boom box/speakers)	Adaptor plugs	Audio tapeplayer
Extension cords	Videotape	Audiotapes
Laptop computer/PowerPoint	CD-ROM drive	

Writing and Note Taking Tools

Pens	Pushpins	Chart paper
#2 Pencils/pencil sharpener location	Scotch tape	Masking tape
3 × 5 index cards	Notepads	Whiteboard pens
Blank transparencies (overhead)	Post-its	Overhead pens
Felt-tip water-based pens	Scissors	

Record Keeping and Attendance

CEU credit forms	Chart paper
Registration materials	Parking permits
Expense forms	Attendance sign-in
Calculator	Class evaluation forms
Roll of poster paper	Certificates of completion

Troubleshooting

String	Hammer	Lens cloth	Pliers/Allen wrenches
Matches	Screwdrivers	Videotapes	Stapler
Extension cords	Overhead transparencies	Box of tissues	Can of orange juice
Safety pins	First-aid kit	Energy bar	

Pocket knife (caution: viewed as a weapon in some places)
Change for vending machines and parking meters

TRAINING SESSION PLANNING WORKSHEET

I. TRAINING SESSION OUTCOMES: After the session, participants will ...

_____

_____

TITLE for Marketing/ Advertising the Training: _____

II. PARTICIPANTS: Description: _____

_____

III. LOGISTICS:

Session Date(s)_____Time(s) _____

Duration _____Number of Participants_____

Location(s) _____

Room Setup Required_____

IV. OBJECTIVES: Participants will demonstrate their knowledge and skill in ...

_____

_____

_____

V: SESSION PLAN:

Time period	Content	Method/activity	Trainer Role	Materials

VI. Evaluation: Tools Used at the Session Close: _____

VII. Follow-Up: Evaluation Planned: _____

TRAINING SESSION EVALUATION

Title of the Training: _____ Date_____

Please use this scale to answer the following questions:

To what extent is the following likely:	No, Not at all	Unlikely	Maybe	Somewhat Likely	Very Likely
1. I needed to learn what is in this session to do my work properly.	1	2	3	4	5
2. The course objectives were clear to me in the beginning.	1	2	3	4	5
3. The course materials, visual aids, and activities kept me interested.	1	2	3	4	5
4. The trainer's style and methods helped me learn.	1	2	3	4	5
5. The session was organized to help me learn the content.	1	2	3	4	5
6. I will apply the skills in this training to my work right away.	1	2	3	4	5
7. The course handouts will help me as a reference in the future.	1	2	3	4	5
8. I will recommend this training to other employees.	1	2	3	4	5

9. Please suggest any changes to the content or presentation methods of this training:

Which parts of the training were most useful or interesting to you?

10. Please comment on the room, location, furniture and the facility in general:

THANK YOU. YOUR COMMENTS HELP US DESIGN THE BEST POSSIBLE TRAINING.

FURTHER READING

Books

Buckley, R. and Caple, J. *The Theory and Practice of Training.* San Diego: University Associates, 1990.

Colin, T. *Strategies for Adult Education.* London: Oxford University Press, 1981.

Daniels, W. *Breakthrough Performance.* Mill Valley, CA: ACT Publishing, 1995.

Gilbert, F. *Power Speaking: How Ordinary People Can Make Extra Ordinary Presentations.* Redwood City, CA: Frederick Gilbert Associates, 1994.

Good, T. W. *Delivering Effective Training.* San Diego, CA: University Associates, 1982.

Hoff, R. *Can See You Naked — A Fearless Guide to Making Great Presentations.* Kansas City, MO: Andrews and McMeel, 1988.

National Research Council. Improving Risk Communication. Washington, D.C.: National Academy Press, 1989.

Pater, R. *Making Successful Safety Presentations.* Portland, OR: Fall Safe, 1987.

Peoples, D. *Presentations Plus.* New York: John Wiley & Sons, 1992.

Pike, R. *Creative Training Techniques Handbook.* Minneapolis: Lakewood Books, 1989.

Other Publications

ANSI Standard Z490.1-2001. *Accepted Practices in Safety, Health and Environmental Training* (New Standard) American National Standards Institute, New York.

Hywel, J. and Race, P. 10 training techniques for safety practitioners. *Health and Safety Practice,* 30(6), 18–21, August 1993.

Merrifield, J. and Bell, B. Don't give us the Grand Canyon to cross. *Adult Learning,* 6(2), 23–24, November/December 1994.

Moeller, M. Video conferencing hits the desktop. *Present Technology: Technology for Better Communication,* 8(5), 16–20, May 1994.

Pike, R. Handouts, a little charity to your audience goes a long way. *Present Technology: Technology for Better Communication,* 8(5), 30–32, May 1994.

## CHAPTER 49

Developing Effective Presentation Skills

Robert S. Fish and Ken Braly

CONTENTS

INTRODUCTION

Giving a terrific presentation is not easy, but it need not be a cause of great anxiety, worry, and dread. One reason it is torturous for so many of us is that we received little training in how to speak to groups at any point in our education. It is bad enough to feel inept, but to have to stand up in front of others and parade our ineptness goes beyond the bounds of decency.

What can one do? First, understand that is not a genetic defect; you really can become an accomplished and confident speaker. Second, understand that the route to that end is to be motivated, persistent, and willing to make mistakes. Remember when you learned a sport? You had to accept that for a time you would fall down, or miss the ball completely, or just look silly. It is part of the process; accept it and get on with it.

Make an earnest attempt to apply as many of the techniques suggested in this chapter as you can over time. As Winston Churchill said in what may be the shortest commencement address ever given: "Never, never, never, never give up."

ENTHUSIASM — THE MAGIC WEAPON

Probably no characteristic of a speaker can do more to keep the attention and interest of an audience than enthusiasm. Yet enthusiasm is often misunderstood. Some points to consider include the following:

Enthusiasm is not volume — We have all heard advertising on radio and television where the announcer shouts at the top of his lungs about what a great deal is being offered and how you had better not miss it. Many presenters, especially technical presenters, fear that showing enthusiasm means coming across like this.

Enthusiasm is not hyperactivity — Your body language is always important when giving a presentation, but showing enthusiasm does not mean racing around and flailing your arms.

Let them know your topic is important — Enthusiasm will show through naturally when you feel that your topic is interesting and important and that it is important for your audience to hear about.

What if you just are not enthusiastic? — What if you do not care about the subject? Or what if you cannot see how to be enthusiastic the tenth time you give the same presentation on why rectabular extrusions will not transmogrify? Well, try. Try to find something about the subject you can be interested in. Challenge yourself to find ways to bring a dull, boring subject to life for your audience. If the subject does not turn you on, be creative and get turned on with giving a terrific presentation.

EXAMPLES AND STORIES

People remember stories, often long after the facts are forgotten. They can add a lot to keeping an audience's interest. They are the best way to add meaning to technical facts and figures and to add humanity and life to dry material.

Stories are not mysterious — You do not have to be a Garrison Keillor or Bill Cosby. Just add some life to the facts by telling about what happened in real life.

Keep them brief — A good story or anecdote can be just 30 to 60 seconds long.

Make them relevant — A story may illustrate a point, or it may provide context for the facts, or it may simply add color, letting the audience know that behind what is being reported are real people, with real concerns and real foibles.

Add details — A good story is rich with details about the people, places, and things that are described. Rather than saying "John didn't think we'd get done on time," add details about how it really happened: "John had been sitting at his computer for hours, trying to get the last tests run. When I walked by, he glanced up with a worried look and said, 'I don't think we're going to make the deadline.'"

Practice and rehearse a story — Even if what you tell about happened to you, do not assume that your words will flow smoothly and succinctly without practice.

ASPECTS OF DELIVERY THAT CONNECT WITH AN AUDIENCE

What do good speakers do during a presentation to demonstrate confidence and to foster a sense of trust and authenticity?

- They have strong eye communication with audience.
- They show enthusiasm for the topic with natural gestures and movement.
- They make their voices expressive, i.e., use fluctuations in volume, pitch, and tone rather than speaking in monotone.
- They smile appropriately.
- They feel comfortable in moving about the room, rather than being tied to one spot.
- They use facial expressions.
- They give the impression that they are enjoying speaking and being with the audience.

DEFINING THE PURPOSE AND MESSAGE

The purpose describes what you want to accomplish with the talk. It is vital for you to know specifically what outcome you want. A statement of your purpose usually begins with the word to. Examples of purpose statements include the following:

- To inform management of the progress made on the back injury control program during the last month.
- To persuade the supervisors that their current inspection system is outdated.
- To demonstrate the functionality of the new record-keeping software.

Sometimes it is helpful to distinguish between your purpose and the purpose you are going to tell the audience. For example, you might define your purpose as "to get the management review committee to fund my project." Yet you probably would not use those words when opening your talk. Instead, you might say, "My purpose today is to explain this project and illustrate why it is a good investment for our company."

The message — sometimes also called the *point* — is the single statement that sums up your whole presentation. If the audience remembers nothing else, you want them to remember this. Using the examples from above, the following message statements might be prepared:

- The back injury control program is nearly complete, but we will need additional resources to finish on schedule.
- You will need to improve the inspection system within a year.
- The new version is substantially more powerful than the previous version.

When you prepare a new presentation, write out the purpose and the message and keep them in front of you as you work. Make every decision with an eye to achieving your purpose and communicating your message.

UNDERSTANDING YOUR AUDIENCE

It is your responsibility to make sure that the audience hears the message you want it to hear. You must analyze your audience to understand its needs and wants. Questions to ask regarding the members of your audience include:

- What is their level of knowledge about your subject? Are they technical? Non-technical? A mixed group?
- What are their expectations? What sort of presentation, and what sort of information, are they looking for?
- What is their attitude? Are they already in favor of what you are going to present, or are you trying to persuade some who may be skeptical or even hostile?
- What are their biggest concerns (hot buttons)? For example, are they most concerned about money? Schedule? Your ability to achieve what you promise?
- What turns them off? Is there anything you should be sure you do not say or do?

OPENING A PRESENTATION

"You never get a second chance to make a first impression." That cliché is especially true when it comes to giving a presentation. In your first few minutes, you can change your audience's thinking from "Oh brother, another safety talk (yawn)" to "Hey, this is going to be interesting." Here is what the opening of your presentation should do:

Capture attention — The minds of those in the audience may be wandering. You want
    them focused on you.
Identify the purpose and message — Get the bottom line out there.
Clarify the agenda — Tell them what you are going to tell them.

Here are some ideas for achieving this:

- Instead of opening with "Good morning, my name is X, and I'd like to tell you about Y," try something more intriguing — a bold statement, a question, dramatic statistics, or a brief story or example. Begin with confidence.
- If the audience does not know you, tell a little about your background to help establish your credibility.
- Your presentation has (or should have) a message — the single most important idea you want your audience to remember. State the message up-front, in the form of a benefit to the audience.
- Explain the agenda of your presentation — what you are going to cover. The audience likes to know where you are going. Show the participants you are interested in them by letting them know you will address their concerns.
- Humor is okay if you know they will think it is funny and if it is relevant to the topic. Do not start with a joke just because you once heard that you should.
- Do not apologize or ever admit you are not as ready as you should be.

BUILDING A STRONG CONCLUSION

The conclusion is arguably the most important part of your presentation. It is often the part that the audience remembers best. So it is a good idea to work to make the ending as strong as possible. Here are some guidelines for a strong conclusion:

- Here is where you summarize your presentation, reiterating the key points. The conclusion should not contain any new material.
- This is your last chance to state (passionately, perhaps) your message.
- This is the place where you "ask for the order," in sales terms. What do you want from the participants? Tell them.
- Consider this sequence:
  - Summarize the main ideas of the body.
  - Conduct a question-and-answer period (we will cover this later).
  - State the message once again.
  - Ask the audience for approval, concurrence, etc.
- Try not to let the talk end with the answer to a question. Have the final say — a restatement of the message, perhaps, or of the main benefit.
- Prepare in advance the last few sentences of the presentation, so you know how you will end before you even begin.

QUESTIONS AND ANSWERS

Some speakers fear the question-and-answer (Q&A) aspect of a presentation: it is unpredictable, it introduces new threats, and it calls for thinking on your feet. Q&A can be the most valuable part of your talk, however — the time when you really sell your ideas best, when you can really connect with your audience over areas it cares most about. Keep these suggestions in mind:

- "Seek first to understand, then to be understood" — Stephen Covey.
- Listen for what is not said: what is the question really asking?
- Think. Ask for clarification if you need it.
- Get to the point quickly, and back it up with examples or data.
- Have eye contact with both the questioner and everyone else.
- Do not get caught up in technical complexity. If it is a long answer, or extremely detailed, ask to put it off for one-on-one discussion later.
- If audience members interrupt you, let them.
- If you are asked a number of questions at once, answer them one at a time.
- Avoid negative words, e.g., can't, won't. Be careful with but — it cancels everything you had said before it.
- Prepare and practice your answers ahead of time. One of life's little pleasures is being asked a killer question you are ready for.
- When you absolutely do not know, admit it with confidence, and if you offer to follow up, do it!
- Conclude your talk with a summary and restatement of your main points, not just an answer to a question.

REHEARSALS

No professional actor would dream of going on stage to perform in front of the public without extensive rehearsal. Why do we resist practicing our presentations beforehand? Well, rehearsing can be time-consuming and boring. However, rehearsals can work magic. Consider the following:

- Rehearsals give you control over your material, so you are smooth instead of stumbling.
- Rehearsals add to your confidence. You present a stronger image, and you handle the unexpected better.
- Rehearsals give you an air of professionalism.
- Rehearsals reduce stage fright and anxiety.

The big drawback is that they take time — and there are always so many other, more important things to take care of. It is hard to find time to do it.

However, you can rehearse your introduction and conclusion while commuting or taking a shower. Use any spare time. Carry your notes and hard copy of visual aids around with you. Look them over while you are standing in line at the post office.

It comes down to priorities. Those who see good speaking skills as career-enhancing will make time to rehearse. How should not rehearse?

- Say the words out loud. We all have a tendency to be brilliant in our minds, so let the words come out to learn where the rough sports are.
- Practice with your visual aids. Again, do not just run through it mentally. Actually flip the pages, or turn the projector on and off.
- If it is an especially important presentation, practice before some colleagues whose feedback you trust.

FURTHER READING

Clancy, J. *The Invisible Powers: The Language of Business.* Lexington, MA: Lexington Books, 1989.

Leech, T. *How to Prepare, Stage, and Deliver Winning Presentations.* New York: Amacom, 1982.

LeRoux, P. *Selling to a Group: Presentation Strategies.* New York: Barnes & Noble, 1984.

Meuse, L. *Succeeding at Business and Technical Presentations.* New York: John Wiley & Sons, 1988.

Perret, G. *Using Humor for Effective Business Speaking.* New York: Sterling Publishing, 1989.

Timm, P. *Functional Business Presentations: Getting Across.* Englewood Cliffs, NJ: Prentice-Hall, 1981.

PART XI

International Developments

CHAPTER **50**

Global Safety and Health Management

Kathy A. Seabrook

CONTENTS

INTRODUCTION

Corporations throughout the world recognize the key to success is increasing their customer base and market share. When domestic markets are saturated, new growth opportunities must be identified. With a population of 1.23 billion in the Peoples Republic of China, 1.74 billion in Brazil, 101 million in Mexico, 10 million in the Czech Republic, and 1.02 billion in India (according to the U.S. Bureau of the Census, International Database), domestic corporations have been quick to capitalize on sustainable growth through global mergers, acquisitions, and joint ventures. In-country investment in plant, equipment, processes, human resources, marketing, etc. creates jobs and begins the cycle to sustain a growing middle class of consumers. Therefore, expect more growth in developing economic markets and, as those markets mature, an increase in occupational workplace safety and health regulations, enforcement, public opinion, and politics.

This chapter will provide you with a high-level, strategic overview of some of the challenges you will face when managing the safety and health of your global

1-56670-370-0/02/$0.00+$1.50
© 2002 by CRC Press LLC

workforce. There are also a number of Internet resources provided for additional, country-specific workplace safety and health regulatory and resource information.

LEADERSHIP RESPONSIBILITY

For a company to manage workplace safety and health globally, its leadership must determine its strategy and commitment to the safety and health of the global workforce. Assessing current corporate culture and leadership direction will allow the safety and health professional to create alignment of the workplace safety and health process throughout the organization.

Leadership should be educated to understand there are local country laws (some noted below) throughout the world that hold it statutorily responsible for worker health and safety. In addition, any major incident, no matter where in the world, can pose a serious public relations challenge for the company, in some situations halting operations or influencing public opinion and company credibility. Leadership responds quickly to the threat of lost market share and consumer confidence. The safety and health professional should bear in mind this is a real driver for global safety and health.

Under the Management of Health and Safety At Work Regulations, 1999, European Union managers and company leadership have a statutory duty to provide a safe and healthful place of work. In addition, a company doing business in any European Union country must provide a competent resource for worker safety and health on site. As of this writing, European Union countries include Germany, Spain, France, Italy, the U.K., Belgium, Denmark, Greece, Ireland, Luxembourg, the Netherlands, Austria, Portugal, Finland, and Sweden.

In countries such as Mexico, China, and Brazil, there are explicit responsibilities for leadership in the area of worker safety and health.

Questions many companies ask is: "How do we really have to comply with the local laws? How well do the countries enforce the laws?" Some thoughts on these questions follow.

One trend seen throughout the world is lack of enforcement of safety and health laws due to insufficient resources for government inspectors. However, some countries may enforce the safety and health laws indiscriminately; in other countries, safety and health inspectors may enforce health and safety laws as a means of setting an example to other in-country multinational companies. Enforcement follows the ability of a company to pay a fee to the government organization. Therefore, it is in the best interest of a company to comply with local regulations as a minimum.

In 1999, the Institute of Chartered Accountants in England and Wales published a document entitled "Internal Control-Guidance for Directors on the Combined Code for Corporate Governance." This document is commonly referred to as the "Turnbull Report," named after Sir Nigel Turnbull, who chaired the committee that drafted the guidance. The significance of the report is that all companies are now required to implement the code in order to be listed on the London Stock Exchange. The code requires a London Stock Exchange company board to maintain a sound system of internal control to safeguard shareholders' investment and the company's assets.

The internal control guidance ensures that:

- A company board is aware of the significant risks (market, credit, liquidity, technology, legal, health and safety, environmental issues, corporate reputation) faced by the company;
- There are procedures in place to manage these risks;
- executive management is responsible for managing risks by maintaining an effective internal control system; and
- The board of directors as a whole is responsible for reporting on it.

The Turnbull Report has had a significant impact on business management for worker health and safety as financial markets are now holding management accountable for worker health and safety.

Leadership must demonstrate commitment to global worker safety and health through good safety and health policies and procedures. It should support financial recommendations for safety and health improvements as well as demonstrate good management of worker safety and health through financial incentives for management. Leadership should set worker safety and health goals and objectives, measure performance, and communicate that performance in the annual report to customers, board members, shareholders, and the public.

A SYSTEM FOR WORKER SAFETY AND HEALTH

To manage a global safety and health process, a management system should be developed and implemented. There are several reference management systems templates available, such as:

- British Standards (BS) 8800: 1996 Guide to Occupational Safety and Health Management Systems.
- Occupational Health and Safety Assessment Series (OHSAS) 18001: 1999 Occupational Health and Safety Management Systems (British).
- International Standards Organization (ISO) 14000 Environmental Management Systems.
- Australian/New Zealand (AS/NZ) 4801: 1999 Occupational Health and Safety Management Systems.

These standards can be used without third-party certification.

The key components of a management system should include, but are not be limited to:

- Leadership commitment.
- Planning and organizing (including a competent person to oversee safety and health issues).
- Implementation.

- Monitoring and measurement.
- Communication.
- Annual leadership review of workplace safety and health performance as well as the upcoming year's goals and objectives.
- Continuously improving the management process and performance outcomes.

The management systems approach is the approach of the United Kingdom and the European Union. Many other countries have followed the U.K. and lead in this approach. They include China, South Korea, Japan, Jamaica, and Thailand.

Using this approach allows local, in-country leadership to own the safety and health process as well as work toward compliance with local standards. It is industry practice to develop corporate worldwide standards for safety and health, including guidelines for lead, noise, lock out–tag out, ergonomics, contractor safety, electricity, etc. In most cases, it is then the responsibility for local, in-country leadership to have a competent person in place (internally or externally, depending on statutory requirements) to understand and implement local standards that are stricter than the corporate standards. If no local standard exists, the corporate standard/guidelines apply where practical or feasible. The key is feasibility; there have been cases where appropriate monitoring equipment or personal protective equipment is not available in some developing countries.

THE IMPACT OF CULTURAL DIFFERENCES

Do your homework when visiting your global sites. Remember, other cultures do not necessarily do business, communicate, or negotiate the same as you do in your own country or culture. In most cultures around the world, relationship development is paramount to doing any business — no relationship, no deal. Also, when visiting an international site, find out something about the country or city you are visiting. This will go a long way in developing the relationships you will need to get your job done. Another tip is to learn a few words, such as good morning, good evening, please, and thank you. It will show your local hosts you appreciate their country, language, and culture. If English is your hosts' second language, they will make an effort to speak in English with you. This will make your job easier, less time-consuming, as well as your stay in the country more pleasurable.

SAFETY AND HEALTH RESOURCES FOR GOVERNMENT SAFETY AND HEALTH AGENCIES AND REGULATIONS

Europe

Directorates General V (Government Agency)
Rue de la Loi, 200
B-1049 Bruxelles
Belgium

European Agency for Safety and Health at Work (Informational Resource for S & H)
Gran Via
33. 48009 Bilbao
Spain
Tel: + 94 479 43 60
Fax: + 94 479 43 83
Web site: http://www. osha.eu.int

United Kingdom

Health and Safety Commission/Health and Safety Executive
Health and Safety Executive
Health and Safety Laboratory
Broad Lane
Sheffield S3 7HQ
Tel: + 44114 289 2342
Fax: 0114 2892333
Web site: www.hse.gov.uk.hsehome.htm

Germany

Bundesministerium für Arbeit und Sozialordnung
Mauerstraße 45-52
D-10117 Berlin
Germany
Fax: +493020071833
E-mail: re.gerber@bma.bund.de
Web site:www.de.osha.eu.int

France

Ministère de l'Emploi et de la Solidarité
DRT/CT
20, bis rue d'Estrées
F-75700 Paris 07 SP
France
Fax: +33144382648
E-mail: robert.mounier-vehier@drt.travail.gouv.fr
Web site: www.fr.osha.eu.int

Ireland

Health and Safety Authority
10 Hogan Place
Dublin 2
Ireland
Fax: +35316147125
E-mail: path@hsa.ie
Web site: www.ie.osha.eu.int

Italy

Istituto Superiore per la Prevenzione e Sicurezza del Lavoro
Head of Documentation, Information and Education Department
Via Alessandria 220 E
I-00198 Roma
Italy
Fax: +390644250972
E-mail: perticaroli.ispesl.doc@infuturo.it
Web site: www.it.osha.eu.int

Spain

Instituto Nacional de Seguridad e Higiene en el Trabajo
c/Torrelaguna 73
E-28027 Madrid
Spain
Fax: +34914030050
E-mail: subdireccioninsht@mtas.es

Sweden

Arbetarskyddsstyrelsen
SE-171 84 Solna
Sweden
Fax: +4687309119
E-mail: arbetarskyddsstyrelsen@arbsky.se
Web site: www.se.osha.eu.int

Portugal

Instituto de Desenvolvimento e Inspecção das Condições de Trabalho
Direcção de Servicos de Prevenção de Riscos Profissionais
Avenida da República 84, 5º Andar
P-1600-205 Lisboa
Portugal
Fax: +351217930515
E-mail: idict@idict.gov.pt
Web site: www.pt.osha.eu.int

Denmark

Arbejdstilsynet
Landskronagade 33
DK-2100 Copenhagen
Denmark
Fax: +4539182062
E-mail: pm@arbejdstilsynet.dk

Russia

Ministry of Labour
All-Russian Center on Occupational Safety and Productivity
4-ja Parkovaja ul. 29
105043 Moscow
Russian Federation
Tel: + 7 95 164 6600
Fax: + 7 95 164 9815

Mexico

Ministry of Labor and Social Welfare
Secretaría del Trabajo y Previsión Social
Web site: http://www.stps.gob.mx/stps1/stpsvp.htm [Spanish only]
New legislation issued April 1997

Chile

Ministerio Del Trabajo Y Prevision Social
Huérfanos 1273 Santiago
Chile
Tel: + 56 2 7530400
Fax: + 56 2 7530467
E-mail: mintrab@mintrab.gob.cl

South Korea

Ministry of Labor
Tel: + (02) 598-2451 8
Web site: http://www.molab.go.kr/English/English.html
Industrial Safety Act 1990

Canada

The Canadian Centre for Occupational Health and Safety (CCOHS)
Web site: www.ccohs.ca

Ministry of Labor (Ontario)
Web site: www.gov.on.ca

FURTHER READING

Axtell, R. E., Parker Pen Company. Do's and Taboos around the World, 3rd ed. New York: John Wiley & Sons, 1993.
European Agency for Safety and Health at Work, Web site: www. osha.eu.int.

Institute of Chartered Accountants. Internal Control — Guidance for Directors on the Com-
    bined Code for Corporate Governance. London: Institute of Chartered Accountants,
    1999.
Seabrook, K. A. Multinational organizations. In: *Safety and Health Management Planning.*
    Kohn, J. P. and Ferry, T. S., Eds. Rockville, MD: Government Institutes, 1999, chap. 7.
Seabrook, K. A. International standards update: occupational safety and health management
    systems. In: *Professional Development Conference Proceedings.* Des Plaines, IL,
    June 2001.

## CHAPTER 51

International Organization for Standards (ISO) 9000:1994 and ISO 9000:2000: The Effect on the Global Safety Community

Mark D. Hansen

CONTENTS

HISTORY AND INTRODUCTION

For those of you who have no idea what ISO 9000 is or its purpose, we will begin with a brief history to illustrate its roots. Today many safety engineers live (and die) through the use of standards, whether they are military standards, Environmental Protection Agency (EPA) standards, Occupational Safety & Health Administration (OSHA) standards, or something very similar. These standards are often inconsistent, and especially when it comes to international trade they are not well, suited for widespread use. In many cases terminology in these standards, as well as in commercial and industrial practice, is inconsistent and confusing.

The publication of the ISO 9000 series and the terminology standard (ISO 8402) in 1987 has attempted to bring some harmonization on an international scale. It has supported the growing impact of quality as a factor in international trade. This is not a fad or a passing buzzword of the day. The ISO 9000 series have been adopted by many nations and regional bodies, and is rapidly supplanting prior national- and industry-based standards.

The ISO 9000-9004 and the American National Standards Institute (ANSI)/American Society for Quality Control (ASQC) Q90-94 series documents contain information relevant to systematic management for product and service development, design, production, and installation activities, including safety, health, and environmental aspects.

These standards represent an international movement to establish worldwide quality system standards for hardware, software, and processed products and services. Although strongly driven by the European community originally, the Total Quality Management (TQM) vision represented by these standards will most probably drive U.S. business to become compliant and require certified practitioners (e.g., certified safety professional (CSP)) in the near future in order to conduct business internationally. The ISO 9000 series documents will directly challenge each of us as safety professionals.

SUMMARY OF ISO 9000

ISO 9000 has set forth four strategic goals: universal acceptance, current compatibility, forward compatibility, and forward flexibility. ISO has set tests for these goals, which are detailed below.

Universal Acceptance — The tests for universal acceptance include the standards are widely adopted and used worldwide, there are few customer complaints in proportion to the volume of use, and there are few sector-specific supplementary or derivative standards under development.

Current Compatibility — The tests for current compatibility include "part number" supplements to existing standards do not change or conflict with requirements in the existing or parent document, the numbering and the clause structure of a supplement facilitate combine use of the parent document and the supplement, and supplements are not stand-alone documents, but are to be used with the parent document.

Forward Compatibility — The tests for forward compatibility include revisions affecting requirements in existing standards are few in number and minor or narrow in scope, and revisions are accepted for existing as well as new contracts.

Forward Flexibility — The tests for forward flexibility include supplements are few in number, but can be combined as needed to meet the needs of virtually any industry/economic sector or generic category of products, and supplement or addendum architecture allows new features or requirements to be consolidated into the parent document at a subsequent revision if the provisions of the supplement are found to be used (almost) universally.

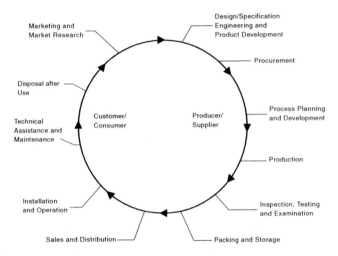

Figure 51.1   A display of final satisfaction and customer expectations.

ISO 9000 Product Categories

Hardware — Hardware is considered to be products consisting of manufactured pieces and parts, or assemblies of parts.

Software — Software is considered to be products, such as computer software, consisting of written, or otherwise recordable information, concepts, transactions, or procedures.

Processed Materials — Processed materials are considered to be products (final to intermediate) consisting of solids, liquids, gases, or combinations thereof, including particulate materials, ingots, filaments, or sheet structures. Processed materials are typically delivered (packaged) in containers such as drums, bags, tanks, cans, pipelines, or rolls.

Services — Services are considered to be intangible products that may be the entire or principal offering, or incorporated features of the offering, relating to activities such as planning, selling, directing, delivering, improving, evaluating, training, operating, or servicing for a tangible product.

For each of the product categories discussed above there is a quality loop that interacts with all activities pertinent to the quality of the product or service. It involves all phases from initial identification to final satisfaction and customer expectations as shown in Figure 51.1. For service functions a more detailed quality loop is provided in Figure 51.2.

Organization of ISO 9000 Documentation

ISO 9000 is organized into four separate documents:

9001: Quality Systems — Model for Assurance in Design/Development, Production, Installation and Servicing
9002: Quality Systems — Model for Assurance in Production and Installation

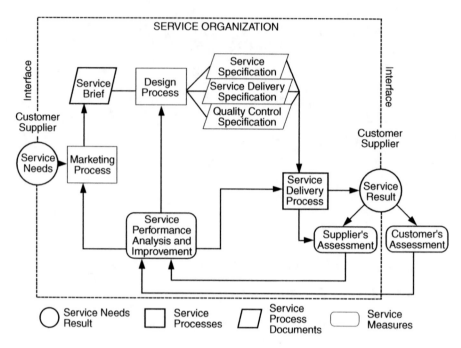

Figure 51.2 A detailed description of service functions.

9003: Quality Systems — Model for Assurance Final Inspection and Test
9004: Quality Management and Quality System Elements — Guidelines

The relationship between ISO 9000 documentation and producer's/supplier's and purchaser's are shown in Figure 51.3.

ISO 9000 Definitions

Some key ISO 9000 concepts are defined below. The relationships between these concepts are illustrated in Figure 51.4.

Quality Policy — Quality policy is the overall quality intentions and direction of an organization as it concerns quality, as formally expressed by top management. The quality policy forms one element of the corporate policy and is authorized by top management.

Quality Management — Quality management is that aspect of the overall management function that determines and implements the quality policy. The attainment of the desired quality requires commitment and participation of all members of the organization, whereas the responsibility for quality management belongs to top management. Quality management includes strategic planning, allocation of resources, and other systematic activities for quality, such as quality planning, operations and evaluations.

Quality System — A quality system is the organization structure, responsibilities, procedures, processes, and resources for implementing quality management.

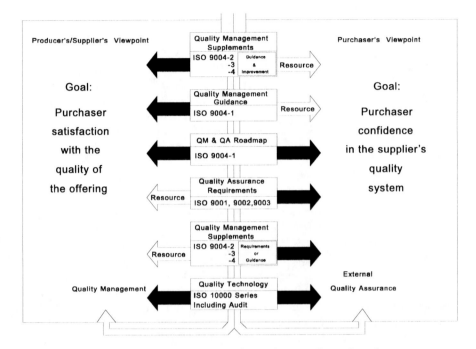

Figure 51.3 Shows the relationship between the product supplier and purchaser.

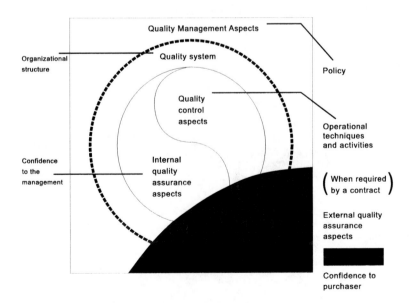

Figure 51.4 An illustration of key ISO 9000 concepts.

Quality Control — Quality control is the operational techniques and activities that are used to fulfill requirements for quality. This includes monitoring processes to eliminate the causes of unsatisfactory performance to improve the processes continually.

Quality Assurance — Quality assurance is all those planned and systematic actions necessary to provide adequate confidence that a product or service will satisfy the given requirements for quality. Some of the related aspects include continuing evaluation of the design or specification and audits of production, installation, and inspection operations. These activities provide confidence to management.

Key Aspects of ISO 9000

There are three key aspects of ISO 9000: management responsibility, the quality system structure, and personnel and material resources. Figure 51.5 illustrates these key aspects and shows that customers are the center of these activities. It also illustrates that customer satisfaction can only be achieved when there is harmony among these aspects. These three aspects are discussed below:

Management Responsibility — Management is responsible for establishing a policy for quality and customer satisfaction. Successful implementation of this policy is dependent on the commitment of management to the development and effective operation of a quality system. This includes having quality objectives, defining clear responsibility and authority guidelines, and conducting periodic reviews.

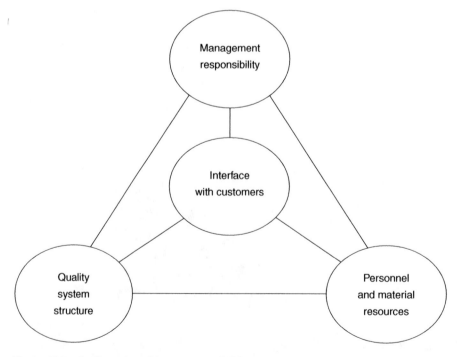

Figure 51.5   An illustration of key aspects of ISO 9000.

Personnel and Material Resources — Personnel is one of the most, if not the most, important resources any organization. Personnel issues include motivation, training and development, and communication. Material resources are also important to any organization, and may include provisioning of equipment, transportation, instrumentation, and documentation.

Quality System Structure — The quality system structure should focus on the output of the organization, which may be different for hardware, software, processed materials, and services. The structure should assure adequate control over all operational processes affecting quality. The quality system structure should also focus on preventive actions that avoid occurrence of problems while not sacrificing the ability to respond to and correct failures, should they occur.

ISO 9000 CONTENT ANALYSIS RELATED TO ENVIRONMENTAL, SAFETY, AND HEALTH PROFESSIONALS

Now that ISO 9000 has been properly introduced, it is time to discuss the safety-related aspects of ISO 9000. ISO 9000-9004 and ANSI/ASQC Q90-Q94 series documents contain information relevant to systematic management for product and service development, design, production, and installation activities, including safety, health, and environmental aspects. The following descriptions provide a summary of key sections of the ISO 9000-9004 and ANSI/ASQC Q90-Q94 that contain the words *safety*, *health*, *liability*, or *environment*. The attached excerpts include references to the established or draft standards. Additionally, many sections not cited contain elements directly applicable to managing safety into products and services, by design.

As you know, these standards represent an international movement to establish worldwide quality system standards for hardware, software, processed materials products, and services. Although strongly driven by the European community originally, the TQM vision represented by these standards will most probably drive U.S. businesses to become compliant and certified practitioners in the near future in order to conduct business internationally.

This summary provides the flavor of the direct challenge to our business that can be best related to traditional Aerospace Industry Product Safety, System Safety, and product liability efforts.

Quality — The totality of features and characteristics of a product or service that bear on its ability to satisfy stated or implied needs.... Needs are usually translated into features and characteristics with specific criteria (which) may include aspects of usability, safety, availability, maintainability, economics, and environment [ISO 8402-1986, Section 3.1, draft ISO/DIS 8402, Section 2.1, and ANSI/ASQC A3-1987, Section 2.1].

Design Review — A formal, documented, comprehensive and systematic examination of a design to evaluate the design requirements and the capability of the design to meet these requirements and to identify problems and propose solutions.... The capability of the design encompasses such things as fitness for purpose, feasibility, manufacturability, measurability, performance, reliability, maintainability, safety, environmental aspects, time scale, and life cycle cost [ISO 8402-1986, Section 3.13 and ANSI/ASQC A3-1987, Section 3.8].

Requirements of Society — Obligations resulting from laws, regulations, rules, codes, statutes, and other considerations [which] include protection of the environment, health, safety, security, conservation of energy and natural resources [draft ISO/DIS 8402, Section 2.4 and ISO 9004:1987, Section 3.3].

Safety — The state in which the risk of harm (to persons) or damage is limited to an acceptable level. NOTE: Safety is one of the aspects of quality [draft ISO/DIS 8402, Section 2.8].

Terms and Definitions ... Product Liability; Service Liability — A generic term used to describe the onus on a producer or others to make restitution for loss related to personal injury, property damage, or other harm caused by a product or service [ISO 8402-1986, Section 3.19, ISO/DIS 8402, Section 2.12, and ANSI/ASQC A3-1987, Section 3.12].

Selection of Model for Quality Assurance ... Selection Factors — In addition to functional criteria detailed in 8.2.1.a) to 8.3.1.c), the following six factors are considered fundamental for selecting the appropriate model for a product or service ... e) Product or service safety. This factor deals with the risk of occurrence of failure and the consequence of such failure [ISO 9000:1987, Section 8.2.3.e and ANSI/ASQC Q90-1987, Section 8.2.3.e].

Selection of Model for Quality Assurance ... Demonstration and Documentation — The nature and degree of demonstration may vary from one situation to another in accordance with such factors as ... e) the safety requirements of the product or service [ISO 9000: 1987, Section 8.3.e and ANSI/ASQC Q90-1987, Section 8.3.e].

Annex ... Cross Reference List of Quality System Elements ... 19 — Product Safety and Liability [ISO 9000: 1987 and ANSI/ASQC Q90-1987].

Design Control ... General — The essential quality aspects, such as safety, performance, and dependability of a product (whether hardware, software, services, or processed materials) are established during design and development phase [draft ISO/DIS 9000-2, Section 4.4.1].

Design Control ... Design and Development Planning — The supplier should establish procedures for design and development planning that include plans for evaluating safety, performance, and dependability incorporated in the product design [draft ISO/DIS 9000-2, Section 4.4.2].

Design Control ... Design Verification — Design reviews for the purpose of design verification can consider questions such as ... are safety considerations covered? [draft ISO/DIS 9000-2, Section 4.4.5].

Design Control ... Design Changes — Design of a product may be changed or modified for a number of reasons. For example: ... safety, regulatory, or other requirements have been changed [draft ISO/DIS 9000-2, Section 4.4.6].

Purchaser's Requirements Specification ... General — In order to proceed with software development, the supplier should have a complete, unambiguous set of functional requirements [which] should include all aspects necessary to satisfy the purchaser's need. These may include, but are not limited to, the following: performance, safety, reliability, security, and privacy [ISO 9000-3:1991, Section 5.3.1].

Development Planning ... Output from development phases — The required output from each [software] development phase should be defined and documented. The output from each development phase should be verified and should ... d) identify those characteristics of the product that are crucial to its safe and proper functioning [ISO 9000-3:1991, Section 5.4.5.d].

Process Control ... General — The supplier shall identify and plan the production and, where applicable, installation processes which directly affect quality and shall ensure that these procedures are carried out under controlled conditions. Controlled conditions shall include the following: a) ... use of suitable production and installation equipment, suitable working environment, compliance with reference standards/codes, and quality plans [ISO 9001: 1987, Section 4.9.1.a) and ANSI/ASQC Q91-1987, Section 4.9.1.a)].

Process Control ... General — The supplier shall identify and plan the production and, where applicable, installation processes which directly affect quality and shall ensure that these procedures are carried out under controlled conditions. Controlled conditions shall include the following: a) ... use of suitable production and installation equipment, suitable working environment, compliance with reference standards/codes, and quality plans [ISO 9002: 1987, Section 4.8.1.a) and ANSI/ASQC Q92-1987, Section 4.8.1.a)].

Risk Considerations ... For the Customer — Consideration has to be given to risks such as those pertaining to the health and safety of people, dissatisfaction with goods and services [ISO 9004: 1987, Section 0.4.2.2 and ANSI/ASQC Q94-1987, Section 0.4.2.2].

Management Responsibility ... Quality Objectives — For the Corporate quality policy, management should define objectives pertaining to key elements of quality, such as fitness for use, performance, safety, and reliability [ISO 9004: 1987, Section 4.3.1, and ANSI/ASQC Q94, Section 4.3.1].

Economics ... Types of Quality-Related Costs ... Operating Quality Costs — Operating Quality costs are those incurred by a business in order to attain and ensure specified quality levels. These include the following: b) Failure costs (or losses) ... external failure: Costs resulting from a product or service failing to meet quality requirements after delivery (e.g., ...liability costs) [ISO 9000.4, Section 6.3.2.b) and ANSI/ASQC Q94-1987, Section 6.3.2.b)].

Quality in Specification and Design ... Design planning and objectives (defining the project) — In addition to customer needs, the designer should give due consideration to the requirements relating to safety, environmental and other regulations, including items in the company's quality policy which go beyond existing statutory requirements [ISO 9004:1987, Section 8.2.4 and ANSI/ASQC Q94-1987, Section 8.2.4].

Quality in Specification and Design ... Design Planning and Objectives (defining the project) — The quality aspects of the design should be unambiguous and adequately define characteristics important to quality, such as the acceptance and rejection criteria. Both fitness for purpose and safeguards against misuse should be considered. Product definition may also include reliability, ... including benign failure and safe disposability [ISO 9004: 1987, Section 8.2.5 and ANSI/ASQC Q94-1987, Section 8.2.5].

Quality in Specification and Design ... Design qualification and validation — The design process should provide periodic evaluation of the design at significant stages. Such evaluation can take the form of analytical methods, such as Failure Mode and Effects Analysis (FMEA), fault tree analysis or risk assessment, as well as inspection or test of prototype models and/or actual production samples.... The tests should include the following activities: a) evaluation of performance, durability, safety, reliability and maintainability under expected storage and operational conditions [ISO 9004:1987, Section 8.4 and ANSI/ASQC Q94-1987, Section 8.4].

Design Review ... Elements of Design Reviews — as appropriate to the design phase and product, the following elements outlined below should be considered: a) Items pertaining to customer needs and satisfaction... 5) safety and environmental compatibility; ... b) Items pertaining to product specification and service requirements... 4) ...disposability; 5) benign failure and fail-safe characteristics; 7) failure modes and effects analyses, and fault tree analysis; 9) Labeling, warnings, ... and user instructions; c) Items pertaining to process specifications and service requirements. 3) specification of materials, components, ... including approved supplies... 4) packaging, handling, and shelf-life requirements, especially safety factors related to incoming and outgoing items [ISO 9004:1987, Section 8.5.2 and ANSI/ASQC Q94-1987, Section 8.5.2].

Corrective Action ... Disposition of Nonconforming Items — Recall decisions are affected by considerations of safety, product liability, and customer satisfaction [ISO 9004:1987, Section 15.8, and ANSI/ASQC Q94-1987, Section 15.8].

Handling and Post-Production Functions ... Installation — Instructional documents should contribute to proper installations and should include provisions which preclude improper installation or factors degrading the quality, reliability, safety, and performance of any product or material [ISO 9004:1987, Section 16.1.5 and ANSI/ASQC Q94-1987, Section 16.1.5].

Product Safety and Liability — The safety aspects of product or service quality should be identified with the aim of enhancing product safety and minimizing product liability. Steps should be taken to both limit the risk of product liability and to minimize the number of cases by a) identifying relevant safety standards in order to make the formulation of product or service specifications more effective; b) carrying out design evaluation tests and prototype (or model) testing for safety and documenting the test results; c) analyzing instructions and warnings to the user, maintenance manuals and labeling and promotional material in order to minimize misinterpretation; d) developing a means of traceability to facilitate product recall if features are discovered which compromise safety and to allow a planned investigation of products or services suspected of having unsafe features (see 15.4, Investigation of possible causes, and 16.1.3, Identification) [ISO 9004:1987, Section 19, and ANSI/ASQC Q94-1987, Section 19].

Use of Statistical Methods ... Statistical Techniques — Specific statistical methods and applications available include, but are not limited to, the following: ... c) safety evaluation/risk analysis [ISO 9004:1987, Section 20.2 and ANSI/ASQC Q94-1987, Section 20.2].

Appendix B — Product Liability and User Safety — This Appendix is not part of ANSI/ASQC Q94-1987..., and is not included in ISO 9004, but is included for information purposes only [ANSI/ASQC Q94-1987, Appendix B].

Characteristics of Service ... Service and Service Delivery Characteristics — Examples of characteristics that might be specified in requirement documents include: ... hygiene, safety, reliability, and security [ISO 9004-2:1991, Section 4.1].

Quality System Operational Elements ... Marketing (Services) Process ... Quality in Market Research and Analysis — Management should establish procedures for planning and implementing market activities. Elements associated with quality in marketing should include: ... review of legislation (e.g. health, safety, and environmental) and relevant national and international standards and codes [ISO 9004-2:1991, Section 6.1.1].

Quality System Operational Elements ... Design Process ... Service Delivery Specification ... General — The service delivery specification should take account of the aims, policies, and capabilities of the service organization, as well as any health, safety, environmental, or other legal requirements [ISO 9004-2:1991, Section 6.2.4.1].

COMPARISON BETWEEN ISO 9000:1994 AND ISO 9000:2000

2000 Version	1994 Version

Foreword

The International Organization for Standardization (ISO) is a worldwide federation of national standards bodies (ISO member bodies). The work of preparing International Standards is carried out through ISO technical committees. Each member body interested in a subject for which a technical committee has been established has the right to be represented on that committee. International organizations, governmental and non-governmental, in liaison with ISO, also take part in the work. ISO collaborates closely with the International Electrotechnical Commission (IEC) on all matters of electrotechnical standardization.

International Standards are drafted in accordance with the rules given in the ISO/IEC Directives, Part 3.

Draft International Standards adopted by the technical committees are circulated to the member bodies for voting. Publication as an International Standard requires approval by at least 75% of the member bodies casting a vote.

Attention is drawn to the possibility that some of the elements of this International Standard may be the subject of patent rights. ISO shall not be held responsible for identifying any or all such patent rights.

International Standard, ISO 9001 was prepared by Technical Committee ISO/TC 176, *Quality management and quality assurance,* Subcommittee SC-2, *Quality systems.*

This third edition of ISO 9001 cancels and replaces the second edition (ISO 9001:1994), which has been technically revised. The provisions of ISO 9002:1994 and ISO 9003:1994 are addressed in this International Standard. ISO 9002:1994 and ISO 9003:1994 will be withdrawn on the publication of ISO 9001:2000. Those organizations which have used ISO 9002:1994 and ISO 9003:1994 in the past may use this International Standard by excluding certain requirements in accordance with 1.2.

This edition of ISO 9001 carried a revised title, which no longer includes the term Quality Assurance. This reflects the fact that the quality management system requirements specified in this edition of ISO 9001 address quality assurance of product as well as customer satisfaction.

Annexes A and B of this International Standard are for information only.

Foreword

ISO (the International Organization for Standardization) is a worldwide federation of national standards bodies (ISO member bodies). The work of preparing International Standards is normally carried out through ISO technical committees. Each member body interested in a subject for which a technical committee has been established has the right to be represented on that committee. International organizations, governmental and non-governmental, in liaison with ISO, can also take part in the work. ISO collaborates closely with the International Electrotechnical Commission (IEC) on all matters of electrotechnical standardization.

Draft International Standards adopted by the technical committees are circulated to the member bodies for voting. Publication as an International Standard requires approval by at least 75% of the member bodies casting a vote.

International Standard ISO 9001 was prepared by Technical Committee ISO/TC 176, *Quality management and quality assurance,* Subcommittee SC2, *Quality systems.*

This second edition cancels and replaces the first edition (ISO 9001:1987), which has been technically revised.

Annex A of this International Standard is for information only.

2000 Version	1994 Version

0 Introduction

0.1 General

This International Standard specifies requirements for a quality management system that can be used by an organization to address customer satisfaction, by meeting customer and applicable regulatory requirements. It can also be used by internal and external parties, including certification bodies, to assess the organization's ability to meet such customer and regulatory requirements.

The adoption of a quality management system needs to be a strategic decision of the organization. The design and implementation of an organization's quality management system is influenced by varying needs, particular objectives, the products provided, the processes employed and the size and structure of the organization. It is not the purpose of this International Standard to imply uniformity in the structure of quality management systems or uniformity of documentation.

It is emphasized that the quality management system requirements specified in this International Standard are complementary to technical requirements for products.

Introduction

This International Standard is one of three International Standards dealing with quality system requirements that can be used for external quality assurance purposes. The quality assurance models, set out in the three International Standards listed below, represent three distinct forms of quality system requirements suitable for the purpose of a supplier demonstrating its capability, and for the assessment of the capability of a supplier by external parties.

a) ISO 9001, *Quality systems — Model for quality assurance in design, development, production, installation and servicing*

 for use when conformance to specified requirements is to be assured by the supplier during design, development, production, installation and servicing.

b) ISO 9002, *Quality systems — Model for quality assurance in production, installation and servicing.*

 for use when conformance to specified requirements is to be assured by the supplier during production, installation and servicing.

c) ISO 9003, *Quality systems — Model for quality assurance in final inspection and test*

 for use when conformance to specified requirements is to be assured by the supplier solely at final inspection and test.

It is emphasized that the quality system requirements specified in this International Standard, ISO 9002 and ISO 9003 are complementary (not alternative) to the technical (product) specified requirements. They specify requirements which determine what elements quality systems have to encompass, but it is not the purpose of these International Standards to enforce uniformity of quality systems. They are generic and independent of any specific industry or economic sector. The design and implementation of a quality system will be influenced by particular objectives, the products and services supplied, and the processes and specific practices employed.

It is intended that these International Standards will be adopted in their present form, but on occasions they may need to be tailored[a] by adding or deleting certain quality system requirements for specific contractual situations. ISO 9000-1 provides guidance on such tailoring as well as on selection of the appropriate quality assurance model, i.e., ISO 9001, ISO 9002 or ISO 9003.

[a] The issue of "tailoring" is addressed in clause 1. Scope.

2000 Version	1994 Version

0.2 Process Approach

This International Standard encourages the adoption of a process approach to quality management.

Any activity that receives inputs and converts them to outputs can be considered as a process. For organizations to function effectively, they have to identify and manage numerous linked processes. Often the output from one process will directly form the input into the next process. The systematic identification and management of the processes employed within an organization and the interactions between such processes, may be referred to as the "process approach."

Figure 51.6 is a conceptual illustration of one model of the process approach presented in clauses 5 to 8. The model recognizes that customers play a significant role in defining requirements as inputs. Monitoring of customer satisfaction is necessary to evaluate and validate whether customer requirements have been met. This model does not reflect processes at a detailed level, but covers all the requirements of this International Standard.

Note: There is no equivalent to this part of the Introduction.

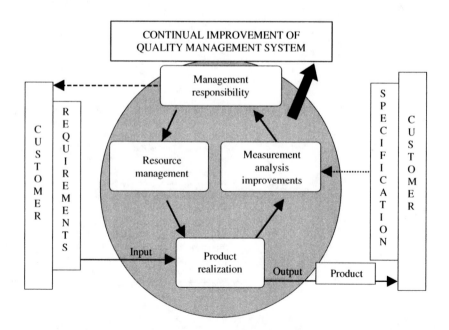

Figure 51.6   Conceptual illustration of a process approach model.

2000 Version	1994 Version

0.3   Relationship with ISO 9004

This edition of ISO 9001 has been developed as one part of a consistent pair of quality management system standards, the other being ISO 9004:2000. The two International Standards are designed to be used together, but can also be used independently. Although the two International Standards have different scopes, they have similar structures for ease of use.

Note: There is no equivalent to this part of the Introduction.

This edition of ISO 9001 specifies requirements for a quality management system that may be used for internal application by organizations, certification, or contractual purposes.

ISO 9004:2000 gives guidance on a wider range of objectives of a quality management system to improve an organization's overall performance. ISO 9004:2000 is not a guideline for implementing ISO 9001:2000 and is not intended for certification or contractual use.

0.4   Compatibility with other management systems

Note: There is no equivalent to this part of the Introduction.

This International Standard is intended to be compatible with other internationally recognized management system standards. It is aligned with ISO 14001:1996 in order to enhance the compatibility of the two standards for the benefit of the user community.

This International Standard does not include requirements specific to other management systems, such as those particular to environmental management, occupational health and safety management, or financial management. However, this International Standard allows an organization to align or integrate its own quality management system with related management system requirements. In some cases, it may be possible for an organization to adapt it existing management system(s) in order to establish a quality management system that complies with the requirements of this International Standard.

2000 Version	1994 Version
Scope	**Scope**

Scope

1.1 General

This International Standard specifies requirements for a quality management system where an organization needs:

a) To demonstrate its ability to provide consistently product that meet customer and applicable regulatory requirements and

b) to address customer satisfaction through the effective application of the system, including processes for continual improvement and the prevention of nonconformity.

NOTE: Monitoring of customer satisfaction, as stated in b), requires the evaluation of information relating to customer perceptions of whether or not the organization has met the customer requirements.

The requirements specified in this International Standard are generic and applicable to all organizations, regardless of type, size and product provided.

It is intended that all requirements of this International Standard be applied. However, certain requirements may be excluded in particular situations (see 1.2).

1.2 Permissible exclusion

The organization may only exclude quality management system requirements that neither affect the organization's ability, nor absolve it from its responsibility, to provide product that meets customer and applicable regulatory requirements. These exclusions are limited to those requirements. These exclusions are limited to those requirements within clause 7 (see also 5.5.5), and may be due to the following:

1. the nature of the organization's product;

2. customer requirements;

3. applicable regulatory requirements.

Where permissible exclusions are exceeded, conformity to this International Standard should not be claimed. This includes situations where the fulfillment of regulatory requirements permits exclusions that exceed those allowed by this International Standard.

Scope

This International Standard specifies quality system requirements for use where a supplier's capability to design and supply conforming product needs to be demonstrated.

The requirements specified are aimed primarily at achieving customer satisfaction by preventing nonconformity at all stages from design through to servicing:

a) Design is required and the product requirements are stated principally in performance terms, or they need to be established, and

b) Confidence in product conformance can be attained by adequate demonstration of a supplier's capabilities in design, development, production, installation and servicing.

Note: There is no equivalent clause in ISO 9001:1994 but the issue of "permissible exclusions" is addressed in the Introduction; the relevant part of which is reproduced here.

It is emphasized that the quality system requirements specified in this International Standard, ISO 9002 and ISO 9003 are complementary (not alternative) to the technical (product) specified requirements. They specify requirements which determine what elements quality systems have to encompass, but it is not the purpose of these International Standards to enforce uniformity of quality systems.. They are generic and independent of any specific industry or economic sector. The design and implementation of a quality system will be influenced by the carrying needs of an organization, its particular objectives, the products and services supplied, and the processes and specific practices employed.

It is intended that these International Standards will be adopted in their present form, but on occasions they may need to be tailored by

2000 Version	1994 Version
	adding or deleting certain quality system requirements for specific contractual situations. ISO 9000-1 provides guidance on such tailoring as well as on selection of the appropriate quality assurance model, *viz* ISO 9001, ISO 9002 or ISO 9003.

2 Normative reference

The following normative document contains provisions which, through reference in this text, constitute provisions of this International Standard. For dated references, subsequent amendments to, or revisions of, this publication does not apply. However, parties to agreements based on this International Standard are encouraged to investigate the possibility of applying the most recent edition of the normative document indicated below. Members of ISO and IEC maintain registers of currently valid International Standards.

ISO 9000:2000, *Quality management systems — Fundamentals and vocabulary.*

3 Terms and definitions

For the purposes of this International Standard, the terms and definitions given in ISO 9000:2000, and the following, apply.

NOTE: The terms used in this edition of this International Standard to describe the supply-chain are as follows:

Supplier   organization   customer

The term organization replaces the previously used term supplier, to mean the unit to which this International Standard applies. The term supplier is now used instead of the previous term "subcontractor." The changes have been introduced to reflect the vocabulary used by organizations.

3.1 Product

Result of a process

Note 1 There are four agreed generic product categories:

– hardware,
– software,
– services,
– processed materials.

Most products are combinations of some of the four generic product categories.

Whether the combined product is then called hardware, processed material, software or service depends on the dominant element.

NOTE 2 Adapted from ISO 9000:2000.

---

2 Normative reference

The following standard contains provisions which, through reference in this text, constitute provisions of this International Standard. At the time of publication, the edition indicated was valid. All standards are subject to revision, and parties to agreements based on this International Standard are encouraged to investigate the possibility of applying the most recent edition of the standard indicated below. Members of IEC and ISO maintain registers of currently valid International Standards.

ISO 8402:1994, *Quality management and quality assurance — Vocabulary.*

3 Definitions

For the purpose of this International Standard, the definitions given in ISO 8402 and the following definitions apply.

3.1 **Product:** Result of activities or processes.

NOTES

2   A product may include service, hardware, processed materials, software or a combination thereof.

3   A product can be tangible (e.g. assemblies or processed materials) or intangible (e.g. knowledge or concepts), or a combination thereof.

4   For the purposes of this International Standard, the term "product" applies to the intended product offering only and not to unintended "by-products" affecting the environment. This differs from the definition given in ISO 8402.

3.2 **Tender:** Offer made by a supplier in response to an invitation to satisfy a contract award to provide product.

3.3 **Contract:** Agreed requirements between a supplier and customer transmitted by any means.

2000 Version	1994 Version

4 Quality management system requirements

4.1 General requirements

The organization shall establish, document, implement, maintain and continually improve a quality management system in accordance with the requirements of this International Standard.

To implement the quality management system, the organization shall:

a) identify the processes needed for the quality management system;

b) determine the sequence and interaction of these processes;

c) determine criteria and methods required to ensure the effective operation and control of these processes;

d) ensure the availability of information necessary to support the operation and monitoring of these processes;

e) measure, monitor and analyze these processes, and implement action necessary to achieve planned results and **continual improvement**.

The organization shall manage these processes in accordance with the requirements of this International Standard.

4.3   General documentation requirements

The quality management system documentation shall include:

a) documented procedures required in this International Standard.

b) Documents required by the organization to ensure the effective operation and control of its processes.

Note 1: Where the term documented procedure appears within this International Standard, this requires the procedure to be established, documented, implemented and maintained.

The extent of the quality management system documentation shall be dependent on the following:

a) size and type of the organization;

b) complexity and interaction of the processes;

c) competence of personnel.

Note 2: The documented procedures and documents may be in any form or type of medium.

---

4   Quality system requirements

4.2.1 General

The supplier shall establish, document and maintain a quality system as a means of ensuring that product conforms to specified requirements. The supplier shall prepare a quality manual covering the requirements of this International Standard. The quality manual shall include or make reference to the quality system procedures and outline the structure of the documentation used in the system.

Note 6: Guidance on quality manuals is give in ISO 10013.

[*Note:* The screened text is not referenced in clause 4 of the new Standard but is addressed in clause 5.6.3.]

4.3.1 Quality system procedures

The supplier shall

a) prepare documented procedures consistent with the requirements of this International Standard and the supplier's stated quality policy, and

b) effectively implement the quality system and its documented procedures.

For the purpose of this International Standard, the range and detail of the procedures that form part of the quality system shall be dependent upon the complexity of the work, the methods used, and the skills and training needed by personnel involved in carrying out the activity.

Note 7: Documented procedures may make reference to work instructions that define how an activity is performed.

2000 Version	1994 Version

5 Management responsibility

5.1 Management commitment

Top management shall provide evidence of its commitment to the development and improvement of the quality management system by:

a) communicating to the organization the importance of meeting customers as well as regulatory and legal requirements;

b) Establishing the quality policy and quality objectives;

c) Conducting management reviews;

d) Ensuring the availability of necessary resources

[*Note:* There is no equivalent of this clause.]

5.2 Customer focus

Top management shall ensure that customer needs and expectations are determined, converted into requirements and fulfilled with the aim of achieving customer satisfaction.

Note: When determining customer needs and expectations, it is important to consider obligations related to product, including regulatory and legal requirements (see 7.2.1)

4.1.1 Quality policy

The supplier's management with executive responsibility shall define and document its policy for quality, including objectives for quality and its commitment to quality. The quality policy shall be relevant to the supplier's organizational goals and the expectations and needs of its customers. The supplier shall ensure that this policy is understood, implemented and maintained at all levels of the organization.

[*Note:* The screened text is dealt with in Clause 5.3 of the 2000 version.]

5.3 Quality policy

Top management shall ensure that the quality policy:

a) is appropriate to the purpose of the organization;

b) includes a commitment to meeting requirements and to continual improvement;

c) provides a framework for establishing and reviewing quality objectives;

d) is communicated and understood at appropriate levels in the organization;

e) is reviewed for continuing suitability.

The quality policy shall be controlled (see 5.5.6)

[*Note:* e) This is not a new requirement and is dealt with again in 5.6 and was covered in clause 4.1.3 of the 1994 version which is mapped against 5.6.]

4.1.1 Quality policy

The supplier's management with executive responsibility shall define and document its policy for quality, including objectives for quality and its commitment to quality. The quality policy shall be relevant to the supplier's organizational goals and the expectations and needs of its customers. The supplier shall ensure that this policy is understood, implemented and maintained at all levels of the organization.

[*Note:* Quality objectives are also dealt with in clause 5.4.]

2000 Version	1994 Version

5.4 Planning

5.4.1 Quality objectives

Top management shall establish quality objectives at relevant functions and levels within the organization. The quality objectives shall be measurable and consistent with the quality policy, including the commitment to continual improvement. Quality objectives shall include those needed to meet requirements for product (See 7.1).

5.4.2 Quality planning

Top management shall ensure that the resources needed to achieve the quality objectives are identified and planned. The output of the planning shall be documented.

Quality planning shall include:

a) the processes of the quality management system, considering permissible exclusions (see 1.2);

b) the resources needed;

c) continual improvement of the quality management system.

Planning shall ensure that change is conducted in a controlled manner and that the integrity of the quality management system is maintained during this change.

4.1.1 Quality policy

The supplier's management with executive responsibility shall define and document its policy for quality, including objectives for quality and its commitment to quality. The quality policy shall be relevant to the supplier's organizational goals and the expectations and needs of its customers. The supplier shall ensure that this policy is understood, implemented and maintained at all levels of the organization.

[*Note:* The screened text is dealt with in 5.3.]

4.2.3 Quality planning

The supplier shall define and document how the requirements for quality will be met. Quality planning shall be consistent with all other requirements of a supplier's quality system and shall be documented in a format to suit the supplier's method of operation.

The supplier shall give consideration to the following activities, as appropriate, in meeting the specified requirements for products, projects or contracts:

a) the preparation of quality plans;

b) the identification and acquisition of any controls, processes, equipment (including inspection and test equipment), fixtures, resources and skills that may be needed to achieve the required quality;

c) ensuring the compatibility of the design, the production process, installation, servicing, inspection and test procedures and the applicable documentation;

d) the updating, as necessary, of quality control, inspection and testing techniques, including the development of new instrumentation;

e) the identification of any measurement requirement involving capability that exceeds the known state of the art, in sufficient time for the needed capability to be developed;

f) the identification of suitable verification at appropriate stages in the realization of product;

g) the clarification of standards of acceptability for all features and requirements, including those which contain a subjective element;

h) the identification and preparation of quality records (see 4.16).

2000 Version	1994 Version
	Note: The quality plans referred to [see 4.2.3a)] may be in the form of a reference to the appropriate documented procedures that form an integral part of the supplier's quality system. [*Note:* Note that e) has not been carried over into the 2000 version.]

5.5   Administration

5.5.1 General

The following clauses describe the administration of the quality management system.

[*Note:* There is no equivalent of this clause.]

5.5.2 Responsibility and authority

Functions and their interrelations within the organization, including responsibilities and authorities, shall be defined and communicated in order to facilitate effective quality management.

4.1.2.1 Responsibility and authority

The responsibility, authority and the interrelation of personnel who manage, perform and verify work affecting quality shall be defined and documented, particularly for personnel who need the organizational freedom and authority to:

a) initiate action to prevent the occurrence of any nonconformities relating to the product, process and quality system;

b) identify and record any problems relating to the product, process and quality system;

c) initiate, recommend or provide solutions through designated channels;

d) verify the implementation of solutions;

e) control further processing, delivery or installation of nonconforming product until the deficiency or unsatisfactory condition has been corrected.

5.5.3 Management representatives

Top management shall appoint member(s) of the management who, irrespective of other responsibilities, shall have responsibility and authority that includes:

a) ensuring that processes of the quality management system are established and maintained;

b) reporting to top management on the performance of the quality management system, including needs for improvement;

c) promoting awareness of customer requirements throughout the organization.

Note: The responsibility of a management representative can include liaison with external parties on matters relating to the quality management system.

4.1.2.2 Management representative

The supplier's management with executive responsibility shall appoint a member of the supplier's own management who, irrespective of other responsibilities, shall have defined authority for:

a) ensuring that a quality system is established, implemented and maintained in accordance with this International Standard, and

b) reporting on the performance of the quality system to the supplier's management for review and as a basis for improvement of the quality system.

Note 5: The responsibility of a management representative may also include liaison with external parties on matters relating to the supplier's quality system.

2000 Version	1994 Version

5.5.4 Internal communication

The organization shall ensure communication between various levels and functions regarding the processes of the quality management system and their effectiveness.

[*Note:* There is no equivalent of this clause.]

5.5.5 Quality Manual

A quality manual shall be established and maintained that includes the following:

a) the scope of the quality management system, including details of, and justification for, any exclusions (see 1.2);

b) documented procedures or reference to them;

c) a description of the processes of the quality management system and their interaction.

The quality manual shall be controlled. (See 5.5.6)

Note: The quality manual can be part of the overall documentation of the organization.

4.2.1 General

The supplier shall establish, document and maintain a quality system as a means of insuring that product conforms to specified requirements. The supplier shall prepare a quality manual covering the requirements of this International Standard. The quality manual shall include or make reference to the quality system procedures and outline the structure of the documentation used in the quality system.

Note 6: Guidance on quality manuals is given in ISO 10013.

5.5.6 Control of documents

Documents required for the quality management system shall be controlled. A documented procedure shall be established:

a) to approve documents for adequacy prior to issue;

b) to review, update as necessary and re-approve documents;

c) to identify the current revision status of documents;

d) to ensure that relevant versions of applicable documents are available at points of use;

e) to ensure that documents remain legible, readily identifiable and retrievable;

f) to ensure that documents of external origin are identified and their distribution controlled;

g) to prevent the unintended use of obsolete documents, and to apply suitable identification to them if they are retained for any purpose.

Documents defined as quality records shall be controlled (see 5.5.7).

[*Note:* Refer to 4.2 Note which indicates that documented procedures and documents may be in any form or any type of media.]

4.5 Document and data control

4.5.1 General

The supplier shall establish and maintain documented procedures to control all documents and data that relate to the requirements of this International Standard including, to the extent applicable, documents of external origin such as standards and customer drawings.

Note 15: Documents and data can be in the form of any type of media, such as hard copy or electronic media.

4.5.2 Document and data approval and issue

The documents and data shall be reviewed and approved for adequacy by authorized personnel prior to use. A master list or equivalent document control procedure identifying the current revision status of documents shall be established and be readily available to preclude the use of invalid and/or obsolete documents.

This control shall ensure that:

a) the pertinent issues of appropriate documents are available at all locations where operations essential to the effective functioning of the quality system are performed;

2000 Version	1994 Version
	b) invalid and/or obsolete documents are promptly removed from all points of issue or use, or otherwise assured against unintended use;
	c) any obsolete documents retained for legal and/or knowledge-preservation purposes are suitably identified.

4.5.3 Document and data changes

Changes to documents and data shall be reviewed and approved by the same functions/organizations that perform the original review and approval, unless specifically designated otherwise. The designated functions/organizations shall have access to pertinent background information upon which to base their review and approval.

Where practicable, the nature of the change shall be identified in the document or the appropriate attachments.

5.5.7 Control of quality records

Records required for the quality management system shall be controlled. Such records shall be maintained to provide evidence of conformance to requirements and of effective operation of the quality management system. A documented procedure shall be established for the identification, storage retrieval, protection, retention time and disposition of quality records.

4.16 Control of quality records

The supplier shall establish and maintain documented procedures for identification, collection, indexing, access, filing, storage, maintenance and disposition of quality records.

Quality records shall be maintained to demonstrate conformance to specified requirements and the effective operation of the quality system. Pertinent quality records from the subcontractor shall be an element of these data.

All quality records shall be legible and shall be stored and retained in such a way that they are readily retrievable in facilities that provide a suitable environment to prevent damage or deterioration and to prevent loss. Retention times of quality records shall be established and recorded. Where agreed contractually, quality records shall be made available for evaluation by the customer or the customer's representative for an agreed period.

NOTE 19 Records may be in the form of any type of media such as hard copy or electronic media.

2000 Version	1994 Version

5.6 Management review

5.6.1 General

Top management shall review the quality management system, at planned intervals, to ensure its continuing suitability, adequacy and effectiveness. The review shall evaluate the need for changes to the organization's quality management system, including quality policy and quality objectives.

5.6.2 Review input

Inputs to management review shall include current performance and improvement opportunities related to the following:

a) results of audits;

b) customer feedback;

c) process performance and product conformance;

d) status of preventive and corrective actions;

e) follow-up actions from earlier management reviews;

f) changes that could affect the quality management system.

5.6.3 Review output

The output from the management review shall include actions related to:

a) improvement of the quality management system and its processes;

b) improvement of product related to customer requirements;

c) resource needs.

Results of management reviews shall be recorded (see 5.5.7).

6 Resource management

6.1 Provision of resources

The organization shall determine and provide, in a timely manner, the resources needed:

a) to implement and improve the processes of the quality management system, and

b) to address customer satisfaction

6.2 Human resources

6.2.1 Assignment of personnel

Personnel who are assigned responsibilities defined in the quality management system shall be competent on the basis of applicable education, training, skills and experience.

4.1.3 Management review

The supplier's management with executive responsibility shall review the quality system at defined intervals sufficient to ensure its continuing suitability and effectiveness in satisfying the requirements of this International Standard and the supplier's stated quality policy and objectives (see 4.1.1). Records of such reviews shall be maintained (see 4.16).

[*Note:* There is no equivalent of this clause.]

4.1.2.2 Resources

The supplier shall identify resource requirements and provide adequate resources, including the assignment of trained personnel (see 4.18), for management, performance of work and verification activities including internal quality audits.

2000 Version	1994 Version

6.2.2 Training, awareness and competency

The organization shall:

a) determine competency needs for personnel performing activities affecting quality;

b) provide training to satisfy these needs;

c) evaluate the effectiveness of the training provided;

d) ensure that its employees are aware of the relevance and importance of their activities and how they contribute to the achievement of the quality objectives;

e) maintain appropriate records of education, experience, training and qualifications (see 5.5.7).

6.3 Facilities

The organization shall determine, provide, and maintain the facilities it needs to achieve the conformity of product, including:

a) workspace and associated facilities;

b) equipment, hardware and software;

c) supporting services.

6.4 Work environment

The organization shall identify and manage the human and physical factors of the work environment needed to achieve conformity of product.

4.17 Training

The supplier shall establish and maintain documented procedures for identifying training needs and provide for the training of all personnel performing activities affecting quality. Personnel performing specific assigned tasks shall be qualified on the basis of appropriate education, training and/or experience, as required. Appropriate records of training shall be maintained (see 4.16).

4.1.2.2 Resources

The supplier shall identify resource requirements and provide adequate resources, including the assignment of trained personnel (see 4.18), for management, performance of work and verification activities including internal quality audits.

[*Note:* The screened text is dealt with in 6.2.]

4.9 Process control

The supplier shall identify and plan the production, installation and servicing processes which directly affect quality and shall ensure that these processes are carried out under controlled conditions. Controlled conditions shall include the following:

a) documented procedures defining the manner of production, installation and servicing, where the absence of such procedures could adversely affect quality;

b) use of suitable production, installation and servicing equipment, and a suitable working environment;

c) compliance with reference standards/codes, quality plans and/or documented procedures;

d) monitoring and control of suitable process parameters and product characteristics;

e) the approval of processes and equipment, as appropriate;

f) criteria and workmanship, which shall be stipulated in the clearest practical manner (e.g. written standards, representative samples or illustrations);

g) suitable maintenance of equipment to ensure continuing process capability.

2000 Version	1994 Version
	[*Note:* The balance of clause 4.9 is not relevant here and is dealt with under 7.5.5 (see page 48).]

7   Product realization

7.1   Planning of realization processes

Product realization is that sequence of processes and sub-processes required to achieve the product. Planning of the realization processes shall be consistent with the other requirements of the organization's quality management system and shall be documented in a form suitable for the organization's method of operation.

In planning the processes for realization of products the organization shall determine the following, as appropriate:

a) quality objectives for the product, project or contract;

b) the need to establish processes and documentation, and provide resources and facilities specific to the product;

c) verification and validation activities, and the criteria for acceptability;

d) the records that are necessary to provide confidence of conformity of the processes and resulting product 175.

Note: Documentation that describes how the processes of the quality management system are applied for a specific product, project or contract may be referred to as a quality plan.

4.2.2   Quality planning

The supplier shall define and document how the requirements for quality will be met. Quality planning shall be consistent with all other requirements of a supplier's quality system and shall be documented in a format to suit the supplier's method of operation.

h) The supplier shall give consideration to the following activities, as appropriate, in meeting the specified requirements for products, project or contracts:

i) the preparation of quality plans;

j) the identification and acquisition of any controls, processes, equipment (including inspection and test equipment), fixtures, resources and skills that may be needed to achieve the required quality;

k) ensuring the compatibility of the design, the production process, installation, servicing, inspection and test procedures and the applicable documentation;

l) the updating, as necessary, of quality control, inspection and testing techniques, including the development of new instrumentation;

m) the identification of any measurement requirement involving capability that exceeds the known state of the art, in sufficient time for the needed capability to be developed;

n) the identification of suitable verification at appropriate stages in the realization of product;

o) the clarification of standards of acceptability for all features and requirements, including those which contain a subjective element;

p) the identification and preparation of quality records (see 4.16).

Note 8 The quality plans referred to [see 4.2.3 a)] may be in the form of a reference to the appropriate documented procedures that form an integral part of the supplier's quality system.

[*Note:* e) has not been carried over into the 2000 version.]

2000 Version	1994 Version

<table>
<tr><td></td><td>

4.9   Process control

The supplier shall identify and plan the production, installation and servicing processes which directly affect quality and shall ensure that these processes are carried out under controlled conditions. Controlled conditions shall include the following:

a) documented procedures defining the manner of production, installation and servicing, where the absence of such procedures could adversely affect quality;

b) use of suitable production, installation and servicing equipment, and a suitable working environment;

c) compliance with reference standards/codes, quality plans and/or documented procedures;

d) monitoring and control of suitable process parameters and product characteristics;

e) the approval of processes and equipment, as appropriate;

f) criteria for workmanship, which shall be stipulated in the clearest practical manner (e.g. written standards, representative samples or illustrations);

g) suitable maintenance of equipment to ensure continuing process capability.

[*Note:* The balance of Clause 4.9 is not relevant here and is addressed in clause 7.5.5 (see page 48).]
</td></tr>
</table>

7.2   Customer-related processes

7.2.1 Identification of customer requirements

The organization shall determine customer requirements including:

a) product requirements specified by the customer, including the requirements for availability, delivery and support;

b) product requirements not specified by the customer but necessary for intended or specified use;

c) obligations related to product, including regulatory and legal requirements.

7.2.2 Review of product requirements

The organization shall review the identified customer requirements together with additional requirements together with additional requirements determined by the organization.

4.3   Contract review

4.3.1 General

The supplier shall establish and maintain documented procedures for contract review and for the coordination of these activities.

4.3.2 Review

Before submission of a tender, or the acceptance of a contract or order (statement of requirement), the tender, contract or order shall be reviewed by the supplier to ensure that:

a) the requirements are adequately defined and documented; where no written statement of requirement is available for an order received by verbal means, the supplier shall ensure that the order requirements are agreed before their acceptance;

b) any differences between the contract or order requirements and those in the tender are resolved;

c) the supplier has the capability to meet the contract or order requirements.

2000 Version	1994 Version

This review shall be conducted prior to the commitment to supply a product to the customer (e.g. submission of a tender, acceptance of a contract or order) and shall ensure that:

a) product requirements are defined;

b) where the customer provides no documented statement of requirement, the customer requirements are confirmed before acceptance;

c) contract or order requirements differing from those previously expressed (e.g. in a tender or quotation) are resolved:

d) the organization has the ability to meet defined requirements.

The results of the review and subsequent follow up actions shall be recorded (see 5.5.7).

Where product requirements are changed, the organization shall ensure that relevant personnel are made award of the changed requirements.

7.2.3 Customer communication

The organization shall identify and implement arrangements for communication with customers relating to:

a) product information;

b) inquiries, contracts or order handling, including amendments;

c) customer feedback, including customer complaints.

7.3 Design and/or development

7.3.1 Design and/or development planning

The organization shall plan and control design and/or development of the product.

Design and/or development planning shall determine:

a) stages of design and/or development processes;

b) review, verification and validation activities appropriate to each design and/or development stage;

c) responsibilities and authorities for design and/or development activities.

4.3.3 Amendment to a contract

The supplier shall identify how an amendment to a contract is made and correctly transferred to the functions concerned within the supplier's organization.

4.3.4 Records

Records of contract reviews shall be maintained (see 4.16).

Note 9: Channels for communication and interfaces with the customer's organization in these contract matters should be established.

4.4 Design control

4.4.1 General

The supplier shall establish and maintain documented procedures to control and verify the design of the product in order to ensure that the specified requirements are met.

4.4.2 Design and development planning

The supplier shall prepare plans for each design and development activity. The plans shall describe or reference these activities, and define responsibility for their implementation. The design and development activities shall be assigned to qualified personnel equipped with adequate resources. The plans shall be updated as the design evolves.

2000 Version	1994 Version

Interfaces between different groups involved in design and/or development shall be managed to ensure effective communication and clarity of responsibilities.

Planning output shall be updated, as appropriate, as the design and/or development progresses.

4.4.3 Organizational and technical interfaces

Organizational and technical interfaces between different groups which input into the design process shall be defined and the necessary information documented, transmitted and regularly reviewed.

7.3.2 Design and/or development inputs

Inputs relating to product requirements shall be defined and documented. These shall include:

a) functional and performance requirements;

b) applicable regulatory and legal requirements;

c) applicable information derived from previous similar designs, and

d) any other requirements essential for design and/or development.

These inputs shall be reviewed for adequacy. Incomplete, ambiguous or conflicting requirements shall be resolved.

4.4.4 Design input

Design input requirements relating to the product, including applicable statutory and regulatory requirements, shall be identified, documented and their selection reviewed by the supplier for adequacy.

Incomplete, ambiguous or conflicting requirements shall be resolved with those responsible for imposing these requirements.

Design input shall take into consideration the results of any contract review activities.

7.3.3 Design and/or development outputs

The outputs of the design and/or development process shall be documented in a manner that enables verification against the design and/or development inputs.

Design and/or development output shall:

a) meet the design and/or development input requirements;

b) provide appropriate information for production and service operations (see 7.5);

c) contain or reference product acceptance criteria;

d) define the characteristics of the product that are essential to its safe and proper use.

Design and/or development output documents shall be approved prior to release.

4.4.5 Design Output

Design output shall be documented and expressed in terms that can be verified and validated against design input requirements.

Design output shall:

a) meet the design input requirements;

b) contain or make reference to acceptance criteria;

c) identify those characteristics of the design that are crucial to the safe and proper functioning of the product (e.g. operating, storage, handling, maintenance and disposal requirements).

Design output documents shall be reviewed before release.

7.3.4 Design and/or development review

At suitable stages, systematic reviews of design and/or development shall be conducted to:

a) evaluate the ability to fulfill requirements;

b) identify problems and propose follow-up actions.

Participants in such reviews shall include representatives of functions concerned with the design and/or development stage(s) being reviewed. The results of the reviews and subsequent follow-up actions shall be recorded (see 5.5.7).

4.4.6 Design review

At appropriate stages of design, formal documented reviews of the design results shall be planned and conducted. Participants at each design review shall include representative of all functions concerned with the design stage being reviewed, as well as other specialist personnel, as required. Records of such reviews shall be maintained (see 4.16).

2000 Version	1994 Version

7.3.5 Design and/or development verification

Design and/or development verification shall be performed to ensure the output meets the design and/or development inputs. The results of the verification and subsequent follow-up actions shall be recorded (see 5.5.7).

4.4.7 Design verification

At appropriate stages of design, design verification shall be performed to ensure that the design stage output meets the design stage input requirements. The design verification measures shall be recorded (see 4.16).

- performing alternative calculations;
- comparing the new design with a similar proven design, if available,
- undertaking tests and demonstrations, and
- reviewing the design stage documents before release.

7.3.6 Design and/or development validation

Design and/or development validation shall be performed to confirm that resulting product is capable of meeting the requirements for the intended use. Wherever applicable, validation shall be completed prior to the delivery or implementation of the product. Where it is impractical to perform full validation prior to delivery or implementation, partial validation shall be performed to the extent applicable.

The results of the validation and subsequent follow-up actions shall be recorded (see 5.5.7).

4.4.8 Design validation

Design validation shall be performed to ensure that product conforms to defined user needs and/or requirements.

NOTES

11 Design validation follows successful design verification (see 4.4.7).

12 Validation is normally performed under defined operating conditions.

13 Validation is normally performed on the final product, but may be necessary in earlier stages prior to product completion.

14 Multiple validations may be performed if there are different intended uses.

7.3.7 Control of design and/or development changes

Design and/or development changes shall be identified, documented and controlled. This includes evaluation of the effect of the changes on constituent parts and delivered products. The changes shall be verified and validated, as appropriate, and approved before implementation.

The results of the review of changes and subsequent follow up actions shall be documented (see 5.5.7).

Note: See ISO 10007 for guidance.

4.4.9 Design changes

All design changes and modifications shall be identified, documented, reviewed and approved by authorized personnel before their implementation.

7.4 Purchasing

7.4.1 Purchasing control

The organization shall control its purchasing processes to ensure purchased product conforms to requirements. The type and extent of control shall be dependent upon the effect on subsequent realization processes and their output.

4.5 Purchasing

4.6.1 General

The supplier shall establish and maintain documented procedures to ensure that purchased product (see 3.1) conforms to specified requirements.

2000 Version	1994 Version

The organization shall evaluate and select suppliers based on their ability to supply product in accordance with the organization's requirements. Criteria for selection and periodic evaluation shall be defined. The results of evaluations and subsequent follow-up actions shall be recorded (see 5.5.7).

4.6.2 Evaluation of subcontractors

The supplier shall:

a) evaluate and select subcontractors on the basis of their ability to meet subcontract requirements including the quality system and any specific quality assurance requirements;

b) define the type and extent of control exercised by the supplier over subcontractors. This shall be dependent upon the type of product, the impact of subcontracted product on the quality of final product and, where applicable, on the quality audit reports and/or quality records of the previously demonstrated capability and performance of subcontractors.

c) Establish and maintain quality records of acceptable subcontractors (see 4.16).

7.4.2 Purchasing information

Purchasing documents shall contain information describing the product to be purchased, including where appropriate:

a) requirements for approval or qualification of

 – product,

 – procedures,

 – processes,

 – equipment and

 – personnel;

b) quality management system requirements.

The organization shall ensure the adequacy of specified requirements contained in the purchasing documents prior to their release.

4.6.3 Purchasing data

Purchasing documents shall contain data clearly describing the product ordered, including where applicable:

a) the type, class, grade or other precise identification;

b) the title or other positive identification, and applicable issues of specifications, drawings, process requirements, inspection instructions and other relevant technical data, including requirements for approval or qualification of product, procedures, process equipment and personnel;

c) the title, number and issue of the quality system standard to be applied.

The supplier shall review and approve purchasing documents for adequacy of the specified requirements prior to release.

7.4.2 Verification of purchased products

The organization shall identify and implement the activities necessary for verification of purchased product.

Where the organization or its customer proposes to perform verification activities at the supplier's premises, the organization shall specify the intended verification arrangements and method of product release in the purchasing information.

4.6.3 Verification of purchased product

4.6.4.1 Supplier verification at subcontractor's premises

Where the supplier proposes to verify purchased product at the subcontractor's premises, the supplier shall specify verification arrangements and the method of product release in the purchasing documents.

2000 Version	1994 Version

4.6.4.2 Customer verification of subcontracted product

Where specified in the contract, the supplier's customer or the customer's representative shall be afforded the right to verify at the subcontractor's premises and the supplier's premises that subcontracted product conforms to specified requirements. Such verification shall not be used by the supplier as evidence of effective control of quality by the subcontractor.

4.10.2 Receiving inspection and testing

The supplier shall ensure that incoming product is not used or processed (except in the circumstances described in 4.10.2.3) until it has been inspected or otherwise verified as conforming to specified requirements. Verification of conformance to the specified requirements shall be in accordance with the quality plan and/or documented procedures.

4.10.2.2

In determining the amount and nature of receiving inspection, consideration shall be given to the amount of control exercised at the subcontractor's premises and the recorded evidence of conformance provided.

4.10.2.3

Where incoming product is released for urgent production purposes prior to verification, it shall be positively identified and recorded (see 4.16) in order to permit immediate recall and replacement in the event of nonconformity to specified requirements.

[*Note:* The screened text has not been carried forward into the 2000 version.]

7.4 Production and service operations

7.5.1 Operations control

The organization shall control production and service operations through:

a) the availability of information that specifies the characteristics of the products;

b) where necessary, the availability of work instructions;

c) the use and maintenance of suitable equipment for production and service operations;

d) the availability and use of measuring and monitoring devices;

e) the implementation of monitoring activities;

4.9 Process control

The supplier shall identify and plan the production, installation and servicing processes which directly affect quality and shall ensure that these processes are carried out under controlled conditions. Controlled conditions shall include the following:

a) documented procedures defining the manner of production, installation and servicing, where the absence of such procedures could adversely affect quality;

b) use of suitable production, installation and servicing equipment, and suitable working environment;

c) compliance with reference standards/codes, quality plans and/or documented procedures;

2000 Version	1994 Version
f) the implementation of defined processes for release, delivery and applicable post-delivery activities.	d) monitoring and control of suitable process parameters and product characteristics;
[*Note:* f) comes from 4.10.4 of the 1994 version.]	e) the approval of processes and equipment, as appropriate;

The balance continues:

2000 Version

2000 Version (continued):

1994 Version (continued):

f) criteria for workmanship, which shall be stipulated in the clearest practical manner (e.g. written standards, representative samples or illustrations);

g) suitable maintenance of equipment to ensure continuing process capability.

[*Note:* The balance of clause 4.9 is addressed in clause 7.5.5.]

4.19   Servicing

Where servicing is a specified requirement, the supplier shall establish and maintain documented procedures for performing, verifying and reporting that servicing meets specified requirements.

7.5.2 Identification and traceability

The organization shall identify, where appropriate, the product by suitable means throughout production and service operations.

The organization shall identify the status of the product with respect to measurement and monitoring requirements.

The organization shall control and record the unique identification of the product, where traceability is a requirement (see 5.5.7).

4.8   Product identification and traceability

Where appropriate, the supplier shall establish and maintain documented procedures for identifying the product by suitable means from receipt and during all stages of production, delivery and installation.

Where and to the extent that traceability is a specified requirement, the supplier shall establish and maintain documented procedures for unique identification of individual product or batches. This identification shall be recorded (see 4.16).

4.12   Inspection and test status

The inspection and test status of product shall be identified by suitable means, which indicate the conformance or nonconformance of product with regard to inspection and tests performed. The identification of inspection and test status shall be maintained, as defined in the quality plan and/or documented procedures, throughout production, installation and servicing of the product to ensure that only product that has passed the required inspections and tests [or released under an authorized concession (see 4.13.2)] is dispatched, used or installed.

7.5.3 Customer property

The organization shall exercise care with customer property while it is under the organization's control or being used by the organization. The organization shall identify, verify, protect and maintain customer property provided for use of incorporation

4.7   Control of customer-supplied product

The supplier shall establish and maintain documented procedures for the control of verification, storage and maintenance of customer-supplied product provided for incorporation into the supplies or for related activities. Any such product that is lost,

2000 Version	1994 Version
into the product. Occurrence of any customer property that is lost, damaged or otherwise found to be unsuitable for use shall be recorded and reported to the customer. Note: Customer property may include intellectual property (e.g. information provided in confidence).	damaged or is otherwise unsuitable for use shall be recorded and reported to the customer (see 4.16). Verification by the supplier does not absolve the customer of the responsibility to provide acceptable product. [*Note:* They screened text has not been carried forward into the 2000 version.]

7.5.4 Handling, packaging, storage, preservation and delivery

The organization shall preserve conformity of product with customer requirements during internal processing and final delivery to the intended destination. This shall include identification, handling, packaging, storage and protection.

This shall also apply to constituent parts of a product.

4.15 Handling, storage, packaging, preservation and delivery

4.15.1 General

The supplier shall establish and maintain documented procedures for handling, storage, packaging, preservation and delivery of product.

4.15.2 Handling

The supplier shall provide methods of handling product that prevent damage or deterioration.

4.15.3 Storage

The supplier shall use designated storage areas of stock rooms to prevent damage or deterioration of product, pending use or delivery. Appropriate methods of authorizing receipt to and dispatch from such areas shall be stipulated.

In order to detect deterioration, the condition of product in stock shall be assessed at appropriate intervals.

4.15.4 Packaging

The supplier shall control packing, packaging and marking processes (including materials used) to the extent necessary to ensure conformance to specified requirements.

4.15.5 Preservation

The supplier shall apply appropriate methods for preservation and segregation of product when the product is under the supplier's control.

4.15.6 Delivery

The supplier shall arrange for the protection of the quality of product after final inspection and test. Where contractually specified, this protection shall be extended to include delivery to destination.

7.5.5 Validation of processes

The organization shall validate any production and service processes where the resulting output cannot be verified by subsequent measurement or monitoring. This includes

4.9 Process control

The supplier shall identify

f) suitable maintenance of equipment to ensure continuing process capability.

2000 Version	1994 Version
any processes where deficiencies may become apparent only after the product is in use or the service has been delivered.	[*Note:* Not all of the preceding text of clause 4.9 is shown as it is not relevant to clause 7.5.5 of the 2000 version.]
Validation shall demonstrate the ability of the processes to achieve planned results.	Where the results of processes cannot be fully verified by subsequent inspection and testing of the product and where, for example, processing deficiencies may become apparent only after the product is in use, the processes shall be carried out by qualified operators and/or shall be carried out by qualified operators and/or shall require continuous monitoring and control of process parameters to ensure that the specified requirements are met.
The organization shall define arrangements for validation and shall include, as applicable:	
a) qualification of processes;	
b) qualification of equipment and personnel;	
c) use of defined methodologies and procedures;	
d) requirements for records;	
e) re-validation.	The requirements for any qualification of process operations, including associated equipment and personnel (see 4.18), shall be specified.
	Note 16: Such processes requiring pre-qualification of their process capability are frequently referred to as special processes.
	Records shall be maintained for qualified processes, equipment and personnel, as appropriate (see 4.16).

7.6 Control of measuring and monitoring devices

4.11 Control of inspection, measuring and test equipment

4.11.1 General

The organization shall identify the measurements to be made and the measuring and monitoring devices required to assure conformity of product to specified requirements,

Measuring and monitoring devices shall be used and controlled to ensure that measurement capability is consistent with the measurement requirements.

Where applicable, measuring and monitoring devices shall:

a) be calibrated and adjusted periodically or prior to use, against devices traceable to international or national standards; where no such standard exist, the basis used for calibration shall be recorded;

b) be safeguarded from adjustments that would invalidate the calibration;

c) be protected from damage and deterioration during handling, maintenance and storage;

d) have the results of their calibration recorded (see 5.5.7);

e) have the validity of previous results reassessed if they are subsequently found to be out of calibration, and corrective action taken.

The supplier shall establish and maintain documented procedures to control, calibrate and maintain inspection, measuring and test equipment (including test software) used by the supplier to demonstrate the conformance of product to the specified requirements. Inspection, measuring and test equipment shall be used in a manner which ensures that the measurement uncertainty in known and is consistent with the required measurement capability.

Where test software or comparative references such as test hardware are used as suitable forms of inspection, they shall be checked to prove that they are capable of verifying the acceptability of product, prior to release for use during production, installation or servicing, and shall be rechecked at prescribed intervals. The supplier shall establish the extent and frequency of such checks and shall maintain records as evidence of control (see 4.16).

Where the availability of technical data pertaining to the inspection, measuring and test equipment is a specified requirement, such data shall be made available, when required by the customer or customer's

2000 Version	1994 Version
Note: See ISO 10012 for guidance.	Note 17: For the purposes of this International Standard, the term "measuring equipment" includes measurement devices.
Software used for measuring and monitoring of specified requirements shall be validated prior to use.	

4.11.2 Control procedure

The supplier shall:

a) determine the measurements to be made and the accuracy required, and select the appropriate inspection, measuring and test equipment that is capable of the necessary accuracy and precision;

b) identify all inspection, measuring and test equipment that can affect product quality, and calibrate and adjust them at prescribed intervals, or prior to use, against certified equipment having a known valid relationship to internationally or nationally recognized standards. Where no such standards exist, the basis used for calibration shall be documented;

c) define the process employed for the calibration of inspection, measuring and test equipment, including details of equipment type, unique identification, location, frequency of checks, check method, acceptance criteria and the action to be taken when results are unsatisfactory;

d) identify inspection, measuring and test equipment with a suitable indicator or approved identification record to show the calibration status;

e) maintain calibration records for inspection, measuring and test equipment (see 4.16);

f) assess and document the validity of previous inspection and test results when inspection, measuring or test equipment is found to be out of calibration;

g) ensure that the environmental conditions are suitable for the calibrations, inspections, measurements and tests being carried out;

h) ensure that the handling, preservation and storage of inspection, measuring and test equipment is such that the accuracy and fitness for use are maintained;

i) safeguard inspection, measuring and test facilities, including both test hardware and test software, from adjustments which would invalidate the calibration setting.

Note 18: The metrological confirmation system for measuring equipment given in ISO 10012 may be used for guidance.

2000 Version	1994 Version

8 Measurement, analysis and improvement

8.1 Planning

The organization shall define, plan and implement the measurement and monitoring activities needed to assure conformity and achieve improvement. This shall include the determination of the need for, and use of, applicable methodologies including statistical techniques.

[*Note:* Clause 4.20 of ISO 9001:1994 covering statistical tools has not been carried over as such into the 2000 version.]

8.2 Measurement and monitoring

8.2.1 Customer satisfaction

The organization shall monitor information on customer satisfaction and/or dissatisfaction as one of the measurements of performance of the quality management system. The methodologies for obtaining and using this information shall be determined.

[*Note:* Clause 4.20 of ISO 9001:1994 covering statistical tools has not been carried over as such into the 2000 version.]

8.2.2 Internal audit

The organization shall conduct periodic internal audits to determine whether the quality management system:

a) conforms to the requirements of this International Standard

b) has been effectively implemented and maintained.

The organization shall plan the audit program taking into consideration the status and importance of the activities and areas to be audited as well as the results of previous audits. The audit scope, frequency and methodologies shall be defined. Audits shall be conducted by personnel other than those who performed the activity being audited.

A documented procedure shall include the responsibilities and requirements for conduction audits, ensuring their independence, recording results and reporting to management.

Management shall take timely corrective action on deficiencies found during the audit.

Follow-up actions shall include the verification of the implementation of corrective action, and the reporting of verification results.

4.17 Internal quality audits

The supplier shall establish and maintain documented procedures for planing and implementing internal quality audits to verify whether quality activities and related results comply with planned arrangements and to determine the effectiveness of the quality system.

Internal quality audits shall be scheduled on the basis of the status and importance of the activity to be audited and shall be carried out by personnel independent of those having direct responsibility for the activity being audited.

The results of the audits shall be recorded (see 4.16) and brought to the attention of the personnel having responsibility in the area audited. The management personnel responsible for the area shall take timely corrective action on deficiencies found during the audit.

Follow-up audit activities shall verify and record the implementation and effectiveness of the corrective action taken (see 4.16).

NOTES

11 The results of internal quality audits form an integral part of the input to management review activities (see 4.1.3).

12 Guidance on quality system audits is given in ISO 10011.

2000 Version	1994 Version

8.2.3 Measurement and monitoring of process

The organization shall apply suitable methods for measurement and monitoring of those realization processes necessary to meet customer requirements.

These methods shall confirm the continuing ability of each process to satisfy its intended purpose.

8.2.4 Measurement and monitoring of product

The organization shall measure and monitor the characteristics of the product to verify that requirements for the product are met. This shall be carried out at appropriate stages of the product realization process.

Evidence of conformity with the acceptance criteria shall be documented. Records shall indicate the authority responsible for release of product (see 5.5.7).

Product release and service delivery shall not proceed until all the specified activities have been satisfactorily completed, unless otherwise approved by the customer.

4.9 Process control

[*Note:* Only Item d) and the text shown of clause 4.9 are relevant here.]

  c) monitoring and control of suitable process parameters and product characteristics;

Where the results of processes cannot be fully verified by subsequent inspection and testing of the product and where, for example, processing deficiencies may become apparent only after the product is in use, the processes shall be carried out by qualified operators and/or shall require continuous monitoring and control of process parameters to ensure that the specified requirements are met.

4.10 Inspection and testing

4.10.1 General

The supplier shall establish and maintain documented procedures for inspection and testing activities in order to verify that the specified requirements for the product are met. The required inspection and testing, and the records to be established, shall be detailed in the quality plan or documented procedures.

4.10.2 Receiving inspection and testing

4.10.2.1

The supplier shall ensure that incoming product is not used or processed (except in the circumstances described in 4.10.2.3) until it has been inspected or otherwise verified as conforming to specified requirements.

Verification of conformance to the specified requirements shall be in accordance with the quality plan and/or documented procedures.

4.10.2.2

In determining the amount and nature of receiving inspection, consideration shall be given to the amount of control exercised at the subcontractor's premises and the recorded evidence of conformance provided.

4.10.2.3

Where incoming product is released for urgent production purposes prior to verification, it shall be positively identified and recorded (see 4.16) in order to permit immediate recall and replacement in the event of nonconformity to specified requirements.

[*Note:* The screened text is dealt with under 7.4.3.]

2000 Version	1994 Version
	4.10.3 In-process inspection and testing
	The supplier shall:
	a) inspect and test the product as required by the quality plan and/or documented procedures;
	b) hold product until the required inspection and tests have been completed or necessary reports have been received and verified, except when product is released under positive-recall procedures (see 4.10.2.3). Release under positive-recall procedures shall not preclude the activities outlined in 4.10.3 a).
	4.10.4 Final inspection and testing
	The supplier shall carry out all final inspection and testing in accordance with the quality plan and/or documented procedures to complete the evidence of conformance of the finished product to the specified requirements.
	The quality plan and/or documented procedures for final inspection and testing shall require that all specified inspection and tests, including those specified either on receipt or product or in-process, have been carried out and that the results meet specified requirements.
	No product shall be dispatched until all the activities specified in the quality plan and/or documented procedures have been satisfactorily completed and the associated data and documentation are available and authorized.
	4.10.5 Inspection and test records
	The supplier shall establish and maintain records which provide evidence that the product has been inspected and/or tested. These records shall show clearly whether the product has passed or failed the inspections and/or tests according to defined acceptance criteria. Where the product fails to pass any inspection and/or test, the procedures for control of nonconforming product shall apply (see 4.13).
	Records shall identify the inspection authority responsible for the release of product (see 4.16).
8.3 Control of nonconformity	**4.13 Control of nonconforming product**
The organization shall ensure that product which does not conform to requirements is identified and controlled to prevent unintended use or delivery. These activities shall be defined in a documented procedure.	**4.13.1 General**
	The supplier shall establish and maintain documented procedures to ensure that product that does not conform to specified requirements is prevented from unintended

2000 Version	1994 Version
Nonconforming product shall be corrected and subject to re-verification after correction to demonstrate conformity.	use of installation. This control shall provide for identification, documentation, evaluation, segregation (when practical), disposition of nonconforming product, and for notification to the functions concerned.
When nonconforming product is detected after delivery or use has started, the organization shall take appropriate action regarding the consequences of the nonconformity.	**4.13.2 Review and disposition of nonconforming product**
It will often be required that the proposed rectification of nonconforming product be reported for concession to the customer, the end-user, regulatory body or other body.	The responsibility for review and authority for the disposition of nonconforming procedures. It may be

The responsibility for review and authority for the disposition of nonconforming procedures. It may be

a) reworked to meet the specified requirements,

b) accepted with or without repair by concession,

c) re-graded for alternative applications, or

d) rejected or scrapped.

Where required by the contract, the proposed use or repair of product [see 4.13.2 b)] which does not conform to specified requirements shall be reported for concession to the customer or customer's representative. The description of the nonconformity that has been accepted, and of repairs, shall be recorded to denote the actual condition (see 4.16).

Repaired and/or reworked product shall be re-inspected in accordance with the quality plan and/or documented procedures.

8.4 Analysis of data

The organization shall collect and analyze appropriate data to determine the suitability and effectiveness of the quality management system and to identify improvements that can be made. This includes data generated by measuring and monitoring activities and other relevant sources.

[*Note:* There is no direct equivalent of this clause.]

[*Note:* The following clause which does not have an equivalent in the 2000 version applies as covering some of the methods of analysis that might be used.]

4.20 Statistical techniques

The organization shall analyze this data to provide information on:

a) customer satisfaction and/or dissatisfaction;

b) conformance to customer requirements;

c) characteristics of processes, product and their trends;

d) suppliers.

4.20.1 Identification of need

The supplier shall identify the need for statistical techniques required for establishing, controlling and verifying process capability and product characteristics.

4.20.2 Procedures

The supplier shall establish and maintain documented procedures to implement and control the application of the statistical techniques identified in 4.20.1.

8.5 Improvement

8.5.1 Planning for continual improvement

The organization shall plan and manage the process necessary for the continual improvement of the quality management system.

[*Note:* There is no equivalent of clause 8.5.1.]

2000 Version	1994 Version

The organization shall facilitate the continual improvement of the quality management system through the use of the quality policy, objectives, audit results, analysis of data, corrective and preventive action and management review.

8.5.2 Corrective action

The organization shall take corrective action to eliminate the cause of nonconformities in order to prevent recurrence. Corrective action shall be appropriate to the impact of the problems encountered.

The documented procedure for corrective action shall define requirements for:

a) identifying nonconformities (including customer complaints);

b) determining the causes of nonconformity;

c) evaluating the need for actions to ensure that nonconformities do not recur;

d) determining and implementing the corrective action needed;

e) recording results of action taken;

f) reviewing of corrective action taken.

8.5.3 Preventive action

The organization shall identify preventive action to eliminate the causes of potential nonconformities to prevent occurrence. Preventative actions taken shall be appropriate to the impact of the potential problems.

The documented procedure for preventive action shall define requirements for:

a) identifying potential nonconformities and their causes;

b) determining and ensuring the implementation of preventative action needed;

c) recording resulting of action taken;

d) reviewing of preventive action taken.

4.14   Corrective and preventive action

4.14.1 General

The supplier shall establish and maintain documented procedures for implementing corrective and preventive action.

Any corrective or preventive action taken to eliminate the causes of actual or potential nonconformities shall be to a degree appropriate to the magnitude of problems and commensurate with the risks encountered.

The supplier shall implement and record any changes to the documented procedures resulting from corrective and preventive action.

4.14.2 Corrective action

The procedures for corrective action shall include:

a) the effective handling of customer complaints and reports of product nonconformities;

b) inveestigation of the cause of nonconformities relating to product, process and quality system, and recording the results of the investigation (see 4.16);

c) determination of the corrective action need to eliminate the cause of nonconformities;

d) application of controls to ensure that corrective action is taken and that it is effective.

4.14.3 Preventive action

The procedures for preventive action shall include:

a) the use of appropriate sources of information such as processes and work operations which affect product quality, concessions, audit results, quality record, service reports and customer complaints to detect, analyze and eliminate potential causes of nonconformities.

b) Determination of the steps needed to deal with any problem requiring preventive action;

c) Initiation of preventive action and application of controls to ensure that it is effective;

d) Ensuring that relevant information on action taken is submitted for management review (see 4.1.3).

CONCLUSION

For companies that are conducting business abroad, or planning to conduct business abroad, ISO 9000 is the worldwide quality defacto standard. It is flexible enough to grow with technology and the times, while ensuring quality and safety by design and repeatability of processes.

FURTHER READING

ISO 9001: Quality Systems — Model for Assurance in Design/Development, Production, Installation and Servicing.
ISO 9002: Quality Systems — Model for Assurance in Production and Installation.
ISO 9003: Quality Systems — Model for Assurance Final Inspection and Test.
ISO 9004: Quality Management and Quality System Elements — Guidelines.
Baum, D. Client/server development tools for Windows, *Computerworld*, April, 73–75, 1993.
O'Lone, E. J. Datapro, document imaging systems, management issues. *Client/Server Computing,* June, 1–8, 1993.

PART XII

Standards of Competence

## CHAPTER 52

Standards of Competence

Roger L. Brauer

CONTENTS

INTRODUCTION

Competency means having the qualifications or capability to perform some task. Standards of competency and occupational credentialing assist employers, the government, and the public in determining who is competent in a particular discipline or area. This chapter reviews competency standards and practices for safety and health professions.

Competency assessment is either voluntary through peer certification boards or mandatory through a license to practice awarded by a government agency. Most assessments of competency involve at least three factors. The first is evaluation of education against some curriculum standard. Frequently, the curriculum standard referenced is that established by an academic accreditation organization and is used to evaluate degree programs that prepare people to enter or advance in the field or discipline. The second factor is experience. The certifying or licensing agency will define what type, level, and breadth of experience is accepted. The third factor is demonstration of knowledge through examinations. The examination contents are based on validation studies of what practitioners actually do in their jobs. More and more disciplines are requiring a fourth factor, demonstrating continued competency through reexamination, continuing education, or similar means, which assures that one is keeping up with changes in practice.

For the safety profession, competency has its basis in several documents. The American Society of Safety Engineers (ASSE) maintains an official definition for professional safety practice entitled *The Scope and Functions of the Professional Safety Position.* This brochure defines the primary functions performed by safety professionals. This task description differentiates the safety profession from other disciplines. A curriculum standard[1] for a baccalaureate safety degree defines the model technical knowledge required for safety profession tasks. Other knowledge and skills may be important to be an effective practitioner. Academic program accreditation for safety degrees derives its curriculum criteria[2] from this standard. The board of certified safety professionals (BCSP) uses the same curriculum standards as the model educational standard for evaluating the academic preparation of individuals who seek the certified safety professional (CSP) designation.

In England, the Occupational Health and Safety Lead Body[3] completed job analyses to define professional, technician, and worker tasks in general safety, radiation protection, and occupational hygiene. The general safety standards form the examination contents for the national diploma and national certificate for safety professionals and safety technicians, respectively, operated by the national examination board in occupational safety and health (NEBOSH).[4]

PROFESSIONAL COMPETENCE

Figure 52.1 illustrates the general process by which organizations and individuals assess competency. The overall goal is to ensure that certain functions or tasks are performed knowledgeably and effectively. The process may address capabilities for general functions or focus on specific tasks. It is difficult to predict actual performance. However, education and training, experience, and knowledge and skills are often measured since they contribute to competence. The evaluation process begins with a definition of the tasks and functions that are to be performed. In the case of an employer trying to select a person for a position, an individual or group may define tasks. For an employer, a job description will contain the specific job tasks, but additional skills and personal style will be considered during placement interviews. For a professional examination, a panel of people representing a cross section of practice defines the job responsibilities in a process called role delineation or job analysis, which is later validated through a survey of practitioners. The results of

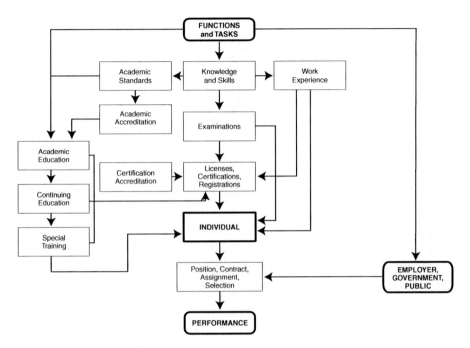

Figure 52.1 Competency process.

the role delineation and survey form the examination blueprint, which defines the examination subjects and their distributions.

Evaluations of education, training, and experience of individuals help predict performance. An employer compares qualifications of job applicants to written job standards. Licensing, registration, and certification boards usually have formal procedures for evaluating the education and experience of a candidate. Professional credentialing procedures usually require particular academic preparation, holding degrees in certain fields from accredited institutions and accredited degree programs. This places a burden on the academic accreditation process to ensure than an education program meets general institutional standards and standards for a professional program for which students are being prepared. Sometimes there are standards for training in lieu of education or combined with education. These standards may define training subjects and courses. The evaluation of experience in particular positions may include the amount of time devoted to a position or major job function within it, the kind of position, the level of job duties, and the breadth of subjects, tasks, and responsibilities for the position.

Most often, employers have final responsibility for competency, since they pay for services of individuals. Sometimes contracts specify particular qualifications to help ensure competency with some degree of certainty. In both of these cases, there has been an increase in reliance on licensing and certification to help evaluate and qualify individuals. With the increasing rate of job change and people changing careers during their lifetime, there is an increasing burden on individuals to provide evidence of independent evaluation of competency in particular functions or responsibilities.

The shift from employer to individual responsibility for competency is one factor that has increased interest in voluntary, peer-operated certifications.

The public may have trouble determining who is qualified to perform some task, particularly when the individual provides services directly to the public rather than through employers or companies. The public may not even know what tasks, knowledge, and skills are required to solve its need. Generally, when the safety, health, and welfare of the public is at stake and a profession offers services directly to the public, government will establish licensing of the profession. In the United States, licensing is the responsibility of states. Governments are reluctant to license professions and practices for which there is not a clear need to protect the public or the potential harm cannot be clearly demonstrated. Governments are more likely to step in when a profession offers its services directly to the public rather than when the professionals work within companies that have the liability for the professional work of their employees or contractors. In many cases, government organizations may depend on peer certification to assess basic competency of practitioners.

CURRICULUM STANDARDS

Four foundation components for any discipline define what constitutes the discipline and its practice, define the body of knowledge essential to be able to practice in the discipline, establish a process for evaluating academic programs that prepare people for entry or advancement in the profession, and establish a process for evaluating the individual competency of those who claim to be practitioners of the discipline.

Since about the 1960s, both the ASSE and the BCSP have maintained guidelines for baccalaureate degrees in safety, the model preparation for the safety profession. In 1991 the two organizations began publishing a joint degree standard. Table 52.1 summarizes the requirements of this standard. Because people enter the safety profession through other academic preparation in addition to the model preparation, additional standards have also been developed and jointly published.[5] These do not cover all of the academic routes of entry. There may be a need for other standards for other routes of entry. In addition, perhaps as many as two thirds or more of those moving into safety positions continue to come from other disciplines. These individuals must learn safety job responsibilities on the job and develop knowledge and skill for them through continuing education or advanced degrees.

To help ensure that academic programs in safety met curriculum and other important standards, the ASSE established a process in the late 1970s to accredit safety degree programs. Because recognition as an accrediting body by academic institutions was important, ASSE moved its accrediting activity under the Related Accreditation Commission of the Accreditation Board for Engineering and Technology (RAC/ABET) in the late 1980s. This gave accreditation of safety degrees recognition under the Council for on Higher Education Accreditation.[6]

In the 1980s, the industrial hygiene community began accreditation of industrial hygiene degrees at the master's level through RAC/ABET. Table 52.2 summarizes these requirements. The American Academy of Industrial Hygiene serves as the lead organization.

Table 52.1 Summary of Baccalaureate Safety Degree Requirements

Category	Course or Subjects
Preparatory	Mathematics
	Through introductory calculus
	Statistics
	Basic computer skills
	Physical, chemical, and other sciences
	Physics with laboratory
	Chemistry (including an introduction to organic) with laboratory
	Physiology, anatomy, or biology
	Psychology
	Other
	Introductory business
	Written and oral communication
	Applied mechanics
	Industrial or manufacturing processes
Professional core courses	Introduction to safety and health
	Safety and health program management
	Design of engineering hazard controls
	Industrial hygiene and toxicology with laboratories
	Fire protection with laboratories
	Ergonomics with laboratories
	Environmental safety and health
	System safety and other analytical methods
Required professional subjects	Measurement of safety performance
	Behavioral aspects of safety
	Product safety
	Construction safety
	Educational and training methods for safety
Electives	Professional electives
	General electives
Experiential Education	Internship or cooperative program with course credit

Table 52.2 Requirements for Industrial Hygiene Master's Degrees

Factor	Criteria
Candidate requirements	120-semester-hour or greater bachelor's degree
	At least 60 hours in science, mathematics, engineering, and technology
Industrial hygiene master's degree	At least 30 semester hours
Curriculum	Industrial hygiene sciences:
	Principles and practices of industrial hygiene
	Principles and practices of environmental sciences
	Epidemiology and biostatistics
	Industrial hygiene practice:
	Control of physical and chemical hazards
	Environmental health
	Occupational safety
Curriculum distribution	Industrial hygiene sciences and practice: 18 hours
	Unspecified hours: 12 hours

COMPETENCY FOR SAFETY AND HEALTH PROFESSIONALS

During the last decade or more, there has been a convergence of safety, health, environment, and ergonomics practice. Most often, this is manifested within an organization by having all these functions located within the same organizational unit. The impact on the professional is that the boundaries among these areas of practice have blurred. Those who wish to advance professionally must either choose to remain a specialist or to develop competency in all areas to advance as a manager.

At the same time, the old model of employment in which one works for the same company an entire career is quickly disappearing. In that model, the employer took care of advancing people in a career path. Today, people must be more fluid to retain employment. This has pushed the burden for demonstrating competency to the individual in order to advance. Some suggest that this change has increased the growth of certifications in the United States.

Those involved in the safety, health, environment, and ergonomics fields can pursue more than 190 titles in the United States. As a result, national accreditation of certification programs has become the means to identify quality certification programs. In the United States, there are two organization that set standards for peer-operated certifications in general and accredit them. One is the National Commission for Certifying Agencies (NCCA)[7] and the other is the Council of Engineering and Scientific Specialty Boards (CESB).[8] Among the 190 titles, there are 12 titles holding national accreditation (Table 52.3). Of the 12 certifications, 2 are at the worker level, 2 at the technician/technologist level, 2 are in medical fields for which there are separate accrediting bodies, and the remaining 6 are professional programs.

Certified Safety Professional

The BCSP awards the CSP designation to individuals who meet an academic standard and an experience standard and pass two examinations. The minimum academic requirement is holding an associate degree in safety and health or a bachelor's degree in any field. The model degree is an RAC/ABET accredited bachelor's degree in safety, which receives full credit (48 points). Other degrees receive varying amounts of credit based on a general comparison to the model degree. Table 52.4 lists the credit awarded. Associate degrees in safety and health, which do not have program accreditation available, receive 18 points (half the credit allowed for an equivalent bachelor's degree).

Candidates for the CSP must have a minimum of 4 years of acceptable professional safety experience, awarded at the rate of one point per month of acceptable experience. More than 4 years of experience is required when someone must make up a deficiency in the academic standard. Graduate degrees can earn up to 2 years of credit toward the experience requirement. Master's degrees receive credit at the rate of one quarter the equivalent bachelor's degree and doctorate degrees receive half the equivalent credit. To qualify, professional safety experience must meet six criteria established by BCSP. The criteria differentiate professional experience from general safety experience.

Table 52.3 Nationally Accredited Safety, Health, Environment, and Ergonomics Titles

Title	Accredited By	Level	Organization
CSP, Certified Safety Professional	NCCA, CESB	Professional	BCSP, Board of Certified Safety Professionals
CIH, Certified Industrial Hygienist	CESB	Professional	ABIH, American Board of Industrial Hygiene
CHP, Certified Health Physicist	CESB	Professional	ABHP, American Board of Health Physics
CHMM, Certified Hazardous Materials Manager	CESB	Professional	IHMM, Institute of Hazardous Materials Management
DEE, Diplomate in Environmental Engineering	CESB	Professional	AAEE, American Academy of Environmental Engineers
QEP, Qualified Environmental Professional	CESB	Professional	IEP, Institute of Professional Environmental Practice
COHN/COHN-S, Certified Occupational Health Nurse	ABNS[a]	Professional	ABOHN, American Board of Occupational Health Nurses
COHP, Certified Occupational Health Physician	ABMS[b]	Professional	ABOHP, American Board of Occupational Health Physicians
OHST, Occupational Health and Safety Technologist	CESB	Technician/ technologist	ABIH/BCSP Joint Committee
CHST, Construction Health and Safety Technician	CESB	Technician/ technologist	ABIH/BCSP Joint Committee
STS-Construction, Safety Trained Supervisor in Construction	CESB	Worker	ABIH/BCSP Joint Committee
CCO, Certified Crane Operator	NCCA	Worker	NCCCO, National Commission for the Certification of Crane Operators

[a] American Board of Nursing Specialties.
[b] American Board of Medical Specialties.

Table 52.4 Schedule of Bachelor's Degree Academic Credit toward the CSP[a]

Baccalaureate Degree	Units of Credit Allowed
ABET-accredited safety, safety technology or safety engineering degree	48
ABET-accredited engineering degree (not safety)	42
Safety or safety technology degree that is not ABET accredited	36
Engineering technology degree (not safety) accredited by ABET	30
Physical and natural science degree	30
Industrial technology degree	24
Engineering technology degree not ABET accredited	24
Business administration, industrial education, or psychology degree	18
Majors not listed above	12

[a] Degrees must be from institutions accredited by a regional accrediting body recognized by CHEA (see Note 6). Foreign degrees are evaluated for U.S. equivalency.

Table 52.5 Subjects Covered on CSP Examinations

Examination Section	Examination Subject
Safety Fundamentals Examination	
1. Basic and Applied Sciences	Mathematics, physics, chemistry, biological sciences, behavioral sciences, ergonomics, engineering and technology, epidemiology
2. Program Management and Evaluation	Organization, planning, and communication; legal and regulatory considerations; program evaluation; disaster and contingency planning; professional conduct and ethics
3. Fire Prevention and Protection	Structural design standards; detection and control systems and procedures; fire prevention
4. Equipment and Facilities	Facilities and equipment design, mechanical hazards, pressures, electrical hazards, transportation, materials handling, illumination
5. Environmental Aspects	Toxic materials, environmental hazards, noise, radiation, thermal hazards, control methods
6. System Safety and Product Safety	Techniques of system safety analysis, design considerations, product liability, reliability, and quality control
Comprehensive Practice Examination	
1. Engineering	Safety engineering, fire protection engineering, occupational health engineering, product and system safety engineering, environmental engineering
2. Management	Applied management fundamentals, business insurance and risk management, industrial and public relations, organizational theory and organizational behavior, quantitative methods for safety management
3. Applied Sciences	Chemistry, physics, life sciences, behavioral sciences
4. Legal/Regulatory Aspects and Professional Conduct and Affairs	Legal aspects, regulatory aspects, professional conduct and affairs

After meeting the academic requirement (48 points), candidates sit for the Safety Fundamentals Examination. Successful candidates receive the associate safety professional (ASP) designation to denote progress toward the CSP. Candidates holding the ASP and also meeting the experience requirement (an additional 48 points) continue by taking the Comprehensive Practice Examination. Successful candidates receive the CSP designation. Table 52.5 outlines the subjects covered on the Safety Fundamentals and Comprehensive Practice Examinations.

Individuals holding the CSP must be recertified every 5 years through the Continuance of Certification program. CSPs earn points through ten activity categories. Retaking and passing the Comprehensive Practice Examination gains all the required points with one activity. Other categories emphasis keeping up with change and continuing education.

Since the BCSP began in 1969, about 16,000 individuals have achieved the CSP. Nearly 10,000 hold the certification currently. BCSP has six sponsoring organizations that participate in the nomination process for many of the 13 positions on the board of directors. The sponsoring organizations include ASSE, American Industrial Hygiene Association, System Safety Society, Society of Fire Protection Engineers,

Institute of Industrial Engineers, and National Safety Council. Other directors come from the profession at large and one represents the public.

Individuals holding the CSP can demonstrate competence in specialty areas. Between 1978 and 1997 candidates for the CSP could choose a specialty examination rather than the Comprehensive Practice Examination. During this period, the specialty examinations contained exactly the same subjects as appeared on the Comprehensive Practice Examination (see Table 52.5), but the distribution of examination items differed among major subjects. Beginning in 1999, specialty examinations were redefined and complement general practice. Validation studies identify that which is unique to the specialty. The new specialty examinations, available to anyone holding the CSP, do not repeat tasks or knowledge and skills covered on the examinations leading to the CSP. Initial specialties under the new structure cover ergonomics, construction safety, and system safety.

Certified Industrial Hygienist

The American Board of Industrial Hygiene (ABIH), begun in 1965, awards the Certified Industrial Hygienist (CIH) to individuals who meet academic and experience requirements and pass two examinations: Core Examination and Comprehensive Practice Examination. An applicant must have an acceptable degree in industrial hygiene, chemistry, physics, medicine, biology, or chemical, mechanical, or sanitary engineering. Recently, ABIH has also accepted accredited bachelor's degrees in safety as meeting the academic standard. Other bachelor's degrees are evaluated for their science content.

An applicant must also have 5 years of acceptable full-time employment experience in industrial hygiene where the position(s) had at least 50% industrial hygiene responsibility. Applicable graduate degrees may substitute for up to 2 years of professional experience.

The Core Examination covers basic knowledge and skills for industrial hygiene. Candidates who pass this examination receive the interim designation of Industrial Hygienist in Training (IHIT). The Comprehensive Practice Examination covers advanced knowledge and skills in industrial hygiene functions. Upon completion of this examination and all other requirements, candidates achieve the CIH designation.

Those holding the CIH must be recertified every 5 years through the Certification Maintenance program involving either retaking and passing the Comprehensive Practice Examination or completing other activities that emphasis continuing education and keeping up with changes in the profession.

Approximately 10,000 individuals have received the CIH since it began, and about 6000 currently hold the title.

Occupational Health and Safety Technologist

The ABIH/BCSP Joint Committee awards the Occupational Health and Safety Technologist (OHST) designation to individuals who have 5 years of health and safety experience with at least 35% of their job duties involved in health and safety. Approximately 1500 individuals hold this designation. Candidates can waive some experience based on academic preparation through certificate programs, safety and health courses, and associate or higher degrees. Candidates must pass a single examination. Many people use this designation as a stepping-stone to the CSP or CIH professional certifications.

The OHST Examination covers seven subject areas:

1. Basic and applied sciences.
2. Laws, regulations, and standards.
3. Control concepts.
4. Investigation (post-event).
5. Survey and inspection techniques (pre-event).
6. Data computation and record keeping.
7. Education, training, and instruction.

Construction Health and Safety Technician

The ABIH/BCSP Joint Committee also awards the Construction Health and Safety Technician (CHST) designation, a program that started in 1994. The program allows for candidates to progress from construction trades into management positions responsible for health and safety or from academic preparation for health and safety jobs. All candidates must have some construction experience and pass the CHST Examination. Individuals holding the CHST typically have responsibility for safety and health at one or more construction sites.

The CHST examination covers the following health and safety functions:

- Inspections.
- General safety training and safety orientation for employees.
- Safety and health record keeping.
- Hazard communication compliance.
- Safety analysis and planning.
- Accident investigations.
- Program management and administration.
- OSHA and other inspections.

Other Certifications

There are other certifications that evaluate competency in safety and health. For example, the American Board of Health Physics[9] awards the Certified Health Physicist (CHP) to specialists in protection of individuals from various forms of radiation and other hazards. One state (Massachusetts) has a safety engineering specialty within its professional engineering licensing programs.

In Canada, the board for Canadian Registered Safety Professionals[10] awards the Canadian Registered Safety Professional (CRSP) designation to candidates meeting education and experience requirements and passing an examination sequence. As noted earlier, NEBOSH offers two designations. At the professional level there is a national diploma and at the technician level there is a national certificate. In Hong Kong and Singapore, government organizations establish examination and other standards for individuals who wish to practice in safety and health.

Table 52.6 Summary of Accreditation Standards for Peer Certifications

Category	Requirement
Agency and operations	Is the agency not-for-profit?
	Is the agency national in scope?
	Is the agency separate and independent from any associated educational body?
	Does at least one member of the public serve as a member of the governing body?
	Are financial statements published? Are they audited using recognized auditing practices?
Eligibility and examinations	Is the certification open to those who are not members of a professional association in the field or not members of the certifying organization?
	Is eligibility logically related to relevant job requirements?
	Is the examination free of bias and nondiscriminatory with regard to published demographic data?
	Has the validity of the examination been established by conducting a national job analysis survey or other psychometrically sound validation survey?
	Are pass/fail cutoff scores established using recognized and psychometrically sound procedures?
	Can anyone be granted the credential without examination and achieving a passing score?
	Are reliability statistics produced regularly or after each examination administration?
Recertification	Is there a recertification program to ensure continued competence?
	What is the program?

NATIONAL STANDARDS FOR CERTIFICATIONS

With the growth of peer-operated certifications in the United States in many disciplines, it has become necessary to establish which programs operate with quality procedures. One objective for accreditation is to ensure fairness to all candidates in the certification process. Another objective is to assure employers, the public, and others relying on a certification that procedures, governance, financial matters, and other aspects are open and reliable. Two organizations that accredit peer-certification programs cited most frequently for their standards are the National Commission for Certifying Agencies (NCCA) and the Council of Engineering and Scientific Specialty Boards (CESB). Accreditation is typically awarded for a period of 5 years. Table 52.6 summarizes the key elements in peer certification accreditation.

National Commission for Certifying Agencies

The NCCA began in the 1970s through a grant from the U.S. Department of Education to provide independent evaluation of certifications in the allied health fields. Later, it expanded its scope so that its standards apply to nearly any peer-operated certification. NCCA standards often serve as a model for standards of other certification accrediting bodies.

Council of Engineering and Scientific Specialty Boards

CESB grew out of a national symposium in the late 1980s on credentialing in engineering and related fields. Member organizations represent engineering, engineering technology, and engineering-related disciplines.

Other Standards

Standards for accreditation of peer certifications have found their way into other standards. For example, ASTM E1929-98[11] draws on NCCA and CESB criteria to establish a means for users of environmental services to identify credible certification programs for environmental professionals. Programs accredited by NCCA or CESB automatically comply with ASTM E1929-98. Another example is found in the Workers' Compensation Commission for the State of Texas. In order for individuals to qualify for field safety representatives with insurance firms selling workers' compensation insurance, they must hold accredited certifications recognized by the commission. The certification programs must apply for recognition and national accreditation is verified by the commission. Use of peer-certification accreditation has been proposed in other national standards and in federal, state, and local government laws, regulations, and contracts.

SUMMARY

Competency among safety and health professionals continues to become more important. More and more employers, government organizations, and the public look for competency in safety and health practice. Contracts involving safety and health services often rely on demonstrated competence. Although academic records and professional experience continue to be a basic measure of professional competence, reliance on certification continues to increase rapidly as an additional means of verifying competence. Certification is becoming a standard used to qualify for safety and health positions and jobs. As a result, individuals holding certifications are becoming more valuable. A recent salary survey[12] showed that individuals with a PE license or a CSP or CIH certification averaged about $16,000 per year more than individuals with no certification. Similarly, another series of surveys[13] showed that the salary for individuals with certifications grew faster than those without certification. Overall, the importance and value for competence in the safety and health profession continues to increase.

NOTES

1. Curriculum Standards for Baccalaureate Degrees in Safety, Joint Report No. 1, American Society of Safety Engineers and Board of Certified Safety Professionals, August 1991.
2. The Related Accreditation Commission of the Accreditation Board for Engineering and Technology (RAC/ABET, 111 Market Place, Suite 1050, Baltimore, MD 21202) publishes curriculum criteria for safety degrees, which are used in accreditation procedures for baccalaureate and master's degrees in safety.

3. Occupational Health and Safety Lead Body, Seventh Floor North, Rose Court, 2 Southwork Bridge, London, SE1 9HS, England.

4. National Examination Board in Occupational Safety and Health, 222 Upingham Road, Leicester LE5 0QG, United Kingdom.

5. Curriculum Standards for Master's Degrees in Safety, Joint Report No. 2, American Society of Safety Engineers and Board of Certified Safety Professionals, March 1994; Curriculum Standards for Safety Engineering Master's Degrees and Safety Engineering Options in Other Engineering Master's Degrees, Joint Report No. 3, American Society of Safety Engineers and Board of Certified Safety Professionals, November 1994; Curriculum Standards for Associate Degrees in Safety, Joint Report No. 4, American Society of Safety Engineers and Board of Certified Safety Professionals, June 1995.

6. Council for Higher Education Accreditation, One Dupont Circle, N.W., Suite 510, Washington, D.C. 20036-1110.

7. National Commission for Certifying Agencies, 1101 Connecticut Avenue, N.W., Suite 700, Washington, D.C. 20036.

8. Council of Engineering and Scientific Specialty Boards, 130 Holiday Court, Suite 100, Annapolis, MD 21041.

9. American Board of Health Physics, 8000 Westpark Drive 400, McLean, VA 22102-3101.

10. Board of Canadian Registered Safety Professionals, 6519B Mississauga Road, Mississauga, Ontario L5N 1A6, Canada.

11. *ASTM E1929-98, Standard Practice for Assessment of Certification Programs for Environmental Professionals: Accreditation Criteria*, American Society for Testing and Materials, 1916 Race Street, Philadelphia, PA 19103.

12. 1997 salary survey, *Industrial Safety and Hygiene News*, Business News Publishing Company, 755 W. Big Beaver Road, Suite 1000, Troy, MI 48084.

13. 1980, 1990, and 1997 salary surveys, American Society of Safety Engineers, 1800 East Oakton St., Des Plaines, IL 60018-2187.

PART XIII

Afterword: The Future

CHAPTER **53**

Where Are We Going? — An Educator's Perspective

Jan Thomas

CONTENTS

INTRODUCTION

It must be considered that there is nothing more difficult to carry out, nor more doubtful of success, nor more dangerous to handle, than to initiate a new order of things.

Machiavelli (1469–1527)

This warning, from the Italian writer and statesman, rings true today as we stand at the threshold of a new millenium and ask the question. "Where is our profession going?" The answer is clear for those who practice futuristic techniques. Change — that new order of things — is afoot. We will explore where these changes are leading us.

The practice of safety and health management has advanced greatly in the past 20 years; it will change even more in the next 10. The education of safety and health professionals, within the U.S. college and university system, is one way to measure this change. Another way is to study current trends in professional continuing education for purposes of career advancement. A third means is to examine the catalytic events and prime agents which appear to influence the practice of safety and health professionals. This third measure may hold the key to the past as well as to the future. We will discuss it first.

A BETTER EXPLANATION OF THE EVOLUTION
OF THE PROFESSION

The common belief is that the safety and health profession grew out of a demand for specialists who could help organizations respond to increasing regulatory scrutiny of hazardous conditions. In fact, the notion of "Specialists" was validated during the 1970s when regulation of hazards became prolific and sophisticated to the point of dividing risks into sectors, such as occupational environmental, consumer, or transportation. However, the belief that the growth in the profession in the 1970s was due solely to the growth in laws and regulations is very misleading.

Instead, we should look at a model of professional state of the practice that demands a higher level of organizational response to hazards and threats. This model shows that, prior to state or federal regulation, there are always critical masses of safety and health professionals operating at a high level of practice. This evolving state of the practice becomes the benchmark for continued standardization and regulation — the models of proaction we prefer in this nation. The OSHAs and EPAs are firmly treading in the footsteps of the professionals — not the other way around!

Figure 53.1 illustrates this model by looking at several key historical events in occupational safety state of the practice. If we can accept this picture, then we see that every time we individually choose to stretch professionally — to strive for continuous quality improvement in safety and health practice — we are leading ourselves into the future. The key here is to strive for improvement — to find better and more effective ways to prevent or control risks.

With this model in mind, let us turn to a discussion of how the education of safety and health professionals has been conducted in the past and how this process is beginning to change.

THE PAST AND THE FUTURE OF PROFESSIONAL EDUCATION

Unfortunately, it must be admitted that most college and university programs for entering practitioners have been unduly influenced by the presence of legislation and regulation. For example, only one or two safety management degrees were available prior to the inception of OSHA. Most professionals, if they had the benefit of a 4-year degree, were graduating with degrees in allied professions such as engineering or science or a traditional specialization such as fire protection. Then, when OSHA and EPA came upon the scene, the educators perked up their ears,

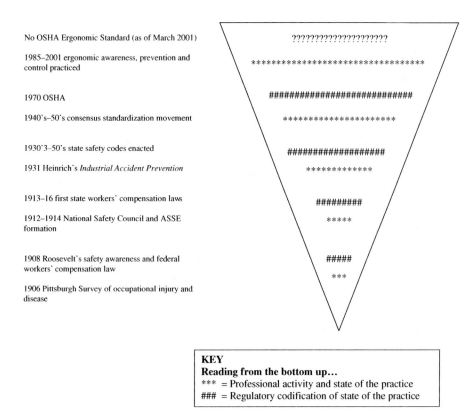

No OSHA Ergonomic Standard (as of March 2001)

1985–2001 ergonomic awareness, prevention and control practiced

1970 OSHA

1940's–50's consensus standardization movement

1930'3–50's state safety codes enacted

1931 Heinrich's *Industrial Accident Prevention*

1913–16 first state workers' compensation laws

1912–1914 National Safety Council and ASSE formation

1908 Roosevelt's safety awareness and federal workers' compensation law

1906 Pittsburgh Survey of occupational injury and disease

KEY
Reading from the bottom up...
*** = Professional activity and state of the practice
= Regulatory codification of state of the practice

Figure 53.1   State of the practice as the catalyst for change (historical examples from occupational safety and health practice).

hired the few credentialed safety and health professionals they could find, and began to offer safety and health degrees of various sorts.

In 1995, the American Society of Safety Engineers (ASSE) has identified 123 colleges or universities that collectively offer 237 degrees in safety and health. Of these institutions, 70 offer undergraduate degrees, 61 offer master's-level work, and 26 universities offer doctoral-level work. These numbers do not represent the many certification programs or 2-year associate degrees offered in the community college system. Correspondence courses and other long-distance study options were not widely available at this time. The actual names and emphases vary greatly among these institutions.

The ASSE directory of college and university safety-related degree programs is now updated regularly on its Web site. In May 2001, ASSE had identified 115 degrees of which 19 were associates, 38 were undergraduate, 46 were at the master's level, and 12 offered doctoral awards. The decline in degree offerings within this 6-year period may, in part, be due to more accurate reporting. Yet, some of the decline must be attributed to a closing of smaller programs as well as an overall reevaluation of professional education, and its modes of delivery. The good news is that several safety degree programs have stood the test of accreditation. Other interesting news is that correspondence courses, distance learning and other forms of professional development are on the rise. ASSE documented ten colleges or universities offering distance learning opportunities in 2001.

Despite a recent movement among colleges and universities to consolidate and eliminate small programs, the number of safety and health degrees is remaining constant. Some growth has actually been observed in the number of master's and doctoral degrees since the early 1990s. This trend is important for several reasons. First, safety and health curriculums offer additional employment opportunities. Women and men may now choose to become safety and health educators, as well as practitioners within the public and private sectors. Second, educators who are truly aligned with the profession will be able to advocate for their peers. They will be able to provide that state-of-the-practice leadership in education, research, and writing that is so necessary to a dynamic profession.

Now that we are beginning to see a dedicated cadre of safety and health educators, we are also beginning to see a debate about the best disciplinary foundations from which to train practitioners. In the past, either the foundation was technical (e.g., engineering) or programs were forced to fit into an existing scheme (e.g., safety education in the physical education department).

The new debate has its origins in the National Institute of Occupational Safety and Health (NIOSH) Minerva Project, which was begun in the 1970s and which continues to promote the teaching of safety and health modules within business schools. Yet, business schools have traditionally been resistant to expanding their already full curriculums to include significant amounts of safety and health education.

The newest school of thought is that management sciences must be combined with technical knowledge. This combination can be achieved through several means. One way is to redesign current undergraduate degrees with a balanced emphasis. Another way is to provide technical training at the undergraduate level with management emphasis at the graduate level. Alternatively, undergraduate and postgraduate safety and health programs could be housed in management schools. Here, technical knowledge could be taught with a management emphasis, running parallel with other management specializations.

The management school model for education of safety and health professionals has not yet been attempted; yet it is one whose time has arrived. After all, it is other managers who the safety and health practitioner must work with, gain respect from, and convince in order to be effective. This process is best begun while everyone is in school.

As this debate begins to take shape, another, very different discussion concerning professional education is already under way. How much education is enough? This question is not unique to the safety and health profession, but it is one which is quickly confronting many colleagues who thought an undergraduate education in safety or health or in an allied field was quite enough.

Now we find that professional certification by examination as well as postgraduate degrees may be needed for advancement. Continuing education in the form of seminars, certificate programs, constant self-study, and participation in distance larning opportunities is all necessary to keep up-to-date, as well as to prepare for the future.

FUTURISTIC TECHNIQUES

What does the future hold for safety and health professionals? We could gaze into a crystal ball, or we could consciously become futurists.

Futurists are interdisciplinary thinkers. They see relationships and change where others do not. They describe the future 5 to 10 years ahead. They try to prescribe the appropriate action, becoming proactive in their risk control and management efforts. They also integrate futuristic techniques into everyday safety and health practice. These techniques include scanning, content analysis, data research, trend analysis, forecasting, cross impact matrix analysis, Delphi surveys, scenario building, decision tree analysis, simulation, and games. These and other tools can be found fully described in future science texts.

The example used in Figure 53.1 illustrates the technique and process. In the early 1980s the issue of ergonomics was creeping into our awareness. This began when some practitioners began to take an in-depth look at a new injury phenomenon of repetitive motion illness and the old problem of back injury. Data were analyzed. Trend analysis was performed and economic and social implications were forecasted. The need for change became obvious to some. At the same time, these professionals became aware of the growing discussion among their colleagues concerning ergonomics. This awareness took the form of scanning and informal content analysis. Early articles began to appear; services and products became available.

Those who decided to do something now about ergonomic problems — not waiting for OSHA to require it — were acting as change agents. Their individual actions developed into a state of the practice that has informed other professionals and organizations. Their actions and comments have also informed the recent OSHA regulatory rulemaking on the issue of ergonomics. And, despite the congressional and presidential disapproval of the OSHA Ergonomics standard on March 20, 2001, safety professionals continue to develop and practice ergonomic prevention strategies. Similar technical issues and opportunities appear on our professional horizons every year.

SEVERAL PREDICTIONS ABOUT THE FUTURE
OF THE PROFESSION

We began by asking, "Where are we going?" We should end by offering several predictions in response. There is no magic to this forecast. Futurists understand that the future is firmly founded in the present. Many will already be aware of their own professional growth in the direction described below.

The safety and health professionals of the new millennium will:

- Be part of a multicultural and diverse population.
- Be well versed in international issues and work across national boundaries.
- Use sophisticated analytical skills.
- Perform survey and data research as a step in risk analysis and decision making.
- Serve as generalists within their organizations.

- Work, in larger number, as consulting specialists to organizations on an as-needed basis.
- Become experts in communication skills of writing, presentations, and training.
- Use electronic-based technology as a daily tool.
- Continue to expand knowledge through long-distance learning.
- Be better educated and credentialed than their predecessors.

Minerva, Roman goddess of wisdom and education, can lead us past Machiavelli's warning and into a predictable future. Have a safe journey!

RESOURCES AND FURTHER READING

American Society of Safety Engineers (ASSE). *Safety and Related Degree Programs.* Des Plaines, IL: ASSE, 1994–1995. The most recent edition may be found at http://www.asse.org/colluni_directory.htm (accessed May, 2001).

Dickson, P. *The Future File: A Guide for People with One Foot in the 21st Century.* New York: Rawson Associates, 1977.

Fowles, J. *Handbook of Futures Research.* Westport, CT: Greenwood Press, 1978.

Holmer, O. *Looking Forward: A Guide to Futures Research.* Beverly Hills, CA: Sage Publications, 1983.

LaConte, R. T. *Technologic Tomorrow Today: A Guide to Futuristics.* New York: Bantam Books, 1975.

Martino, J. P. *Technologic Forecasting for Decision Making,* New York: McGraw-Hill, 1993.

Oregon OSHA. Online safety courses at http://www.cbs.state.or.us/external/osha/educate/training/pages/course.htm (accessed May, 2001).

World Future Society. *The Futurist: A Journal of Forecasts, Trends, and Ideas about the Future.* Bethesda, MD: World Future Society.

CHAPTER **54**

Safety and Health: Managing Using Computer Technology

David L. Fender

CONTENTS

INTRODUCTION

Whether responsible for occupational-related safety, health, environment, or some combination of those areas, professionals will find that proper use of computers and associated equipment can make them more efficient. Professionals spend a great deal of time finding out and managing information, and proper use of computer-based resources can make this process simpler and more useful. To do this, professionals must know what is possible and what resources are available. One of the best sources of information is to discover what others are doing. If someone else has done it, not only is it possible, but you can find out how it was done and its advantages and disadvantages. When it comes to developing computer applications there is nothing like letting other people spend the time, money, and make the mistakes and then apply their lessons learned to your application.

HARDWARE

There are four common types of computer systems in use today: stand-alone systems (commonly called desktop systems), notebook, handheld, and networked systems. These systems do not necessarily exist totally separately, but can interrelate to each other.

Desktop Systems

Desktop systems are computers capable of connecting with a variety of peripheral devices such as a printer, modem, and scanner and are commonly used by a single user. They consist of a central processing unit that usually contains the actual workings of the computer and input, monitor, and storage devices. Desktop system designs vary and new designs are always being developed. Desktop systems often work in a stand-alone mode; that is, all work is done within the desktop and any peripheral devices connected to it.

It is difficult to give specific guidance about what to look for in a desktop system because of rapid technology changes, but some general advice is possible. When selecting a computer first consider what you want to use the computer for and what peripheral devices you will need. If there is specific software you need to run, find out what specifications the computer needs to meet in order to run the software. Generally speaking, it is advisable to buy a little bit more computer than you currently need because in the long run you are more likely to be happy with the purchase. Purchase as much memory as you can afford, a reasonably fast processor, and more hard drive space than you think you will need. You will also likely need to transfer large files from time to time, so it is wise to think about some type of large-capacity portable file storage. There are many such devices on the market and you will need to research to discover what type of device is available and best for you.

A monitor is necessary, and since you will spend virtually all the time you work on the desktop system by looking at the monitor, it is very important. Purchase as good and as large a monitor as you can reasonably afford. The recommendation is to buy a monitor with a screen that is 15 inches viewable or larger. Some monitors are physically quite large, so be careful that it will fit the intended space.

Notebooks

Notebooks, laptops and their variations are relatively small, lightweight, portable versions of desktop systems. They will usually fit in a standard briefcase, although with some models, nothing else will fit in the briefcase with the notebook. Many individuals prefer using a notebook instead of a traditional desktop system. Notebooks essentially have the same capabilities as desktops with the added advantage of being readily portable.

When looking for a notebook consider the same features as previously mentioned regarding desktops. Due largely to size limitations there will be fewer options available for a specific notebook and upgrading options are more limited on notebooks as compared with desktop systems. Especially for individuals who travel or need to take computer-based work from one location to another, notebooks can be very useful and efficient.

Handheld Devices

Handheld devices, sometimes called personal digital assistants (PDA), are small, fit-in-the-hand computing devices. Basic functions of most devices include address book, date book, expense reporting, calculator, memo pad, to-do lists, and appointment calendar. The features of these devices vary and most of the better devices are capable of transferring information to and from desktop or notebook computers and some can receive and send e-mail as well as access information on the Internet. Many of the devices let you write on the screen and save what you write; they also have handwriting recognition capability. Some models also have the ability to download and run programs. Safety-related applications includes inspection checklists and recording audit findings, which can later be uploaded to a desktop computer and integrated into a report. Some individuals are finding these devices useful enough that they use them instead of a notebook. Uses for these devices will only grow in the future.

Networks

Businesses are increasingly using computer networks. As indicated in Figure 54.1 these systems consist of a server that is capable of connecting to multiple desktop or notebook computers and various peripherals. Servers are large-capacity computers that usually have very large data storage capability along with large memories and fast central processors.

Networks have several advantages, including the ability to share files readily among users, and allow access to large databases among multiple users. Additionally, many software programs can be run through the network so that the program is installed only on the server and not on each individual computer that needs access. This eases installation and makes better use of hard drive space on individual computers. Networks also have the ability to share printers and mass storage devices among users, thus cutting down the need to buy multiple devices. Networks usually

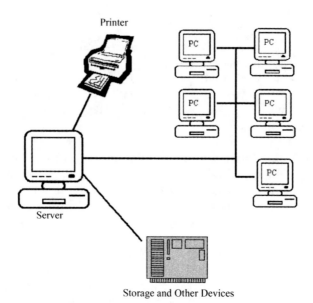

Figure 54.1 Computer network.

require an experienced individual(s) to install and maintain them. Reliability becomes critical in a network because if the network fails to work many individuals may also be unable to work.

Computer networks usually use various types of wire or fiber-optic cable to connect devices physically but networks can also be wireless. Wireless networks do not require the installation of the large amount of cable as with conventional networks, and desktop and notebook computers can "talk" to the network anywhere within range of the system.

When networks exist, communication such as Internet access and e-mail normally functions through the network. This makes these capabilities available to anyone who is connected to the network. Additionally, computer-based training can be conducted right at the user's desk using network resources.

Purchasing

In addition to the comments previously made about purchasing equipment, another concern is who to buy from. Local resellers have an advantage of having someone to assist in the purchase directly, and the store may provide technical support. Mass merchandisers and office superstores may be able to provide some advice but very little, if any, technical support. It is common today to purchase systems via mail order. Major companies have Internet sites at which you can configure your system and order online. As long as you are working with a major vendor, there is little risk to purchasing systems this way.

When purchasing from other than a local computer vendor, technical support is normally provided via telephone and the Internet. Whomever you buy from, consider more than the final price. Assume that you will, at least once, have need for technical support, and ready availability and quality of support will quickly make up for the small amount you might save by buying a cheaper system from a vendor that does not provide good support.

SOFTWARE

Software for the safety office can be divided into two general types: off-the-shelf software that is used for general office purposes and specialized software that has very narrow, specific purposes. Examples of the most common types and uses for this software will be discussed.

Office Suites

Having the proper software or computer programs is essential to get the most out of computer resources. Basic software would include what has become known as an office suite. Features of this suite will vary with the exact version of the program purchased but normally consists of word processing, spreadsheet, database, and presentation software. Publishing, graphics, and Web page development software is also included in some packages.

Word processing software can be used for more than basic reports, its real value comes from developing standardized formats and reports that can save time as they are used as the basis for each new document. Spreadsheets are very useful for any number-intensive operation such as budgeting. They can also be used as a basic database. Databases are specifically designed to store and allow access to large amounts of information. Most people find that getting the most out of database programs requires training on the specific program in use. Databases can be customized for the needs of the safety office and proper design can make the recording and extracting of information simpler.

Presentation software is principally used to make overhead transparencies and computer-based presentations. These programs are not restricted to just one type of presentation, but after a presentation is finished it can be printed on paper and transparencies or used in a computer-based presentation with projection equipment. These programs have graphic design features and pictures, graphs, charts, drawings, and text can be incorporated. In addition, notes can be typed with each slide, and the slides and notes may then be printed and used as a reference for the actual presentation.

Publishing software is used for making calendars, newsletters, forms, cards, brochures, etc. Word processing software is capable of doing some of these same functions but the specialized publishing software provides additional options to the process. Graphics software is used to develop and edit graphic images and photographs. These may then be used to enhance documents and presentations.

Web page development software is gaining increasing use as the Internet gains importance as an information source. This specialized software takes much of the programming out of Web page design and makes it so that almost anyone can develop workable Web pages.

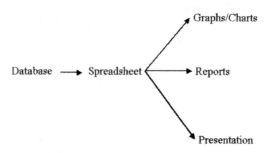

Figure 54.2 Database program.

The strength of these suites for the safety office is the ability to link the various programs within a suite to speed the development of standard reports that are routinely produced. For example, you can set up the programs so that information is extracted from the database program and inserted into a spreadsheet. Graphs and charts are automatically updated (Figure 54.2). These graphs and charts as well as spreadsheet data can then be automatically exported to word processing documents and presentations. All that is left to do is to open the final documents and review and update any wording. This linking of programs is particularly useful for repetitive reports that require data updating. Doing it this way saves time, and provides consistency in the report format.

Specialized Applications

Specialized applications vary widely in their purpose and scope. Common applications are Material Safety Data Sheets (MSDS) management, incident recording, workers' compensation data, OSHA data recording, SARA management, employee training, regulatory information, and others. MSDS management programs are intended to simplify management and control of MSDS sheets. They vary from a simple database that allows important information to be inputted from each MSDS to programs that allow MSDS to be scanned and automatically indexed. There are also companies that will put MSDS information into a database and update it as necessary. Whether the company does the work in-house or hires an outside vender to do it, with the right program the resulting MSDS data can be made available at any networked computer within the company and some programs can make the information available on the Internet, which can be useful for extended sites.

Programs that record details about incidents can assist in OSHA recording and reporting requirements as well as help in incident analysis. These programs frequently have chart and graph features that can assist in trend analysis. Some programs can be linked to human resources databases and make data entry easier. There are also specialized programs that record workers' compensation data and print first report of injury reports. Some programs are designed to record incidents and workers' compensation data and these programs can minimize the retyping of information in separate programs.

Computer databases can be very useful for SARA management and other required EPA-related records. MSDS programs are sometimes designed to include

SARA management as part of their function. For small numbers of chemicals, such a program probably is not necessary but for large numbers of chemicals such a program may be indispensable for compliance purposes and accurate reporting.

A company with many employees may find that keeping up with employee training is an arduous task. Employee training programs can make this task much easier, especially if the program is linked to the human resources database to reduce data entry and to better track new and ex-employees. These programs manipulate data and print reports such as what employees have completed certain training as well as employees who still need training. They let the manager know when refresher training is necessary, and many programs can track scores and other quantitative and qualitative data regarding employee training. There is also computer-based training available from several vendors. These training programs, using multimedia, can provide very effective training and keep excellent records of student performance and comprehension. Some companies provide software that can track training needs and automatically update the records as the computer-based training is completed.

It can be much easier to find the many government regulations that affect safety, health, and the environment using regulatory software. The strength of this software lies in their search capabilities. By using a single word or phrase, the software almost instantly shows every place in the standard where that word or phrase exists. This can save much time in researching what regulatory requirements exist for a particular area of concern. This software ranges from just one regulation such as OSHA or EPA to packages that combine many regulations. This type of software is available from many different companies as well as from the Government Printing Office. Better packages offer enhancements to the basic regulations, for example, by adding such things as training guidance, highlighting recent regulation changes, providing other government-produced documents such as OSHA letters of interpretation, and some include specific state requirements.

Purchasing Considerations

How can the average professional know what software is the best to use? In some cases the choice will be simple. For example, if you need to share word processing documents with others in the company, with major clients, or if your organization has standardized on certain software, you will most likely have to obtain specific packages. Beyond this, research is essential and will be eased if you know what you need to accomplish.

Before looking for software, define exactly what you need the program to do. Do you want to keep track of workers' compensation claims and cost? Do you need to track incidents for later analysis? Or do you want a program to do all the above? If you have multiple sites, especially if they are in different states, you will want the program must be able to identify sites and print reports that will meet the regulatory requirements of the different states.

Research is essential in purchasing specialized software. Many of the companies that offer specialized programs are small and may not advertise widely. Read trade publications, including the advertising, for information. Software is occasionally reviewed in these publications and the reviews can be quite useful. The Internet can also provide information on available software and the features of specific packages. Trade shows can also be very useful. Exhibitors at the shows will usually demonstrate the software and nothing takes the place of actually seeing and working with the software. In this vein, many vendors will provide demonstration copies of their software so that you can work with them in your office. These demos usually will work for a limited time or have other restrictions that make the software unsuitable for continuous use, until you purchase it.

When evaluating software, look for ease of data entry and appropriate report generation. Once you get your data into a program you can find yourself "locked" into the program because of the time and difficulty it would take to get the data into another program. Because of this ensure that the program will do what you need it to do, before you purchase it. Also, it is a plus if the program has the capability to export the data into standard file formats that other programs can import. This feature will make it much easier, if in the future you wish to use a different program.

Other professionals can be a valuable source of information. Find out what software they are using and how they are using it. Ask if they are happy with the software and what problems they have. You may be able to go and look at what they use and get a firsthand report. Find out the purchase price, any additional cost, their experience with the software company technical support, and how long it took to learn the program.

At some time or another you will likely need support from the company, so look at its support polices very carefully. Support ranges from toll-free numbers to call and unlimited support calls to a very limited free support time and fees and toll calls after that. Do not underestimate the need for support. A slightly inferior package with good support is probably a better selection, in the long run, than the perfect package with expensive support policies.

INTERNET

The major strength of the Internet is the ability to find and transfer information. One way to think of the Internet is as a giant worldwide library. All types of information is readily available including safety, health, and environmental data. There is so much available that it can be difficult to sort through everything for the specific information needed. Much of the information available is free and Internet sources may be able to replace some paid or subscribed publications.

To use the Internet you will need a subscription to an Internet Service Provider (ISP) and a computer with either a modem or a network connection if the network is connected to the Internet. The only other tool needed is browser software that provides the interface between the computer and the Internet. This software is readily available and very good versions are available for free.

Searching

There is a specific address for each Internet site. Most addresses are in the format: http://www.osha.gov. The http://www is a standard prefix for most addresses and the osha.gov is the specific address, in this case the OSHA Web site. Computers are very exact, so any misspelling or typos will be interpreted literally. If you type an address and do not get to the site you want, check the address very carefully for spelling, punctuation, and spaces.

To search for information, if you don't know the address of a specific site you can use a search engine to search for key words. There are many search sites that index content on the Web but no search engine indexes everything. Make the search as specific as possible to restrict the search and use Boolean operators to narrow the search. Search engines will have a help feature that explains how to use the Boolean operators and the best way to search using that particular engine. When the search is completed, the search engine will give you "hits" that meet the criteria you stipulate. You then click on the hits to go to the various pages indicated.

Information Available

Web sites can be a wealth of information, but cautions are in order. It is very easy to establish a Web site and Web sites are established for a variety of purposes that in some cases may not be obvious. For example, the sites may have a marketing focus and should be evaluated the same as you would any marketing or promotional material; they may promote a particular belief or cause and should be viewed as you would any information from a special interest group; they may provide information like a library, may be a company Web site for company communications and public relations, or may exist for many other reasons. The bottom line is to evaluate the validity of the information and its source as knowledgeable professionals would from any other conventional source.

The U.S. government puts a great deal of effort into their Web sites, and many documents published by various agencies are readily available on the agency Web site. For example, NIOSH makes most of its publications available, as does the Consumer Product Safety Commission. OSHA and EPA publish their standards as well as letters of interpretation and many other helpful documents regarding their areas of enforcement. The OSHA site also has specialized software packages available for download to help with compliance. Many companies make MSDS for their products downloadable from a Web site. There are also sites that allow you to search for specific products and provide MSDS from their library. This makes getting MSDS much easier and quicker. Professional organizations such as the American Society of Safety Engineers, the National Safety Council, and others have Web sites that provide much information about the organization. Safety statistics are available from several sites including OSHA and the Bureau of Labor Statistics. The appendix to this chapter lists recommended sites that may have information you can use, and there are many others.

Companies can use the Internet to publish company documents and make the documents more widely available to employees. This is particularly valuable for companies with multiple locations. By using the Internet, companies can avoid the expense and delay of printing information, the information is much easier to update, and the information is readily available to employees when they want it. Many safety departments have put their safety manuals and procedures on Web pages for just such a purpose.

E-Mail

An e-mail address is a mailing address that is similar in concept to a regular postal address in that a specific name and address gets a message to a particular person. The format is "name"@"location." The information before the @ identifies the person it is going to and the information after the @ indicates the location or computer network where this person can be found. If an address is incorrect, the mail will usually be returned to the sender, with an explanation of why the message could not be delivered. E-mail is usually very fast, taking only seconds to arrive at its destination. Another advantage of e-mail is the ability to attach computer-generated files to the message. This can be a very quick and easy way to pass along information to others and to ensure that they get exactly what is sent.

Most browsers have a built-in e-mail function and separate programs are available that do e-mail. The better programs have advanced features that check spelling of outgoing messages, maintain address books, and help filter and sort messages in order to better manage incoming messages.

E-Mail Lists

There are many listserve or e-mail groups built around safety, health, and environmental issues. Each group has a specific audience that it serves. Examples of subjects covered by specific lists include accident and emergency news, agricultural health and safety, biohazards, general industrial hygiene, emergency preparedness, chemistry education, construction safety and health, consumer product safety, campus safety, cumulative trauma disorders, environmental health and safety, ergonomics and human factors in medicine, radiation safety, and many others. The list automatically sends any message sent to it by the list members to all list members via e-mail. Members can ask questions of the list members, make comments, or just monitor what goes on. Monitoring a list can be a good way to keep informed about topics of interest and is a good way to network with others who have similar interests and concerns. To subscribe to most lists you send an e-mail message to the specified address and request to join.

One of the largest of the safety-related lists is called SAFETY. This mailing list discusses worldwide safety issues including laboratory safety, hazardous waste disposal, safety management, and the electronic resources available to assist with these topics. Subscribers include representatives from academia, industry, the military, and government agencies. See the appendix for SAFETY subscription address.

CONCLUSION

Computers and associated functions can be very useful to the working safety and health professional. The secret is to use computers for what they do best and to use other means for what they do best. For example, computer databases are very useful when you have large amounts of data to compile, but if you have very small amounts of data a pencil and paper may be quicker and more efficient. The number-crunching ability of spreadsheet programs is phenomenal, but in some cases a calculator may be more efficient. In other words, processes should not be "computerized" just because they can be but because it saves time or adds value to the previous way of performing the function. Computers and the Internet can be very useful tools for the professional, but ensure that you run the tools and do not let the tools run you.

FURTHER READING

Brooks, R. Help for hazardous moves, *Occupational Health & Safety*, 70, 24, April 2001.
Grimaldi, J. V. and Simonds, R. H. *Safety Management*. Des Plaines, IL: American Society of Safety Engineers, 1993.
Lang, R. and Park, S. Speed up your Internet searches. *Occupational Hazards*, 63, 44, April 2001.
Stuart, R. B. and Moore, C. *Safety & Health on the Internet*, Rockville, MD: Government Institutes, Inc, 1999.

APPENDIX: RECOMMENDED WEB SITES

Government-Related Sites

FedWorld — Entry portal into federal government information
http://www.fedworld.gov

Canadian Centre for Occupational Health & Safety (CCOHS)
http://www.ccohs.ca

National Institute for Occupational Safety & Health (NIOSH)
http://www.cdc.gov/niosh/homepage.html

U.S. Centers for Disease Control
http://cdc/gov

Environmental Protection Agency
http://www.epa.gov

National Institutes of Health
http://www.nih.gov

Bureau of Labor Statistics
http://www.bls.gov

Federal Register
http://www.access.gpo.gov/su_docs/aces/aces140.html

Occupational Safety & Health Administration (OSHA)
http://www.osha.gov

Brookhaven National Laboratory, Environmental, Safety, Health & Quality Division
http://www.bnl.gov/bnlweb/ESH.html

Private and Commercial Sites

Firenet — British fire information site
http://www.fire.org.uk/

DuPont Safety Site
http://www.dupont.com/safety/

Safety Online — A safety marketplace for professionals
http://www.safetyonline.com/

University Sites

Vermont SIRI
http://siri.org/library/library.html

SIRI MSDS Search Site
http://hazard.com/msds/

E-Mail Lists

Site for Safety-Related E-Mail Lists
http://www.ccohs.ca/resources/listserv.htm

SAFETY subscription address
listserv@list.uvm.edu

Safety Organizations

ASSE
http://www.asse.org/

National Safety Council
http://www.nsc.org

## CHAPTER 55

The Safety and Health Professional of the Future*

Richard W. Lack

CONTENTS

INTRODUCTION

In March 1991, I presented a paper at the Annual Conference of the Safety Executives of New York entitled "The Role of the Safety Professional as We Approach the Year 2000." The research conducted for this paper, plus my ongoing interest and research on this subject, will form the foundation of this chapter.

---

* I am grateful to Tillinghast Publications for publishing this material in part. The details on this publication may be found in the Further Reading section at the end of this chapter under Kloman.

As we all recognize, nothing in this great world of ours stays the same. We change as we get older and so does our environment. The process is constant, but the cycles are usually gradual so that it really becomes a progression or evolution. Every creature on Earth, including the human race, has to adjust continuously to change.

In spite of this almost relentless universal law, many people resist change, and instead of recognizing it as an opportunity for improvement they cling to their memories of the good old days. Because of this reluctance to embrace change, history has shown many times that few people seem to be able to identify trends or shifts in the making. For an inquiring open mind the signs are there, but they are either ignored or misunderstood. Eventually, some dramatic series of events may occur to introduce sweeping change. For example, many of us brought up in the Cold War years could not envisage the collapse of communism in Russia and its satellite states.

While attending the American Society of Safety Engineers Annual Professional Development Conference in Las Vegas, Nevada in June 1994, I listened to a fascinating keynote speech by Dan Burrus, a leading futurist. Dan pointed out that technically the next 10 years has already been invented and that what we need to do is change the way we think. He said that we must be opportunity managers and use the new tools. This takes new thinking.

Other trends Dan mentioned were that time would be the currency of the late 1990s. We cannot continue to do things the old way. We need to "leverage our time with technology," as he put it. "Change is opportunity," said Dan. We need to open our minds to the creative application of technology. This is the communication age, and one trend Dan foresaw is that we are going to automate and humanize training so that it is more interactive, self-directive, and self-diagnostic. His final words were that we must think in terms of idealism, think beyond to the future problems that are about to happen, and what can we do about them. We must take the new tools and become the craftspeople of the 21st century — in other words, be proactive!

Six years later at the American Society for Industrial Security annual seminar in Orlando, Florida, Dan Burrus was again a keynote speaker, and building on his earlier comments, I noted the following key points in his address:

- In the next 5 years we will go through more change than in the last 20.
- Anticipation is like taking time back. The more you look, the more you see. Take 1 hour a week to consider.... Look into the visible future. We must be opportunity managers not crisis managers.
- The new role of technology is if it works, its obsolete! Obsolescence is a nonissue. The point is what are you doing with it! It's not the tool, it's what you are doing with the tool. Application is the key.
- We need to think communication age not information age and become opportunity managers of our profession.

These are inspiring concepts that I hope will provide food for thought while you are reading this chapter.

FUTURE TRENDS

The following is a list of trends based on my research:

- Population explosion — China and India are the most populous nations.
- Megacities — By the year 2000, Mexico City, São Paulo, Shanghai, and Calcutta were all forecast to have populations exceeding 20 million.
- Worldwide free trade is coming — NAFTA and GATT are just the beginning.
- The Pacific Rim, South America, India, and Africa all have huge growth potential, not to mention the former communist-controlled countries.
- Travel and tourism is forecast to be a huge growth area.
- In the world of business, entrepreneurs are creating the global economy. Intense competition will force the big corporations to break up in order to survive.
- Environmental contamination will become a worldwide problem.
- Terrorism and civil unrest will probably increase.
- Trend forecasts for the United States and other developed countries include:
  - Population moves from the urban areas to smaller rural industrial/high-technology centers — the so-called fifth migration.
  - Renaissance of the arts.
  - A more caring age, concerns for elderly people and the less fortunate.
  - Health care becoming a huge growth industry due to the aging population.
  - Increasing community activity to help combat crime and other uncivil behavior.

In the world of business, most of us have already felt the effects of downsizing, reengineering, and outsourcing to reduce costs and become more competitive.

The burden of costs to business arising from employee injuries and illnesses is already providing creative opportunities for safety and health professionals. Government resources are already totally unable to solve this problem, and increased legislation can only deal with the symptoms — not the root causes. In fact, government officials are already seeking ways to promote a more cooperative approach, and more and more that key word *prevention* is coming up in these debates.

With all these anticipated issues and trends, my prediction is that the safety and health professional of the future will, in general, require a much broader technical knowledge and must gain increased management and communications skills to help the organizations they serve to set up and continuously improve effective prevention systems.

KNOWLEDGE AND SKILLS FOR THE FUTURE

Expanding on these attributes, here is a list of knowledge and skills that I believe will form the job requirements for the safety and health professional of the future:

Personal Abilities and Characteristics

- Excellent education in language and communication skills, written and oral
- A wide area of interests
- A high level of professional and ethical conduct

Professional Technical Knowledge

- State-of-the-art safety management techniques
- Process safety management
- Industrial hygiene and toxicology
- Medicine/health (at least the preventive aspects)
- Ergonomics
- Security
- Fire protection
- All related laws, codes, and standards

Professional Management and Related Knowledge and Skills

- Advanced professional management
- Computer skills
- Communicating
- Training
- Psychology/human behavior
- Civil and criminal law (at least the fundamentals)
- Risk management and insurance
- Human resource management

Certifications

This aspect will most likely continue expanding in the years ahead, so the *minimum* recommended certifications are as follows:

- Certified Safety Professional (CSP)
- Certified Safety and Health Manager (CSHM)
- Certified Industrial Hygienist (CIH)

Additional certifications will be needed as position responsibilities require. Examples:

- Certified Hazard Control Manager (CHCM)
- Certified Hazardous Materials Manager (CHMM)
- Master of Public Health (MPH)
- Registered Nurse (RN)
- Certified Protection Professional (CPP)
- Fire Protection Engineer (FPE)
- System Safety Engineer (SSE)
- Professional Engineer (P.E.) in Safety Engineering or other disciplines
- Global responsibilities will include international certifications and memberships.

Needless to say, there are many examples of other technical certifications. Some of those are related to certain industries such as construction (Crane Instructor) or mining (MHSA Instructor) or to specific materials and chemicals such as asbestos or lead. This list is continuously growing, and needs will depend on your specific job responsibilities.

For more information on how to obtain these certification, please refer to Dr. Roger Brauer's chapter (Chapter 52) entitled Standards of Competence.

SUMMARY AND CONCLUSIONS

Continuous professional development will be absolutely vital for the safety and health professional of the future. Here is my three-step prescription for success:

- **Step I. Get out on the professional edge** — We will need to constantly update, improve, and broaden the scope of our professional knowledge and skills. This will involve taking courses, attending seminars, joining professional organizations, obtaining reference materials, and gaining certifications. At all costs, avoid "floating along with the crowd." This is the sure path to mediocrity and eventual disillusionment with your situation in life. Stay flexible, and be ready to accept and master challenges as they face you throughout your professional career. At all costs, never stop learning!
- **Step II. Constantly develop your skills in the art and science of professional management** — Prevention is the name of the game in terms of helping our organizations to control and avoid unintentional and intentional losses of their assets, both human and physical. We need to improve our management skills continuously by taking courses, reading professional books and journals, and becoming involved in management-related organizations such as the AMA (American Management Association), ASTD (American Society for Training and Development), and SHRM (Society for Human Resource Management).
- **Step III. Find Ways to Help Promote Collective Cooperation and Coordination among All Related Professional Societies and Organizations** — The old saying "united we stand, divided we fall!" has never been truer for all our professions in the years ahead. Each of us must do our part to help bring our various professional societies closer together. This will not happen until we, the membership, make our desires known to the leadership of the societies. How do we do this? To me the key is involvement — we must get involved in the process. Take an active role in your local chapter, serve on a committee, volunteer for projects, write, give presentations, travel to see how others have done it. There are many ways to do this, but ACTION — DOING — is the key. By getting involved in this way you will automatically be improving your skills and knowledge in many different areas, besides gaining valuable recognition from your peers.

One thing is for sure, by going *singly* on their own way, some of our societies will not survive the years ahead. They will be merged into another organization or they will just slowly fade into oblivion. *Collectively* they can become a powerful force for good in the service of the peoples of this world. It is up to us to see that this happens.

PROFESSIONAL ORGANIZATIONS IN THE FIELD OF SAFETY AND HEALTH MANAGEMENT AND RELATED FIELDS

American Board of Industrial Hygiene
6015 West St. Joseph, Suite 102
Lansing, MI 48917-3980
(517) 321-2638

American Conference of Governmental Industrial Hygienists
Kemper Woods Center
1330 Kemper Meadow Drive
Cincinnati, OH 45240
(513) 742-2020

American Industrial Hygiene Association
2700 Prosperity Avenue, Suite 250
Fairfax, VA 22031
www.aiha.org

American Management Association
135 West 50th Street
New York, NY 10020-1201
1-800-262-9699

American Management Association Extension Institute
P.O. Box 1026
Saranac Lake, NY 12983-9986
1-800-225-3215
(518) 891-0065

American Public Heath Association
1015 15th St., N.W.
Washington, D.C. 20005
(202) 789-5600
www.apha.org

American Society for Industrial Security
1625 Prince Street
Alexandria, VA 22314
(703) 519-6200
www.asisonline.org

American Society for Quality
611 E. Wisconsin Avenue
P.O. Box 3005
Milwaukee, WI 53203
1-800-248-1946/(414) 272-8575
www.asq.org

American Society of Safety Engineers
1800 East Oakton St.
Des Plaines, IL 60018-2187
(847) 699-2929
www.asse.org

Note: Readers interested in management aspects are strongly advised to join the Management Practice Specialty and possibly other practice specialties depending on your specialty interests.

American Society for Training and Development
1640 King Street
Box 1443
Alexandria, VA 22313-2043
(703) 683-8129
www.astd.org

Association for Quality and Participation
801-B West 8th Street
Cincinnati, OH 45203-1067
1-800-733-3310/(513) 381-1959
www.aqp.org

Canadian Society of Safety Engineering
P.O. Box 294
Kleinburg
Ontario, LOJ 1CO
Canada
(905) 893-1689
www.csse.org

Human Factors and Ergonomics Society
P.O. Box 1369
Santa Monica, CA 90406-1369
(310) 394-1811

Chartered Institute of Personnel and Development
IPD House
Camp Road
London SW 19 4UX
United Kingdom
020-8971-9000
www.cipd.co.uk

Institution of Occupational Safety and Health
The Grange
Highfield Drive
Wigston
Leicestershire LE18 INN
United Kingdom
0116-257-3100
www.iosh.co.uk

International Commission on Occupation Health (ICOH)
Department of Community, Occupational and Family Medicine
National University Hospital
Lower Kent Road S. (0511)
Republic of Singapore 119074
(65) 772-4290

International Institute of Risk and Safety Management
British Safety Council
70 Chancellors Road
London W6 9RS
United Kingdom
0181-741-1231
www.britishsafetycouncil.co.uk

International Professional Security Association
IPSA House
3 Dendy Road
Paignton
S. Devon TQ4 5DB
United Kingdom
01803 554849
www.ipsa.uk.com

National Fire Protection Association
Batterymarch Park
P.O. Box 9101
Quincy, MA 02269-9101
1-800-344-3555
www.nfpa.org

National Safety Council
1121 Spring Lake Drive
Itasca, IL 60143-3201
(708) 285-1121
www.nsc.org

National Safety Management Society
Business Manager, Phillip Mueller
2004 Halton Court
Columbia, MO 65203
(800) 321-2910
www.nsms.ws

International Society for Performance and Improvement
1400 Spring Street, Suite 206
Silver Spring, MD 20910
(202) 408-7967
www.ispi.org

Risk Insurance Management Society, Inc.
655 Third Avenue
New York, NY 10017-5637
(212) 286-9292/(800) 713-RIMS

Royal Society for the Prevention of Accidents
Edgbaston Park
353 Bristol Road
Birmingham B5 7ST
United Kingdom
0121-248-2000
www.rospa.co.uk

Society for Human Resource Management
1800 Duke Street
Alexandria, VA 22314-3499
(703) 548-3440
www.shrm.og

System Safety Society
P.O. Box 70
Unionville, VA 22567-0070

World Future Society
7910 Woodmont Avenue, Suite 450
Bethesda, MD 20814
1-800-989-8274
(301) 656-8274
www.wfs.org/wfs

For information on international safety and health organizations, readers can consult the International Directory of Occupational Safety and Health Institutions, which is published by the International Labour Office (ILO). ILO Publications can be obtained through major booksellers, from ILO local offices in many countries, or directly from ILO Publications, International Labor Office, CH-1211 Geneva 22, Switzerland.

FURTHER READING

Books

Boyett, J. H. and Conn, H. P. *Workplace 2000*. New York: Plume Penguin Books, 1992.
Boylston, R. P. et al. *The Safety Profession Year 2000*. Des Plaines, IL: American Society of Safety Engineers, 1991.
Burrus, D. and Gittines, R. *Technotrends*. New York: HarperCollins Publishers, 1993.
Cornish, E. et al. *The 1990's and Beyond*. Bethesda, MD: World Future Society, 1990.

Drucker, P. F. *Managing for the Future — The 1990's and Beyond.* New York: Truman Talley
    Books/Dutton, 1992.
Drucker, P. F. *Post Capitalist Society,* New York: HarperCollins Publishers, 1993.
Fennelly, L. et al. *Security in the Year 2000 and Beyond.* Palm Springs, CA: ETC Publications,
    1987.
Holt, A. St. J. et al. *Health and Safety towards the Millennium.* Leicester, U.K.: IOSH
    Publishing, 1987.
Knowdell, R. L., Branstead, E., and Moravec, M. *From Downsizing to Recovery: Strategic
    Transition Options for Organizations and Individuals,* Palo Alto, CA: CPP Books,
    1994.
Manuele, F. A. *On the Practice of Safety,* 2nd ed. New York: Van Nostrand Reinhold, 1997.
Naisbitt, J. *Global Paradox.* New York: William Morrow, 1994.
Olesen, E. *12 Steps to Mastering the Winds of Change.* New York: Rawson Associates/Mac-
    millan, 1993.
Peck, M. S. *A World Waiting to Be Born: Civility Rediscovered.* New York: Bantam Books,
    1994.
Peters, T. *The Tom Peters Seminar.* New York: Random House, 1994.
Scott, C. D. and Jaffe, D. T. *Take This Job and Love It.* New York: Simon & Schuster, 1988.
Scott, C. D. and Jaffe, D. T. *Managing Personal Change.* Los Altos, CA: Crisp Publications,
    1989.
Thomas, H. G. *Safety, Work and Life — An International View.* Des Plaines, IL: American
    Society of Safety Engineers, 186–212, 1991.

Other Publications

Lack, R. W. The safety–HR connection, *Personnel Journal,* 71(7), 18, July 1992.
Kloman, H. F., Ed. Risk Management Reports — The Safety Professional and the Year 2000
    (Author Lack, R. W.) 20(2), 15–27, Stanford, CT: Tillinghast Publications, March/April
    1993.
National Safety Council. Tomorrow's safety solutions. *Safety & Health,* October, 57–65, 2000.

Index

A

Q

R